ENGINEERING
VIBRATIONS

ENGINEERING VIBRATIONS

WILLIAM J. BOTTEGA

CRC Taylor & Francis
Taylor & Francis Group
Boca Raton London New York

A CRC title, part of the Taylor & Francis imprint, a member of the
Taylor & Francis Group, the academic division of T&F Informa plc.

Published in 2006 by
CRC Press
Taylor & Francis Group
6000 Broken Sound Parkway NW, Suite 300
Boca Raton, FL 33487-2742

International Standard Book Number-10: 0-8493-3420-9 (Hardcover)
International Standard Book Number-13: 978-0-8493-3420-7 (Hardcover)
Library of Congress Card Number 2005054946

Library of Congress Cataloging-in-Publication Data
Bottega, William J.
Engineering vibrations / William J. Bottega.
p. cm.
Includes bibliographical references and index.
ISBN 0-8493-3420-9 (alk. paper)
1. Vibration. I. Title.
TA355.B635 2006
620.1'1248--dc22   2005054946

informa

Taylor & Francis Group
is the Academic Division of Informa plc.

Visit the Taylor & Francis Web site at
http://www.taylorandfrancis.com

and the CRC Press Web site at
http://www.crcpress.com

To my mother, *Marie Bottega*

Preface

The effects of vibrations on the behavior of mechanical and structural systems are often of critical importance to their design, performance, and survival. For this reason the subject of mechanical vibrations is offered at both the advanced undergraduate level and graduate level at most engineering schools. I have taught vibrations to mechanical and aerospace engineering students, primarily seniors, for a number of years and have used a variety of textbooks in the process. As with many books of this type, the emphasis is often a matter of taste. Some texts emphasize mathematics, but generally fall short on physical interpretation and demonstrative examples, while others emphasize methodology and application but tend to oversimplify the mathematical development and fail to stress the fundamental principles. Moreover, both types fail to stress the underlying mechanics and physics to a satisfactory degree, if at all. For these reasons, there appeared to be a need for a textbook that couples thorough mathematical development and physical interpretation, and that emphasizes the mechanics and physics of the phenomena. The book would need to be readable for students with the background afforded by a typical university engineering curriculum, and would have to be self-contained to the extent that concepts are developed, advanced and abstracted using that background as a base. The present volume has been written to meet these goals and fill the apparent void.

Engineering Vibrations provides a systematic and unified presentation of the subject of mechanical and structural vibrations, emphasizing physical interpretation, fundamental principles and problem solving, coupled with rigorous mathematical development in a form that is readable to advanced undergraduate and graduate university students majoring in engineering and related fields. Abstract concepts are developed and advanced from principles familiar to the student, and the interaction of theory, numerous illustrative examples and discussion form the basic pedagogical approach. The text, which is extensively illustrated, gives the student a thorough understanding of the basic concepts of the subject, and enables him or her to apply these principles and techniques to any problem of interest. In addition, the pedagogy encourages the reader's physical sense and intuition, as well as analytical skills. The text also provides the student with a solid background for further formal study and research, as well as for self study of specialized techniques and more advanced topics.

Particular emphasis is placed on developing a connected string of ideas, concepts and techniques that are sequentially advanced and generalized throughout the text. In this way, the reader is provided with a thorough background in the vibration of single degree of freedom systems, discrete multi-degree of freedom systems, one-dimensional continua, and the relations between each, with the subject viewed as a whole. Some distinctive features are as follows. The concept of mathematical modeling is introduced in the first chapter and the question of validity of such models is emphasized throughout. An extensive review of elementary dynamics is presented as part of the introductory chapter. A discussion and demonstration of the underlying physics accompany the introduction of the phenomenon of resonance. A distinctive approach incorporating generalized functions and elementary dynamics is used to develop the general impulse response. Structural damping is introduced and developed from first principle as a phenomenological theory, not as a heuristic empirical result as presented in many other texts. Continuity between basic vector operations including the scalar product and normalization in three-dimensions and their extensions to N-dimensional space is clearly established. General (linear) viscous damping, as well as Rayleigh (proportional) damping, of discrete multi-degree of freedom systems is discussed, and represented in state space. Correspondence between discrete and continuous systems is established and the concepts of linear differential operators are introduced. A thorough development of the mechanics of pertinent 1-D continua is presented, and the dynamics and vibrations of various structures are studied in depth. These include axial and torsional motion of rods and transverse motion of strings, transverse motion of Euler-Bernoulli Beams and beam-columns, beams on elastic foundations, Rayleigh Beams and Timoshenko Beams. Unlike in other texts, the Timoshenko Beam problem is stated and solved in matrix form. Operator notation is introduced throughout. In this way, all continua discussed are viewed from a unified perspective. Case studies provide a basis for comparison of the various beam theories with one another and demonstrate quantitatively the limitations of single degree of freedom approximations. Such studies are examined both as examples and as exercises for the student.

The background assumed is typical of that provided in engineering curricula at U.S. universities. The requisite background includes standard topics in differential and integral calculus, linear differential equations, linear algebra, boundary value problems and separation of variables as pertains to linear partial differential equations of two variables, sophomore level dynamics and mechanics of materials. MATLAB is used for root solving and related computations, but is not required. A certain degree of computational skill is, however, desirable.

The text can basically be partitioned into preliminary material and three major parts: single degree of freedom systems, discrete multi-degree of freedom systems, and one-dimensional continua. For each class of system the fundamental dynamics is discussed and free and forced vibrations under various conditions are studied. A breakdown of the eleven chapters that comprise the text is provided below.

The first chapter provides introductory material and includes discussions of degrees of freedom, mathematical modeling and equivalent systems, a review of complex numbers and an extensive review of elementary dynamics. Chapters 2 through 4 are devoted to free and forced vibration of single degree of freedom systems. Chapter 2 examines free vibrations and includes undamped, viscously damped and Coulomb damped systems. An extensive discussion of the linear and nonlinear pendulum is also included. In Chapter 3 the response to harmonic loading is established and extended to various applications including support excitation, rotating imbalance and whirling of shafts. The mathematical model for structural damping is developed from first principle based on local representation of the body as comprised of linear hereditary material. The chapter closes with a general Fourier Series solution for systems subjected to general periodic loading and its application. The responses of systems to nonperiodic loading, including impulse, step and ramp loading and others, as well as general loading, are discussed in Chapter 4. The Dirac Delta Function and the Heaviside Step Function are first introduced as generalized functions. The relation and a discussion of impulsive and nonimpulsive forces follow. The general impulse response is then established based on application of these concepts with basic dynamics. The responses to other types of loading are discussed throughout the remainder of the chapter. Chapter 5, which is optional and does not affect continuity, covers Laplace transforms and their application as an alternate, less physical/nonphysical, approach to problems of vibration of single degree of freedom systems.

The dynamics of multi-degree of freedom systems is studied in Chapter 6. The first part of the chapter addresses Newtonian mechanics and the derivation of the equations of motion of representative systems in this context. It has been my experience (and I know I'm not alone in this) that many students often have difficulty and can become preoccupied or despondent with setting up the equations of motion for a given system. As a result of this they often lose site of or never get to the vibrations problem itself. To help overcome this difficulty, Lagrange's equations are developed in the second part of Chapter 6, and a methodology and corresponding outline are established to derive the equations of motion for multi-degree of freedom systems. Once mastered, this approach provides the student a direct means of deriving the equations of motion of complex multi-degree of freedom systems. The instructor, who chooses not to cover Lagrange's equations may bypass these sections. The chapter closes with a fundamental discussion of the symmetry of the mass, stiffness and damping matrices with appropriate coordinates.

The free vibration problem for multi-degree of freedom systems with applications to various systems and conditions including semi-definite systems is presented in Chapter 7. The physical meanings of the modal vectors for undamped systems are emphasized and the properties of the modal vectors are discussed. The concepts of the scalar product, orthogonality and normalization of three-dimensional vectors are restated in matrix form and abstracted to N-dimensional space, where they are then discussed in the context of the modal vectors. The chapter closes with extensive discussions of the free vibration of discrete systems with viscous damping. The problem is examined in both N-dimensional space and in the corresponding state space. Analogies to the properties of the modal vectors for undamped systems are then abstracted to the complex eigenvectors for the problem of damped systems viewed in state space. Forced vibration of discrete multi-degree of freedom systems is studied in Chapter 8. A simple matrix inversion approach is first introduced for systems subjected to harmonic excitation. The introductory section concludes with a discussion of the simple vibration absorber. The concepts of coordinate transformations, principal coordinates and modal coordinates are next established. The bulk of the chapter is concerned with modal analysis of undamped and proportionally damped systems. The chapter concludes with these procedures abstracted to systems with general (linear) viscous damping in both N-dimensional space and in state space.

The dynamics of one-dimensional continua is discussed in Chapter 9. Correlation between discrete and continuous systems is first established, and the concept of differential operators is introduced. The correspondence between vectors and functions is made evident as is that of matrix operators and differential operators. This enables the reader to identify the dynamics of continua as an abstraction of the dynamics of discrete systems. The scalar product and orthogonality in function space then follow directly. The kinematics of deforming media is then developed for both linear and geometrically nonlinear situations. The equations governing various one-dimensional continua are established, along with corresponding possibilities for boundary conditions. It has been my experience that students have difficulty in stating all but the simplest boundary conditions when approaching vibrations problems. This discussion will enlighten the reader in this regard and aid in alleviating that problem. Second order systems that are studied include longitudinal and torsional motion of elastic rods and transverse motion of strings. Various beam theories are developed from a general, first principle, point of view with the limitations of each evident from the discussion. Euler-Bernoulli Beams and beam-columns, Rayleigh Beams and Timoshenko Beams are discussed in great detail, as is the dynamics of accelerating beam-columns. The various operators pertinent to each system are summarized in a table at the end of the chapter.

The general free vibration of one-dimensional continua is established in Chapter 10 and applied to the various continua discussed in Chapter 9. The operator notation introduced earlier permits the student to perceive the vibrations problem for continua as merely an extension of that discussed for discrete systems. Case studies are presented for various rods and beams, allowing for a direct quantitative evaluation of the one degree of freedom approximation assumed in the first five chapters. It further allows for direct comparison of the effectiveness and validity of the various beam theories. Properties of the modal functions, including the scalar product, normalization and orthogonality are established. The latter is then used in the evaluation of amplitudes and phase angles. Forced vibration of one-dimensional continua is discussed in Chapter 11. The justification for generalized Fourier Series representation of the response is established and modal analysis is applied to the structures of interest under various loading conditions.

The material covered in this text is suitable for a two-semester sequence or a one-semester course. The instructor can choose appropriate chapters and/or sections to suit the level, breadth and length of the particular course being taught.

To close, I would like to thank Professor Haim Baruh, Professor Andrew Norris, Ms. Pamela Carabetta, Mr. Lucian Iorga and Ms. Meghan Suchorsky, all of Rutgers University, for reading various portions of the manuscript and offering helpful comments and valuable suggestions. I would also like to express my gratitude to Ms. Carabetta for preparing the index. I wish to thank Glen and Maria Hurd for their time, effort and patience in producing the many excellent drawings for this volume. Finally, I wish to thank all of those students, past and present, who encouraged me to write this book.

William J. Bottega

Contents

6. DYNAMICS OF MULTI-DEGREE OF FREEDOM SYSTEMS 287

7. FREE VIBRATION OF MULTI-DEGREE OF FREEDOM SYSTEMS 341

10. FREE VIBRATION OF ONE-DIMENSIONAL CONTINUA   579

11. FORCED VIBRATION OF ONE-DIMENSIONAL CONTINUA   683

1

Preliminaries

The subject of mechanical vibrations is primarily concerned with the study of repeated, or nearly repeated, motion of mechanical systems. As engineers, we may be interested in avoiding excessive vibration in a structure, machine or vehicle, or we may wish to induce certain types of vibrations in a very precise manner. Stealth of a submarine is intimately connected to vibration suppression, and earthquakes can have dramatic effects on engineering structures. The response and durability of an engineering system to short duration, high intensity, loading is a function of the vibration characteristics of the system as well. Most of us have experienced the effects of vibrations in our everyday lives. We might feel undesirable vibrations in an automobile, or similarly while riding a bicycle. Likewise we might observe the vibration of an airplane wing while flying to or from a vacation, on our way to visiting friends or relatives, or while traveling on business. We all enjoy the benefit of vibrations when we have a conversation on a telephone or when we listen to music coming from our stereo speakers. Even our ability to speak stems from the vibrations of our vocal chords.

The earliest modern scientific studies of vibrations are generally attributed to Galileo, who examined the motion of the simple pendulum and the motion of strings. Based on his observations, Galileo arrived at a relationship between the length of the pendulum and its frequency and described the phenomenon of resonance, whereby a system exhibits large amplitude vibrations when excited at or near its natural frequency. Galileo also observed the dependence of the frequencies of a string on its length, mass density and tension, and made comparisons with the behavior of the pendulum. The fundamental understanding of mechanical vibrations was advanced in the centuries that followed, with the development and advancement of mechanics and the calculus. Investigations toward this end continue to the present day.

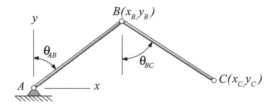

Figure 1.1 A two bar mechanism.

To study vibrations properly we must first understand and bring into context certain preliminary material that will be used throughout this text. Much of this material is presented in the present chapter, while other material of this type is introduced and discussed in subsequent chapters of this book as needed. The preliminary material presented in this chapter includes a discussion of the concepts of *degrees of freedom, mathematical modeling and equivalent systems*, and a review of *complex numbers*. The chapter finishes with an extensive review of *elementary dynamics*.

1.1 DEGREES OF FREEDOM

When we study the behavior of a system we need to choose parameters that describe the motion of that system and we must make sure that we are employing enough parameters to characterize the motion of interest completely. That is to say, if we know the values of these variables at a particular instant in time then we know the configuration of the system at that time. Consider, for example, the two (rigid) bar mechanism shown in Figure 1.1. Note that if we know the location of pins B and C at any time, then we know the configuration of the entire system at that time, since the lengths of the rigid rods are specified. That is, we know the location of every particle (e.g., point) of the system. It may be noted that the location of pins B and C may be characterized in many ways, some more efficient than others. We may, for example, describe their locations by their Cartesian coordinates (x_B, y_B) and (x_C, y_C), or we may describe their locations by the angular coordinates θ_{AB} and θ_{BC}, as indicated. Both sets of coordinates describe the configuration of the mechanism completely. A combination of the two sets of coordinates, say (x_B, y_B) and θ_{BC}, also describes the configuration of the system. It may be seen, however, that if we choose the angular coordinates then we only need two coordinates to describe the configuration of the system, while if we choose the Cartesian coordinates we need four, and if we choose the mixed set of coordinates we need three. We see that, for this particular system, the minimum number of coordinates needed to characterize its configuration completely is two. This minimum number of coordinates is referred to as the *degrees of freedom* of the system. We also note that the two angular coordinates may not be expressed in terms of one another. They are said to be *independent* in this regard. In general then, *the number of degrees of freedom of a system refers to the number of independent coord-*

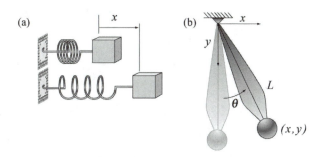

Figure 1.2 Sample single degree of freedom systems: (a) mass-spring system, (b) simple pendulum.

inates needed to describe its configuration at any time. Examples of one degree of freedom (1 d.o.f.) systems, two degree of freedom systems (2 d.o.f.), 'N' degree of freedom systems (N d.o.f. — where N is any integer) and continuous (infinite degree of freedom) systems are discussed in the remainder of this section.

Single Degree of Freedom Systems

Single degree of freedom systems are the simplest systems as they require only one independent coordinate to describe their configuration. The simplest example of a single degree of freedom system is the mass-spring system shown in Figure 1.2a. For the system shown, the coordinate x indicates the position of the mass measured relative to its position when the massless elastic spring is unstretched. If x is known as a function of time t, that is $x = x(t)$ is known, then the motion of the entire system is known as a function of time. Similarly, the simple pendulum shown in Figure 1.2b is also a one degree of freedom system since the motion of the entire system is known if the angular coordinate θ is known as a function of time. Note that while the position of the bob may be described by the two Cartesian coordinates, $x(t)$ and $y(t)$, these coordinates are not independent. That is, the Cartesian coordinates (x,y) of the bob are related by the *constraint equation*, $x^2 + y^2 = L^2$. Thus, if x is known then y is known and vice versa. Further, both $x(t)$ and $y(t)$ are known if $\theta(t)$ is known. In either case, only one coordinate is needed to characterize the configuration of the system. The system therefore has one degree of freedom.

Two Degree of Freedom Systems

The two bar mechanism described in the introduction of this section was identified as a two degree of freedom system. Two other examples include the two mass-spring system shown in Figure 1.3a and the double pendulum depicted in Figure 1.3b. In the first case, the configuration of the entire system is known if the position of mass m_1 is

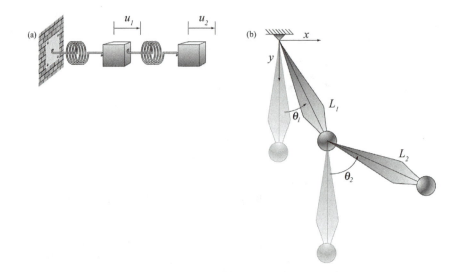

Figure 1.3 Sample two degree of freedom systems: (a) two-mass two-spring system, (b) double pendulum.

known and the position of mass m_2 is known. The positions are known if the coordinates u_1 and u_2 are known, where u_1 and u_2 represent the displacements of the respective masses from their equilibrium configurations. Likewise, the motion of the double pendulum is known if the angular displacements, θ_1 and θ_2, measured from the vertical equilibrium configurations of the masses, are known functions of time.

General Discrete Multi-Degree of Freedom Systems

Two degree of freedom systems are a special case of multi-degree of freedom systems (systems with more than 1 d.o.f.). Thus, let us consider general N degree of freedom systems, where N can take on any integer value as large as we like. Examples of such systems are the system comprised of N masses and $N+1$ springs shown in Figure 1.4a, and the compound pendulum consisting of N rods and N bobs shown in Figure 1.4b and the discrete model of an aircraft structure depicted in Figure 1.4c.

Continuous Systems

To this point we have been discussing discrete systems — systems that have a finite (or even infinite) number of masses separated by a finite distance. Continuous systems are systems whose mass is distributed continuously, typically over a finite domain. An example of a continuous system is the elastic beam shown in Figure 1.5. For the case of a linear beam (one for which the strain-displacement relation contains only first order terms of the displacement gradient), the transverse motion of the beam is known if the transverse deflection, $w(x,t)$, of each particle located at the coordinates $0 \le x \le L$ along the axis of the beam is known.

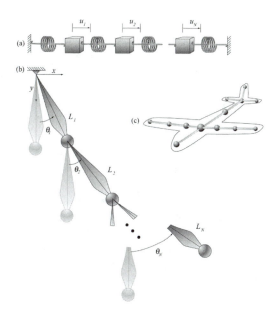

Figure 1.4 Sample N-degree of freedom systems: (a) N-mass $N+1$-spring system, (b) compound pendulum, (c) discrete model of aircraft structure.

The systems we described above are all examples of mathematical models that may represent actual systems. Each has its place depending, of course, on the particular system and the degree of accuracy required for the given application. In most cases there is a tradeoff between accuracy and facility of solution. Too simple a model may not capture the desired behavior at all. Too complex a model may not be practical to solve, or may yield results that are difficult to interpret. The modeler must choose the most suitable representation for the task at hand. In the next section we shall discuss how some complicated systems may be modeled as much simpler systems. Such simplifications can often capture dominant behavior for certain situations. We shall examine the vibrations of single degree of freedom systems in the next three chapters. The behavior of discrete multi-degree of freedom systems and continuous systems will then be examined in subsequent chapters. The richness of the behavior of such systems and the restrictions imposed by simplified representations will also be discussed.

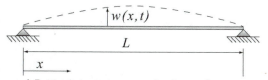

Figure 1.5 Elastic beam: an example of a continuous system.

1.2 EQUIVALENT SYSTEMS

In many applications the motion of a certain point of the system is of primary concern, and a single type of motion is dominant. For such cases certain simplifications may be made that allow us to approximate a higher degree of freedom system by a lower degree of freedom system, say a single degree of freedom system. Such simplifications shall be demonstrated in this section. Simplifications of this type approximate one type of motion (the lowest *mode*) of the many possible motions of discrete multi-degree of freedom systems and continuous systems. Thus, even if such a representation adequately represents a particular mode, it cannot capture all possible motion. Therefore, such approximations are only suitable for applications where the motion that is captured by the simplified model is dominant. Results of simplified models may be compared with those of multi-degree of freedom and continuous systems as they are studied in full in subsequent chapters of this text. The concept of equivalent systems will be introduced via several examples. In these examples, an equivalent stiffness is determined from a static deflection of a continuous system such as an elastic beam or rod. Since the inertia of the structure is neglected, such models are justifiable only when the mass of the beam or rod is much smaller than other masses of the system.

1.2.1 Extension/Contraction of Elastic Rods

Elastic rods possess an infinite number of degrees of freedom. Nevertheless, if the mass of the rod is small compared with other masses to which it is attached, and if we are interested only in the motion of a single point, say the unsupported end, the elastic rod may be modeled as an equivalent elastic spring as discussed below.

Consider a uniform elastic rod of length L, cross-sectional area A, and elastic modulus E. Let x correspond to the axial coordinate, and let the rod be fixed at the end $x = 0$ as shown in Figure 1.6. Further, let the rod be subjected to a tensile force of magnitude F_0 applied at the end $x = L$, as indicated. If $u(x)$ corresponds to the axial displacement of the cross section originally located at x then, for small axial strains $\varepsilon(x)$, the strain and displacement are related by

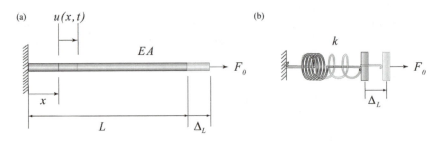

Figure 1.6 (a) Elastic rod subjected to axial load, (b) equivalent single degree of freedom system.

$$\varepsilon(x) = \frac{du}{dx} \tag{1.1}$$

The constitutive relation for an elastic rod in uniaxial tension/compression is

$$\sigma(x) = E\varepsilon(x) \tag{1.2}$$

where σ is the axial stress in the rod. It follows from Eqs. (1.1) and (1.2) that the resultant membrane force, $N(x)$, acting over the cross section at x is given by

$$N(x) = \sigma(x)A = EA\frac{du}{dx} \tag{1.3}$$

Consideration of the equilibrium of a differential volume element of the rod yields its governing equation as

$$EA\frac{d^2u}{dx^2} = n(x) \tag{1.4}$$

where $n(x)$ represents a distributed axial load. For the present problem $n(x) = 0$, and the boundary conditions for the rod of Figure 1.6 are stated mathematically as

$$u(0) = 0, \qquad EA\frac{du}{dx}\bigg|_{x=L} = F_0 \tag{1.5}$$

Integrating Eq. (1.4), with $n(x) = 0$, imposing the boundary conditions (1.5), and evaluating the resulting expression at $x = L$ gives the axial deflection of the loaded end, Δ_L, as

$$\Delta_L = \frac{F_0 L}{EA} \tag{1.6}$$

Rearranging Eq. (1.6) then gives the relation

$$F_0 = k\Delta_L \tag{1.7}$$

where

$$k = \frac{EA}{L} \tag{1.8}$$

Equation (1.7) may be seen to be the form of the constitutive relation for a linear spring. Thus, if we are only interested in the motion of the free end of the rod, and if the mass of the rod is negligible, then the elastic rod may be modeled as an equivalent spring whose stiffness is given by Eq. (1.8). In this way, the continuous system (the elastic rod) is modeled as an equivalent single degree of freedom system.

1.2.2 Bending of Elastic Beams

As discussed earlier, continuous systems such as elastic beams have an infinite number of degrees of freedom. Yet, under appropriate circumstances (loading type, kinematical constraints, mass ratios, etc.) a certain type of motion may be dominant. Further, as a simple model may be desirable and still capture important behavior, we next consider several examples of elastic beams modeled as equivalent single degree of freedom systems.

The Cantilever Beam

Consider a uniform elastic beam of length L, cross-sectional area moment of inertia I and elastic modulus E that is supported as shown in Figure 1.7a. Let the beam be subjected to a transverse point load of magnitude P_0 applied on its free end, and let Δ_L correspond to the deflection of that point as indicated. Suppose now that we are only interested in the motion of the point of the beam under the load, and that the inertia of the beam is negligible compared with other masses that the beam will ultimately be connected to. If we wish to construct an equivalent single degree of freedom system for the beam then we must seek a relation between the applied load and the load point deflection of the form

$$P_0 = k\Delta_L \tag{1.9}$$

where the parameter k is an *equivalent stiffness*. That is, we wish to treat the beam as an equivalent elastic spring of stiffness k as shown in Figure 1.7b. To find k, let us consider the static deflection of the beam due to the applied point load. If $w(x)$ corresponds to the deflection of the centerline of the beam at the axial coordinate x, then we know from elementary beam theory that the governing equation for the transverse motion of an elastic beam subjected to a distributed transverse load of intensity $q(x)$ is of the form

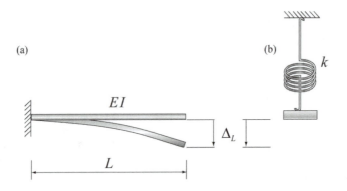

Figure 1.7 (a) Cantilever beam, (b) equivalent single degree of freedom system.

$$EI\frac{d^4w}{dx^4} = q(x) \tag{1.10}$$

where $q(x) = 0$ for the case under consideration. The boundary conditions for a beam that is clamped at the origin and loaded by a point load at its free end are

$$w(0) = \frac{dw}{dx}\bigg|_{x=0} = 0, \quad EI\frac{d^2w}{dx^2}\bigg|_{x=L} = 0, \quad EI\frac{d^3w}{dx^3}\bigg|_{x=L} = -P_0 \tag{1.11}$$

Integrating Eq. (1.10) with $q(x) = 0$, imposing the boundary conditions of Eq. (1.11) and evaluating the resulting solution at $x = L$ gives the load point deflection

$$\Delta_L \equiv w(L) = \frac{P_0}{EI}\frac{L^3}{3} \tag{1.12}$$

Solving Eq. (1.12) for P_0 gives the relation

$$P_0 = k\Delta_L \tag{1.13}$$

where

$$k = \frac{3EI}{L^3} \tag{1.14}$$

We have thus found the equivalent stiffness (i.e., the stiffness of an equivalent spring) for a cantilever beam loaded at its free edge by a transverse point load. We shall next use this result to establish mathematical models for selected sample structures.

Side-Sway of Structures

In the previous section we found the equivalent stiffness of a cantilever beam as pertains to the motion of its free end. In this section we shall employ that stiffness in the construction of a dynamic single degree of freedom model of a one-story structure undergoing side-sway motion as may occur, for example, during an earthquake.

Consider a structure consisting of four identical elastic columns supporting an effectively rigid roof of mass m, as shown in Figure 1.8a. Let the columns, each of length L and bending stiffness EI, be embedded in a rigid foundation as indicated. Further, let the mass of the roof be much larger than the mass of the columns. We shall consider two types of connections of the columns with the roof, pinned and clamped/embedded.

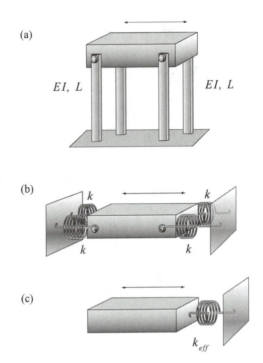

Figure 1.8 Side-sway of one-story structure with pinned connections at roof: (a) representative structure, (b) roof with columns represented as equivalent springs, (c) equivalent system.

Pinned Connections

Let the columns be connected to the roof of the structure as shown in Figure 1.8a. If we are interested in side-sway motion of the structure as may occur during earthquakes, and if the mass of the columns is negligible compared with the mass of the roof, then the columns may be treated as cantilever beams as discussed earlier. For this purpose, the structure may be modeled as four equivalent springs, each of stiffness k as given by Eq. (1.14) and shown in Figure 1.8b. This, in turn, is equivalent to a mass attached to a single effective spring of stiffness k_{eff} (see Section 1.3), given by

$$k_{eff} = 4k = \frac{12EI}{L^3}$$

(1.15)

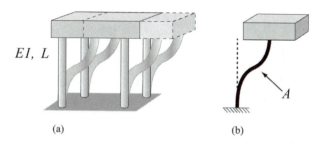

EI, L

(a)

(b)

Figure 1.9 Side-sway of one-story structure with clamped connections at roof: (a) structure in motion, (b) deflection of column showing inflection point A.

Clamped Connections

If the columns are embedded (framed) into the roof structure, as shown in Figure 1.9a, the deflections differ from those for the pinned case. One way to determine the equivalent stiffness of a beam that is clamped-fixed at one end and clamped-free at the other is to solve Eq. (1.10) with $q(x) = 0$ subject to the boundary conditions

$$w(0) = \frac{dw}{dx}\bigg|_{x=0} = 0, \quad \frac{dw}{dx}\bigg|_{x=L} = 0, \quad EI\frac{d^3 w}{dx^3}\bigg|_{x=L} = -P_0 \qquad (1.16)$$

in lieu of the boundary conditions of Eq. (1.11). It may be seen that only the last condition differs from the previous case. This approach, however, will be left as an exercise (Problem 1.6). Instead, we shall use the results for the cantilever beam to obtain the desired result. This may be done if we realize that, due to the anti-symmetry of the deformation, the deflection of the column for the present case possesses an inflection point at the center of the span (point A, Figure 1.9b). Since, by definition, the curvature and hence the bending moment vanishes at an inflection point such a point is equivalent to a pin joint. Thus, each of the columns for the structure under consideration may be viewed as two cantilever beams of length $L/2$ that are connected by a pin at the center of the span. The total deflection of the roof will then be twice that of the inflection point, as indicated. Therefore, letting $L \to L/2$ and $\Delta_L \to \Delta_L/2$ in Eq. (1.13) gives, for a single clamped-fixed/clamped-free column, that

$$k = \frac{12EI}{L^3} \qquad (1.17)$$

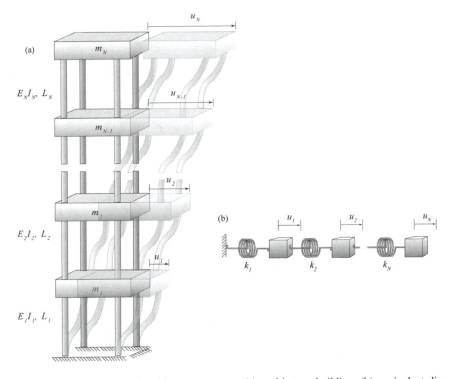

Figure 1.10 Side-sway of multi-story structure: (a) multi-story building, (b) equivalent discrete system.

As for the pinned roof structure considered earlier, the four equivalent springs for the present structure act in parallel (see Section 1.3) and are thus equivalent to a single effective spring of stiffness

$$k_{eff} = 4k = \frac{48EI}{L^3} \tag{1.18}$$

Note that since, for this case, the columns are embedded in the roof and hence provide greater resistance to bending and therefore to lateral translation of the roof than for the pinned case, the effective stiffness is higher (by a factor of 4) than the stiffness for the pinned case.

Multi-Story Buildings

Consider the N-story building shown in Figure 1.10a. Let each floor of the building be connected by four columns below it and four columns above it, with the obvious exception that the roof (floor number N) has no columns above it. Let each floor, numbered $j = 1, 2, ..., N$ from bottom to top, possess mass m_j and let the ends of the columns be embedded into the floors. The ground floor, $j = 0$, is fixed to the ground. Further, let each column that connects floor j with floor $j - 1$ possess bending stiffness $E_j I_j$, as indicated, where E_j and I_j respectively correspond to the elastic modulus and area moment of inertia of the column. If we are interested in side-sway motion of the building, and if the masses of the columns are negligible compared to those of the floors, then the building may be represented by the equivalent discrete N – degree of freedom system shown in Figure 1.10b. It follows from our discussions of a single story building with end-embedded columns that the equivalent stiffness of the jth spring may be obtained directly from Eq. (1.18). Hence,

$$k_j = \frac{48 E_j I_j}{L^3} \quad (j = 1, 2, ... N) \tag{1.19}$$

The Simply Supported Beam

We next construct an equivalent single degree of freedom system for a simply supported beam subjected to a transverse point load applied at the midpoint of the span. The equivalent stiffness of this structure can, of course, be found by solving Eq. (1.10) subject to the appropriate boundary conditions. However, we shall use the equivalent stiffness of the cantilever beam, Eq. (1.15), as a shortcut to establish the equivalent stiffness of the present structure, as was done earlier for the modeling of side-sway of a single story building. Toward this end, let us consider a simply supported beam of length $\tilde{L} = 2L$ and bending stiffness EI, and let the beam be subjected to a transverse point load of magnitude $Q_0 = 2P_0$ applied at the center of the span as shown in Figure 1.11a. Consideration of the differential beam element on the interval $-dx/2 \leq x \leq dx/2$ (Figure 1.11b) shows that the problem is equivalent to that of half of the structure on $0 \leq x \leq L$ subjected to a transverse point load of magnitude P_0 acting at the edge $x = 0$ (Figure 1.11c). This, in turn may be seen to be equivalent to the problem of the cantilever beam shown in Figure 1.11d. Next, let Δ_0 correspond to the deflection of the cantilever beam under the point load P_0. It may be seen that Δ_0 also corresponds to the center-span deflection of the beam of Figure 1.11a. It then follows form Eq. (1.12) that

$$\Delta_0 = \frac{P_0 L^3}{3EI} = \frac{Q_0 L^3}{6EI} \tag{1.20}$$

and hence that

$$Q_0 = k \Delta_0 \tag{1.21}$$

where

$$k = \frac{6EI}{L^3} = \frac{48EI}{\tilde{L}^3} \tag{1.22}$$

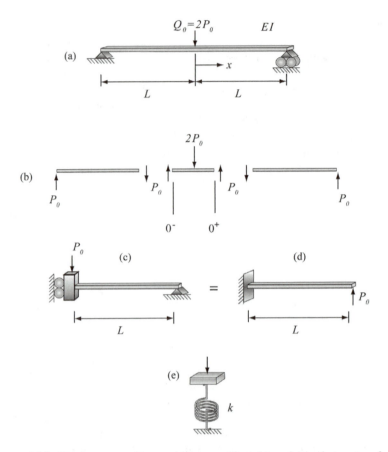

Figure 1.11 Simply supported beam: (a) beam subjected to point load at center of span, (b) free-body diagram of segmented beam, (c) half span problem, (d) equivalent cantilever beam, (e) equivalent single degree of freedom system.

Compound Systems

In many applications a beam may be attached to another structure, or to compliant supports. The effect of the second structure, or the compliance of the supports, may often be represented as a linear elastic spring, in the manner discussed throughout this section. As before, and under similar circumstances, we may be interested in representing the primary beam as an equivalent linear spring, and ultimately the combined structure of the beam and spring as a single equivalent spring. We shall do this for two related cases as examples.

We next consider and compare the two related systems shown in Figures 1.12a and 1.12b. In each case the system consists of a simply supported elastic beam to which a spring of stiffness k_s is attached at the center of the span. In the first case the other end of the spring is attached to a rigid foundation while a point load is applied to the beam at center span (Figure 1.12a), while in the second case the bottom edge of

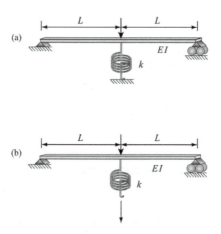

Figure 1.12 Compound system of elastic beam and spring: (a) fixed spring, (b) loaded spring.

the spring is free to translate and a point load is applied to that edge (Figure 1.12b).

Simply Supported Beam Attached to a Fixed Spring

Since we are interested in the vertical motion of the center-span of the beam we may model the beam as an equivalent linear spring. It follows that the effective stiffness, k_{beam}, of the equivalent spring for the beam is given by Eq. (1.22). The stiffness of the compound system consisting of the two springs may then be obtained by superposition, as shown in Figure 1.13. For this case, the springs are seen to act in parallel and thus to act as a single equivalent spring whose stiffness, k_{eq}, is the sum of the stiffnesses of the two parallel springs (see Section 1.3). We therefore have that

$$k_{eq} = k_{beam} + k_s = \frac{6EI}{L^3} + k_s \qquad (1.23)$$

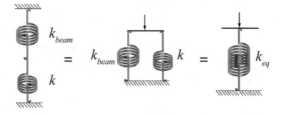

Figure 1.13 Equivalent system for beam and spring of Figure 1.12a.

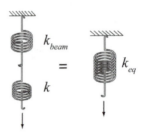

Figure 1.14 Equivalent system for beam and spring of Figure 1.12b.

Simply Supported Beam Attached to a Loaded Spring
Let us again consider a simply supported elastic beam attached to a linear
spring of stiffness k_s. In this case, however, a point load is applied to the free
edge of the spring (Figure 1.12b). Once again, if we are only interested in the
motion of the point of the beam that lies directly over the point load (the center-
span of the beam), we may model the beam as an equivalent linear spring as we
did for the previous case. Using superposition, as shown in Figure 1.14, it may
be seen that the two springs act in series and hence that the effect of the two
springs is equivalent to that of a single equivalent spring. As shown in Section
1.3, the stiffness of the equivalent spring representing the compound system of
the two springs in series is given by

$$k_{eq} = \frac{1}{\left(1/k_{beam}\right)+\left(1/k_s\right)} = \frac{1}{\left(L^3/6EI\right)+\left(1/k_s\right)} \qquad (1.24)$$

1.2.3 Torsion of Elastic Rods

In Section 1.2.1 we examined axial motion of elastic rods and the bending of elastic
beams. In each case we found the stiffness of an equivalent elastic spring for situations where we would be concerned with axial or transverse motion of a single point
of the structure. This stiffness could then be used in the construction of a simpler,
single degree of freedom system representation for situations where the mass of the
rod or beam is much smaller than other masses of the system. An example of the use
of such a representation was in the side-sway motion of a roof structure. In this section we shall determine the analogous stiffness of an equivalent *torsional* spring representing the rotational resistance of an elastic rod of circular cross section. In this
regard, such a model will be applicable in situations where we are interested in small
rotational motion of a single cross section at some point along the axis of the rod, say
at its free end, and when the mass moment of inertia of the rod is small compared
with other mass moments of the system.

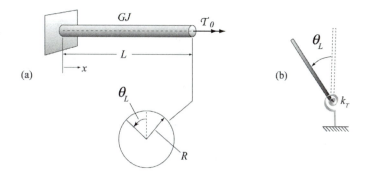

Figure 1.15 Torsion of elastic rod: (a) elastic rod, (b) equivalent 1 d.o.f. system.

Single Rod

Consider a long thin elastic rod of circular cross section. Let the rod be of length L, shear modulus G, and polar moment of inertia J. Further, let the rod be fixed at one end and be subjected to a twisting moment (torque) of magnitude T_0 at its free end, as shown in Figure 1.15a. Let a coordinate x originate at the fixed end of the rod and run along the axis of the rod, and let $\theta(x)$ correspond to the rotation of the cross section located at coordinate x as indicated. The governing equilibrium equation for torsion of a uniform elastic rod subjected to a distributed torque (torque per unit length) $\mu(x)$ is given by

$$GJ\frac{d^2\theta}{dx^2} = \mu(x) \tag{1.25}$$

where $\mu(x) = 0$ for the present case. The boundary conditions for the case under consideration are

$$\theta(L) = 0, \quad GJ\frac{d\theta}{dx}\bigg|_{\theta=L} = T_0 \tag{1.26}$$

Integrating Eq. (1.25), with $\mu = 0$, imposing the boundary conditions defined in Eq. (1.26), and evaluating the resulting expression at $x = L$ gives the rotation at the free end of the rod as

$$\theta_L \equiv \theta(L) = T_0\frac{L}{GJ} \tag{1.27}$$

or

$$T_0 = k_T\theta_L \tag{1.28}$$

where

$$k_T = \frac{GJ}{L} \tag{1.29}$$

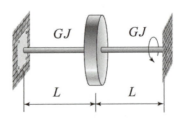

Figure 1.16 Rigid disk at junction of two elastic rods.

The parameter k_T is the stiffness of an equivalent torsional spring (Figure 1.15b) simulating the motion of the free edge of an elastic rod subjected to a torque at that edge and fixed at the other edge (Figure 1.15b). This model will be used in Chapters 2 – 8 for applications where the moment of inertia of the rod is small compared with other mass moments of the system.

Compound Systems

In practice, the supports to which an elastic rod is secured have a certain degree of compliance. If we wish to include this effect, the support may be modeled as an equivalent torsional spring. In addition, many mechanical systems are comprised of several connected elastic rods. If we are interested in the motion of a single point, and if the masses of the rods are small compared with other masses of the system, then we may model the system as an equivalent single degree of freedom system in a manner similar to that which was done for beams. We do this for two sample systems in this section.

As a first example, suppose we are interested in the motion of the rigid disk connected to the junction of two elastic rods such that all axes of revolution are coincident (Figure 1.16). If the masses of the rods are small compared to that of the disk we may treat the resistance (restoring moment) imparted by the two elastic rods as that due to equivalent torsional springs. The effect of the two rods fixed at their far ends is then equivalent to a single torsional spring whose stiffness is the sum of the stiffnesses of the individual rods as given by Eq. (1.29). (See also the discussion of parallel springs in Section 1.3.1.) Hence, the two rods may be represented as a single torsional spring of stiffness

$$k_T^{(eq)} = k_{T_1} + k_{T_2} = \frac{G_1 J_1}{L_1} + \frac{G_2 J_2}{L_2} \tag{1.30}$$

As another example, let us consider the effect of a compliant support of torsional stiffness k_{Ts}, on the rotation of the rigid disk at the free end of an elastic rod of torsional stiffness GJ and length L (Figure 1.17). The equivalent stiffness for this system is found from an analogous argument with that of the beam attached to a load-

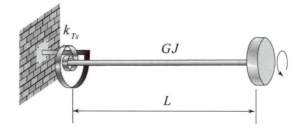

Figure 1.17 Elastic rod with compliant support.

ed spring. (See also the discussion of springs connected in series in Section 1.3.2.) The combined effect of the rod and compliant support on the motion of a rigid disk at the free end of the rod is then that of a single torsional spring of stiffness

$$k_T^{(eq)} = \frac{1}{(GJ/L) + (1/k_{Ts})} \tag{1.31}$$

Similar expressions may be found for the effect of two elastic rods connected in series as shown in Figure 1.18.

Finally, consider the multi-component shaft of Figure 1.19. The torsional motion of the discrete system comprised of N rigid disks connected to N elastic shafts aligned sequentially, as shown, is directly analogous to the side-sway motion of a multi-story building considered in Section 1.2.2. Thus, each rod may be modeled as an equivalent torsional spring, with the corresponding stiffnesses given by

$$k_T^{(j)} = \frac{G_j J_j}{L_j} \quad (j = 1, 2, ..., N) \tag{1.32}$$

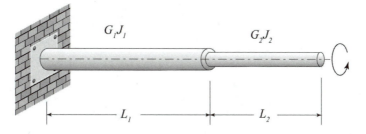

Figure 1.18 Elastic rods in series.

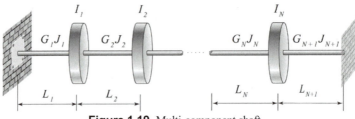

Figure 1.19 Multi-component shaft.

1.2.4 Floating Bodies

If we push down on a floating body we observe that the body deflects into the fluid. We also observe that the fluid exerts a resistance to the applied force that restricts the extent of the deflection of the floating body. If we subsequently release the body we will observe that it returns to its original position, first bobbing about that position before eventually coming to rest. The fluid thus exerts a restoring force on the floating body and may, under appropriate circumstances, be treated as an equivalent elastic spring. We next compute the stiffness of that equivalent spring.

Consider the vertical motion of a rigid body of mass m that floats in a fluid of mass density ρ_f, as shown in Figure 1.20. We shall not consider wobbling of the body here. That will be left to the chapters concerned with multi-degree of freedom systems (Chapters 6–9). We wish to model the system shown in the figure as an equivalent mass-spring system. We thus wish to determine the stiffness provided by the buoyant effects of the fluid, say water.

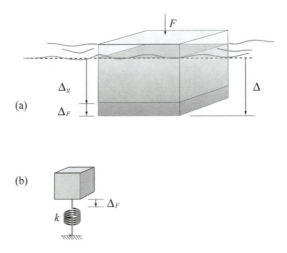

Figure 1.20 (a) Floating body, (b) equivalent system.

Let Δ_g be the deflection of the body due to gravity and thus correspond to the initial equilibrium configuration of the bottom surface of the body relative to the free surface of the fluid as indicated. Let Δ_F represent the additional deflection of the body due to a force F that is subsequently applied along a vertical axis through the centroid of the body (and thus does not cause any rotation of the body). Let us first determine Δ_g.

Archimedes Principle tells us that, at equilibrium, the weight of the displaced water is equal to the weight of the body. We also know, from fluid statics, that the pressure acting on the surface of the body varies linearly with depth from the free surface. Given this, the free-body diagram for the floating body under its own weight alone is as shown in Figure 1.21. Letting y correspond to the depth coordinate measured from the stationary surface of the fluid, and g represent the gravitational acceleration, then the (gage) pressure, p, is given by

$$p = \rho_f g\, y \qquad (1.33)$$

The buoyant force, F_{bg}, the resultant force acting on the bottom surface ($y = \Delta_g$) of the body is thus given by

$$F_{bg} = p\,A = \rho_f g\, \Delta_g A \qquad (1.34)$$

where A is the area of the bottom surface of the body. Now, the balance of forces in the vertical direction, $\Sigma F_y = 0$, gives

$$\rho_f g\, \Delta_g A - mg = 0 \qquad (1.35)$$

which is seen to be a statement of Archimedes Principle. Solving for the deflection, Δ_g, gives

$$\Delta_g = \frac{m}{\rho_f A} \qquad (1.36)$$

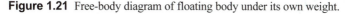

Figure 1.21 Free-body diagram of floating body under its own weight.

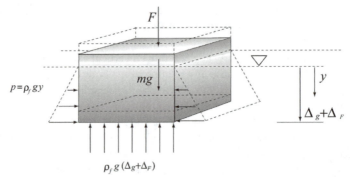

Figure 1.22 Free-body diagram of floating body subjected to an applied force.

Let us next determine the additional deflection due to the applied force F. The free-body diagram for this case is shown in Figure 1.22. For this case, the pressure exerted on the bottom surface of the body is given by

$$p = \rho_f g \left(\Delta_g + \Delta_F \right) \tag{1.37}$$

where Δ_F is the additional deflection due to the applied force F, as indicated. The resultant force acting on the bottom surface of the body is then given by

$$F_{buoy} = F_{bg} + F_{bF} \tag{1.38}$$

where F_{bg} is given by Eq. (1.34), and

$$F_{bF} = \rho_f g A \Delta_F \tag{1.39}$$

The force F_{bF} is evidently the *restoring force* exerted by the fluid on the body as it is moved away from its initial equilibrium configuration. The effective stiffness of the fluid, k, is then given by the coefficient of the associated defection appearing in Eq. (1.39). Hence,

$$k = \rho_f g A \tag{1.40}$$

1.2.5 The Viscous Damper

A simple type of dissipation mechanism typically considered in vibrations studies is that of viscous damping. Though damping may be introduced in a variety of ways, the following model captures the characteristics of a standard viscous damper.

Consider a long cylindrical rod of radius R_i that is immersed in a Newtonian fluid of viscosity μ that is contained within a cylinder of radius R_o possessing rigid walls. Let the axis of the rod and that of the cylinder be coincident, as shown in Figure 1.23, and let the rod be moving through the fluid with velocity v_0 in the axial di-

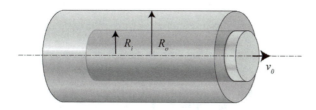

Figure 1.23 Rod moving through viscous fluid contained within cylinder.

rection, as indicated. For such a fluid the shear stress, τ, is proportional to the rate of deformation. If we define the z-axis to be coincident with the axes of the cylinder and the rod, and let r be the radial coordinate measured from this axis, then the shear stress may be expressed as

$$\tau = \frac{\mu}{2}\left(\frac{\partial v_r}{\partial z} + \frac{\partial v_z}{\partial r}\right)$$

where v_r and v_z represent the radial and axial components of the velocity of the fluid. If no slip conditions are imposed on the fluid at the rod and cylinder walls, the fluid velocity profile varies logarithmically, as indicated in Figure 1.24, such that

$$v_z(r) = v_0 \frac{\left(\ln R_o - \ln r\right)}{\ln\left(R_0 / R_i\right)}, \quad v_r = 0$$

The shear stress acting on the surface of the rod is then seen to be given by

$$\tau = \frac{-\mu/R_i}{\ln\left(R_o/R_i\right)} v_0$$

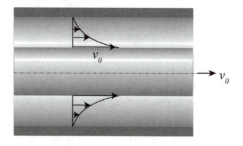

Figure 1.24 Flow field of damper fluid.

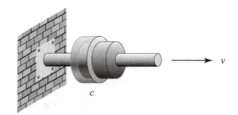

Figure 1.25 Representation of viscous damper.

It follows that the resultant force, F_d, applied to the rod by the viscous fluid is given by

$$F_d = \tau A = -cv_0$$

where

$$c = \frac{\mu A / R_i}{\ln(R_o / R_i)}$$

and A is the surface area of the rod.

As demonstrated by the above example, the force applied to the body by the linear viscous fluid damper opposes the motion of the body and is linearly proportional to the speed, v, at which the body travels *relative* to the damper. Hence, in general, the damping force is

$$F_d = -cv \tag{1.41}$$

where the constant c is referred to as the *damping coefficient*. A viscous damper is typically represented schematically as a piston or dashpot (Figure 1.25).

1.2.6 Aero/Hydrodynamic Damping (Drag)

Drag is a retarding force exerted on a body as it moves through a fluid medium such as air or water, as shown in Figure 1.26. It is generally comprised of both viscous and pressure effects. However, for incompressible flows of classical fluids at very low Reynolds numbers,

$$Re \equiv \frac{\rho v L}{\mu} \leq 1$$

where ρ and μ are respectively the (constant) mass density and (constant) viscosity of the fluid, v is the magnitude of the velocity of the fluid relative to the body and L is a characteristic length of the body, the drag force exerted on the body is predominantly due to friction and is linearly proportional to the velocity. Thus, for such flows,

$$F_D = -cv \tag{1.42}$$

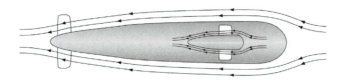

Figure 1.26 Body moving through fluid medium.

where, for a sphere,

$$c = 6\pi \mu R$$

and R is the diameter of the sphere. Equation (1.42) is seen to be of identical form to Eq. (1.41). Thus, from a vibrations perspective, the low *Re* drag force and the viscous force affect the system in the same way. This is not surprising since, for low Reynolds numbers, the drag force is predominantly frictional. For larger Reynolds numbers the drag force depends on the velocity in a nonlinear manner, with the specific form depending on the range of Reynolds number, and Eq. (1.42) is no longer valid.

1.3 SPRINGS CONNECTED IN PARALLEL AND IN SERIES

When linear springs are connected to one another and viewed collectively, the displacement of the outermost points is related to the applied load in a manner identical to that of a single spring. That is, when viewed collectively, the system of linear springs behaves as a single equivalent linear spring. There are two fundamental ways in which linear elastic springs may be connected: (a) in parallel (Figure 1.27a), and (b) in series (Figure 1.27b). Other arrangements correspond to combinations of these two fundamental configurations. In this section we shall obtain the effective stiffness of the equivalent springs corresponding to these two fundamental configurations. We begin with a discussion of parallel springs.

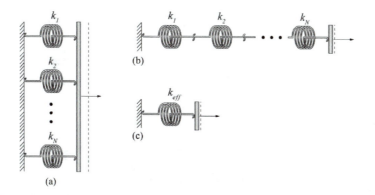

Figure 1.27 Compound springs: (a) springs in parallel, (b) springs in series, (c) equivalent system.

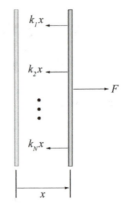

Figure 1.28 Free-body diagram for springs in parallel.

1.3.1 Springs in Parallel

Consider a rigid plate attached to any number of elastic springs, say N, with the other end of the springs connected to a fixed rigid wall as shown in Figure 1.27a. Let the stiffnesses of the springs that comprise the system be respectively designated k_j ($j = 1$, 2, ..., N) as indicated. If the plate is displaced a distance Δ to the right (or left) then each spring exerts a restoring force of the form $F_j = k_j \Delta$ ($j = 1, 2, ..., N$) acting on the plate, as shown in the free-body diagram depicted in Figure 1.28. The total restoring force, that is the resultant of all the forces exerted by the springs on the plate, is then the sum of the individual restoring forces. Thus,

$$F = \sum_{j=1}^{N} F_j = \sum_{j=1}^{N} k_j \Delta = k_{\mathit{eff}} \Delta \qquad (1.43)$$

where

$$k_{\mathit{eff}} = \sum_{j=1}^{N} k_j \qquad (1.44)$$

The system of parallel springs therefore behaves as a single spring whose stiffness is equal to the sum of the stiffnesses of the individual springs that comprise the system.

1.3.2 Springs in Series

Consider a system of N springs connected end to end (i.e., in series), and let one end of spring number 1 be attached to a rigid wall as shown in Figure 1.27b. In addition, let an external force P be applied to the free end of spring number N. Further, let k_j ($j = 1, 2, ..., N$) correspond to the stiffness of spring number j, and let Δ_j represent the "stretch" (the relative displacement between the two ends) in that spring. Note that since spring 1 is fixed at one end, the stretch in that particular spring, Δ_1, is also the absolute displacement of the joint connecting spring 1 and spring 2. Let Δ^* represent the absolute displacement of the free end of the system (i.e., the displacement of joint number N measured with respect to its rest position), and thus the displacement of the applied force P. The displacement Δ^* then also represents the total stretch in the system, or the stretch of an equivalent spring with effective stiffness k_{eff}. We wish to determine k_{eff} such that the relationship between the applied force and its displacement is of the form

$$P = k_{eff} \Delta^* \tag{1.45}$$

To do this, let us first isolate each spring in the system and indicate the forces that act on them as shown in Figure 1.29. It then follows from Newton's Third Law applied at each joint, and the implicit assumption that the springs are massless, that

$$k_1 \Delta_1 = k_2 \Delta_2 = \ldots = k_N \Delta_N = P \tag{1.46}$$

Dividing through by the stiffness of each individual spring then gives the relations

$$\Delta_j = \frac{P}{k_j} \quad (j = 1, 2, \ldots, N) \tag{1.47}$$

Now, as discussed earlier, the deflection of the load is equal to the total stretch in the system. Further, the total stretch of the system is equal to the sum of the individual stretches. Hence,

$$\Delta^* = \Delta_1 + \Delta_2 + \ldots + \Delta_N = \sum_{j=1}^{N} \Delta_j \tag{1.48}$$

Substitution of each of Eqs. (1.47) into Eq. (1.48) gives the relation

$$\Delta^* = \frac{P}{k_1} + \frac{P}{k_2} + \ldots + \frac{P}{k_N} = P \sum_{j=1}^{N} \frac{1}{k_j} \tag{1.49}$$

or

$$\Delta^* = \frac{P}{k_{eff}} \tag{1.50}$$

where

Figure 1.29 Free-body diagram for springs in series.

$$\frac{1}{k_{eff}} = \frac{1}{k_1} + \frac{1}{k_2} + ... + \frac{1}{k_N} = \sum_{j=1}^{N} \frac{1}{k_j} \tag{1.51}$$

Equation (1.51) gives the relation between the effective stiffness of the single equivalent spring and the stiffnesses of the springs that comprise the system.

1.4 A BRIEF REVIEW OF COMPLEX NUMBERS

During the course of our study of vibrations we shall find that many pertinent functions and solutions may be expressed more generally and more compactly using complex representation. Likewise, solutions to many vibrations problems are facilitated by the use of complex numbers. In this section we briefly review complex numbers and derive certain identities that will be used throughout this text.

Let us consider numbers of the form

$$z = x + iy \tag{1.52}$$

where $i \equiv \sqrt{-1}$. The number x is said to be the *real part* of the *complex number z*, and y is said to be the *imaginary part* of z. Alternatively, we may write

$$x = \text{Re}(z), \quad y = \text{Im}(z) \tag{1.53}$$

The *complex conjugate* of z, which we shall denote as z^c, is defined as

$$z^c \equiv x - iy \tag{1.54}$$

The product of a complex number and its conjugate may be seen to have the property

$$z\,z^c = x^2 + y^2 = \|z\|^2 \tag{1.55}$$

where $\|z\|$ is called the *magnitude* of the complex number z. Alternatively, we may write

$$\text{mag}(z) = \|z\| = \sqrt{x^2 + y^2} = \sqrt{z\,z^c} \tag{1.56}$$

The complex number z may be expressed in vector form as $z = (x, y)$, and may be represented graphically in the *complex plane* as shown in Figure 1.30. We then define the magnitude and argument of z as the radius or length, r, of the line from the origin to the point (x, y) and the angle, ψ, that this line makes with the x-axis, respectively. Hence,

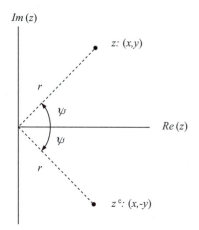

Figure 1.30 Graphical representation of a complex number and its conjugate.

$$r = \sqrt{x^2 + y^2} = \text{mag}(z) = \| z \| = \sqrt{z\,z^c} \qquad (1.57)$$

and

$$\psi = \tan^{-1}\left(y/x\right) = \arg(z) \qquad (1.58)$$

Both z and z^c are displayed in Figure 1.30 where it is seen that z^c is the reflection of z through the real axis. It is also seen from Figure 1.30 that a complex number and its conjugate may be expressed in terms of its magnitude, r, and its argument, ψ, as

$$\begin{aligned} z &= r(\cos\psi + i\sin\psi) \\ z^c &= r(\cos\psi - i\sin\psi) \end{aligned} \qquad (1.59)$$

Note that if $\psi = \omega\,t$, where the parameter t is the time, then ω corresponds to *angular frequency*, a quantity that will be central to our studies of vibrating systems. In this case, ω is the angular rate (angular velocity — see Section 1.5) at which the radial line segment connecting the origin and point z (i.e., the "vector" z) rotates about an axis through the origin and perpendicular to the complex plane.

The forms given by Eqs. (1.59) will lead us to further identities that will be useful to us in our study of vibrations. Toward this end, let us first recall the series representation for $\cos\psi$ and $\sin\psi$,

$$\begin{aligned} \cos\psi &= 1 - \frac{\psi^2}{2!} + \frac{\psi^4}{4!} - \cdots \\ \sin\psi &= \psi - \frac{\psi^3}{3!} + \frac{\psi^5}{5} - \cdots \end{aligned} \qquad (1.60)$$

Let us next take the complex sum of the two series as follows,

$$\cos\psi + i\sin\psi = \left(1 - \frac{\psi^2}{2!} + \cdots\right) + i\left(\psi - \frac{\psi^3}{3!} + \cdots\right)$$

$$= 1 + \frac{(i\psi)}{1!} + \frac{(i\psi)^2}{2!} + \frac{(i\psi)^3}{3!} + \cdots$$

$$= e^{i\psi}$$

Similarly, letting $\psi \to -\psi$ in the above expressions gives the identity

$$\cos\psi - i\sin\psi = e^{-i\psi}$$

Combining the above two results gives *Euler's Formula,*

$$e^{\pm i\psi} = \cos\psi \pm \sin\psi \tag{1.61}$$

Complex numbers and their conjugates may be written in useful forms using Euler's Formula. Substitution of Eq. (1.61) into Eqs. (1.59) gives the summary of the various forms for a complex number and its conjugate,

$$z = x + iy = r(\cos\psi + i\sin\psi) = re^{i\psi}$$
$$z^c = x - iy = r(\cos\psi - i\sin\psi) = re^{-i\psi} \tag{1.62}$$

Lastly, letting $\psi \to -i\psi$ in Eq. (1.60) and paralleling the development of Eq. (1.61) gives the analog of Euler's Formula for hyperbolic functions,

$$e^{\pm\psi} = \cosh\psi \pm \sinh\psi \tag{1.63}$$

The complex forms of functions, and Euler's Formula in particular, will greatly facilitate our analyses throughout this text.

1.5 A REVIEW OF ELEMENTARY DYNAMICS

Dynamics is the study of motion. As such, the principles of dynamics are central to our study of vibrations. In fact, vibrations may be viewed as a subset of dynamics, focusing on certain types of motions. For the study of mechanical and structural vibrations, which constitutes the scope of this book, we are interested in classical mechanics. In this section we shall review some of the basic principles of *Newtonian Mechanics,* while certain concepts and principles of the subject known as *Analytical*

Mechanics will be introduced in Chapter 6. (*The reader who is well grounded in elementary dynamics may proceed to Chapter 2 without loss of continuity.*) We shall first discuss the dynamics of single particles, and then extend these ideas to particle systems. These concepts will then be abstracted to a continuum, viewed as a continuous distribution of matter or particles, with the dynamics of rigid bodies presented as a special case at the close of this section. The dynamics of deformable bodies is introduced in Chapter 9.

The study of dynamics can be separated into two sub-areas, kinematics and kinetics. Kinematics is the study of the geometry of motion. That is, it is the study of how we describe a given motion mathematically. Kinetics, on the other hand, deals with the forces imparted on bodies and the response (motion) of the bodies to these forces. The notion of a particle is an idealization. A particle is a body that has mass but no volume. It is thus a point that moves through space. We shall see that, for many situations, the motion of a finite body may be adequately described by that of a particle. The consequences of such an idealization for finite bodies will be examined in subsequent sections. More generally, a body may be viewed as an assemblage of particles. We first review the kinematics of particles.

1.5.1 Kinematics of Particles

As stated in the introduction to this section, kinematics is the study of the geometry of motion. In this section we introduce fundamental mathematical measures that characterize the motion of a particle.

Basic Kinematic Measures

In order to locate a particle, we must specify its location with respect to some reference. Therefore, let us define a coordinate system with origin at point "O." All quantities are then measured with respect to this point. Alternatively, we may view such quantities as those "seen by an observer standing at O." In this context, the location of a particle at a particular time is defined as the *position* of the particle at that time. We thus introduce the *position vector*

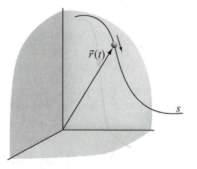

Figure 1.31 A particle and its trajectory.

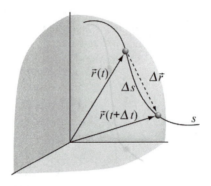

Figure 1.32 Displacement of a particle.

$$\vec{r} = \vec{r}(t) \tag{1.64}$$

which is represented as the directed line segment between the origin O and the location of the particle at time t, as shown in Figure 1.31.

The path that the particle follows during its motion is called the particle's *trajectory*. Let us consider the particle at two points along its trajectory at two instants in time, t and $t + \Delta t$, as shown in Figure 1.32. The change in position of the particle between these two points is called the displacement of the particle and is defined by the *displacement vector*

$$\Delta \vec{r} \equiv \vec{r}(t+\Delta t) - \vec{r}(t) \tag{1.65}$$

If we wish to characterize how quickly the particle is changing its location we must continue our development by quantifying the rate at which the position of the particle is changing. The time rate of change of the position vector is called the velocity vector. The average velocity over a given time interval, Δt, is simply the ratio of the change of position to the duration of the interval. The *average velocity* is thus

$$\vec{v}_{avg} \equiv \frac{\Delta \vec{r}}{\Delta t} = \frac{\vec{r}(t+\Delta t) - \vec{r}(t)}{\Delta t} \tag{1.66}$$

The instantaneous velocity, or simply the velocity, at a given time t is established by letting the time interval approach zero. Thus, the *instantaneous velocity* at time t is given by

$$\vec{v}(t) \equiv \frac{d\vec{r}}{dt} = \lim_{\Delta t \to 0} \frac{\Delta \vec{r}}{\Delta t} = \lim_{\Delta t \to 0} \frac{\vec{r}(t+\Delta t) - \vec{r}(t)}{\Delta t} \tag{1.67}$$

If one considers the displacement vector between two positions of the particle, and lets this vector get smaller and smaller as shown in Figure 1.32, it is seen that as $\Delta t \to 0$ the vector $\Delta \vec{r} \to d\vec{r}$ and becomes tangent to the path at time t. It follows from Eq. (1.67) that the velocity vector is always tangent to the path traversed by the particle.

To characterize how the velocity changes as a function of time we introduce its rate of change. The time rate of change of the velocity vector is referred to as the acceleration vector, or simply the acceleration. Paralleling our discussion of velocity we first introduce the *average acceleration*,

$$\vec{a}_{avg} \equiv \frac{\Delta \vec{v}}{\Delta t} = \frac{\vec{v}(t + \Delta t) - \vec{v}(t)}{\Delta t} \tag{1.68}$$

The *instantaneous acceleration* is then

$$\vec{a}(t) \equiv \frac{d\vec{v}}{dt} = \lim_{\Delta t \to 0} \frac{\Delta \vec{v}}{\Delta t} = \lim_{\Delta t \to 0} \frac{\vec{v}(t + \Delta t) - \vec{v}(t)}{\Delta t} \tag{1.69}$$

Relative Motion

Consider the motions of two particles, A and B, and let $\vec{r}_A(t)$ and $\vec{r}_B(t)$ be the corresponding position vectors of the particles with respect to a common origin O. Further, let $\vec{r}_{B/A}(t)$ correspond to the position vector of particle B as seen by an observer translating (but not rotating) with particle A, as indicated in Figure 1.33. It may be seen from the figure that, through vector addition, the relative position of particle B with respect to particle A may be expressed in terms of the positions of the two particles with respect to the origin O by the relation

$$\vec{r}_{B/A}(t) = \vec{r}_B(t) - \vec{r}_A(t) \tag{1.70}$$

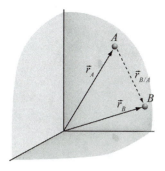

Figure 1.33 Two particles in motion.

Differentiation of Eq. (1.70) with respect to time gives the *relative velocity* of particle B with respect to particle A,

$$\vec{v}_{B/A}(t) = \vec{v}_B(t) - \vec{v}_A(t) \tag{1.71}$$

where $\vec{v}_A(t)$ and $\vec{v}_B(t)$ are, respectively, the velocities of particles A and B with respect to O. Differentiating Eq. (1.71) gives the corresponding *relative acceleration*,

$$\vec{a}_{B/A}(t) = \vec{a}_B(t) - \vec{a}_A(t) \tag{1.72}$$

where $\vec{a}_A(t)$ and $\vec{a}_B(t)$ are the accelerations of the indicated particles with respect to the origin. The relative velocity $\vec{v}_{B/A}(t)$ is interpreted as the velocity of particle B as seen by an observer that is translating (but not rotating) with particle A. The relative acceleration $\vec{a}_{B/A}(t)$ is interpreted similarly.

Coordinate Systems

It is often expedient to use a particular coordinate system for a particular problem or application. We next consider Cartesian, path, cylindrical-polar and spherical coordinates, and express the position, velocity and acceleration vectors in terms of their components with respect to these coordinate systems.

Cartesian Coordinates

Let $\vec{i}, \vec{j}, \vec{k}$ represent unit base vectors oriented along the x, y, z coordinate axes, respectively, as indicated in Figure 1.34. As the basis vectors are constant in direction as well as magnitude for this case, it follows that their derivatives with respect to time vanish. It then follows that the position, velocity and acceleration vectors expressed in terms of their Cartesian components, are respectively

$$\vec{r}(t) = x(t)\vec{i} + y(t)\vec{j} + z(t)\vec{k}$$
$$\vec{v}(t) = v_x(t)\vec{i} + v_y(t)\vec{j} + v_z(t)\vec{k} = \dot{x}(t)\vec{i} + \dot{y}(t)\vec{j} + \dot{z}(t)\vec{k} \tag{1.73}$$
$$\vec{a}(t) = a_x(t)\vec{i} + a_y(t)\vec{j} + a_z(t)\vec{k} = \ddot{x}(t)\vec{i} + \ddot{y}(t)\vec{j} + \ddot{z}(t)\vec{k}$$

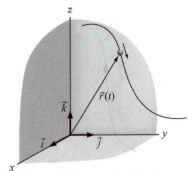

Figure 1.34 Cartesian Coordinates.

Path Coordinates

Let s represent a coordinate along the path traversed by a particle, as indicated in Figure 1.35. Let $\vec{e}_t$ represent the unit vector that is tangent to the path in the direction of increasing s at a given point, let $\vec{e}_n$ represent the unit normal to the path directed toward the center of curvature at that point, and let $\vec{e}_b \equiv \vec{e}_t \times \vec{e}_n$ be the corresponding unit binormal vector that completes the triad of basis vectors, as indicated. We note that, though the basis vectors are of unit magnitude, their directions are constantly changing as the particle proceeds along its trajectory. In fact, it is easily shown that

$$\dot{\vec{e}}_t = \frac{\dot{s}}{\rho}\vec{e}_n \tag{1.74}$$

where ρ is the radius of curvature of the path at the point in question. Since $s(t)$ measures the distance along the path, and hence locates the particle at a given time, it follows that the speed is given by

$$v(t) = \dot{s}(t)$$

Since the velocity vector is always tangent to the path, we have that

$$\vec{v}(t) = v(t)\vec{e}_t(t) = \dot{s}(t)\vec{e}_t \tag{1.75}$$

Differentiating Eq. (1.75) and incorporating the identity stated by Eq. (1.74) gives the acceleration in terms of its normal and tangential components. Hence,

$$\vec{a}(t) = a_t(t)\vec{e}_t + a_n(t)\vec{e}_n = \ddot{s}(t)\vec{e}_t + \frac{\dot{s}^2}{\rho}\vec{e}_n \tag{1.76}$$

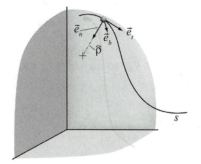

Figure 1.35 Path coordinates.

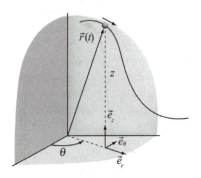

Figure 1.36 Cylindrical-polar coordinates.

Cylindrical Polar Coordinates
Let R, θ, z, represent the radial, angular and axial coordinates of a particle at a given instant, as indicated in Figure 1.36. Let $\vec{e}_R, \vec{e}_\theta, \vec{e}_z$ represent the corresponding unit vectors. Though the magnitude of all three basis vectors remains constant, the directions associated with the first two are constantly changing as the particles moves along its trajectory. The relation between the time derivatives of the first two unit vectors is similar to that for the basis vectors associated with path coordinates. The position vector expressed in terms of its components in cylindrical-polar coordinates takes the form

$$\vec{r}(t) = R(t)\vec{e}_R + z(t)\vec{e}_z \tag{1.77}$$

Differentiating Eq. (1.77) with respect to time, and noting that

$$\dot{\vec{e}}_R = \dot{\theta}\vec{e}_\theta \quad \text{and} \quad \dot{\vec{e}}_\theta = -\dot{\theta}\vec{e}_R$$

gives the corresponding velocity vector

$$\vec{v}(t) = v_R(t)\vec{e}_R + v_\theta(t)\vec{e}_\theta + v_z(t)\vec{e}_z = \dot{R}\vec{e}_R + R\dot{\theta}\vec{e}_\theta + \dot{z}\vec{e}_z \tag{1.78}$$

Differentiating again gives the acceleration vector in terms of its cylindrical-polar components as

$$\vec{a}(t) = a_R(t)\vec{e}_R + a_\theta(t)\vec{e}_\theta + a_z(t)\vec{e}_z = \left(\ddot{R} - r\dot{\theta}^2\right)\vec{e}_r + \left(R\ddot{\theta} + 2\dot{R}\dot{\theta}\right)\vec{e}_\theta + \ddot{z}\vec{e}_z \tag{1.79}$$

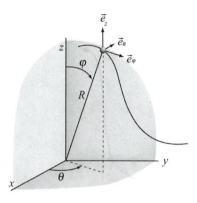

Figure 1.37 Spherical coordinates.

Spherical Coordinates

Let r, θ, φ represent the radial, polar angle and azimuth coordinates and let $\vec{e}_r, \vec{e}_\theta, \vec{e}_\varphi$ represent the corresponding unit vectors, as indicated in Figure 1.37. As for polar and path coordinates, the unit vectors associated with spherical coordinates have constant magnitude but constantly change direction throughout the motion of the particle. Therefore, their time derivatives do not vanish. Proceeding as we did for path and polar coordinates, we first express the position vector in terms of its spherical components. This is simply

$$\vec{r}(t) = r(t)\vec{e}_r \tag{1.80}$$

Differentiating Eq. (1.80) gives the velocity vector in terms of its spherical components. Hence,

$$\vec{v}(t) = v_r(t)\vec{e}_r + v_\theta(t)\vec{e}_\theta + v_\varphi(t)\vec{e}_\varphi = \dot{r}\vec{e}_r + r\dot{\theta}\cos\varphi\,\vec{e}_\theta + r\dot{\varphi}\vec{e}_\varphi \tag{1.81}$$

Differentiating again gives the corresponding expression for the acceleration vector as

$$\vec{a}(t) = a_r(t)\vec{e}_r + a_\theta(t)\vec{e}_\theta + a_\varphi(t)\vec{e}_\varphi$$
$$= \left[\ddot{r} - r\dot{\varphi}^2 - r\dot{\theta}^2\cos^2\varphi\right]\vec{e}_r + \left[\frac{\cos\varphi}{r}\frac{d}{dt}\left(r^2\dot{\theta}\right) - 2r\dot{\theta}\dot{\varphi}\sin\varphi\right]\vec{e}_\theta \tag{1.82}$$
$$+ \left[\frac{1}{r}\frac{d}{dt}\left(r^2\dot{\varphi}\right) + r\dot{\theta}^2\sin\varphi\cos\varphi\right]\vec{e}_\varphi$$

1.5.2 Kinetics of a Single Particle

Classical mechanics is based on the three fundamental laws posed by Newton, and the integrals of one of them. We first discuss Newton's Laws of Motion.

Newton's Laws of Motion

Newton's three laws of motion form the basis for our study of dynamics. They are paraphrased below.

Newton's First Law
A body at rest, or in motion at constant velocity, remains in that state unless acted upon by an unbalanced force.

Newton's Second Law
If a body is acted upon by an unbalanced force, its velocity changes at a rate proportional to that force. This is stated mathematically by the well-known relation

$$\vec{F} = m\vec{a} \tag{1.83}$$

where $\vec{F}$ is the force acting on the particle, $\vec{a}$ is the time rate of change of the velocity of the particle and m is the mass of the particle. The mass (or inertia) of the particle is seen to be a measure of the resistance of the particle to changes in its velocity. The larger the mass, the larger the force required to produce the same rate of change of velocity.

Newton's Third Law
If a body exerts a force on a second body, the second body exerts an equal and opposite force on the first body.

In principle, the motion of a particle is completely defined by these laws. However, it is often convenient to approach a problem from an alternate perspective. Certain integrals of Newton's Second Law accomplish this, and lead to other principles of classical mechanics. These principles are discussed in the following sections.

Work and Kinetic Energy

If we take the scalar dot product of the mathematical statement of Newton's Second Law, Eq. (1.83), with the increment of the position vector, $d\vec{r}$, multiply and divide the right hand side by dt, and integrate the resulting expression between two points on the particle's trajectory we arrive at the *Principle of Work-Energy*,

$$\mathcal{W} = \Delta T = T_2 - T_1 \tag{1.84}$$

where

$$\mathcal{W} \equiv \int_{\vec{r}_1}^{\vec{r}_2} \vec{F} \cdot d\vec{r} \tag{1.85}$$

is the work done by the applied force in moving the particle form position $\vec{r}_1 \equiv \vec{r}(t_1)$ to position $\vec{r}_2 \equiv \vec{r}(t_2)$, t_1 and t_2 are the times at which the particle is at these positions,

$$T \equiv \tfrac{1}{2}mv^2 \qquad (1.86)$$

is the kinetic energy of the particle, and $v_j = v(t_j)$. It is instructive to write Eq. (1.85) in terms of path coordinates. Hence, expressing the resultant force in terms of its tangential, normal and binormal components, noting that $d\vec{r} = ds\,\vec{e}_t$, substituting into Eq. (1.85) and carrying through the dot product gives

$$\mathcal{W} = \int_{s_1}^{s_2} \left[F_t \vec{e}_t + F_n \vec{e}_n + F_b \vec{e}_b \right] \cdot \left(ds\,\vec{e}_t \right) = \int_{s_1}^{s_2} F_t\, ds \qquad (1.87)$$

where $s_1 = s(t_1)$ and $s_2 = s(t_2)$. It is seen from Eq. (1.87) that only the tangential component of the force does work.

Path Dependence, Conservative Forces and Potential Energy

Let us consider a particular type of force for which the work done by that force in moving the particle from position 1 to position 2 is independent of the particular path along which the particle moves. Let us denote this force as $\vec{F}^{(C)}$. The work done by such a force,

$$\mathcal{W}^{(C)} \equiv \int_{\vec{r}_1}^{\vec{r}_2} \vec{F}^{(C)} \cdot d\vec{r} \qquad (1.88)$$

is thus a function of the coordinates of the end points of the path only. If we denote this function as $-\mathcal{U}$, where we adopt the minus sign for convention, then

$$\begin{aligned}
\int_{\vec{r}_1}^{\vec{r}_2} \vec{F}^{(C)} \cdot d\vec{r} &= -\left[\mathcal{U}(s_2) - \mathcal{U}(s_1) \right] = -\Delta \mathcal{U} \\
&= -\int_{\vec{r}_1}^{\vec{r}_2} \nabla \mathcal{U} \cdot d\vec{r}
\end{aligned} \qquad (1.89)$$

where ∇ is the gradient operator. Comparison of the integrals on the right and left hand sides of Eq. (1.89) gives the relation

$$\vec{F}^{(C)} = -\nabla \mathcal{U} \qquad (1.90)$$

It is seen from Eq. (1.90) that a force for which the work done is independent of the path traversed is derivable from a scalar potential. Such a force is referred to as a *conservative force*, and the corresponding potential function as the *potential energy*.

Forces that do not fall into this category, that is forces for which the work done is dependent on the path traversed, are referred to as nonconservative forces. It is seen from Eq. (1.89) that only the *difference* in potential energy between positions, or its gradient, enters the formulation and thus the potential energy is defined to within an arbitrary constant. It is often convenient to introduce a 'datum' in order to assign a definite value to the potential energy. The potential energy is defined through its change. Hence, the change in potential energy is the negative of the work done by a conservative force in moving a particle between two positions. The potential energy is thus seen to be the work that would be done if the process were reversed. That is, it is the work that would be done by the conservative force if the particle were to move from the latter position to the former position. The potential energy may therefore be viewed as "stored energy" or "the ability to do work". Examples of conservative forces are the gravitational force and the force of an elastic spring. Examples of non-conservative forces are friction forces, damping forces, and follower forces such as the thrust of a rocket.

    If we partition our forces into conservative and nonconservative then the work-energy principle, Eq. (1.84), may be written in an alternative form. Toward this end, let the resultant force acting on a particle be comprised of a resultant conservative force and a resultant nonconservative force, $\vec{F}^{(C)}$ and $\vec{F}^{(NC)}$, respectively. Hence,

$$\vec{F} = \vec{F}^{(C)} + \vec{F}^{(NC)} = -\nabla \mathcal{U} + \vec{F}^{(NC)} \qquad (1.91)$$

where we have incorporated Eq. (1.90). Substituting Eqs. (1.91) and (1.88) into Eqs. (1.85) and (1.84), and rearranging terms, gives the *alternate form of the work energy principle*

$$\mathcal{W}^{(NC)} = \Delta \mathcal{T} + \Delta \mathcal{U} \qquad (1.92)$$

where

$$\mathcal{W}^{(NC)} \equiv \int_{\vec{r}_1}^{\vec{r}_2} \vec{F}^{(NC)} \cdot d\vec{r} \qquad (1.93)$$

is the work of the nonconservative force. Note that the work of the conservative force is already taken into account as the change in potential energy. Thus, $\mathcal{W}^{(NC)}$ represents the work of the remaining forces (those not included in $\Delta \mathcal{U}$) acting on the particle.

Example 1.1 – Work done by the weight of a body

A car travels between two points, A and B, along the road shown. Evaluate the work done by the weight of the car as it travels between these two points.

Figure E1.1-1

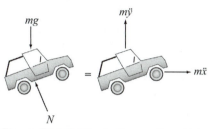

Figure E1.1-2 Kinetic diagram for vehicle.

Solution
The kinetic diagram corresponding to the car at a generic point along the path is shown in Figure E1.1-2. The weight and the increment in position may be expressed in terms of the Cartesian coordinates shown as

$$\vec{W} = -mg\,\vec{j} \tag{a}$$

and

$$d\vec{r} = dx\,\vec{i} + dy\,\vec{j} \tag{b}$$

where the relation between the coordinates x and y depends on the specific equation that describes the road (not given). We next evaluate the work done by the weight by substituting Eqs. (a) and (b) into Eq. (1.85). Thus,

$$\int_{\vec{r}_A}^{\vec{r}_B} \vec{F} \cdot d\vec{r} = \int_{s_A}^{s_B} \left[-mg\,\vec{j} \right] \cdot \left[dx\,\vec{i} + dy\,\vec{j} \right] = -\int_{y_A}^{y_B} mg\,dy \tag{c}$$

Hence,

$$\mathcal{W}_W = -mg(y_B - y_A) \tag{◁ (d)}$$

It may be seen that the work done by the weight depends only on the coordinates of the end points of the path. The particular road on which the car travels between the two points A and B is thus immaterial as far as the work of the weight is concerned. Since the work done is independent of path, the weight is then a *conservative force*. The change in potential energy is then, by definition,

$$\Delta\mathcal{U} = mg\,\Delta y \tag{e}$$

If we choose our datum (the level of zero potential energy) to be at A, we recover the elementary formula

$$\mathcal{U} = mgh$$

where $h = y - y_A$ is the height above the datum.

Example 1.2 – Work done by a follower force

Consider the motion of a rocket car as it moves along a straight track or along a circular track between two points A and B, as shown. For simplicity, let us assume that the magnitude of the thrust is constant throughout the motion. The thrust, $\vec{T} = T\,\vec{e}_t$, which is always tangent to the path, is an example of what is referred to as a "follower force," since it follows the direction of the path of the particle.

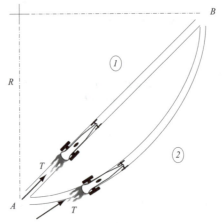

Figure E1.2 Rocket car traversing two different tracks.

Solution

For $T = T_0 =$ constant, the work done by the thrust as the rocket car moves from A to B along the straight track (Path 1) clearly differs from that done along the circular track (Path 2). Specifically, using Eq. (1.87), we have that

$$\mathcal{W}^{(1)} = T_0 R\sqrt{2} \quad \neq \quad \mathcal{W}^{(2)} = T_0\,\pi R/2$$

Since the work done clearly depends on the particular path traversed by the car, the thrust is then a *nonconservative* force.

Example 1.3 – Potential energy of elastic springs

Determine the potential energy of (*a*) a deformed linear spring of stiffness k and (*b*) a deformed torsional spring of stiffness k_T.

Solution

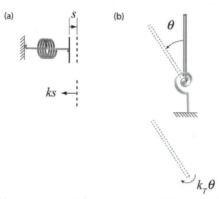

Figure E1.3 Displacement and restoring action: (a) linear spring, (b) torsional spring.

(*a*)

The work done by the restoring force of a linear spring as it is stretched from the reference (unstretched) configuration to the current configuration is readily evaluated as

$$\mathcal{W}^{(s)} = \int_0^s -k\tilde{s}\,d\tilde{s} = -\tfrac{1}{2}ks^2 \tag{a}$$

where s is the stretch in the spring (Figure E3.1a). The corresponding potential energy of the deformed spring is then, from Eq. (1.89),

$$\mathcal{U}^{(s)} = \tfrac{1}{2}s^2 \tag{b}$$

Note that it is implicit in the above expression that the datum is chosen as the undeformed state of the spring [as per the lower limit of integration of Eq. (a)].

(*b*)

The potential energy of the deformed torsional spring is similarly determined by first calculating the work done by the restoring torque and then using Eq. (1.89). Hence,

$$\mathcal{W}^{(TS)} = \int_0^\theta -k_T\tilde{\theta}\,d\tilde{\theta} = -\tfrac{1}{2}k_T\theta^2 \tag{c}$$

and

$$\mathcal{U}^{(TS)} = \tfrac{1}{2}k_T\theta^2 \tag{d}$$

where it is implicit that that the datum is taken as the undeformed state of the torsional spring.

Conservation of Mechanical Energy

A system for which only conservative forces do work is said to be a *conservative system*. If this is the case, that is if

$$\int_{\bar{r}_1}^{\bar{r}_2} \bar{F}^{(NC)} \cdot d\bar{r} = 0 \tag{1.94}$$

then Eq. (1.92) reduces to the statement that

$$\Delta \mathcal{T} + \Delta \mathcal{U} = 0$$

This may also be expressed in the alternate form

$$\mathcal{E} = \mathcal{T} + \mathcal{U} = \text{constant} \tag{1.95}$$

where $\mathcal{E}$ is the total mechanical energy of the system. Equation (1.95) is the statement of *conservation of mechanical energy* of the system. It is thus seen that conservative forces conserve mechanical energy.

Example 1.4

A coaster traveling with speed v_0 enters a vertical loop of radius R and proceeds around the loop as shown in Fig. E1.4-1. (*a*) If the total mass of the coaster and its passengers is m, determine the force exerted by the track on the coaster as it moves around the loop (i.e., as a function of the angular coordinate θ). (*b*) What is the minimum entry speed for the coaster to successfully traverse the loop?

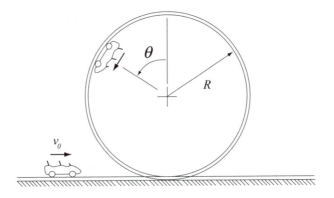

Figure E1.4-1 Roller coaster and loop.

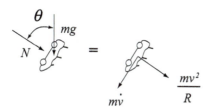

Figure E1.4-2 Kinetic diagram for coaster.

Solution
We first draw the kinetic diagram (dynamic free-body diagram) of the coaster at a generic location, as depicted in Figure E1.4-2. This displays the forces that act on the "particle" (in this case the coaster) on one figure, and the inertia "forces" (the response of the particle) on another. The kinetic diagram is basically a pictorial statement of Newton's Second (and Third) Law.

(*a*)
For this particular problem it is convenient to work in either path of polar coordinates. It is, however, somewhat more informative if we choose the former. We shall therefore solve the problem using path coordinates.

With the help of the kinetic diagram, and the incorporation of Eq. (1.76), the component of the statement of Newton's Second Law along the normal direction is written as

$$N + mg \cos \theta = \frac{mv^2}{R}$$

which, when solved for the normal force N, gives

$$N = \frac{mv^2}{R} - mg \cos \theta \qquad\qquad (a)$$

In order to find $N(\theta)$ for a given v_0 we must first determine $v(\theta)$. That we require the velocity as a function of position suggests that we should employ the work-energy principle. It is evident form Eq. (1.87) that the normal force does no work. Since the only other force acting on the coaster is the weight, which is a conservative force, we know that the energy of the system is conserved throughout its motion. Evaluating Eq. (1.95) at the entrance and at the current point of the loop gives the relation

$$\tfrac{1}{2}mv_0^2 = mgR(1 + \cos \theta) + \tfrac{1}{2}mv^2 \qquad\qquad (b)$$

where we have chosen the entrance level of the loop as our datum. Solving Eq. (b) for mv^2 and substituting the resulting expression into Eq. (a) gives the normal force as a function of location around the loop. Hence,

$$N(\theta) = \frac{mv_0^2}{R} - mg(2 + 3\cos\theta) \qquad\qquad \triangleleft (c)$$

(b)
It may be seen upon inspection of Eq. (c) that the normal force, N, achieves its minimum value when $\theta = 0$. Thus, the critical entry speed (the minimum speed at which the coaster can round the loop without leaving the track) is determined from conditions at the top of the loop. Further, when the coaster is about to fall away from the track, $N \to 0$. Substituting these values of θ and N into Eq. (c) gives the critical entry speed

$$v_{0cr} = \sqrt{5gR} \qquad\qquad \triangleleft (d)$$

Linear Impulse and Momentum
We obtained the Principle of Work-Energy as an integral of Newton's Second Law over space. We shall next consider an integral of Newton's Second Law over time that is generally concerned with translational motion.

Let us multiply Newton's Second Law, Eq. (1.83), by the differential time increment dt and integrate between two instants in time, t_1 and t_2, during the particle's motion. Doing this results in the *Principle of Linear Impulse-Momentum*,

$$\int_{t_1}^{t_2} \vec{F}\, dt = m\vec{v}(t_2) - m\vec{v}(t_1) \qquad\qquad (1.96)$$

or, equivalently,

$$\vec{\mathcal{I}} = \Delta\vec{\wp} \qquad\qquad (1.97)$$

where

$$\vec{\mathcal{I}} \equiv \int_{t_1}^{t_2} \vec{F}\, dt \qquad\qquad (1.98)$$

is the *linear impulse* imparted by the force $\vec{F}$ over the time interval $\Delta t = t_2 - t_1$, and

$$\vec{\wp}(t) \equiv m\vec{v}(t) \qquad\qquad (1.99)$$

is the *linear momentum* of the particle at time t. Thus, a linear impulse that acts on a particle for a given duration produces a change in linear momentum of that particle during that time period.

Conservation of Linear Momentum

If the linear impulse vanishes over a given time interval, Eqs. (1.96) and (1.97) reduce to the statements

$$m\vec{v}(t_2) = m\vec{v}(t_1) \qquad (1.100)$$

or

$$\vec{\wp} = m\vec{v} = \text{constant} \qquad (1.101)$$

When this occurs, the linear momentum is said to be *conserved* over the given time interval.

Angular Impulse and Momentum

In the previous section we established an integral, over time, of Newton's Second Law that led to the principle of linear impulse-momentum. We next establish the rotational analogue of that principle.

Let us form the vector cross product of Newton's Second Law, Eq. (1.83), with the position vector of a particle at a given instant. Doing this results in the relation

$$\vec{M}_O = \dot{\vec{H}}_O \qquad (1.102)$$

where

$$\vec{M}_O = \vec{r} \times \vec{F} \qquad (1.103)$$

is the moment of the applied force about an axis through the origin, and

$$\vec{H}_O = \vec{r} \times m\vec{v} \qquad (1.104)$$

is referred to as the *angular momentum*, or moment of momentum, of the particle about O. Let us next multiply Eq. (1.102) by the differential time increment dt, and integrate the resulting expression between two instants in time, t_1 and t_2, during the particle's motion. This results in the statement of the *Principle of Angular Impulse-Momentum*,

$$\int_{t_1}^{t_2} \vec{r} \times \vec{F}\, dt = \left[\vec{r} \times m\vec{v}\right]_{t=t_2} - \left[\vec{r} \times m\vec{v}\right]_{t=t_1} \qquad (1.105)$$

or, equivalently,

$$\vec{\mathcal{J}}_O = \Delta \vec{H}_O \qquad (1.106)$$

where

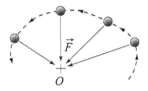

Figure 1.38 Central force motion.

$$\vec{\mathcal{J}}_O \equiv \int_{t_1}^{t_2} \vec{M}_O \, dt = \int_{t_1}^{t_2} \vec{r} \times \vec{F} \, dt \qquad (1.107)$$

is the *angular impulse* about an axis through O, imparted by the force $\vec{F}$, or simply the angular impulse about O. The angular impulse is seen to be the impulse of the moment of the applied force about the origin.

Conservation of Angular Momentum

If the angular impulse about an axis vanishes, then Eqs. (1.105) and (1.106) reduce to the equivalent statements

$$\left[\vec{r} \times m\vec{v}\right]_{t=t_2} = \left[\vec{r} \times m\vec{v}\right]_{t=t_1} \qquad (1.108)$$

or

$$\vec{H}_O = \vec{r} \times m\vec{v} = \text{constant} \qquad (1.109)$$

over the given time interval. When this is so, the angular momentum is said to be *conserved about an axis through O*. It should be noted that angular momentum may be conserved about an axis through one point and not another. An example of this is when a particle undergoes *central force motion*, where the line of action of the applied force is always directed through the same point (Figure 1.38).

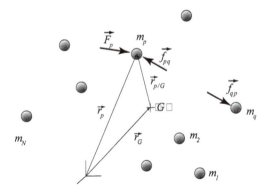

Figure 1.39 System of N particles.

1.5.3 Dynamics of Particle Systems

Mechanical systems are typically comprised of many particles. In fact, rigid bodies and deformable bodies may each be considered as an assemblage of a continuous distribution of particles with certain characteristic constraints. In this section we extend the concepts discussed for a single particle to general particle systems. These results may then be abstracted to more complex systems, as needed. We begin by examining the equations of motion for a system particles.

Equations of Motion

Consider the N particle system shown in Figure 1.39. Let m_p ($p = 1, 2, ..., N$) represent the mass of particle p, and let $\vec{r}_p$ represent its position vector with respect to a fixed reference frame, as shown. In addition, let $\vec{F}_p$ be the *resultant external force* acting on particle p, and let $\vec{f}_{pq}$ ($p,q = 1, 2, ..., N$) be the *internal force* exerted on particle p by particle q. (We assume that the resultant internal force that a particle exerts on itself vanishes. Thus, $\vec{f}_{11} = \vec{f}_{22} = ... = \vec{f}_{NN} = \vec{0}$). The *resultant internal force* acting on particle p by all other particles of the system is then

$$\vec{f}_p^* \equiv \sum_{q=1}^{N} \vec{f}_{pq}$$

It follows from Newton's Third Law that

$$\vec{f}_{qp} = -\vec{f}_{pq} \quad (p,q = 1,2,...,N)$$

and hence that

$$\sum_{p=1}^{N} \vec{f}_p^* = \sum_{p=1}^{N}\sum_{q=1}^{N} \vec{f}_{pq} = \vec{0} \tag{1.110}$$

Applying Newton's Second Law to each particle individually gives the set of equations

$$\vec{F}_p + \vec{f}_p^* = m_p \vec{a}_p \quad (p = 1,2,...,N) \tag{1.111}$$

Adding the equations for all of the particles, and incorporating Eq. (1.110) gives the relation

$$\vec{F} = \sum_{p=1}^{N} m_p \vec{a}_p \tag{1.112}$$

where

$$\vec{F} \equiv \sum_{p=1}^{N} \vec{F}_p \tag{1.113}$$

is the resultant external force acting on the system.

Let us next consider some point G that moves with the system of particles and let $\vec{r}_G(t)$ be the position vector of that point measured with respect to the fixed reference frame defined earlier. Further, let $\vec{r}_{p/G}(t)$ correspond to the position of particle p as seen by an observer translating with point G. It then follows from Eqs. (1.70)–(1.72) and Figure 1.33 that

$$\vec{r}_p = \vec{r}_G + \vec{r}_{p/G} \quad (p = 1,2,...,N) \tag{1.114}$$

$$\vec{v}_p = \vec{v}_G + \vec{v}_{p/G} \quad (p = 1,2,...,N) \tag{1.115}$$

$$\vec{a}_p = \vec{a}_G + \vec{a}_{p/G} \quad (p = 1,2,...,N) \tag{1.116}$$

where $\vec{v}_{p/G}$ and $\vec{a}_{p/G}$ are, respectively the velocity and acceleration of particle p as seen by an observer translating with point G. Substitution of Eq. (1.116) into Eq. (1.112) and regrouping terms gives

$$\vec{F} = m\vec{a}_G + \frac{d^2}{dt^2}\left(m_1\vec{r}_{1/G} + m_2\vec{r}_{2/G} + ... + m_N\vec{r}_{N/G}\right) \tag{1.117}$$

where

$$m \equiv \sum_{p=1}^{N} m_p \tag{1.118}$$

is the total mass of the particle system. If we now define the point G such that

$$\sum_{p=1}^{N} m_p \vec{r}_{p/G} = 0 \tag{1.119}$$

or, equivalently, that

$$\vec{r}_G \equiv \frac{1}{m}\sum_{p=1}^{N} \vec{r}_p \tag{1.120}$$

then Eq. (1.117) reduces to the familiar form

$$\vec{F} = m\vec{a}_G \tag{1.121}$$

The point G defined by Eq. (1.120) is referred to as the *center of mass* of the system. It is seen from Eq. (1.121) that the motion of the center of mass is governed by Newton's Second Law of Motion. Thus, the center of mass of the system behaves as a single particle whose mass is equal to the total mass of the system. It follows that the Principles of Work-Energy and Impulse-Momentum for a single particle also hold for the center of mass of a particle system. The motion of a system of particles can be described by the motion of the center of mass acting as a particle and the motion of the system relative to the center of mass. If the motion relative to the center of mass is negligible for a given application, then Eq. (1.121) and its integrals adequately describe the motion of the system.

Work and Energy

The total kinetic energy of the system is the sum of the kinetic energies of the individual particles that comprise the system. Summing the kinetic energies and incorporating Eq. (1.115) gives the total kinetic energy of the system in the form

$$T = \sum_{p=1}^{N} \tfrac{1}{2} m_p \vec{v}_p \cdot \vec{v}_p = \tfrac{1}{2} m v_G^2 + \sum_{p=1}^{N} \tfrac{1}{2} m_p v_{p/G}^2 \tag{1.122}$$

It may be seen that the total kinetic energy of the system may be partitioned into the sum of two kinetic energies: the kinetic energy of motion of the center of mass of the system acting as a single particle, and the kinetic energy of motion of the system relative to the center of mass. The total work done on the system may be similarly partitioned by adding the work done by the external and internal forces acting on the individual particles, and incorporating Eq. (1.114). Hence,

$$W = \sum_{p=1}^{N} \int_{\vec{r}_p^{(1)}}^{\vec{r}_p^{(2)}} \left[\vec{F}_p + \vec{f}_p^* \right] \cdot d\vec{r}_p = \int_{\vec{r}_G^{(1)}}^{\vec{r}_G^{(2)}} \vec{F} \cdot d\vec{r}_G + \sum_{p=1}^{N} \int_{\vec{r}_{p/G}^{(1)}}^{\vec{r}_{p/G}^{(2)}} \left[\vec{F}_p + \vec{f}_p^* \right] \cdot d\vec{r}_{p/G} \tag{1.123}$$

The first integral on the right hand side of Eq. (1.123) is seen to be the work done by the resultant external force moving along the trajectory of the center of mass, while the second term may be seen to be the work done by the forces acting on the individual particles in moving them along their trajectories relative to the center of mass.

Summing the work-energy relations of the individual particles gives the *Work-Kinetic Energy Principle for a Particle System*,

$$\sum_{p=1}^{N} \int_{\vec{r}_p^{(1)}}^{\vec{r}_p^{(2)}} \left[\vec{F}_p + \vec{f}_p^* \right] \cdot d\vec{r}_p = \sum_{p=1}^{N} \tfrac{1}{2} m_p v_p^{(2)2} - \sum_{p=1}^{N} \tfrac{1}{2} m_p v_p^{(1)2} \tag{1.124}$$

Paralleling the development of Eq. (1.84) with Eq. (1.121) replacing Eq. (1.83), or simply applying Eq. (1.83) for the center of mass directly, gives the *Work-Kinetic Energy Principle for the Center of Mass of a Particle System*,

$$\int_{\vec{r}_G^{(1)}}^{\vec{r}_G^{(2)}} \vec{F} \cdot d\vec{r}_G = \tfrac{1}{2} m v_G^{(2)2} - \tfrac{1}{2} m v_G^{(1)2} \tag{1.125}$$

Substitution of Eqs. (1.122), (1.123) and (1.125) into Eq. (1.124), and incorporating Eq. (1.119), gives the *Work-Kinetic Energy Principle for motion of a system relative to its center of mass*,

$$\sum_{p=1}^{N} \int_{\vec{r}_{p/G}^{(1)}}^{\vec{r}_{p/G}^{(2)}} \left[\vec{F}_p + \vec{f}_p^* \right] \cdot d\vec{r}_{p/G} = \sum_{p=1}^{N} \frac{1}{2} m_p v_{p/G}^{(2)\ 2} - \sum_{p=1}^{N} \frac{1}{2} m_p v_{p/G}^{(1)\ 2} \tag{1.126}$$

Equations (1.124)–(1.126) hold for all particle systems. When a subset of the forces that act on the system are conservative, these work-energy relations can be written in alternative forms, replacing the work done by the conservative forces by corresponding changes in potential energy, as discussed for single particles in Section 1.5.2. Doing this results in the alternative forms of Eqs. (1.124)–(1.126), respectively, as

$$\mathcal{W}^{(NC)} = \Delta \mathcal{T} + \Delta \mathcal{U} \tag{1.127}$$

$$\mathcal{W}_G^{(NC)} = \Delta \mathcal{T}_G + \Delta \mathcal{U}_G \tag{1.128}$$

$$\mathcal{W}_{rel}^{(NC)} = \Delta \mathcal{T}_{rel} + \Delta \mathcal{U}_{rel} \tag{1.129}$$

where the superscript *NC* indicates work done by nonconservative forces, a subscript *G* indicates work, kinetic energy and potential energy measured following the center of mass, and a subscript *rel* indicates work, kinetic energy and potential energy measured relative to the center of mass.

Linear Impulse and Momentum

The impulse-momentum principles for particle systems may be obtained in a manner similar to that employed to obtain the work-energy principles. We first consider linear impulse and momentum.

The linear impulse-momentum relation for each particle may be obtained by multiplying Eq. (1.111) by dt and integrating between two instants in time. Alternatively, we could simply apply Eq. (1.96) for each particle of the system. Either approach gives the relation

$$\int_{t_1}^{t_2} \left[\vec{F}_p + \vec{f}_p^* \right] dt = m_p \vec{v}_p(t_2) - m_p \vec{v}_p(t_1) \quad (p = 1, 2, ..., N) \tag{1.130}$$

The total linear momentum of a system of particles is the sum of the momenta of the individual particles. If we add the linear-momentum relations for the individual particles, and recall Eq. (1.110), we obtain the *Linear Impulse-Momentum Principle for a particle system* given by

$$\int_{t_1}^{t_2} \vec{F} \, dt = \sum_{p=1}^{N} m_p \vec{v}_p(t_2) - \sum_{p=1}^{N} m_p \vec{v}_p(t_1) \tag{1.131}$$

where $\vec{F}$ is the resultant external force acting on the system, as defined by Eq. (1.113).

Multiplying Eq. (1.121) by dt, and integrating over the given time interval, gives the *Principle of Linear Impulse-Momentum for the center of mass of a particle system*,

$$\int_{t_1}^{t_2} \vec{F} \, dt = m\vec{v}_G(t_2) - m\vec{v}_G(t_1) \tag{1.132}$$

Substituting Eq. (1.115) into Eq. (1.131), and subtracting Eq. (1.132) from the resulting expression gives the *Principle of Linear Impulse-Momentum for motion relative to the center of mass*,

$$\sum_{p=1}^{N} m_p \vec{v}_{p/G}(t_2) = \sum_{p=1}^{N} m_p \vec{v}_{p/G}(t_1) \tag{1.133}$$

It is seen form Eq. (1.133) that the linear momentum for motion relative to the center of mass is always conserved. Further, it may be seen from Eqs. (1.131) and (1.132) that if the resultant external impulse acting on the system vanishes over a given time interval then the total momentum of the system is conserved, and the velocity of the center of mass is constant during this time interval. Thus, in such situations, the momentum of the individual particles may be altered but the total momentum of the system is unchanged. An example of this phenomenon may be seen when a cue ball collides with a set of billiard balls.

Angular Impulse and Momentum

Let us next take the vector cross product of the position vector of particle p with Eq. (1.111). This gives the relation

$$\vec{M}_p = \dot{\vec{H}}_p \quad (p = 1, 2, ..., N) \tag{1.134}$$

where

$$\vec{M}_p \equiv \vec{r}_p \times \left[\vec{F}_p + \vec{f}_p^* \right] \tag{1.135}$$

and

$$\vec{H}_p \equiv \vec{r}_p \times m_p \vec{v}_p \tag{1.136}$$

are, respectively, the moment of the forces acting on particle p about an axis through the origin O, and the angular momentum of particle p about O. Summing Eq. (1.134) over all particles of the system, noting that it is implicitly assumed that each external force is collinear with its reciprocal, and recalling Eq. (1.110) gives the relation

$$\vec{M}_O = \dot{\vec{H}}_O \tag{1.137}$$

where

$$\vec{M}_O \equiv \sum_{p=1}^{N} \vec{r}_p \times \vec{F}_p \tag{1.138}$$

and

$$\vec{H}_O \equiv \sum_{p=1}^{N} \vec{r}_p \times m_p \vec{v}_p \tag{1.139}$$

respectively correspond to the resultant moment of the external forces about O, and total angular momentum of the system about an axis through O.

If we next multiply Eq. (1.137) by the differential time increment dt, and integrate between two instants in time, we obtain the *Principle of Angular Impulse-Momentum for the particle system,*

$$\int_{t_1}^{t_2} \vec{M}_O dt = \vec{H}_O(t_2) - \vec{H}_O(t_1) \tag{1.140}$$

or, in expanded form,

$$\int_{t_1}^{t_2} \sum_{p=1}^{N} \vec{r}_p \times \vec{F}_p \, dt = \sum_{p=1}^{N} \left[\vec{r}_p \times m_p \vec{v}_p \right]_{t=t_2} - \sum_{p=1}^{N} \left[\vec{r}_p \times m_p \vec{v}_p \right]_{t=t_1} \tag{1.141}$$

Proceeding for the center of mass as for a single particle gives the *Principle of Angular Impulse-Momentum for the center of mass of a particle system,*

$$\int_{t_1}^{t_2} \vec{r}_G \times \vec{F} \, dt = \left[\vec{r}_G \times m\vec{v}_G \right]_{t=t_2} - \left[\vec{r}_G \times m\vec{v}_G \right]_{t=t_1} \tag{1.142}$$

Substitution of Eqs. (1.114) and (1.115) into Eq. (1.141), subtracting Eq. (1.142) from the resulting expression and incorporating Eq. (1.119) gives the *Principle of Angular Impulse-Momentum for motion relative to the center of mass,*

$$\sum_{p=1}^{N} \int_{t_1}^{t_2} \vec{r}_{p/G} \times \vec{F}_p \, dt = \sum_{p=1}^{N} \left[\vec{r}_{p/G} \times m_p \vec{v}_{p/G} \right]_{t=t_2} - \sum_{p=1}^{N} \left[\vec{r}_{p/G} \times m_p \vec{v}_{p/G} \right]_{t=t_1} \tag{1.143}$$

Equations (1.140)–(1.143) are the statements of angular impulse-momentum for general particle systems.

Example 1.5

Consider a circular disk comprised of a skin of mass m_s, a relatively rigid core of mass m_c and a compliant weave modeled as n identical linear springs of stiffness k and negligible mass, as shown. The length of each spring when unstretched is $R_0 = R_s - R_c - \varepsilon$. The disk is translating at a speed v_0 in the direction indicated when the skin fragments into four identical pieces and separates from the remaining structure. If the fragments travel at the same speed relative to the core, and at equal angles as viewed from the core, determine the absolute velocity of the fragments if the energy loss due to the fragmentation is negligible. Assume that the fragmentation occurs instantaneously.

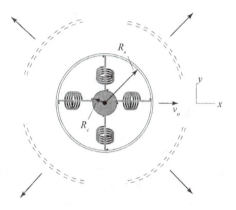

Figure E1.5 Fragmented disk.

Solution

If the fragmentation occurs instantaneously, then the impulse imposed by the weight of the ball is vanishingly small. (See the discussion of nonimpulsive forces in Chapter 4.) Therefore, since there are no external impulses acting on the system, the total linear momentum of the system is conserved throughout the interval of interest. For these conditions, Eq. (1.142) tells us that the velocity of the center of mass is unchanged. Thus,

$$\vec{v}_G(t) = \vec{v}_0 \tag{a}$$

If the energy of fragmentation is negligible, then the work done by the nonconservative forces vanishes and the energy of the system is conserved. We next apply Eq. (1.129) with $\mathcal{W}_{rel}^{(NC)} = 0$. This gives

$$0 = \left[\tfrac{1}{2} m_s v_{rel}^2 - 0 \right] + \left[0 - \tfrac{1}{2} n k \varepsilon^2 \right] \tag{b}$$

where v_{rel} is the speed of the fragments relative to the center of mass. Solving for v_{rel} gives

$$v_{rel} = \varepsilon \sqrt{\frac{n k}{m_s}} \tag{c}$$

Substitution of Eqs. (a) and (c) into Eq. (1.115) gives the absolute velocities of the fragments,

$$\vec{v}_{1,2} = \vec{v}_0 + \varepsilon \sqrt{\frac{nk}{2m_s}} \left[\vec{i} \pm \vec{j}\right] \qquad \triangleleft \text{(d-1,2)}$$

$$\vec{v}_{3,4} = \vec{v}_0 - \varepsilon \sqrt{\frac{nk}{2m_s}} \left[\vec{i} \pm \vec{j}\right] \qquad \triangleleft \text{(d-3,4)}$$

1.5.4 Kinematics of Rigid Bodies

A rigid body is an idealization that, in certain applications, may capture the dominant motion of the body. Alternatively, we may be interested in motions such as vibrations where the circumstances are such that the rigid body portion of the motion is unimportant to us, for example in predicting structural or material failure in aircraft or other vehicular structures. In such cases it may be necessary to identify the rigid body portion of the response and subtract it out. In any event rigid body motion, and hence the dynamics of rigid bodies, is of interest.

A rigid body may be considered to be a continuous distribution of particles, and hence a particle system, for which the relative distances and orientations of the constituent particles remains fixed. Therefore, the principles pertaining to the dynamics of particle systems discussed in Section 1.5.3 may be applied to this particular class of particle systems. As the relative motions of the particles that comprise a rigid body are restricted, it is expedient to incorporate these constraints into the description of the motion. In this regard, we first discuss the kinematics of rigid body motion. We shall restrict the overall discussion herein to planar motion of rigid bodies.

The motion of rigid bodies is comprised of two basic motions — translation and rotation. We shall first consider each type of motion separately and then together.

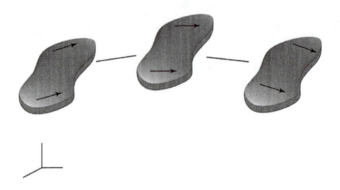

Figure 1.40 Rigid body in pure translation.

Pure Translation

Translation is a motion for which the velocity vectors are the same for each and every point of the body throughout the motion. As a result of this, the orientation of the body with regard to a given reference frame is preserved throughout the motion, as demonstrated in Figure 1.40.

Pure Rotation

As the relative distances between particles or points of a rigid body remain fixed, it is evident that if one point on the body is fixed with regard to translation (say pinned) as in Figure 1.41, then each point on the body traverses a circular path about the axis through the fixed point. The most general motion of a rigid body with one point fixed is therefore equivalent to a rotation about an axis through the fixed point.

It is apparent that if the rotation of a rigid body with a fixed point is known, then the displacement of each and every particle or point of the body is known. Further, the velocity and acceleration of each point is known if we know the first and second time rates of change of this angle, Letting the z-axis correspond to the axis of rotation, we introduce the *angular displacement,* $\vec{\theta}$, the *angular velocity,* $\vec{\varpi}$, and the *angular acceleration,* $\vec{\alpha}$, respectively defined as

$$\vec{\theta} = \theta(t)\,\vec{k} \tag{1.144}$$

$$\vec{\varpi} \equiv \frac{d\vec{\theta}}{dt} = \varpi(t)\,\vec{k} = \dot{\theta}(t)\,\vec{k} \tag{1.145}$$

$$\vec{\alpha} \equiv \frac{d\vec{\varpi}}{dt} = \alpha(t)\,\vec{k} = \dot{\varpi}(t)\,\vec{k} = \ddot{\theta}(t)\,\vec{k} \tag{1.146}$$

where $\vec{k}$ is the corresponding unit vector parallel to the axis.

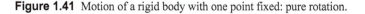

Figure 1.41 Motion of a rigid body with one point fixed: pure rotation.

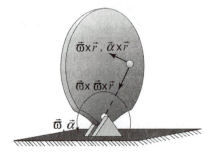

Figure 1.42 Velocity and acceleration of generic point of a rigid body in pure rotation.

As each particle of the body traverses a circular path about the axis of rotation, it follows from Eqs. (1.78), (1.79), (1.145) and (1.146) that the velocity and acceleration of a point on the body located a radial distance r from the axis of rotation are respectively given by

$$\vec{v}(t) = r\,\dot{\theta}(t)\vec{e}_\theta = \vec{\omega} \times \vec{r} \tag{1.147}$$

and

$$\vec{a}(t) = r\,\ddot{\theta}\,\vec{e}_\theta - r\,\dot{\theta}^2\,\vec{e}_r = (\vec{\alpha} \times \vec{r}) + (\vec{\omega} \times \vec{\omega} \times \vec{r}) \tag{1.148}$$

where $\vec{r}(t)$ is the position vector of the point on the body in question, and $\vec{e}_r$, $\vec{e}_\theta$ are unit vectors in the directions indicated (Figure 1.42).

General Motion

Consider the representative body shown in Figure 1.43 and two generic points, A and B. Suppose that you are translating but not rotating with point A and that you are observing the motion of point B which is painted on the body. Then, since the body is rigid, and therefore $\|\vec{r}_{B/A}\| = \text{constant}$, the motion of point B that you would observe is simply that point moving in a circular path around you at distance $\|\vec{r}_{B/A}\|$. It is evident that, for a rigid body in general motion (i.e., no point on the body is fixed) the relative motion between two points on the body is purely rotational. Therefore, from Eqs. (1.147) and (1.148), the relative velocity and relative acceleration of particle B with respect to particle A are, respectively,

$$\vec{v}_{B/A}(t) = \vec{\omega} \times \vec{r}_{B/A} \tag{1.149}$$

$$\vec{a}_{B/A}(t) = \vec{\alpha} \times \vec{r}_{B/A} + \vec{\omega} \times \vec{\omega} \times \vec{r}_{B/A} \tag{1.150}$$

Substitution of Eqs. (1.149) and (1.150) into Eqs. (1.71) and (1.72) gives the velocity and acceleration of point B with respect to the fixed reference frame at O. Hence,

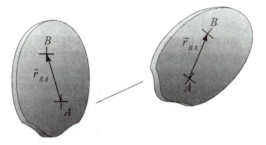

Figure 1.43 General motion of a rigid body.

$$\vec{v}_B(t) = \vec{v}_A(t) + \vec{\omega} \times \vec{r}_{B/A} \tag{1.151}$$

$$\vec{a}_B(t) = \vec{a}_A(t) + \vec{\alpha} \times \vec{r}_{B/A} + \vec{\omega} \times \vec{\omega} \times \vec{r}_{B/A} \tag{1.152}$$

Equations (1.151) and (1.152) are statements of *Euler's Theorem*, which says that *the most general motion of a rigid body is equivalent to the translation of one point on the body and a rotation about an axis through that point.*

Rolling Motion

Consider a rigid wheel that rolls along the surface of a track such that no slip occurs between the surface of the wheel and the surface of the track. Such motion is referred to as *rolling without slip*. If the wheel rolls without slipping, then the velocity of the point of the wheel in instantaneous contact with the track must have the same velocity as the track. Thus, the relative velocity of the contact point with respect to the track must vanish. Consider a wheel of radius R that is rolling on a stationary track, as shown in Figure 1.44. Let the velocity and tangential component of the acceleration of the hub, or geometric center, of the wheel be designated as v_C and a_{C_t} respectively, and let the angular velocity and angular acceleration of the wheel be $\omega(t)$ and $\alpha(t)$, respectively. Let us next consider a small time interval Δt during the motion of the wheel, and let Δs be the displacement of the hub during this time. Further, let $\Delta \theta$ represent the corresponding angle through which the wheel rotates. If we imagine a piece of tape attached to the circumference of the wheel peels during rolling and adheres point by point to the surface of the track as shown, it is evident from the figure that

$$\Delta s = R \Delta \theta \tag{1.153}$$

Dividing both sides of the above equation by the time increment and taking the limit as $\Delta t \to 0$ gives the relation

$$v_C(t) = R \omega(t) \tag{1.154}$$

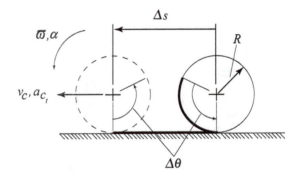

Figure 1.44 Wheel rolling without slip.

Differentiating with respect to time gives the companion relation

$$a_{C_t} = R\alpha(t) \tag{1.155}$$

An extension of the above arguments that accounts for rocking of the wheel shows that the relations stated in Eqs. (1.154) and (1.155) hold for rolling on curved tracks as well. For rolling on curved tracks the normal component of the acceleration of the hub does not vanish identically ($a_{C_n} \neq 0$) as it does for flat tracks. Formally, regardless of the curvature of the track, the velocity of the point of the wheel (say, point P) that is instantaneously in contact with the surface of the track is found from Eq. (1.151) as

$$\vec{v}_P(t) = \vec{v}_C(t) + \vec{\omega} \times \vec{r}_{P/C} = \vec{v}_C(t) - \vec{\omega} \times \vec{r}_{C/P} \tag{1.156}$$

Recognizing that for no slip $\vec{v}_P = \vec{0}$, the above expression gives

$$\vec{v}_C = \vec{\omega} \times \vec{r}_{C/P} \leftrightarrow v_C(t) = R\varpi(t) \tag{1.157}$$

regardless of the curvature of the track.

To this point we have established how the motion of a rigid body may be described. We next discuss the physical laws that govern this motion.

1.5.5 (Planar) Kinetics of Rigid Bodies

A rigid body may be considered as an assemblage of particles that are subject to certain kinematical constraints. We can therefore apply the principles established in Section 1.5.3 and exploit the constraints and the kinematics of rigid body motion developed in Section 1.5.4.

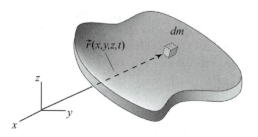

Figure 1.45 Rigid body, showing generic mass element.

Let us consider a rigid body as a continuous distribution of mass. Let dm be the mass of a differential volume element as shown in Figure 1.45. We have thus replaced the discrete particle system of Section 1.5.3 by a continuous distribution of mass. The associated position, velocity and acceleration vectors for the discrete particles are similarly replaced by corresponding vector functions as follows;

$$m_p \to dm, \quad \vec{r}_p(t) \to \vec{r}(x,y,z,t), \quad \vec{v}_p(t) \to \vec{v}(x,y,z,t), \quad \vec{a}_p(t) \to \vec{a}(x,y,z,t)$$

$$(1.158)$$

where the Cartesian Coordinates (x, y, z) are indicated simply to emphasize the spatial dependence of the vectors. The actual coordinate system employed may be any convenient system, such as polar or spherical. Further, since all quantities involved are now continuous functions of the spatial coordinates, summations over discrete masses are replaced by integrals over the entire mass, and hence volume, of the body. Thus, for any quantity $\vec{\lambda}$,

$$\sum_{p=1}^{N} m_p \vec{\lambda}_p(t) \to \int_m \vec{\lambda}(x,y,x,t)\, dm \qquad (1.159)$$

Equations of Motion

The equation governing the motion of the center of mass, Eq. (1.121), may be directly applied to any particle system including the system that comprises a rigid body. Hence,

$$\vec{F} = m\, \vec{a}_G \qquad (1.160)$$

however, the total mass of the system is now given by

$$m = \int_m dm = \int_V \rho(x,y,z)\, dV \qquad (1.161)$$

where ρ is the mass density (mass per unit volume) of the body.

Let us next express Eqs. (1.137)–(1.139) in terms of continuous functions using Eqs. (1.158) and (1.159). Doing this gives the equations governing rotational motion about the center of mass, and about an arbitrary point P. Hence,

$$\vec{M}_G = I_G \, \vec{\alpha} \tag{1.162}$$

$$\vec{M}_P = I_G \, \vec{\alpha} + \left[\vec{r}_{P/G} \times m \, \vec{a}_G \right] \tag{1.163}$$

where $\vec{M}_G$ and $\vec{M}_P$ are, respectively, the resultant moment about an axis through the center of mass and the resultant moment about an axis through P of the external forces acting on the body. Further, the parameter

$$I_G = \int_m r_{rel}^2 \, dm \tag{1.164}$$

is the mass moment of inertia (the second moment of the mass) about the axis though the center of mass of the body, and

$$r_{rel}(x, y, z) \equiv \left\| \vec{r}_{rel}(x, y, z) \right\|$$

is the distance of the mass element, dm, from that axis.

If one point of the body is fixed with regard to translation, say point P as in Figures 1.41 and 1.42, then the acceleration of the center of mass is obtained from Eq. (1.148) which, when substituted into Eq. (1.163), gives *the equation of motion for pure rotation*,

$$\vec{M}_P = I_P \, \vec{\alpha} \tag{1.165}$$

where

$$I_P = I_G + \left\| \vec{r}_{P/G} \right\|^2 m \tag{1.166}$$

is the moment of inertia about the axis through P. Equation (1.166) is a statement of *the Parallel Axis Theorem*. Equations (1.160), (1.162) and (1.163) govern the motion of a rigid body. For the special case of a rigid body with one point fixed with regard to translation, these equations reduce to Eq. (1.165).

Work and Energy

The relations that govern work and energy for a rigid body may be found by applying the corresponding relations for particle systems in a manner similar to that which was done to obtain the equations of motion. Alternatively, we can operate directly on the equations of motion that were established above for a rigid body.

Equation (1.125) can be applied directly, with the total mass m interpreted as given by Eq. (1.161). This gives the *work-kinetic energy relation for translational motion of the center of mass of a rigid body*,

$$\int_{\vec{r}_{G_1}}^{\vec{r}_{G_2}} \vec{F} \cdot d\vec{r}_G = \tfrac{1}{2}m{v_{G_2}}^2 - \tfrac{1}{2}m{v_{G_1}}^2 \tag{1.167}$$

The left hand side of Eq. (1.167) corresponds to the work done by the resultant external force acting on the body as it follows the trajectory of the center of mass, while the right hand side is the change in *kinetic energy of translational motion* of the body. The work-energy relation for motion relative to the center of mass may be found in an analogous fashion. We, therefore, first form the scalar dot product of the governing equation for rotational motion about the center of mass, Eq. (1.162), with incremental rotational displacement and then integrate the resulting expression between two configurations. We thus obtain the *work-kinetic energy relation for motion of a rigid body about the center of mass*,

$$\int_{\vec{\theta}_1}^{\vec{\theta}_2} \vec{M}_G \cdot d\vec{\theta} = \tfrac{1}{2}I_G\dot{\theta}_2^{\,2} - \tfrac{1}{2}I_G\dot{\theta}_1^{\,2} \tag{1.168}$$

The expression on the left-hand side of Eq. (1.168) corresponds to the work done by the resultant moment of the external forces about the axis through the center of mass as it rotates between the two configurations. The right-hand side is the corresponding change in kinetic energy of rotational motion (i.e., motion relative to the center of mass). Performing the same operation on Eq. (1.165) gives the *work-kinetic energy principle for a rigid body with one point fixed*,

$$\int_{\vec{\theta}_1}^{\vec{\theta}_2} \vec{M}_P \cdot d\vec{\theta} = \tfrac{1}{2}I_P\dot{\theta}_2^{\,2} - \tfrac{1}{2}I_P\dot{\theta}_1^{\,2} \tag{1.169}$$

The kinetic energy for this case is, of course, observed to be purely rotational.

Let us next partition the forces and moments acting on the body into conservative and nonconservative forces and moments as follows,

$$\vec{F} = \vec{F}^{(C)} + \vec{F}^{(NC)} \tag{1.170}$$

$$\vec{M} = \vec{M}^{(C)} + \vec{M}^{(NC)} \tag{1.171}$$

where the superscript C and superscript NC denote conservative and nonconservative, respectively. Proceeding as in Section 1.5.2, the work done by the conservative forces are each, by definition, independent of path and thus may be expressed as (the negative of) a change in potential energy. Hence,

$$\int_{\vec{r}_1}^{\vec{r}_2} \vec{F}^{(C)} \cdot d\vec{r} = -\Delta \mathcal{U}^{(F)} \tag{1.172}$$

$$\int_{\vec{\theta}_1}^{\vec{\theta}_2} \vec{M}^{(C)} \cdot d\vec{\theta} = -\Delta \mathcal{U}^{(M)} \tag{1.173}$$

Substitution of Eqs. (1.170)–(1.173) into Eqs. (1.167)–(1.169) gives the *alternate forms of the work-energy relations for a rigid body*,

$$\int_{\vec{r}_1}^{\vec{r}_2} \vec{F} \cdot d\vec{r} = \Delta \mathcal{U}^{(F)} + \Delta \tfrac{1}{2} m v_G^{\;2} \tag{1.174}$$

$$\int_{\vec{\theta}_1}^{\vec{\theta}_2} \vec{M}_G^{(NC)} \cdot d\vec{\theta} = \Delta \mathcal{U}^{(M)} + \Delta \tfrac{1}{2} I_G \dot{\theta}^2 \tag{1.175}$$

$$\int_{\vec{\theta}_1}^{\vec{\theta}_2} \vec{M}_P^{(NC)} \cdot d\vec{\theta} = \Delta \mathcal{U}^{(M)} + \Delta \tfrac{1}{2} I_P \dot{\theta}^2 \tag{1.176}$$

If the work done by the nonconservative forces and moments vanish, then the total mechanical energy of the body is conserved.

Impulse and Momentum

The relations that govern impulse and momentum for a rigid body may be found by integrating the corresponding equations of motion for translation and rotation over time. Hence, multiplying Eqs. (1.160), (1.162), (1.163) and (1.165) by dt and integrating between two instants in time gives the linear and angular impulse-momentum relations for a rigid body. We thus have the *impulse-momentum principle for a rigid body*,

$$\int_{t_1}^{t_2} \vec{F} \, dt = m \vec{v}_G(t_2) - m \vec{v}_G(t_1) \tag{1.177}$$

We likewise obtain the *angular impulse-momentum principle for rotation of a rigid body about the center of mass*,

$$\int_{t_1}^{t_2} \vec{M}_G \, dt = I_G \vec{\omega}(t_2) - I_G \vec{\omega}(t_1) \tag{1.178}$$

and the *angular impulse-momentum principle about an arbitrary point P*

$$\int_{t_1}^{t_2} \vec{M}_P \, dt = \left[I_G \vec{\omega} + \left(\vec{r}_{P/G} \times m\vec{v}_{P/G} \right) \right]_{t=t_2} - \left[I_G \vec{\omega} + \left(\vec{r}_{P/G} \times m\vec{v}_{P/G} \right) \right]_{t=t_1} \quad (1.179)$$

Equations (1.177)–(1.179) pertain to a rigid body in general motion. If point P is fixed with regard to translation, the above equations may be replaced by the *angular impulse-momentum principle about a fixed point*,

$$\int_{t_1}^{t_2} \vec{M}_P \, dt = I_P \vec{\omega}(t_2) - I_P \vec{\omega}(t_1) \quad (1.180)$$

If an impulse in one of the equations vanishes then the corresponding momentum is said to be conserved. It may be seen that, for a given force system, angular momentum may be conserved about one point while not about another.

Example 1.6

A uniform rigid rod of mass m and length L is pinned at one end as shown. If the rod is released from rest when horizontal, determine the velocity of the tip when the rod is vertical. Also determine the reactions at the pin.

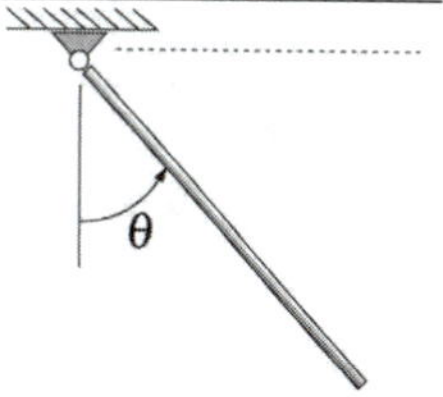

Figure E1.6-1

Solution

Let us first consider the kinetic diagram for the system (Figure E1.6-2), where it may be observed that only conservative forces and moments act on the body.

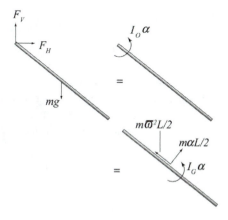

Figure E1.6-2 Kinetic diagram for swinging rod.

Application of Eq. (1.176) with $\vec{M}_O^{(NC)} = \vec{0}$ gives

$$0 = \left[\tfrac{1}{2}I_O\dot{\theta}^2 - 0\right] + \left[-mg(L/2)\cos\theta - 0\right] \tag{a}$$

where we have chosen the datum to be at the horizontal configuration. Solving for the angular speed, and using Eq. (1.78) and the relation $I_O = mL^3/3$, gives the speed of the tip of the rod as

$$v_{tip}\big|_{\theta=0} = L\dot{\theta}\big|_{\theta=0} = \sqrt{3gL} \qquad \triangleleft\text{(b)}$$

directed to the left.

We next apply Eq. (1.165) and solve for α to get

$$\alpha_2 = \alpha\big|_{\theta=0} = -\frac{mg(L/2)\sin\theta}{I_O}\bigg|_{\theta=0} = 0 \qquad \text{(c)}$$

Then, applying Eq. (1.160) in component form along the horizontal and vertical directions when $\theta = 0$, and incorporating the results of Eqs. (b) and (c) and the relation $I_O = mL^3/3$, gives the reactions at the pin,

$$F_H = m(L/2)\alpha_2 = 0 \qquad \triangleleft\text{(d-1)}$$

$$F_V - mg = m(L/2)\dot{\theta}_2^2 \implies F_V = m\left[g + \tfrac{3}{2}g\right] = 2mg \qquad \triangleleft\text{(d-2)}$$

1.6 CONCLUDING REMARKS

In this chapter we discussed and reviewed some fundamental issues pertinent to the study of vibrations. These included the concepts of degrees of freedom and the modeling of complex systems as equivalent lower degree of freedom systems under appropriate circumstances. We also reviewed complex numbers and the basic principles of elementary dynamics. With this basic background we are now ready to begin our study of vibrations. Additional background material will be introduced in subsequent chapters, as needed.

BIBLIOGRAPHY

Baruh, H. *Analytical Dynamics*, McGraw-Hill, Boston, 1999.

Beer, F.P., Johnston, E.R., and Clausen, W.E., *Vector Mechanics for Engineers - Dynamics*, 7th ed., McGraw-Hill, New York, 2004.

Churchill, R.V. and Brown, J.W., *Complex Variables with Applications*, 4th ed., McGraw-Hill, New York, 1984.

Fox, R.W. and McDonald, A.T., *Introduction to Fluid Mechanics*, 5th ed., Wiley, New York, 1998.

Greenwood, D.T., *Principles of Dynamics*, 2nd ed., Prentice-Hall, Englewood Cliffs, 1988.

Landau, L.D. and Lifschitz, E.M., *Course of Theoretical Physics – Vol. 6: Fluid Mechanics*, Pergamon, Oxford, 1975.

Meriam, J.L. and Kraig, L.G., *Engineering Mechanics - Dynamics*, 5th ed., Wiley, New York, 2002.

Rayleigh, J.W.S., *The Theory of Sound*, Vol. 1, Dover, New York, 1941.

Shames, I.H., *Mechanics of Fluids*, McGraw-Hill, New York, 1962.

Geer, J.M. and Timoshenko, S.P., *Mechanics of Materials*, 4th ed., PWS, Boston, 1997.

PROBLEMS

1.1 Assess the number of degrees of freedom for the system shown in Figure P1.1.

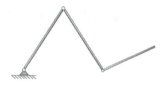

Fig. P1.1

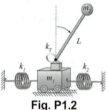

Fig. P1.2

1.2 Assess the number of degrees of freedom for the system shown in Figure P1.2.

1.3 Assess the number of degrees of freedom for the system shown in Figure P1.3.

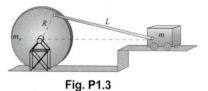

Fig. P1.3

1.4 A 200 lb weight is placed at the free end of a cantilevered beam that is 10 ft in length and has a 2" × 4" rectangular cross section (Figure P1.4). Determine the elastic modulus of the beam if the weight deflects ½ inch.

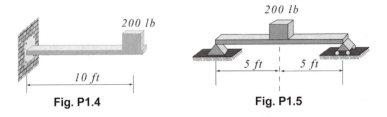

| Fig. P1.4 | Fig. P1.5 |

1.5   Determine the elastic modulus of the beam of Problem 1.4 if it is simply supported and the weight is placed at the center of the span (Figure P1.5).

1.6   Determine the effective stiffness of an equivalent single degree of freedom system for the cantilever beam whose free end is embedded in a rigid block that is free to move transversely as indicated. Do this by solving the elementary beam equation with the appropriate boundary conditions.

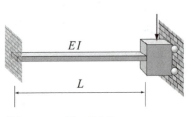

Fig. P1.6

1.7   A flat raft with a 6 ft × 6 ft surface floats in a fresh water lake. Determine the deflection of the raft if a 190 lb man stands at the geometric center.

Fig. P1.7

1.8   Determine the stiffness of a single equivalent spring that represents the three-spring system shown in Figure P1.8.

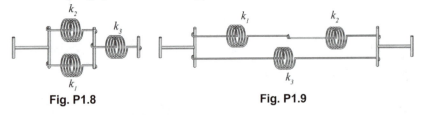

Fig. P1.8   **Fig. P1.9**

1.9   Determine the stiffness of a single equivalent spring that represents the three-spring system shown in Figure P1.9.

1.10 A 2m aluminum rod of circular cross section and 2cm radius is welded end to end to a 3m steel rod of circular cross section and radius 3cm as shown. If the steel rod is fixed at one end and the axes of the two rods are coincident, determine the deflection of the free end of the composite rod if it is pulled axially by a force of 10N at the free end as indicated.

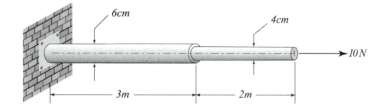

Fig. P1.10

1.11 Determine the rotation of the free end of the rod of Problem 1.10 if a torque of 200N-m is applied at that end (Figure P1.11).

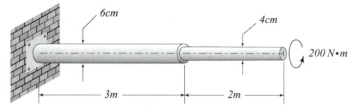

Fig. P1.11

1.12 Determine the effective stiffness of an equivalent single degree of freedom system that models side-sway motion of the frame shown in Figure P1.12.

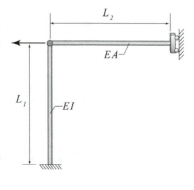

Fig. P1.12

1.13 Determine the effective stiffness of an equivalent single degree of freedom system that models the beam structure shown in Figure P1.13.

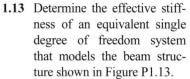

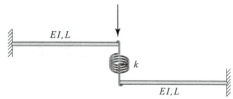

Fig. P1.13

1.14 Determine the effective stiffness of an equivalent single degree of freedom system that models the beam shown in Figure P1.14.

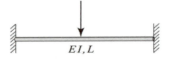

Fig. P1.14

1.15 A raft consists of a board sitting on four cylindrical floats, each of 2 ft radius and oriented vertically as shown. If the raft sits in sea water, determine the deflection of the raft if a 125 lb woman sits at its geometric center.

Fig. P1.15

1.16 Determine the effective stiffness of the multi-rod system shown in Figure P1.16.

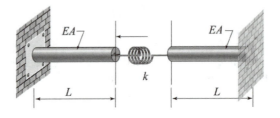

Fig. P1.16

1.17 Determine the effective torsional stiffness of the multi-rod system shown in Figure P1.17 when it is twisted at the center of the span.

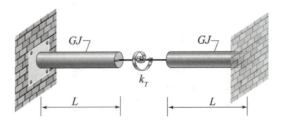

Fig. P1.17

1.18 Determine the effective stiffness of the multi-beam system shown in Figure P1.18. (Neglect twisting.)

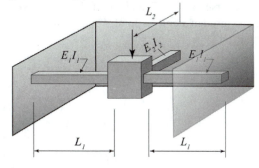

Fig. P1.18

1.19 Use Euler's Formula to establish the identities

$$\cos\psi = \frac{e^{i\psi} + e^{-i\psi}}{2} \quad \text{and} \quad \sin\psi = \frac{e^{i\psi} - e^{-i\psi}}{2i}$$

1.20 Use Eq. (1.63) to show that

$$\cosh\psi = \frac{e^{\psi} + e^{-\psi}}{2} \quad \text{and} \quad \sinh\psi = \frac{e^{\psi} - e^{-\psi}}{2}$$

1.21 Consider the function

$$f(\theta) = \tfrac{1}{2}(a + ib)e^{i\theta} + \tfrac{1}{2}(a - ib)e^{-i\theta}$$

where a and b are real numbers. Show that f can be written in the form

$$f(\theta) = C_1 \cos\theta + C_2 \sin\theta$$

and determine the values of C_1 and C_2.

1.22 Consider the complex number

$$z = \frac{a + ib}{c + id}$$

where a, b, c and d are real. Determine Re(z) and Im(z).

1.23 A cart is attached to an elastic tether of stiffness k as shown. Determine the work done by the tether force as the cart moves from point A to point C; **(a)** if the cart traverses path AC, and **(b)** if the cart traverses path ABC. The tether is fixed through a hole in point A and is unstretched when the cart is at that location.

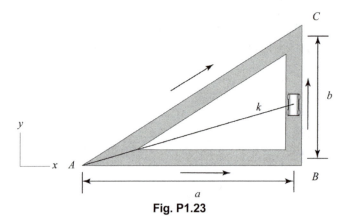

Fig. P1.23

1.24 A block of mass m moves in the horizontal plane. **(a)** Determine the work done by the friction force if the block moves from A to B and back to A along path 1. **(b)** Determine the work done by the friction force as the block moves from A to B to C to D to A along paths 1, 2, 3 and 4. What do your results show?

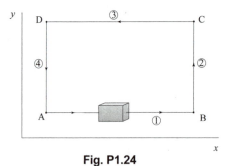

Fig. P1.24

1.25 The timing device shown is deflected by an angle θ_0 and released from rest. Determine the velocity of the mass m when the rod passes through the vertical. The mass of the rod may be considered negligible and the spring is untorqued when $\theta = 0$.

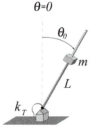

Fig. P1.25

1.26 Derive the equations of motion for the mass-spring-damper system shown, (**a**) using Newton's Second Law and (**b**) using the Principle of Work – Energy.

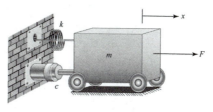

Fig. P1.26

1.27 A tire of mass m and radius of gyration r_G and outer radius R rolls without slipping down the hill as shown. If the tire started from rest at the top of the hill determine the linear and angular velocities of the tire when it reaches the flat road.

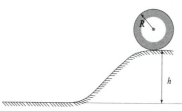

Fig. P1.27

2

Free Vibration of Single Degree of Freedom Systems

The most fundamental system germane to the study of vibrations is the single degree of freedom system. By definition (see Section 1.1), a single degree of freedom system is one for which only a single independent coordinate is needed to describe the motion of the system completely. It was seen in Chapter 1 that, under appropriate circumstances, many complex systems may be adequately represented by an equivalent single degree of freedom system. Further, it will be shown in later chapters that, under a certain type of transformation, the motion of discrete multi-degree of freedom systems and continuous systems can be decomposed into the motion of a series of independent single degree of freedom systems. Thus, the behavior of single degree of freedom systems is of interest in this context as well as in its own right. In the next few chapters the behavior of these fundamental systems will be studied and basic concepts of vibration will be introduced.

2.1 FREE VIBRATION OF UNDAMPED SYSTEMS

The oscillatory motion of a mechanical system may be generally characterized as one of two types, free vibration or forced vibration. Vibratory motions that occur without the action of external dynamic forces are referred to as *free vibrations*, while those resulting from dynamic external forces are referred to as forced vibrations. In this chapter we shall study free vibrations of single degree of freedom systems. We first establish the general form of the equation of motion and its associated solution.

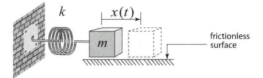

Figure 2.1 Mass-spring system.

2.1.1 Governing Equation and System Response

It was seen in Chapter 1 that many mechanical systems can be represented as equivalent single degree of freedom systems and, in particular, as equivalent mass-spring systems. Let us therefore consider the system comprised of a mass m attached to a linear spring of stiffness k that is fixed at one end, as shown in Figure 2.1. Let the mass be constrained so as to move over a horizontal frictionless surface, and let the coordinate x measure the position of the mass with respect to its rest configuration, as indicated. Thus, $x = 0$ corresponds to the configuration of the horizontally oriented system when the spring is unstretched. We wish to determine the motion of the mass as a function of time, given the displacement and velocity of the mass at the instant it is released. If we let the parameter t represent time, we will know the motion of a given system if we know $x(t)$ for all times of interest. To accomplish this, we must first derive the equation of motion that governs the given system. This is expedited by examination of the *dynamic free-body diagram* (DFBD) for the system, also known as the *kinetic diagram*, depicted in Figure 2.2. In that figure, the applied force acting on the system (the cause) is shown on the left hand side of the figure, and the inertia force (the response) is shown on the right hand side of the figure. The kinetic diagram is simply a pictorial representation of Newton's Second Law of Motion and, as was seen in Section 1.5, greatly aids in the proper derivation of the governing equations for complex systems. In Figure 2.2, and throughout this text, we employ the notation that superposed dots imply (total) differentiation with respect to time (i.e., $\dot{x} \equiv dx/dt$, etc.). Stating Newton's Second Law mathematically for the one-dimensional problem at hand, we have that

$$-kx = m\ddot{x}$$

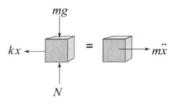

Figure 2.2 Kinetic diagram.

Upon rearranging terms and dividing through by m, we obtain the governing equation

$$\ddot{x} + \omega^2 x = 0 \tag{2.1}$$

where

$$\omega = \sqrt{\frac{k}{m}} \tag{2.2}$$

Equation (2.1) is known as the *harmonic equation*, and the parameter ω is referred to as the *natural* (circular) *frequency*. It may be seen that the natural frequency defines the undamped system in the sense that all the parameters that characterize the system are contained in the single parameter ω. Thus, systems with the same stiffness to mass ratio will respond in the same way to a given set of initial conditions. The physical meaning and importance of the natural frequency of the system will be es-tablished once we determine the response of the system. As the solution of Eq. (2.1) will give the motion of the system as a function of time, we shall next determine this solution.

Suppose the mass is pulled some distance away from its equilibrium position and subsequently released with a given velocity. We wish to determine the motion of the mass-spring system when it is released from such an initial configuration. That is, we wish to obtain the solution to Eq. (2.1) subject to a general set of initial condi-tions. To do this, let us assume a solution of the form

$$x(t) \sim C e^{st} \tag{2.3}$$

where e is the exponential function, and the parameters C and s are constants that are yet to be determined. In order for Eq. (2.3) to be a solution to Eq. (2.1) it must, by definition, satisfy that equation. That is it must yield zero when substituted into the left-hand side of Eq. (2.1). Upon substitution of Eq. (2.3) into Eq. (2.1) we have that

$$\left(s^2 + \omega^2\right) C e^{st} = 0 \tag{2.4}$$

Thus, for Eq. (2.3) to be a solution of Eq. (2.1), the parameters C and s must satisfy Eq. (2.4). One obvious solution is $C = 0$, which gives $x(t) \equiv 0$ (the trivial solution). This corresponds to the equilibrium configuration, where the mass does not move. This solution is, of course, uninteresting to us as we are concerned with the dynamic response of the system. For nontrivial solutions [$x(t) \neq 0$ identically], it is required that $C \neq 0$. If this is so, then the bracketed term in Eq. (2.4) must vanish. Setting the bracketed expression to zero and solving for s we obtain

$$s = \pm i\omega \tag{2.5}$$

where $i \equiv \sqrt{-1}$. Equation (2.5) suggests two values of the parameter s, and hence two solutions of the form of Eq. (2.3), that satisfy Eq. (2.1). Since Eq. (2.1) is a linear differential equation, a linear combination of these solutions is also a solution. Thus, the general solution to Eq. (2.1) is given by

$$x(t) = C_1 e^{i\omega t} + C_2 e^{-i\omega t} \tag{2.6}$$

where C_1 and C_2 are complex constants. While Eq. (2.6) is a form of the general solution to Eq. (2.1), the solution may be written in alternative forms that lend themselves to physical interpretation. If we apply Euler's Formula, Eq. (1.61), to Eq. (2.6) we find that the solution may be expressed in the alternate form

$$x(t) = A_1 \cos \omega t + A_2 \sin \omega t \tag{2.7}$$

where

$$A_1 = C_1 + C_2 \quad \text{and} \quad A_2 = i(C_1 - C_2) \tag{2.8}$$

are real constants. Let us further define two additional constants, A and ϕ, such that

$$A = \sqrt{A_1^2 + A_2^2} \quad \text{and} \quad \phi = \text{Tan}^{-1}\left(A_2 / A_1\right) \tag{2.9}$$

or, equivalently,

$$A_1 = A \cos \phi \quad \text{and} \quad A_2 = A \sin \phi \tag{2.10}$$

Substitution of Eqs. (2.9) and (2.10) into Eq. (2.7) provides the alternate, and physically interpretable, form of the solution given by

$$x(t) = A \cos(\omega t - \phi) \tag{2.11}$$

The constant A is referred to as the *amplitude* of the oscillation, and the constant ϕ is called the *phase angle*. The reason for this terminology will become apparent shortly. Equations (2.6), (2.7) and (2.11) are different forms of the same solution. Each has its place. The last form, Eq. (2.11), allows for the most direct physical interpretation. The constants of integration are evaluated by imposing the initial conditions. Clearly, once one pair of integration constants is determined then all integration constants are determined.

The system is put into motion by displacing the mass and releasing it at some instant in time. We shall take the instant of release as our reference time, $t = 0$. Thus, if at the instant of release the mass is at position x_0 and is moving at velocity v_0, the initial conditions may be formally stated as

$$x(0) = x_0, \quad \dot{x}(0) = v(0) = v_0 \tag{2.12}$$

Imposition of the initial conditions, Eq. (2.12), on the solution given by Eq. (2.7) yields

$$A_1 = x_0 \quad \text{and} \quad A_2 = v_0 / \omega \tag{2.13}$$

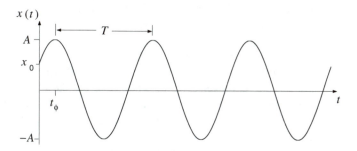

Figure 2.3 Time history of a harmonic oscillator in free vibration.

It follows from Eqs. (2.7) and (2.13) that the response is given by

$$x(t) = x_0 \cos \omega t + \frac{v_0}{\omega} \sin \omega t \qquad (2.14)$$

It follows from Eqs. (2.9), (2.11) and (2.13) that the response is also given by the equivalent form

$$x(t) = A \cos(\omega t - \phi) \qquad (2.15)$$

where

$$A = \sqrt{x_0^2 + (v_0/\omega)^2} \qquad (2.16)$$

and

$$\phi = \tan^{-1}\left(\frac{v_0}{\omega x_0}\right) \qquad (2.17)$$

A plot of the response described by Eq. (2.15) is displayed in Figure 2.3. It may be seen from the figure that A is indeed the *amplitude* of the oscillations. It corresponds to the magnitude of the maximum displacement of the mass during its motion and therefore represents a bound on the displacement of the mass from its equilibrium position. It is also seen that the cosine function is shifted by an amount

$$t_\phi = \phi/\omega \qquad (2.18)$$

along the time axis. This time difference is referred to as the *phase shift*, or *phase lag*, of the response in that the response is shifted from, or lags behind, a pure cosine re-sponse by this amount of time. It is seen from Eq. (2.17) that the pure cosine response ($\phi = 0$) corresponds to the case of vanishing initial velocity ($v_0 = 0$). From a physical perspective, for such initial conditions, the mass is initially displaced (held) away from its equilibrium configuration, say $x_0 > 0$, and released from rest. The mass then immediately begins to move toward the equilibrium configuration, eventually passes it and moves away from it, and so on, as described by the cosine function. For the case of nonvanishing initial velocity the mass is initially displaced away from the equilibrium configuration, say $x_0 > 0$, and "launched" with an initial velocity, say $v_0 > 0$, as if the mass is thrown or hit with a baseball bat or a golf club. For such initial conditions, the mass first moves further away from the equilibrium configuration, until it stops for an instant and reverses its direction. The time to this first direction reversal is the phase lag, t_ϕ. For subsequent times the mass follows the pure cosine function as if it were released from rest. The initial velocity thus causes the system to lag behind the initially quiescent system by t_ϕ.

It may be seen from Figure 2.3, as well as from Eq. (2.15), that the response repeats itself (i.e., it is periodic) at time intervals of $2\pi/\omega$. In this context, the quantity

$$T = \frac{2\pi}{\omega} \tag{2.19}$$

is referred to as the *period* of the vibratory response. If the mass is at a certain position x at time t, the mass will be at the same location at times $t = t + nT$, where n is any integer. The response therefore has the property that

$$x(t + nT) = x(t) \quad (n = 1, 2, ...) \tag{2.20}$$

Equation (2.20) is characteristic of all functions of period T and is, in fact, the formal definition of a periodic function. The inverse of the period is the natural frequency v. It corresponds to the rate at which the vibrations are occurring and hence to the number of oscillations, or *cycles*, that the system goes through per unit of time. The frequency v differs slightly from the natural circular frequency ω, which is related to v, and represents an angular rate in the complex plane as discussed in Section 1.4. A summary of the parameters introduced in this section (along with sample units) is given in Table 2.1.

Table 2.1 Free Vibration Parameters

SYMBOL	DEFINITION	SAMPLE UNITS
A	amplitude	meters, feet
$T = 2\pi/\omega = 1/v$	period	seconds
$v = 1/T = \omega/2\pi$	natural frequency	cycles per second
$\omega = 2\pi v = 2\pi/T$	natural (circular) frequency	radians per second
ϕ	phase angle	radians
$t_\phi = \phi/\omega$	phase shift	seconds

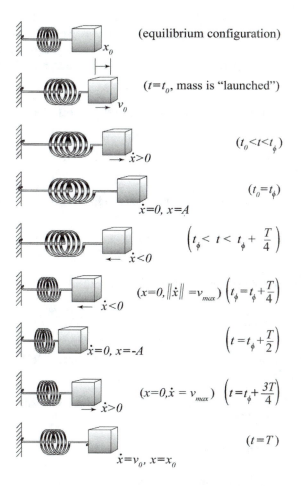

Figure 2.4 Freeze-frame depiction of motion of mass-spring system.

The corresponding physical depiction of the motion described by Eq. (2.15) is sketched, at selected instants in time, in Figure 2.4. Recognizing that the slope of the x vs. t plot of Figure 2.3 corresponds to the velocity of the mass at any instant in time, the solution given by Eq. (2.15) and its plot (Figure 2.3) offer the scenario depicted in Figure 2.4. It may be seen that the mass moves first in one direction, slows, then stops and subsequently reverses its direction when $t = t_\phi = \phi/\omega$. The mass then continues to move in the opposite direction and eventually slows, stops and reverses direction again. The mass then moves in its original direction and eventually reaches its initial position and achieves its initial velocity at the same instant. The process then starts all over again and repeats itself continuously.

Example 2.1

A 2 kg block sits on a frictionless table and is connected to a coil of stiffness 4.935 N/m. The mass is displaced 1 m and released with a velocity of 2.721 m/sec. Determine the natural frequency of the system and the amplitude and phase lag of the response. Sketch and label the time history of the response.

Solution

The natural frequency of the mass-spring system is calculated from Eq. (2.2) to give

$$\omega = \sqrt{\frac{k}{m}} = \sqrt{\frac{4.935}{2}} = 1.571 \text{ rad/sec} \tag{a}$$

The natural period of the motion is then, from Eq. (2.19),

$$T = \frac{2\pi}{\omega} = \frac{2\pi}{1.571} = 4.0 \text{ secs} \tag{b}$$

The amplitude of the motion is found using Eq. (2.16). Hence,

$$A = \sqrt{x_0^2 + (v_0/\omega)^2} = \sqrt{(1)^2 + (2.721/1.571)^2} = 2 \text{ m} \tag{c}$$

The phase angle is calculated using Eq. (2.17). We thus obtain

$$\phi = \tan^{-1}\left(\frac{v_0}{\omega x_0}\right) = \tan^{-1}\left(\frac{3}{(1.571)(1)}\right) = 63.36° = 1.088 \text{ rads} \tag{d}$$

The phase lag is then, from Eq. (2.18),

$$t_\phi = \phi/\omega = 1.088/1.571 = 0.693 \text{ secs} \tag{e}$$

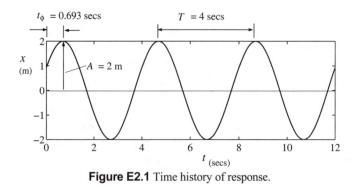

Figure E2.1 Time history of response.

Example 2.2

Consider side-sway motion of the elastic column of length L and bending stiffness EI, which is pinned to a rigid mass m as shown (Figure E2.2a), where the total mass of the column is much smaller than that of the supported mass. If ρ is the mass density of the column and A is its cross-sectional area, determine the response of the structure when the supported mass is displaced a distance x_0 from the equilibrium position and then released from rest at that position.

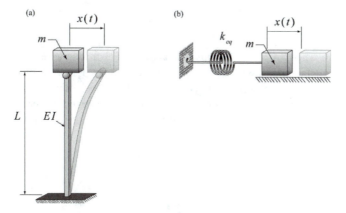

Figure E2.2 (a) Column-mass structure, (b) equivalent system.

Solution

If $\rho A L \ll m$ we may treat the column-point mass system as the equivalent single degree of freedom system shown in Figure E2.2b, as discussed in Section 1.2.2. For the particular structure under consideration we have, from Eq. (1.14), that

$$k_{eq} = 3EI/L^3$$

It then follows from Eqs. (2.2) and (2.19) that the natural frequency of the system is

$$\omega = \sqrt{\frac{3EI}{mL^3}} \tag{a}$$

and the natural period is

$$T = 2\pi \sqrt{\frac{mL^3}{3EI}} \tag{b}$$

If the mass is released from rest from the initial position x_0 then $v_0 = 0$. The response is then found from Eqs. (2.15)–(2.17) to be

$$x(t) = x_0 \cos \sqrt{\frac{3EI}{mL^3}} t \qquad \triangleleft (c)$$

The structure therefore oscillates from side to side at the frequency given by Eq. (a) with amplitude $A = x_0$, as described by Eq. (c).

Example 2.3

A 20 lb wheel of 15 inch radius is attached to a 3 ft long axle that is supported by a fixed frictionless collar, as shown. The axle is 1 inch in diameter and its mass is negligible compared with that of the wheel. If the wheel is rotated slightly and released, a period of 0.1 seconds is observed. Determine the shear modulus of the axle. (Assume that the mass of the wheel is uniformly distributed.)

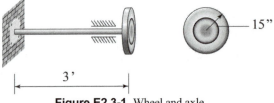

3'

Figure E2.3-1 Wheel and axle.

Solution

Since the mass of the axle is small compared with that of the wheel, the axle may be modeled as an equivalent torsional spring as discussed in Section 1.2.3. This renders the equivalent single degree of freedom system shown in Figure E2.3-2.

Let us first derive the equation of motion for the equivalent system, with the z-axis chosen to coincide with the axis of the shaft. In this regard, we apply Eq. (1.165) to the present problem with the help of the kinetic diagram of Figure E2.3-3.

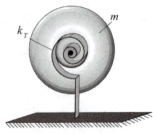

Figure E2.3-2 Equivalent system.

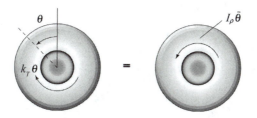

Figure E2.3-3 Kinetic diagram.

Hence,

$$-k_T\theta = I_{zz}\ddot{\theta}$$

or

$$\ddot{\theta} + \omega^2\theta = 0 \tag{a}$$

where

$$\omega = \sqrt{\frac{k_T}{I_{zz}}} \tag{b}$$

The stiffness of the equivalent system is given by Eq. (1.29) as

$$k_T = GJ/L \tag{c}$$

where G, J, and L are respectively the shear modulus, area polar moment of inertia, and length of the axle. Further, the mass moment of inertia for a disk of mass m and radius R is

$$I_{zz} = \tfrac{1}{2}mR^2 \tag{d}$$

Substitution of Eqs. (c) and (d) into Eq. (b) and solving for the shear modulus gives

$$G = \frac{k_T L}{J} = \frac{\omega^2 I_{zz} L}{J} = \left(\frac{2\pi}{T}\right)^2 \frac{\left(mR^2/2\right)L}{\left(\pi D^4/32\right)} = 64\pi \frac{mR^2 L}{T^2 D^4} \tag{e}$$

Substitution of the given values of the various parameters into Eq. (e) gives the value

$$G = 64\pi \frac{[20/(32.2\times12)](15)^2(3\times12)}{(0.1)^2(1)^4} = 8.43\times10^6 \text{ psi} \qquad \triangleleft \text{ (f)}$$

Example 2.4

A 200 lb man floats on a 6 ft by 2 ft inflatable raft in a quiescent swimming pool. Estimate the period of vertical bobbing of the man and raft system should the man be disturbed. Assume that the weight of the man is distributed uniformly over the raft and that the mass of the raft is negligible.

Solution

The effect of the stiffness of the water, as pertains to the man and the raft, is modeled as an equivalent spring as discussed in Section 1.2.4. The corresponding equivalent system is shown in the adjacent figure.

We thus have from Eq. (1.40) that

$$k_{eq} = (\rho_{water}g)A = 62.4[(2)(6)] = 749 \text{ lb/ft} \qquad \text{(a)}$$

The natural frequency of the system is then obtained from Eq. (2.2) as

$$\omega = \sqrt{\frac{749}{200/32.2}} = 11.0 \text{ rad/sec} \qquad \text{(b)}$$

Equation (2.19) then gives the period as

$$T = \frac{2\pi}{11.0} = 0.571 \text{ sec} \qquad \text{(c)}$$

Now, from Table 2.1,

$$v = \frac{1}{0.571} = 1.75 \text{ cps} \qquad \triangleleft \text{(d)}$$

Thus, the man will oscillate through a little less than 2 cycles in a second. (How does this compare with your own experience?)

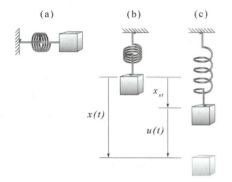

Figure 2.5 Mass-spring system: (a) in horizontal equilibrium, (b) vertical configuration with spring unstretched, (c) vertical equilibrium configuration and generic position of mass in motion.

2.1.2 The Effect of Gravity

In the previous section we considered systems modeled as an equivalent mass-spring system. In those models the equivalent mass-spring system was aligned horizontally along a frictionless surface, and so gravity was not a factor. Suppose now that we take the horizontally configured mass-spring system in its undeformed configuration as depicted in Figure 2.5a and hold the mass so that the spring remains unstretched as we rotate the system until it is aligned vertically as shown in Figure 2.5b. Let us next allow the mass to deflect very slowly until the system comes to equilibrium as shown in Figure 2.5c. Let x measure the position of the mass with respect to it position when the spring is unstretched, and let u measure the deflection of the mass from its equilibrium position in the vertical configuration as indicated.

Static Deflection

Let x_{st} correspond to the static deflection of the mass as depicted in Figure 2.5, and let us consider the associated free body diagram of the equilibrated system shown in Figure 2.6. For equilibrium the forces must sum to zero and hence the gravitational force and the spring force must balance. Thus,

Figure 2.6 Free-body diagram for mass in equilibrium.

$$kx_{st} = mg \tag{2.21}$$

where g is the gravitational acceleration.

Dynamic Response

Suppose now that the mass of the vertically oriented system is deflected away from its equilibrium position and released. The corresponding dynamic free body diagram of the mass is shown in Figure 2.7. Application of Newton's Second Law gives

$$mg - kx = m\ddot{x} \tag{2.22}$$

which, after substitution of Eq. (2.21) and rearranging terms, takes the form

$$m\ddot{x} + k(x - x_{st}) = 0 \tag{2.23}$$

Let

$$u = x - x_{st} \tag{2.24}$$

represent the displacement of the mass from its equilibrium position. Substituting Eq. (2.24) into Eq. (2.23) and noting that x_{st} is a constant gives the equation of motion of the system expressed in terms of the position relative to the equilibrium configuration. Hence,

$$\ddot{u} + \omega^2 u = 0 \tag{2.25}$$

where ω is given by Eq. (2.2). It follows from Eqs. (2.15) - (2.17) that

$$u(t) = A\cos(\omega t - \phi) \tag{2.26}$$

where

$$A = \sqrt{u_0^2 + (v_0/\omega)^2} \tag{2.27}$$

$$\phi = \tan^{-1}\left(\frac{v_0}{\omega u_0}\right) \tag{2.28}$$

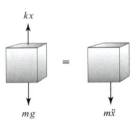

Figure 2.7 Kinetic diagram for vertical motion.

and u_0 and v_0 are, respectively, the initial displacement and initial velocity of the mass. Consideration of Eq. (2.26) shows that the mass oscillates about its equilibrium position in precisely the same manner that the mass of the horizontally oriented system oscillates about its (unstretched) equilibrium position. Thus, from a vibrations perspective, gravity has no effect on the behavior of the system except to change the position about which the mass oscillates.

Example 2.5

A block of mass m is attached to a spring of stiffness k_s that is suspended from a long cantilever beam of length L and bending stiffness EI as shown in Figure E2.5a. If the mass of the block is much greater than the mass of the beam, determine the natural frequency of the system.

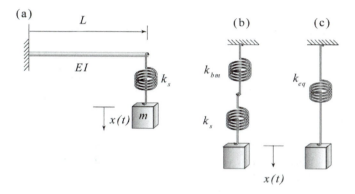

Figure E2.5 Elastic beam with suspended mass: (a) physical system, (b) representation of compound system with beam as equivalent 1 d.o.f. system, (c) equivalent system for beam and spring combination.

Solution

The natural frequency of an equivalent single degree of freedom system of stiffness k_{eq} and mass m is given by Eq. (2.2) as

$$\omega = \sqrt{\frac{k_{eq}}{m}} \tag{a}$$

We therefore need to determine the equivalent stiffness k_{eq}.

The equivalent single degree of freedom system is realized by first considering the equivalent system for the beam alone. As discussed in Section 1.2.2, a cantilever beam may be modeled as a spring of stiffness k_{bm}, where, from Eq. (1.14),

$$k_{bm} = 3EI/L^3 \tag{b}$$

The composite beam-spring structure is thus represented as two springs connected in series as shown in Figure E2.5b. The two spring structure may in
turn be represented as an equivalent single spring of stiffness k_{eq} as shown in
Figure E2.5c and discussed in Section 1.3.2. We then have, from Eq. (1.51),
that

$$k_{eq} = \frac{k_s}{1+\left(k_s/k_{bm}\right)} \tag{c}$$

Substituting Eq. (b) in to Eq. (c) gives the equivalent stiffness of the beam-
spring-mass system as

$$k_{eq} = \frac{k_s}{1+\left(k_s L^3/3EI\right)} \tag{d}$$

Substitution of Eq. (d) into Eq. (a) gives the natural frequency sought. Hence,

$$\omega = \frac{\sqrt{k_s/m}}{\sqrt{1+\left(k_s L^3/3EI\right)}} \tag{$\triangleleft$ (e)}$$

It may be seen that the expression in the numerator of Eq. (e) corresponds to
the natural frequency of the mass-spring system attached to a rigid support, and
that this frequency is recovered if the beam has infinite bending stiffness. That
is, it may be seen that

$$\omega \to \sqrt{k_s/m} \text{ as } EI \to \infty$$

We thus see that the effect of adding compliance to the support of the mass-
spring system, in this case the flexibility of an elastic beam, is to lower the
natural frequency of the system. Such behavior is characteristic of mechanical
systems in general.

Example 2.6

Consider the small two wheel trailer of *known* mass m_A (shown in end view in
Figure E2.6-1). Let the tires of the trailer be modeled as elastic springs, each of
unknown stiffness $k/2$, as indicated, and let a removable package of unknown
mass, m_B, be secured to the trailer as shown. How might we measure (***a***) the
stiffness of the spring and (***b***) the mass of the package if there is no scale available in our laboratory? (***c***) If the natural period of oscillation of the trailer is increased by a factor of 1.05 when hauling the package, determine the weight of
the package as a percentage of the weight of the unloaded trailer.

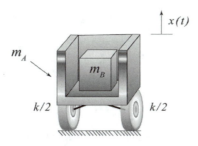

Figure E2.6-1 Two wheel trailer carrying package.

Solution

If we displace and release the known mass, m_A, with or without the unknown mass, m_B, attached we may measure the period of vibration (or equivalently the corresponding natural frequency), as well as the amplitude and phase of the motion. The latter two are immaterial for the purposes of the present problem. Thus, we may determine the stiffness of the springs and the unknown mass of the package in terms of the known mass of the trailer and the measured period for both the one mass and two mass systems.

(a)

Let us first remove the package. Let us next push down on and then release the trailer (mass m_A), and trace the subsequent motion of the lone mass on a recording device. The trace of the response will be similar in appearance to that shown in Figure E2.6-2. We may then read off the period, T_A, of the motion of the single mass from this trace. Equations (2.2) and (2.19) then give, for the trailer alone,

$$\omega_A = \frac{2\pi}{T_A} = \sqrt{\frac{k}{m_A}} \qquad (a)$$

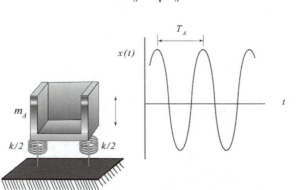

Figure E2.6-2 Free vibration of trailer alone.

Equation (a) may be solved for k to give the spring stiffness in terms of the known mass and measured period,

$$k = \frac{4\pi^2 m_A}{T_A}$$

◁ (b)

(b)

Let us next secure the package in the trailer and then pull and release the combined mass system (Figure E2.6-1). We then trace and record the resulting motion, which would appear as in Fig. E2.6-3, and read off the corresponding period, T_{A+B}, from the plot. Once again, we substitute the measured value of the period into Eqs. (2.2) and (2.19) to determine the natural frequency for the combined mass system. Thus,

$$\omega_{A+B} = \frac{2\pi}{T_{A+B}} = \sqrt{\frac{k}{m_A + m_B}}$$

(c)

Equation (c) may now be solved for m_B. This gives the unknown mass in terms of the known mass and measured spring stiffness as

$$m_B = \frac{k T_{A+B}^2}{4\pi^2} - m_A$$

(d)

Substitution of Eq. (b) into Eq. (d) gives the unknown mass of the package in terms of the known mass of the trailer and the measured periods of the single and combined mass systems. Hence,

$$m_B = m_A \left[(T_{A+B}/T_A)^2 - 1 \right]$$

◁ (e)

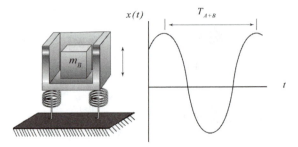

Figure E2.6-3 Free vibration of trailer with package.

(*c*)

From Eq. (e),

$$\frac{m_B g}{m_A g} = (1.05)^2 - 1 = 0.1025 \qquad \lhd \text{(f)}$$

Thus, the weight of the package is 10.25% of the weight of the unloaded trailer.

2.1.3 Work and Energy

Consider a system for which there are no dissipation mechanisms. Since there are no mechanisms for energy dissipation in this ideal system, the total mechanical energy of the system is conserved. As discussed in Section 1.5, the total mechanical energy of the system, $\mathcal{E}$, consists of the sum of the kinetic energy of the system, T, and the potential energy of the system, $\mathcal{U}$. Thus, for such a conservative system,

$$\mathcal{E} = T + \mathcal{U} = \text{constant} \qquad (2.29)$$

It is evident from Eq. (2.29) that during any motion of the system, the kinetic energy increases as the potential energy decreases, and vice versa. It then follows that

$$\begin{aligned} \mathcal{U} &= \mathcal{U}_{max} \quad \text{when} \quad T = T_{min} \\ T &= T_{max} \quad \text{when} \quad \mathcal{U} = \mathcal{U}_{min} \end{aligned} \qquad (2.30)$$

Let us now focus our attention on the simple mass-spring system of Figure 2.1. For this particular system Eq. (2.29) may be stated explicitly as

$$\tfrac{1}{2} m\dot{x}^2 + \tfrac{1}{2} kx^2 = \text{constant} \qquad (2.31)$$

Substitution of Eqs. (2.30) into Eq. (2.31), and equating the resulting expressions, gives the identity

$$\tfrac{1}{2} m\dot{x}_{max}^2 + 0 = 0 + \tfrac{1}{2} kx_{max}^2 \qquad (2.32)$$

Thus, for this simple system, the maximum potential energy is achieved when the spring achieves its maximum extension (contraction). This occurs at the instant the motion is about to reverse itself, at which point the mass stops momentarily. Conversely, the potential energy of the system vanishes when the stretch in the spring vanishes. Thus, the mass achieves its maximum kinetic energy, and hence its maximum speed, as the mass passes through this configuration. Solving Eq. (2.32) for k/m and evaluating Eq. (2.15) at the instants of maximum potential energy and maximum kinetic energy gives the natural frequency of the system in terms of the amplitude of the motion and the maximum speed of the mass. Hence,

$$\omega^2 = \frac{k}{m} = \frac{v_{max}^2}{A^2} \tag{2.33}$$

where $v(t) \equiv \dot{x}(t)$ is the velocity of the mass and thus corresponds to the slope of the x vs. t trace (Figure 2.3), and $x_{max} = A$ (the amplitude of the oscillation). Finally, note that when we differentiate Eq. (2.31) with respect to t, we obtain the equality

$$\dot{x}\left(m\ddot{x} + kx\right) = 0$$

We thus recover the equation of motion for the mass-spring system, Eq. (2.1), for $\dot{x} \neq 0$. This should not surprise us since the work-energy principle is an integral of Newton's Second Law (Section 1.5.1). Such an approach provides an alternative method for the derivation of equations of motion for conservative systems.

2.1.4 The Simple Pendulum

A classic problem in mechanics in general, and vibrations in particular, is the problem of the simple pendulum. The pendulum consists of a "bob" of mass m that is attached to the end of a rod or cord of length L and negligible mass. The rod is pinned to a rigid support at its opposite end, as shown in Figure 2.8. In that figure, the angular coordinate θ measures the position of the rod with respect to its (downward) vertical configuration, and the path coordinate s locates the position of the bob with respect to its position in that configuration, as indicated.

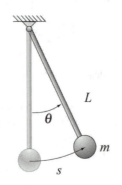

Figure 2.8 The simple pendulum.

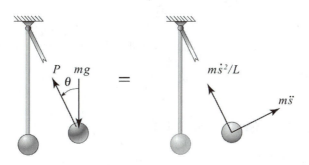

Figure 2.9 Kinetic diagram for bob of simple pendulum.

The Equations of Motion

The equations of motion for the simple pendulum may be derived in several ways. In the context of the discussions of this and the preceding chapter, we may use Newton's laws or we may form the total energy of the system, as per Eq. (2.31), and then differentiate with respect to time. The latter approach will be demonstrated and employed in a subsequent section. We shall here apply Newton's laws directly. We therefore first consider the dynamic free-body diagram of the system (Figure 2.9). In that figure the parameter P corresponds to the tension in the rod, g is the gravitational acceleration, and the inertial force has been resolved into its tangential and normal components. Writing Newton's Second Law in component form along the normal and tangential directions of the path of the bob, we have

$$-mg \sin \theta = m\ddot{s}$$

$$-mg \cos \theta + P = m\frac{\dot{s}^2}{L}$$

Rearranging terms gives

$$\ddot{s} + g \sin \theta = 0 \tag{2.34}$$

and

$$P(t) = m\left[\frac{\dot{s}^2}{L} + g \cos \theta\right] \tag{2.35}$$

It may be seen that the mass drops out of the equation that governs the tangential components since both the restoring force (the weight) and the inertia force are proportional to the mass. It is thus seen that the motion along the path is independent of the mass. Once the motion $s(t)$ is known, the time history of the tension in the cord may be computed using Eq. (2.35). We are therefore interested in solving Eq. (2.34). To do this, it is convenient to express the above equations solely in terms of the angular coordinate θ. In this regard, we substitute the kinematical relationship

$$s = L\theta \tag{2.36}$$

into Eq. (2.34) and divide by L. Doing this renders the equation of motion to the form

$$\ddot{\theta} + \frac{g}{L}\sin\theta = 0 \tag{2.37}$$

Similarly, Eq. (2.35) takes the form

$$P(t) = mg\left[\frac{\dot{\theta}^2}{g/L} + \sin\theta\right] \tag{2.38}$$

The solution of Eq. (2.37) determines the motion of the pendulum as a function of time. It may be seen that the only system parameter that affects the motion of the pendulum is the length of the rod (or cord). It is seen from Eq. (2.38), however, that the tension in the rod (or cord) depends on the mass of the bob, as well as the length of the rod (or cord), as one might expect.

Linearization and the "Small Angle Response"
Equation (2.37) is seen to be a nonlinear differential equation and is valid for all values of the angular displacement θ. Its solution is nontrivial and will be examined in detail in the next section. However, if we here restrict our attention to *small motions* about the vertical (we shall specify what is meant by "small" momentarily), an approximate, yet very applicable, solution is easily established. To obtain this solution we shall first *linearize* Eq. (2.37) and then determine the solution to the resulting equation. To do this, let us first consider the series representation of the sine function given by

$$\sin\theta = \theta - \frac{\theta^3}{3!} + \frac{\theta^5}{5!} - \ldots = \theta\left[1 - \frac{\theta^2}{3!} + \frac{\theta^4}{5!} - \ldots\right] \tag{2.39}$$

If we limit ourselves to motions for which $\theta^2 \ll 1$ (where θ is, of course, measured in radians) then all terms involving θ within the brackets on the right-hand side of Eq. (2.39) may be neglected. We thus make the approximation

$$\sin\theta \approx \theta \tag{2.40}$$

In doing this, we have *linearized* the sine function by approximating it by its dominant linear term. Substitution of Eq. (2.40) into Eq. (2.37) gives the *linearized* equation of motion of the pendulum,

$$\ddot{\theta} + \omega^2\theta = 0 \tag{2.41}$$

where

$$\omega = \sqrt{\frac{g}{L}} \tag{2.42}$$

is the natural frequency of the system. Equation (2.41) is seen to correspond to the harmonic equation, Eq. (2.1), with x replaced by θ. It therefore follows from Eqs. (2.15)–(2.17) that the free-vibration response of the pendulum is given by

$$\theta(t) = A\cos(\omega t - \phi) \tag{2.43}$$

where

$$A = \sqrt{\theta_0^2 + (\chi_0/\omega)^2} \tag{2.44}$$

$$\phi = \tan^{-1}\left\{\frac{\chi_0}{\omega\theta_0}\right\} \tag{2.45}$$

and

$$\theta(0) = \theta_0 \quad \text{and} \quad \dot{\theta}(0) = \chi_0 \tag{2.46}$$

are respectively the initial angular displacement and angular velocity of the pendulum. The response is thus seen to be harmonic with period

$$T = \frac{2\pi}{\omega} = \frac{2\pi}{\sqrt{g/L}} \tag{2.47}$$

The natural frequency and natural period of the harmonic oscillator, and hence of the linear model of the pendulum, is a constant that depends solely on the parameters of the system (in this case the length of the rod or cord). It will be seen that the exact period of the free vibration response, the response of the nonlinear pendulum described by Eq. (2.37), is dependent on the initial conditions described by Eq. (2.46) as well. It will also be seen that this exact period is asymptotic to the approximate constant period given by Eq. (2.47). It will be demonstrated in an example at the end of the next subsection, and in Problem P2.24, that the constant period predicted by the linear model is a reasonable approximation of the actual period, even for moderate initial displacements.

Example 2.7

A child's swing consists of a seat hanging from four chains. A child is sitting quietly in a swing when a parent gently pulls and releases the seat. If the parent observes that it takes approximately 2 seconds for the child and swing to return, what is the length of the swing's hoist?

Solution
From Eq. (2.47),

$$T = \frac{2\pi}{\sqrt{g/L}}$$ (a)

Solving for L and substituting the observed period and the value of g gives the length of the hoist as

$$L = g\left(\frac{T}{2\pi}\right)^2 = 32.2\left(\frac{2}{2\pi}\right)^2 = 3.26' \cong 3'\text{-}3''$$ ◁ (b)

Example 2.8

A uniform rigid disk of mass m and radius r_D is released from rest from some initial position along a circular track segment of radius R as shown. Determine the resulting *small amplitude motion* if the disk rolls without slipping.

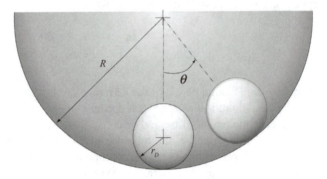

Figure E2.8-1 Disk rolling on circular track.

Solution
Kinematics: Let s represent the path coordinate following the motion of the geometric center of the disk, and let $\rho = R - r_D$ be the corresponding radius of curvature of that path. Further, let v_G be the (linear) speed of the geometric center of the disk and let a_{Gt} and a_{Gn} represent the tangential and normal components of the (linear) acceleration of that point, respectively, as indicated. We then have the following kinematical relations between the path coordinate s and the angular displacement θ (see Section 1.5):

$$s = \rho\theta = (R - r_D)\theta$$ (a)

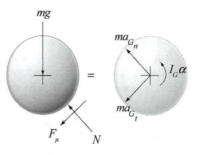

Figure E2.8-2 Kinetic diagram for rolling disk.

$$v_G = \dot{s} = (R - r_D)\dot{\theta} \tag{b}$$

$$a_{Gt} = \ddot{s} = (R - r_D)\ddot{\theta} \tag{c}$$

$$a_{Gn} = \frac{\dot{s}^2}{\rho} = (R - r_D)\dot{\theta}^2 \tag{d}$$

Furthermore, since the disk rolls without slipping we have, from Eqs. (1.154) and (1.155), the following *no slip conditions* between the linear speed and the tangential component of the linear acceleration of the center of the disk, and the angular velocity, ϖ, and the angular acceleration, α :

$$v_G = \varpi\, r_D \tag{e}$$

$$a_{Gt} = \alpha\, r_D \tag{f}$$

Substituting Eq. (b) into Eq. (e), and Eq. (c) into Eq. (f), gives the relations

$$\varpi = \frac{R - r_D}{r_D}\dot{\theta} \tag{g}$$

$$\alpha = \frac{R - r_D}{r_D}\ddot{\theta} \tag{h}$$

Kinetics: Having established the kinematics of the system, we are now ready to derive the associated equations of motion. To do this, we first consider the kinetic diagram for the system (Figure E2.8-2). In that figure, the instantaneous point of contact of the disk with the track is labeled P. The rotational equation of motion of a rigid body about some point P, Eq. (1.163), is rewritten for the present problem as

$$\sum \vec{M}_P = I_G\, \vec{\alpha} + \vec{r}_{G/P} \times m\vec{a}_G \tag{i}$$

where $\vec{r}_{G/P}$ is the position of point G as seen by an observer translating with point P. Since the disk rolls in a single plane, the only nontrivial component of Eq. (i) is the component in the direction of the normal to that plane. Taking moments about an axis perpendicular to the plane and through the contact point P, and substituting these moments into Eq. (i), gives

$$mgr_D \sin\theta = -I_G\alpha - mr_{G/P}^2\alpha = -I_P\alpha \qquad (j)$$

where

$$I_P = I_G + mr_{G/P}^2 \qquad (k)$$

is the moment of inertia of the disk about an axis through the contact point. Further, for a uniform disk, the geometric center and the center of mass coincide. The moment of inertia of a uniform disk of radius r_D about an axis through its geometric center G is given by

$$I_G = I_{cm} = \tfrac{1}{2}mr_D^2 \qquad (l)$$

Substitution of Eq. (l) into Eq. (k) gives the moment of inertia about the axis through P as

$$I_P = \tfrac{3}{2}mr_D^2 \qquad (m)$$

Substitution of Eq. (h) into Eq. (j), and rearranging terms, gives the equation of motion of the disk in terms of the angular displacement θ. Hence,

$$\ddot{\theta} + \frac{g}{L_{eff}}\sin\theta = 0 \qquad (n)$$

where

$$L_{eff} = \tfrac{3}{2}(R - r_D) \qquad (o)$$

Comparison of Eq. (n) with Eq. (2.37) shows that the rigid disk rolling on a circular segment of track behaves as a pendulum with a rod of length $L = L_{eff}$. The parameter L_{eff} therefore represents an effective rod length. As for the pendulum, we may *linearize* Eq. (n) if we restrict the motions of the disk to those for which $\theta^2 \ll 1$. Hence, employing the approximation given by Eq. (2.40), Eq. (o) simplifies to the harmonic equation in θ,

$$\ddot{\theta} + \omega^2\theta = 0 \qquad \triangleleft (p)$$

with the natural frequency identified as

$$\omega = \sqrt{\frac{g}{L_{\textit{eff}}}} = \sqrt{\frac{2g}{3(R - r_D)}} \qquad \text{(q)}$$

The response is then given by Eqs. (2.43) – (2.46), with ω and T interpreted accordingly.

Example 2.9

A glider of mass m and radius of gyration r_G teeters on its belly just before lifting off the ground. The center of mass of the craft lies a distance ℓ from the geometric center of the fuselage of radius R, as indicated. Determine the period of the small angle motion if no slip occurs.

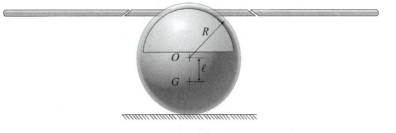

Figure E2.9-1 Teetering glider.

Solution

The kinetic diagram for the system is shown in Figure E2.9-2. We next derive the equation of motion of the system by taking moments about the contact point P. To do this, we must first establish the acceleration of the center of mass. For no slip conditions, we have from Eq. (1.155) that the acceleration of the geometric center is

$$\vec{a}_O = R\ddot{\theta}\,\vec{i} \qquad \text{(a)}$$

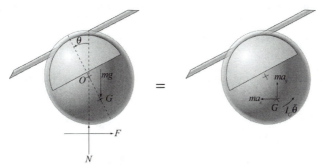

Figure E2.9-2 Kinetic diagram for glider.

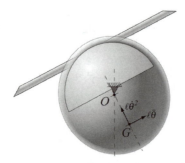

Figure E2.9-3 Relative motion of glider.

The acceleration of the center of mass relative to the geometric center is seen to be pure rotation (Figure E2.9-3). From Eqs. (1.72), (1.148) and (a), the acceleration of the center of mass with respect to the ground is

$$\vec{a}_G = \vec{a}_O + \vec{a}_{G/O}$$
$$= R\ddot{\theta}\vec{i} + \left[\ell\left(\dot{\theta}^2\sin\theta - \ddot{\theta}\cos\theta\right)\vec{i} + \ell\left(\dot{\theta}^2\cos\theta + \ddot{\theta}\sin\theta\right)\vec{j}\right] \tag{b}$$

For small angle motion, let us linearize the above expression. Thus, neglecting terms of second order and above gives the approximate representation of the acceleration of the center of mass as

$$\vec{a}_G \cong (R - \ell)\ddot{\theta}\vec{i} \tag{c}$$

We next take moments about the contact point. Hence,

$$-mg\ell\sin\theta = I_G\ddot{\theta} + \left[\vec{r}_{G/P} \times m\vec{a}_G\right]_z \tag{d}$$

Linearizing the left hand side and substituting Eq. (c) gives the equation of motion as

$$-mg\ell\theta = m\left[r_G^2 + (R - \ell)^2\right]\ddot{\theta} \tag{e}$$

Bringing nonvanishing terms to one side and dividing by the coefficient of $\ddot{\theta}$ gives the equation of motion as

$$\ddot{\theta} + \frac{g}{L_{eff}}\theta = 0 \tag{f}$$

where

$$L_{eff} = \left[r_G^2 + (R - \ell)^2\right]/\ell \tag{g}$$

Thus,

$$\omega^2 = g/L_{eff} \tag{h}$$

The period for small angle motion of the glider is then

$$T = \frac{2\pi}{\omega} = \frac{2\pi}{\sqrt{g/L_{eff}}} = 2\pi\sqrt{\frac{r_G^2 + (R-\ell)^2}{g\ell}} \qquad \triangleleft (i)$$

Finite Angle Response

In the previous subsection we approximated the equation of motion for the simple pendulum, Eq. (2.37), by its linear counterpart, Eq. (2.41). The latter was seen to be the harmonic equation and therefore yielded sinusoidal solutions with constant period. The methods used to obtain the solution to the approximate equation are limited to linear differential equations. Solving the exact nonlinear equation of motion, Eq. (2.37), requires a different approach. We shall here examine the nonlinear response and assess how well the linear approximation represents the motion of the pendulum. We first consider the static response.

Static Response

For the static case, Eq. (2.37) reduces to the equilibrium equation given by

$$mgL\sin\theta = 0 \tag{2.48}$$

The roots of Eq. (2.48) yield the solution, and hence the corresponding equilibrium configurations, defined by

$$\theta = n\pi \qquad (n = 0,1,2,...) \tag{2.49}$$

Thus, the exact equation of motion gives the two distinct equilibrium configurations, $\theta = 0, \pi$, shown in Figure 2.10. These obviously correspond to alignment of the rod vertically downward and vertically upward, respectively. If we parallel the above calculation with Eq. (2.41), it is seen that the linear counterpart to Eq. (2.48) only gives the configuration $\theta = 0$. Evidently the upward configuration is not predicted by the linearized equation of motion since we restricted our attention to angles that are small compared to unity in that approximation. In other words, when we linearized Eq. (2.37), we restricted our attention to angles close to the first equilibrium configuration ($n = 0$).

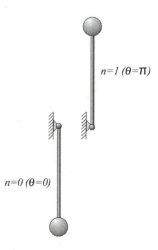

Figure 2.10 Equilibrium configurations of the simple pendulum.

The omission of equilibrium configurations is typical of what may happen when an equation, or system of equations, is linearized.

If we focus our attention on the second (upward) equilibrium configuration, we intuitively know that this configuration is *unstable*. That is, we expect that the pendulum will move away from this configuration if it is given even the smallest perturbation ("nudge"). It is instructive to analyze this instability. To do this we must establish the potential energy of the system, which for the case under consideration is solely due to gravity. If we choose the lower equilibrium position of the bob as our datum (i.e., the level of zero potential energy), then the potential energy of the system at any other point, within allowable motions of the mass, is given by

$$\mathcal{U} = mgL(1 - \cos\theta) \tag{2.50}$$

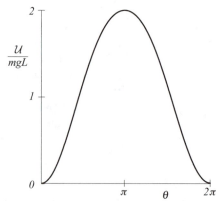

Figure 2.11 Potential energy of the simple pendulum as a function of angular displacement.

A sketch of this function is displayed in Figure 2.11. Let us next take the derivative of the potential energy with respect to the angular displacement θ, and set the resulting expression to zero. This results in the equation

$$\frac{d\mathcal{U}}{d\theta} = mgL\sin\theta = 0 \qquad (2.51)$$

which is seen to be the equilibrium equation, Eq. (2.48), the solution of which yields the roots that correspond to the equilibrium configurations of the system as established previously. What we have effectively done is apply the statement of *conservation of mechanical energy*, Eq. (1.95), for the particular (conservative) single particle system under consideration. It may be seen that the equilibrium configurations correspond to points on the potential energy curve for which the function is stationary (i.e., points where the derivative of the potential energy vanishes, and hence where the potential energy is a relative minimum or maximum). Additional information regarding the behavior of the system at these points can be attained by examination of the second derivative of the potential energy of the system. Thus, differentiating Eq. (2.51) with respect to θ gives

$$\frac{d^2\mathcal{U}}{d\theta^2} = mgL\cos\theta \qquad (2.52)$$

Evaluation of this expression at each of the equilibrium configurations shows that

$$\begin{aligned} \left.\frac{d^2\mathcal{U}}{d\theta^2}\right|_{\theta=0} &= +mgL > 0 \\ \left.\frac{d^2\mathcal{U}}{d\theta^2}\right|_{\theta=\pi} &= -mgL < 0 \end{aligned} \qquad (2.53)$$

It may be observed that the second derivative of the potential energy is positive for the stable configuration ($\theta = 0$) and negative for the unstable configuration ($\theta = \pi$). This is also seen by examination of Figure 2.11. These characteristics may be understood by analogy with a marble that is perched either at the bottom of a well or at the top of a hill whose profile is similar to that of the potential energy function shown in Figure 2.11. When in the well, the potential energy of the marble will increase ($d^2\mathcal{U}/d\theta^2 > 0$) as the marble is perturbed away from its equilibrium configuration. Further, the marble will return to, or remain in the vicinity of, that configuration when the perturbation is removed. In contrast, when the marble is at the top of the hill its potential energy will decrease ($d^2\mathcal{U}/d\theta^2 < 0$) as it is perturbed away from its equilibrium configuration, and it will continue to move away from the equilibrium configuration after the perturbation is removed. In general for any system, whether single degree of freedom or multi-degree of freedom, the convexity of the potential energy of that system at a given equilibrium configuration establishes the stability or instability of that configuration.

Dynamic Response

In this section we shall obtain a solution to the equation of motion, Eq. (2.37), without any restriction on the size of the angular displacement. Because of the nonlinear nature of the problem, we shall obtain the response in the form $t = t(\theta)$ rather than the conventional form $\theta = \theta(t)$.

It may be seen from the dynamic free-body diagram for the system (Figure 2.9) that the tension in rod, P, is always perpendicular to the path of the bob and therefore does no work throughout the motion of the particle. Further, the only other force acting on the bob is its own weight, which is a conservative force. Finally, the pin at O is assumed to be frictionless and thus exerts no transverse shear or moment on the rod and hence on the bob. The total energy of the system is therefore conserved throughout the motion. If we take $\theta = 0$ as our datum for the gravitational potential energy, then the statement of conservation of mechanical energy for the pendulum may be written in the explicit form

$$mgL(1-\cos\theta) + \tfrac{1}{2}mL^2\dot{\theta}^2 = mgL(1-\cos\theta_0) + \tfrac{1}{2}mL^2\chi_0^2 \qquad (2.54)$$

where $\theta(0) = \theta_0$ and $\dot{\theta}(0) = \chi_0$. Solving Eq. (2.54) for $\dot{\theta}$ gives

$$\frac{d\theta}{dt} = \sqrt{\frac{2g}{L}\left(\cos\theta - \cos\theta_0\right) + \chi_0^2} \qquad (2.55)$$

which may be rearranged to the form

$$dt = \frac{d\theta}{\sqrt{\dfrac{2g}{L}\left(\cos\theta - \cos\theta_0\right) + \chi_0^2}} \qquad (2.56)$$

We remark that Eq. (2.55) could also be obtained by first multiplying the equation of motion, Eq. (2.37), by $\dot{\theta}$, exploiting the identity

$$\dot{\theta}\ddot{\theta} = d\left(\tfrac{1}{2}\dot{\theta}^2\right)/dt$$

and then integrating the resulting expression and solving for $\dot{\theta}$. We bypassed these calculations by starting from the statement of conservation of energy, since the work-energy principle is, in fact, an integral of the equation of motion (see Section 1.5).

Let us next integrate Eq. (2.56) over the time interval $[0, t]$ and divide the resulting expression by the period of oscillation of the linear response,

$$T_0 \equiv \frac{2\pi}{\sqrt{g/L}} \tag{2.57}$$

Doing this gives the response as

$$\bar{t} \equiv \frac{t}{T_0} = \frac{1}{2\pi} \int_{\theta_0}^{\theta} \frac{d\theta}{\sqrt{2(\cos\theta - \cos\theta_0) + \left(\dfrac{\chi_0^2}{g/L}\right)}} \tag{2.58}$$

where $\bar{t}$ is the *normalized time*. Equation (2.58) is the solution we are seeking and may be integrated for given initial conditions to give the response.

Let us next consider motions for which the bob is released from rest ($\chi_0 = 0$). In this way we exclude possible motions for which the bob orbits the pin. We also know, from conservation of energy, that the motion is periodic. Further, by virtue of the same arguments, we know that the time for the bob to move between the positions $\theta = \theta_0$ and $\theta = 0$ is one quarter of a period. Likewise, the reverse motion, where the bob moves from the position $\theta = 0$ to $\theta = \theta_0$, takes a quarter of a period to traverse as well. Taking this into account and utilizing the equivalence of the cosine function in the first and fourth quadrants in Eq. (2.58) gives the exact period, T, of an oscillation of the simple pendulum,

$$\bar{T} \equiv \frac{T}{T_0} = 4 \cdot \frac{1}{2\pi} \int_0^{\theta_0} \frac{d\theta}{\sqrt{2(\cos\theta - \cos\theta_0)}} \tag{2.59}$$

where $\bar{T}$ is the *normalized period*. It is seen that the "true" period depends on the initial conditions, that is $T = T(\theta_0)$, while the period predicted by the linear approximation is independent of the initial conditions. We might expect, however, that the two solutions will converge as $\theta_0 \to 0$. It is left as an exercise (Problem P2.24) to demonstrate that this is so. The integral of Eq. (2.59) may be applied directly, or it may be put in a standard form by introducing the change of variable

$$\varphi \equiv \sin^{-1}\left(\frac{1}{q}\sin(\theta/2)\right) \tag{2.60}$$

where

$$q \equiv \sin(\theta_0/2) \tag{2.61}$$

After making these substitutions, Eq. (2.59) takes the equivalent form

$$\bar{T} = \frac{2}{\pi} \mathbb{F}(q, \pi/2) \tag{2.62}$$

where

$$\mathbb{F}(q, \beta) \equiv \int_0^\beta \frac{d\varphi}{\sqrt{1 - q^2 \sin^2 \varphi}} \tag{2.63}$$

is referred to as an *Elliptic Integral of the First Kind*, and $\mathbb{F}(q, \pi/2)$ is called a *Complete Elliptic Integral of the First Kind.* Elliptic integrals are tabulated in tables in much the same way as trigonometric functions and, in an analogous way, are also available in certain mathematical software. Equation (2.62), or equivalently Eq. (2.59), is normalized by the period that is predicted by the linear approximation. It thus allows for a direct comparison of the response predicted by the linear approximation with that predicted by the exact nonlinear model.

Example 2.10

Determine the percent error in approximating the true period of a simple pendulum by the period predicted by the linear approximation for an initial angular displacement of 30 degrees. Do the same for initial angular displacements of 15 degrees and 10 degrees.

Solution

We must first evaluate the normalized period for $\theta_0 = 30°$. For this initial angular displacement, Eq. (2.61) gives

$$q = \sin(30°/2) = 0.2588 \tag{a}$$

The corresponding value of the complete elliptic integral is found from a table, using software, or from numerical integration of Eq. (2.63). We here use the MATLAB function "ellipke" to compute

$$\mathbb{F}(0.2588, \pi/2) = 1.598 \tag{b}$$

Substitution of Eq. (a) into Eq. (2.62) gives the normalized period

$$\bar{T} = \frac{2}{\pi}(1.598) = 1.017 \tag{c}$$

Finally,

$$\frac{T - T_0}{T} = 1 - \frac{1}{\bar{T}} = 1 - \frac{1}{1.017} = 0.01672 \tag{d}$$

Hence,

$$\% \text{ Error} = 1.67\% \qquad\qquad \triangleleft (e)$$

Carrying out similar calculations gives, for $\theta_0 = 15°$, $\overline{T} = 1.005$ and

$$\% \text{ Error} = 0.498\% \qquad\qquad \triangleleft (f)$$

and, for $\theta_0 = 10°$, $\overline{T} = 1.002$ and

$$\% \text{ Error} = 0.200\% \qquad\qquad \triangleleft (g)$$

It is seen that the linear approximation gives quite accurate values of the period for small, and even for moderate, initial angular displacements.

2.2 FREE VIBRATION OF SYSTEMS WITH VISCOUS DAMPING

In all physical systems, a certain amount of energy is dissipated through various mechanisms. Hence, the vibrations considered in the previous section, though important, applicable and representative for many applications, conserve energy and are therefore idealistic in that sense. It is therefore of interest to examine and to characterize to the extent possible, the effects of damping on the vibratory behavior of mechanical systems. The inclusion of some damping and its influence on the vibratory behavior of single degree of freedom systems is the subject of the next two sections. In particular, we shall discuss the effects of viscous damping in the present section and damping due to Coulomb friction in Section 2.3. Structural (material) damping and its effects on vibrating systems will be introduced in Chapter 3 (Section 3.4).

2.2.1 Equation of Motion and General System Response

A simple type of dissipation mechanism often considered in vibrations studies is that of viscous damping. It was demonstrated in Section 1.2.5 that the force applied to a body by a linear viscous damper opposes the motion of the body and is linearly proportional to the speed at which the body travels *relative* to the damper. In this section we derive the equations of motion and obtain the general solution for systems possessing this type of damping.

Consider the system comprised of a mass, m, attached to a spring of stiffness k, and a viscous damper with damping coefficient c (see Section 1.2.5), where the opposite end of the spring and the damper are fixed, as shown in Figure 2.12. Let $x(t)$ denote the displacement of the mass, measured from its equilibrium configuration. We shall be interested in the response of the system after it is initially displaced and released. The corresponding kinetic diagram for the system is shown in Figure 2.13.

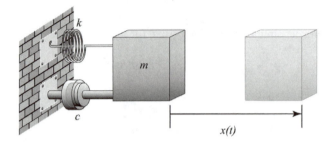

Figure 2.12 Mass-spring-damper system.

Applying Newton's Second Law along the horizontal direction gives

$$-kx - c\dot{x} = m\ddot{x}$$

which, after rearranging terms, takes the form

$$\ddot{x} + 2\omega\zeta\dot{x} + \omega^2 x = 0 \qquad (2.64)$$

where

$$\zeta = \frac{c}{2\omega m} = \frac{c}{2\sqrt{mk}} \qquad (2.65)$$

and ω is given by Eq. (2.2). The parameter ζ is referred to as the *damping factor*, while ω is seen to correspond to the natural frequency when no damping is present. (It will be seen shortly that, when damping is present, ω no longer corresponds to the frequency of oscillation.)

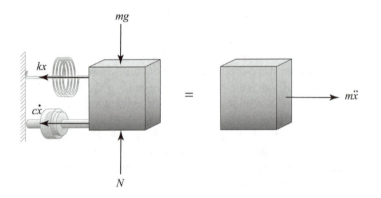

Figure 2.13 Kinetic diagram for mass-spring-damper system.

To determine the response of the mass-spring-damper system we must solve Eq. (2.64). To do this we shall proceed as we did for the undamped system. Hence, we first assume a solution of the form

$$x(t) \sim Ce^{st} \tag{2.66}$$

where C and s are (complex) parameters that are yet to be determined. Substitution of this expression into the governing equation (2.64) results in the characteristic equation

$$s^2 + 2\omega\zeta s + \omega^2 = 0 \tag{2.67}$$

which yields the roots

$$s = -\omega\zeta \pm \omega\sqrt{\zeta^2 - 1} \tag{2.68}$$

The response of the system is thus comprised of the sum of two solutions of the form of Eq. (2.66) corresponding to the two roots of Eq. (2.68). It is evident from Eqs. (2.66) and (2.68) that the behavior of the system is characterized by whether the damping factor is less than, greater than, or equal to unity. We shall consider each case separately.

2.2.2 Underdamped Systems ($\zeta^2 < 1$)

Systems for which $\zeta^2 < 1$ are referred to as *underdamped systems*. When the square of the damping factor is less than unity, the characteristic roots given by Eq. (2.68) may be written as

$$s = -\omega\zeta \pm i\omega_d \tag{2.69}$$

where

$$\omega_d = \omega\sqrt{1 - \zeta^2} \tag{2.70}$$

It will be seen that ω_d corresponds to the frequency of oscillation of the damped system (the *damped natural frequency*). Substitution of each of the characteristic roots defined by Eq. (2.69) into Eq. (2.66) gives two solutions of the form

$$x(t) \sim Ce^{-\zeta\omega t}e^{\pm i\omega_d t} \tag{2.71}$$

The general solution to Eq. (2.64) consists of a linear combination of these two solutions. Hence,

$$x(t) = e^{-\zeta\omega t}\left[C_1 e^{i\omega_d t} + C_2 e^{-i\omega_d t} \right] \tag{2.72}$$

The terms within the brackets of Eq. (2.72) are seen to be identical in form to the right-hand side of Eq. (2.6). It therefore follows from Eqs. (2.6)–(2.11) that the solution for the underdamped case ($\zeta^2 < 1$) may also be expressed in the equivalent forms

$$x(t) = e^{-\zeta \omega t} \left[A_1 \cos \omega_d t + A_2 \sin \omega_d t \right] \tag{2.73}$$

and

$$x(t) = Ae^{-\zeta \omega t} \cos(\omega_d t - \phi) = A_d(t)\cos(\omega_d t - \phi) \tag{2.74}$$

where

$$A_d(t) = Ae^{-\zeta \omega t} \tag{2.75}$$

and the pairs of constants (A, ϕ), (A_1, A_2) and (C_1, C_2) are related by Eqs. (2.8), (2.16) and (2.17). The specific values of these constants are determined by imposing the initial conditions

$$x(0) = x_0 \quad \text{and} \quad \dot{x}(0) = v_0$$

on the above solutions. Doing this gives the relations

$$A_1 = x_0, \quad A_2 = \frac{(v_0/\omega) + \zeta x_0}{\sqrt{1-\zeta^2}} \tag{2.76}$$

and

$$A = x_0\sqrt{1 + \frac{\left[(v_0/\omega x_0) + \zeta\right]^2}{1-\zeta^2}}, \quad \phi = \tan^{-1}\left(\frac{(v_0/\omega x_0) + \zeta}{\sqrt{1-\zeta^2}}\right) \tag{2.77}$$

It may be seen from Eqs. (2.74) and (2.75) that the response of the system corresponds to harmonic oscillations whose amplitudes, $A_d(t)$, decay with time at the rate $\zeta\omega$ ($= c/2m$). It may be noted that the frequency of the oscillation for the underdamped case is ω_d, as defined by Eq. (2.70). It follows that the corresponding period is

$$T_d = \frac{2\pi}{\omega_d} = \frac{2\pi}{\omega\sqrt{1-\zeta^2}} \tag{2.78}$$

and that the associated phase lag is

$$t_\phi = \omega_d/\phi \tag{2.79}$$

Note that the frequency of the damped oscillations is lower, and hence the corresponding period is longer, than that for the undamped system ($\zeta = 0$). It is thus seen that damping tends to slow the system down, as might be anticipated.

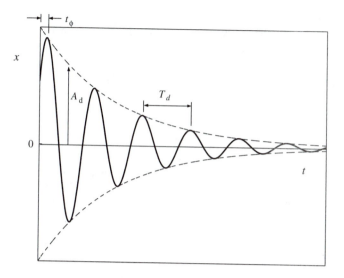

Figure 2.14 Characteristic response history of an underdamped system.

A typical response history of an underdamped system is shown in Figure 2.14. It may be noted that, unlike the undamped case, the underdamped system will generally exhibit a nonvanishing phase shift for vanishing initial velocity ($v_0 = 0$). The phase shift corresponds to the instant at which the time dependent amplitude and the displacement first become equal. Inclusion of the damping force in the development presented in Section 2.1.2 shows that the effect of gravity is the same for damped systems. That is, the damped system will exhibit decaying oscillations about the equilibrium configuration corresponding to the statically stretched spring due to gravity alone, with $x(t)$ and x_0 measured relative to the equilibrium position of the mass.

Example 2.11

A 4 kg mass is attached to a spring of stiffness 400 N/m and a viscous damper of coefficient 16 N-sec/m as shown. If the mass is displaced 0.5 m from its equilibrium position and released from rest, determine the position of the mass after it has oscillated through 3 cycles from the point of release. Evaluate the ratio of the magnitude of the displacement to the initial displacement at this instant.

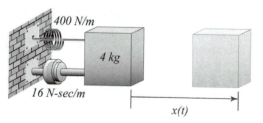

Solution

The undamped natural frequency and the damping factor may be calculated from the given parameters as

$$\omega = \sqrt{k/m} = \sqrt{400/4} = 10 \text{ rad/sec} \tag{a}$$

and

$$\zeta = \frac{c}{2m\omega} = \frac{16}{2(4)(10)} = 0.2 \tag{b}$$

The damped natural frequency and the associated damped natural period are then obtained by substituting Eqs. (a) and (b) into Eqs. (2.70) and (2.78), respectively. Hence,

$$\omega_d = \omega\sqrt{1-\zeta^2} = 10\sqrt{1-(0.2)^2} = 9.798 \text{ rad/sec} \tag{c}$$

$$T_d = \frac{2\pi}{\omega_d} = \frac{2\pi}{9.798} = 0.6413 \text{ secs} \tag{d}$$

The initial conditions for the problem are

$$x(0) = x_0 = 0.5\text{m} \quad \text{and} \quad \dot{x}(0) = v_0 = 0 \tag{e}$$

The amplitude coefficient A and phase angle ϕ may be evaluated by substituting Eqs. (b) and (e) into Eqs. (2.77). We thus have that

$$A \rightarrow \frac{x_0}{\sqrt{1-\zeta^2}} = \frac{0.5}{\sqrt{1-(0.2)^2}} = 0.5103 \text{ m} \tag{f}$$

and

$$\phi \rightarrow \tan^{-1}\left(\frac{\zeta}{\sqrt{1-\zeta^2}}\right) = \tan^{-1}\left(\frac{0.2}{\sqrt{1-(0.2)^2}}\right) = 11.54° = 0.2014 \text{ rads} \tag{g}$$

Now, the time to the completion of 3 cycles is

$$t_3 = 3T_d = 3(0.6413) = 1.924 \text{ secs} \tag{h}$$

The amplitude at this instant is then computed using Eq. (2.75). This gives

$$A_d(t_3) = A_d(3T_d) = Ae^{-\zeta\omega(3T_d)} = (0.5103)e^{-0.2(10)(1.924)} = 0.01088 \text{ m} \tag{i}$$

Next,

$$\cos(\omega_d t_3 - \phi) = \cos(\omega_d \cdot 3T_d - \phi) = \cos\{\omega_d \cdot 3(2\pi/\omega_d) - \phi\}$$
$$= \cos(6\pi - 0.2014) = 0.9798 \tag{j}$$

Finally, substitution of Eqs. (i) and (j) into Eq. (2.74) gives the desired response

$$x\big|_{t=3T_d} = A_d(3T_d)\cos(\omega_d \cdot 3T_d - \phi) = 0.1066 \text{ m} \qquad \triangleleft \text{(k)}$$

The ratio of the displacement after three cycles to the initial displacement is given by

$$\frac{x(3T_d)}{x_0} = \frac{0.01066}{0.5} = 0.02132 \qquad \triangleleft \text{(l)}$$

It is seen that the displacement of the damped oscillation reduces to 2.1% of its original value after three cycles.

2.2.3 Logarithmic Decrement (measurement of ζ)

Suppose we have an underdamped system ($\zeta^2 < 1$) and we wish to measure the damping coefficient, ζ. How might we accomplish this? Let us imagine that we displace the mass of Figure 2.12 an initial distance, release it, and then plot a trace of the ensuing motion as depicted in Figure 2.15. Let us consider the displacements measured at two instants in time occurring one period apart, say at two consecutive peaks, as indicated. Let the displacement at time $t = t_1$ be denoted as $x(t_1) = x_1$, and the displacement measured at $t = t_2$ be denoted as $x(t_2) = x_2$. It follows from Eq. (2.74) that the displacements at these instants are given by

$$x_1 = Ae^{-\zeta\omega t_1}\cos(\omega_d t_1 - \phi)$$
$$x_2 = Ae^{-\zeta\omega t_2}\cos(\omega_d t_2 - \phi) \tag{2.80}$$

Since the measurements are taken at times separated by one period, we have the relation

$$t_2 = t_1 + T_d \tag{2.81}$$

It then follows that

$$\cos(\omega_d t_2 - \phi) = \cos\{\omega_d(t_1 + T_d) - \phi\} = \cos(\omega_d t_1 - \phi) \tag{2.82}$$

Next, let us take the ratio of the displacements measured a period apart. This gives

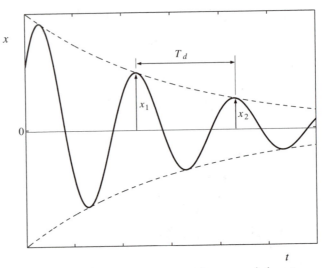

Figure 2.15 Two measurements taken one period apart.

$$\frac{x_1}{x_2} = \frac{Ae^{-\zeta\omega t_1}\cos(\omega_d t_1 - \phi)}{Ae^{-\zeta\omega t_2}\cos(\omega_d t_2 - \phi)} = \frac{Ae^{-\zeta\omega t_1}\cos(\omega_d t_1 - \phi)}{Ae^{-\zeta\omega t_2}\cos(\omega_d t_1 - \phi)} = e^{\zeta\omega(t_2 - t_1)} = e^{\zeta\omega T_d} \quad (2.83)$$

The *logarithmic decrement*, $\bar{\delta}$, is defined as the natural log of this ratio. Hence,

$$\bar{\delta} \equiv \ln\left(\frac{x_1}{x_2}\right) = \zeta\omega T_d = \frac{2\pi\zeta}{\sqrt{1-\zeta^2}} \quad (2.84)$$

where we have used Eq. (2.78) to eliminate the period. Solving for ζ gives the inverse relation

$$\zeta = \frac{\bar{\delta}}{\sqrt{4\pi^2 + \bar{\delta}^2}} \quad (2.85)$$

It is seen that $\bar{\delta}$ completely determines ζ. We may therefore measure successive peaks off of an *x-t* plot, take the natural log of the ratio, and substitute the logarithmic decrement into Eq. (2.85) to determine ζ. It may be seen from Eqs. (2.84) and (2.85) that for light damping ($\zeta^2 \ll 1$, or equivalently $\bar{\delta}^2 \ll 4\pi^2$),

$$\zeta \approx \frac{\bar{\delta}}{2\pi} \quad (2.86)$$

Let us next consider $n+1$ successive instants in time, such that adjacent instants are separated by a natural period on the displacement plot for an underdamped system (Figure 2.15). If we let x_j and x_{j+1} represent the jth and $j+1^{st}$ measurement, then we have from Eq. (2.84) that

$$\bar{\delta} = \ln\left(\frac{x_j}{x_{j+1}}\right) = \zeta\omega T_d \tag{2.87}$$

Let us next take the ratio of the first and last displacement measurements. Using the above identity in the resulting expression gives

$$\frac{x_1}{x_{n+1}} = \frac{x_1}{x_2}\frac{x_2}{x_3}\cdots\frac{x_n}{x_{n+1}} = \left(e^{\zeta\omega T_d}\right)^n = e^{n\zeta\omega T_d} \tag{2.88}$$

Evaluating the natural log of this expression gives the relation

$$\bar{\delta} = \frac{1}{n}\ln\left(\frac{x_1}{x_{n+1}}\right) \tag{2.89}$$

Thus, we may measure two displacements separated by n periods ($\Delta t = nT_d$) and calculate the logarithmic decrement using Eq. (2.89). We may then determine the damping coefficient ζ using Eq. (2.85) or, under appropriate circumstances, Eq. (2.86).

Finally, another useful relation is obtained if we take the inverse natural log of Eq. (2.89), rearrange terms, and substitute Eq. (2.84) into the resulting expression. This yields the relation

$$x_{n+1} = x_1 e^{-n\bar{\delta}} = x_1 e^{-n\zeta\omega T_d} = x_1 e^{-2\pi n\zeta/\sqrt{1-\zeta^2}} \tag{2.90}$$

Example 2.12

A portion of an automobile suspension system consists of an elastic spring and a viscous damper, as shown. If the spring is chosen such that $k/m = 40$ sec^{-2}, determine the allowable range of the ratio c/m so that any oscillations that occur will decay by a factor of 95% within 1 cycle.

Solution

It is required that $x_2 = 0.05 \, x_1$. Thus, from Eq. (2.84),

$$\bar{\delta} = \ln\left(\frac{1}{0.05}\right) = 2.996 \qquad\qquad (a)$$

Using Eq. (2.85), we calculate the damping factor to be

$$\zeta = \frac{2.996}{\sqrt{4\pi^2 + (2.296)^2}} = 0.4479 \qquad\qquad (b)$$

Substituting Eq. (b) into Eq. (2.65) and solving for c/m gives the value for the desired ratio as

$$c/m = 2(0.4479)\sqrt{40} = 5.67 \text{ sec}^{-1} \qquad\qquad (c)$$

Thus, in order for an oscillation to decay at the desired rate, the damper must be chosen such that

$$c/m \geq 5.67 \text{ sec}^{-1} \qquad\qquad \triangleleft (d)$$

Example 2.13 (Ex. 2.11 revisited)

A 4 kg mass is attached to a spring of stiffness 400 N/m and a viscous damper of coefficient 16 N-sec/m as shown in Figure E2.11. If the mass is displaced 0.5 m from its equilibrium position, and released from rest, determine the position of the mass after it has oscillated through 3 cycles.

Solution

The given the initial conditions are

$$x(0) = x_0 = 0.5 \text{ m} \quad\text{and}\quad \dot{x}(0) = v_0 = 0 \qquad\qquad (a)$$

As in our prior analysis of this problem (Example 2.11), we first calculate the undamped natural frequency and the damping factor. We thus have

$$\omega = \sqrt{k/m} = \sqrt{400/4} = 10 \text{ rad/sec} \qquad\qquad (b)$$

and

$$\zeta = \frac{c}{2m\omega} = \frac{16}{2(4)(10)} = 0.2 \qquad\qquad (c)$$

We next employ Eq. (2.90) where, for the present problem, $n = 3$ and $x_1 = 0.5$ m. This gives the desired position directly as

$$x(3T_d) = 0.5e^{-3(0.2)(10)(0.6413)} = 0.01066 \text{ m} \qquad \triangleleft \text{(d-1)}$$

Equivalently,

$$x(3T_d) = 0.5e^{-6\pi(0.2)/\sqrt{1-(0.2)^2}} = 0.01066 \text{ m} \qquad \triangleleft \text{(d-2)}$$

2.2.4 Overdamped Systems ($\zeta^2 > 1$)

Systems for which $\zeta^2 > 1$ are referred to as *overdamped* systems. For such systems, the characteristic roots given by Eq. (2.68) are all real. Substitution of these roots into Eq. (2.66) gives the solution for the overdamped case as

$$x(t) = C_1 e^{-(\zeta - \jmath)\omega t} + C_2 e^{-(\zeta + \jmath)\omega t} \qquad (2.91)$$

where

$$\jmath = \sqrt{\zeta^2 - 1} \qquad (2.92)$$

An equivalent form of the solution is easily obtained with the aid of Eq. (1.63) as

$$x(t) = e^{-\zeta \omega t} \left[A_1 \cosh \jmath \omega t + A_2 \sinh \jmath \omega t \right] \qquad (2.93)$$

and

$$A_1 = C_1 + C_2 \ , \ A_2 = C_1 - C_2 \qquad (2.94)$$

Upon imposing the initial conditions $x(0) = x_0$ and $\dot{x}(0) = v_0$, Eq. (2.93) gives the integration constants as

$$A_1 = x_0 \ , \ A_2 = \frac{(v_0/\omega) + \zeta x_0}{\sqrt{\zeta^2 - 1}} \qquad (2.95)$$

The solution for the overdamped case then takes the form

$$x(t) = e^{-\zeta \omega t} \left[x_0 \cosh \jmath \omega t + \frac{(v_0 + \zeta \omega x_0)}{\omega \jmath} \sinh \jmath \omega t \right] \qquad (2.96)$$

where $\jmath$ is given by Eq. (2.92). Consideration of the exponential form of the solution, Eq. (2.91), shows that both terms of the solution decay exponentially. A typical response is depicted in Figure 2.16.

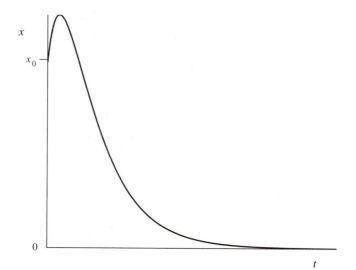

Figure 2.16 Typical response history of an overdamped system.

Example 2.14

The automobile system of Figure E2.12 is initially at rest when it is impacted, imparting an initial vertical velocity of 0.5 m/sec. Determine the maximum vertical displacement of the system if $c/m = 19$ sec^{-1}.

Solution
For the given system, $k/m = 40$ sec^{-2}, $c/m = 19$ sec^{-1}, and the initial conditions are $x_0 = 0$ and $v_0 = 2$ m/sec. Thus, from Eqs. (2.2), (2.65) and (2.92),

$$\omega = \sqrt{40} = 6.32 \text{ rad/sec} \tag{a}$$

$$\zeta = \frac{c/m}{2\omega} = \frac{19}{2(6.32)} = 1.50 \tag{b}$$

and

$$\mathfrak{z} = \sqrt{(1.5)^2 - 1} = 1.12 \tag{c}$$

Differentiating Eq. (2.96) once with respect to time, setting the resulting expression to zero, and imposing the condition of vanishing initial displacement gives the time to the maximum displacement, t^*, as

$$\mathfrak{z}\omega t^* = \tanh^{-1}\left(\frac{1.12}{1.50}\right) = 0.965 \tag{d}$$

Hence,

$$t^* = 0.641/[(1.12)(6.32)] = 0.136 \text{ secs} \tag{e}$$

Substitution of the nondimensional time, Eq. (d), or the dimensional time, Eq. (e), into Eq. (2.96) gives the maximum deflection. Thus,

$$x_{max} = x(t^*) = \frac{v_0}{\omega_{\mathfrak{z}}} e^{-\zeta\omega t^*} \sinh(\mathfrak{z}\omega t^*)$$

$$= \frac{0.5}{(6.32)(1.12)} e^{-1.50(0.136)} \sinh(0.965) \qquad \triangleleft (f)$$

$$= 0.0647 \text{ m} = 6.47 \text{ cm}$$

2.2.5 Critically Damped Systems ($\zeta^2 = 1$)

It was seen in the prior two sections that the response of an underdamped system ($\zeta^2 < 1$) is a decaying oscillation, while the response of an overdamped system ($\zeta^2 > 1$) is an exponential decay with no oscillatory behavior at all. A system for which $\zeta^2 = 1$ is referred to as *critically damped* since it lies at the boundary between the underdamped and overdamped cases and therefore separates oscillatory behavior from nonoscillatory behavior. We examine this case next.

For critically damped systems, the characteristic roots given by Eq. (2.68) reduce to

$$s = -\omega, -\omega \tag{2.97}$$

When substituted into Eq. (2.66), these roots yield the solution for the critically damped case in the form

$$x(t) = \left(A_1 + A_2 t \right) e^{-\omega t} \tag{2.98}$$

where the factor t occurs because the roots are repeated. Imposition of the initial conditions, $x(0) = x_0$ and $\dot{x}(0) = v_0$, renders the response given by Eq. (2.98) to the form

$$x(t) = \left[x_0 + \left(v_0 + \omega x_0 \right) t \right] e^{-\omega t} \tag{2.99}$$

Representative characteristic responses are displayed in Figure 2.17. It is seen from Eq. (2.99) that when the initial displacement and the initial velocity are of opposite sign and the latter is of sufficient magnitude, such that $v_0/\omega x_0 < -1$, then the displacement passes through zero once and occurs at time

$$t = t_a = \frac{1}{\left\| v_0/x_0 \right\| - \omega} \tag{2.100}$$

Thus, for this situation, the displacement is initially positive and then becomes negative for $t > t_a$. The mass then continues to move in the opposite direction, and eventually achieves a displacement of maximum magnitude, after which it changes direction. The displacement then decays exponentially. The change in direction occurs when $\dot{x} = 0$. Imposing this condition on Eq. (2.99) and solving for the time gives the time to maximum overshoot as

$$t = t_{os} = t_a + \frac{1}{\omega} \tag{2.101}$$

This maximum displacement following reversal of direction is referred to as the "overshoot." Thus, from Eq. (2.99), we have that the *overshoot* is given by

$$x_{os} = x(t_{os}) = -\frac{x_0}{\omega t_{os}} e^{(t_{os}+1)} = -2.718 \frac{x_0}{\omega t_{os}} e^{t_{os}} \tag{2.102}$$

To visualize this phenomenon let us imagine that we pull the mass in one direction, hold it, and then "throw" the mass toward its equilibrium configuration. The mass then moves in this direction, and eventually passes through the equilibrium configuration. At some later time, the mass stops for an instant (the displacement at this instant is the "overshoot"). It then reverses its direction and travels back toward the equilibrium position asymptotically.

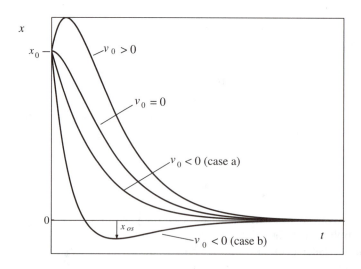

Figure 2.17 Characteristic response histories of critically damped systems.

Example 2.15

A plank that bridges a small segment of a body of still water is pinned at one end and sits on a pontoon at a distance L from the pin as shown in Fig. E2.15-1. The plank is of length $L\sqrt{3}$, the cross-sectional area of the pontoon is 1 m^2 and the total mass of the plank and the pontoon together is 100 kg. It is desired to attach a viscous damper of coefficient 1200 N-sec/m at some distance l along the plank to minimize the effects of vibration. (**a**) Assume "small angle" motion of the plank and determine the ratio of the attachment length of the damper to that of the pontoon which will cause the system to achieve critical damping, and in so doing establish the allowable range of l/L that will prohibit (free) oscillatory motion. (**b**) If the pontoon is raised, held, and then pushed downward and released at a clockwise angular speed of 1.5 rad/sec at an angle of 0.1 rads counterclockwise from its level at equilibrium, determine the overshoot (if any) of the float if the system is critically damped. The density of water is $\rho_w = 1000$ kg/m^3.

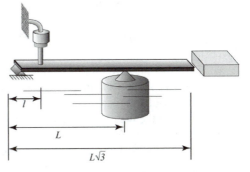

Figure E2.15-1 Plank and float system.

Solution

(**a**)

To solve this problem we must (**i**) establish an equivalent system (recognizing that the buoyant force exerted on the pontoon by the water acts as a restoring force, (**ii**) derive the equation of motion for the pontoon using this model, (**iii**) put the governing equation in the form of Eq. (2.64) and establish the effective damping factor, ζ, and the undamped natural frequency, ω. (**iv**) We can then set the damping factor to unity and solve for the desired value of the length ratio in question.

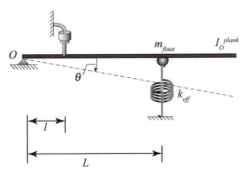

Figure E2.15-2 Equivalent system.

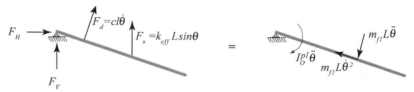

Figure E2.15-3 Kinetic diagram.

(*i*) The equivalent system is shown in Figure E2.15-2 where, using Eq. (1.22),

$$k_{eff} = \rho_w gA = (1000)(9.81)(1) = 9810 \text{ N/m} \tag{a}$$

(*ii*) To derive the governing equation of motion for the equivalent system we first draw the kinetic diagram shown in Fig. E2.15-3. As we need the rotational acceleration of the plank and the velocity of the damper, as well as the displacement of the mass, it is convenient to express our equations in terms of the angular coordinate $\theta(t)$.

The damping force, F_d, and the spring force, F_s, are respectively

$$F_d = c\dot{s}_l = cl\dot{\theta} \tag{b}$$

and

$$F_s = k_{eff} y_L = k_{eff} L \sin\theta \tag{c}$$

The inertia force of the pontoon may be resolved into its normal and tangential components, $m_{fl}a_n$ and $m_{fl}a_t$, with regard to the circular path described by the float, where a_n and a_t are respectively the normal and tangential components of the acceleration of the pontoon, and m_{fl} is its mass. Thus, if $s = L\theta$ is the path coordinate for the pontoon, we have from Eq. (1.76) that

$$m_{fl}a_n = m_{fl}\frac{\dot{s}^2}{L} = m_{fl}L\dot{\theta}^2 \tag{d}$$

$$m_{fl}a_t = m_{fl}\ddot{s} = m_{fl}L\ddot{\theta} \tag{e}$$

[Note that we could have just as easily used polar coordinates, Eq. (1.79), with $r = L$.] The moment of inertia of the plank about hinge O is given by

$$I_O^{(pl)} = \tfrac{1}{3}m_{pl}L_{pl}^2 \tag{f}$$

where m_{pl} and L_{pl} respectively correspond to the mass and length of the plank, and we have modeled the plank as a (uniform) rod. For the present problem the length of the plank is given as

$$L_{pl} = L\sqrt{3}$$

Upon substituting this into Eq. (f), the moment of inertia of the plank is given by

$$I_O^{(pl)} = \tfrac{1}{3} m_{pl} L_{pl}^2 = m_{pl} L^2 \tag{g}$$

We are now ready to write the equation of motion for the plank-pontoon-damper system.

(***iii***) Let F_H and F_V represent the horizontal and vertical reactions at pin O, respectively. Taking moments about pin O, and using Eqs. (a)–(e) and (1.163), gives the rotational equation of motion for the plank. Thus,

$$\vec{M}_O = I_O^{(pl)} \vec{\alpha} + \vec{r}_{fl/O} \times m_{fl} \vec{a}_{fl}:$$

$$\left(cl\dot{\theta}\right)l + (k_{eff} L \sin\theta)(L\cos\theta) = -m_{pl} L^2 \ddot{\theta} - \left(m_{fl} L\ddot{\theta}\right)L$$
$$= -\left(m_{pl} + m_{fl}\right)L^2\ddot{\theta} \tag{h}$$

(Note that the coefficient of the angular acceleration in Eq. (h) is simply the moment of inertia, about O, of an equivalent rod with a singularity at $r = L$ in the otherwise uniform mass distribution.) Upon rearranging terms, the equation of angular motion takes the form

$$mL^2\ddot{\theta} + cl^2\dot{\theta} + k_{eff}L^2 \sin\theta\cos\theta = 0 \tag{i}$$

where

$$m = m_{pl} + m_{fl} \tag{j}$$

is the total mass of the system. If we divide through by mL^2 and employ the small angle assumption and linearize Eq. (i) by setting $\sin\theta\cos\theta \approx \theta$, the equation of motion of the plank-pontoon-damper system reduces to the familiar form

$$\ddot{\theta} + 2\omega\zeta\dot{\theta} + \omega^2\theta = 0 \tag{k}$$

where

$$\omega = \sqrt{\frac{k_{eff}}{m}} = \sqrt{\frac{\rho_w g \pi A}{m}} = \sqrt{\frac{(1000)(9.81)(1)}{100}} = 9.90 \text{ rad/sec} \tag{l}$$

and

$$\zeta = \frac{c}{2\omega m}(l/L)^2 \tag{m}$$

(*iv*) For *critical damping*, we set $\zeta = 1$ in Eq. (m) and solve for the critical length ratio. Doing this gives

$$\left(\frac{l}{L}\right)_{cr} = \left[\frac{2\omega m}{c}\right]^{1/2} = \sqrt{\frac{2(9.90)(100)}{1200}} = 1.28 \tag{n}$$

Thus, for the system to avoid free vibratory behavior, the length ratio must be such that

$$\frac{l}{L} \geq \left[\frac{2\sqrt{mk}}{c}\right]^{1/2} = 1.28 \tag{◁ (o)}$$

(*b*) Let us first check to see that there will be an overshoot. Hence,

$$v_0/\omega\theta_0 = 1.5/[9.9(-0.1)] = -16.67 < -1$$

Since the ratio satisfies the requisite inequality, we anticipate that the plank will pass through its equilibrium position one time before returning to it asymptotically. We now proceed to calculate by how much. To determine the overshoot, we first calculate the time to the overshoot. Substitution of the given initial conditions and the value of the undamped natural frequency given by Eq. (l) into Eq. (2.100) gives the time to crossing the equilibrium position as

$$t_a = \frac{1}{\|(1.5)/(-0.1)\| - 9.9} = 0.196 \text{ secs} \tag{p}$$

Substitution of Eqs. (l) and (p) into Eq. (2.101) then gives the time to overshoot as

$$t_{os} = 0.196 + \frac{1}{9.90} = 0.297 \text{ secs} \tag{q}$$

The overshoot may now be determined using Eq. (2.102). Thus,

$$\theta_{os} = -2.718\frac{(-0.10)}{(9.90)(0.297)}e^{0.297} = 0.0458 \text{ rads} \tag{◁ (r)}$$

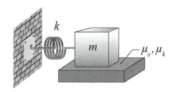

Figure 2.18 Mass-spring system on rough surface.

2.3 COULOMB (DRY FRICTION) DAMPING

The mass of the single degree of freedom system considered in Section 2.1 was assumed to move along a frictionless surface. Suppose that we relax the assumption of
the frictionless surface (Figure 2.18). How will the presence of friction affect the response of the mass-spring system? Since friction is a dissipative (nonconservative)
force we know that the energy of the system will not be conserved (Section 1.5.2).
We therefore anticipate that friction will have a damping effect on the motion of the
mass. Let's examine how.

2.3.1 Stick-Slip Condition

Consider a stationary block that sits on a frictional surface and is acted upon by a
force F, as shown in Figure 2.19a. The corresponding free-body diagram is shown in
Figure 2.19b. It is seen from the free-body diagram that the applied force, F, is bal-
anced by the friction force, F_μ, provided that

$$\left\| F_\mu \right\| < \mu_s N = \mu_s mg \tag{2.103}$$

where μ_s is the *coefficient of static friction* for the pair of surfaces in mutual contact,
and N is the (compressive) normal force exerted on the block by the support. Once
the magnitude of the friction force, and hence that of the applied force, achieves the
critical level $\left\| F \right\|_{cr}$ where

$$\left\| F \right\|_{cr} = \mu_s N = \mu_s mg \tag{2.104}$$

sliding begins. The friction force then maintains the magnitude

$$\left\| F_\mu \right\| = \mu_k N = \mu_k mg \tag{2.105}$$

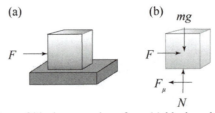

Figure 2.19 Equilibrium of block on rough surface: (a) block under applied load, (b) free-body diagram.

throughout the motion, where μ_k is the *coefficient of kinetic friction* and $\mu_k \leq \mu_s < 1$. Conversely, if during motion the mass should stop momentarily, and if the applied force drops below the critical value while the mass is stationary, then the motion of the mass will cease. Thus, if $\|F\|$ falls below $\|F\|_{cr}$ as defined by Eq. (2.104), then the motion of the mass is arrested (i.e., the mass "sticks"). For the mass-spring system of interest, we anticipate that the mass will first move in one direction, stop for an instant, then move in the opposite direction, stop for an instant, then move in the original direction, stop for an instant and reverse direction, and so on. That this is the case will be confirmed shortly. In this regard, for the mass-spring system, the force appearing in Eqs. (2.103) and (2.104) is replaced by the restoring force of the spring, $F_s = -kx$, at the instants when the velocity of the mass vanishes. Substituting F_s for F in Eq. (2.104) and solving for x gives the associated critical displacement, $\|x\|_{cr}$, as

$$\| x \|_{cr} = \bar{\mu} f_\mu \tag{2.106}$$

where

$$f_\mu \equiv \frac{\|F_\mu\|}{k} = \mu_k \frac{mg}{k} \tag{2.107}$$

and

$$\bar{\mu} \equiv \frac{\mu_s}{\mu_k} \geq 1 \tag{2.108}$$

The parameter f_μ is seen to be the static displacement of the mass if subjected to a constant force having the magnitude of the friction force. After incorporating Eqs. (2.106)–(2.108) into Eqs. (2.103)–(2.105), the *stick-slip criterion* for the mass-spring system may be stated as follows;

If, at time $t = t_s$, $\dot{x}(t_s) = 0$

then $\|\dot{x}(t_s^+)\| > 0$ provided that $\|x(t_s)\| \geq \bar{\mu} f_\mu$. $\qquad$ (2.109)

If not, $\dot{x} \equiv 0$ for $t > t_s$.

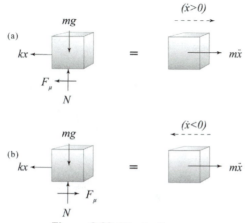

Figure 2.20 Kinetic diagram.

2.3.2 System Response

Let us consider once again the mass-spring system of Section 2.1, however, let us now relax the assumption that the surface on which the mass slides is frictionless (Figure 2.18). The kinetic diagram for the present case is shown in Figure 2.20. When motion occurs, the magnitude of the friction force is given by Eq. (2.105) and its direction is always opposite to that of the velocity of the mass. Thus, the friction force reverses direction whenever the mass reverses direction. To solve the problem we therefore write the equations of motion for each case, $\dot{x} > 0$ and $\dot{x} < 0$, separately. Applying Newton's Second Law as depicted in Figures 2.20a and 2.20b, respectively, and rearranging terms gives

$$\begin{aligned}
\ddot{x} + \omega^2 x &= -\omega^2 f_\mu \quad (\dot{x} > 0) \\
\ddot{x} + \omega^2 x &= +\omega^2 f_\mu \quad (\dot{x} < 0)
\end{aligned} \tag{2.110}$$

where ω and f_μ are given by Eqs. (2.2) and (2.107), respectively. The solution for each case is easily found by adding the *particular solution* corresponding to the appropriate right-hand side of Eq. (2.110) to the *complementary solution* given by Eq. (2.11). Thus, in each case, the response is of the form of a harmonic oscillation with a constant shift in displacement, $\mp f_\mu$. Hence,

$$\begin{aligned}
x(t) &= A^+ \cos(\omega t - \phi^+) - f_\mu \quad (\dot{x} > 0) \\
x(t) &= A^- \cos(\omega t - \phi^-) + f_\mu \quad (\dot{x} < 0)
\end{aligned} \tag{2.111}$$

Since each solution is only valid for the sign of the velocity indicated, each of the above forms yields the response for only half of a cycle.

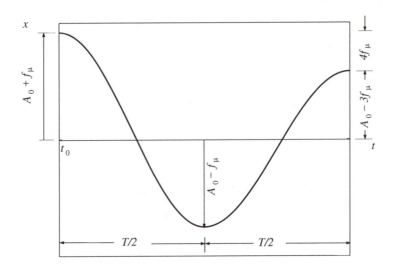

Figure 2.21 Typical response history over one cycle.

As the solutions described by Eqs. (2.111) are valid on intervals bounded by
the vanishing of the velocity, and hence when the mass is about to reverse its direction, let us consider the motion between the corresponding instants in time. If we start
our "measurements" in this way, then the initial velocity of each half cycle is zero
and the phase angle appearing in Eqs. (2.111) vanishes ($\phi^\pm = 0$). Likewise, the initial
displacement for each half cycle is given by the final displacement of the prior half
cycle, and the amplitude of the half cycle is adjusted accordingly (see Figure 2.21).
One can increment between half cycles in this way to obtain the full response as a
function of time. It will, however, prove sufficient to examine a single cycle in detail.
Let us now consider the motion through one cycle, beginning at time $t = t_0$ and
concluding at time $t = t_0 + T$, where the period T is related to the frequency ω
through Eq. (2.19). We shall take the reference displacement as positive, therefore the
mass moves to the left ($\dot{x} < 0$) for the first half of the cycle as indicated in Figure
2.21. Let $A^- = A_0$ correspond to the amplitude of the half cycle that begins at time
$t = t_0$. The second of Eqs. (2.111) then gives

$$x(t_0) = A_0 + f_\mu \tag{2.112}$$

which can, of course, be solved for the amplitude A_0 for given $x(t_0) = x_0$. Similarly,

$$x\left(t_0 + T/2\right) = -\left(A_0 - f_\mu\right) \tag{2.113}$$

We next perform a similar calculation for the second half of the cycle ($t_0 + T/2 \le t \le t_0 + T$) using the first of Eqs. (2.111) together with Eq. (2.113) as the initial condition. This gives

$$A^+ = 2f_\mu - A_0 \tag{2.114}$$

Substituting Eq. (2.114) into the first of Eqs. (2.111) and evaluating the resulting expression at $t = t_0 + T$ gives the displacement at that particular instant as

$$x(t_0 + T) = A_0 - 3f_\mu \tag{2.115}$$

Let Δ correspond to the reduction in deflection over the cycle. Then, subtracting Eq. (2.115) from Eq. (2.112) gives the reduction

$$\Delta = x(t_0 + T) - x(t_0) = 4f_\mu = \frac{4\mu_k mg}{k} \tag{2.116}$$

It is similarly seen, by subtracting Eq. (2.113) from Eq. (2.112), that the reduction in displacement over half of the cycle is $\Delta/2$. The displacement reduction is clearly the same for each and every cycle. Thus, the effect of dry friction is to damp the free vibration response of the simple harmonic oscillator. However, the damping rate is seen to be constant causing the oscillations to decay in a linear manner for this case, as shown in Figure 2.22, rather than in an exponential manner as for the viscous damper discussed in Section 2.2.2. It is also seen that, unlike for systems with viscous damping, the frequency of oscillation and hence the period is unaltered by the presence of friction. Further, the motion terminates when $\dot{x} = 0$ and $\|x\| < \|x\|_{cr}$. At this point, the spring force can no longer overcome the friction force.

To determine the time at which the mass sticks and motion subsides, let us consider the number of half-cycles the mass goes through to that point. Let $n = 1, 2, \ldots$, represent the half-cycle number and let x_n represent the corresponding peak displacement at the end of that half-cycle. It is seen from Figures 2.21 and 2.22 that

$$\|x_n\| = A_0 - (2n-1)f_\mu = x_0 - 2n f_\mu \tag{2.117}$$

The stick-slip condition, Eq. (2.109), tells us that motion will continue beyond a given peak provided that

$$x_0 - 2n f_\mu \ge \bar{\mu} f_\mu \tag{2.118}$$

The value of n for which sticking occurs is thus the first value of n for which the inequality of Eq. (2.118) is violated. Solving the corresponding equality for n gives the critical parameter

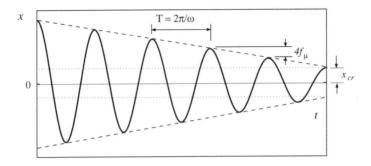

Figure 2.22 Response history of system with Coulomb damping from onset to cessation of motion.

$$n_{cr} = \frac{1}{2}\left[\frac{x_0}{f_\mu} - \bar{\mu}\right] \tag{2.119}$$

In general, the parameter n_{cr} will not be an integer. Thus, according to the criterion of Eq. (2.109) together with Eq. (2.118), the value of the integer n at which sticking occurs is described as follows:

$$n_{stick} \text{ is the first value of } n \text{ that is greater than or equal to } n_{cr} \tag{2.120}$$

The time, t_{stick}, at which sticking of the mass occurs is then

$$t_{stick} = n_{stick} T/2 \tag{2.121}$$

The mass is thus seen to stick at the first peak that lies between the dotted horizontal envelope, $\|x\| = \|x\|_{cr}$, shown in Figure 2.22.

Example 2.16

A 2 kg mass attached to a linear elastic spring of stiffness $k = 200$ N/m is released from rest when the spring is stretched 10 cm. If the coefficients of static and kinetic friction between the mass and the surface that it moves on are $\mu_s = \mu_k = 0.1$, determine the time after release at which the mass sticks and the corresponding displacement of the mass.

Solution

We first calculate the natural frequency and period using Eqs. (2.2) and (2.19). Hence,

$$\omega = \sqrt{\frac{200}{2}} = 10 \text{ rad/sec} \tag{a}$$

$$T = \frac{2\pi}{10} = 0.628 \text{ secs} \tag{b}$$

We next evaluate the displacement parameter defined by Eq. (2.107), giving

$$f_\mu = \frac{(0.1)(2)(9.81)}{200} = 0.00981 \text{ m} \tag{c}$$

The half-cycle number for which stick occurs is next obtained by evaluating Eq. (2.108), and substituting this result, along with Eq. (c), into Eq. (2.119). This gives

$$n_{cr} = 0.5\left[\frac{0.1}{0.00981} - \frac{0.1}{0.1}\right] = 4.60 \tag{d}$$

Thus, from Eq. (2.120),

$$n_{stick} = 5 \tag{e}$$

Therefore, the mass comes to rest after going through 2½ cycles. The time at which the motion of the mass is arrested is found by substituting Eqs. (b) and (e) into Eq. (2.121), giving

$$t_{stick} = t_5 = (5)\frac{(0.628)}{2} = 1.57 \text{ secs} \tag{f}$$

Finally, we have from the given initial conditions that $x_0 = 0.10$ m. The displacement of the mass at the sticking point is then found using Eq. (2.117). We thus find that

$$\|x_{stick}\| = \|x_5\| = \|0.10 - 2(5)(0.00981)\| = 0.0019 \text{ m} = 0.19 \text{ cm} \qquad \triangleleft \text{(g)}$$

2.4 CONCLUDING REMARKS

In this chapter we have examined the behavior of single degree of freedom systems when they are free from forces from outside of the system. That is, no dynamic external forces were considered. Rather, only the forces exerted by one part of the system on another part of the system (internal forces) entered into the problem. In this regard, the spring, the viscous damper and, in the case of Coulomb friction, the surface upon which the mass moved were all considered to be part of the system. We remark that the springs and dampers were considered to be massless. That is, the models considered apply to systems for which the mass, represented as a block or bob etc., is much greater than the masses of the objects represented as springs and dampers such as elastic beams. The only external force that was considered was the static gravitational force, the effect of which was simply to shift the equilibrium configuration about which the mass oscillated for the case of the mass-spring-damper system. For the simple pendulum the gravitational force was seen to act as a restoring force in the spirit of that of an elastic spring for a mass-spring system. Gravity was seen to also play a role in the buoyant force of a Newtonian fluid acting on a floating body. This resulted in the representation of the fluid as an equivalent elastic spring.

The mathematical process of linearization was demonstrated for the problem of the simple pendulum. The complex nonlinear equation of motion for the pendulum was simplified and took the form of the simple harmonic oscillator for situations where the amplitudes of the angular motions were suitably restricted. It was seen, however, that when this was done the second (vertical) equilibrium configuration was not predicted. This is typical of what occurs when linearization is performed; some information is lost. It was, however, seen that the oscillatory behavior of the pendulum was well represented by the linearized equation for a fairly wide range of amplitudes. It is instructive to remark here that the Euler-Bernoulli beam equations, used in this and the prior chapter to represent the members of certain structural systems as equivalent linear springs, are also obtained by linearization of more complex models. Such beam equations are valid for a restricted range of strains (infinitesimal), rotations and deflections. More correct models include the small strain and moderate rotation model typically used to examine buckling behavior (both static and dynamic). These issues are discussed in greater detail in Chapter 9. For truly finite deflections the structure must be modeled, more correctly, as an *elastica*. It is interesting to note that the equation governing the tangent angle as a function of distance along the axis for the (quasi-static) elastica is identical to the nonlinear equation of motion (angle of rotation as a function of time) for the simple pendulum, with the parameters suitably interpreted.

The effects of two types of damping were considered. These included viscous and Coulomb damping. Other types of damping that influence the vibratory behavior of mechanical systems include structural/internal/material or hysteresis damping, and aerodynamic damping, though these are often treated as effective viscous damping in practice. The effects of structural damping will be discussed in Chapter 3.

At this point we have an understanding of the vibratory behavior of single degree of freedom systems when free of external dynamic forces. We are now ready to examine the motion of such systems due to time dependent external excitation.

BIBLIOGRAPHY

Abramowitz, M. and Stegun, I.A., *Handbook of Mathematical Functions*, Dover, Mineola, 1972.

Zwillinger, D., *CRC Standard Mathematical Tables and Formulae*, 31st ed., Chapman and Hall/CRC, Boca Raton, 2003.

Fox, R.W. and McDonald, A.T., *Introduction to Fluid Mechanics*, 5th ed., Wiley, New York, 1998.

Love, A.E.H., *A Treatise on the Mathematical Theory of Elasticity*, 4th ed., Dover, Mineola, 1944.

Meirovitch, L., *Elements of Vibration Analysis*, 2nd ed., McGraw-Hill, New York, 1986.

Meirovitch, L., *Fundamentals of Vibrations*, McGraw-Hill, Boston, 2001.

Thomson, W.T., *Theory of Vibration with Applications*, 4th ed., Prentice-Hall, Englewood Cliffs, 1993.

PROBLEMS

2.1 The mass of a mass-spring system is displaced and released from rest. If the 20 gm mass is observed to return to the release point every 2 seconds, determine the stiffness of the spring.

2.2 Two packages are placed on a spring scale whose plate weighs 10 lb and whose stiffness is 50 lb/in. When one package is accidentally knocked off the scale the remaining package is observed to oscillate through 3 cycles per second. What is the weight of the remaining package?

2.3 Determine the natural frequency for side-sway motion of the one story structure shown in Figure 1.8. Do the same for the structure shown in Figure 1.9. What is the relation between the two and hence what is the effect of embedding the columns in the roof on the motion of the structure?

2.4 A single degree of freedom system is represented as a 4 kg mass attached to a spring possessing a stiffness of 6 N/m. Determine the response of the horizontally configured system if the mass is displaced 2 meters to the right and released with a velocity of 4 m/sec. What is the amplitude, period and phase lag for the motion? Sketch and label the response history of the system.

2.5 A single degree of freedom system is represented as a 2 kg mass attached to a spring possessing a stiffness of 4 N/m. Determine the response of the vertically configured system if the mass is displaced 1 meter downward and released from rest. What is the amplitude, period and phase lag for the motion? Sketch and label the response history of the system.

2.6   A 30 cm aluminum rod possessing a circular cross
section of 1.25 cm radius is inserted into a testing
machine where it is fixed at one end and attached
to a load cell at the other end. At some point during
a tensile test the clamp at the load cell slips, releasing that end of the rod. If the 20 kg clamp remains
attached to the end of the rod, determine the frequency of the oscillations of the rod-clamp system?

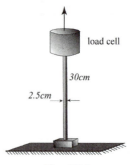

load cell

30cm

2.5cm

Fig. P2.6

2.7   A 30 cm aluminum rod possessing a circular cross section of 1.25 cm radius is
inserted into a testing machine where it is fixed at one end and attached to a
load cell at the other end. At some point
during a torsion test the clamp at the load
cell slips, releasing that end of the rod. If
the 20 kg clamp remains attached to the
end of the rod, determine the frequency of
the oscillations of the rod-clamp system.
The radius of gyration of the clamp is 5 cm.

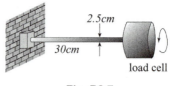

2.5cm

30cm

load cell

Fig. P2.7

2.8   Determine the natural period of a typical ice cube floating in water. Measure
the dimensions of a typical cube from your refrigerator and calculate its natural
frequency in water. (The dimensions may vary depending on your particular ice
tray.) Confirm your "experiment." Place an ice cube in water, displace it
slightly and release it. Make an approximate measure of the period of an oscillation with your wrist watch, or a
stop watch if available. Repeat
this operation several times and
compare the average measured
value with the calculated value.

Fig. P2.8

2.9   The manometer shown is used to measure the
pressure in a pipe. If the pipe pressure is suddenly
reduced to atmospheric (the gage pressure reduced to zero), determine the frequency of the oscillations of the manometer fluid about its equilibrium configuration if the total length of the
manometer fluid is L and the tube is uniform.

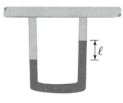

ℓ

Fig. P2.9

2.10 The tip of cantilever beam of Problem 1.4 is displaced 1″ and released from
rest. Determine the response of the weight if it has been welded to the beam.
Sketch and label the response.

2.11 The center span of the simply supported beam of Problem 1.5 is displaced 1″ and released at a speed of 13.9 in/sec. Determine the response of the weight if it has been welded to the beam. Sketch and label the response.

2.12 Determine the response of the movable support for the system of Problem 1.6 if the support is displaced 1″ and released from rest. The beam and moveable support have the properties of the beam and weight of Problem 1.4.

2.13 A railroad car of mass m is attached to a stop in a railroad yard. The stop consists of four identical metal rods of length L, radius R and elastic modulus E that are arranged symmetrically and are fixed to a rigid wall at one end and welded to a rigid plate at the other. The plate is hooked to the stationary railroad car as shown. In a docking maneuver, a second car of mass m approaches the first at speed v_1. If the second car locks onto the first upon contact, determine the response of the two car system after docking.

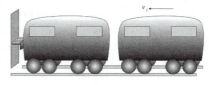

Fig. P2.13

2.14 Determine the response of the 200 lb. raft of Problem 1.7 if the man suddenly dives off.

2.15 Determine the response of the 200 lb. raft of Problem 1.15 if the woman suddenly dives off.

2.16 The line of action of the resultant buoyant force exerted on a ship by the water passes through a point Q along the centerplane as indicated. If the point Q lies a distance ℓ above the ship's centroid G, and the ship possesses mass m and mass moment of inertia I_G, determine the frequency of small angle rolling.

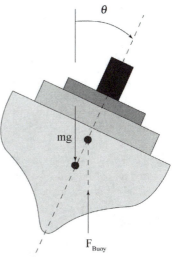

Fig. P2.16

2.17 A woman observes that the chandelier hanging in the hallway of her home is swinging to and fro about every three seconds. What is the length of the chandelier's cable?

2.18 The bob of a simple pendulum is
immersed in a medium whose resistance may be represented as a
spring of stiffness k connected at
distance a from the base of the rod
as shown. Determine the natural
frequency of the system.

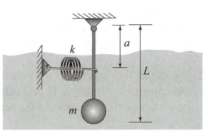

Fig. P2.18

2.19 The timing device shown consists of a movable cylinder of known mass m that
is attached to a rod of negligible mass supported by a torsional spring at its
base. If the stiffness of the spring is k_T, where $k_T/mgL > 1$, determine the period of small angle motion of the device as a function of the attachment length,
L, if the spring is untorqued when $\theta = 0$.

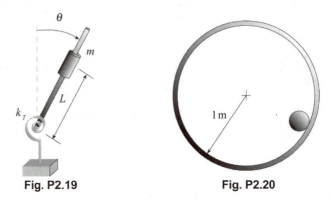

Fig. P2.19　　　　　　　　　　　　　　**Fig. P2.20**

2.20 A circular tube of 1 m inner diameter stands in the vertical plane. A 6 gm marble of 1.5 cm diameter is placed in the tube, held at a certain height and then released. Determine the time it takes for the marble to reach its maximum height
on the opposite rise. What is the frequency of the marble's motion?

2.21 The system shown consists of a rigid rod, a flywheel of radius R and mass m,
and an elastic belt of stiffness k. Determine the natural frequency of the system.
The belt is unstretched when $\theta = 0$.

Fig. P2.21

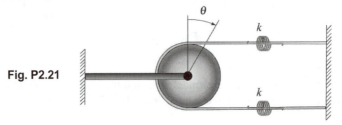

2.22 Determine the response of the flywheel of Problem 2.21 if it is slowly rotated through angle θ_0 and then released from rest.

2.23 The cranking device shown consists of a mass-spring system of stiffness k and mass m that is pin-connected to a massless rod which, in turn, is pin-connected to a wheel at radius R, as indicated. If the mass moment of inertia of the wheel about an axis through the hub is I_O, determine the natural frequency of the system. (The spring is unstretched when connecting pin is directly over hub 'O'.)

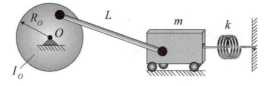

Fig. P2.23

2.24 Use the software of your choice (or, if this is not option, any suitable mathematical tables at your disposal) to evaluate the period of the finite motion of a pendulum when it is released from rest at a series of initial angles within the range $0 < \theta_0 < \pi$. (*a*) Plot the normalized period as a function of the initial angle. (*b*) Determine the initial angle at which the percent error for the linear approximation is (*i*) less than 5%, (*ii*) less than 1%.

2.25 A single degree of freedom system is represented as a 4 kg mass attached to a spring possessing a stiffness of 6 N/m and a viscous damper whose coefficient is 1 N-sec/m. (*a*) Determine the response of the horizontally configured system if the mass is displaced 2 meters to the right and released with a velocity of 4 m/sec. Plot and label the response history of the system. (*b*) Determine the response and plot its history if the damping coefficient is 5 N-sec/m. (*c*) Determine the response and plot its history if the damping coefficient is 10 N-sec/m.

2.26 A single degree of freedom system is represented as a 2 kg mass attached to a spring possessing a stiffness of 4 N/m and a viscous damper whose coefficient is 2 N-sec/m. (*a*) Determine the response of the horizontally configured system if the mass is displaced 1 meter to the right and released from rest. Plot and label the response history of the system. (*b*) Determine the response and plot its history if the damping coefficient is 8 N-sec/m.

2.27 A 12 kg spool that is 1 m in radius is pinned to a viscoelastic rod of negligible mass with effective properties $k = 10$ N/m and $c = 8$ N-sec/m. The end of the rod is attached to a rigid support as shown. Determine the natural frequency of the system if the spool rolls without slipping.

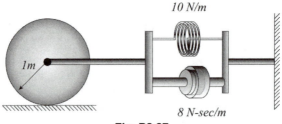

Fig. P2.27

2.28 A viscous damper with coefficient c is fixed at one end and is attached to the cylindrical mass of the timing device of Problem 2.19 at the other end. Determine the natural frequency of the system.

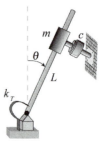

Fig. P2.28

2.29 (*a*) A mass-spring system oscillates freely with a period of 2.6 seconds. After a viscous damper is attached, the resulting mass-spring-damper system is observed to oscillate with a period of 3.0 seconds. Determine the damping factor for the system. (*b*) A mass-spring-damper system is displaced and released from rest. The time history of the motion of the mass is recorded and displayed, as shown (Figure P2.29). Estimate the damping factor from the measured response.

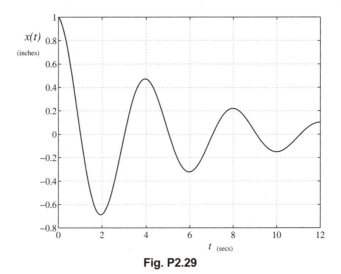

Fig. P2.29

2.30 A diving board that is 7 feet in length, has a cross-sectional area of 2 ft² and a specific weight of 48 lb/ft³ is supported by a torsional spring of stiffness $k_T = 13.45 \times 10^3$ ft-lb/rad and a viscous damper of coefficient c that is located 2 feet ahead of the spring's pivot, as shown. Assuming that the flexure of the board is accounted for in the model of the spring, and hence that the board may be treated as rigid, determine the value of c so that the oscillations will decay to 2% of their initial amplitude within 5 cycles.

Fig. P2.30

2.31 A simple pendulum is immersed in a viscous fluid. If the resistance of the fluid can be modeled by the linear spring and viscous damper attached a distance a from the support, determine the critical value of c.

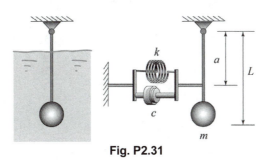

Fig. P2.31

2.32 Determine the critical value of the damping coefficient for the diving board of Problem 2.30.

2.33 A screen door of mass m, height L and width ℓ is attached to a door frame as indicated. A torsional spring of stiffness k_T is attached as a closer at the top of the door as indicated, and a damper is to be installed near the bottom of the door to keep the door from slamming. Determine the limiting value of the damping coefficient so that the door closes gently, if the damper is to be attached a distance a from the hinge.

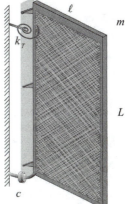

Fig. P2.33

2.34 Determine the overshoot if the system of Problem 2.25 is critically damped and $v_0 = -4$ m/sec.

2.35 Determine the overshoot if the system of Problem 2.26 is critically damped and $v_0 = -2$ m/sec.

2.36 A single degree of freedom system is represented as a 4 kg mass attached to a spring possessing a stiffness of 6 N/m. If the coefficients of static and kinetic friction between the mass and the surface it moves on are $\mu_s = \mu_k = 0.1$, and the mass is displaced 2 meters to the right and released with a velocity of 4 m/sec, determine the time after release at which the mass sticks and the corresponding displacement of the mass.

2.37 A single degree of freedom system is represented as a 2 kg mass attached to a spring possessing a stiffness of 4 N/m. If the coefficients of static and kinetic friction between the block and the surface it moves on are respectively $\mu_s = 0.12$ and $\mu_k = 0.10$, determine the drop in amplitude between successive periods during free vibration. What is the frequency of the oscillations?

2.38 Use the work-energy principle, Eq. (1.84) or Eq. (1.92), to arrive at Eq. (2.116). That is show, by way of the Principle of Work-Energy, that the decay in displacement per cycle for a mass-spring system with coulomb friction is $\Delta = 4\mu_k mg / k$.

3

Forced Vibration of Single Degree of Freedom Systems – 1: *Periodic Excitation*

In Chapter 2 we studied the response of single degree of freedom systems that were free from external loading. In doing so we established parameters that characterize the system and discussed the motion that would ensue if the system was disturbed from equilibrium and then moved under its own volition. In this and the next chapter we shall study the response of single degree of freedom systems that are subjected to time dependent external forcing, that is forcing which is applied to the system from an outside source. In the present chapter we consider forcing that is continuously applied and repeats itself over time. Specifically, we now consider forcing that varies harmonically with time, and forcing that is periodic but is otherwise of a general nature. We begin with a discussion of the general form of the equation of motion for forced single degree of freedom systems, followed by a discussion of superposition.

3.1 STANDARD FORM OF THE EQUATION OF MOTION

It was seen in the previous two chapters that a mass-spring damper system is representative of a variety of complex systems. It was also seen that the equations of motion of various single degree of freedom systems took on a common general form. We therefore consider the externally forced mass-spring-damper system to initiate our discussion.

Consider the system comprised of a mass m attached to a spring of stiffness k and a viscous damper of damping coefficient c that is subjected to a time dependent force $F(t)$ as shown in Figure 3.1. The mass moves along a frictionless surface, or hangs vertically, as indicated. In either case, $x(t)$ measures the displacement of the

143

mass from its equilibrium configuration. The corresponding dynamic free body diagram (DFBD) is depicted in Figure 3.2. Applying Newton's Second Law along the x direction gives

$$-kx - c\dot{x} + F(t) = m\ddot{x}$$

which, when rearranged, takes the form

$$\ddot{x} + 2\omega\zeta\dot{x} + \omega^2 x = \omega^2 f(t) \qquad (3.1)$$

where

$$f(t) = \frac{F(t)}{k} \qquad (3.2)$$

and

$$\zeta = \frac{c}{2\sqrt{mk}} \quad \text{and} \quad \omega = \sqrt{\frac{k}{m}}$$

respectively correspond to the viscous damping factor and undamped natural frequency introduced in Chapter 2. An equation of motion expressed in the form of Eq. (3.2) will be said to be in *standard form*. Equation (3.2) is seen to be of the same general form as Eq. (2.64), but with nonvanishing right hand side. It may be seen that, for the mass-spring-damper system, the quantity $f(t)$ has units of length.

3.2 SUPERPOSITION

In this section we shall establish the principle of superposition for the specific class of linear systems under consideration. This principle shall be of great importance in evaluating the response of the many types of systems and loading considered throughout our study of linear vibrations. We consider the mass-spring-damper system to represent the class of systems of interest.

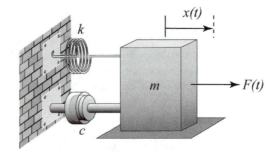

Figure 3.1 Mass-spring-damper system subjected to external forcing.

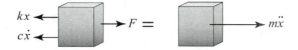

Figure 3.2 Kinetic diagram for mass-spring-damper system.

Suppose the mass of the system of Figure 3.1 is simultaneously subjected to two different forces $F_1(t)$ and $F_2(t)$. Let $x_1(t)$ correspond to the response of the system when it is subjected to $F_1(t)$ alone, and let $x_2(t)$ correspond to the response of the system when it is subjected to $F_2(t)$ alone. Further, let

$$f_1(t) = F_1(t)/k \quad \text{and} \quad f_2(t) = F_2(t)/k$$

If this is so then each force-displacement pair, $\{F_1, x_1\}$ and $\{F_2, x_2\}$, must separately satisfy Eq. (3.1) and equivalently Eq. (3.2). Hence,

$$\ddot{x}_1 + 2\omega\zeta\dot{x}_1 + \omega^2 x_1 = \omega^2 f_1(t) \tag{3.3}$$

$$\ddot{x}_2 + 2\omega\zeta\dot{x}_2 + \omega^2 x_2 = \omega^2 f_2(t) \tag{3.4}$$

Adding Eqs. (3.3) and (3.4), and exploiting the fact that the differential operators are linear, gives the relation

$$\frac{d^2}{dt^2}\{x_1 + x_2\} + 2\omega\zeta\frac{d}{dt}\{x_1 + x_2\} + \omega^2\{x_1 + x_2\} = \omega^2\{f_1 + f_2\}$$

or, equivalently,

$$\ddot{x} + 2\omega\zeta\dot{x} + \omega^2 x = \omega^2 f(t) \tag{3.5}$$

where

$$x(t) = x_1(t) + x_2(t) \tag{3.6}$$

and

$$f(t) = f_1(t) + f_2(t) = \frac{F_1(t) + F_2(t)}{k} \tag{3.7}$$

Equation (3.7) is seen to be identical to Eq. (3.2), and the response is seen to be the sum of the responses to the individual forces. We thus see that the response of the system subjected to two forces acting simultaneously is equal to the sum of the responses to the two forces acting individually. This process can be extended to any number of forces and is referred to as the *principle of superposition*. What permits this convenient property to occur when we add Eqs. (3.3) and (3.4) to get Eq. (3.6) is the fact that the differential operators in Eq. (3.1), and hence in Eqs. (3.3) and (3.4), appear to first power only as does the displacement. If this was not the case (say, for example, one or more of the operators was squared) then it is evident that the equation resulting from the sum of the individual equations of motion would not be of the same form as the original two equations.

Example 3.1

Determine the response of an undamped mass-spring system to an applied force of the form $F(t) = F_0(1 + bt)$, where F_0 and b are constants.

Solution

Let us consider the applied force as the sum of two forces $F_1(t) = F_0$ and $F_2(t) = F_0 bt$. Let us further determine the response of the system (the solution) to each force acting separately. Thus, let us solve the following two problems:

$$\ddot{x} + \omega^2 x = \omega^2 \frac{F_0}{k} \tag{a}$$

$$\ddot{x} + \omega^2 x = \omega^2 \frac{F_0}{k} bt \tag{b}$$

The solutions to Eqs. (a) and (b) are readily obtained as

$$x_1(t) = A_1 \cos \omega t + A_2 \sin \omega t + \frac{F_0}{k} \tag{c}$$

$$x_2(t) = B_1 \cos \omega t + B_2 \sin \omega t + \frac{F_0}{k} bt \tag{d}$$

respectively. Based on the principle of superposition, the response of the system in question to the given applied force $F(t) = F_0(1 + bt)$ is obtained by adding Eqs. (c) and (d). Hence,

$$x(t) = C_1 \cos \omega t + C_2 \sin \omega t + \frac{F_0(1 + bt)}{k} \qquad \triangleleft (e)$$

where

$$C_1 = A_1 + B_1 \quad \text{and} \quad C_2 = A_2 + B_2$$

and the constants of integration, C_1 and C_2, are determined from the specific initial conditions for a given problem.

In preparation for the development and analyses of the next section, we finish the present discussion by establishing a result for complex forces and displacements (forces and displacements possessing both real and imaginary parts). To accomplish this, let us multiply Eq. (3.4) by the imaginary number $i \equiv \sqrt{-1}$ and add Eq. (3.3) to the resulting expression. This gives the differential equation

$$\ddot{\hat{x}} + 2\omega\zeta\dot{\hat{x}} + \omega^2\hat{x} = \omega^2\hat{f}(t) \tag{3.8}$$

where

$$\hat{x}(t) = x_1(t) + ix_2(t) \tag{3.9}$$

and

$$\hat{f}(t) = f_1(t) + if_2(t) = \frac{F_1(t) + iF_2(t)}{k} \tag{3.10}$$

It is seen from the above superposition that the real part of the complex response is the response to the real part of the complex force and the imaginary part of the complex response is the response to the imaginary part of the complex force. We shall use this important property in the developments and analyses of the next section, and throughout our study of vibrations.

3.3 HARMONIC FORCING

An important class of forcing in the study of vibrations, both fundamentally and with regard to applications, is harmonic excitation. In this section we shall consider the specific class of forces whose time dependence is harmonic. That is, we shall consider forces that vary temporally in the form of sine and cosine functions.

3.3.1 Formulation

Let us consider forcing functions of the form

$$F(t) = F_0 \cos\Omega t \quad \text{and} \quad F(t) = F_0 \sin\Omega t \tag{3.11}$$

where F_0 = constant is the *amplitude* of the applied force and Ω is the frequency of the applied force. The latter is referred to as the *forcing frequency* or the *excitation frequency*.

We can solve the problem of harmonic forcing for time dependence in the form of cosine or sine functions individually, however that will be left as an exercise. Instead we shall solve both problems simultaneously by using the principle of superposition (Section 3.2) together with Euler's Formula, Eq. (1.61). Let us combine the two forcing functions described by Eqs. (3.11) by defining the complex forcing function

$$\hat{F}(t) = F_0 \cos \Omega t + iF_0 \sin \Omega t = F_0 e^{i\Omega t} \qquad (3.12)$$

It follows from Eqs. (3.8)–(3.10) that once the response to the complex force is determined then the response to the cosine function will be the real part of the complex response and the response to the sine function will be the imaginary part of the complex response. We shall therefore solve the problem

$$\ddot{x} + 2\omega\zeta\dot{x} + \omega^2 x = \omega^2 \hat{f}(t) = \omega^2 f_0 e^{i\Omega t} \qquad (3.13)$$

where

$$f_0 = F_0 / k \qquad (3.14)$$

It may be seen that the parameter f_0 corresponds to the deflection that the mass of the system would undergo if it was subjected to a static force of the same magnitude, F_0, as that of the dynamic load.

The general solution of Eq. (3.13) consists of the sum of the complementary solution and the particular solution associated with the specific form of the forcing function considered. Hence,

$$\hat{x}(t) = \hat{x}_c(t) + \hat{x}_p(t) \qquad (3.15)$$

where subscripts c and p indicate the complementary and particular solution, respectively. The former corresponds to the solution to the associated homogeneous equation, as discussed in Chapter 2. Incorporating Eqs. (2.15), (2.74), (2.96) and (2.98) gives the general solutions for undamped and viscously damped systems as

$$x(t) = Ae^{-\zeta\omega t} \cos(\omega_d t - \phi) + x_p(t) \qquad (0 \le \zeta < 1) \qquad (3.16)$$

$$x(t) = e^{-\zeta\omega t} \left[A_1 \cosh \jmath\omega t + A_2 \sinh \jmath\omega t \right] + x_p(t) \quad (\zeta > 1) \qquad (3.17)$$

$$x(t) = \left(A_1 + A_2 t \right) e^{-\omega t} + x_p(t) \qquad (\zeta = 1) \qquad (3.18)$$

where ω_d and $\jmath$ are defined by Eqs. (2.70 and 2.92), respectively. The constants of integration are evaluated by imposition of the initial conditions on the pertinent form of the response, Eq. (3.16), (3.17) or (3.18), after the specific form of the particular solution is determined.

It is seen that the complementary solution damps out and becomes negligible with respect to the particular solution after a sufficient amount of time. (Note that the undamped case, $\zeta = 0$, is an idealization for very light damping. Since all systems possess some damping we shall, for the purposes of the present discussion, consider the complementary solution to decay for vanishing damping as well.) Since we are presently considering forces that act continuously over very long time intervals, the particular solution for these cases corresponds to the *steady state response* of the sys-

tem. As an example, imagine that we are modeling the behavior of a machine or vehicle that is switched on and then runs for the entire day. The normal operating state
will be achieved a short time after the system is turned on, after the transients have
died out. This state is referred to as the *steady state*, and the response during this time
is referred to as the *steady state response*. Thus, for any loading which repeats itself
over long intervals of time, including the present harmonic loading, we denote the
particular solution as x_{ss}. Thus, $x_p \leftrightarrow x_{ss}$ for the class of loading considered in this
chapter. We first examine the steady state response for undamped systems and then
study the effects of damping on forced vibrations.

3.3.2 Steady State Response of Undamped Systems

Though all systems possess some degree of damping it is instructive, as well as practical, to examine the response of undamped systems. We therefore first consider the
special case of vanishing damping.

    To establish the steady state response we seek to determine the complex func-
tion $\hat{x}_{ss}$ that, when substituted into the left hand side of Eq. (3.13) with $\zeta = 0$, results
in the right hand side of that equation. Given that the time dependence of the forcing
function is exponential, and given that differentiation of exponentials results in exponentials, let us assume a particular solution of the form

$$\hat{x}_p(t) = \hat{x}_{ss}(t) = \hat{X}e^{i\Omega t} \tag{3.19}$$

where $\hat{X}$ is a (possibly complex) constant to be determined. Substitution of the assumed form of the solution, Eq. (3.19), into the governing equation, Eq. (3.13), results in the algebraic equation

$$\left[(i\Omega)^2 + \omega^2 \right] \hat{X}e^{i\Omega t} = \omega^2 f_0 e^{i\Omega t}$$

which may be solved for $\hat{X}$ to give

$$\hat{X} = \frac{f_0}{1 - \bar{\Omega}^2} \tag{3.20}$$

where

$$\bar{\Omega} \equiv \Omega / \omega \tag{3.21}$$

Equivalently,

$$\hat{X} = Xe^{-i\Phi} \tag{3.22}$$

where

$$X = f_0 \Gamma_0(\bar{\Omega}) \tag{3.23}$$

$$\Gamma_0\left(\bar{\Omega}\right) = \frac{1}{\left\|1 - \bar{\Omega}^2\right\|} \tag{3.24}$$

and

$$\begin{aligned}\Phi &= 0 \quad \text{when} \quad \bar{\Omega} < 1 \\ \Phi &= \pi \quad \text{when} \quad \bar{\Omega} > 1\end{aligned} \tag{3.25}$$

Substituting Eq. (3.20) into Eq. (3.19) and using Euler's Formula gives the particular solution

$$\hat{x}_{ss}(t) = X\left[\cos(\Omega t - \Phi) + i\sin(\Omega t - \Phi)\right] = \frac{f_0}{1 - \bar{\Omega}^2}\left[\cos\Omega t + i\sin\Omega t\right] \tag{3.26}$$

As discussed at the end of Section 3.2, the response to the real part of a complex forcing function is the real part of the complex response. Likewise, the response to the imaginary part of a complex forcing function is the imaginary part of the complex response. Thus, pairing Eq. (3.26) with Eq. (3.12) in this sense, we find the following responses:

if $F(t) = F_0 \cos\Omega t$, then

$$x_{ss}(t) = X\cos(\Omega t - \Phi) = \frac{f_0}{1 - \bar{\Omega}^2}\cos\Omega t \tag{3.27}$$

if $F(t) = F_0 \sin\Omega t$, then

$$x_{ss}(t) = X\sin(\Omega t - \Phi) = \frac{f_0}{1 - \bar{\Omega}^2}\sin\Omega t \tag{3.28}$$

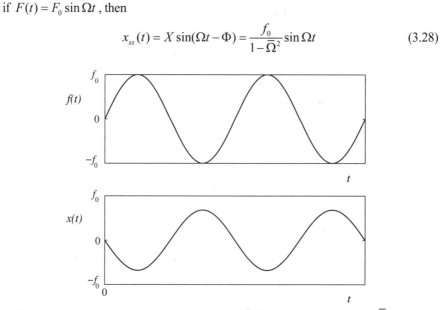

Figure 3.3 Typical time histories of excitation, $f(t)$, and response, $x(t)$, for $\bar{\Omega} > 1$.

The parameter X is seen to be the *amplitude* of the steady state response. Consequently, the parameter Γ_0 is referred to as the *magnification factor*, and the angle Φ is seen to be the *phase angle* of the steady state response measured with respect to the excitation.

It may be seen that the steady state response is in phase with the force when the excitation frequency is less than the natural frequency of the system and that the response is π radians out of phase with the force, and hence lags the force by $t_{lag} = \pi / \omega$, when the excitation frequency is greater than the natural frequency. Thus, when in steady state motion, the mass oscillates about its equilibrium position with frequency Ω and constant amplitude X, completely in phase or 180 degrees out of phase with the force according to Eq. (3.25). Typical time histories of the excitation and the corresponding response of the system are displayed in Figure 3.3 for the case when $\bar{\Omega} > 1$. In addition, it may be seen that the magnification factor corresponds to the ratio of the amplitude of the deflection induced by the applied harmonic force to the static deflection that would be produced by an applied static force having the same magnitude as the dynamic force. The magnification factor therefore measures the magnification of the static response due to the dynamic nature of the applied force and mechanical system and, as will be demonstrated in subsequent sections, is an important parameter in design considerations where vibratory behavior is pertinent.

Example 3.2

A 4 kg mass is attached to a spring of stiffness 2 N/m. If the mass is excited by the external force $F(t) = 5\cos 2t$ N, determine the amplitude and phase of the steady state response. Write down the response. Plot the steady state response of the system and label pertinent measures.

Solution

For the mass-spring system, the natural frequency is

$$\omega = \sqrt{\frac{2}{4}} = .7071 \text{ rad/sec} \tag{a}$$

The frequency of the given excitation is $\Omega = 2$ rad/sec. Thus,

$$\bar{\Omega} = \frac{2}{0.7071} = 2.828 \ (>1)$$

It then follows from Eq. (3.25) that

$$\Phi = \pi \tag{◁ (b)}$$

The amplitude of the steady state response is next computed using Eq. (3.23). Hence,

$$X = \frac{5/2}{\left\|1-(2.828)^2\right\|} = 0.357 \text{ m} = 35.7 \text{ cm} \qquad \triangleleft \text{(c)}$$

The steady state response of the mass-spring system is now calculated using Eq. (3.27). Thus,

$$x_{ss}(t) = 0.357\cos(2t - \pi) = -0.357\cos 2t \text{ m} = -35.7\cos 2t \text{ cm} \qquad \triangleleft \text{(d)}$$

The force and response histories are displayed below (Figure E3.2).

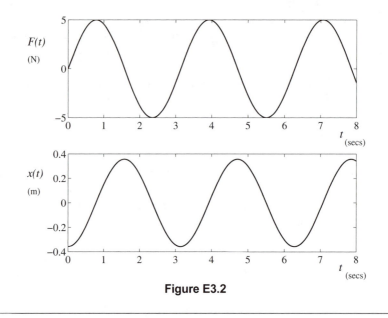

Figure E3.2

Example 3.3

The teetering glider of Example 2.9 undergoes a sustained breeze while on the ground. The force imposed by the breeze consists of a constant uniform lift force of magnitude $\overline{F}_0$ ($< mg$) and a nonuniform time dependent perturbation. The perturbation of the wind force manifests itself as the equivalent of two harmonic forces, each of magnitude F_ε and frequency Ω, mutually out of phase and acting at the half length points of the wings as indicated. If F_L lags F_R by $180°$, determine the steady state response of the glider.

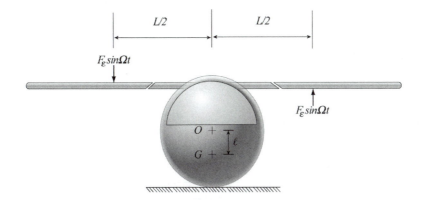

Figure E3.3 Glider with wind load perturbation.

Solution

The perturbed portion of the wind forces may be expressed as follows:

$$F_R(t) = F_\varepsilon \sin \Omega t$$

$$F_L(t) = F_\varepsilon \sin(\Omega t - \pi) = -F_\varepsilon \sin \Omega t$$

To derive the equations of motion for the present case, we augment the kinetic diagram of Example 2.9 by including the wind forces. The inertia portion of the kinetic diagram remains the same. Adding the moments of the wind forces to the development of Example 2.9, and linearizing as for that problem, gives the equation of motion for the forced system as

$$LF_\varepsilon \sin \Omega t - mg\, \ell\theta = m\left[r_G^2 + (R - \ell)^2 \right]\ddot{\theta} \qquad (a)$$

Rearranging terms gives the equation of motion for the wind excited glider as

$$\ddot{\theta} + \frac{g}{L_{eff}}\theta = \frac{g}{L_{eff}}\frac{LF_\varepsilon}{\ell mg}\sin \Omega t \qquad (b)$$

where

$$L_{eff} = \left[r_G^2 + (R - \ell)^2 \right]/\ell \qquad (c)$$

Equation (b) may be written in the "standard form"

$$\ddot{\theta} + \omega^2 \theta = \omega^2 f_\varepsilon \sin \Omega t \qquad (d)$$

where

$$\omega^2 = \frac{g}{L_{eff}} \tag{e}$$

and

$$f_\varepsilon = \frac{LF_\varepsilon}{\ell mg} \tag{f}$$

With Eqs. (d)–(f) identified, the response can now be written directly from Eq. (3.28). Doing this we find that the steady state response of the glider is

$$\theta(t) = \frac{f_\varepsilon}{1-\left(\Omega^2 L_{eff}/g\right)}\sin\Omega t \qquad\qquad \triangleleft (g)$$

The discussion to this point has pertained to excitation frequencies for which $\Omega \neq \omega$. However, upon examination of Eqs. (3.23), (3.24), (3.27) and (3.28), it is seen that the amplitude, and hence the solutions, become singular when $\overline{\Omega}^2 = 1$ ($\Omega = \omega$). Though this tells us that something interesting occurs when the forcing frequency takes on the value of the natural frequency of the system, the solutions given by Eqs. (3.26) – (3.28) are actually invalid when $\overline{\Omega} = 1$. This becomes evident when we go back and examine the equation of motion for this special case ($\zeta = 0$). We must therefore re-solve the problem for the case where the forcing frequency equals the natural frequency of the system.

The Phenomenon of Resonance

When $\zeta = 0$ and $\overline{\Omega} = 1$ ($\Omega = \omega$), Eq. (3.13) reduces to the form

$$\ddot{\hat{x}} + \omega^2 \hat{x} = \omega^2 f_0 e^{i\omega t} \tag{3.29}$$

It is seen that, for this case, the time dependence of the forcing function is of precisely the same form as the solution to the corresponding homogenous equation studied in Section 2.1. Thus, any solution of the form $\hat{x}(t) = Ce^{i\omega t}$ will yield zero when substituted into the left-hand side of Eq. (3.29). The particular solution for this special case is therefore found by seeking a solution of the form

$$\hat{x}_{ss}(t) = \hat{C}te^{i\omega t} \tag{3.30}$$

where $\hat{C}$ is a (complex) constant. Substituting Eq. (3.30) into Eq. (3.29) and solving for the constant gives

$$\hat{C} = -\tfrac{1}{2} i \omega f_0 \tag{3.31}$$

Substituting Eq. (3.31) back into Eq. (3.30) and using Euler's Formula gives the particular solution of Eq. (3.29) as

$$x_{ss}(t) = \tfrac{1}{2} f_0 \omega t \left[\sin \omega t - i \cos \omega t\right]$$

or   (3.32)

$$x_{ss}(t) = \tfrac{1}{2} f_0 \omega t \left[\cos(\omega t - \pi/2) + i \sin(\omega t - \pi/2)\right]$$

As per earlier discussions, the real part of the solution is the response to the real part of the complex forcing function, and the imaginary part of the solution is the response to the imaginary part of the complex forcing function. Thus,

when $F(t) = F_0 \cos \omega t$,

$$x_{ss}(t) = \tfrac{1}{2} f_0 \omega t \sin \omega t = X(t) \cos(\omega t - \pi/2) \tag{3.33}$$

when $F(t) = F_0 \sin \omega t$,

$$x_{ss}(t) = -\tfrac{1}{2} f_0 \omega t \cos \omega t = X(t) \sin(\omega t - \pi/2) \tag{3.34}$$

where

$$X = X(t) = \tfrac{1}{2} f_0 \omega t \tag{3.35}$$

is the (*time dependent*) amplitude of the steady state response. It is seen that, when the forcing frequency is numerically equal to the natural frequency of the undamped system, the steady state response of the system is out of phase with the force by $\Phi = \pi/2$ radians and hence lags the force by $t_{lag} = \pi/2\omega$. The amplitude, X, is seen to grow linearly with time. A typical response is displayed in Figure 3.4. We see that, when the undamped system is excited by a harmonically varying force whose frequency is identical in value to the natural frequency of the system, the steady state response increases linearly in time without bound. This phenomenon is called *resonance*. Clearly the energy supplied by the applied force is being used in an optimum manner in this case. We certainly know, from our discussions of free vibrations in Chapter 2, that the system moves naturally at this rate (the natural frequency) when disturbed and then left on its own. The system is now being forced at this very same rate. Let's examine what is taking place more closely.

To better understand the mechanics of resonance let us examine the work done by the applied force over one cycle of motion, $nT < t < (n+1)T$, where T is the period of the excitation (and thus of the steady state response as well) and n is any integer. We shall compare the work done by the applied force, $F(t)$, for three cases; (i) $\bar{\Omega} < 1$, (ii) $\bar{\Omega} > 1$ and (iii) $\bar{\Omega} = 1$.

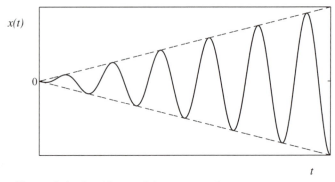

Figure 3.4 Time history of the response of a system at resonance.

From the definition of work given in Section 1.5.2 [see Eqs. (1.85) and (1.87)], and utilization of the chain rule of differentiation, the work done by the applied force for the single degree of freedom system under consideration is seen to be given by the relation

$$W = \int_{x_1}^{x_2} F \, dx = \int_{t_1}^{t_2} F \dot{x} \, dt \tag{3.36}$$

where $x_1 = x(t_1)$, $x_2 = x(t_2)$ and, for the interval under consideration, $t_1 = nT$ and $T_2 = (n+1)T$ for any given n.

For the sake of the present discussion we will assume, without loss of generality, that the forcing function is of the form of a cosine function. The response is then given by Eq. (3.33). Typical plots of the applied force and the resulting steady state response for cases (i)–(iii) are displayed as functions of time over one cycle in Figures 3.5a–3.5d, respectively. Noting that the slope of the response plot corresponds to the velocity of the mass, we can examine the work done by the applied force qualitatively during each quarter of the representative period considered for each case.

Case (i): $\overline{\Omega} < 1$
Consideration of Figures 3.5a and 3.5b shows that $F > 0$ and $\dot{x}_{ss} < 0$ in the first quadrant. Thus, it may be concluded from Eq. (3.36) that $W < 0$ over the first quarter of the period. If we examine the second quadrant, it is seen that $F < 0$ and $\dot{x}_{ss} < 0$ during this interval. Hence, $W > 0$ during the second quarter of the period. Proceeding in a similar manner, it is seen that $F < 0$ and $\dot{x}_{ss} > 0$ during the third quarter of the period. Hence, $W < 0$ during this interval. Finally, it may be observed that $F > 0$ and $\dot{x}_{ss} > 0$ during the fourth quarter of the period. Therefore, $W > 0$ during the last interval. It is thus seen that the applied force does positive work on the system during half of the period and negative work during half of the period. Therefore, the applied force reinforces the motion of the mass during half of the period and opposes the motion of the mass during half the period. In fact, the total work done by the applied force over a cycle

vanishes for this case. Hence, the motion of the mass remains bounded for situations when $\bar{\Omega} < 1$.

Case (ii): $\bar{\Omega} > 1$

Proceeding as for case (i), it is seen from Figures 3.5a and 3.5c that $\mathcal{W} > 0$ during the first and third quarters of the period, and that $\mathcal{W} < 0$ for the second and fourth quarters of the period. Thus, as for case (i), the applied force does positive work on the system during half of the period and negative work during half of the period. It therefore reinforces the motion of the mass for half of the cycle and opposes it during half of the cycle. As for case (i), the total work done by the applied force during a cycle vanishes and the motion of the mass remains bounded.

Case (iii): $\bar{\Omega} = 1$

For this case it may be seen, upon consideration of Figures 3.5a and 3.5d together with Eq. (3.36), that $F > 0$ and $\dot{x}_{ss} > 0$ during the first and fourth quarters of the period. Hence, that $\mathcal{W} > 0$ during these intervals. It may be similarly observed that $F < 0$ and $\dot{x}_{ss} < 0$ during the second and third quarters of the period. Therefore $\mathcal{W} > 0$ during these intervals as well. Thus, for the case when $\Omega = \omega$, the applied force does positive work in moving the mass during the entire period. Hence, during resonance, the phase relationship between the applied force and the response of the system is such that the force continuously reinforces the motion of the mass (does positive work). The amplitude of the displacement of the mass thus increases continuously. It is seen that, during resonance, the system uses the work imparted by the applied force in the most optimum manner possible.

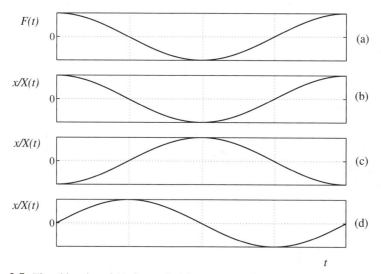

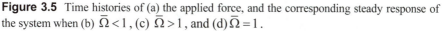

Figure 3.5 Time histories of (a) the applied force, and the corresponding steady response of the system when (b) $\bar{\Omega} < 1$, (c) $\bar{\Omega} > 1$, and (d) $\bar{\Omega} = 1$.

Example 3.4

A 10 kg mass is attached to the end of a 1 m long cantilever beam of rectangular cross section as shown. If the mass attached to the 30 mm by 5 mm beam is driven by a harmonic point force it is found that the beam vibrates violently when the forcing frequency approaches 3 cps. Determine Young's Modulus for the beam.

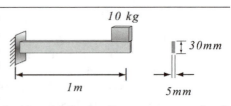

Solution

The beam has clearly achieved resonance at this frequency. Hence,

$$\omega = \Omega = 2\pi\nu = 2\pi(3) = 6\pi \text{ rad/sec} \tag{a}$$

From Eqs. (2.2) and (1.14),

$$\omega^2 = \frac{k}{m} = \frac{3EI}{mL^3} \tag{b}$$

Solving Eq. (b) for Young's Modulus, substituting the resulting expression into Eq. (a) and evaluating the final form in terms of the given values of the system parameters yields

$$E = \frac{\omega^2 mL^3}{3I} = \frac{(6\pi)^2 (10)(1)^3}{3\left[0.005(0.030)^3/12\right]} = 1.05 \times 10^{11} \text{ N/m}^2 \qquad \triangleleft (c)$$

Example 3.5

A system consisting of a 1 kg mass attached to a spring of stiffness $k = 900$ N/m is initially at rest. It is subsequently excited by the force $F(t) = 5\cos(30t)$ N, where t is measured in seconds. (*a*) Determine the displacement of the mass due to a static force of equivalent magnitude. (*b*) Determine the displacement of the mass 25 seconds after the given time dependent force is applied.

Solution

(*a*)

Applying Eq. (3.14):

$$f_0 = \frac{F_0}{k} = \frac{5}{900} = 0.00556 \text{ m} \qquad \triangleleft (a)$$

(b)
Applying Eq. (2.2):

$$\omega = \sqrt{\frac{k}{m}} = \sqrt{\frac{900}{1}} = 30.0 \text{ rad/sec} \tag{b}$$

Now, for the given excitation, $\Omega = 30$ rad/sec. Hence,

$$\bar{\Omega} = \frac{\Omega}{\omega} = \frac{30}{30} = 1 \tag{c}$$

This is evidently a resonance condition for the undamped system under consideration.

Though we argued that the complementary solution damps out after some time for real systems, it is instructive to include it for the present problem since the system is initially at rest and since the time of interest after application of the load is finite. Therefore, we have from Eqs. (3.16) and (3.34) that the forced response is given by

$$x(t) = A\cos(\omega t - \phi) + \tfrac{1}{2} f_0 \omega t \sin \omega t \tag{d}$$

To evaluate the constants of integration, A and ϕ, we must impose the given initial conditions on Eq. (d). Doing so we find that

$$x(0) = 0 = A\cos(-\phi)$$
$$\Rightarrow \quad \text{either} \quad A = 0 \quad \text{or} \quad \cos(-\phi) = 0 \tag{e}$$

and

$$\dot{x}(0) = 0 = -\omega A \sin(-\phi)$$
$$\Rightarrow \quad \text{either} \quad A = 0 \quad \text{or} \quad \sin(-\phi) = 0 \tag{f}$$

Since both $\cos(-\phi)$ and $\sin(-\phi)$ cannot vanish simultaneously, we conclude from Eqs. (e) and (f), that

$$A = 0 \tag{g}$$

Upon substitution of Eq. (g) into Eq. (d) it is seen that the response of the system which is initially at rest is simply the steady state response

$$x(t) = \tfrac{1}{2} f_0 \omega t \sin \omega t \quad (t \geq 0) \tag{h}$$

Substitution of the given values of the system parameters into Eq. (h), or equivalently Eq. (3.33), gives the displacement in question. Hence,

$$x\big|_{t=25} = \frac{(0.00556)}{2}(30)(25)\sin\{30(25)\} = 1.55 \text{ m} \qquad \triangleleft \text{(i)}$$

It is interesting to note that

$$\frac{x\big|_{t=25}}{f_0} = \frac{1.55}{0.00556} = 279 \qquad \text{(j)}$$

Thus, within the first 25 seconds of application, the dynamic force deflects the initially resting mass by a factor of 279 times greater than the deflection that would be imparted by a static force of equal magnitude. (We remark that, at this stage, the system may have passed beyond the critical damage state or beyond the range of validity of the linear spring model employed, depending on the actual system being represented by this model and its dimensions and material properties.)

The Phenomenon of Beating

It was seen that resonance of an undamped system occurs when the mass is harmonically forced at a frequency equal to the natural frequency of the system. Another interesting phenomenon occurs when such a system is forced harmonically at a frequency very near, but not equal to, the natural frequency of the system. We examine this situation next.

Consider a single degree of freedom system subjected to sinusoidal forcing $F(t) = F_0 \sin \Omega t$. If the system is initially undisturbed when the force is applied, i.e.; if $x(0) = 0$ and $\dot{x}(0) = 0$, then the response is found by incorporating Eq. (3.34) into Eq. (3.16) and then imposing the stated initial conditions. Doing this, we find that the response takes the form

$$x(t) = \frac{f_0}{1-\bar{\Omega}^2}\left[\sin \Omega t - \bar{\Omega}\sin \omega t\right] \qquad (3.37)$$

The solution described by Eq. (3.37) is valid for all $\Omega \neq \omega$. However, let us now restrict our attention to the situation where the forcing frequency is very near, but not equal to, the natural frequency of the system ($\Omega \approx \omega$). For this case, the solution can be simplified somewhat and, more importantly, it can be put into a form that has a clear physical interpretation.

If the forcing frequency is very close in value to the natural frequency of the system, the quotient appearing in Eq. (3.37) can be simplified as follows:

$$\frac{f_0}{1-\bar{\Omega}^2} = \frac{f_0}{(1-\bar{\Omega})(1+\bar{\Omega})} = \frac{f_0}{\varepsilon[1+(1-\varepsilon)]} \cong \frac{f_0}{2\varepsilon} \qquad (3.38)$$

where

$$\varepsilon \equiv 1-\bar{\Omega} \quad (\|\varepsilon\| \ll 1) \qquad (3.39)$$

Let us next introduce the average between the excitation and natural frequencies, ω_a, and its conjugate frequency, ω_b. Hence,

$$\omega_a \equiv \frac{\omega+\Omega}{2} \qquad (3.40)$$

$$\omega_b \equiv \frac{\omega-\Omega}{2} = \tfrac{1}{2}\varepsilon\omega \qquad (3.41)$$

Next, let us use the first of the identities

$$\sin\psi = \frac{e^{i\psi}-e^{-i\psi}}{2i} \quad \text{and} \quad \cos\psi = \frac{e^{i\psi}+e^{-i\psi}}{2}$$

(see Problem 1.19) in the trigonometric terms of Eq. (3.37). Doing this, then incorporating Eqs. (3.40) and (3.41) and regrouping terms gives

$$\sin\Omega t - \bar{\Omega}\sin\omega t = -2\left(\frac{e^{i\omega_b t}-e^{-i\omega_b t}}{2i}\right)\left(\frac{e^{i\omega_a t}+e^{-i\omega_a t}}{2}\right)+\varepsilon\sin\omega t$$

Using the aforementioned identities (Problem 1.19) once again simplifies the above equality to the convenient form

$$\sin\Omega t - \bar{\Omega}\sin\omega t = -2\sin\omega_b t\cos\omega_a t + \varepsilon\sin\omega t$$
$$\cong -2\sin(\tfrac{1}{2}\varepsilon\omega t)\cos\omega_a t \qquad (3.42)$$

Finally, substituting Eqs. (3.38) and (3.42) into Eq. (3.37) gives the desired physically interpretable form of the response as

$$x(t) \cong X(t)\sin(\omega_a t - \pi/2) \qquad (3.43)$$

where

$$X(t) = \frac{f_0}{\varepsilon}\sin(\tfrac{1}{2}\varepsilon\omega t) \qquad (3.44)$$

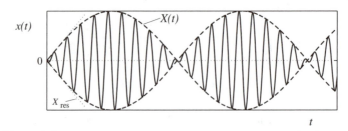

Figure 3.6 Time history of response when $\Omega \approx \omega$, demonstrating beating.

The response described by Eqs. (3.43) and (3.44) is sketched in Figure 3.6. It may be seen from the solution, and with the aid of the figure, that the system oscillates at the average value of the excitation and natural frequencies and within an envelope corresponding to a time dependent amplitude that oscillates at the much slower frequency $\omega_b = \varepsilon\omega / 2$. This phenomenon is referred to as *beating*. It is also seen that the response is out of phase with the excitation by $\Phi \cong \pi / 2$ radians and hence that the displacement lags the force by $t_{lag} \cong \pi / 2\omega_a$ for beating as for resonance. Thus, when the forcing frequency is very near but not equal to the natural frequency of the system, the system nearly achieves resonance. However, in this case, the response does not "run away," but rather is "captured" and remains bounded with the amplitude oscillating at relatively slow rate. In fact, if we make the small angle approximation for the sine function in Eq. (3.44),

$$X(t) = \frac{f_0}{\varepsilon}\sin(\tfrac{1}{2}\varepsilon\omega t) \approx \tfrac{1}{2}f_0\omega t$$

and let $\omega_a \approx \omega$ in the sine function of Eq. (3.43), then the solution takes the form

$$x(t) \approx \tfrac{1}{2}f_0\omega t \sin(\omega t - \pi / 2)$$

which is identical to the resonance solution given by Eqs. (3.34) and (3.35). Thus, as indicated in Figure 3.6, beating parallels resonance for times small compared with the natural period of the system.

Our study of the behavior of undamped systems subjected to harmonic forcing has revealed some interesting and important characteristics. We shall next examine how damping alters this behavior.

3.3.3 Steady State Response of Systems with Viscous Damping

In this section we consider the response of viscously damped systems to harmonic excitation. It was seen in Section 3.3.1 that the complimentary solution damps out over time. This is not the same for the particular solution associated with the steady state response. We shall first obtain the corresponding solution and then examine the associated response under various conditions. As for the case of vanishing damping, we shall obtain the response for both cosine and sine excitation functions simultaneously by solving the corresponding problem for complex excitation. The governing equation is given by Eq. (3.13).

We seek to determine the function which, when substituted into the left-hand side of Eq. (3.13) results in the exponential function on the right-hand side of that equation. As for the case of vanishing damping considered in Section 3.3.2, we assume a particular solution of the form

$$\hat{x}_{ss}(t) = \hat{X}e^{i\Omega t} \tag{3.45}$$

where $\hat{X}$ is a complex constant that is yet to be determined. Substitution of Eq. (3.45) into Eq. (3.13) results in the algebraic equality

$$\left[(i\Omega)^2 + 2\omega\zeta(i\Omega) + \omega^2 \right] \hat{X}e^{i\Omega t} = \omega^2 f_0 e^{i\Omega t}$$

which when solved for $\hat{X}$ gives

$$\hat{X} = \frac{f_0}{\left(1 - \bar{\Omega}^2\right) + i2\zeta\bar{\Omega}} \tag{3.46}$$

where $\bar{\Omega}$ is defined by Eq. (3.21). Let us next multiply both the numerator and the denominator of Eq. (3.46) by the complex conjugate of the denominator. This will put the complex amplitude in the usual complex form $\hat{X} = \operatorname{Re}\hat{X} + i\operatorname{Im}\hat{X}$. Hence, multiplying the right hand side of Eq. (3.46) by the unit expression

$$\frac{\left(1 - \bar{\Omega}^2\right) - i2\zeta\bar{\Omega}}{\left(1 - \bar{\Omega}^2\right) - i2\zeta\bar{\Omega}}$$

gives the alternate form of the complex amplitude

$$\hat{X} = \frac{f_0}{\left(1 - \bar{\Omega}^2\right)^2 + \left(2\zeta\bar{\Omega}\right)^2} \left[\left(1 - \bar{\Omega}^2\right) - i2\zeta\bar{\Omega}\right] \tag{3.47}$$

Equation (3.47) can be expressed in exponential form with the aid of Eq. (1.62). This gives

$$\hat{X} = Xe^{-i\Phi} \tag{3.48}$$

where

$$X = \left\|\hat{X}\right\| = f_0\,\Gamma\left(\bar{\Omega};\zeta\right) \tag{3.49}$$

$$\Gamma \equiv \frac{X}{f_0} = \frac{1}{\sqrt{\left(1-\bar{\Omega}^2\right)^2 + \left(2\zeta\bar{\Omega}\right)^2}} \qquad (3.50)$$

and

$$\Phi = \tan^{-1}\left(\frac{2\zeta\bar{\Omega}}{1-\bar{\Omega}^2}\right) \qquad (3.51)$$

Substitution of Eq. (3.48) into Eq. (3.45) gives the particular solution to Eq. (3.13). Hence,

$$\hat{x}_{ss}(t) = Xe^{i(\Omega t - \Phi)} = X\left[\cos(\Omega t - \Phi) + i\sin(\Omega t - \Phi)\right] \qquad (3.52)$$

From our discussion of superposition in Section 3.2 we see that if the force is of the form of a cosine function then the response is given by the real part of Eq. (3.52). Likewise, if the force is of the form of a sine function then the corresponding response is given by the imaginary part of Eq. (3.52). It follows that

if $F(t) = F_0 \cos\Omega t$, then

$$x_{ss}(t) = X\cos(\Omega t - \Phi) = f_0\Gamma(\bar{\Omega};\zeta)\cos(\Omega t - \Phi) \qquad (3.53)$$

and if $F(t) = F_0 \sin\Omega t$, then

$$x_{ss}(t) = X\sin(\Omega t - \Phi) = f_0\Gamma(\bar{\Omega};\zeta)\sin(\Omega t - \Phi) \qquad (3.54)$$

It is seen that, after the transients die out, the system oscillates with the frequency of the excitation, but that the displacement of the mass lags the force by the time t_{lag} = Φ / Ω. The angle Φ is thus the phase angle of the steady state response and characterizes the extent to which the response lags the excitation. (The phase angle Φ should not be confused with the phase angle ϕ associated with the transient or free vibration response appearing in Eqs. (3.16) or (2.15), respectively.) The parameter X, defined by Eq. (3.49), is seen to be the *amplitude* of the steady state response of the system to the applied harmonic force. The amplitude of the response is seen to depend on the effective static deflection, f_0, the damping factor, ζ, and the frequency ratio $\bar{\Omega}$. Representative plots of the applied force and the resulting steady state response are displayed as functions of time in Figure 3.7.

It may be seen from Eq.(3.48) that the parameter Γ corresponds to the ratio of the amplitude of the dynamic response to that of the effective static response. For this reason, Γ is referred to as the *magnification factor* as it is a measure of the magnification of the response of the system to the harmonic force above the response of the system to a static force having the same magnitude. Note that Eq. (3.23) is a special case of Eq. (3.50). That is,

$$\Gamma_0\left(\bar{\Omega}\right) = \Gamma\left(\bar{\Omega},0\right)$$

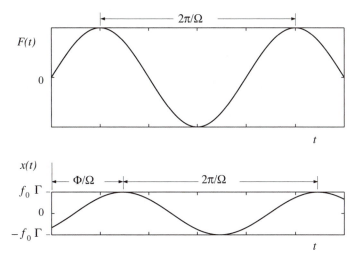

Figure 3.7 Time histories of excitation and corresponding steady state response.

It may be seen from Eqs. (3.50) and (3.51) that, for a given system, the parameters Γ and Φ are solely dependent on the frequency ratio $\overline{\Omega}$. Plots of the magnification factor and the corresponding phase angle of the steady state response are displayed in Figures 3.8 and 3.9 as functions of the ratio of the excitation frequency to the undamped natural frequency of the system for a range of values of the damping factor.

Consideration of Figures 3.8 and 3.9 reveals several important features. It is seen that the magnification factor, and hence the response of the system, achieves a maximum when the frequency ratio is at or near $\overline{\Omega} = 1$, depending on the value of the damping factor for the particular system of interest. (We shall discuss this optimum response shortly.) We note that in all cases, except zero damping, the response is bounded. We also note that $\Gamma \to 1$ and $\Phi \to 0$ as $\overline{\Omega} \to 0$, regardless of the value of the damping factor. This is the *static limit*. That is, as $\overline{\Omega} \to 0$, the magnitude of the dynamic response approaches that of the static response, and the motion of the mass becomes synchronized with the applied force as the frequency of the excitation becomes small compared with the natural frequency of the system. The system therefore behaves *quasi-statically* when the forcing frequency is low enough. We have all had the experience of trying to carry a delicate object very slowly so as not to disturb it. When we do this, we are attempting to achieve the static limit for the object we are carrying. In the opposite limit, we see that $\Gamma \to 0$ and $\Phi \to \pi$ as $\overline{\Omega} \to \infty$, regardless of the damping factor. That is, the system is essentially unaffected by the applied harmonic force when the frequency of the excitation is very large compared with the natural frequency of the system. Hence, for this case, the motion of the mass is completely out of phase with the force. Under these conditions, the period of the excitation is very small compared with the natural time scale of the system, the period of free vibration, so the system effectively does not "sense" the excitation for large enough forcing frequencies. As this limit is approached, the force is essentially moving too fast for the system to react to it, so the system remains almost stationary.

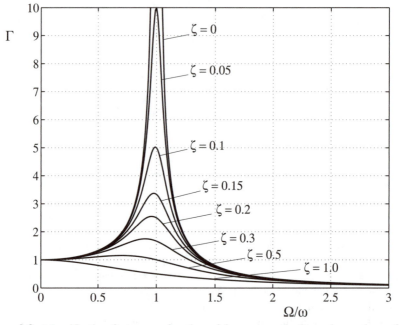

Figure 3.8 Magnification factor as a function of frequency ratio for various values of the damping factor.

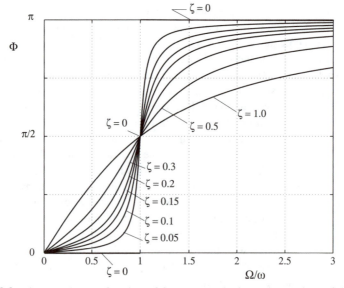

Figure 3.9 Phase angle as a function of frequency ratio for various values of the damping factor.

It is seen from Figure 3.8 that for a large range of values of ζ the amplitude of the dynamic response of the system achieves its maximum when $\bar{\Omega} \approx 1$. It may be observed that the maximum shifts left, to lower values of the frequency ratio, as the damping factor increases. In fact, if the damping of the system is large enough, no maximum is seen at all. To determine the frequency ratio at which the peak response occurs, $\bar{\Omega} = \bar{\Omega}_{pk}$, we simply differentiate the magnification factor with respect to the frequency ratio, set the resulting expression to zero and solve the corresponding equation for $\bar{\Omega}_{pk}$. Doing this, we find that $\Gamma = \Gamma_{max}$ when

$$\bar{\Omega} = \bar{\Omega}_{pk} \equiv \sqrt{1 - 2\zeta^2} \quad \left(0 < \zeta < 1/\sqrt{2}\right) \tag{3.55}$$

This is the resonance condition for the damped system. It is seen that $\bar{\Omega}_{pk} \to 1$ as $\zeta \to 0$, which recovers the resonance condition for the case of vanishing damping discussed in Section 3.3.2. It may be further seen from Figure 3.8 and Eq. (3.55) that for $\zeta > 1/\sqrt{2}$ no peak is achieved in the Γ vs. $\bar{\Omega}$ curve. In fact, for these cases, $\Gamma \leq 1$ with the equality being achieved when $\bar{\Omega} = 0$. Damping is thus seen to retard the motion of the system and therefore to slow things down. It should be emphasized that in the above measures the forcing frequency is divided by the natural frequency for vanishing damping. Recall that the natural frequency, ω_d, for an underdamped system ($\zeta^2 < 1$) is given by Eq. (2.70), from which it follows that

$$\tilde{\Omega}_{pk} \equiv \frac{\Omega_{pk}}{\omega_d} = \sqrt{\frac{1 - 2\zeta^2}{1 - \zeta^2}} < 1 \tag{3.56}$$

Thus, for nonvanishing damping, the peak response is seen to occur when the forcing frequency is less than the damped natural frequency as well, though it may be noted that $\bar{\Omega}_{pk} < \tilde{\Omega}_{pk}$. It may be seen from Figure 3.9 that $\Phi \approx \pi/2$ at resonance.

Example 3.6

A mechanical system represented as a mass-spring-damper system has the properties $m = 4$ kg, $k = 2$ N/m and $c = 3$ N-sec/m. Determine the amplitude of the steady state response of the system, (*a*) if the mass is subjected to the external force $F(t) = 5\sin 2t$ N, (*b*) if it is subjected to the force $F(t) = 5\sin \omega t$ N, (*c*) if it is subjected to the force $F(t) = 5\sin(\omega_d t)$ N, and (*d*) if it is subjected to the force $F(t) = 5\sin(\Omega_{pk} t)$ N. (In each case, t is in seconds.)

Solution
Let us first calculate the system parameters. Hence,

$$\omega = \sqrt{\frac{k}{m}} = \sqrt{\frac{2}{4}} = .7071 \text{ rad/sec} \tag{a}$$

$$\zeta = \frac{c}{2\omega m} = \frac{3}{2(0.7071)(4)} = 0.5303 \tag{b}$$

$$\omega_d = \omega\sqrt{1-\zeta^2} = 0.7071\sqrt{1-(0.5303)^2} = 0.5995 \text{ rad/sec} \tag{c}$$

Further, for each case,

$$f_0 = \frac{F_0}{k} = \frac{5}{2} = 2.5 \text{ m} \tag{d}$$

(a)
For this case,

$$\bar{\Omega} = \frac{\Omega}{\omega} = \frac{2}{0.7071} = 2.828 \tag{e}$$

For this excitation frequency the magnification factor is calculated as

$$\Gamma = \frac{1}{\sqrt{\left(1-\bar{\Omega}^2\right)^2 + \left(2\zeta\bar{\Omega}\right)^2}}$$

$$= \frac{1}{\sqrt{\left[1-(2.828)^2\right]^2 + \left[2(0.5303)(2.828)\right]^2}} = 0.1313 \tag{f}$$

We now calculate the corresponding amplitude of the steady state response,

$$X = f_0\Gamma = (2.5)(0.1313) = 0.3283 \text{ m} = 32.83 \text{ cm} \qquad \triangleleft \text{(g)}$$

Let us compare the amplitude of the steady state response just calculated with the amplitude calculated in Example 3.2 for the same system without damping subjected to the same excitation. For the undamped system the amplitude of the steady state response was calculated to be $X = 35.7$ cm. We see that the damping has reduced the amplitude of the response by about 3 cm at this excitation frequency.

(b)
For the case where the forcing frequency has the same value as the undamped natural frequency,

$$\bar{\Omega} = 1$$

and hence,

$$\Gamma \rightarrow \frac{1}{2\zeta} = \frac{1}{2(0.5303)} = 0.9429 \tag{h}$$

Thus, the amplitude of the steady state response when the excitation frequency has the same value as the undamped natural frequency is

$$X = (2.5)(0.9429) = 2.355 \text{ m} \tag{i}$$

(*c*)
For the case where the excitation frequency has the same value as the damped natural frequency,

$$\bar{\Omega} = \frac{\omega\sqrt{1-\zeta^2}}{\omega} = \sqrt{1-(0.5303)^2} = 0.8478 \tag{j}$$

The magnification factor is then

$$\Gamma = \frac{1}{\sqrt{\left[1-(0.8478)^2\right]^2 + \left[2(0.5303)(0.8478)\right]^2}} = 1.061 \tag{k}$$

Thus, the magnitude of the steady state response for this excitation frequency is

$$X = (2.5)(1.061) = 2.653 \text{ m} \tag{l}$$

(*d*)
For the peak (resonance) response, the normalized excitation frequency is calculated using Eq. (3.55). Hence,

$$\bar{\Omega}_{pk} = \sqrt{1-2\zeta^2} = \sqrt{1-2(0.5303)^2} = 0.6615 \tag{m}$$

The magnification factor for the peak (resonance) response is next calculated to be

$$\Gamma = \frac{1}{\sqrt{\left[1-(0.6615)^2\right]^2 + \left[2(0.5303)(0.6615)\right]^2}} = 1.112 \tag{n}$$

The amplitude of the steady state response at resonance is then

$$X = (2.5)(1.112) = 2.780 \text{ m} \tag{o}$$

Example 3.7

A sensor and actuator are attached to a mechanical system in an effort to control any vibrations that may occur. The total mass of the system is 6 kg, its effective stiffness is 2 N/m and its coefficient of viscous damping is 3.7 N-sec/m.
If the sensor detects a sustained harmonic vibration of amplitude 60 cm and
frequency 2 rad/sec, what force must be applied by the actuator to counter the
observed motion?

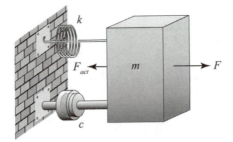

Figure E3.7-1 Mass-spring system with applied force and actuator force.

Solution

To counter the observed motion, the actuator must apply a force that would
produce a motion that is equal in magnitude and 180° out of phase with the applied force. That is, using superposition, we wish the actuator to apply a force
such that

$$x(t) = x_{obs}(t) + x_{act}(t) = 0 \tag{a}$$

Since the observed motion is sustained (i.e., is steady state) our objective can
be accomplished by applying a force that is of equal amplitude and 180° out of
phase with the applied force. (See Figures E3.7-1 and E3.7-2.) We must therefore determine the characteristics of the applied force from the motion of the
system detected by the sensor.

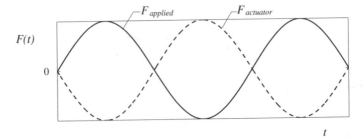

Figure E3.7-2 Time history of applied force and actuator force.

We may determine the amplitude of the applied force from the amplitude of the observed response as follows. The observed response may be expressed mathematically as

$$x_{obs}(t) = X_{obs} \sin(\Omega_{obs}t - \Phi) = 0.6\sin(2t - \Phi) \text{ m}$$ (b)

(The phase angle is unknown, but is merely a reference in this case. Note that a force that is 180° out of phase with the applied force will produce a response that is 180° out of phase with the observed motion and Eq. (a) will be satisfied.) For the modified system,

$$\omega = \sqrt{\frac{k}{m}} = \sqrt{\frac{2}{6}} = 0.5774 \text{ rad/sec}$$ (c)

and

$$\zeta = \frac{c}{2\omega m} = \frac{3.7}{2(0.5774)(6)} = 0.5340$$ (d)

Hence, for the observed steady state motion,

$$\bar{\Omega} = \frac{2}{0.5774} = 3.464$$ (e)

Now, from Eq. (3.49),

$$X_{obs} = f_0 \Gamma = \frac{F_0}{k} \Gamma$$ (f)

Hence,

$$F_0 = \frac{k X_{obs}}{\Gamma} = \frac{(2)(0.6)}{\sqrt{\left[1-(3.464)^2\right]^2 + \left[2(3.464)(0.5340)\right]^2}} = 0.103 \text{ N}$$ (g)

If we reference the external force as the sine function

$$F(t) = F_0 \sin \Omega t = 0.103 \sin 2t$$ (h)

then the actuator must apply the force

$$F_{act}(t) = 0.103\sin(2t - \pi) = -0.103\sin(2t) \text{ N}$$ ◁ (i)

to counter the effects of the excitation.

Example 3.8

A 20 lb rigid baffle hangs in the vertical plane. The
baffle is 5 ft in length and is restrained by a viscous
damper of coefficient 1 lb-sec/ft attached at mid-
span as shown. A motor exerts a harmonic torque
at the support causing the baffle to waffle at a prescribed rate. Determine the range of allowable excitation frequencies if the magnitude of the applied
torque is 2.5 ft-lbs and the maximum allowable deflection of the baffle is 6 inches.

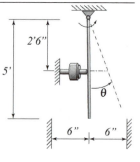

Solution

We first derive the equation of motion. To do this, we draw the corresponding
kinetic diagram and then take moments about the support at O. Hence,

$$M(t) - c v_c \frac{L}{2} \cos\theta - mg\frac{L}{2}\sin\theta = I_O \ddot{\theta} \tag{a}$$

where

$$v_c = \frac{d}{dt}\left(\frac{L}{2}\sin\theta\right) = \frac{L}{2}\dot{\theta}\cos\theta \tag{b}$$

and

$$I_O = \tfrac{1}{3}mL^2 \tag{c}$$

for a uniform baffle. Next, we note that

$$\Theta_{max} \equiv \|\theta\|_{max} = \frac{\|\Delta\|_{max}}{L} = \frac{6"\times(1'/12")}{5'} = 0.1 \text{ rads} \tag{d}$$

Making the small angle approximation in Eq. (a) and rearranging terms gives
the equation of motion for the baffle as

$$I_O \ddot{\theta} + \tfrac{1}{4}cL^2\dot{\theta} + \tfrac{1}{2}mgL\,\theta = M(t) \tag{e}$$

We next put Eq. (d) in standard form by dividing through by I_O and grouping
terms accordingly. Doing this we arrive at the equation

$$\ddot{\theta} + 2\omega\zeta\dot{\theta} + \omega^2\theta = \omega^2 f(t) \tag{f}$$

where

$$\omega^2 = \frac{mgL/2}{I_O} = \frac{3}{2}\frac{g}{L} \tag{g}$$

$$\zeta = \frac{1}{2\omega}\frac{cL^2}{4I_O} = \frac{3}{8}\frac{c}{\omega m} \tag{h}$$

and

$$f(t) = \frac{M(t)}{mgL/2} \tag{i}$$

Since phase is unimportant for the present analysis let us take the excitation as the sine function

$$M(t) = M_0 \sin\Omega t \tag{j}$$

Substituting Eq. (j) into Eq. (i) gives

$$f(t) = f_0 \sin\Omega t \tag{k}$$

where

$$f_0 = \frac{M_0}{mgL/2} \tag{l}$$

We next compute the values of the system parameters using the given system properties. Hence,

$$\omega = \sqrt{\frac{3}{2}\frac{g}{L}} = \sqrt{1.5\frac{(32.2)}{5}} = 3.108 \text{ rad/sec} \tag{m}$$

and

$$\zeta = \frac{3}{8}\frac{(1)}{(3.108)(20/32.2)} = 0.1943 \tag{n}$$

The amplitude of the excitation function is similarly computed from Eq. (l) and the given amplitude of the applied moment. Hence,

$$f_0 = \frac{(2.5)}{(20)(5)/2} = 0.05 \tag{o}$$

From Eq. (3.54), we know that the response of the baffle will be of the form

$$\theta(t) = \Theta\sin(\Omega t - \Phi)$$

where

$$\Theta = f_0\Gamma \tag{p}$$

and the magnification factor, Γ, is given by Eq. (3.50). The constraints on the system require that

$$\Theta < \Theta_{max} = 0.1 \text{ rads} \tag{q}$$

Substitution of Eq. (q) into Eq. (p) gives the relation

$$\Gamma - \frac{\Theta_{max}}{f_0} = 0 \tag{r-1}$$

Expanding Γ in the above relation gives the polynomial

$$\left(\overline{\Omega}^2\right)^2 - 2\left(1 - 2\zeta^2\right)\left(\overline{\Omega}^2\right) + \left[1 - \left(\frac{f_0}{\Theta_{max}}\right)^2\right] = 0 \tag{r-2}$$

which defines the bounds on the allowable excitation frequencies. Substituting Eqs. (d), (n) and (o) into Eq. (r-2) and solving for $\overline{\Omega}$ gives the bounds on the allowable frequency ratios as

$$\overline{\Omega} = 0.7752, \ 1.117 \tag{s}$$

Now,

$$\Omega = \omega \overline{\Omega} = 3.108 \overline{\Omega}$$

Upon substituting the values listed in Eq. (s) into the above expression we find that the frequencies of the applied torque must lie in the ranges

$$\Omega < 2.41 \text{ rad/sec} \quad \text{and} \quad \Omega > 3.47 \text{ rad/sec}$$

to satisfy the constraints imposed on the baffle.

Sharpness of the Resonance Peak

In many applications it is desired to have the maximum performance of the system achieved within a narrow range of excitation frequencies. The performance of acoustic speakers, transducers, and telephone receivers fall in this category. The necessity of a clean peak response requires that the resonance peak of the associated Γ vs. $\overline{\Omega}$ plot, Figure 3.8, be as sharp as possible $(\zeta \ll 1)$. In this way, a relatively small deviation in the driving frequency will not excite vibrations of appreciable amplitude. It also insures that a relatively large value of the amplitude will be achieved when the driving frequency is such that the frequency ratio lies within a narrow band in the vicinity of the peak. It is useful to characterize the sharpness of the peak for such situations.

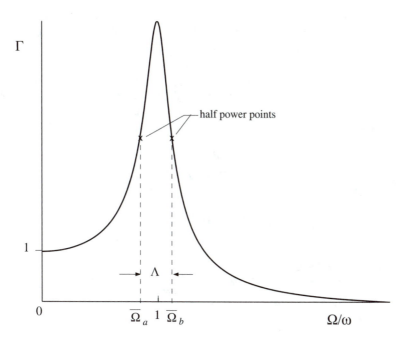

Figure 3.10 Typical plot of magnification factor versus frequency ratio for lightly damped system, showing resonance peak, half-power points and band width.

To characterize the sharpness of the peak of a typical Γ vs. $\bar{\Omega}$ curve (see Figure 3.10) we must establish some measure of the width of the peak as a function of its height. A standard approach is to define these measures in terms of work and energy, or average power, over one cycle of the response of the system. In particular, we shall employ as our measure the average power imparted by the applied force acting on the system over a cycle of the response. Specifically, let $\mathcal{P}_Q$ represent the average power of the applied force when operating at the excitation frequency associated with the peak response (i.e., when $\bar{\Omega} = \bar{\Omega}_{pk}$) as defined by Eq. (3.55). Further, let $\bar{\Omega}_a$ and $\bar{\Omega}_b$ correspond to the excitation frequencies at which the average power imparted by the applied force over a cycle is half that associated with the peak response (i.e., $\mathcal{P}_a = \mathcal{P}_b = \mathcal{P}_Q/2$). The corresponding points on the associated Γ vs. $\bar{\Omega}$ curve are referred to as the *half power points*. The width of the peak at this level of the average power is referred to as the *band width*, Λ, and is seen to characterize the sharpness of the peak. (See Figure 3.10.) The bandwidth is thus defined as

$$\Lambda \equiv \bar{\Omega}_b - \bar{\Omega}_a \tag{3.57}$$

To evaluate the sharpness of the peak explicitly, we must express these quantities in terms of the parameters of the system. We do this next.

As discussed in Section 1.5, the work done by the forces acting on a mass may be partitioned into two parts; the work done by the conservative forces acting on the body and the work done by the nonconservative forces acting on the body. As shown in the aforementioned section, the work of the conservative forces may be expressed as the negative of the change in potential energy of the system. For a mass-spring-damper system, the potential energy is simply the elastic energy of the spring. We further partition the remaining work into that done by the applied force, $F(t)$, and that done by the viscous damping force, $F_d = -c\dot{x}$. When this is done Eq. (1.92) takes the form

$$W_{ext} + W_d = \Delta T + \Delta U \tag{3.58}$$

where

$$W_{ext} = \int_{x_1}^{x_2} F(t)\,dx = \int_{t_1}^{t_2} F\dot{x}\,dt \tag{3.59}$$

is the work done by the applied force,

$$W_d = \int_{x_1}^{x_2} F_d\,dx = -\int_{t_1}^{t_2} c\dot{x}^2\,dt \tag{3.60}$$

is the work done by the viscous damping force, U and T are respectively the potential and kinetic energy of the system, t_1 and t_2 are two instants in time, $x_1 = x(t_1)$ and $x_2 = x(t_2)$. Recall that the steady state response of the system given by Eqs. (3.53) and (3.54) is purely harmonic. Therefore, if we consider the motion of the system over one cycle (i.e., $t_2 = t_1 + T_{ss}$, where $T_{ss} = 2\pi/\Omega$), then the change in potential energy and the change in kinetic energy vanish over this interval. For this situation, Eq. (3.58) gives the equality

$$W_{ext} = -W_d \quad (\Delta t = T_{ss}) \tag{3.61}$$

Evaluating Eq. (3.60) for an applied force of the form of either of Eq. (3.11), together with the corresponding steady state response given by Eq. (3.53) or Eq. (3.54), gives the work done by the viscous damper and thus by the applied force over a cycle as

$$W_{ext} = -W_d = c\Omega^2 X^2 \tag{3.62}$$

It is seen that the integrand of the right most integral of Eq. (3.60) corresponds to the *power*. The *average power* over a cycle is evaluated with the aid of Eqs. (3.62) and (3.49) to give

$$P_{avg} = \frac{1}{T}\int_{t_1}^{t_2} F\dot{x}\,dt = \frac{\Omega}{2\pi}W_{ext} = \frac{c\Omega^3}{2\pi}X^2 = \frac{c\Omega^3 f_0^2}{2\pi}\Gamma^2 \tag{3.63}$$

As stated at the outset of the present discussion, the sharpness of the resonance peak will be characterized by the average power.

It may be seen from Eq. (3.63) that the average power imparted by the applied force over a cycle achieves a maximum when the amplitude of the response, X, achieves a maximum. It follows from Eq. (3.49) that the amplitude, and hence the average power, achieves a maximum when the magnification factor, Γ, achieves a maximum. For systems possessing a resonance peak (systems for which $\zeta < 1/\sqrt{2}$) and, in particular, those systems possessing very light damping ($\zeta \ll 1$) the maximum value of the magnification factor, Γ_{pk}, is referred to as the *quality factor, Q_f*. The quality factor may be expressed in terms of the damping factor alone by substituting Eq. (3.55) into Eq. (3.50). Thus,

$$Q_f \equiv \Gamma_{pk} = \Gamma\left(\bar{\Omega}_{pk};\zeta\right) = \frac{1}{2\zeta\sqrt{1-\zeta^2}} \approx \frac{1}{2\zeta} \tag{3.64}$$

It follows from Eq. (3.63) that

$$P_Q \equiv P_{avg}\big|_{max} = \frac{c\,\Omega^3 f_0^2}{2\pi}Q_f^{\,2} \tag{3.65}$$

Equation (3.64) allows for the evaluation of the damping factor by measurement of the quality factor from the Γ vs. $\bar{\Omega}$ curve for a given system.

To determine the bandwidth we must first evaluate the excitation frequencies at the half power points. It may be seen from Eqs. (3.65) and (3.63) that the frequency ratios at which $P_{avg} = P_Q/2$ is achieved occur when the equality

$$\Gamma\left(\bar{\Omega};\zeta\right) = \frac{Q_f}{\sqrt{2}} \tag{3.66}$$

is satisfied. The frequency ratios associated with the half power points will be designated as $\bar{\Omega}_a$ and $\bar{\Omega}_b$, and may be found by solving Eq. (3.66) for $\bar{\Omega}$. Substitution of Eqs. (3.50) and (3.64) into Eq. (3.66), and rearranging terms, gives the fourth order polynomial equation for the frequency ratios associated with the half power points. Hence,

$$\left(1-\bar{\Omega}^2\right)^2 + \left(2\zeta\bar{\Omega}\right)^2 - 8\zeta^2 = 0 \tag{3.67}$$

When expanded, Eq. (3.67) may be solved for the square of the frequency ratio to give

$$\bar{\Omega}_a^{\,2} \cong 1-2\zeta , \quad \bar{\Omega}_b^{\,2} \cong 1+2\zeta \tag{3.68}$$

for $\zeta \ll 1$. It follows from Eqs. (3.68), and the definition of bandwidth as stated by Eq. (3.57), that

$$\bar{\Omega}_b{}^2 - \bar{\Omega}_a{}^2 = \left(\bar{\Omega}_b + \bar{\Omega}_a\right)\left(\bar{\Omega}_b - \bar{\Omega}_a\right) \cong 2\Lambda = 4\zeta \tag{3.69}$$

Solving Eq. (3.69) for Λ gives the bandwidth in terms of the damping factor as

$$\Lambda \cong 2\zeta \tag{3.70}$$

The bandwidth, as given by Eq. (3.70), characterizes the sharpness of the resonance for a given system. Finally, substitution of Eq. (3.70) into Eq. (3.64) gives the relation between the quality factor and the bandwidth,

$$Q_f \cong \frac{1}{\Lambda} \tag{3.71}$$

Thus, measurement of the quality factor determines the bandwidth directly.

Example 3.9

A harmonic force is applied to the mass of a lightly damped system. The excitation frequency is slowly varied, and the magnitude of the displacement of the mass is measured as a function of the excitation frequency giving the record shown in Figure E3.9. Determine the damping factor of the system. What is the bandwidth?

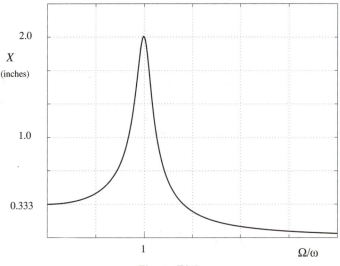

Figure E3.9

Solution

The values of the displacement parameters X_{pk} and f_0 may be read directly from the plot of the test data. Then, using Eqs. (3.49) and (3.64) along with these values, we calculate the quality factor

$$Q_f \equiv \Gamma_{pk} = \frac{X_{pk}}{f_0} = \frac{2}{0.333} = 6 \tag{a}$$

The damping factor may now be determined by substituting Eq. (a) into Eq. (3.64) and solving for ζ. This gives

$$\zeta = \frac{1}{2Q_f} = \frac{1}{2(6)} = 0.0833 \tag{◁ (b)}$$

Finally, using Eq. (3.71), the bandwidth is calculated to be

$$\Lambda \cong \frac{1}{Q_f} = \frac{1}{6} = 0.1667 \tag{◁ (c)}$$

3.3.4 Force Transmission and Vibration Isolation

In many applications it is of interest to determine the force transmitted to the support by the oscillating system. We may wish, for example, to minimize the magnitude of the transmitted force so as not to damage the support or what is beyond or attached to it. Similarly, we may wish to minimize the magnitude of the transmitted force to avoid sound transmission through the boundary so as to keep the operation of the system quiet. In contrast, we may wish to maximize, or at least optimize in some sense, the amount of information transmitted via this force.

Let us consider the mass-spring-damper system that is subjected to a harmonically varying force applied to the mass (Figure 3.1) as our representative system. From Newton's Third Law, and the associated dynamic free body diagram displayed in Figure 3.11, it is seen that the force transmitted to the support by the vibrating system is comprised of the elastic spring force and the viscous damping force. Hence,

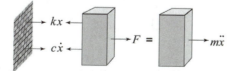

Figure 3.11 Kinetic diagram for mass-spring-damper system and support.

$$F_{tr} = c\dot{x} + kx = m\left[2\omega\zeta\dot{x} + \omega^2 x\right] \tag{3.72}$$

For the sake of the present discussion, it is convenient to express the applied force in the exponential form of Eq. (3.12). Substituting Eq. (3.12) and the corresponding complex form of the steady state solution, Eq. (3.52), into Eq. (3.72) gives the complex form of the transmitted force as

$$\hat{F}_{tr} = F_0 e^{i\Omega t}\left[(1 + i2\zeta\bar{\Omega})\Gamma e^{-i\Phi}\right] \tag{3.73}$$

where Γ and Φ are given by Eqs. (3.50) and (3.51), respectively. It may be seen from Eq. (3.51), with the aid of Figure 3.12, that

$$\cos\Phi = (1 - \bar{\Omega}^2)\Gamma \quad\text{and}\quad \sin\Phi = 2\zeta\bar{\Omega}\Gamma \tag{3.74}$$

Applying Eq. (1.61) to the exponential term within the brackets, and employing the identities of Eq. (3.74), renders the complex representation of the transmitted force to the form

$$\hat{F}_{tr} = \|F_{tr}\|e^{i(\Omega t - \Psi)} \tag{3.75}$$

where

$$\|F_{tr}\| = F_0\Upsilon(\bar{\Omega};\zeta) \tag{3.76}$$

is the magnitude of the transmitted force,

$$\Upsilon(\bar{\Omega};\zeta) = \Gamma(\bar{\Omega};\zeta)\sqrt{1 + (2\zeta\bar{\Omega})^2} \tag{3.77}$$

and

$$\Psi = \tan^{-1}\left\{\frac{2\zeta\bar{\Omega}^3}{1 - \bar{\Omega}^2 + (2\zeta\bar{\Omega})^2}\right\} \tag{3.78}$$

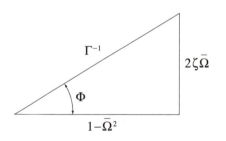

Figure 3.12 Geometric relation between Γ and Φ.

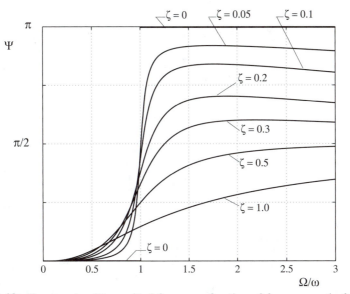

Figure 3.13 Phase angle of transmitted force as a function of frequency ratio for various values of the damping factor.

The latter is the phase angle that measures the extent that the transmitted force lags behind the applied force. The corresponding lag time is thus $t_{lag} = \Psi/\Omega$. Plots of the phase angle as a function of the frequency ratio are displayed in Figure 3.13 for various values of the damping factor. Finally, it follows from Eq. (3.75) that

if $F(t) = F_0 \cos \Omega t$ then

$$F_{tr} = F_0 \, \Upsilon\!\left(\bar{\Omega}; \zeta\right) \cos(\Omega t - \Psi) \tag{3.79}$$

if $F(t) = F_0 \sin \Omega t$ then

$$F_{tr} = F_0 \, \Upsilon\!\left(\bar{\Omega}; \zeta\right) \sin(\Omega t - \Psi) \tag{3.80}$$

It may be seen from Eqs. (3.76) and (3.77) that the magnitude of the force transmitted to the support is the product of the magnification factor and a nonlinear function of the damping factor and frequency ratio, as well as of the magnitude of the applied force. We shall characterize the frequency dependence of the transmitted force in a manner analogous to that for the steady state displacement. To do this we define the *transmissibility*,

$$\mathbb{T}_{\mathrm{R}} \equiv \frac{\left\| F_{tr} \right\|}{F_0} = \Upsilon\!\left(\bar{\Omega}; \zeta\right) \tag{3.81}$$

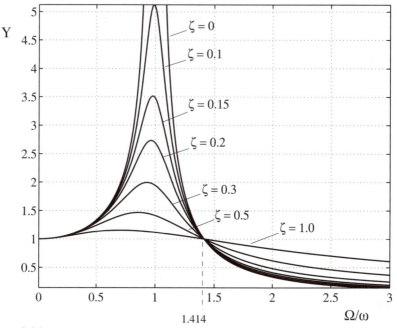

Figure 3.14 Transmissibility as a function of frequency ratio for various values of the damping factor.

where Y is given by Eq. (3.77). The transmissibility is thus the ratio of the magnitude of the transmitted force to the magnitude of the applied force, and is seen to give all pertinent information concerning the force transmitted to the support by the system. Plots of the transmissibility as a function of the frequency ratio are displayed in Figure 3.14 for various values of the damping factor. It is seen from the figure that the maximum force transmitted to the support occurs when the maximum displacement occurs (i.e., at resonance conditions), which is to be expected. Setting Y to unity and solving for the frequency ratio, we find that

$$\left\| F_{tr} \right\| = F_0 \text{ when}$$

$$\bar{\Omega} = 0 \quad \text{and} \quad \bar{\Omega} = \sqrt{2} \tag{3.82}$$

regardless of the value of ζ, as demonstrated in Figure 3.14. It may be noted from the figure that the magnitude of the transmitted force reduces to less than the magnitude of the applied force (i.e., to less than that which would occur for a static force of the same magnitude) when $\bar{\Omega} > \sqrt{2}$. We may also note that increasing the damping factor increases the transmitted force in this frequency range. Thus, if the force transmitted to the support is a consideration in the design of a system, then it would be desirable to operate the system in the aforementioned frequency range, and with minimum damping, provided that the desired frequency range can be achieved without passing through resonance. The desired frequency range may also be achieved if the system can be restrained during start-up, or a large amount of damping can be temporarily imposed until the excitation frequency is sufficiently beyond the resonance frequency. To close, it is seen that if the system is operated in the desired frequency range and possesses low damping, then the vibrations of the system become *isolated* from the surroundings. This is a very desirable result in many situations.

Example 3.10

Consider the system of Example 3.6. (*a*) Determine the magnitude of the force transmitted to the support if the system is excited by the external force $F(t) =$ 5sin2t N, where t is measured in seconds. Also determine the lag time of the reaction force with respect to the applied force. (*b*) Determine the magnitude of the transmitted force at resonance.

Solution

(*a*)

From Part (*a*) of Example 3.6, $\zeta = 0.5303$, $\bar{\Omega} = 2.828$ and $\Gamma = 0.1313$. Substituting these values into Eq. (3.77) gives the transmissibility as

$$\mathbb{T}_{\mathbb{R}} = \Upsilon = 0.1313\sqrt{1+[2(0.5303)(2.828)]^2} = 0.4151 \tag{a}$$

Thus, 42% of the applied force is transmitted to the support. The magnitude of the force transmitted to the support is then, from Eq. (3.81),

$$\left\|F_{tr}\right\| = F_0 \mathbb{T}_{\mathbb{R}} = (5)(0.4151) = 2.076 \text{ N} \qquad \triangleleft \text{(b)}$$

The corresponding phase angle is calculated using Eq. (3.78). Hence,

$$\Psi = \tan^{-1}\left\{\frac{2(.5303)(2.828)^3}{1-(2.828)^2+[2(.5303)(2.828)]^2}\right\} = 85.24° = 1.488 \text{ rads} \qquad \triangleleft \text{(c)}$$

It follows that

$$t_{lag} = \Psi/\Omega = 1.488/2 = 0.7440 \text{ secs} \qquad \triangleleft \text{(d)}$$

(*b*)

For this case we have from Part (*d*) of Example 3.6 that $\bar{\Omega} = 0.6615$ and $\Gamma = 1.112$. Thus,

$$\mathbb{T}_{\mathbb{R}} = 1.112\sqrt{1+[2(.5303)(0.6615)]^2} = 1.358 \tag{e}$$

The magnitude of the force transmitted to the support is then

$$\left\| F_{tr} \right\| = (5)(1.358) = 6.790 \text{ N} \qquad \qquad \triangleleft (f)$$

Example 3.11

The beam of Example 3.4 was designed on the basis of a static analysis. If the structure is to support harmonic loads at or below the static design level, determine the range of allowable frequencies of the applied load.

Solution

A static analysis predicts that the maximum transverse shear occurs at the support. This is then the force that is transmitted to the support since the resultant axial force in the beam vanishes. The problem is therefore to determine the frequency range for the applied dynamic load at which the magnitude of the force transmitted to the support is equal to the magnitude of the applied force (which we take to be equal to the level of the static design load). This corresponds to the situation where

$$\mathbb{T}_{\mathbb{R}} = 1 \qquad \qquad (a)$$

and hence, from Eq. (3.82) and Figure 3.14, that

$$\bar{\Omega} = \sqrt{2} \qquad \qquad (b)$$

From Example 3.4, the natural frequency of the system is 6π rad/sec. Substituting this value into Eq. (b) gives

$$\Omega = \omega\sqrt{2} = (6\pi)\sqrt{2} = 26.7 \text{ rad/sec} \left(= 4.24 \text{ cps} \right) \qquad (c)$$

The allowable operating range is thus

$$\Omega > 26.7 \text{ rad/sec} \qquad \qquad \triangleleft (d)$$

3.4 STRUCTURAL DAMPING

In Chapter 1, and thereafter, we employed models of equivalent single degree of freedom systems based on stiffnesses derived from the quasi-static behavior of linear elastic structures. Such models assume that the elastic moduli are constants that can be measured form simple quasi-static experiments. It is observed that elastic structures such as rods, beams, plates and shells, and their simplified 1 d.o.f. counterparts, do oscillate at natural frequencies often adequately predicted based on these moduli. It is, however, also observed (even in our every day experiences) that the vibrations of such structures decay with time. Such damping is seen when structures oscillate in a vacuum, as well. The damping cannot then be a result of some external viscous medium, but rather must be a function of *internal friction* of the material comprising the structure itself. In this section we discuss and develop a model of internal friction for the case of structures subjected to harmonic excitation.

3.4.1 Linear Hereditary Materials

The constitutive relations (the stress-strain relations) for the elastic materials used in the structural models relate the current stress and the current strain. They implicitly assume that the history of loading does not affect the current strain, and thus that only the current level of the load is influential. In order to account for the effects of internal friction (which, as discussed in Section 1.5, is by its very nature nonconservative) we shall relax this implicit restriction and allow the current value of the strain to depend on the history of the stress as well as on its current level. Materials whose behavior depends on the history of the loading are referred to as *hereditary materials*, or *materials with memory*. Materials for which this relation is linear are referred to as *linear hereditary materials*. We shall be interested in the behavior of structures comprised of linear hereditary materials.

Consider a material for which the current state of strain is dependent, not only on the current state of stress, but on the entire history of stress to the present time. For the purposes of the present discussion, let us consider a specimen of such a material loaded in a simple manner, say in uniaxial tension or in pure shear, as would occur during a tension test or a torsion test. Let $\sigma(t)$ represent the stress at some point in the body at time t and let $\varepsilon(t)$ represent the corresponding strain at that time. Consider a generic stress history as depicted in Figure 3.15a and let the increment in strain at time t, due to the increment in stress at some prior time τ (Figure 3.165) be related to that increment in stress in a linear fashion. Hence, let

$$d\varepsilon(t) = \mathcal{J}(t-\tau)\dot{\sigma}(\tau)d\tau \qquad (3.83)$$

where $\mathcal{J}(t-\tau)$ is a material property referred to as the *creep function*. A material that obeys the constitutive relation given by Eq. (3.83) is referred to as a *linear hereditary material*. Let us sum the strains due to the stress states at all prior times to obtain the strain at the current time. Thus,

$$\varepsilon(t) = \int_{t_0}^{t} \mathcal{J}(t-\tau)\dot{\sigma}(\tau)d\tau \qquad (3.84)$$

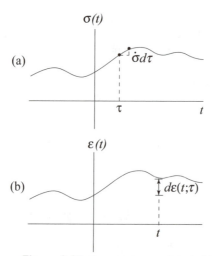

Figure 3.15 Generic stress and strain histories.

where t_0 is the initial (reference) time. An integral of the type that appears in Eq. (3.84) is called a *convolution integral*. Such integrals will be introduced and employed in a different context in Chapter 4. In a similar manner, the stress at the current time t is related to the strain history by the convolution integral

$$\sigma(t) = \int_{t_0}^{t} \Xi(t-\tau)\dot{\varepsilon}\,d\tau \tag{3.85}$$

where $\Xi(t-\tau)$ is referred to as the *relaxation function* for the material. In this section we shall be interested in the behavior of structures comprised of linear hereditary materials when they are subjected to harmonic excitation.

3.4.2 Steady State Response of Linear Hereditary Materials

Let us consider the response of systems after they are operating for times long enough for all transients to have died out. That is, let us consider the steady state response of these systems. For such situations, we extend the reference time back through negative infinity to capture the entire history of loading. In this case Eq. (3.84) assumes the form

$$\varepsilon(t) = \int_{-\infty}^{t} \mathcal{J}(t-\tau)\dot{\sigma}(\tau)\,d\tau \tag{3.86}$$

To discuss the response of the linear hereditary materials to harmonic excitation it is convenient to introduce the time shift

$$\xi = t - \tau \tag{3.87}$$

into Eq. (3.86). The stress-strain relation then takes the form

$$\varepsilon(t) = \int_0^\infty J(\xi)\dot{\sigma}(t - \xi)d\xi \tag{3.88}$$

where we are now integrating back in time. Let us next consider the situation where the stress history varies harmonically in time. Hence, let us consider the stress history of the form

$$\sigma(t) = \sigma_0 e^{i\Omega t} \tag{3.89}$$

where σ_0 = constant. Substitution of the harmonic stress form, Eq. (3.89), into the constitutive relation, Eq. (3.88), gives the corresponding strain history

$$\varepsilon(t) = \varepsilon_0 e^{i\Omega t} \tag{3.90}$$

where

$$\varepsilon_0 = i\Omega \hat{J}(\Omega)\sigma_0 \tag{3.91}$$

and

$$\hat{J}(\Omega) = \int_0^\infty J(\xi)e^{-i\Omega\xi}d\xi \tag{3.92}$$

The parameter $\hat{J}(\Omega)$ is known as the *complex compliance* of the material and may be recognized as the Fourier transform of its counterpart J. As $\hat{J}(\Omega)$ is generally a complex function of the excitation frequency, it may be expressed as the sum of its real and imaginary parts. Hence, we may express the complex compliance in the form

$$\hat{J}(\Omega) = \hat{J}_R(\Omega) + i\hat{J}_I(\Omega) \tag{3.93}$$

where $\hat{J}_R(\Omega)$ is referred to as the *storage compliance* and $\hat{J}_I(\Omega)$ is called the *loss compliance*. Substitution of Eq. (3.93) into Eq. (3.91) and solving for σ_0 renders the relation between the magnitudes of the stress and strain to the form

$$\sigma_0 = \hat{E}(\Omega)\varepsilon_0 \tag{3.94}$$

where $\hat{E}(\Omega)$ is defined as the *complex elastic modulus* of the material given by

$$\hat{E}(\Omega) = E(\Omega) + i\tilde{E}(\Omega) \tag{3.95}$$

with

$$E(\Omega) = \frac{\hat{\mathcal{J}}_I(\Omega)}{\Omega \left\| \hat{\mathcal{J}}(\Omega) \right\|^2} \quad \text{and} \quad \tilde{E}(\Omega) = \frac{-\hat{\mathcal{J}}_R(\Omega)}{\Omega \left\| \hat{\mathcal{J}}(\Omega) \right\|^2} \tag{3.96}$$

The parameters $E(\Omega)$ and $\tilde{E}(\Omega)$ are referred to as the *storage modulus* and the *loss modulus*, respectively. It is seen that the components of the complex modulus are, in general, dependent on the excitation frequency. An analogous development may be used to establish the corresponding shear modulus of the material during a state of pure shear imposed, say, during a torsion test. If this is done, the corresponding *complex shear modulus* takes the form

$$\hat{G}(\Omega) = G(\Omega) + i\tilde{G}(\Omega) \tag{3.97}$$

where the associated shear storage modulus and shear loss modulus are generally dependent on the excitation frequency.

   If we restrict our attention to materials for which the real part of the complex elastic modulus (the elastic storage modulus) and the real part of the complex shear modulus (the shear storage modulus) are independent of the excitation frequency and hence are equal to the corresponding elastic modulus, E, and shear modulus, G, that would be measured during quasi-static tests, then the frequency dependence of the material lies solely in the imaginary part of the corresponding modulus (the elastic loss modulus or the shear loss modulus). For such materials, Eqs. (3.95) and (3.97) take the respective forms

$$\hat{E}(\Omega) = E + i\tilde{E}(\Omega) \tag{3.98}$$

and

$$\hat{G}(\Omega) = G + i\tilde{G}(\Omega) \tag{3.99}$$

In this way, the standard elastic constants, E and G, together with the frequency dependent loss moduli, $\tilde{E}(\Omega)$ and $\tilde{G}(\Omega)$ characterize the material behavior for the case of harmonic loading. These parameters may be measured directly, and therefore provide an equivalent characterization of the behavior of the material to that provided by the creep function, $\mathcal{J}$, for the case of harmonic excitation.

   The complex moduli defined by Eqs. (3.98) and (3.99) may be substituted for the standard elastic constants in any structural model of interest, for the loading type under consideration. This may be done, for example, for the continuous dynamic rod and beam models discussed in Chapters 9–11, or for the simplified representations of such structures considered thus far. We shall next incorporate the model of hereditary material behavior introduced above into the approximate representations for structural systems employed to this point.

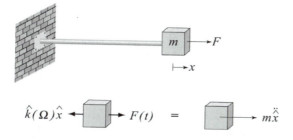

Figure 3.16 Kinetic diagram for structure.

3.4.3 Steady State Response of Single Degree of Freedom Systems

When the mass of a structural member, such as a beam or rod, is small compared with the other mass measures of a system, and we are primarily interested in the motion of a single point on that structure, we often approximate that system as an equivalent single degree of freedom system. We next incorporate the effects of internal friction into such models.

If, in the equivalent single degree of freedom models of the elastic systems discussed in Chapter 1 and employed to this point, we replace Young's modulus, E, and the shear modulus, G, by the complex elastic modulus, $\hat{E}(\Omega)$, and complex shear modulus, $\hat{G}(\Omega)$, respectively defined by Eqs. (3.98) and (3.99), then the equivalent stiffness in each case will be replaced by an equivalent complex stiffness, $\hat{k}(\Omega)$, of the form

$$\hat{k}(\Omega) = k + i\tilde{k}(\Omega) = k\left[1 + i\tilde{\gamma}(\Omega)\right] \tag{3.100}$$

where k = constant is the equivalent *elastic stiffness* as computed in Section 1.3 and

$$\tilde{\gamma}(\Omega) = \tilde{k}(\Omega)/k \tag{3.101}$$

is the *structural loss factor*. It follows from the corresponding dynamic free body diagram (Figure 3.16) that the equation of motion for a harmonically excited system with material damping takes the form of that for an undamped system with the spring stiffness replaced by the complex stiffness defined by Eq. (3.100). If we introduce the harmonic excitation in complex form, then the equation of motion for the corresponding equivalent single degree of freedom system takes the form

$$m\ddot{\hat{x}} + \hat{k}(\Omega)\hat{x} = F_0 e^{i\Omega t} \tag{3.102}$$

or equivalently

$$\ddot{\hat{x}} + \omega^2\left[1 + i\tilde{\gamma}(\Omega)\right]\hat{x} = \omega^2 f_0 e^{i\Omega t} \tag{3.103}$$

where the familiar parameters f_0 and ω are defined by Eqs. (3.14) and (2.2), respectively. We next parallel the approach of Sections 3.3.2 and 3.3.3 to obtain the particular solution, and hence the steady state response of the structurally damped system described by Eq. (3.103). We thus assume a solution of the form

$$\hat{x}(t) = \hat{X} e^{i\Omega t} \tag{3.104}$$

and substitute it into Eq. (3.99). Solving the resulting algebraic equation for $\hat{X}$ and substituting the corresponding expression back into Eq. (3.104) gives the particular solution

$$\hat{x}(t) = f_0 \, \tilde{\Gamma}(\Omega) e^{i(\Omega t - \tilde{\Phi})} \tag{3.105}$$

where

$$f_0 = \frac{F_0}{k} \tag{3.106}$$

$$\tilde{\Gamma}(\Omega) = \frac{1}{\sqrt{\left(1 - \bar{\Omega}^2\right)^2 + \tilde{\gamma}^2}} \tag{3.107}$$

and

$$\tilde{\Phi} = \tan^{-1} \left\{ \frac{\tilde{\gamma}}{1 - \bar{\Omega}^2} \right\} \tag{3.108}$$

are, respectively, the magnification factor and phase angle of the structurally damped response. Employing Euler's Formula, Eq. (1.61), in Eq. (3.105) and equating real and imaginary parts gives the steady state response for the explicit harmonic excitations as follows:

if $F(t) = F_0 \cos \Omega t$ then

$$x_{ss}(t) = f_0 \, \tilde{\Gamma}(\Omega) \cos\left(\Omega t - \tilde{\Phi}\right) \tag{3.109}$$

if $F(t) = F_0 \sin \Omega t$ then

$$x_{ss}(t) = f_0 \, \tilde{\Gamma}(\Omega) \sin\left(\Omega t - \tilde{\Phi}\right) \tag{3.110}$$

A comparison of Eqs. (3.107)–(3.110) with Eqs.(3.50), (3.51), (3.53) and (3.54) suggests the definition of the *effective damping factor*

$$\zeta_{eff}(\Omega) \equiv \frac{\tilde{\gamma}(\Omega)}{2\bar{\Omega}} \tag{3.111}$$

which is seen to be dependent on the excitation frequency. Thus, the loss factor $\tilde{\gamma}(\Omega)$ is often referred to as the structural loss factor. Equations (3.111) and (3.101) together with Eq. (2.65) suggest the definition of an *effective damping coefficient* of an equivalent, but frequency dependent, viscous damper in the form

$$c_{\text{eff}}(\Omega) \equiv \frac{\tilde{k}(\Omega)}{\Omega} = \frac{k\,\tilde{\gamma}(\Omega)}{\Omega} \tag{3.112}$$

Hereditary materials are also referred to as *viscoelastic materials*.

Example 3.12

A $1' \times 1'' \times 1/16''$ metal strip is supported by rollers at its edges. Young's modulus of the strip is measured in a quasi-static test as 3×10^7 psi. A 3 lb weight is bonded to the beam at center-span and the lower edge of the beam is excited by a harmonic force as indicated. When the excitation frequency is set at 6 cps, the time history of the displacement of the weight is observed to be out of phase with the force by 30°. Determine the loss factor, the effective damping factor and the damping coefficient of the beam
for the given frequency.

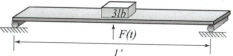

Solution

We first determine the effective stiffness of the equivalent single degree of freedom system. The area moment of inertia of a cross section is

$$I = \frac{(1)(1/16)^3}{12} = 2.035 \times 10^{-5}\ \text{in}^4 \tag{a}$$

The effective stiffness is next found using Eq. (1.22). Hence,

$$k = \frac{48(3 \times 10^7)(2.035 \times 10^{-5})}{(12)^3} = 16.95\ \text{lb/in} = 203.4\ \text{lb/ft} \tag{b}$$

It then follows that

$$\omega = \sqrt{\frac{203.4}{3/32.2}} = 46.72\ \text{rad/sec} \tag{c}$$

The excitation frequency is

$$\Omega = 2\pi(6) = 37.70\ \text{rad/sec} \tag{d}$$

Hence,

$$\bar{\Omega} = 37.70/46.72 = 0.8069 \qquad\qquad (e)$$

From Eq. (3.108), the structural loss factor is computed as

$$
\begin{aligned}
\tilde{\gamma} &= \left[1 - \bar{\Omega}^2\right] \tan(30^\circ) \\
&= \left[1 - (0.8069)^2\right](0.5774) = 0.2015
\end{aligned}
\qquad \triangleleft (f)
$$

Now, the effective damping factor is computed using Eq. (3.111) giving

$$\zeta_{eff} = \frac{\tilde{\gamma}}{2\bar{\Omega}} = \frac{(0.2015)}{2(0.8069)} = 0.1249 \qquad \triangleleft (g)$$

Finally, the effective damping coefficient is found using Eq. (3.112) to give

$$c_{eff} = \frac{k\tilde{\gamma}}{\Omega} = \frac{(203.4)(0.2015)}{37.70} = 1.087 \text{ lb-sec/ft} \qquad \triangleleft (h)$$

3.5 SELECTED APPLICATIONS

In this section, we examine three representative applications that involve harmonic excitation. These include harmonic motion of the support, the unbalanced motor, and synchronous whirling of rotating shafts.

3.5.1 Harmonic Motion of the Support

In many applications the support or foundation of the system is driven and undergoes motion. Examples include earthquake loadings on buildings and devices attached to machinery. In this section we consider the response of single degree of freedom systems that are excited in this manner.

Consider the system shown in Figure 3.17, and let the foundation undergo the prescribed displacement $x_F(t)$ as indicated. The associated kinetic diagram for the system is displayed in Figure 3.18. Since the spring force is proportional to the relative displacement of the mass with respect to the support, and the damping force is proportional to the relative velocity of the mass with respect to the support, the equation of motion is found from Newton's Second Law as follows;

$$-k(x - x_F) - c(\dot{x} - \dot{x}_F) = m\ddot{x}$$

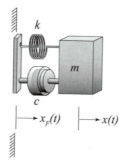

Figure 3.17 Mass-spring-damper system with movable support.

or

$$\ddot{x} + 2\omega\zeta(\dot{x} - \dot{x}_F) + \omega^2(x - x_F) = 0 \tag{3.113}$$

Relative Motion

Let $u(t)$ measure the displacement of the mass relative to the moving foundation, and hence measure the stretch in the spring. Hence,

$$u(t) = x(t) - x_F(t) \tag{3.114}$$

It then follows that $\dot{u}(t)$ represents the relative velocity of the mass with respect to the foundation and therefore measures the rate at which the viscous damper is being separated. Substitution of Eq. (3.114) into Eq. (3.113) gives the governing equation in terms of the relative motion of the mass as

$$\ddot{u} + 2\omega\zeta\dot{u} + \omega^2 u = -\ddot{x}_F(t) \tag{3.115}$$

Equation (3.115) is valid for any form of support excitation and is seen to have the form of the governing equation for the damped harmonic oscillator subjected to a time dependent force. As we are presently interested in harmonic motion of the support let us consider support motion in the form

$$x_F(t) = h_0 \sin\Omega t \tag{3.116}$$

Figure 3.18 Kinetic diagram for system with support motion.

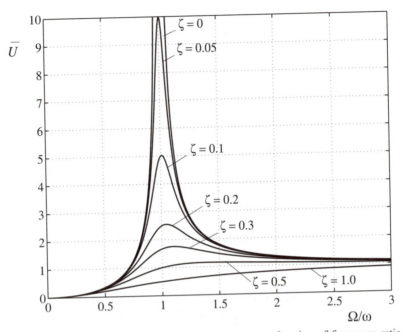

Figure 3.19 Normalized amplitude of relative motion as a function of frequency ratio for various values of the damping factor.

where h_0 is the amplitude of the displacement of the foundation. Substitution of the prescribed support displacement given by Eq. (3.116) into Eq. (3.115) gives the equation of motion in the form

$$\ddot{u} + 2\omega\zeta\dot{u} + \omega^2 u = \omega^2 f_0 \sin\Omega t \tag{3.117}$$

where

$$f_0 = h_0\bar{\Omega}^2 \tag{3.118}$$

and, as before, $\bar{\Omega} = \Omega/\omega$. Equation (3.117) is seen to be in standard form. Therefore, the steady state response may be written directly from Eq. (3.54), with appropriate change of variables. Thus,

$$u_{ss}(t) = U_0 \sin(\Omega t - \Phi) \tag{3.119}$$

where

$$U_0 = h_0\bar{\Omega}^2\Gamma(\bar{\Omega};\zeta) \tag{3.120}$$

is the amplitude of the steady state response, and Γ and Φ are given by Eqs. (3.50) and Eq. (3.51) respectively. The phase angle Φ measures the degree that the relative motion of the mass lags the motion of the support.

It may be seen that the ratio of the amplitude of the relative motion of the mass to the amplitude of the motion of the support is given by

$$\bar{U} \equiv \frac{U_0}{h_0} = \bar{\Omega}^2 \Gamma\left(\bar{\Omega}; \zeta\right) \tag{3.121}$$

The normalized amplitude, $\bar{U}$, characterizes the relative motion of the support excited system in the same sense that the magnification factor, Γ, characterizes the absolute motion of a system excited by a harmonic force applied to the mass. It may be noted from Eqs. (3.121) and (3.50) that when $\bar{\Omega}^2 \gg 1$, $\bar{U} \approx 1$ and hence $U_0 \approx h_0$ regardless of the excitation frequency. Plots of the normalized amplitude are displayed as functions of the frequency ratio for various values of the damping factor in Figure 3.19. The phase angle is seen to be of the same form as that for the case of the harmonically forced mass. Hence the plots of the phase angle displayed in Figure 3.9 pertain to the present case as well, but with the current interpretation.

Absolute Motion

The solution given by Eq. (3.119) gives the motion of the mass of the support excited system as seen by an observer moving with the support. We next determine the motion of the mass as seen by an observer attached to a fixed reference frame.

The absolute steady state motion of the mass, that is the motion with regard to a fixed reference frame, is obtained by first rearranging Eq. (3.113) so that all terms associated with the prescribed motion of the support appear on the right-hand side. Hence,

$$\ddot{x} + 2\omega\zeta\dot{x} + \omega^2 x = 2\omega\zeta\dot{x}_F + \omega^2 x_F \tag{3.122}$$

Equation (3.122) is valid for any excitation of the foundation. As for relative motion, we are here interested in harmonic motion of the foundation in the form of Eq. (3.116). It is, however, expedient to solve for the absolute motion of the mass in complex form. Hence, let us consider the complex form of the support motion

$$\hat{x}_F(t) = h_0 e^{i\Omega t} \tag{3.123}$$

After solving, we shall extract the imaginary part of the solution, as it will correspond to the response to forcing in the form of the sine function of interest. Substitution of Eq. (3.123) into Eq. (3.122) gives the differential equation governing the complex response as

$$\ddot{\hat{x}} + 2\omega\zeta\dot{\hat{x}} + \omega^2\hat{x} = \omega^2 h_0 \left(1 + i2\zeta\bar{\Omega}\right)e^{i\Omega t} \tag{3.124}$$

The right-hand side of Eq. (3.124) may be viewed as the superposition of two harmonic forces,

$$f^{(1)}(t) = f_0^{(1)}e^{i\Omega t} = h_0 e^{i\Omega t}$$

and

$$f^{(2)}(t) = f_0^{(2)}e^{i\Omega t} = ih_0\, 2\zeta\, \overline{\Omega}e^{i\Omega t}$$

It follows form the Principle of Superposition (Section 3.1) that the response of the system to the two forces is the sum of the responses to each of the excitations applied individually. In addition, Eq. (3.124) is in standard form. Applying Eqs. (3.47)–(3.51) for the two forces of Eq. (3.124) and summing gives the complex form of the steady state response as

$$\hat{x}_{ss}(t) = \hat{X}e^{i\Omega t} \tag{3.125}$$

where

$$\hat{X} = h_0\Gamma e^{-i\Phi}\left(1 + i2\zeta\,\overline{\Omega}\right) \tag{3.126}$$

and Γ and Φ are given by Eqs. (3.50) and Eq. (3.51), respectively. Next, paralleling the development of Eqs. (3.73)–(3.74) gives the steady state response for the absolute motion of the mass as

$$\hat{x}_{ss}(t) = Xe^{i(\Omega t - \Psi)} \tag{3.127}$$

where

$$X = h_0\Upsilon\left(\overline{\Omega};\zeta\right) = h_0\,\Gamma\left(\overline{\Omega};\zeta\right)\sqrt{1 + \left(2\zeta\,\overline{\Omega}\right)^2} \tag{3.128}$$

and

$$\Psi = \tan^{-1}\left\{\frac{2\zeta\,\overline{\Omega}^3}{\left[\left(1-\overline{\Omega}^2\right) + \left(2\zeta\,\overline{\Omega}\right)^2\right]}\right\} \tag{3.129}$$

If the motion of the foundation is given by Eq. (3.116) then the steady state response of the support is given by the imaginary part of the complex response, Eq. (3.127). Hence,

$$x_{ss}(t) = h_0\Upsilon\sin(\Omega t - \Psi) \tag{3.130}$$

The absolute motion of the mass is seen to be characterized by the ratio of the amplitude of the motion of the system to the amplitude of the motion of the foundation

$$\overline{X} \equiv \frac{X}{h_0} = \Upsilon\left(\overline{\Omega};\zeta\right) \tag{3.131}$$

Comparison of Eqs. (3.131) and (3.81) shows that the same functional form governs both the response for the present case and the force transmitted to a fixed support, though the parameters they represent have, of course, very different interpretations. Thus, the plots displayed in Figures 3.15 and 3.14 characterize the absolute motion of support excited systems with appropriate interpretation of the parameters.

Force Transmitted to the Moving Support

The kinetic diagram of the system (Figure 3.19) together with Eq. (3.113) shows that the force transmitted to the support by the system is

$$F_{tr} = c\dot{u} + ku = -m\ddot{x} \tag{3.132}$$

Substitution of the steady state response, Eq. (3.130), into the right-hand side of Eq. (3.132) gives the transmitted force as

$$F_{tr} = \left\|F_{tr}\right\| \sin(\Omega t - \Psi) \tag{3.133}$$

where

$$\left\|F_{tr}\right\| = k\,h_0\,\widehat{\mathbb{T}} \tag{3.134}$$

and

$$\widehat{\mathbb{T}} = \overline{\Omega}^2 \Upsilon = \frac{\overline{\Omega}^2\sqrt{1+\left(2\zeta\overline{\Omega}\right)^2}}{\sqrt{\left(1-\overline{\Omega}^2\right)^2+\left(2\zeta\overline{\Omega}\right)^2}} \tag{3.135}$$

Example 3.13

The foundation of a one-story building undergoes harmonic ground motion of magnitude 2 inches and frequency 4 cps. If the roof structure weighs one ton, the bending stiffness of each of the four identical 12 ft columns is 4×10^6 lb-ft² and the structural damping factor at this frequency is estimated to be 0.1, determine the steady state response of the structure. Also determine the magnitude of the shear force within each column.

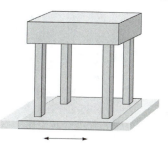

Solution

The effective stiffness of the structure may be computed using Eq. (1.18) to give

$$k_{eff} = \frac{48EI}{L^3} = \frac{48(4\times10^6)}{(12)^3} = 1.11\times10^5 \text{ lb/ft} \tag{a}$$

Hence,

$$\omega = \sqrt{\frac{k_{eff}}{m}} = \sqrt{\frac{48EI}{mL^3}} = \sqrt{\frac{1.11 \times 10^5}{(2000/32.2)}} = 42.3 \text{ rad/sec} \tag{b}$$

The excitation frequency is given as

$$\Omega = 4 \text{ cps} \times 2\pi \text{ rads/cycle} = 8\pi \text{ rad/sec} \tag{c}$$

Thus,

$$\bar{\Omega} = \frac{8\pi}{42.3} = 0.594 \tag{d}$$

Now, to calculate the amplitude of the response using Eq. (3.128) we first compute Υ, giving

$$\Upsilon = \frac{\sqrt{1+\left(2\zeta\bar{\Omega}\right)^2}}{\sqrt{\left(1-\bar{\Omega}^2\right)^2+\left(2\zeta\bar{\Omega}\right)^2}} = \frac{\sqrt{1+\left[2(0.1)(0.594)\right]^2}}{\sqrt{\left[1-(0.594)^2\right]^2+\left[2(0.1)(0.594)\right]^2}} = 1.53 \tag{e}$$

The amplitude of the steady state response is then

$$X = h_0\Upsilon = 2(1.53) = 3.06'' \tag{f}$$

The phase angle is next computed using Eq. (3.129) giving

$$\Psi = \tan^{-1}\left\{\frac{2\zeta\bar{\Omega}^3}{1-\bar{\Omega}^2+\left(2\zeta\bar{\Omega}\right)^2}\right\}$$

$$= \tan^{-1}\left\{\frac{2(0.10)(0.594)^3}{1-(0.594)^2+\left[2(0.1)(0.594)\right]^2}\right\} = 0.0633 \text{ rads} \tag{g}$$

Substituting Eqs. (c), (f) and (g) into Eq. (3.130) gives the steady state response of the building,

$$x_{ss}(t) = 3.06\sin(8\pi t - 0.0633) \text{ (inches)} \qquad \triangleleft \text{(h)}$$

The force transmitted to the support is computed from Eqs. (3.134) and (3.135). For $\zeta^2 \ll 1$ we have

$$\|F_{tr}\| = k\,h_0\widehat{\mathbb{T}} = (1.11 \times 10^5)(2/12)\left[(0.594)^2(1.53)\right] = 9.99 \times 10^3 \text{ lb} \tag{i}$$

Finally, the shear force in each of the four columns is one quarter of the total transmitted force. Hence,

$$V = \|F_{tr}\|/4 = 2.50 \times 10^3 \text{ lb} \qquad \lhd \text{(j)}$$

Example 3.14 Vehicle Traveling on a Buckled Road

During very hot weather roads often suffer thermal buckling. If the buckle is sinusoidal, as shown, the rise of the buckle, y, may be expressed as a function of the distance along the road, η, in the form $y = y_0 \sin(2\pi\eta/\lambda)$, where y_0 is the amplitude of the buckle (the maximum rise) and λ is the wavelength (the spatial period) of the buckle. Those of us that drive, or even just ride, all know that the ride over a bumpy road is often worse at particular speeds. If a vehicle modeled as the equivalent, lightly damped, single degree of freedom system shown rides along the buckled road at constant speed, v_0, determine the value of the speed at which the amplitude of the vehicle vibration achieves a maximum. Also determine the magnitude of the force transmitted to the axle at this speed.

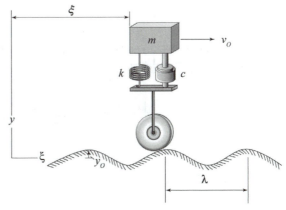

Solution

The vehicle moves to the right at constant speed v_0 over the buckled road, which causes the vehicle to rise and dip as it follows the contour of the road. The first thing we need to do is determine the vertical motion of the wheel as a function of time. Since the vehicle travels at constant speed, the horizontal component of the velocity is constant. The distance traveled in the horizontal direction is then given by

$$\eta = v_0 t \qquad \text{(a)}$$

as indicated in the figure. Substitution of Eq. (a) into the given equation for the road deflection gives the vertical motion of the wheel as

$$y = y_0 \sin \Omega t \qquad \text{(b)}$$

where

$$\Omega = 2\pi v_0 / \lambda \tag{c}$$

With the vertical motion of the wheel hub given by Eq. (b), it may be seen that the present problem corresponds to one of (vertical) excitation of the support, with $y(t)$ identified with $x_F(t)$ and y_0 identified with h_0. Since the damping in the vehicle's suspension system is "light" ($\zeta^2 \ll 1$), the maximum vertical response of the vehicle will occur when $\overline{\Omega} \cong 1$. Substituting Eq. (c) into the resonance condition, $\Omega \cong \omega$, and solving for the velocity gives the critical speed of the vehicle,

$$v_{cr} = \frac{\lambda}{2\pi} \sqrt{\frac{k}{m}} \tag{$\triangleleft$ (d)}$$

The magnitude of the force transmitted to the axle is found using Eq. (3.134) and (3.135) with $\overline{\Omega} \cong 1$ and $\zeta^2 \ll 1$. This gives

$$\|F_{tr}\| \cong \frac{k\, y_0}{2\zeta} \tag{e}$$

Substituting Eqs. (c) and (d) into Eq. (e) gives the magnitude of the transmitted force at the critical speed as

$$\|F_{tr}\| = \frac{k\, y_0}{2(c/2\omega m)} = 2\pi v_0 \frac{y_0}{\lambda} \frac{k\, m}{c} \tag{$\triangleleft$ (f)}$$

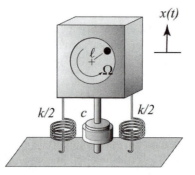

Figure 3.20 Rotor system with imbalance.

3.5.2 Unbalanced Motor

In many practical systems the situation exists where the mass of a supposedly axisymmetric rotating body is not evenly distributed. (In reality, this will likely be the case to some extent in all systems.) If, for example, a motor is slightly unbalanced in the sense that the mass of the rotor is not quite symmetrically distributed throughout its cross section, unwanted vibrations of the motor or its support system can occur. We examine this phenomenon in the present section.

Equation of Motion

Consider a circular wheel or shaft rotating within the frictionless sleeve of the representative system shown in Figure 3.20. Let the total mass of the system, including the shaft, be m, and let the distribution of the mass of the rotor be slightly nonuniform. A simple way to model this situation is to represent the uneven distribution of mass as an eccentric point mass embedded in a uniform distribution of mass. We shall thus consider an eccentric point mass, m_e, to be embedded in the rotor and located a distance ℓ from the axis of rotation, as indicated. If the shaft rotates at a constant angular rate Ω, the angular displacement, θ, of the radial generator to the eccentric mass is then $\theta = \Omega t$. Finally, let the vertical displacement of the axis of rotation relative to its equilibrium position be x. We are interested in the motion $x(t)$ of the system that results from the rotation of the nonuniform rotor.

   To derive the equation of motion of the composite system it is useful to first express the Cartesian coordinates (x_e, y_e) that establish the position of the eccentric mass, in terms of the vertical displacement of the axis of rotation and the angular displacement of the rotor or flywheel. The desired relations are found, with the aid of Figure 3.21, as

$$x_e(t) = x(t) + \ell \sin \Omega t \tag{3.136}$$

$$y_e(t) = -\ell \cos \Omega t \tag{3.137}$$

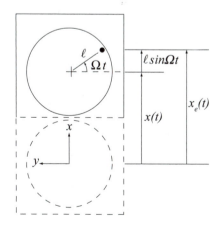

Figure 3.21 Description of motion of the system.

We next isolate each of the bodies that comprise the idealized system. The corresponding dynamic free-body diagram is shown in Figure 3.22. In those diagrams the mass of the rotor is labeled m_w, and the resultant internal forces acting on the wheel, the eccentric mass and the block are labeled $\vec{F}^{(w)}$, $\vec{F}^{(e)}$ and $\vec{F}^{(b)}$, respectively. A subscript x indicates the corresponding vertical component of that force. With the aid of the kinetic diagrams, the vertical component of Newton's Second Law may be written for each body as

$$F_x^{(b)} - kx - c\dot{x} = \left[m - m_e - m_w \right] \ddot{x}$$
$$F_x^{(e)} = m_e \ddot{x} = m_e \left[\ddot{x} - \Omega^2 \ell \sin \Omega t \right] \qquad (3.138)$$
$$F_x^{(w)} = m_w \ddot{x}$$

We next sum Eqs. (3.138) and note that the internal forces sum to zero via Newton's Third Law. This results in the governing equation for the vertical motion of the entire system

$$m\ddot{x} + c\dot{x} + kx = \Omega^2 \ell m_e \sin \Omega t \qquad (3.139)$$

When rearranged, Eq. (3.139) takes the standard form

$$\ddot{x} + 2\omega\zeta\dot{x} + \omega^2 x = \omega^2 f_0 \sin \Omega t \qquad (3.140)$$

where

$$f_0 = f_0\left(\bar{\Omega}\right) = \frac{m_e \ell}{m} \bar{\Omega}^2 \qquad (3.141)$$

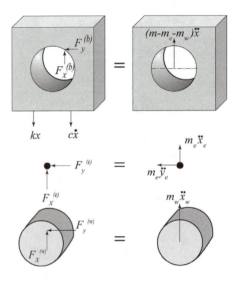

Figure 3.22 Kinetic diagram for each component of the system.

Steady State Response

The steady state solution of Eq. (3.140) is given by Eq. (3.54) with appropriate inter-pretation of the parameters. Hence, the steady state response of the unbalanced motor is given by

$$x_{ss}(t) = X \sin(\Omega t - \Phi) = \frac{m_e \ell}{m} \bar{\Omega}^2 \Gamma\left(\bar{\Omega}; \zeta\right) \sin(\Omega t - \Phi) \qquad (3.142)$$

where Γ and Φ are given by Eqs. (3.50) and Eq. (3.51), respectively. The ratio of the first moments of the mass,

$$Q \equiv \frac{mX}{m_e \ell} = \bar{\Omega}^2 \Gamma\left(\bar{\Omega}; \zeta\right) \qquad (3.143)$$

is seen to characterize the response of the unbalanced motor. We note that when $\bar{\Omega}^2 \gg 1$, $Q \approx 1$ and hence

$$X \approx m_e \ell / m \qquad (3.144)$$

Thus, for large frequency ratios the amplitude of the response is effectively independ-ent of the excitation frequency. It may be seen that the normalized (first) mass mo-ment defined by Eq. (3.143) is characterized by the same function as the normalized amplitude of the relative motion associated with harmonic motion of the support, Eq. (3.121). The corresponding plots of that function displayed in Figure 3.19 then de-scribe the characteristics of the response for the unbalanced motor, as well. Plots of the associated phase angle are shown in Figure 3.9. Upon consideration of Figure 3.19 it may be seen that $Q \to Q_{max}$ in the vicinity of $\bar{\Omega} = 1$ as may be expected. It may also be seen that $Q \to 0$ and $\Phi \to 0$ as $\bar{\Omega} \to 0$, and also that $Q \to 1$ and $\Phi \to \pi$ as $\bar{\Omega} \to \infty$. These trends may be put into context by examination of the mo-tion of the center of mass of the system.

The location of the center of mass of the block-rotor-eccentric mass system shifts as the block moves up and down and the eccentric mass rotates with the rotor about the corresponding axis of rotation. The vertical location of the center of mass is found, as a function of time, by applying Eqs. (3.136) and (3.143) to Eq. (1.120). This gives

$$x_{cm}(t) = X \left[\sin(\Omega t - \Phi) + \frac{m_e \ell}{mX} \sin \Omega t \right] \qquad (3.145)$$

Recall, from our earlier discussion, that $Q \to 1$ and $\Phi \to 0$ as $\bar{\Omega} \to 0$. It is seen from Eq. (3.145) that, in this extreme,

$$x_{cm} \to \frac{m_e \ell}{m} \sin \Omega t$$

Thus, in the static limit, the excitation is so slow compared with the natural motion of the system that the block essentially remains stationary ($X = 0$) and the center of mass slowly moves up and down as the eccentric mass rotates very slowly about the axis of rotation. In the opposite extreme, recall that $\mathbb{Q} \to 0$ and $\Phi \to \pi$ as $\bar{\Omega} \to \infty$. It is seen from Eq. (3.145) that, in this extreme,

$$x_{cm} \to X\left[-1 + \frac{m_e \ell}{mX}\right] \sin \Omega t \to 0$$

Thus, in this limit, the center of mass of the system remains essentially stationary while the eccentric mass rotates rapidly about the axis of rotation, and the block moves rapidly up and down to compensate. This scenario is in keeping with that for the high (excitation) frequency response found for the cases discussed in previous sections in that, in this extreme, the excitation is so rapid that the system as a whole essentially does not respond to the excitation.

Force Transmitted to the Support

The force transmitted to the support can be obtained by adopting the development of Section 3.3.4 to the present case. To do this, we substitute the amplitude of the steady state response given by Eq. (3.142) into Eq. (3.72) and parallel the development leading to Eq. (3.80). This gives the force transmitted to the support of the unbalanced motor as

$$F_{tr} = \|F_{tr}\| \sin(\Omega t - \Psi) \tag{3.146}$$

where

$$\|F_{tr}\| = kX\sqrt{1 + \left(2\zeta\bar{\Omega}\right)^2} = k\left(m_e\ell/m\right)\bar{\Omega}^2\Upsilon \tag{3.147}$$

Υ is defined by Eq. (3.77) and Ψ is given by Eq. (3.78).

Example 3.15

A 4 kg motor sits on an isolation mount possessing an equivalent stiffness of 64 N/m and damping coefficient 6.4 N-sec/m. It is observed that the motor oscillates at an amplitude of 5 cm when the rotor turns at a rate of 6 rad/sec. (*a*) Determine the offset moment of the motor. (*b*) What is the magnitude of the force transmitted to the support at this rate of rotation?

Solution

(*a*)
Let us first determine the system parameters ω and ζ. Hence,

$$\omega = \sqrt{\frac{k}{m}} = \sqrt{\frac{64}{4}} = 4 \text{ rad/sec} \tag{a}$$

and

$$\zeta = \frac{c}{2\omega m} = \frac{6.4}{2(4)(4)} = 0.2 \tag{b}$$

Therefore,

$$\bar{\Omega} = \frac{\Omega}{\omega} = \frac{6}{4} = 1.5 \tag{c}$$

Substituting Eqs. (b) and (c) into Eq. (3.143) gives the ratio of the first inertial moments as

$$Q = \bar{\Omega}^2 \Gamma = \frac{(1.5)^2}{\sqrt{\left[1-(1.5)^2\right]^2 + \left[2(1.5)(0.2)\right]^2}} = 1.623 \tag{d}$$

Rearranging Eq. (3.143) and substituting Eq. (d) gives the offset moment

$$m_e \ell = \frac{mX}{Q} = \frac{(4)(0.05)}{1.623} = 0.1232 \text{ kg-m} \qquad \triangleleft \text{(e)}$$

(b)
To determine the magnitude of the transmitted force, we simply substitute the given displacement and effective stiffness, as well as the computed damping factor and frequency ratio, into Eq. (3.147). Carrying through the computation gives the magnitude of the force transmitted to the support as

$$\|F_{tr}\| = (64)(0.05)\sqrt{1 + \left[2(0.2)(1.5)\right]^2} = 3.732 \text{ N} \qquad \triangleleft \text{(f)}$$

Example 3.16
Determine the amplitude of the response of the system of Example 3.15 when the rotation rate is increased to 60 rad/sec.

Solution
For this case,

$$\bar{\Omega} = \frac{60}{4} = 15 \tag{a}$$

and hence

$$\bar{\Omega}^2 = 225 \gg 1 \qquad\qquad (b)$$

Therefore $Q \approx 1$ and

$$X \approx \frac{m_e \ell}{m} = \frac{0.1232}{4} = 0.03080 \text{ m} = 3.08 \text{ cm} \qquad\qquad \triangleleft (c)$$

Let us next calculate the amplitude using the "exact" solution. Hence,

$$Q = \frac{(15)^2}{\sqrt{\left[1-(15)^2\right]^2 + \left[2(0.2)(15)\right]^2}} = 1.004 \qquad\qquad (d)$$

The amplitude of the steady state response is then

$$X = \frac{m_e \ell}{m} Q = \frac{0.1232}{4}(1.004) = 0.03092 \text{ m} = 3.09 \text{ cm} \qquad\qquad \triangleleft (e)$$

Comparing the two answers we see that the error in using the large frequency approximation is

$$\% \text{ error} = \frac{3.09 - 3.08}{3.09} \times 100\% = 0.324\% \qquad\qquad (f)$$

3.5.3 Synchronous Whirling of Rotating Shafts

Consider a circular disk of mass m that is coaxially attached at center span of an elastic shaft as shown in Figure 3.23. Let the center of mass of the disk be offset a distance ℓ from the axis of the shaft as shown, and let the shaft-disk system be spinning about its axis at the angular rate Ω. Further let the mass of the shaft be very small compared with the mass of the disk and thus be considered negligible. The system may then be modeled as an equivalent single degree of freedom system as shown in Figure 3.24. In this regard, let k represent the equivalent elastic stiffness of the shaft in bending (Section 1.2.2), and let c represent some effective viscous damping of the system. The latter may be provided, for example, by oil in the bearings, by structural damping (Section 3.4) or by aerodynamic damping (Section 1.2.6). To examine this problem, we first derive the equations of motion for the disk during whirl.

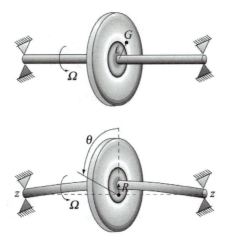

Figure 3.23 Whirling shaft.

To derive the equations of motion, we must apply Newton's Second Law of Motion to the system of interest. In order to this we must first obtain expressions for the acceleration of the center of mass during this relatively complex motion. If point O refers to the geometric center of the disk then we have, from Eq. (1.72), that the acceleration of the center of mass of the disk is related to the acceleration of the geometric center of the disk, and hence of the axis of the shaft, by the relation

$$\vec{a}_G = \vec{a}_O + \vec{a}_{G/O} \tag{3.148}$$

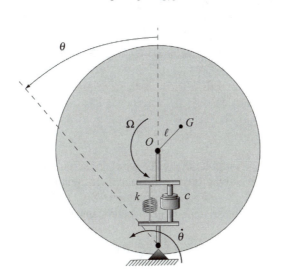

Figure 3.24 Equivalent system.

where $\vec{a}_G$ and $\vec{a}_O$ correspond to the absolute acceleration of the center of mass and the absolute acceleration of the geometric center, respectively, and $\vec{a}_{G/O}$ is the relative acceleration of the center of mass with respect to the geometric center (i.e., the acceleration of the center of mass as seen by an observer translating, but not rotating, with point O). For the present problem, it is convenient to work in polar coordinates. The acceleration of the geometric center is then given by Eq. (1.79) as

$$\vec{a}_O = \left(\ddot{R} - R\dot{\theta}^2\right)\vec{e}_R + \left(R\ddot{\theta} + 2\dot{R}\dot{\theta}\right)\vec{e}_\theta \tag{3.149}$$

where R is the radial distance of the geometric center of the disk from the axis between the supports (the z-axis) as shown in Figure 3.23. Thus, $R(t)$ represents the amplitude of the whirling motion. The relative acceleration of the center of mass with respect to the geometric center is similarly obtained, with the aid of Figure 3.25. We thus find that

$$\vec{a}_{G/O} = \left[-\Omega^2 \ell \cos(\Omega t - \theta) + \ell\dot{\Omega}\sin(\Omega t - \theta)\right]\vec{e}_R \\ + \left[\ell\dot{\Omega}\cos(\Omega t - \theta) - \Omega^2 \ell \sin(\Omega t - \theta)\right]\vec{e}_\theta \tag{3.150}$$

where θ is the angular displacement of the geometric center about the axis connecting the supports, as indicated in Figure 3.24, and we have employed the fact that the offset, ℓ, of the center of mass is constant for a rigid disk. The kinematic relations described by Eqs. (3.148)–(3.150) are valid for general whirling of the system.

For *synchronous whirl*, we restrict our attention to motions for which the whirling rate is numerically equal to the rotational rate of the shaft, and for which the whirling motion remains steady. We therefore, at this juncture, restrict our attention to motions for which

$$\dot{\theta} = \Omega = \text{constant} \quad \text{and} \quad R = \text{constant} \tag{3.151}$$

Integration of the first of Eqs. (3.151) with respect to time gives the angular displacement of the whirl as

$$\theta = \Omega t - \Phi \tag{3.152}$$

where Φ is seen to measure the lag of the whirl angle with respect to the spin angle (see Figure 3.25). Substituting Eqs. (3.151) and (3.152) into Eqs. (3.149) and (3.150), and then substituting the resulting expressions into Eq. (3.148), gives the acceleration of the center of mass of the system during synchronous whirl as

$$\vec{a}_G = -\Omega^2 \ell \left[\left(\bar{R} + \cos\Phi\right)\vec{e}_R + \sin\Phi\,\vec{e}_\theta\right] \tag{3.153}$$

where

$$\bar{R} \equiv R/\ell \tag{3.154}$$

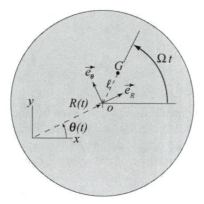

Figure 3.25 Inertial and axial reference frames.

is the normalized amplitude of the whirling motion. With the acceleration evaluated, we are now ready to apply Newton's Second Law of Motion to the system of interest. We thus write the equations of motion for our system in terms of polar coordinates, substitute Eqs. (3.151)–(3.153) into these equations, and equate components. This gives the governing equations for the system along the radial and circumferential directions, respectively, as

$$-kR = -m\Omega^2\ell\left(\bar{R} + \cos\Phi\right)$$
$$-c\Omega R = -m\Omega^2\ell\sin\Phi \tag{3.155}$$

Solving the first of Eqs. (3.155) for $\cos\Phi$, the second for $\sin\Phi$, and dividing the latter by the former gives the phase angle in terms of the damping factor and the spin rate of the shaft as

$$\Phi = \tan^{-1}\left\{\frac{2\zeta\bar{\Omega}}{1-\bar{\Omega}^2}\right\} \tag{3.156}$$

where $\bar{\Omega}$ is the ratio of the spin rate of the shaft to the undamped natural frequency of the system and ζ is the damping factor. It may be seen that Eq. (3.156) is identical to Eq. (3.51). Thus, the plots displayed in Figure 3.9 pertain to the current problem as well, with suitable interpretation. However, recall that for structural damping the damping factor is dependent on the excitation frequency. (See Section 3.4) For the present case, the angle Φ represents the phase difference between the rotation of the eccentric mass about the geometric center of the disk (see Figure 3.26) and the motion of the geometric center about the axis between the supports. Substitution of the first (second) of Eqs. (3.74) into the first (second) of Eqs. (3.155) gives the normalized amplitude of whirling as a function of the normalized spin rate and the damping factor. Hence,

$$\bar{R} \equiv \frac{R}{\ell} = \bar{\Omega}^2 \Gamma\left(\bar{\Omega}; \zeta\right) \tag{3.157}$$

It may be seen that the normalized amplitude given by Eq. (3.157) is of the identical functional form as Eq. (3.121). The associated plots displayed in Figure 3.19 therefore describe the normalized amplitude for the problem of synchronized whirl, as well. Recall, however, that for structural damping the damping factor is dependent on the excitation frequency.

Example 3.17

A 50 lb rotor is mounted at center span of a 6 ft shaft having a bending stiffness of 20×10^6 lb-in^2. The shaft is supported by rigid bearings at each end. When operating at 2000 rpm the shaft is strobed and the rotor is observed to displace off-axis by ½ inch. If the damping factor for the system is 0.1 at this frequency, determine the offset of the rotor.

Solution

The effective stiffness for transverse motion of the shaft is determined using Eq. (1.17) to give

$$k = 2 \times \frac{12EI}{(L/2)^3} = 192\frac{EI}{L^3} = 192\frac{(20 \times 10^6)}{(6 \times 12)^3} = 10,290 \text{ lb/in} \tag{a}$$

(See Problem 1.14.) The natural frequency of the system is then computed to be

$$\omega = \sqrt{\frac{10,290}{50/(32.2 \times 12)}} = 282 \text{ rad/sec} \tag{b}$$

The rotation rate of the rod is given as

$$\Omega = 2000 \text{ rev/min} \times 2\pi \text{ rad/rev} \times 1 \text{ min/60sec} = 209 \text{ rad/sec} \tag{c}$$

Hence,

$$\bar{\Omega} = \frac{209}{282} = 0.741 \tag{d}$$

Now,

$$\bar{\Omega}^2 \Gamma = \frac{(0.741)^2}{\sqrt{\left[1 - (0.741)^2\right]^2 + \left[2(0.10)(0.741)\right]^2}} = 1.16 \tag{e}$$

Rearranging Eq. (3.157) and substituting the observed displacement and Eq. (d) gives the offset

$$\ell = \frac{R}{\overline{\Omega}^2 \Gamma} = \frac{0.5}{1.16} = 0.432 \text{ inches} \qquad \triangleleft \text{(f)}$$

3.6 RESPONSE TO GENERAL PERIODIC LOADING

To this point we have studied the response of single degree of freedom systems subjected to excitations that vary harmonically with time. That is the forcing functions considered were of the form of cosine or sine functions. We next consider excitations that are periodic, but not necessarily harmonic, in time. Such excitations are prevalent in common everyday systems as well as in engineering systems. A ratcheting action, a uneven cam driving a mechanism or a parent pushing a child on a swing are just three examples. In this section we study the response of single degree of freedom systems that are subjected to this broader class of excitations, namely general periodic forcing of which harmonic loading is a special case. We shall see that the response to harmonic loading is intimately related to the response to other types of periodic loading.

3.6.1 General Periodic Excitation

We shall consider the class of time dependent excitations that are applied to systems over a very long span of time and whose temporal form repeats itself over specified intervals (see Figure 3.26). The forces and hence the associated functions that describe them, may be discontinuous as in the case of periodically applied pulses such as those imparted by a parent pushing a child on a swing. Such periodic forces are represented mathematically by periodic functions which are defined as functions possessing the property

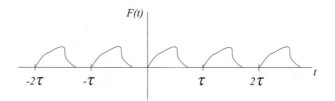

Figure 3.26 Generic periodic function.

$$F(t+\tau) = F(t) \tag{3.158}$$

where t is the time and τ is the period of application of the force.

The standard form of the equation of motion for single degree of freedom systems is given by Eq. (3.1), and is repeated here for convenience. Hence,

$$\ddot{x} + 2\omega\zeta\dot{x} + \omega^2 x = \omega^2 f(t)$$

where now

$$f(t+\tau) = f(t) \tag{3.159}$$

which is related to $F(t)$ by Eq. (3.2) for a mass-spring-damper system, or by analogous relations for other single degree of freedom systems. It is evident from Eq. (3.2) that $f(t)$ is periodic and has the same period as $F(t)$. We wish to determine the general form of the steady state response of single degree of freedom systems (or equivalently establish an algorithm to determine the steady state response of systems) subjected to excitations that vary periodically with time. To do this we shall first resolve the applied force into its projections (components) onto a set of periodic basis functions. That is we shall express $f(t)$ in terms of its Fourier series.

Functions such as $f(t)$ are basically vectors in infinite dimensional space (see Chapters 9 and 10). We can therefore express a function as the sum of the products of its components with respect to a set of mutually orthogonal basis functions just as we can express a vector in three dimensional space as the sum of the products of its components with respect to three mutually orthogonal basis vectors. In three dimensional space there must be three basis vectors. In function space there must be an infinite number of basis functions. For the class of periodic functions under consideration, the set of functions

$$\left\{ \cos(2p\pi t/\tau), \sin(2p\pi t/\tau) \mid p = 0, 1, 2, \ldots \right\}$$

forms such a basis. These functions have the property that

$$\int_{-\tau/2}^{\tau/2} \cos(2p\pi t/\tau)\sin(2q\pi t/\tau)\,dt = 0 \quad (p, q = 0, 1, 2, \ldots) \tag{3.160}$$

$$\frac{2}{\tau}\int_{-\tau/2}^{\tau/2} \cos(2p\pi t/\tau)\cos(2q\pi t/\tau)\,dt = \begin{cases} 0 & (p \neq q) \\ 1 & (p = q) \end{cases} \quad (p, q = 0, 1, 2, \ldots) \tag{3.161}$$

$$\frac{2}{\tau}\int_{-\tau/2}^{\tau/2} \sin(2p\pi t/\tau)\sin(2q\pi t/\tau)\,dt = \begin{cases} 0 & (p \neq q) \\ 1 & (p = q) \end{cases} \quad (p, q = 0, 1, 2, \ldots) \tag{3.162}$$

The functions that comprise the set are thus mutually orthogonal in this sense (Chapters 9 and 10).

Let us express the excitation function $f(t)$ in terms of its Fourier series. Hence,

$$f(t) = f_0 + \sum_{p=1}^{\infty} f_p^{(c)} \cos\Omega_p t + \sum_{p=1}^{\infty} f_p^{(s)} \sin\Omega_p t \qquad (3.163)$$

where

$$\Omega_p = \frac{2p\pi}{\tau} \qquad (p = 1, 2, ...) \qquad (3.164)$$

and the associated Fourier components are given by

$$f_0 = \frac{1}{\tau} \int_{-\tau/2}^{\tau/2} f(t)\, dt \qquad (3.165)$$

which is seen to be the average value of the excitation $f(t)$ over a period, and

$$f_p^{(c)} = \frac{2}{\tau} \int_{-\tau/2}^{\tau/2} f(t) \cos\Omega_p t\, dt \qquad (p = 1, 2, ...) \qquad (3.166)$$

and

$$f_p^{(s)} = \frac{2}{\tau} \int_{-\tau/2}^{\tau/2} f(t) \sin\Omega_p t\, dt \qquad (p = 1, 2, ...) \qquad (3.167)$$

which are the Fourier coefficients (components) of the cosine and sine basis functions, respectively. These coefficients are thus the components of the given forcing function with respect to the harmonic basis functions. [The equations for the Fourier components arise from multiplying Eq. (3.163) by $\cos\Omega_p t$ or $\sin\Omega_p t$, then integrating over the period τ and incorporating Eqs. (3.160)–(3.162).] At this point we are reminded of certain characteristics regarding convergence of a Fourier series representation, particularly for functions that possess discontinuities. Notably, that the Fourier series of a discontinuous function converges to the average value of the function at points of discontinuity, and that the series does not converge uniformly. Rather, if the series for such a function is truncated, the partial sums exhibit Gibbs Phenomenon whereby overshoots (bounded oscillatory spikes) relative to the mean occur in the vicinity of the discontinuity. With the Fourier series of an arbitrary periodic excitation established, we now proceed to obtain the general form of the corresponding steady state response.

3.6.2 Steady State Response

We next substitute the Fourier series for the excitation into the equation of motion for the system. The equation of motion then takes the form

$$\ddot{x} + 2\omega\zeta\dot{x} + \omega^2 x = \omega^2 f_0 + \sum_{p=1}^{\infty} \omega^2 f_p^{(c)} \cos\Omega_p t + \sum_{p=1}^{\infty} \omega^2 f_p^{(s)} \sin\Omega_p t \quad (3.168)$$

We see that by incorporating the expansion defined by Eq. (3.163) in the right hand side of Eq. (3.1) we are, in effect, treating the system as if it is subjected to an equivalent system of harmonic forces and a constant force. Recall that the constant force is the average value of $f(t)$. From the superposition principle discussed in Section 3.2 we know that the response of the system to the system of forces is the sum of the responses to the individual forces applied separately. The steady state response to $F(t)$ therefore corresponds to the sum of the responses to the individual Fourier components of $F(t)$. The steady state response for the general periodic force may thus be found directly from Eqs. (3.53) and (3.54) for each Ω_p. Hence,

$$x_{ss}(t) = x_0 + \sum_{p=1}^{\infty} \left[x_p^{(c)}(t) + x_p^{(s)}(t) \right] \quad (3.169)$$

where

$$x_0 = f_0 \quad (3.170)$$

$$x_p^{(c)}(t) = f_p^{(c)} \Gamma_p \cos\left(\Omega_p t - \Phi_p\right) \quad (3.171)$$

$$x_p^{(s)}(t) = f_p^{(s)} \Gamma_p \sin\left(\Omega_p t - \Phi_p\right) \quad (3.172)$$

$$\Gamma_p \equiv \Gamma\left(\overline{\Omega}_p; \zeta\right) = \frac{1}{\sqrt{\left(1 - \overline{\Omega}_p^2\right)^2 + \left(2\zeta\overline{\Omega}_p\right)^2}} \quad (p = 1, 2, \ldots) \quad (3.173)$$

$$\Phi_p \equiv \Phi\left(\overline{\Omega}; \zeta\right) = \operatorname{Tan}^{-1}\left\{ \frac{2\zeta\overline{\Omega}_p}{1 - \overline{\Omega}_p^2} \right\} \quad (p = 1, 2, \ldots) \quad (3.174)$$

$$\overline{\Omega}_p \equiv \frac{\Omega_p}{\omega} \quad (p = 1, 2, \ldots) \quad (3.175)$$

For vanishing damping

$$x_p^{(c)} = \frac{f_p^{(c)}}{1 - \overline{\Omega}_p^2} \cos\Omega t \quad (3.176)$$

and

$$x_p^{(s)} = \frac{f_p^{(s)}}{1 - \overline{\Omega}_p^2} \sin\Omega_p t \quad (3.177)$$

as per Eqs. (3.27) and (3.28). Note that for the case of structural damping the damping factor will generally have a different value for each harmonic, Ω_p, of the excitation. That is, for the case of structural damping,

$$\zeta \to \zeta_p \equiv \zeta(\Omega_p) \quad (p = 1, 2, \ldots) \tag{3.178}$$

Thus, for practical purposes, one must have an extensive knowledge of the structural damping factor if this approach is to be used for systems with appreciable structural damping.

Example 3.18

A parent pushes a child on a swing whose chains are 6 ft long, as shown. The force history exerted by the parent is approximated by the sequence of periodically applied step pulses of 0.75 second duration applied every 3 secs as indicated. Determine the motion of the child if the magnitude of the applied force is $1/10^{th}$ the combined weight of the child and the seat. Assume that the system is free from damping and that the mass of the chain is negligible.

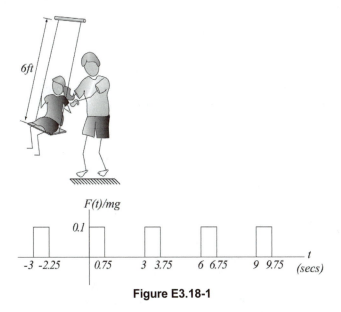

Figure E3.18-1

Solution

The equation of motion is derived by following the procedure for the simple pendulum of Section 2.1.4. The kinetic diagram for the system is shown in Figure E3.18-2, where we have assumed that the pushing force is applied tangent to the path. Writing Newton's Second Law along the tangential direction, rearranging terms and linearizing the resulting expression gives the equation of motion for the child on the swing as

$$\ddot{\theta} + \omega^2 \theta = \omega^2 f(t) \tag{a}$$

where

$$f(t) = \frac{F(t)}{mg} \tag{b}$$

and

$$\omega = \sqrt{\frac{g}{L}} = \sqrt{\frac{32.2}{6}} = 2.317 \text{ rad/sec} \tag{c}$$

Since the swing is pushed periodically we may express the applied force in terms of its Fourier Series. The frequencies of the Fourier basis functions are then computed as

$$\Omega_p = \frac{2\pi p}{\tau} = \frac{2\pi p}{3} \tag{d}$$

from which it follows that

$$\bar{\Omega}_p = \frac{2\pi p}{\omega \tau} = \frac{2\pi p}{(2.317)(3)} = 0.904 p \tag{e}$$

We next compute the Fourier components of the applied force. The force is described over the interval $-3 \le t \le 3$ seconds by

$$F(t) = \begin{cases} 0 & (-3 \le t < 0) \\ F_0 & (0 \le t < 0.75) \\ 0 & (0.75 \le t < 3) \end{cases} \tag{f}$$

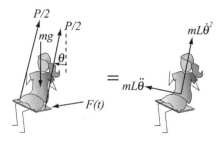

Figure E3.18-2 Kinetic Diagram.

Substituting Eq. (f) into Eqs. (3.165)–(3.167) and carrying through the integration gives

$$f_0 = \frac{1}{\tau} \int_{-\tau/2}^{\tau/2} f(t)\,dt = \frac{1}{3} \int_{-1.5}^{0} 0\,dt + \frac{1}{3} \int_{0}^{0.75} \frac{F_0}{mg}\,dt + \frac{1}{3} \int_{0.75}^{1.5} 0\,dt$$

$$= \frac{F_0/mg}{4} = \frac{(0.1)}{4} = 0.025 \tag{g}$$

(note that f_0 is the average value of f over a period),

$$f_p^{(c)} = \frac{2}{\tau} \int_{-\tau/2}^{\tau/2} f(t) \cos\Omega_p t\,dt$$

$$= \frac{2}{3} \int_{-1.5}^{0} 0\cdot\cos\Omega_p t\,dt + \frac{2}{3} \int_{0}^{0.75} \frac{F_0}{mg}\cos\Omega_p t\,dt + \frac{2}{3} \int_{0.75}^{1.5} 0\cdot\cos\Omega_p t\,dt \tag{h}$$

$$= 0.1\frac{\sin(p\pi/2)}{p\pi}$$

and

$$f_p^{(s)} = \frac{2}{\tau} \int_{-\tau/2}^{\tau/2} f(t) \sin\Omega_p t\,dt$$

$$= \frac{2}{3} \int_{-1.5}^{0} 0\cdot\sin\Omega_p t\,dt + \frac{2}{3} \int_{0}^{0.75} \frac{F_0}{mg}\sin\Omega_p t\,dt + \frac{2}{3} \int_{-1.5}^{0} 0\cdot\sin\Omega_p t\,dt \tag{i}$$

$$= 0.1\frac{[1-\cos(p\pi/2)]}{p\pi}$$

Substituting Eqs. (d)–(i) into Eqs. (3.169), (3.170), (3.176) and (3.177) gives the steady state response of the child and swing,

$$\theta_{ss}(t) = 0.025$$

$$+ \frac{0.1}{\pi} \sum_{p=1}^{\infty} \frac{1}{p(1-0.816p^2)}\left\{\sin(p\pi/2)\cos(2p\pi t/3) + [1-\cos(p\pi/2)]\sin(2p\pi t/3)\right\}$$

$$\text{radians} \triangleleft \text{(j)}$$

Example 3.19

A motor is configured to drive a system at the constant rate $\dot{F}_0$. A controller alters the sense of the applied force over specified time intervals of duration $\tilde{\imath}$ as indicated by the force history shown. Determine the response of the mass-spring-damper system.

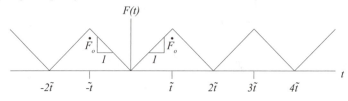

Solution

The force applied to the single degree of freedom system is clearly periodic with period

$$\tau = 2\tilde{\imath} \qquad (a)$$

and may be described over the interval $-\tilde{\imath} \le t \le \tilde{\imath}$ by the relations

$$F(t) = -\dot{F}_0\, t \quad (-\tilde{\imath} \le t \le 0)$$
$$F(t) = \dot{F}_0\, t \quad (0 \le t \le \tilde{\imath}) \qquad (b)$$

We may therefore represent the force imparted by the motor by its Fourier Series, where the frequencies of the Fourier basis functions are

$$\Omega_p = \frac{2p\pi}{\tau} = \frac{p\pi}{\tilde{\imath}} \qquad (c)$$

Hence,

$$\bar{\Omega}_p = \frac{p\pi}{\omega \tilde{\imath}} \qquad (d)$$

We next evaluate the Fourier components of the applied force. Substituting Eq. (b) into Eqs. (3.165)–(3.167) and carrying through the integration gives

$$f_0 = \frac{1}{\tau}\int_{-\tau/2}^{\tau/2} f(t)\, dt = -\frac{1}{2\tilde{\imath}}\int_{-\tilde{\imath}}^{0}\frac{\dot{F}_0\, t}{k}\, dt + \frac{1}{2\tilde{\imath}}\int_{0}^{\tilde{\imath}}\frac{\dot{F}_0}{k}\, t\, dt = \frac{1}{\tilde{\imath}}\int_{0}^{\tilde{\imath}}\frac{\dot{F}_0}{k}\, t\, dt = \frac{\dot{F}_0}{2k}\tilde{\imath} \qquad (e)$$

$$f_p^{(c)} = -\frac{2}{2\tilde{\imath}}\int_{-\tilde{\imath}}^{0}\frac{\dot{F}_0}{k}\, t\cos\Omega_p t\, dt + \frac{2}{2\tilde{\imath}}\int_{0}^{\tilde{\imath}}\frac{\dot{F}_0}{k}\, t\cos\Omega_p t\, dt = 0 \qquad (f)$$

$$f_p^{(s)} = -\frac{2}{2\tilde{t}} \int_{-\tilde{t}}^{0} \frac{\dot{F}_0}{k} t \sin \Omega_p t\, dt + \frac{2}{2\tilde{t}} \int_{0}^{\tilde{t}} \frac{\dot{F}_0}{k} t \sin \Omega_p t\, dt = \frac{2}{\tilde{t}} \int_{0}^{\tilde{t}} \frac{\dot{F}_0}{k} t \sin \Omega_p t\, dt$$

$$= -\frac{8\dot{F}_0\tilde{t}}{\pi pk}\cos p\pi = \frac{8\dot{F}_0\tilde{t}}{\pi pk}(-1)^{p+1}$$

(g)

With the Fourier components of the applied force calculated, we now determine the steady state response of the system by substituting Eqs. (d) – (g) into Eqs. (3.169) – (3.175). In doing so we obtain the steady state response of the system as

$$x_{ss}(t) = \frac{\dot{F}_0\tilde{t}}{2k}\left[1 + \frac{16}{\pi}\sum_{p=1}^{\infty}\frac{(-1)^{p+1}}{p}\Gamma_p \sin(p\pi t/\tilde{t})\right]$$

◁ (h)

where

$$\Gamma_p = \frac{1}{\sqrt{\left[1-\left(p\pi/\omega\tilde{t}\right)^2\right]^2 + \left[2\varsigma\left(p\pi/\omega\tilde{t}\right)\right]^2}}$$

◁ (i)

3.7 CONCLUDING REMARKS

In this chapter we have considered the motion of single degree of freedom systems that are subjected to periodic excitation, that is excitation that repeats itself at regular intervals over long periods of time. We began by studying systems which were subjected to external forces that vary harmonically in time. We also considered classes of applications corresponding to motion of the support, unbalanced motors and synchronous whirling of rotating shafts. In each case the steady state response was seen to be strongly influenced by the ratio of the excitation frequency to the natural frequency undamped motion, and the viscous damping factor. A notable feature is the phenomenon of resonance whereby the mass of the system undergoes large amplitude motion when the excitation frequency achieves a critical value. For undamped systems this occurs when the forcing frequency is equal to the excitation frequency. It was seen that at resonance the work of the external force is used in the optimal manner. For undamped systems the amplitude of the steady state response was seen to grow linearly with time. The related phenomenon of beating was seen to occur, for systems with vanishing damping, when the excitation frequency was very close to but not equal to the natural frequency. In this case the system is seen to oscillate at the average between the excitation and natural frequency, with the amplitude oscillating harmonically with very large period. For viscously damped systems the peak response was seen to occur at values of the excitation frequency away from the undamped natural frequency, and at lower frequencies for a force excited system, the damping slowing the system down. We remark that in some literature the term resonance is associated solely with undamped systems, and the term resonance frequency with the excitation frequency equal to the undamped natural frequency. We here use the terms to mean the peak/maximum response and the frequency at which it occurs. In all cases the amplitude of the steady state response is characterized by a magnification factor, or related expression, which measures the effect of the dynamics on the amplitude of the response. In addition, the steady state response lags the excitation with the phase lag dependent upon the same parameters as the magnification factor. The phasing was seen to be central to the description of the resonance phenomenon. The force transmitted to the support during excitation was also examined and a critical frequency ratio, beyond which the system is essentially isolated form the effects of vibration, was identified. A model for structural damping was presented, based on the mechanics of linear hereditary materials. The model was adapted to single degree of freedom systems and took the form of an effective viscous damping factor that is generally dependent on the frequency of the excitation. We finished by studying the motion of systems subjected to general periodic loading based on the principle of superposition and using Fourier Series. In the next chapter we shall study the response of single degree of freedom systems subjected to excitations that are of finite duration in time.

BIBLIOGRAPHY

Churchill, R.V. and Brown, J.W., *Fourier Series and Boundary Value Problems*, 3rd ed., McGraw-Hill, New York, 1978.

Fung, Y.C., *Foundations of Solid Mechanics*, Prentice Hall, Englewood Cliffs, 1965.

Malvern, L.E., *Introduction to the Mechanics of a Continuous Medium*, Prentice Hall, Englewood Cliffs, 1969.

Meirovitch, L., *Elements of Vibration Analysis*, 2nd ed., McGraw-Hill, New York, 1986.

Meirovitch, L., *Fundamentals of Vibrations*, McGraw-Hill, Boston, 2001.

Pinsky, M.A., *Introduction to Partial Differential Equations with Applications*, McGraw-Hill, New York, 1984.

Thomson, W.T., *Theory of Vibration with Applications*, 4th ed., Prentice-Hall, Englewood Cliffs, 1993.

PROBLEMS

3.1   The mass of the system of Problem 2.1 is subjected to the force $F(t) = 250 \cos 4t$ dynes, where t is in seconds. Determine the response of the system. Plot the response and identify the amplitude, period and lag time.

3.2   The load cell clamp attached to the end of the rod of Problem 2.6 is observed to oscillate with an amplitude of 0.1 cm when attached to a driving mechanism operating at the rate of 1 cps. Determine the magnitude of the force imparted by the mechanism.

3.3   The wheel of Example 2.3 is subjected to a harmonic torque of frequency 15 rad/sec and magnitude 10 lb-ft. Determine the amplitude of the steady state response of the system.

3.4   A 3000 pound cylindrical pontoon having a radius of 6 feet floats in a body of fluid. A driver exerts a harmonic force of magnitude 500 lb at a rate of 200 cycles per minute at the center of the upper surface of the float as indicated. (**a**) Determine the density of the fluid if the pontoon is observed to bob with an amplitude of 1 foot. (**b**) What is the magnitude of the bobbing motion of the pontoon when the excitation frequency is reduced to 5 rad/sec?

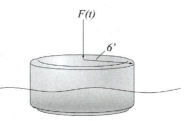

Fig. P3.4

3.5   A horizontally directed harmonic force $F(t) = F_0 \sin \Omega t$ acts on the bob of mass m of a simple pendulum of length L. Determine the steady state response of the pendulum.

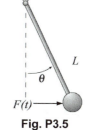

Fig. P3.5

3.6   The left wheel of a conveyor belt is locked into position when the motor is accidentally switched on, exerting a harmonic torque $M(t) = M_0 \sin \Omega t$ about the hub of the right wheel, as indicated. The mass and radius of the flywheel are m and R, respectively. If no slipping occurs between the belt and wheels the effective stiffness of each leg of the elastic belt may be represented as $k/2$, as shown. Determine the response of the system at resonance. What is the amplitude of the response at a time of 4 natural periods after the motor is switched on?

$k/2$

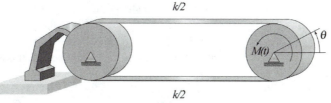

$k/2$

Fig. P3.6

3.7   An aircraft mock-up for ground testing is modeled as shown, where a brace represented as a torsional spring of stiffness k_T has been attached at the pivot support to restrain excessive yawing. It is assumed that during a test, both thrusters will exert a constant force of magnitude F_0. However, during a simulation, one of the thrusters deviates from the prescribed force by adding a perturbation in the form of a harmonic thrust of small amplitude. Thus, the force applied by this thruster takes the form $F(t) = F_0(1 + \varepsilon \sin \Omega t)$, as indicated, where $\varepsilon \ll 1$. Determine the steady state yawing motion of the mock-up. What is the amplitude of the response at resonance?

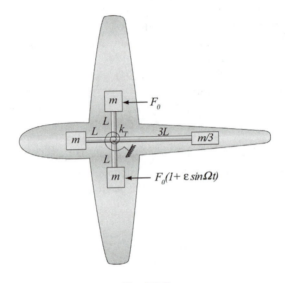

Fig. P3.7

3.8   The tip of the beam of Problem 1.4 is subjected to a sinusoidal force whose magnitude is 1% of the supported weight. Determine the motion of the supported mass if the excitation frequency is tuned to 98% of the natural frequency of the system. Plot and label the response.

3.9   The disk of Example 2.8 is driven by the torque $M(t) = 2mgr\cos(gt/R)$ (positive clockwise). Determine the small angle motion of the disk.

3.10   A horizontally directed harmonic force $F(t) = F_0 \sin \Omega t$ acts on the mass of the timing device of Problem 2.28. Determine the steady state response of the device.

3.11   Determine the torque transmitted to the support of the system of Problem 3.3.

3.12 The mount for the system of Problem 3.1 is to be replaced by a mount having the same stiffness as the original but possessing appreciable damping. If the system is to be driven by the same force during operation, determine the minimum value of the damping coefficient needed by the mount so that the force transmitted to the base never exceeds 150% of the applied force.

3.13 A viscoelastic beam is configured as in Problem 1.4 and is subjected to a harmonic edge load of magnitude 1000 lb. (**a**) Determine the deflection of the beam due to a static load of the same magnitude as the dynamic load. (**b**) When the frequency of the excitation approaches the natural frequency of an elastic beam possessing the same geometry and elastic modulus (Problem 3.8) the amplitude of the deflection of the supported weight is measured to be 12.5 inches. Determine the structural loss factor of the system for this excitation frequency.

3.14 The base of the inverted pendulum shown is attached to a cranking mechanism causing the base motion described by $x(t) = X_0(1 - \cos \Omega t)$. Determine the steady state motion of the pendulum if $k_T > mgL$.

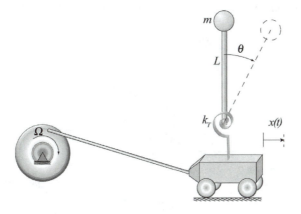

Fig. P3.14

3.15 Determine the range of excitation frequencies for which the amplitude of the (absolute) steady state motion of the roof of the structure of Example 3.13 will always be smaller than the amplitude of the ground motion.

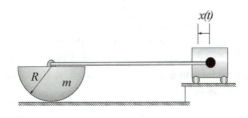

Fig. P3.16

3.16 An automatic slicer consists of an elastic rod of length L and axial stiffness EA attached to the hub of a semi-circular blade of radius R and mass m, as shown. If the base of the rod is controlled so that its horizontal motion is described by $x(t) = h_0 \sin \Omega t$, determine the motion of the blade during the cutting process. What force must be applied to the rod by the motor in order to produce this motion?

3.17 A structure identical to the building of Example 3.13, but with unknown damping factor, undergoes the same ground motion. If the magnitude of the steady state response is 2.75 inches, determine the structural loss factor for that excitation frequency.

3.18 The pitching motion of a certain vehicle is studied using the model shown. A controller is embedded in the system and can provide an effective moment $M(t)$ about the pivot, as indicated. If the base at the right undergoes the prescribed harmonic motion $y(t) = y_0 \sin \Omega t$ while the base at the left remains fixed, determine the moment that must be applied by the actuator to compensate for the base motion so that a passenger within the vehicle will experience a smooth ride?

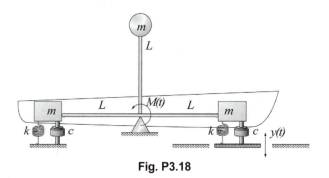

Fig. P3.18

3.19 A 150 pound boy stands at the edge of the diving board of Problem 2.30, preparing to execute a dive. During this time, he shifts his weight in a leaping motion, moving up and down at the rate of 1 cps. If, at the apogee of each bob his feet just touch the board so that they are nearly losing contact, determine the steady state motion of the boy.

3.20 A controller is attached to the pendulum of Problem 3.5. If a sensor measures the amplitude of the motion of the pendulum driven by the applied force to be X_1, determine the moment that must be exerted by the actuator to maintain the quiescence of the system.

3.21 Suppose the block sitting at the center of the simply supported beam of Problem 1.5 corresponds to a motor whose rotor spins at a constant rate. If the beam is observed to deflect with an amplitude of 3 inches (about the static deflection) when the motor operates at 50% of the critical rate of rotation determine the offset moment of the rotor.

Fig. P3.21

3.22 A flywheel of radius R and mass m is supported by an elastic rod of length L and bending stiffness EI, as shown. The wheel possesses an offset moment $m_e\ell$ and is connected to the rest of the system by a fan belt, as indicated. If the flywheel rotates such that the belt moves with constant speed v_0, determine the vertical motion of the wheel. What is the critical speed of the belt? (Assume that no slip of the belt occurs and that its stiffness is negligible.)

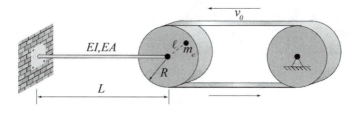

Fig. P3.22/P3.23

3.23 Determine the horizontal motion of the flywheel of Problem 3.22, and the corresponding critical speed of the fanbelt, if the cross-sectional area of the rod is $\mathcal{A}$.

3.24 Several bolts on the propeller of a fanboat detach, resulting in an offset moment of 5 lb-ft. Determine the amplitude of bobbing of the boat when the fan rotates at 200 rpm, if the total weight of the boat and passengers is 1000 lbs and the wet area projection is approximately 30 sq ft. What is the amplitude at 1000 rpm?

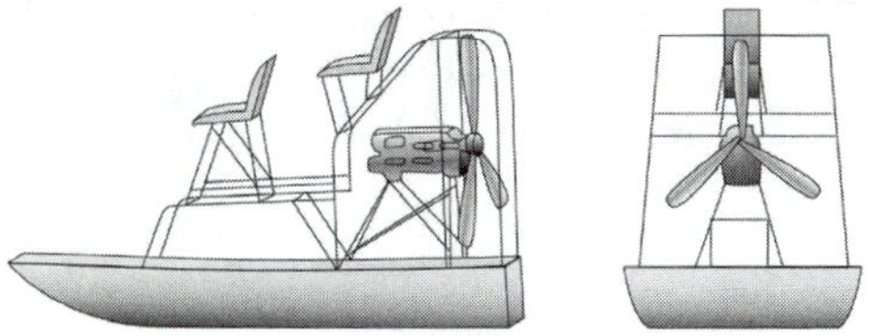

Fig. P3.24

3.25 The damping factor for a shaft-turbine system is measured to be 0.15. When the turbine rotates at a rate equal to 120% of the undamped natural frequency of the shaft, the system is observed to whirl with an amplitude equal to the radius of the shaft. What will be the amplitude of whirling when the rotation rate of the turbine is reduced to 80% of the undamped natural frequency of the shaft if the radius of the shaft is 1 inch?

3.26 Determine the amplitude of whirling motion of the system of Example 3.17 if the supports permit in-plane rotation as well as spin.

3.27 Determine the response of the system of Problem 3.1 when it is subjected to the sawtooth force history shown in Figure P3.27.

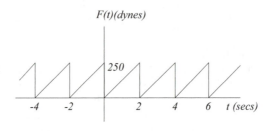

Fig. P3.27

3.28 Determine the response of the right wheel of the conveyor belt of Problem 3.6 if the left wheel is locked and the alternating step torque shown in Figure P3.28 is applied to the right wheel by the motor.

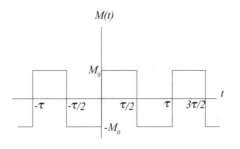

Fig. P3.28

3.29 Determine the response of the structure of Example 3.13 if the base motion is described by the stuttering sinusoid shown in Figure P3.29. Neglect the effects of damping.

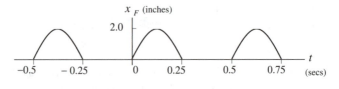

Fig. P3.29

3.30 Determine the response of the vehicle of Problem 3.18 when the right base undergoes the step motion shown in Figure P3.30 and damping is negligible.

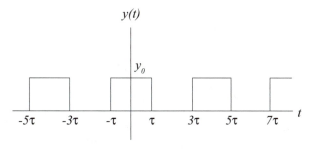

Fig. P3.30

3.31 A vehicle of mass m, with suspension of stiffness k and negligible damping, is traveling at constant speed v_0 when it encounters the series of equally spaced semi-circular speed bumps shown. Assuming that no slip occurs between the wheel and the road, determine (**a**) the motion of the engine block while it is traveling over this part of the roadway and (**b**) the force transmitted to the block.

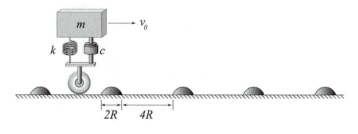

Fig. P3.31

4

Forced Vibration of Single Degree of Freedom Systems – 2: *Nonperiodic Excitation*

To this point we have considered the response of systems to excitations that are continuously applied over a very long period of time. Moreover, we restricted our attention to excitations that vary periodically in time. Though such excitations are clearly important in the study of vibrations, engineering systems may obviously be subjected to other forms of excitation. A system may be at rest when it is first excited by a periodic or other continuously varying force. Alternatively, an engineering system may be subjected to a short duration load, a *pulse*, which can clearly affect its response and performance. Examples of such *transient excitation* include the start-up of an automobile engine, the behavior of a machine before it achieves steady state operation, the loads on an aircraft during take-off or landing, or the impact of a foreign object on a structure. In addition, a system may be subjected to loading that does not fall into any particular category. In this chapter we shall examine the behavior of single degree of freedom systems that are subjected to transient loading and to the general case of arbitrary excitation.

4.1 TWO GENERALIZED FUNCTIONS

To facilitate our study of the response of single degree of freedom systems to transient and general excitation it is expedient to introduce two *generalized functions*, the *Dirac Delta Function* and the *Heaviside Step Function*. The qualification "generalized" is employed because these mathematical entities do not follow the traditional definitions and rules of ordinary functions. Rather, they fall into the category of "distributions."

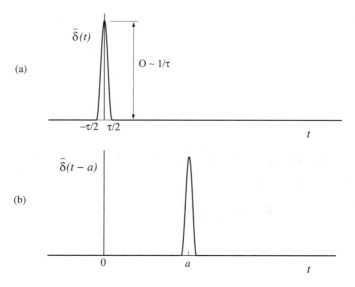

Figure 4.1 The Dirac Delta Function (unit impulse).

4.1.1 The Dirac Delta Function (Unit Impulse)

Consider a function, $\widehat{\delta}(t)$, which is of the form shown in Figure 4.1a. The function in question is effectively a spike acting over the very small time interval $-\tau/2 \leq t \leq \tau/2$, and the maximum height of the function is of the order of the inverse of the duration, as indicated. The specific form of the function is not specified. However, its integral is. The generalized function just described is referred to as the *Dirac Delta Function*, also known as the *unit impulse function*, and is defined in terms of its integral as follows:

$$\lim_{\tau \to 0} \int_{-\tau/2}^{\tau/2} \widehat{\delta}(t)\, dt = 1 \tag{4.1}$$

The integral of the unit impulse is dimensionless, hence the dimension of the unit impulse corresponds to that of the inverse of the independent variable. Thus, for the case under discussion, the Dirac Delta Function has units of $1/t$. It may be seen that the Delta Function has the property that

$$\int_{-\infty}^{t} \widehat{\delta}(t)\, dt = \begin{cases} 0 & (t < 0) \\ 1 & (t > 0) \end{cases} \tag{4.2}$$

Likewise,

$$\int_{t}^{\infty} \widehat{\delta}(t)\, dt = \begin{cases} 1 & (t < 0) \\ 0 & (t > 0) \end{cases} \tag{4.3}$$

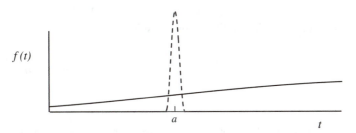

Figure 4.2 Generic function $f(t)$, with unit impulse in background.

If we introduce a time shift so that the delta function acts in the vicinity of $t = a$, as indicated in Figure 4.1b, we see that the unit impulse has the following properties:

$$\lim_{\tau \to 0} \int_{a-\tau/2}^{a+\tau/2} \widehat{\delta}(t-a)\, dt = 1 \tag{4.4}$$

$$\int_{-\infty}^{t} \widehat{\delta}(t-a)\, dt = \begin{cases} 0 & (t < a) \\ 1 & (t > a) \end{cases} \tag{4.5}$$

$$\int_{t}^{\infty} \widehat{\delta}(t-a)\, dt = \begin{cases} 1 & (t < a) \\ 0 & (t > a) \end{cases} \tag{4.6}$$

Another useful property of the Dirac Delta Function is found when we consider the integral of its product with some regular well behaved function, $f(t)$. (See Figure 4.2.) Piecewise evaluation of that integral takes the form

$$\int_{-\infty}^{\infty} f(t)\,\widehat{\delta}(t-a)\, dt = \lim_{\tau \to 0} \int_{-\infty}^{a-\tau/2} f(t)\,\widehat{\delta}(t-a)\, dt$$
$$+ \lim_{\tau \to 0} \int_{a-\tau/2}^{a+\tau/2} f(t)\,\widehat{\delta}(t-a)\, dt + \lim_{\tau \to 0} \int_{a+\tau/2}^{\infty} f(t)\,\widehat{\delta}(t-a)\, dt$$

It may be seen from Eqs. (4.5) and (4.6) that the first and third integrals on the right-hand side of the above expression vanish. Further, over the infinitesimal time interval bounded by the limits of integration for the second integral on the right-hand side, the function $f(t)$ effectively maintains the constant value $f(a)$ and can be taken out of the corresponding integrand. Application of Eq. (4.4) on the resulting expression then gives the identity

$$\int_{-\infty}^{\infty} f(t)\,\widehat{\delta}(t-a)\, dt = f(a) \tag{4.7}$$

We next define the derivative of the unit impulse.

As for the delta function itself, the derivative of the Dirac Delta Function is defined through its integral. To motivate this, consider the integral

$$\int_{-\infty}^{\infty} f(t)\,\dot{\hat{\delta}}(t-a)\,dt = \left[f(t)\hat{\delta}(t-a) \right]_{-\infty}^{+\infty} - \int_{-\infty}^{\infty} \dot{f}(t)\,\hat{\delta}(t-a)\,dt$$

where we have assumed that we may perform integration by parts with the Delta function as we do for regular functions. We may argue that the term in brackets vanishes at the limits and thus that

$$\int_{-\infty}^{\infty} f(t)\,\dot{\hat{\delta}}(t-a)\,dt = -\int_{-\infty}^{\infty} \dot{f}(t)\,\hat{\delta}(t-a)\,dt = -\dot{f}(a) \qquad (4.8)$$

Regardless, we take Eq. (4.8) as the definition of the derivative of the Dirac Delta Function. The above procedure may be extended to give the jth derivative of the delta function as

$$\int_{-\infty}^{\infty} f(t)\frac{d^{(j)}}{dt^{(j)}}\hat{\delta}(t-a)\,dt = (-1)^j \int_{-\infty}^{\infty} \frac{d^{(j)}f(t)}{dt^{(j)}}\hat{\delta}(t-a)\,dt = (-1)^j \left.\frac{d^{(j)}f}{dt^{(j)}}\right|_{t=a} \qquad (4.9)$$

We next introduce a (generalized) companion function to the Dirac Delta Function.

4.1.2 The Heaviside Step Function (Unit Step)

Consider a function $\mathcal{H}(t)$ of the form shown in Figure 4.3a. Such a function can represent, say, a sudden start-up of a system to a constant operating level if the times of interest are large relative to the start-up time. The function in question is seen to be discontinuous at the origin, jumping from zero for $t < 0$ to unity for $t > 0$. This generalized function is referred to as the *Heaviside Step Function*, also known as the *unit step function*. It is formally defined as follows,

$$\mathcal{H}(t) = \begin{cases} 0 & (t < 0) \\ 1 & (t > 0) \end{cases} \qquad (4.10)$$

The unit step function is dimensionless. If we introduce a time shift so that the jump occurs at $t = a$ (Figure 4.3b), we have that

$$\mathcal{H}(t-a) = \begin{cases} 0 & (t < a) \\ 1 & (t > a) \end{cases} \qquad (4.11)$$

The Heaviside Step Function is inherently related the Dirac Delta Function. This relation is discussed next.

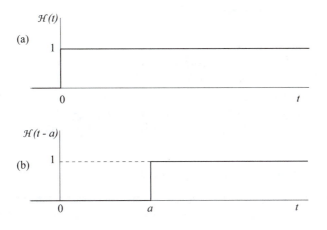

Figure 4.3 The Heaviside Step Function (unit step).

4.1.3 Relation Between the Unit Step and the Unit Impulse

When Eqs. (4.2) and (4.10) are compared, it is evident that

$$\int_{-\infty}^{t} \widehat{\delta}(t)\,dt = \mathcal{H}(t) \tag{4.12}$$

Similarly, comparison of Eqs. (4.2) and (4.12) gives the relation

$$\int_{-\infty}^{t} \widehat{\delta}(t-a)\,dt = \mathcal{H}(t-a) \tag{4.13}$$

Since the unit step and unit impulse functions are related in this way, the derivatives of the Heaviside Step Function may be defined as the inverse operation. Equations (4.12) and (4.13) therefore suggest the definition of the *derivative* of the unit step function as

$$\frac{d}{dt}\mathcal{H}(t) = \widehat{\delta}(t) \tag{4.14}$$

$$\frac{d}{dt}\mathcal{H}(t-a) = \widehat{\delta}(t-a) \tag{4.15}$$

Both the Dirac Delta Function and the Heaviside Step Function will prove to be very useful in describing the loading and response for various types of transient excitation, as well as for the development of the response for arbitrary excitation.

4.2 IMPULSE RESPONSE

Engineering systems are often subjected to forces of very large magnitude that act over very short periods of time. Examples include forces produced by impact, explosions or shock. In this section we examine the response of single degree of freedom systems to such loading. It will be seen that the response to impulse loading provides a fundamental solution from which the response to more general loading types may be based. Toward these ends, we first classify forces into two fundamental types.

4.2.1 Impulsive and Nonimpulsive Forces

Time dependent forces may be classified as impulsive or nonimpulsive. Impulsive forces are those that act over very short periods of time but possess very large magnitudes, such as the forces associated with explosions and impact. Nonimpulsive forces are those that are well behaved over time, such as the gravitational force, the elastic spring force and the viscous damping force. Impulsive and nonimpulsive forces are defined formally in what follows. With the qualitative descriptions of these two types of forces established we now proceed to formally define impulsive and nonimpulsive forces mathematically.

Impulsive Forces

Forces that act over very short periods of time, such as those due to explosions or impact, may be difficult to measure directly or to quantify mathematically. However, their impulses can be measured and quantified. In this light, an impulsive force is defined as a force that imparts a finite (nonvanishing) impulse over an infinitesimal time interval. Formally, an impulsive force $F(t)$ is a force for which

$$\lim_{\Delta t \to 0} \int_0^{\Delta t} F(t)\,dt \to \mathcal{I} \neq 0 \qquad (4.16)$$

where $\mathcal{I}$ is the impulse imparted by the impulsive force F. Consequently, and in light of Eqs. (4.7) and (4.16), an impulsive force may be expressed in the form

$$F(t) = \mathcal{I}\,\widehat{\delta}(t) \qquad (4.17)$$

where $\widehat{\delta}(t)$ is the Dirac Delta Function discussed Section 4.1.

Nonimpulsive Forces

Nonimpulsive forces, such as spring forces, damping forces and the gravitational force, are forces that are well behaved over time. Since such forces are finite, the impulse they produce over infinitesimal time intervals is vanishingly small. Formally, a nonimpulsive force is a force $F(t)$ for which

$$\lim_{\Delta t \to 0} \int_0^{\Delta t} F(t)\, dt \to 0 \tag{4.18}$$

With the mathematical description of impulsive and nonimpulsive forces established, we next determine the general response of a standard single degree of freedom system to an arbitrary impulsive force.

4.2.2 Response to an Applied Impulse

Consider a mass-spring-damper system that is initially at rest when it is subjected to an impulsive force $F(t)$, as shown in Figure 4.4. Let us next apply the principle of linear impulse-momentum, Eq. (1.96), to the initially quiescent system over the time interval when the impulsive force defined by Eq. (4.17) is applied. Hence,

$$\int_{0^-}^{0^+} \mathcal{I}\,\hat{\delta}(t)\, dt - \int_{0^-}^{0^+} \left[kx + c\dot{x} \right] dt = m \left[\dot{x}(0^+) - \dot{x}(0^-) \right] \tag{4.19}$$

where m, k and c correspond to the mass, spring constant and damping coefficient of the system, respectively. Since the spring force and damping force are nonimpulsive forces, the second integral on the left-hand side of Eq. (4.19) vanishes. Since the system is initially at rest, the corresponding initial velocity is zero as well. The impulse-momentum balance, Eq. (4.19), then reduces to the relation

$$\mathcal{I} = m\dot{x}(0^+) \tag{4.20}$$

If we consider times after the impulsive force has acted ($t > 0$) then, using Eq. (4.20) to define the initial velocity, the problem of interest becomes equivalent to the problem of free vibrations with the initial conditions

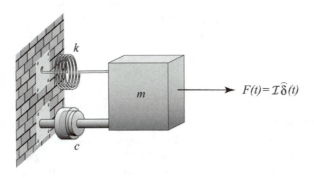

Figure 4.4 Mass-spring-damper system subjected to an impulsive force.

$$x_0 \equiv x(0) = 0 \quad \text{and} \quad v_0 \equiv \dot{x}(0) = \mathcal{I}/m \tag{4.21}$$

Recognizing this, the response can be written directly by incorporating the initial conditions stated in Eq. (4.21) into the solution given by Eqs. (2.73), (2.76), (2.77) and (2.10). The motion of an underdamped system is then described as follows:

$$x(t) = 0 \quad (t < 0)$$

$$x(t) = \frac{\mathcal{I}}{m\omega_d} e^{-\zeta\omega t} \sin \omega_d t \quad (t > 0)$$

Let us next incorporate Eq. (4.10) into the above solution. The response of the under-damped system to the impulse $\mathcal{I}$ can then be restated in the compact form

$$x(t) = \mathcal{I}\,\mathcal{G}(t) \tag{4.22}$$

where

$$\mathcal{G}(t) \equiv \frac{1}{m\omega_d} e^{-\zeta\omega t} \sin \omega_d t\, \mathcal{H}(t) \tag{4.23}$$

is the *unit impulse response* (the response of a single degree of freedom system to an impulse of unit magnitude applied at $t = 0$). In Eq. (4.23), $\mathcal{H}(t)$ is the Heaviside Step Function and ω, ζ and ω_d are given by Eqs. (2.2), (2.65) and (2.70) respectively.

The response to an impulse applied at $t = \tau$ is found by incorporating a time shift into Eq. (4.22). The motion of the system for this case is then

$$x(t) = \mathcal{I}\,\mathcal{G}(t - \tau) \tag{4.24}$$

The response of a system to an impulse is evidently of fundamental, as well as of practical, importance in its own right. In addition, the response of engineering systems to impulse loading is fundamental to their response to loading of any type. This relation is discussed in the next section.

Example 4.1

A system consisting of 4 kg mass, a spring of stiffness 400 N/m and a damper of coefficient 16 N-sec/m is initially at rest when it is struck by a hammer. (**a**) If the hammer imparts an impulse of magnitude 2 N-sec, determine the motion of the system. (**b**) Determine the response if the hammer again strikes the mass with the same impulse 10 seconds later. (**c**) Determine the response of the system if instead, the second strike occurs 6.413 seconds after the first.

Solution

The natural damping factor, natural frequency and natural period are easily computed to be (see Example 2.11)

$$\omega = \sqrt{400/4} = 10 \text{ rad/sec}, \quad \zeta = 16/[2(10)(4)] = 0.2,$$
$$\omega_d = 10\sqrt{1-(0.2)^2} = 9.798 \text{ rad/sec}, \quad T_d = 2\pi/9.798 = 0.6413 \text{ secs}$$

(a)

(a)

The unit step response for the system is computed from Eq. (4.23) giving

$$\mathcal{G}(t) = \frac{1}{(4)(9.798)} e^{-(0.2)(10)t} \sin(9.798t)\,\mathcal{H}(t)$$
$$= 0.02552 e^{-2t} \sin(9.798t)\,\mathcal{H}(t) \text{ (m/N-sec)}$$

(b)

Substitution of the given impulse and Eq. (b) into Eq. (4.22) gives the response of the system as

$$x(t) = (2)\left[0.02552 e^{-2t} \sin(9.798t)\,\mathcal{H}(t)\right]$$
$$= 0.05104 e^{-2t} \sin(9.798t)\,\mathcal{H}(t) \text{ (meters)}$$

(c)

Hence,

$$x(t) = 0 \quad (t < 0)$$
$$x(t) = 0.05104 e^{-2t} \sin(9.798t) \quad (t > 0)$$

◁ (d)

(b)

The response of the system to the two impulses is the sum of the response to each impulse applied individually. Hence,

$$x(t) = 0.05104\left[e^{-2t} \sin(9.798t)\,\mathcal{H}(t)\right.$$
$$\left. + e^{-2(t-10)} \sin\{9.798(t-10)\}\,\mathcal{H}(t-10)\right] \text{ (m)}$$

(e)

Thus,

$$x(t) = 0 \text{ (m)} \quad (t < 0)$$
$$x(t) = 0.05104 e^{-2t} \sin(9.798t) \text{ (m)} \quad (0 < t < 10 \text{ secs})$$
$$x(t) = 0.05104\left[e^{-2t} \sin(9.798t) + e^{-2(t-10)} \sin\{9.798(t-10)\}\right] \text{ (m)} \quad (t > 10 \text{ secs})$$

◁ (f)

(*c*)

Paralleling the computation of Eq. (d) gives

$$x(t) = 0.05104\left[e^{-2t} \sin(9.798t)\,\mathcal{H}(t) \right.$$
$$\left. + e^{-2(t-6.413)} \sin\left\{9.798(t-6.413)\right\} \mathcal{H}(t-6.413) \right] \text{(m)} \tag{g}$$

The response of the system before the second impulse is applied $(0 < t < 6.413)$ follows as

$$x(t) = 0 \text{ (m)} \quad (t < 0)$$
$$\lhd \text{ (h-i,ii)}$$
$$x(t) = 0.05104e^{-2t} \sin(9.798t) \text{ (m)} \quad (0 < t < 6.413 \text{ secs)}$$

However, for later times, Equation (g) can be simplified by noting that the application of the second pulse occurs at precisely ten natural periods after the first pulse is imparted. Thus, for these times,

$$x(t) = 0.05104\left[e^{-2t}\left(1 + e^{12.83}\right)\right] \sin 9.798t$$
$$\lhd \text{ (h-iii)}$$
$$= 1.905 \times 10^4 e^{-2t} \sin(9.798t) \text{ (m)} \quad (t > 6.413 \text{ secs)}$$

Example 4.2

A tethered 1 pound ball hangs in the vertical plane when it is tapped with a racket. Following the tap the ball is observed to exhibit oscillatory motion of amplitude 0.2 radians with a period of 2 seconds. Determine the impulse imparted by the racket.

Solution

From Eq. (4.22), the path of the ball as a function of time takes the form

$$s(t) = L\theta(t) = \mathcal{I}\,\mathcal{G}(t) \tag{a}$$

The ball and tether are evidently equivalent to a pendulum of length L. The natural frequency for small angle motion of the ball is then, from Eq. (2.42),

$$\omega = \sqrt{g/L} \tag{b}$$

Substituting the frequency into Eq. (a) gives the small angle response of the ball as

$$\theta(t) = \frac{\mathcal{I}}{m\omega L}\sin\left(\sqrt{g/L}\,t\right)\mathcal{H}(t) \tag{c}$$

where $\mathcal{I}$ is the unknown impulse imparted by the racket and L is the unknown length of the tether. The length can be determined from the period of the observed motion as follows. The natural period for small angle motion of a pendulum, and hence of the tethered ball, is from Eq. (2.47) or Eq. (b)

$$T = \frac{2\pi}{\sqrt{g/L}} \tag{d}$$

The length of the cord may now be computed using Eq. (d), giving

$$L = \frac{gT^2}{4\pi^2} = \frac{32.2(2)^2}{4\pi^2} = 3.263 \text{ ft} \tag{d}$$

Next, it follows from Eq. (c) that the amplitude of the observed motion is related to the magnitude of the applied impulse by

$$\|\theta\| = \mathcal{I}/m\omega L \tag{e}$$

Solving Eq. (e) for $\mathcal{I}$ and substituting Eq. (d) and the observed amplitude gives the impact imparted by the racket on the ball as

$$\mathcal{I} = \|\theta\| Lm\omega = (0.2)(3.263)(1/32.2)(2\pi/2) = 0.06367 \text{ lb-sec} \qquad \triangleleft (f)$$

4.3 RESPONSE TO ARBITRARY EXCITATION

In this section we develop the response of an initially quiescent single degree of freedom system to a force of arbitrary form, based on the response to impulse loading. Toward this end, consider the generic time dependent force, $F(t)$, shown in Figure 4.5a. Let us further consider the impulse of the force over the particular time interval $\tau \le t \le \tau + d\tau$, as indicated. The impulse, $d\mathcal{I}$, imparted by the force on the mass during this particular differential time interval is thus given by

$$d\mathcal{I} = F(\tau)\,d\tau \tag{4.25}$$

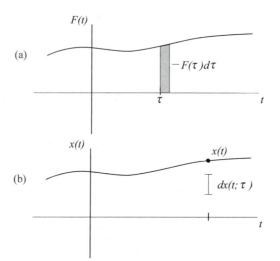

Figure 4.5 Generic time histories of force and corresponding system response.

The increment in the response of the system at some later time t to the particular increment of the impulse acting during the interval $\tau \leq t \leq \tau + d\tau$, as depicted in Figure 4.5b, is obtained using Eq. (4.24) and takes the form

$$dx(t;\tau) = d\mathcal{I}\, G(t-\tau) = F(\tau)G(t-\tau)\,d\tau \tag{4.26}$$

The response at time t to the impulses imparted by the force F during all times $\tau < t$ is obtained by superposing the responses of each of the corresponding impulses. Hence, "summing" all such increments gives the general form of the total response of the system as

$$x(t) = \int_0^t F(\tau)G(t-\tau)\,d\tau \tag{4.27}$$

Equation (4.27) is referred to as the *convolution integral* and it gives the response of an initially quiescent system to any force history $F(t)$.

An alternate form of the convolution integral is obtained by introducing the coordinate shift $\bar{\tau} = t - \tau$ into Eq. (4.27). This transforms the corresponding integral to the form

$$x(t) = -\int_t^0 F(t-\bar{\tau})G(\bar{\tau})\,d\bar{\tau}$$

which, after reversing the direction of integration, gives the response as

$$x(t) = \int_0^t F(t-\bar{\tau})G(\bar{\tau})\,d\bar{\tau} \tag{4.28}$$

Either convolution integral, Eq. (4.27) or Eq. (4.28), may be used to obtain the response of an initially quiescent system to any force $F(t)$. In the next two sections we employ the convolution integral to obtain the response of single degree of freedom systems subjected to a number of fundamental loading types. The response to other types of loading are then obtained, using these results, in Section 4.6. Finally, in Section 4.7, the convolution integral is used to assess the maximum peak response for systems subjected to transient loading.

4.4 RESPONSE TO STEP LOADING

We next consider the situation where a system is loaded by a force $F(t)$ that is applied very rapidly to a certain level F_0 and is then maintained at that level thereafter (Figure 4.6). If the rise time is small compared with the times of interest then the loading may be treated as a step function. Such loading is referred to as *step loading* and may be represented mathematically in the form

$$F(t) = F_0 \mathcal{H}(t) \tag{4.29}$$

where $\mathcal{H}(t)$ is the unit step function. The response is easily evaluated by substituting Eq. (4.29) into the convolution integral defined by Eq. (4.27). Hence, for an underdamped system,

$$x(t) = \int_0^t F_0 \mathcal{H}(t) G(t-\tau) d\tau = \frac{F_0}{m\omega_d} \int_0^t e^{-\zeta\omega t} \sin\omega_d(t-\tau) d\tau \, \mathcal{H}(t) \tag{4.30}$$

Performing the required integration gives the *step response*

$$x(t) = f_0 \, S(t) \tag{4.31}$$

where

$$S(t) \equiv \left[1 - \frac{e^{-\zeta\omega t}}{\sqrt{1-\zeta^2}} \cos(\omega_d t - \phi) \right] \mathcal{H}(t) \tag{4.32}$$

is the *unit step response,*

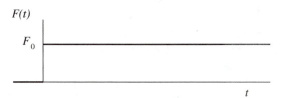

$F(t)$

F_0

t

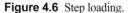

Figure 4.6 Step loading.

$$f_0 = F_0/k \qquad (4.33)$$

for a mass-spring-damper system,

$$\omega_d = \omega\sqrt{1-\zeta^2} \qquad (4.34)$$

and

$$\phi = \tan^{-1}\left\{\frac{\zeta}{\sqrt{1-\zeta^2}}\right\} \qquad (4.35)$$

Example 4.3

The beam and mass system of Example 3.4 is initially at rest when a 5 kg block is suddenly placed on the 10 kg mass of the system. Determine the motion of the beam and block. Plot the time history of the response.

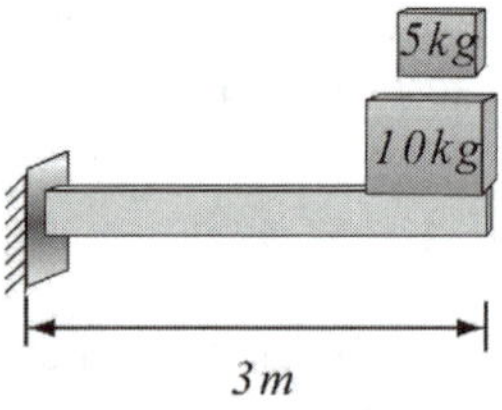

3 m

Figure E4.3-1

Solution
From Example 3.4, $m = 10$ kg and $\omega = 6\pi$ rad/sec. Hence,

$$k = (6\pi)^2(10) = 360\pi^2 \text{ N/m} \qquad (a)$$

We shall take the time at which the block is placed on the structure as $t = 0$. Now, the magnitude of the applied force is simply the weight of the block. Further, since the block is placed suddenly, the time dependence of the applied force may be taken as a step function. Thus,

$$F(t) = F_0 \mathcal{H}(t) = -mg\mathcal{H}(t) = -(5)(9.81)\mathcal{H}(t) = -49.1\mathcal{H}(t) \text{ N} \qquad (b)$$

It follows that

$$f_0 = \frac{F_0}{k} = \frac{-49.1}{360\pi^2} = -0.0138 \text{ m} \qquad (c)$$

Substitution of the Eq. (c) and the known natural frequency into Eq. (4.31), with $\zeta = 0$, gives the motion of the system as

$$x(t) = -0.0138[1 - \cos 6\pi t]\,\mathcal{H}(t) \text{ m}$$
$$= -1.38[1 - \cos 6\pi t]\,\mathcal{H}(t) \text{ cm} \qquad \triangleleft (d)$$

Recalling that the deflection $x(t)$ is measured from the equilibrium configuration of the 10 kg mass, it may be seen that the system oscillates about the equilibrium configuration of the combined mass system. A plot of the response is shown in Figure E4.3-2.

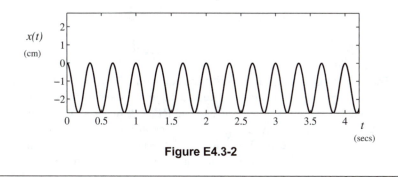

Figure E4.3-2

Example 4.4

A package of mass m is being shipped in the (rigid) crate shown, where k and c represent the stiffness and damping coefficient of the packing material. If the system is initially at rest, determine the response of the package if the crate is suddenly displaced a distance h_0 to the right, (**a**) when damping is negligible and (**b**) when the system is underdamped.

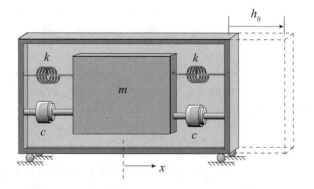

Figure E4.4-1 Package inside crate.

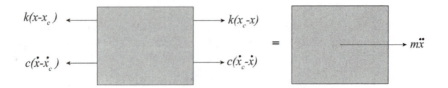

Figure E4.4-2 Kinetic diagram of package.

Solution

The first thing we must do is derive the equation of motion for the system. To accomplish this we first draw the kinetic diagram for the package. Let $x(t)$ correspond to the motion of the package and let $x_C(t)$ represent the motion of the crate. To establish a convention, let us assume that the package displaces to the right relative to the crate. When this is so the left spring is in tension and the right spring is in compression, as indicated. Similarly, if the relative velocity of the package with respect to the crate is assumed positive (i.e., to the right) both the left and right dashpots act in the direction indicated.

We next apply Newton's Second Law, based on the forces shown in the kinetic diagram (dynamic free-body diagram). Doing this gives

$$k\left(x_C - x\right) + c\left(\dot{x}_C - \dot{x}\right) - k\left(x - x_C\right) - c\left(\dot{x} - \dot{x}_C\right) = m\ddot{x}$$

which upon rearranging takes the form

$$m\ddot{x} + 2c\dot{x} + 2kx = 2c\dot{x}_C(t) + 2kx_C(t) \tag{a}$$

Note that we have brought the expressions for the displacement and velocity of the crate to the right-hand side of the equation since the excitation enters the problem through the *prescribed* motion of the crate. Equation (a) is valid for any prescribed motion of the crate. Let us next consider the specific excitation under consideration.

Since the crate is *suddenly* moved a given distance, the displacement of the crate is readily expressed as a function of time with the aid of the Heaviside Step Function. Hence,

$$x_C(t) = h_0 \mathcal{H}(t) \tag{b}$$

The corresponding velocity of the crate is then found by direct application of Eq. (4.14) to the expression for the displacement just established. This gives

$$\dot{x}_C(t) = h_0 \hat{\delta}(t) \tag{c}$$

Substitution of Eqs. (b) and (c) into Eq. (a) gives the explicit form of the equation of motion for the problem at hand. Thus,

$$m\ddot{x} + 2c\dot{x} + 2kx = 2c\,h_0\,\hat{\delta}(t) + 2k\,h_0\mathcal{H}(t) \tag{d}$$

It is useful to put the above equation of motion in standard form. Doing so gives

$$\ddot{x} + 2\omega\zeta\,\dot{x} + \omega^2 x = \omega^2\,\underbrace{\hat{\mathcal{I}}\hat{\delta}(t)}_{f_1(t)} + \omega^2\,\underbrace{h_0\mathcal{H}(t)}_{f_2(t)} \tag{e}$$

where

$$\hat{\mathcal{I}} = \mathcal{I}/k = h_0\,c/k \tag{f}$$

$$\omega^2 = 2k/m \quad \text{and} \quad \zeta = c/\omega m \tag{g,h}$$

(a)
For negligible damping, $c = \zeta = 0$ and the governing equation, Eq. (e), reduces to the form

$$\ddot{x} + \omega^2 x = \omega^2\,\underbrace{h_0\mathcal{H}(t)}_{f(t)} \tag{i}$$

The response is obtained directly from Eq. (4.31) as

$$x(t) = h_0\left[1 - \cos\left(\sqrt{2k/m}\ t\right)\right]\mathcal{H}(t) \tag{j}$$

and is sketched in Figure E4.4-3. It is seen that the package oscillates about the new equilibrium configuration of the crate.

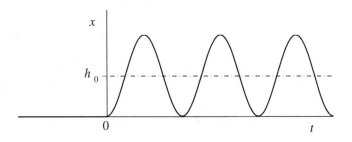

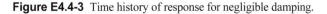

Figure E4.4-3 Time history of response for negligible damping.

(**b**)
When damping is taken into account, the response of the package to the combined effects of the damping and stiffness forces, f_1 and f_2 of Eq. (e), may be obtained using superposition. In this context, the response of the package is simply the sum of the responses to the two forces acting individually. Direct application of Eqs. (4.22) and (4.31) then gives the response of the package as

$$x(t) = \frac{h_0 c}{m \omega_d} e^{-\zeta \omega t} \sin \omega_d t \, \mathcal{H}(t) + h_0 \left[1 - \frac{e^{-\zeta \omega t}}{\sqrt{1 - \zeta^2}} \cos(\omega_d t - \phi) \right] \mathcal{H}(t) \qquad \lhd \text{(k)}$$

or

$$x(t) = h_0 \mathcal{H}(t) + h_0 \frac{e^{-\zeta \omega t}}{\sqrt{1 - \zeta^2}} \left[\zeta \sin \omega_d t - \cos(\omega_d t - \phi) \right] \mathcal{H}(t) \qquad \lhd \text{(l)}$$

where ω_d and ϕ are given by Eqs. (4.34) and (4.35), respectively. It may be seen from Eq. (l) that the package oscillates about the new equilibrium position of the crate, with the amplitude of the oscillation decaying exponentially with time.

4.5 RESPONSE TO RAMP LOADING

We next consider an initially quiescent single degree of freedom system that is subsequently excited by a force that increases linearly with time. If we consider the load to be activated at time $t = 0$, then the excitation is such that the magnitude of the load is zero for $t < 0$ and increases linearly with time for $t > 0$, as depicted in Figure 4.7. Such loading is referred to as *ramp loading*, and may be expressed mathematically as

$$F(t) = \dot{F} t \, \mathcal{H}(t) \tag{4.36}$$

where $\dot{F}$ is the (constant) rate at which the loading is applied to the system. It may be noted that the ramp function with unit loading rate ($\dot{F} = 1$) is simply the integral of the unit step function and is a generalized function. It follows that the derivative of the unit ramp function is the unit step function.

The response to ramp loading may be determined by direct substitution of the forcing function defined by Eq. (4.36) into the convolution integral, Eq. (4.27). Hence,

$$x(t) = \frac{1}{m \omega_d} \int_0^t \dot{F} t \left[e^{-\zeta \omega(t - \tau)} \sin \omega_d (t - \tau) \right] d\tau \, \mathcal{H}(t)$$

Carrying out the integration gives the response to the ramp loading as

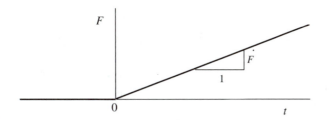

Figure 4.7 Ramp loading.

$$x(t) = \dot{f} \mathcal{R}(t) \tag{4.37}$$

where

$$\dot{f} = \dot{F}/k \tag{4.38}$$

for a mass-spring-damper system, and

$$\mathcal{R}(t) \equiv \left\{ t - \frac{2\zeta}{\omega} + e^{-\zeta\omega t} \left[\frac{2\zeta}{\omega} \cos \omega_d t - \frac{\left(1 - 2\zeta^2\right)}{\omega_d} \sin \omega_d t \right] \right\} \mathcal{H}(t) \tag{4.39}$$

is the *unit ramp response*. The interpretation of the parameter $\dot{f}$ in Eq. (4.37) is adjusted accordingly for systems other than mass-spring-damper systems.

Example 4.5

The man and the raft of Example 2.4 are at rest when a long rope is lowered to the raft. If the man guides the rope so that it forms a coil whose weight increases at the rate of 0.5 lb/sec, determine the vertical motion of the raft during this process.

Figure E4.5-1 Rope lowered to man on raft.

Solution

From Example 2.4, $k_{eq} = 749$ lb/ft, $\zeta = 0$ and the natural frequency of the un-damped system is $\omega = 11.0$ rad/sec. It is evident, from the given loading, that the problem at hand corresponds to ramp loading of the man and raft, with $\dot{F} = 0.5$ lb/sec. The applied force may be expressed mathematically as

$$F(t) = 0.5t\,\mathcal{H}(t) \tag{a}$$

where $t = 0$ represents the instant that the rope first comes into contact with the man and raft. It follows that

$$\dot{f} = \frac{0.5}{749} = 6.68 \times 10^{-4} \text{ ft/sec} \tag{b}$$

Substituting Eq. (b), $\zeta = 0$ and $\omega = 11.0$ rad/sec into Eqs. (4.37) and (4.39) gives the motion of the raft during the loading process as

$$x(t) = 6.68 \times 10^{-4} \left[t - \tfrac{1}{11}\sin(11t) \right] \mathcal{H}(t) = 6.07 \times 10^{-5} \left[11t - \sin(11t) \right] \mathcal{H}(t) \text{ ft} \quad \triangleleft (c)$$

The time history of the response is displayed in Figure E4.5-2.

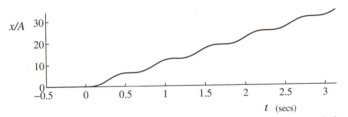

Figure E4.5-2 Time history of vertical motion of man and raft as rope is lowered ($A = 6.07 \times 10^{-5}$ ft).

4.6 TRANSIENT RESPONSE BY SUPERPOSITION

Many transient loads and pulses may be constructed from combinations of the basic step and ramp functions, as well as other functions, by way of superposition. The corresponding response is then simply the sum of the responses to the participating functions. To demonstrate this approach, we consider the response of an equivalent mass-spring-damper system to two basic forcing functions, the rectangular pulse and the linear transition to constant load.

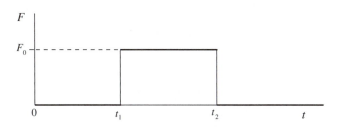

Figure 4.8 The rectangular pulse.

4.6.1 The Rectangular Pulse

Consider a rectangular pulse of magnitude F_0 applied to a single degree of freedom system over the time interval $t_1 < t < t_2$ as depicted in Figure 4.8. The pulse can be expressed as the sum of two step loads, as indicated. The pulse may thus be represented mathematically in the form

$$F(t) = F_0 \left[\mathcal{H}(t - t_1) - \mathcal{H}(t - t_2) \right] \tag{4.40}$$

From the superposition principle established in Section 3.1, the response of the system to this pulse is then given by the sum of the responses to the individual step loadings. Hence, upon application of Eq. (4.31), the motion of the system is given by

$$x(t) = f_0 \left[S(t - t_1) - S(t - t_2) \right] \tag{4.41}$$

When expanded, Eq. (4.41) takes the form

$$x(t) = f_0 \left\{ \left[1 - \frac{e^{-\zeta\omega(t-t_1)}}{\sqrt{1-\zeta^2}} \cos\left\{ \omega_d (t - t_1) - \phi \right\} \right] \mathcal{H}(t - t_1) \right.$$
$$\left. - \left[1 - \frac{e^{-\zeta\omega(t-t_2)}}{\sqrt{1-\zeta^2}} \cos\left\{ \omega_d (t - t_2) - \phi \right\} \right] \mathcal{H}(t - t_2) \right\} \tag{4.42}$$

Thus,

$$x(t) = 0 \quad (t < t_1)$$

$$x(t) = f_0 \left[1 - \frac{e^{-\zeta\omega(t-t_1)}}{\sqrt{1-\zeta^2}} \cos\{\omega_d(t-t_1) - \phi\} \right] \quad (t_1 < t < t_2)$$

$$x(t) = f_0 \left[\frac{e^{-\zeta\omega(t-t_2)}}{\sqrt{1-\zeta^2}} \cos\{\omega_d(t-t_2) - \phi\} \right. \tag{4.43}$$

$$\left. - \frac{e^{-\zeta\omega(t-t_1)}}{\sqrt{1-\zeta^2}} \cos\{\omega_d(t-t_1) - \phi\} \right] \quad (t > t_2)$$

where f_0 and ϕ are given by Eqs. (4.33) and (4.35), respectively.

Example 4.6

Determine the motion of the system of Example 4.3, (**a**) if the 5 kg block is knocked off the beam 10.5 seconds after it has been placed on the beam, (**b**) if the 5 kg block is knocked off the beam 10 seconds after it has been placed on the beam.

Solution
(**a**)
For this situation, the force may be expressed as the step loading

$$F(t) = -49.1 \left[\mathcal{H}(t) - \mathcal{H}(t-10.5) \right] \text{ N} \tag{a}$$

The response then follows by direct application of Eqs. (4.41)–(4.43), for an undamped system. We thus have that

$$x(t) = -1.38 \left\{ \left[1 - \cos(6\pi t) \right] \mathcal{H}(t) - \left[1 - \cos\{6\pi(t-10.5)\} \right] \mathcal{H}(t-10.5) \right\} \text{ cm} \tag{b}$$

Expanding equation (b) gives the motion of the system as

$$x(t) = 0 \quad (t < 0)$$
$$x(t) = -1.38 \left[1 - \cos(6\pi t) \right] \text{ cm} \quad (0 \le t < 10.5 \text{ secs}) \qquad \triangleleft \text{(c)}$$
$$x(t) = 1.38 \left[\cos(6\pi t) - \cos\{6\pi(t-10.5)\} \right] \text{ cm} \quad (t \ge 10.5 \text{ secs})$$

It may be seen that the motion of the system between the time the block is placed on the beam and when it is knocked off is identical to that predicted in Example 4.3, as it should be. It is also seen that, after the 5 kg block is knocked off the beam, the system oscillates about the original equilibrium configuration, $x = 0$. The time history of the motion of the beam is displayed in Figure E4.6-1.

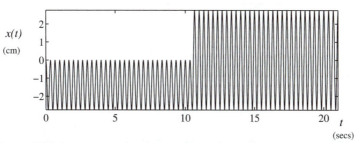

Figure E4.6-1 Time history of the motion of the beam-mass system when the 5 kg mass is removed 10.5 seconds after it is placed.

(b)

The forcing function for this case may be expressed in the form

$$F(t) = -49.1\left[\mathcal{H}(t) - \mathcal{H}(t-10)\right] \text{ N} \tag{d}$$

Proceeding as for the previous case yields the response

$$x(t) = -1.38\left\{\left[1 - \cos(6\pi t)\right]\mathcal{H}(t) - \left[1 - \cos\{6\pi(t-10)\}\right]\mathcal{H}(t-10)\right\} \text{ cm} \tag{e}$$

Expanding Eq. (e) gives the explicit form

$$x(t) = 0 \quad (t < 0)$$
$$x(t) = -1.38\left[1 - \cos(6\pi t)\right] \text{ cm} \quad (0 \le t < 10 \text{ secs}) \qquad \triangleleft \text{(f)}$$
$$x(t) = 1.38\left[\cos\{6\pi(t-10)\} - \cos(6\pi t)\right] = 0 \text{ cm} \quad (t \ge 10 \text{ secs})$$

The time history of the motion is displayed in Figure E4.6-2. Note that, for this case, the system comes to rest after the 5 kg block is removed. This is because the amplitude of the oscillation to that point corresponds to the static deflection due to the 5 kg block, and because the time between the instant the block is placed on the beam and the time it is knocked off the beam is precisely an integer multiple of the natural period of the system. The block is therefore removed at the exact instant that the system passes through the original equilibrium position and is about to reverse direction. At this instant the position and velocity are both zero. This situation is equivalent to a free vibration problem in which the initial conditions vanish identically. The system thus remains in the equilibrium configuration, $x = 0$, for all later times.

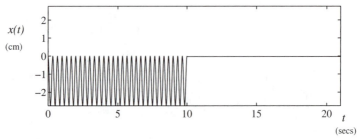

Figure E4.6-2 Time history of the motion of the beam-mass system when the 5 kg mass is removed 10.0 seconds after it is placed.

Example 4.7

A vehicle travels at constant speed over a flat road when it encounters a bump of rise h_0 and length L, having the shape shown. Determine the response of the vehicle to the disturbance in the road if the shape may be approximated as harmonic and, assuming that the horizontal speed of the vehicle is maintained, that the wheel rolls over the bump without leaving the surface of the road. Damping is negligible.

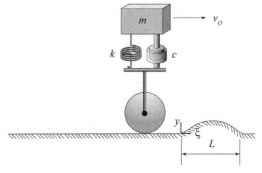

Solution

The equation for the path of the vehicle when it traverses the bump is given, in terms of the indicated coordinates, as

$$y(\xi) = h_0 \left[1 - \cos \frac{2\pi\xi}{L} \right] \qquad (0 < \xi < L) \qquad \text{(a)}$$

where h_0 is the rise of the bump and L is its length. Since the vehicle travels at constant speed, the horizontal position of the vehicle may be expressed as a function of time as

$$\xi = v_0 t \qquad \qquad \text{(b)}$$

where v_0 is the speed and ξ is measured from the start of the bump, as indicated. Substitution of Eq. (b) into Eq. (a) gives the vertical motion of the wheel as a function of time. Hence,

$$
\begin{aligned}
y(t) &= 0 & (t < 0) \\
y(t) &= h_0 \left[1 - \cos \Omega t\right] & (0 < t < t_L) \\
y(t) &= 0 & (t > t_L)
\end{aligned}
$$

or

$$
\begin{aligned}
y(t) &= h_0 \left[1 - \cos \Omega t\right] \mathcal{H}(t) - h_0 \left[1 - \cos \Omega (t - t_L)\right] \mathcal{H}(t - t_L) \\
&= h_0 \left[1 - \cos \Omega t\right] \left[\mathcal{H}(t) - \mathcal{H}(t - t_L)\right]
\end{aligned}
\tag{c}
$$

where

$$
\Omega = 2\pi / t_L \tag{d}
$$

and

$$
t_L = L / v_0 \tag{e}
$$

is the time that it takes for the vehicle to traverse the bump. The problem is thus equivalent to that of a single degree of freedom system subjected to prescribed motion of its support. The equation of motion is then given by Eq. (3.122) which, for vanishing damping and support motion given by Eq. (c), takes the form

$$
\ddot{x} + \omega^2 x = \omega^2 y
$$

or

$$
\ddot{x} + \omega^2 x = \omega^2 h_0 \left[\mathcal{H}(t) - \mathcal{H}(t - t_L)\right] - \omega^2 h_0 \cos \Omega t \left[\mathcal{H}(t) - \mathcal{H}(t - t_L)\right] \tag{f}
$$

It may be seen from Eqs. (c) and (f) that the system is excited by the superposition of a rectangular pulse and a cosine pulse. We shall solve Eq. (f) two different ways.

The response to the rectangular pulse is found directly from Eq. (4.42), while the response to the cosine pulse can be evaluated using the convolution integral, Eq. (4.27). Thus,

$$
\begin{aligned}
x(t) &= h_0 \left[1 - \cos \omega t\right] \mathcal{H}(t) - h_0 \left[1 - \cos \omega (t - t_L)\right] \mathcal{H}(t - t_L) \\
&\quad - k h_0 \int_0^t \cos \Omega \tau \, \frac{1}{m\omega} \sin \omega (t - \tau) d\tau \, \mathcal{H}(t) \\
&\quad + k h_0 \int_0^t \cos \Omega \tau \, \frac{1}{m\omega} \sin \omega (t - t_L - \tau) d\tau \, \mathcal{H}(t - t_L)
\end{aligned}
$$

which, after carrying out the integration and using trigonometric identities and lengthy algebraic operations, takes the form

$$x(t) = h_0 \left[1 + (\Omega/\omega)^2 \beta \cos \omega t - \beta \cos \Omega t \right] \mathcal{H}(t)$$
$$- h_0 \left[1 + (\Omega/\omega)^2 \beta \cos \omega (t - t_L) - \beta \cos \Omega (t - t_L) \right] \mathcal{H}(t - t_L) \qquad \triangleleft \text{(g)}$$

where

$$\beta = \frac{1}{1 - (\Omega/\omega)^2} = \frac{1}{1 - (2\pi/\omega t_L)^2} \qquad \text{(h)}$$

    Though the use of the convolution integral to obtain the response is, in principle, straightforward it required a great deal of algebraic and trigonometric manipulation for the current loading type. [The reader is invited to evaluate the corresponding convolution integrals to obtain the solution (g).] An alternative, and perhaps simpler, approach to solving Eq. (f) is to use the solutions pertaining to harmonic excitation developed in Chapter 3. To do this, we shall first obtain the response to the first forcing on the right hand side of Eq. (c). The solution to the second forcing function will then be the same as the first but with a time shift (t replaced by $t - t_L$). The actual response will be obtained by superposing the two solutions.

    The general solution to Eq. (f) is given by the sum of the complementary and particular solutions. Hence, with aid of Eqs. (2.7) and (3.27), the general form of the response $x^{(1)}(t)$ to the first forcing function $f^{(1)}(t) = h_0 \left[1 - \cos \Omega t \right] \mathcal{H}(t)$ is given by

$$x^{(1)}(t) = x_c^{(1)} + x_p^{(1)} = A_1 \cos \omega t + A_2 \sin \omega t + h_0 \left[1 - \beta \cos \Omega t \right] \qquad \text{(i)}$$

Imposing the quiescent initial conditions $x(0) = \dot{x}(0) = 0$ gives

$$A_1 = h_0 (\Omega/\omega)^2 \beta, \qquad A_2 = 0$$

Hence,

$$x^{(1)}(t) = h_0 \left[1 + (\Omega/\omega)^2 \beta \cos \omega t - \beta \cos \Omega t \right] \mathcal{H}(t) \qquad \text{(j)}$$

The solution $x^{(2)}(t)$ associated with the second forcing function, $f^{(2)}(t) = -h_0 \left[1 - \cos \Omega (t - t_L) \right] \mathcal{H}(t - t_L)$, is then simply given by

$$x^{(2)}(t) = -h_0 \left[1 + (\Omega/\omega)^2 \beta \cos \omega (t - t_L) - \beta \cos \Omega (t - t_L) \right] \mathcal{H}(t - t_L) \quad \text{(k)}$$

The total response to the bump is then obtained by superposition. Hence, adding Eqs. (j) and (k) gives

$$x(t) = x^{(1)}(t) + x^{(2)}(t)$$

$$= h_0 \left[1 + (\Omega/\omega)^2 \, \beta \cos \omega t - \beta \cos \Omega t \right] \mathcal{H}(t) \tag{l}$$

$$-h_0 \left[1 + (\Omega/\omega)^2 \, \beta \cos \omega (t - t_L) - \beta \cos \Omega (t - t_L) \right] \mathcal{H}(t - t_L)$$

which is identical to Eq. (g) obtained using the convolution integral. The explicit form of the response is then

$$x(t) = 0 \quad (t < 0)$$

$$x(t) = h_0 \left[1 + (\Omega/\omega)^2 \, \beta \cos \omega t - \beta \cos \Omega t \right] \quad (0 < t < t_L)$$

$$x(t) = h_0 \beta \left[(\Omega/\omega)^2 \left\{ \cos \omega t - \cos \omega (t - t_L) \right\} \right. \tag{m}$$

$$\left. - \left\{ \cos \Omega t - \cos \Omega (t - t_L) \right\} \right] \quad (t > t_L)$$

4.6.2 Linear Transition to a Constant Load Level

We next consider loading that consists of a linear transition over time τ to a constant load level, F_0, as depicted in Figure 4.9. For this case, the loading function may be constructed as the sum of two ramp functions, as shown in the figure. Thus,

$$F(t) = \frac{F_0}{\tau} t \, \mathcal{H}(t) - \frac{F_0}{\tau} (t - \tau) \, \mathcal{H}(t - \tau) \tag{4.44}$$

The response of a single degree of freedom system to this loading is then the sum of the responses to the individual ramp functions. Hence,

$$x(t) = \frac{f_0}{\tau} \left[\mathcal{R}(t) - \mathcal{R}(t - \tau) \right] \tag{4.45}$$

where $\mathcal{R}(t)$ is given by Eq. (4.39) and, for a mass-spring-damper system,

$$f_0 = F_0/k \tag{4.46}$$

Expanding Eq. (4.45) gives the explicit form of the response of the system as

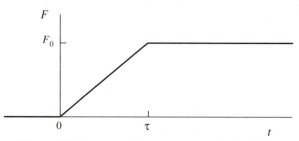

Figure 4.9 Linear transition to a constant level of force.

$$x(t) = 0 \quad (t < 0)$$

$$x(t) = \frac{f_0}{\tau}\left\{t - \frac{2\zeta}{\omega} + e^{-\zeta\omega t}\left[\frac{2\zeta}{\omega}\cos\omega_d t - \frac{\left(1-2\zeta^2\right)}{\omega_d}\sin\omega_d t\right]\right\} \quad (0 < t < \tau)$$

$$x(t) = f_0 + f_0\frac{2\zeta}{\omega\tau}\left[e^{-\zeta\omega t}\cos\omega_d t - e^{-\zeta\omega(t-\tau)}\cos\omega_d(t-\tau)\right]$$

$$\qquad - f_0\frac{\left(1-2\zeta^2\right)}{\omega_d\tau}\left[e^{-\zeta\omega t}\sin\omega_d t - e^{-\zeta\omega(t-\tau)}\sin\omega_d(t-\tau)\right] \quad (t > \tau)$$

(4.47)

Example 4.8

Determine the motion of the raft of Problem 4.5 if the loading of the rope is completed 20 seconds after the first of it touches down.

Solution

For the present problem $\tau = 20$ seconds and, from Eq. (b) of Example 4.5,

$$f_0/\tau = \dot{f} = 6.68 \times 10^{-4} \text{ ft/sec}^2 \qquad (a)$$

Substituting these values into Eqs. (4.45) and (4.47) gives the response

$$x(t) = 6.07 \times 10^{-5}\left\{\left[11t - \sin(11t)\right]\mathcal{H}(t)\right.$$
$$\left. - \left[11(t-20) - \sin 11(t-20)\right]\mathcal{H}(t-20)\right\} \text{ ft} \qquad (b)$$

It is seen from Eq. (b) that during the 20 second drop of the rope, the motion of the raft is identical to that described by Eq. (c) of Example 4.5. After the drop is completed, that is for $t \geq 20$ seconds, the vertical motion of the raft is described by

$$x(t) = 0.0134 + 6.07 \times 10^{-5}\left[\sin 11(t-20) - \sin(11t)\right] \text{ ft} \quad (t \geq 20 \text{ secs}) \qquad \triangleleft (c)$$

Note that the static deflection due to the weight of the entire rope is

$$\Delta_{static} = \frac{W_{rope}}{k_{eff}} = \frac{\dot{F}\tau}{k_{eff}} = \dot{f}\tau = (6.68\times10^{-4})(20) = 0.0134 \text{ ft} \qquad \text{(d)}$$

Since $x(t)$ is measured from the equilibrium position of the man and raft it may be seen from Eq. (c) that, after the rope has completed it's drop, the system oscillates about the new equilibrium position of the combined raft, man and rope system.

In this section we have demonstrated the use of superposition to obtain the response of linear systems to pulses that can be constructed by adding together other pulses for which we already know, or can easily obtain, the response to. This technique was used to obtain the response of single degree of freedom systems to two sample pulse forms: the rectangular pulse and the linear transition to a constant load level. It should be emphasized that this technique is not restricted to these sample pulse forms, but rather can be applied to many other pulse forms as well. In the next section we study a procedure to characterize and compare the response of systems to various forms of short duration pulses, and the sensitivities of different systems to a given pulse.

4.7 SHOCK SPECTRA

Engineering systems are often subjected to loads possessing large magnitudes acting over short periods of time. Excitations for which the duration is on the order of, or shorter than, the natural period of the system upon which they act are referred to as *shocks*. As shocks may have detrimental effects on engineering systems, compromising their effectiveness or causing significant damage, the sensitivity of a system to shocks must generally be taken into account in its design. It is therefore of interest to characterize the response of a given system to shocks of various forms, and also to compare the sensitivities of different systems (or the same system with different values of the system parameters) to a given type of shock. This is generally accomplished by determining and interpreting the *shock spectrum* for a given type of pulse and system. This is the subject of the present section.

In the present section we consider the response of single degree of freedom systems to shock, and establish a measure for comparison. To do this we must first identify the parameters that characterize a shock and the parameters that describe the system of interest. In this context, a shock of a given type may be distinguished by its magnitude and duration. Alternatively, the shock may be characterized by its impulse. Since the effects of damping generally accrue over time, and since damping tends to retard motion, damping is typically neglected when considering the severity of the response of systems to shock. The system may thus be defined by its natural frequency for vanishing damping, or equivalently by the corresponding natural period. The severity of the response of a system, and hence the sensitivity of that system, to a given shock may be characterized by the magnitude of the maximum deflection of the mass. Evaluations can then be made by comparing the *maximum peak response* for different values of the shock and system parameters.

For the purposes of comparison, a natural timescale (normalized time) is obtained by dividing the time by the natural period of the system in question. Similarly, a natural length scale (normalized response) is obtained by dividing the dynamic deflection of the mass by the characteristic static deflection (the ratio of the magnitude of the applied force to the stiffness of the system). The sensitivity of various systems to a given type of shock may then be characterized by plotting the normalized *maximum peak response* (the *maximax*) of the system as a function of the normalized duration of the pulse. In this way a "universal" plot is created. Such a plot is referred to as the *Shock Spectrum* or *Shock Response Spectrum*.

In general, the *maximum peak response* for an initially quiescent undamped single degree of freedom system subjected to a shock, F, of duration t^*, may be obtained using the convolution integral of Section 4.3. When this is done the maximax may generally be found from the convolution integral

$$\|x\|_{\max} = \frac{1}{m\omega}\left\|\int_0^t F(\tau)\sin\omega(t-\tau)d\tau\right\|_{\max} \tag{4.48}$$

The plot of $\|x\|_{\max}/f_0$ vs. $t^*/T = \omega t^*/2\pi$ is the *Shock Response Spectrum* for pulses of the form of F. It should be noted that the peak response as defined above is not unique to any particular shock, but rather may be the same for various cases.

We shall next generate the shock spectra for two sample shock types; those in the form of a rectangular pulse, and those in the form of a half-sine pulse. The spectra for other pulse types may be found in a similar manner.

Example 4.9 – Shock Spectrum for a Rectangular Pulse

Consider a single degree of freedom system subjected to a rectangular pulse of magnitude F_0 and duration t^*, as shown. Determine the shock spectrum for this type of pulse.

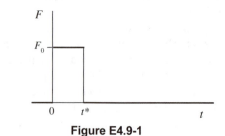

Figure E4.9-1

Solution

The response of a single degree of freedom system to a rectangular pulse was established in Section 4.6.1. If we neglect the effects of damping, the response of the system to the rectangular pulse is obtained from Eq. (4.43) as

$$x(t) = 0 \quad (t < 0) \tag{a}$$

$$x(t) = f_0 \left[1 - \cos \omega t\right] \quad (0 < t \le t^*) \tag{b}$$

$$x(t) = f_0 \left[\cos \omega(t - t^*) - \cos \omega t\right] \quad (t > t^*) \tag{c}$$

where $\omega = 2\pi/T$ is the natural frequency of the system and T is the corresponding natural period. The peak response may occur either (i) during the time interval that the applied pulse is active $(0 < t < t^*)$, or (ii) after the pulse subsides $(t > t^*)$.

(i) $0 < t < t^*$: *Initial Spectrum*

The maximum response of the system, while the applied pulse is active is determined by consideration of Eq. (b). For short duration pulses $(t^* < 0.5T)$ it is seen from Eq. (b) that

$$\frac{x_{max}}{f_0} = 1 - \cos \omega t^* = 2\sin^2(\pi t^*/T) \quad (t^* < 0.5T) \qquad \lhd \text{(d-1)}$$

For long duration pulses $(t^* \ge 0.5T)$, it is seen that

$$\frac{\|x\|_{max}}{f_0} = 2 \quad (t^* > 0.5T) \qquad \lhd \text{(d-2)}$$

It may be seen from Eqs. (d-1) and (d-2) that the initial shock spectrum for the rectangular pulse increases monotonically as the duration of the pulse increases, and reaches a plateau when the duration of the pulse is at least half the natural period of the system (Figure E4.9-2).

(ii) $t > t^*$: *Residual Spectrum*

We next consider the maximum response achieved by the system after the pulse subsides. To find the time $(t = t_{pk})$ at which the peak response occurs for this case, and hence to find the peak response when $t > t^*$, we take the derivative of Eq. (c), set it to zero and solve for the time. Hence,

$$\left. \frac{dx}{dt} \right|_{t=t_{pk}} = 0 = -f_0 \omega \left[\sin \omega (t - t^*) - \sin \omega t \right]_{t=t_{pk}} \tag{e}$$

With the aid of the identity $sin(a) - sin(b) = 2\ cos\{(a+b)/2\}\ sin\{(a-b)/2\}$, Eq. (e) takes the form

$$2 \cos \left\{ \tfrac{1}{2} \omega (2t_{pk} - t^*) \right\} \cos \left\{ \tfrac{1}{2} \omega t^* \right\} = 0 \tag{f}$$

Hence,

$$\omega t_{pk} = \frac{\omega t^*}{2} + (2n - 1) \frac{\pi}{2} \tag{g}$$

Evaluating Eq. (c) at $t = t_{pk}$ and substituting Eq. (g) into the resulting expression gives the maximum response

$$\frac{\left\| x \right\|_{max}}{f_0} = \left\| 2 \sin \left(\frac{\pi t^*}{T} \right) \right\| \qquad \triangleleft \text{(h)}$$

A plot of Eq. (h) (the residual shock spectrum) is displayed in Figure 4.9-2 along with the initial shock spectrum described by Eq. (d). It may be seen that the maximum displacements are larger and hence the residual spectrum dominates for short duration pulses while the initial spectrum dominates for large duration pulses. It may also be seen that the most sensitive systems are those for which the duration of the pulse is at least as large as half the natural period of the system (i.e., those for which $t^*/T \geq 0.5$). The complete shock spectrum is indicated by solid lines in Figure E4.9-2.

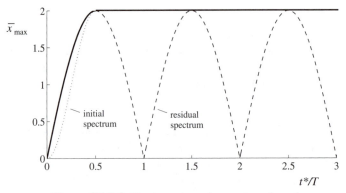

Figure E4.9-2 Shock spectrum for a rectangular pulse.

Example 4.10 – Response Spectrum for Rectangular Support Motion

Determine the shock spectrum for a single degree of freedom system when the time history of the motion of the support is in the form of a rectangular pulse of amplitude h_0 and duration t^*.

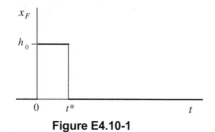

Figure E4.10-1

Solution

The motion of the support shown in Figure E4.10-1 may be expressed in the form

$$x_F(t) = h_0 \left[\mathcal{H}(t) - \mathcal{H}(t - t^*) \right] \tag{a}$$

Substitution of Eq. (a) into Eq. (3.122) gives the equation governing absolute motion as

$$\ddot{x} + \omega^2 x = \omega^2 h_0 \left[\mathcal{H}(t) - \mathcal{H}(t - t^*) \right] \tag{b}$$

Thus, the present problem is identical in form with that of Example 4.9, with f_0 replaced by h_0. The absolute motion of the mass then follows from Eq. (4.43) or Eqs. (a) – (c) of Example 4.9 with f_0 replaced by h_0. For the present problem, the extension/compression of the equivalent elastic spring is described by the motion of the mass relative to the support,

$$u(t) = x(t) - x_F(t) \tag{c}$$

Hence,

$$u(t) = x(t) = 0 \quad (t < 0) \tag{d}$$

$$u(t) = -h_0 \cos \omega t \quad (0 < t \le t^*) \tag{e}$$

$$u(t) = x(t) = -h_0 \left[\cos \omega (t - t^*) - \cos \omega t \right] \quad (t > t^*) \tag{f}$$

We see from Eq. (e) that the maximum relative deflection of the mass while the pulse is active is

$$\frac{\left\| u \right\|_{\max}}{h_0} = 1 \qquad \triangleleft \text{(g)}$$

Equation (g) describes the initial shock spectrum. Since the deflection of the support vanishes for $t > t^*$ the relative motion and absolute motion are the same during this interval. The residual spectrum is thus described by Eq. (h) of Example 4.9 with f_0 replaced by h_0. Hence,

$$\frac{\left\| u \right\|_{\max}}{h_0} = \left\| 2\sin\left(\frac{\pi t^*}{T}\right) \right\| \qquad \triangleleft \text{(h)}$$

The complete shock spectrum is formed by taking the larger of the values computed from Eqs. (g) and (h) for a given value of the pulse duration. The shock spectrum is displayed in Figure 4.10-2.

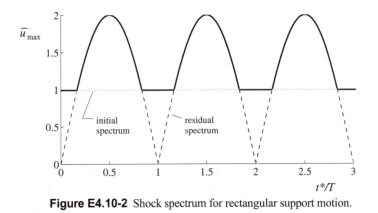

Figure E4.10-2 Shock spectrum for rectangular support motion.

Example 4.11 – Shock Spectrum for a Half Sine Pulse

Consider a single degree of freedom subjected to a pulse in the form of a half sine wave, as indicated. For this case the duration of the pulse is $t^* = T^*/2$, where T^* is the period of the sine function. Determine the shock spectrum for this type of pulse.

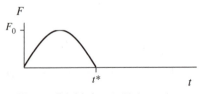

Figure E4.11-1 A half sine pulse.

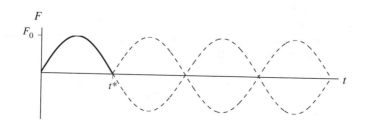

Figure E4.11-2 Pulse represented as superposition of sine waves.

Solution

The forcing function for this pulse takes the form

$$F(t) = F_0 \sin \Omega t \left[\mathcal{H}(t) - \mathcal{H}(t - t^*) \right]$$
$$= F_0 \sin \Omega t \, \mathcal{H}(t) + F_0 \sin \Omega(t - t^*) \mathcal{H}(t - t^*) \qquad \text{(a)}$$

where

$$\Omega = 2\pi/T^* = \pi/t^* \qquad \text{(b)}$$

(see Figure E4.11-2). Substitution of Eq. (a) into Eq. (4.48) gives

$$x(t) = \frac{F_0}{m\omega} \int_0^t \sin \Omega \tau \, \sin \omega(t - \tau) d\tau \, \mathcal{H}(t)$$

$$+ \frac{F_0}{m\omega} \int_0^t \sin \Omega(\tau - t^*) \, \sin \omega(t + t^* - \tau) d\tau \, \mathcal{H}(t - t^*) \qquad \text{(c)}$$

Carrying through the integration gives the response to the half-sine pulse as

$$\frac{x(t)}{f_0} = \frac{1}{1 - (T/2t^*)^2} \left\{ \left[\sin\left(\pi t/t^* \right) - \frac{T}{2t^*} \sin \omega t \right] \mathcal{H}(t) \right.$$

$$\left. + \left[\sin\{\pi(t - t^*)/t^*\} - \frac{T}{2t^*} \sin \omega(t - t^*) \right] \mathcal{H}(t - t^*) \right\} \qquad \text{(d)}$$

We must next determine the maximum values of the response and when they occur. We shall consider two intervals, (i) the time interval during which the pulse is still active ($t < t^*$) and (ii) the interval after the pulse subsides ($t > t^*$). The maximum displacements will then be plotted as a function of the duration of the pulse, constructing the shock spectrum for the half-sine pulse.

(i) $0 < t < t^*$: Initial Spectrum

To determine the time at which the maximum deflection occurs we take the derivative of the response, set it to zero and solve for the time. The response of the system while the pulse is active is, from Eq. (d),

$$\frac{x(t)}{f_0} = \frac{1}{1 - (T/2t^*)^2} \left[\sin(\pi t/t^*) - \frac{T}{2t^*} \sin \omega t \right] \tag{e}$$

Differentiating Eq. (e) and setting the resulting expression to zero gives

$$\cos \Omega t - \cos \omega t = 0$$

Now, using the identity $\cos(a) - \cos(b) = -2 \sin\{(a + b)/2\} \sin\{(a - b)/2\}$ in the above expression gives the equivalent statement

$$\sin\left\{\tfrac{1}{2}(\Omega + \omega)t\right\} \sin\left\{\tfrac{1}{2}(\Omega - \omega)t\right\} = 0 \tag{f}$$

It may be seen that Eq. (f) is satisfied if either

$$\left(\frac{\Omega + \omega}{2}\right) t = n\pi \quad \text{or} \quad \left(\frac{\Omega - \omega}{2}\right) t = n\pi \quad (n = 0, 1, 2, \ldots) \tag{g}$$

Equation (g) thus yields two sets of peak times (times that render the response an extremum),

$$t_{pk}^{(a)} = \frac{2n\pi}{\Omega + \omega} \quad (n = 0, 1, 2, \ldots) \tag{h-1}$$

and

$$t_{pk}^{(b)} = \frac{2n\pi}{\Omega - \omega} \quad (n = 0, 1, 2, \ldots) \tag{h-2}$$

It remains to establish which set renders the response the true maximum.

Substituting Eq. (h-1) into Eq. (e) and noting that $\sin(\omega t \mp 2n\pi) = \sin \omega t$ and that

$$\omega t_{pk}^{(a)} - 2n\pi = \frac{2n\pi \omega}{\Omega + \omega} - 2n\pi = -\frac{2n\pi \Omega}{\Omega + \omega}$$

gives the responses at the first set of extrema as

$$\| x \|_{t = t_{pk}^{(a)}} = f_0 \left\| \frac{1}{1 - (\Omega/\omega)} \sin \Omega t_{pk}^{(a)} \right\| \tag{i}$$

Similarly, substituting Eq. (h-2) into Eq. (e) and noting that

$$\omega t_{pk}^{(b)} + 2n\pi = \frac{2n\pi\,\omega}{\Omega-\omega} + 2n\pi = \frac{2n\pi\,\Omega}{\Omega-\omega}$$

gives the second set of extrema as

$$\left\| x \right\|_{t=t_{pk}^{(b)}} = f_0 \left\| \frac{1}{\left[1+(\Omega/\omega) \right]} \sin \Omega t_{pk}^{(b)} \right\|$$ (j)

It may be seen by comparing Eqs. (i) and (j) that

$$\left\| x \right\|_{t=t_{pk}^{(a)}} > \left\| x \right\|_{t=t_{pk}^{(b)}}$$

and hence that

$$\left\| x \right\|_{max} = \left\| x \right\|_{t=t_{pk}^{(a)}}$$ (k)

Substituting Eqs. (h-1) and (i) into Eq. (e) gives the maximum response during the interval for which the pulse is active as

$$\frac{\left\| x \right\|_{max}}{f_0} = \left\| \frac{1}{\left[1-(T/2t*) \right]} \sin\left\{ \frac{2n\pi}{1+(2t*/T)} \right\} \right\|$$ ◁ (l)

where $t_{pk} < t*$ and, from Eq. (h-1),

$$n < \frac{1}{2} + \frac{t*}{T}$$ ◁ (m)

Since n must be an integer, it is seen from Eq. (m) that no maxima occur when the duration of the pulse is less than half the natural period of the system.

(ii) $t > t*$: *Residual Spectrum*
Evaluating Eq. (e) for $t > t*$ gives the response of the system after the pulse has subsided as

$$\frac{x(t)}{f_0} = \frac{T/2t*}{\left[(T/2t*)^2 - 1 \right]} \left[\sin \omega t + \sin \omega(t - t*) \right]$$ (n)

Differentiating Eq. (n) and setting the resulting expression to zero gives

$$\left[\cos \omega t + \cos \omega(t - t*)\right]_{t=t_{pk}} = 0$$

which, after using the identity $\cos(a) + \cos(b) = 2\cos\{(a + b)/2\}\cos\{(a - b)/2\}$, takes the alternate form

$$\cos\left\{\omega\left(t - \tfrac{1}{2}t *\right)\right\}\cos\left\{\tfrac{1}{2}\omega t *\right\} = 0 \tag{o}$$

It follows that

$$\omega t_{pk} = \frac{\omega t *}{2} + (2n - 1)\frac{\pi}{2} \tag{p}$$

Substituting Eq. (p) into Eq. (n) gives the maximum response after the pulse has subsided as

$$\frac{\|x\|_{max}}{f_0} = 2\left\|\frac{(2t */T)}{\left[(2t */T)^2 - 1\right]}\cos\left(\pi t */T\right)\right\| \qquad \lhd (q)$$

A plot of the residual spectrum is displayed in Figure E4.11-3, along with the initial spectrum. For a given range of the pulse duration ratio $t*/T$, the greater of the initial or residual maxima form the shock spectrum. It may be seen that the residual spectrum dominates when the duration of the pulse is less than half the natural period of the system while the initial spectrum governs for $t* > T/2$

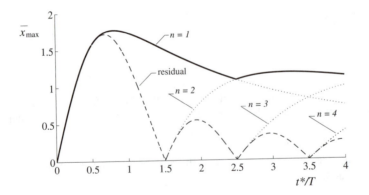

Figure E4.11-3 Shock spectrum for a half-sine pulse.

Example 4.12 – Motion of the Support in Form of a Half-Sine Wave

Consider a shock in the form of a sudden motion of the support of a mass-spring-damper system (such as in Example 4.1). Set up the equations that describe the shock spectrum of the system if the time history of the displacement of the support is in the form of a half-sine pulse.

Solution

The motion of the support is of the form

$$x_F(t) = \begin{cases} 0 & (t < 0) \\ h_0 \sin \Omega t & (0 < t < t^*) \\ 0 & (t > t^*) \end{cases} \tag{a}$$

where

$$\Omega = 2\pi/T^* = \pi/t^* \tag{b}$$

The motion can be described formally by the equation

$$\begin{aligned} x_F(t) &= h_0 \sin \Omega t \left[\mathcal{H}(t) - \mathcal{H}(t-t^*) \right] \\ &= h_0 \left[\sin \Omega t \, \mathcal{H}(t) + \sin \Omega(t-t^*) \, \mathcal{H}(t-t^*) \right] \end{aligned} \tag{a'}$$

The equation that governs the absolute motion of the mass is, from Eq. (3.122),

$$m \ddot{x}(t) + k \, x(t) = k \, x_F(t) \tag{c}$$

(i) $0 < t < t^*$: *Initial Spectrum*

During the interval that the pulse is active the equation of motion takes the form

$$m \ddot{x}(t) + k \, x(t) = k h_0 \sin \Omega t \tag{d}$$

The absolute response may be determined directly from Eq. (e) of Example 4.11 by letting $F_0 = kh_0$. Hence,

$$\frac{x(t)}{h_0} = \frac{1}{1-(T/2t^*)^2} \left[\sin\left(\pi t/t^*\right) - \frac{T}{2t^*} \sin \omega t \right] \tag{e}$$

The relative displacement of the mass with respect to the support is then

$$\frac{u(t)}{h_0} = \frac{x(t) - x_F(t)}{h_0} = \frac{(T/2t^*)}{1-(T/2t^*)^2} \left[(T/2t^*) \sin\left(\pi t/t^*\right) - \sin \omega t \right] \quad \triangleleft \text{(f)}$$

$$\dot{u} = 0 = \left(T/2t*\right)^2 \cos\left(\pi t_{pk}/t*\right) - \cos\left(2\pi t_{pk}/T\right) \qquad \lhd \text{(g)}$$

The roots of Eq. (g) yield the times $t_{pk} < t*$ at which the relative displacement is an extremum for a given value of the pulse duration. These values can be substituted into Eq. (f) to determine the corresponding maximum displacement. The plot of the corresponding maximum displacement as a function of the pulse duration then yields the initial shock spectrum.

(ii) $t > t*$: *Residual Spectrum*
For this time interval, the absolute displacement of the mass may be obtained directly from Eq. (n) of Example 4.11 by letting $F_0 = kh_0$. Since after the pulse subsides $x_F = 0$ it follows that $u(t) = x(t)$ and hence, from Eq. (q) of Example 4.11, that

$$\frac{\|u\|_{max}}{h_0} = \frac{\|x\|_{max}}{h_0} = 2 \left\| \frac{\left(2t*/T\right)}{\left[\left(2t*/T\right)^2 - 1\right]} \cos\left(\pi t*/T\right) \right\| \qquad \lhd \text{(h)}$$

The larger of the values of the initial and residual spectra form the shock spectrum for the system under support shock. The generation of the shock spectrum for this case is left as a project for the reader (Problem 4.20).

4.8 CONCLUDING REMARKS

In this chapter we have focused our attention on single degree of freedom systems subjected to transient loads of various types. We began by introducing two generalized functions and their properties, operations and relations. These functions allowed us to formally obtain the response of single degree of freedom systems to impulse loading. A representation of the response of initially quiescent systems to arbitrary loading was developed based on the response to impulse loading and took the form of the convolution integral. The convolution integral can be used to determine the response of a system to any loading, given the time history of the load. This formulation was used to determine the response of systems to the specific forms of step loading and ramp loading. For linear systems superposition may be used to construct solutions of complex loading types from the known responses of loads that, together, comprise the loading of interest. This was done explicitly for two loading types, step loading and linear transition to constant load level. The method may, of course, be used to obtain solutions of other load types. The chapter finished with a discussion of shock spectra, a characterization of the severity of the response of a system to a given type of pulse or shock. Specific spectra were presented for rectangular pulses and half sine pulses. The methodology outlined could be used to develop the response spectra for other types of pulses as well.

BIBLIOGRAPHY

Greenberg, M.D., *Applications of Green's Functions in Science and Engineering*, Prentice-Hall, 1971.

Meirovitch, L., *Elements of Vibration Analysis*, 2nd ed., McGraw-Hill, New York, 1986.

Meirovitch, L., *Fundamentals of Vibrations*, McGraw-Hill, Boston, 2001.

Thomson, W.T., *Theory of Vibration with Applications*, 4th ed., Prentice-Hall, Englewood Cliffs, 1993.

Tse, F.S., Morse, I.E. and Hinkle, R.T., *Mechanical Vibrations, Theory and Applications*, 2nd ed., Prentice Hall, Englewood Cliffs, 1978.

PROBLEMS

4.1 Consider the function $f(t) = 2t^4$. Evaluate the following integrals:

(a) $\displaystyle\int_0^{10} f(t)\,\widehat{\delta}(t-3)\,dt$

(b) $\displaystyle\int_0^{10} f(t)\,\dot{\widehat{\delta}}(t-3)\,dt$

(c) $\displaystyle\int_0^{10} f(t)\,\ddot{\widehat{\delta}}(t-3)\,dt$

(d) $\displaystyle\int_0^{10} \ddot{f}(t)\,\widehat{\delta}(t-3)\,dt$

4.2 Consider the function $f(t) = 2t^4$. Evaluate the following integrals:

(a) $\displaystyle\int_0^{10} f(t)\,\mathcal{H}(t-3)\,dt$

(b) $\displaystyle\int_0^{10} f(t)\,\dot{\mathcal{H}}(t-3)\,dt$

(c) $\displaystyle\int_0^{10} f(t)\,\ddot{\mathcal{H}}(t-3)\,dt$

(d) $\displaystyle\int_0^{10} \ddot{f}(t)\,\mathcal{H}(t-3)\,dt$

4.3 The mass of the system of Problem 2.25 is subjected to an impulse of amplitude 20 N-sec. Determine the response history of the system.

4.4 A 12 inch wrench is attached to the hub of the wheel of Example 2.3 and extends horizontally, as indicated. Determine the rotational history of the wheel if a tool suddenly falls on the free end of the wrench, imparting an impulse of magnitude 10 lb-sec at that point. (The wrench may be treated as rigid for the purposes of the present problem.)

Fig. P4.4/P4.5

4.5   A 12 inch wrench is attached to the hub of the wheel of Example 2.3 and extends horizontally, as indicated. After a tool suddenly falls and strikes the free end of the wrench, the wheel is observed to oscillate at its natural frequency with an amplitude of 0.15 radians. Determine the impulse imparted by the tool.

4.6   The timing device of Problem 2.19 is tapped to initiate motion. Determine the magnitude of the impulse required so that the motion of the device has an amplitude Θ_0.

4.7   A 10 N dead load is suddenly applied to the mass of the system of Problem 2.26a. Determine the response history of the system.

4.8   A flat 225 lb raft with a 6 ft × 6 ft surface floats in a fresh water lake. Determine the response history for vertical motion of the system when a 120 lb boy suddenly jumps onto the float. Assume that the jump is primarily horizontal so that the vertical component of the boy's velocity is negligible as he lands on the raft. Also assume that the boy doesn't "bounce" on the raft after landing.

Fig. P4.8

4.9   Determine the motion of the one-story building of Example 3.13 if the base suddenly moves 2 inches to the right.

4.10 The diving board of Problem 2.30 is at rest when a 110 lb boy stands at its edge. Determine the response of the diving board after the boy jumps into the pool.

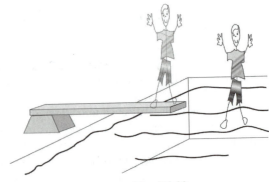

Fig. P4.10

4.11 The spool of the system of Problem 2.27 is initially at rest when it is pulled on by a force whose magnitude increases linearly with time at the rate of 0.1 N/sec. Determine the motion of the spool assuming that no slipping occurs between the spool and the ground.

4.12 Solve Problem 4.11 if damping is negligible and the magnitude of the force increases parabolically with time such that the rate of application of the load increases at the rate of 0.1 N/sec².

4.13 Determine the yawing motion of the aircraft mock-up of Problem 3.7 if the thrust supplied by one of the engines suddenly deviates from the norm by a factor of ε for a time interval of duration τ.

4.14 Determine the time history of the response of the system of Problem 2.21 when, over the interval $0 < t < \tau$, the flywheel is loaded at a constant rate to the ultimate level M_0 and then maintained at that level.

4.15 Differential settlement of a segment of roadway causes a small drop of magnitude h. Repairs are performed effecting a transition ramp of length L between the two segments of the roadway. A vehicle travels at constant speed v_0 along a flat road when it encounters the drop. Assuming that the driver maintains the same horizontal speed throughout his motion, determine the vertical motion of the vehicle during and after it encounters the drop. The vehicle never leaves the roadway and the vertical speed due to gravity is negligible.

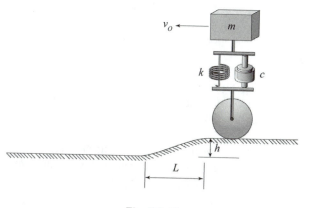

Fig. P4.15

4.16 Determine the response of the system of Problem 3.1 when it is subjected to a half-sine pulse of 250 dyne amplitude and 0.25π seconds duration.

4.17 Determine the response of a standard mass-spring-damper system when it is subjected to a sinusoidal transition of duration $t*$ to a constant load of magnitude F_0.

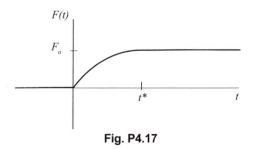

Fig. P4.17

4.18 Determine the time history of the response of a standard mass-spring-damper system when it is subjected to a triangular pulse of magnitude F_0 and total duration $t*$, when the ramp-up and ramp-down times are of the same duration.

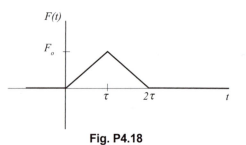

Fig. P4.18

4.19 Determine the shock spectrum for a linear system subjected to the transition loading of Section 4.6.2.

4.20 (Project) Use Eqs. (f), (g) and (h) of Example 4.12 to generate the shock spectrum for a system whose base undergoes a motion in the form of a half-sine pulse.

4.21 (Project) Determine the shock spectrum for a linear system subjected to a triangular pulse of magnitude F_0 and total duration $t*$, when the ramp-up and ramp-down times are of the same duration.

5

Operational Methods

In this chapter we shall study an alternate approach to solving vibration problems for mechanical systems. This approach is used in various situations, particularly with regard to systems analysis and vibration control. (The reader may bypass this chapter and continue on to subsequent chapters without any loss of continuity.)

Operational methods, in the present context, refer to techniques associated with integral operators and are also known as integral transform techniques. We shall here be interested in the particular transform attributed to Laplace. When applied, an integral transform maps an ordinary differential equation (o.d.e.) to an algebraic equation in terms of a transformed dependent variable. The algebraic equation may then be solved directly for the transformed dependent variable, and that variable may then be inverted (mapped back) to give the solution to the original problem. We first review the Laplace transform and some of its properties.

5.1 THE LAPLACE TRANSFORM

Consider some real function ξ of a real variable t. The *Laplace transform* of $\xi(t)$, which we shall denote as $\breve{\xi}(s)$, is defined by the relation

$$\breve{\xi}(s) = \mathcal{L}\{\xi(t)\} \equiv \int_0^\infty e^{-st}\xi(t)\,dt \qquad (5.1)$$

where s is a complex variable such that $\mathrm{Re}(s) > s_0 > 0$, and $\mathcal{L}\{\xi\}$ is read "$\mathcal{L}$ operating on ξ." Equation (5.1) maps the function $\xi(t)$ to the function $\breve{\xi}(s)$. In all applications of the Laplace transform considered in this text, the inversions (i.e., the inverse mappings) may be accomplished by algebraic manipulation and inspection, as will be discussed shortly. In this regard, many pertinent Laplace transforms are evaluated below as examples. The Laplace transforms of many other functions may be found in published mathematical tables.

Though we shall have no need to employ it in our present study of vibrations, the formal *inverse Laplace transform* is presented for completeness, and for situations where functions are encountered such that their inverses are not readily available. For these situations, the associated inverse transform that maps $\bar{\xi}(s) \to \xi(t)$ is given by

$$\xi(t) = \mathcal{L}^{-1}\left\{\bar{\xi}(s)\right\} \equiv \frac{1}{2\pi i} \int_{\beta - i\infty}^{\beta + i\infty} e^{st}\bar{\xi}(s)\,ds \tag{5.2}$$

where the integration is performed in the complex plane and $\beta > s_0$ defines what is referred to as the *Bromwich line*. The reader is referred to the variety of applied mathematics texts for discussion of the evaluation of the inversion integral for arbitrary functions. In the remainder of this section we shall evaluate the Laplace transforms of a variety of pertinent functions. We will then apply this technique to vibration analysis in subsequent sections.

5.1.1 Laplace Transforms of Basic Functions

We next evaluate the Laplace transform of some basic functions of interest. These functions include the generalized functions introduced in Chapter 4 (the unit impulse function, the unit step function and the unit ramp function), as well as the exponential and harmonic functions. All are easily evaluated by straightforward integration and may be verified by the reader.

The Unit Impulse

The Laplace transform of the Dirac Delta Function (the unit impulse) is obtained by setting $\xi(t) = \hat{\delta}(t)$ in Eq. (5.1) and using Eq. (4.7). Doing this, we find that

$$\mathcal{L}\left\{\hat{\delta}(t)\right\} = \int_0^\infty e^{-st}\,\hat{\delta}(t)\,dt = 1 \tag{5.3}$$

Similarly,

$$\mathcal{L}\left\{\hat{\delta}(t-\tau)\right\} = \int_0^\infty e^{-st}\,\hat{\delta}(t-\tau)\,dt = e^{-s\tau} \tag{5.4}$$

The Unit Step Function

The Laplace transform of the Heaviside Step Function (the unit step function) is obtained by setting $\xi(t) = \mathcal{H}(t)$ in Eq. (5.1) and evaluating the resulting integral. This gives

$$\mathcal{L}\{\mathcal{H}(t)\} = \int_0^\infty e^{-st}\,dt = -\frac{1}{s}e^{-st}\bigg|_0^\infty = \frac{1}{s} \tag{5.5}$$

Similarly,

$$\mathcal{L}\{\mathcal{H}(t-\tau)\} = \int_0^\infty e^{-st}\mathcal{H}(t-\tau)\,dt = \int_\tau^\infty e^{-st}\,dt = \frac{e^{-s\tau}}{s} \tag{5.6}$$

The Unit Ramp Function

The Laplace transform of the unit ramp function is obtained by substituting the function $\xi(t) = t\mathcal{H}(t)$ (the unit ramp function) into Eq. (5.1) and evaluating the resulting integral. Carrying out this calculation gives

$$\mathcal{L}\{t\mathcal{H}(t)\} = \int_0^\infty te^{-st}\,dt = -\frac{te^{-st}}{s}\bigg|_0^\infty + \frac{1}{s}\int_0^\infty e^{-st}\,dt = 0 + \frac{1}{s^2}$$

Hence,

$$\mathcal{L}\{t\mathcal{H}(t)\} = \frac{1}{s^2} \tag{5.7}$$

The Exponential Function

The Laplace transform of the exponential function $\xi(t) = e^{at}$ is obtained by direct evaluation as

$$\tilde{\xi}(s) = \mathcal{L}\{e^{at}\} = \int_0^\infty e^{-(s-a)t}\,dt = \frac{1}{s-a} \tag{5.8}$$

The Harmonic Functions

The Laplace transform of the harmonic functions $\cos \omega t$ and $\sin \omega t$ can be evaluated with the aid of Eq. (5.8) and the two identities

$$\cos\psi = \frac{e^{i\psi} + e^{-i\psi}}{2} \quad \text{and} \quad \sin\psi = \frac{e^{i\psi} - e^{-i\psi}}{2i} \tag{5.9}$$

(see Problem 1.19). Now, from Eqs. (5.8) and (5.9),

$$\mathcal{L}\{\cos\omega t\} = \frac{1}{2}\left[\mathcal{L}\{e^{i\omega t}\} + \mathcal{L}\{e^{-i\omega t}\}\right] = \frac{1}{2(s-i\omega)} + \frac{1}{2(s+i\omega)} = \frac{s}{s^2+\omega^2}$$

Hence,

$$\mathcal{L}\{\cos\omega t\} = \frac{s}{s^2+\omega^2} \tag{5.10}$$

Similarly,

$$\mathcal{L}\{\sin\omega t\} = \frac{\omega}{s^2+\omega^2} \tag{5.11}$$

Harmonic Functions with Exponential Amplitudes

Two more functions of evident interest for our study of vibrations are the combinations $\xi(t) = e^{at}\cos\omega t$ and $\xi(t) = e^{at}\sin\omega t$. The Laplace transform of these *composite functions* are evaluated in a manner similar to that for the harmonic functions alone. Hence,

$$\mathcal{L}\{e^{at}\cos\omega t\} = \frac{1}{2}\left[\mathcal{L}\{e^{(a+i\omega)t}\} + \mathcal{L}\{e^{(a-i\omega)t}\}\right] = \frac{1}{2(s-a-i\omega)} + \frac{1}{2(s-a+i\omega)}$$

Combining the two fractions gives the desired transform

$$\mathcal{L}\{e^{at}\cos\omega t\} = \frac{s-a}{(s-a)^2+\omega^2} \tag{5.12}$$

A similar calculation gives the companion transform

$$\mathcal{L}\{e^{at}\sin\omega t\} = \frac{\omega}{(s-a)^2+\omega^2} \tag{5.13}$$

5.1.2 Shifting Theorem

Consider a function $f(t)$ that is shifted by time τ, as shown in Figure 5.1, but is otherwise arbitrary. Let us evaluate the Laplace transform of such a function. The shifted function may be represented as

$$f_\tau(t) = f(t-\tau)\mathcal{H}(t-\tau) \tag{5.14}$$

where $\mathcal{H}(t)$ is the Heaviside Step Function defined in Chapter 4. We wish to evaluate $\mathcal{L}\{f_\tau(t)\}$. Thus, letting $\xi(t) = f_\tau(t)$ in Eq. (5.1) gives

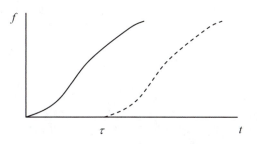

Figure 5.1 Generic function shifted by time τ.

$$\mathcal{L}\{f_\tau(t)\} = \int_0^\infty e^{-st} f(t-\tau)\mathcal{H}(t-\tau)dt = \int_\tau^\infty e^{-st} f(t-\tau)dt \qquad (5.15)$$

Next, let us introduce the time shift

$$\hat{t} = t - \tau \qquad (5.16)$$

from which it follows that $dt = d\hat{t}$. Incorporating Eq. (5.16) into Eq. (5.15) gives

$$\breve{f}_\tau(s) = \int_0^\infty e^{-s(\hat{t}+\tau)} f(\hat{t})d\hat{t} = e^{-s\tau}\int_0^\infty e^{-s\hat{t}} f(\hat{t})d\hat{t} = e^{-s\tau}\breve{f}(s)$$

Hence,

$$\mathcal{L}\{f(t-\tau)\mathcal{H}(t-\tau)\} = e^{-s\tau}\mathcal{L}\{f(t)\} \qquad (5.17)$$

5.1.3 Laplace Transforms of the Derivatives of Functions

We have seen that establishing the dynamic response of a mechanical system gener-ally involves solving a differential equation. It is therefore of interest to evaluate the Laplace transform of the derivatives of a function. The Laplace transform of the de-rivative of a function is determined by substituting that derivative into Eq. (5.1) and performing integration by parts. Doing this we find that

$$\mathcal{L}\{\dot{\xi}(t)\} = \int_0^\infty \frac{d\xi}{dt}e^{-st}dt = \left[e^{-st}\xi\right]_0^\infty + \int_0^\infty s\,e^{-st}\xi\,dt = -\xi(0) + s\,\breve{\xi}(s)$$

Hence,

$$\mathcal{L}\{\dot{\xi}\} = -\xi(0) + s\,\breve{\xi}(s) \qquad (5.18)$$

The Laplace transform of the second derivative of a function is similarly obtained by defining a second function equal to the derivative of the first and applying Eq. (5.18). This gives the Laplace transform of the second derivative as

$$L\{\ddot{\xi}(t)\} = -\dot{\xi}(0) - s\xi(0) + s^2\breve{\xi}(s) \tag{5.19}$$

This process can be continued to obtain the Laplace transform of any order derivative. It is seen that the Laplace transforms of the derivatives of a function are simply functions of the transform of the function itself and of the associated initial conditions. We can therefore anticipate that taking the Laplace transform of an ordinary differential equation results in an algebraic equation involving these parameters.

5.1.4 Convolution

In the previous section we evaluated the Laplace transforms of the product of harmonic functions and exponential functions directly, with the aid of Euler's Formula. In this section we develop a general expression for the determination of the Laplace transforms and the associated inverses of the products of arbitrary functions.

Consider two functions, $f(t)$ and $g(t)$, and their respective Laplace transforms, $\breve{f}(s)$ and $\breve{g}(s)$. It follows from Eq. (5.1) that

$$\breve{g}(s)\breve{f}(s) = \breve{g}(s)\int_0^\infty e^{-s\tau}f(\tau)d\tau = \int_0^\infty f(\tau)e^{-s\tau}\breve{g}(s)d\tau \tag{5.20}$$

Expressing $\breve{g}(s)$ in terms of its definition in the right most integral of Eq. (5.20) and employing the shifting theorem, Eq. (5.17), gives

$$\breve{g}(s)\breve{f}(s) = \int_0^\infty f(\tau)e^{-s\tau}L\{g(t)\}d\tau = \int_0^\infty f(\tau)L\{g(t-\tau)\mathcal{H}(t-\tau)\}d\tau$$

and thus,

$$\breve{f}(s)\breve{g}(s) = \int_0^\infty f(\tau)\int_0^\infty e^{-st}g(t-\tau)\mathcal{H}(t-\tau)dt\,d\tau$$

Now, interchanging the order of integration gives

$$\breve{f}(s)\breve{g}(s) = \int_0^\infty e^{-st}\int_0^\infty f(\tau)g(t-\tau)\mathcal{H}(t-\tau)d\tau\,dt$$

which may be written as

$$\breve{f}(s)\breve{g}(s) = \lim_{T\to\infty} \int_0^T e^{-st} \int_0^T f(\tau)g(t-\tau)\mathcal{H}(t-\tau)\,d\tau\,dt$$

$$+ \lim_{T\to\infty} \int_T^\infty e^{-st} \int_T^\infty f(\tau)g(t-\tau)\mathcal{H}(t-\tau)\,d\tau\,dt$$

The second term on the right-hand side is seen to vanish in the limit. Finally, from the definition of the Heaviside Step Function, Eq. (4.11), it is noted that $\mathcal{H}(t-\tau) = 0$ for $\tau > t$. Hence, we finally have the relation

$$\breve{f}(s)\breve{g}(s) = \int_0^\infty e^{-st} \int_0^t f(\tau)g(t-\tau)\,d\tau\,dt \tag{5.21}$$

Equation (5.21) may be expressed in the form

$$\breve{f}(s)\breve{g}(s) = \mathcal{L}\{f(t) * g(t)\} \tag{5.22}$$

where

$$f(t) * g(t) \equiv \int_0^t f(\tau)g(t-\tau)\,d\tau \tag{5.23}$$

is referred to as the *convolution* of f and g, and the integral on the right-hand side as the *convolution integral*, also known as the Faltung Integral. It follows that the inverse Laplace transform of a product of two functions is given by

$$\mathcal{L}^{-1}\{\breve{f}(s)\breve{g}(s)\} = f(t) * g(t) \tag{5.24}$$

Equation (5.24) is very useful in a variety of applications.

5.2 FREE VIBRATIONS

Recall the standard form of the equation of motion pertaining to single degree of freedom systems, Eq. (2.64),

$$\ddot{x} + 2\omega\zeta\dot{x} + \omega^2 x = 0$$

and the associated initial conditions

$$x(0) = x_0 \,, \quad \dot{x}(0) = v_0$$

Taking the Laplace transform of the governing equation, exploiting the property that Laplace transformation is a linear operation and using Eqs. (5.18) and (5.19), results in the algebraic equation for the transformed displacement

$$\left[-\dot{x}(0) - s\,x(0) + s^2 \tilde{x}(s)\right] + 2\omega\zeta\left[-x(0) + s\,\tilde{x}(s)\right] + \omega^2 \tilde{x}(s) = 0 \qquad (5.25)$$

Equation (5.25) may be solved for $\tilde{x}(s)$ to give

$$\tilde{x}(s) = \frac{v_0 + (s + 2\omega\zeta)x_0}{s^2 + 2\omega\zeta s + \omega^2} \qquad (5.26)$$

If we compare the denominator of Eq. (5.26) with the characteristic function $Z(s)$, the left-hand side of Eq. (2.67), it is seen that the denominator corresponds to the characteristic function. Further, let us recall that the roots of the characteristic equation $Z(s) = 0$ yield the exponents of Eq. (2.66) and hence the general form of the solution. Let us next factor the denominator of Eq. (5.26) to give the form

$$\tilde{x}(s) = \frac{v_0 + (2\omega\zeta + s)x_0}{\left[s + (\zeta\omega + i\omega_d)\right]\left[s + (\zeta\omega - i\omega_d)\right]} \qquad (5.27)$$

where ω_d is defined by Eq. (2.70) for underdamped systems. It may be seen that the singularities of the transformed displacement, the values of s for which the bracketed expressions in the denominator of Eq. (5.27) vanish, correspond to the roots of the characteristic equation $Z(s) = 0$. In order to invert the transformed displacement, and hence to obtain the response $x(t)$, it is desirable to express Eq. (5.27) in a form, or combination of forms, that corresponds to the Laplace transform of functions that we have already established or are listed elsewhere. This may be accomplished by multiplying the respective factors in the denominator back again, and regrouping terms. The transformed displacement then takes the form

$$\tilde{x}(s) = x_0 \frac{(s + \zeta\omega)}{(s + \zeta\omega)^2 + \omega_d^2} + \frac{(v_0/\omega)\zeta x_0}{\sqrt{1-\zeta^2}} \frac{\omega}{(s + \zeta\omega)^2 + \omega_d^2} \qquad (5.28)$$

Equation (5.28) may now be inverted by direct application of Eqs. (5.12) and (5.13). This gives the free vibration response as

$$x(t) = \mathcal{L}^{-1}\{\tilde{x}(s)\} = x_0 e^{-\zeta\omega t} \cos\omega_d t + \left[\frac{(v_0/\omega) + \zeta x_0}{\sqrt{1-\zeta^2}}\right] e^{-\zeta\omega t} \sin\omega_d t \qquad (5.29)$$

Substituting Eqs. (2.6)–(2.11) puts the response in the form identical to Eq. (2.74).

5.3 FORCED VIBRATIONS

We next consider the response of single degree of freedom systems when they are subjected to externally applied forces.

5.3.1 The Governing Equations

Recall the general equation of motion in standard form, Eq. (3.1),

$$\ddot{x} + 2\omega\zeta\dot{x} + \omega^2 x = \omega^2 f(t) \tag{5.30}$$

The right-hand side of Eq. (5.30), $\omega^2 f(t)$, is seen to correspond to the specific applied force (the force per unit mass acting on the system). Taking the Laplace transform of Eq. (5.30) results in the algebraic equation

$$\mathcal{Z}(s)\breve{x}(s) - \mathcal{Z}_0(s) = \omega^2 \breve{f}(s) \tag{5.31}$$

where

$$\mathcal{Z}(s) = s^2 + 2\omega\zeta s + \omega^2 \tag{5.32}$$

$$\mathcal{Z}_0(s) = v_0 + \left(2\omega\zeta + s\right)x_0 \tag{5.33}$$

and $\breve{f}(s) \equiv \mathcal{L}\{f(t)\}$. The function $\mathcal{Z}(s)$ is referred to as the *mechanical impedance* of the system. Equation (5.31) may be solved for $\breve{x}(s)$ to give

$$\breve{x}(s) = \frac{\omega^2 \breve{f}(s) + \mathcal{Z}_0(s)}{\mathcal{Z}(s)} \tag{5.34}$$

which may then be inverted for given $\breve{f}(s)$. It may be noted that the inverse transform of $\mathcal{Z}_0(s)/\mathcal{Z}(s)$ is given by Eq. (5.28). Also note that Eq. (5.34) reduces to Eq. (5.26) when $f = 0$. If the system is initially at rest it may be seen that $\mathcal{Z}_0(s) = 0$, and hence that

$$\mathcal{Z}(s) = \omega^2 \breve{f}(s)/\breve{x}(s) \tag{5.35}$$

In this context, the mechanical impedance is seen to correspond to the (specific) force per unit displacement in the space of the transformed variables. The inverse of the impedance,

$$\mathcal{Q}(s) = \mathcal{Z}^{-1}(s) = \breve{x}(s)/\omega^2 \breve{f}(s) \tag{5.36}$$

Figure 5.2 Block diagram for forced vibration problem.

is referred to as the *mechanical admittance* (transfer function) and, in the present context, is seen to correspond to the displacement per unit (specific) force in the transformed space. From a systems perspective, $\tilde{f}(s)$ and $\tilde{x}(s)$ may be viewed as input-output pairs, with $Q(s)$ the corresponding transfer function as depicted in the block diagram of Figure 5.2. Both $Z(s)$ and $Q(s)$ characterize the system and are important parameters in certain applications (such as control theory).

5.3.2 Steady State Response

Let us next consider the response of single degree of freedom systems to external time dependent forces. In particular, let us first consider the response to harmonic forcing. Thus, let us consider the external force of the form

$$f(t) = f_0 e^{i\Omega t} = f_0 \left[\cos \Omega t + i \sin \Omega t\right] \tag{5.37}$$

It follows from Eq. (5.8), or equivalently from Eqs. (5.10) and (5.11), that

$$\tilde{f}(s) = \frac{f_0}{s - i\Omega} \tag{5.38}$$

The transform of the response is then readily obtained by substituting Eq. (5.38) into Eq. (5.34). This gives

$$\tilde{x}(s) = \frac{\omega^2 f_0 + Z_0(s)}{(s - i\Omega)(s^2 + 2\omega\zeta s + \omega^2)} = \frac{\omega^2 f_0 + Z_0(s)}{(s - i\Omega)\left[(s + \zeta\omega)^2 + \omega_d^2\right]} \tag{5.39}$$

The physical response is found by inverting Eq. (5.39). This can be done by expanding the denominator and using partial fractions. However, it is expedient as well as instructive to perform the inversion using the convolution theorem. To do this we first partition Eq. (5.39) using Eqs. (5.8), (5.12) and (5.13). We then incorporate the resulting expression into Eq. (5.23) to get

$$x(t) = \mathcal{L}^{-1}\left\{\tilde{x}(s)\right\} = \omega^2 f_0 \int_0^t e^{i\Omega(t-\tau)}\left[e^{-\zeta\omega\tau} \sin \omega_d \tau + x_a(\tau)\right] d\tau \tag{5.40}$$

where $x_a(t)$ is given by Eq. (5.29). Evaluating the integral in Eq. (5.40) gives the response of the system as

$$x(t) = \frac{f_0 e^{i\Omega t}}{\left(1-\bar{\Omega}^2\right)+i\,2\zeta\,\bar{\Omega}} + e^{-\zeta\omega t}\left[x_a(t)+x_b(t)\right] \qquad (5.41)$$

where

$$\bar{\Omega} \equiv \Omega/\omega \qquad (5.42)$$

$$x_a(t) = \Gamma^2\left[\left(1-\bar{\Omega}^2\right)-i\,2\zeta\,\bar{\Omega}\right]\left[\cos\omega_d t + \frac{\zeta+i\bar{\Omega}}{\sqrt{1-\zeta^2}}\sin\omega_d t\right] \qquad (5.43)$$

$$x_b(t) = \cos\omega_d t + \frac{\left(v_0/\omega\right)+\zeta x_0}{\sqrt{1-\zeta^2}}\sin\omega_d t \qquad (5.44)$$

and Γ is given by Eq. (3.50). The second term in Eq. (5.41) may be seen to tend to zero for large t, and thus will become negligible after sufficient time has elapsed following the start-up of the system. Therefore, neglecting the decaying term in that expression gives the *steady state response* as

$$x_{ss}(t) = f_0\Gamma\left(\bar{\Omega};\zeta\right)e^{i(\Omega t-\Phi)} \qquad (5.45)$$

where Γ and Φ respectively correspond to the *magnification factor* and *phase angle* of the steady state response, as discussed in Section 3.3.3. The response given by Eq. (5.45) is seen to be identical with that described by Eq. (3.52) as, of course, it should be.

5.3.3 Transient Response

We next consider the response of initially quiescent systems to short duration loads (pulses). It is seen from Eq. (5.33) that $\mathcal{Z}_0 = 0$ for systems that are initially at rest.

Impulse Response

Consider a single degree of free system, say a mass-spring-damper system that is subjected to an impulse of the form

$$f(t) = \frac{\mathcal{I}}{k}\hat{\delta}(t) \qquad (5.46)$$

where k is the stiffness of the spring, $\mathcal{I}$ is the magnitude of the impulse, and $\hat{\delta}(t)$ is the Dirac Delta Function. The Laplace transform of the excitation is found directly from Eq. (5.3) to be

$$\breve{f}(s) = \mathcal{I}/k \tag{5.47}$$

The response is found by substituting Eq. (5.47) into Eq. (5.34) and inverting the resulting expression. Then, using the inversion results of the free vibration analysis performed earlier in this chapter, with mv_0 replaced by $\mathcal{I}$, gives the response to an impulse as

$$x(t) = \mathcal{I}\,G(t) \tag{5.48}$$

where

$$G(t) = \frac{1}{m\omega_d} e^{-\zeta\omega t} \sin\omega_d t\, \mathcal{H}(t) \tag{5.49}$$

It is seen that Eq. (5.48) compares directly with Eq. (4.22). The response to step loading may be similarly obtained using the Laplace transform approach. The calculation is left as an exercise for the reader.

Ramp Loading

Consider next a transient load that varies linearly with time. We thus consider an excitation of the form

$$f(t) = \dot{f}\,t\,\mathcal{H}(t) \tag{5.50}$$

where $\dot{f}$ is a constant. The Laplace transform of this forcing function is given by Eq. (5.7). Hence, for the given loading function,

$$\breve{f}(s) = \frac{\dot{f}}{s^2} \tag{5.51}$$

The transform of the displacement is then obtained by substituting Eq. (5.51) into Eq. (5.34). This gives

$$\breve{x}(s) = \frac{\omega^2 \dot{f}}{s^2 Z(s)} \tag{5.52}$$

The response may be obtained by inverting Eq. (5.52) using the convolution theorem. Thus,

$$x(t) = \omega^2 \dot{f} \int_0^t (t-\tau) \frac{e^{-\zeta\omega\tau} \sin\omega_d\tau}{\omega_d} d\tau \tag{5.53}$$

Evaluating the integral, we obtain the response to the ramp loading as

$$x(t) = \dot{f} \, \mathcal{R}(t) \tag{5.54}$$

where

$$\mathcal{R}(t) = \left\{ t - \frac{2\zeta}{\omega} + e^{-\zeta\omega t} \left[\frac{2\zeta}{\omega} \cos \omega_d t - \frac{\left(1 - 2\zeta^2\right)}{\omega_d} \sin \omega_d t \right] \right\} \mathcal{H}(t) \tag{5.55}$$

is the unit ramp response. Equation (5.55) compares directly with Eq. (4.37) as, of course, it should.

5.4 CONCLUDING REMARKS

In this chapter we introduced an alternate approach to vibration problems using Laplace transforms. The basic definition was established, the shifting theorem established and the convolution integral derived. The transforms of fundamental relevant functions were determined and the solutions for free and forced vibration problems were developed. The concepts of mechanical impedance and admittance were also introduced.

BIBLIOGRAPHY

Benaroya, H., *Mechanical Vibration: Analysis, Uncertainties and Control*, 2nd ed., Dekker/CRC, Boca Raton, 2004.

Boyce, W.E. and DiPrima, R.C., *Elementary Differential Equations*, 5th ed., Wiley, New York, 1992.

Churchill, R.V., *Operational Mathematics*, 3rd ed., McGraw-Hill, New York, 1972.

Graff, K.F., *Wave Motion in Elastic Solids*, Dover, New York, 1991.

Greenberg, M.D., *Advanced Engineering Mathematics*, 2nd ed., Prentice-Hall, Upper Saddle River, 1998.

Meirovitch, L., *Fundamentals of Vibrations*, McGraw-Hill, Boston, 2001.

Meirovitch, L., *Principles and Techniques of Vibrations*, Prentice-Hall, Upper Saddle River, 1996.

PROBLEMS

5.1   Evaluate the Laplace transform of the step function $F_0 \mathcal{H}(t)$.

5.2   Evaluate the Laplace transform of the step function $F_0 \mathcal{H}(t - \tau)$.

5.3 Evaluate the Laplace transform of the ramp function $f(t) = \dot{f}_0 \cdot (t - \tau) \cdot \mathcal{H}(t - \tau)$ where $\dot{f}_0 = \text{constant}$.

5.4 Verify Eq. (5.11) by evaluating $\mathcal{L}\{\sin \omega t\}$.

5.5 Verify Eq. (5.13) by evaluating $\mathcal{L}\{e^{at} \sin \omega t\}$.

5.6 Evaluate the mechanical impedance and the mechanical admittance of the simple pendulum.

5.7 Evaluate the mechanical impedance and the mechanical admittance of the system of Problem 3.6 when one wheel is fixed.

5.8 Evaluate the mechanical impedance and the mechanical admittance of the structure of Example 3.13.

5.9 Use the Laplace transform approach to obtain the solution to Example 3.1 if the system is initially at rest.

5.10 Use the Laplace transform approach to determine the motion of the mass of Example 3.5 if it is subjected to the applied force $F(t) = 0.05t^2$ N.

6

Dynamics of Multi-Degree of Freedom Systems

Actual mechanical systems generally possess many, if not an infinite number, of degrees of freedom. Further, many continuous systems may be adequately modeled as discrete systems with many degrees of freedom. It is therefore of practical, as well as fundamental, interest to understand the nature of such systems. The next three chapters are devoted to the study of vibrations of discrete multi-degree of freedom systems. Examples of such systems include the multi-story building shown in Figure 6.1a, the motorcycle frame shown in Figure 6.1b and the compound pendulum displayed in Figure 6.1c. The analysis of multi-degree of freedom systems requires the introduction of many new concepts that will be used in conjunction with the ideas and concepts presented thus far. Before we can examine the vibratory behavior of these rather complex systems, we must first develop the facility to derive the equations of motion that describe them. Acquisition of this capability will also allow for the representation and characterization of the various systems, as well. Toward this end, we first introduce multi-degree of freedom systems by example and derive the governing equations for selected systems using Newtonian Mechanics in Section 6.1. Though a useful approach, and most certainly a fundamentally correct one, the derivation of the equations of motion by direct application of Newton's laws is often cumbersome and tedious for systems other than relatively simple assemblages. To facilitate the derivation of equations of motion for all multi-degree of freedom systems we introduce aspects of a subject area known as analytical dynamics in Section 6.2. In particular, we focus our attention on the development and application of *Lagrange's Equations* as an alternate, and often more convenient, approach for deriving the governing equations of complex systems.

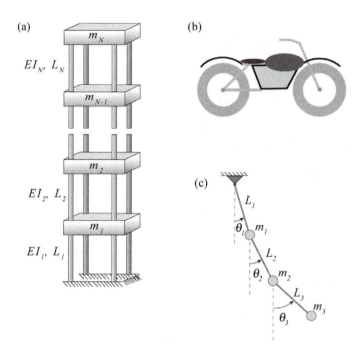

Figure 6.1 Examples of multi-degree of freedom systems.

6.1 NEWTONIAN MECHANICS OF DISCRETE SYSTEMS

In this chapter, and in the two chapters that follow, we shall be interested in the behavior of systems possessing many degrees of freedom. We will introduce the topic through several representative examples using the principles of elementary dynamics discussed in Section 1.5. Sample systems include mass-spring systems, multiple pendulum systems and rigid frames. We begin with a discussion of a simple two degree of freedom system, the results of which are then generalized to systems with any number of degrees of freedom.

6.1.1 Mass-Spring Systems

Mass-spring systems, with or without damping, are appropriate representations for many physical systems such as the multi-story building shown in Figure 1.10 and other elastic structures from submarines to aircraft. The derivation of the equations of motion for simple systems, such as masses and springs in series, is easily accomplished by direct application of Newton's Laws of Motion. In this section, we demonstrate how this is done by first introducing two examples; a two degree of freedom system consisting of two masses subjected to external forces and three springs, and a three degree of freedom system consisting of three masses and three springs excited by the motion of its base. We then extend these results to the formulation for analogous N-degree of freedom systems, where N is any integer.

Example 6.1

Consider a system consisting of two masses, m_1 and m_2, connected by three elastic springs with stiffnesses k_1, k_2 and k_3 respectively, as indicated in Figure E6.1-1. Let the masses be subjected to the externally applied forces F_1 and F_2, respectively, and let u_1 and u_2 represent the corresponding displacements from equilibrium of the masses, as shown. We wish to derive the equations of motion for this system using Newton's Laws of Motion.

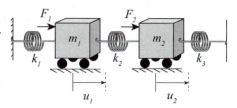

Figure E6.1-1 Two-mass three-spring system.

Solution

To derive the equations of motion for the system, we must derive the equation of motion for each mass individually. To do this we first isolate each mass and draw its kinetic diagram as shown in Figure E6.1-2. Note that Newton's Third Law is implied in the figure. For the sake of convention, we take $u_2 > u_1 > 0$ throughout our derivation. This renders all but the third spring to be in tension as a reference. Results to the contrary in subsequent analyses may then be interpreted accordingly.

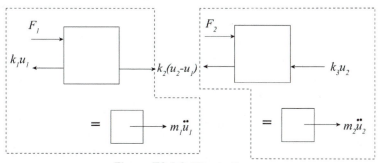

Figure E6.1-2 Kinetic diagram.

With the individual kinetic diagrams drawn, we may directly apply Newton's Second Law to each mass of the system. We then have

$$F_1 - k_1 u_1 + k_2 (u_2 - u_1) = m_1 \ddot{u}_1$$
$$F_2 - k_2 (u_2 - u_1) - k_3 u_2 = m_2 \ddot{u}_2$$

which when rearranged gives the equations of motion for mass 1 and mass 2, respectively, as

$$m_1\ddot{u}_1 + (k_1 + k_2)u_1 - k_2 u_2 = F_1$$
$$m_2\ddot{u}_2 - k_2 u_1 + (k_2 + k_3)u_2 = F_2$$

(a)

Note that the equation governing m_1 depends on the displacement of m_2 as well as the displacement of m_1, while the equation governing the second mass also involves the displacements of both masses. The equations, and hence the motions of each mass, are said to be *coupled*. This, of course, makes physical sense since we might anticipate that the motion of one mass is dependent on the motion of the other since they are connected through the springs. This particular type of coupling is referred to as *stiffness coupling* since the coupling is manifested through the stiffness elements of the system (in this case the elastic springs). We shall elaborate on this and other types of coupling later in this section.

It is seen that this two degree of freedom system is governed by two (coupled) equations of motion. This is the case for two degree of freedom systems, in general. A three degree of freedom system will be governed by three equations of motion, and so on. Thus, an N-degree of freedom system will be governed N equations of motion. While it is cumbersome to work with many equations individually, the governing equations, and hence any analysis and interpretation is made easier and cleaner by writing the equations in matrix form. Analyses may then be performed using the methods of matrix analysis, and of linear algebra in general. Toward this end, Eqs. (a) may be easily written in matrix form as

$$\begin{bmatrix} m_1 & 0 \\ 0 & m_2 \end{bmatrix}\begin{Bmatrix} \ddot{u}_1 \\ \ddot{u}_2 \end{Bmatrix} + \begin{bmatrix} k_1 + k_2 & -k_2 \\ -k_2 & k_2 + k_3 \end{bmatrix}\begin{Bmatrix} u_1 \\ u_2 \end{Bmatrix} = \begin{Bmatrix} F_1 \\ F_2 \end{Bmatrix}$$

(b)

or, in the compact form

$$\mathbf{m\ddot{u} + ku = F}$$

(c)

where

$$\mathbf{m} = \begin{bmatrix} m_1 & 0 \\ 0 & m_2 \end{bmatrix} = \mathbf{m}^\mathsf{T}$$

(d)

is referred to as the *mass matrix* of the system,

$$\mathbf{k} = \begin{bmatrix} k_1 + k_2 & -k_2 \\ -k_2 & k_2 + k_3 \end{bmatrix} = \mathbf{k}^\mathsf{T}$$

(e)

is referred to as the *stiffness matrix* of the system, and a superposed "T" implies the transpose of the matrix. The matrices **m** and **k** are seen to contain the properties of the system. Similarly,

$$\mathbf{F} = \begin{Bmatrix} F_1(t) \\ F_2(t) \end{Bmatrix}$$ (f)

is the *force matrix* for the system,

$$\mathbf{u} = \begin{Bmatrix} u_1(t) \\ u_2(t) \end{Bmatrix}$$ (g)

is the *displacement matrix*, and

$$\ddot{\mathbf{u}} = \begin{Bmatrix} \ddot{u}_1(t) \\ \ddot{u}_2(t) \end{Bmatrix}$$ (h)

is the *acceleration matrix*. In general, we wish to solve Eq. (b), or equivalently Eq. (c), for **u**(t).

Consideration of Eqs. (d) and (e) indicates that both the mass matrix and the stiffness matrix for this system are symmetric. This is typical of such systems and will be found to apply generally. It may also be seen that, for this particular system, the mass matrix is diagonal while the stiffness matrix is not. This is because, as discussed earlier, this particular type of system is coupled through the stiffnesses but not through the masses. As we will see, coupling can occur through the masses or through the stiffnesses, or through both.

Example 6.2

Consider the system comprised of three masses and three springs shown in Figure E6.2-1. Further, let the support/base of the system be subjected to a "prescribed" motion $u_0(t)$. Derive the equations of motion for this system.

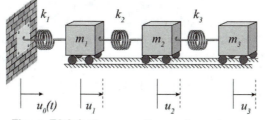

Figure E6.2-1 Three-mass three-spring system.

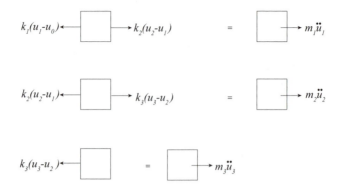

Figure E6.2-2 Kinetic diagram.

Solution

The procedure will be similar to that of Example 6.1. However, for this case, the masses of the system are not subjected to explicit applied forces. Rather, the forcing enters the problem through the motion of the support.

As in the previous example, we first isolate each mass and draw the corresponding kinetic diagram for each (Figure E6.2-2). We note that the displacement of the support, u_0, does not vanish identically for this case, and hence the stretch in spring 1 and the associated spring force are then $u_1 - u_0$ and $k_1(u_1 - u_0)$, respectively. We next write express Newton's Second Law for each mass as follows

$$
\begin{aligned}
m_1\ddot{u}_1 &= k_2(u_2 - u_1) - k_1(u_1 - u_0) \\
m_2\ddot{u}_2 &= k_3(u_3 - u_2) - k_2(u_2 - u_1) \\
m_3\ddot{u}_3 &= -k_3(u_3 - u_2)
\end{aligned}
\tag{a}
$$

Regrouping terms in each of Eqs. (a) and bringing all terms except that associated with the base motion to the left hand-side of the equations results in

$$
\begin{aligned}
m_1\ddot{u}_1 + (k_1 + k_2)u_1 - k_2 u_2 &= k_1 u_0(t) \\
m_2\ddot{u}_2 - k_2 u_1 + (k_2 + k_3)u_2 - k_3 u_3 &= 0 \\
m_3\ddot{u}_3 - k_3 u_2 + k_3 u_3 &= 0
\end{aligned}
\tag{b}
$$

The above equations governing the displacements of the masses of the system may be written in matrix form as

$$
\mathbf{m}\ddot{\mathbf{u}} + \mathbf{k}\mathbf{u} = \mathbf{F}
\tag{c}
$$

where

$$\mathbf{m} = \begin{bmatrix} m_1 & 0 & 0 \\ 0 & m_2 & 0 \\ 0 & 0 & m_3 \end{bmatrix} \tag{d}$$

$$\mathbf{k} = \begin{bmatrix} (k_1 + k_2) & -k_2 & 0 \\ -k_2 & (k_2 + k_3) & -k_3 \\ 0 & -k_3 & k_3 \end{bmatrix} \tag{e}$$

$$\mathbf{F} = \begin{Bmatrix} k_1 u_0(t) \\ 0 \\ 0 \end{Bmatrix} \tag{f}$$

$$\ddot{\mathbf{u}} = \begin{Bmatrix} \ddot{u}_1 \\ \ddot{u}_2 \\ \ddot{u}_3 \end{Bmatrix}, \quad \mathbf{u} = \begin{Bmatrix} u_1 \\ u_2 \\ u_3 \end{Bmatrix} \tag{g, h}$$

N-Degree of Freedom Mass-Spring-Damper Systems

The formulations of Examples 6.1 and 6.2 can be generalized to a system comprised of any number of masses, say N, connected in series by a system of springs and dampers as shown in Figure 6.2. For this particular system the state of the system is known if the position, or equivalently the displacement, of each of the N masses is known as a function of time. By definition (see Section 1.1), a system that requires N "coordinates" to describe its state, possesses N degrees of freedom. The system presently under consideration is, therefore, an N-degree of freedom system. The equations of motion for this system are relatively simple to derive using Newton's Laws of Motion. (This, however, will not always be the case.)

The kinetic diagram for mass m_j ($j = 1, 2, \ldots, N$) is depicted in Figure 6.3. Writing Newton's Second Law for this generic mass gives

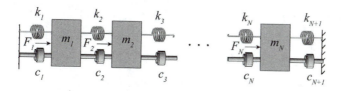

Figure 6.2 *N*-mass, *N*+1-spring, *N*+1-damper system.

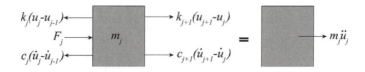

Figure 6.3 Kinetic diagram for *N*-mass system.

$$F_j - k_j\left(u_j - u_{j-1}\right) - c_j\left(\dot{u}_j - \dot{u}_{j-1}\right) + k_{j+1}\left(u_{j+1} - u_j\right) + c_{j+1}\left(\dot{u}_{j+1} - \dot{u}_j\right) = m_j\ddot{u}_j$$

which, after rearranging terms, takes the form

$$m_j\ddot{u}_j - c_j\dot{u}_{j-1} + (c_j + c_{j+1})\dot{u}_j - c_{j+1}\dot{u}_{j+1} - k_j u_{j-1} + (k_j + k_{j+1})u_j - k_{j+1}u_{j+1} = F_j \quad (6.1)$$
$$(j = 1, 2, \ldots, N)$$

The motion of the system is therefore described by *N* equations of the form of Eq. (6.1). We here present a general case, so we shall consider the displacement of the left support to be a prescribed function, $u_0 = u_0(t)$, and we shall consider the right support to be fixed ($u_{N+1} = \dot{u}_{N+1} = 0$). The equations for each mass, and hence for the entire system can be arranged in matrix form as

$$\mathbf{m}\ddot{\mathbf{u}} + \mathbf{c}\dot{\mathbf{u}} + \mathbf{k}\mathbf{u} = \mathbf{F} \tag{6.2}$$

where

$$\mathbf{m} = \begin{bmatrix} m_1 & 0 & \cdots & 0 \\ 0 & m_2 & \cdots & 0 \\ \vdots & \vdots & \ddots & \vdots \\ 0 & 0 & \cdots & m_N \end{bmatrix} = \mathbf{m}^\mathsf{T} \tag{6.3}$$

is the mass matrix,

$$\mathbf{k} = \begin{bmatrix} (k_1 + k_2) & -k_2 & 0 & \cdots & 0 \\ -k_2 & (k_2 + k_3) & -k_3 & \ddots & \vdots \\ 0 & -k_3 & (k_3 + k_4) & \ddots & 0 \\ \vdots & \ddots & \ddots & \ddots & -k_N \\ 0 & 0 & \cdots & -k_N & (k_N + k_{N+1}) \end{bmatrix} = \mathbf{k}^\mathsf{T} \tag{6.4}$$

is the stiffness matrix,

$$\mathbf{c} = \begin{bmatrix} (c_1 + c_2) & -c_2 & 0 & \cdots & & 0 \\ -c_2 & (c_2 + c_3) & -c_3 & \ddots & & \vdots \\ 0 & -c_3 & (c_3 + c_4) & \ddots & & 0 \\ \vdots & \ddots & \ddots & \ddots & & -c_N \\ 0 & 0 & \cdots & & -c_N & (c_N + c_{N+1}) \end{bmatrix} = \mathbf{c}^{\mathsf{T}} \quad (6.5)$$

is the damping matrix, and

$$\ddot{\mathbf{u}} = \begin{Bmatrix} \ddot{u}_1(t) \\ \ddot{u}_2(t) \\ \vdots \\ \ddot{u}_N(t) \end{Bmatrix}, \quad \mathbf{u} = \begin{Bmatrix} u_1(t) \\ u_2(t) \\ \vdots \\ u_N(t) \end{Bmatrix}, \quad (6.6)$$

are the acceleration and displacement matrices, respectively, and

$$\mathbf{F} = \begin{Bmatrix} F_1(t) + k_1 u_0(t) + c_1 \dot{u}_0(t) \\ F_2(t) \\ \vdots \\ F_N(t) \end{Bmatrix} \quad (6.7)$$

is the force matrix. Note that if the left base is fixed, then we set $u_0 = \dot{u}_0 = 0$ in the force matrix, Eq. (6.7). If the right end is free (there is no spring or damper at the right end of the system), then we respectively set $k_{N+1} = 0$ in the stiffness matrix, Eq. (6.4), and $c_{N+1} = 0$ in the damping matrix, Eq. (6.5). The reader may wish to verify that the equations for the system of Example 6.1 may be obtained by setting $\mathbf{c} = \mathbf{0}$, $N = 2$ and $u_0 = \dot{u}_0 = 0$ in the above formulation. Similarly, the equations for the system of Example 6.2 may be obtained by setting $\mathbf{c} = \mathbf{0}$, $N = 3$ and $k_4 = 0$ in the above formulation.

It may be noted that for this class of system, the stiffness and damping matrices are banded. That is, nonzero elements occur along or near the diagonal of the matrix. This is so because, for this system, the springs and dashpots are connected in series and therefore a given mass is only directly acted upon by the springs and dashpots connected to its neighboring masses. The matrices would not be banded if nonadjacent masses were connected (for example if a spring connected m_1 with m_N, etc., as well as m_2). Banding of matrices provides a computational advantage for the analysis of large systems. However, for general systems, such convenient banding does not always occur directly.

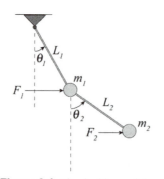

Figure 6.4 The double pendulum.

6.1.2 The Double Pendulum

Another exemplary problem is that of a multiple pendulum. We here consider the double pendulum (Figure 6.4) for simplicity. The pendulum is comprised of two masses m_1 and m_2 that are pinned to rigid rods (or cords) of length L_1 and L_2, respectively. One end of the first rod is pinned to a rigid support, as indicated. For this system the coordinates θ_1 and θ_2 measure the displacement of the two masses. The system therefore possesses two degrees of freedom. In this section we will derive the equations of motion for the double pendulum using Newton's Laws of Motion. It will be seen that, even for this relatively simple two degree of freedom system, the derivation is quite complex and must be carefully implemented. (The difficulty in deriving the governing equations for systems of this type possessing more than two degrees of freedom, say the triple or quadruple pendulum, is compounded accordingly.) The governing equations for the double pendulum are considered again in Section 6.2, where the advantage of Lagrange's Equations will become apparent. For the moment, however, let us derive the corresponding equations of motion using traditional vector mechanics. To do this, we must first evaluate the acceleration of each mass with respect to a common, and convenient, reference frame.

Kinematics

Consider the coordinates (ξ, η) aligned with the deflected position of the first rod at the instant of observation. Since mass m_1 moves along a circular path, its acceleration is easily expressed in terms of polar coordinates (see Section 1.5.1), and is thus of the same form as that for the mass of the simple pendulum of Section 2.1.4. Hence,

$$\vec{a}^{(1)} = L_1 \ddot{\theta}_1 \vec{e}_\xi + L_1 \dot{\theta}^2 \vec{e}_\eta \tag{6.8}$$

where $\vec{e}_\xi$ and $\vec{e}_\eta$ correspond to unit vectors in the axial and normal directions of the first rod, respectively.

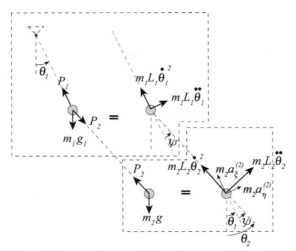

Figure 6.5 Kinetic diagram for the double pendulum.

The acceleration of mass m_2 may be expressed as the sum of the acceleration of mass m_1 and the acceleration of mass m_2 as seen by an observer translating, but not rotating, with mass m_1 (see Section 1.5.1). Thus,

$$\vec{a}^{(2)} = \vec{a}^{(1)} + \vec{a}^{(2/1)}$$

The above form may be decomposed into its components along the (ξ, η) directions, and may be expressed in terms of the angular coordinates θ_1 and θ_2 as follows:

$$\vec{a}^{(2)} = a_\xi^{(2)} \vec{e}_\xi + a_\eta^{(2)} \vec{e}_\eta \tag{6.9}$$

where

$$a_\xi^{(2)} = L_1 \ddot{\theta} + L_2 \ddot{\theta}_2 \cos\psi - L_2 \dot{\theta}_2^{\,2} \sin\psi$$
$$a_\eta^{(2)} = L_1 \dot{\theta}_1^{\,2} + L_2 \ddot{\theta}_2 \sin\psi + L_2 \dot{\theta}_2^{\,2} \cos\psi \tag{6.10}$$

and

$$\psi = \theta_2 - \theta_1 \tag{6.11}$$

Kinetics

With the expressions for the acceleration established, we may now derive the equations of motion for the system. To do this we first isolate each mass and draw the associated kinetic diagram (Figure 6.5). Once this is done we can express Newton's Second Law for each mass. For this system, it is more expedient to take the moment of Newton's Second Law about the common origin, O, for each mass individually.

We then have, for the first mass,

$$\sum \vec{M}_0^{(1)} = \vec{r}^{(1)} \times m_1 \vec{a}^{(1)} :$$

$$-L_1 m_1 g \sin \theta_1 + L_1 F_1 \cos \theta_1 + L_1 P_2 \sin \psi = L_1 m_1 L_1 \ddot{\theta}_1$$

or, after rearranging,

$$m_1 L_1^2 \ddot{\theta}_1 + m_1 g L_1 \sin \theta_1 - L_1 P_2 \sin \psi = F_1 L_1 \cos \theta_1 \qquad (6.12)$$

It may be seen that the unknown internal force P_2 appears in Eq. (6.12). To find this force in terms of the chosen coordinates and their derivatives, we write out the ξ component of Newton's Second Law for mass m_2. Thus,

$$\sum F_\xi^{(2)} = m_2 a_\xi^{(2)} :$$

$$-P_2 \sin \psi - m_2 g \sin \theta_1 + F_2 \cos \theta_1 = m_2 (L_1 \ddot{\theta}_1 + L_2 \ddot{\theta}_2 \cos \psi - L_2 \dot{\theta}_2^2 \sin \psi) \quad (6.13)$$

Solving Eq. (6.13) for P_2 and substituting the resulting expression into Eq. (6.12) eliminates the unknown internal force and gives the equation of motion for mass m_1 as

$$(m_1 + m_2) L_1^2 \ddot{\theta}_1 + m_2 L_1 L_2 \ddot{\theta}_2 \cos \psi - m_2 L_1 L_2 \dot{\theta}_2^2 \sin \psi$$
$$+ (m_1 + m_2) g L_1 \sin \theta_1 = (F_1 + F_2) L_1 \cos \theta_1 \qquad (6.14)$$

Linearization of Eq. (6.14) about $\theta_1 = \theta_2 = 0$, as discussed in Section 2.1.4, renders the corresponding equation of motion for mass m_1 to the form

$$(m_1 + m_2) L_1^2 \ddot{\theta}_1 + m_2 L_1 L_2 \ddot{\theta}_2 + (m_1 + m_2) g L_1 \theta_1 = (F_1 + F_2) L_1 \qquad (6.15)$$

To obtain the equation of motion for mass m_2 we proceed in a similar manner. We thus take the moment about the origin of the statement of Newton's Second Law for mass m_2. Hence,

$$\sum \vec{M}_0^{(2)} = \vec{r}^{(2)} \times m_2 \vec{a}^{(2)} :$$

$$P_2 \cos \psi L_2 \sin \psi - P_2 \sin \psi (L_1 + L_2 \cos \psi) + F_2 (L_1 \cos \theta_1 + L_2 \cos \theta_2)$$
$$-m_2 g (L_1 \sin \theta_1 + L_2 \sin \theta_2) = m_2 a_\xi^{(2)} (L_1 + L_2 \cos \psi) + m_2 a_\eta^{(2)} L_2 \sin \psi$$

Substituting Eqs. (6.10) and (6.13) into the above expression then gives the desired form of the governing equation as

$$m_2 L_1 L_2 \left[\ddot{\theta}_1 \cos \psi + \dot{\theta}_1^2 \sin \psi \right] + m_2 L_2^2 \left[\ddot{\theta}_2 - \dot{\theta}_2^2 \sin \psi (1 - \cos \psi) \right]$$
$$+ m_2 g L_2 \sin \theta_2 = F_2 L_2 \cos \theta_2 \qquad (6.16)$$

Linearization of Eq. (6.16) about $\theta_1 = \theta_2 = 0$ renders the corresponding equation of motion for mass m_2 as

$$m_2 L_1 L_2 \ddot{\theta}_1 + m_2 L_2^2 \ddot{\theta}_2 + m_2 g\, L_2 \theta_2 = F_2 L_2 \qquad (6.17)$$

Equations (6.15) and (6.17) describe the coupled motion of the double pendulum. As for the class of mass-spring systems discussed previously, it is of interest to express these equations in matrix form. When this is done, the equation of (small angle) motion for the double pendulum takes the form

$$\begin{bmatrix} (m_1 + m_2)L_1^2 & m_2 L_1 L_2 \\ m_2 L_1 L_2 & m_2 L_2^2 \end{bmatrix} \begin{Bmatrix} \ddot{\theta}_1 \\ \ddot{\theta}_2 \end{Bmatrix} + \begin{bmatrix} (m_1 + m_2)gL_1 & 0 \\ 0 & m_2 gL_2 \end{bmatrix} \begin{Bmatrix} \theta_1 \\ \theta_2 \end{Bmatrix}$$
$$= \begin{Bmatrix} (F_1 + F_2)L_1 \\ F_2 L_2 \end{Bmatrix} \qquad (6.18)$$

or, equivalently,

$$\mathbf{m}\ddot{\mathbf{u}} + \mathbf{k}\mathbf{u} = \mathbf{F}$$

where

$$\mathbf{m} = \begin{bmatrix} (m_1 + m_2)L_1^2 & m_2 L_1 L_2 \\ m_2 L_1 L_2 & m_2 L_2^2 \end{bmatrix} \qquad (6.19)$$

$$\mathbf{k} = \begin{bmatrix} (m_1 + m_2)gL_1 & 0 \\ 0 & m_2 gL_2 \end{bmatrix} \qquad (6.20)$$

$$\mathbf{F} = \begin{Bmatrix} (F_1 + F_2)L_1 \\ F_2 L_2 \end{Bmatrix} \qquad (6.21)$$

$$\mathbf{u} = \begin{Bmatrix} \theta_1 \\ \theta_2 \end{Bmatrix} \qquad (6.22)$$

Note that, for this system, the elements of the displacement matrix are actually angular displacements and the elements of the force matrix are actually moments. Thus, we speak of the "displacement matrix" and of the "force matrix" in the most general sense, with their elements interpreted accordingly. It may be seen that this system is coupled through the mass matrix. Such coupling is referred to as *inertial coupling*.

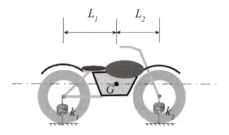

Figure 6.6 Motorcycle frame.

6.1.3 Two-Dimensional Motion of a Rigid Frame

Consider the motorcycle frame shown in Figure 6.6. The frame has mass m and length $L = L_1 + L_2$, as indicated. Its center of mass G is located a horizontal distance L_1 from the rear of the bike, as shown, and its moment of inertia about an axis through the center of mass is I_G. The tires are modeled as elastic springs, with the respective stiffnesses k_1 and k_2, that resist vertical motion as shown. We shall derive the equations of motion for this twice, using different sets of coordinates. It will be seen that the type of coupling is dependent of the coordinates chosen to describe the motion of the system.

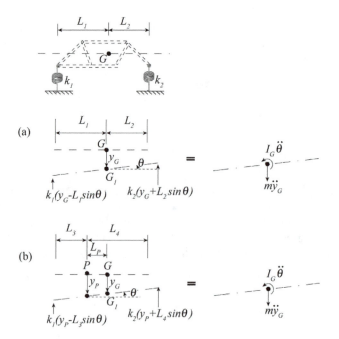

Figure 6.7 Kinetic diagram for motorcycle frame.

Case 1: Translation of, and Rotation About, the Center of Mass

A natural set of coordinates to describe the motion of the frame is the coordinate that describes the lateral position of the center of mass, and the coordinate that measures the rotation about an axis through that point (both measured with respect to the equilibrium configuration of the system). These measures are denoted as y_G and θ, respectively, as shown on the kinetic diagram of Figure 6.7a. Writing Newton's Second Law for the vertical motion of the frame gives

$$\sum F_y = ma_y :$$

$$-k_1(y_G - L_1 \sin\theta) - k_2(y_G + L_2 \sin\theta) = m\ddot{y}_G$$

which for small angle motion simplifies to

$$m\ddot{y}_G + (k_1 + k_2)y_G + (k_2 L_2 - k_1 L_1)\theta = 0 \tag{6.23}$$

We next write the equation of rotational motion of the frame about its center of mass. Hence, applying Eq. (1.162),

$$\sum M_G = I_G \alpha :$$

$$k_2(y_G + L_2 \sin\theta)L_2 - k_1(y_G - L_1 \sin\theta)L_1 = -I_G \ddot{\theta}$$

which for small angle motion simplifies to

$$I_G \ddot{\theta} + (k_2 L_2 - k_1 L_1)y_G + (k_1 L_1^2 + k_2 L_2^2)\theta = 0 \tag{6.24}$$

Eqs. (6.23) and (6.24) can be combined and represented in matrix form as

$$\begin{bmatrix} m & 0 \\ 0 & I_G \end{bmatrix} \begin{Bmatrix} \ddot{y}_G \\ \ddot{\theta} \end{Bmatrix} + \begin{bmatrix} (k_1 + k_2) & (k_2 L_2 - k_1 L_1) \\ (k_2 L_2 - k_1 L_1) & (k_1 L_1^2 + k_2 L_2^2) \end{bmatrix} \begin{Bmatrix} y_G \\ \theta \end{Bmatrix} = \begin{Bmatrix} 0 \\ 0 \end{Bmatrix} \tag{6.25}$$

or

$$\mathbf{m\ddot{u} + ku = 0}$$

where

$$\mathbf{m} = \begin{bmatrix} m & 0 \\ 0 & I_G \end{bmatrix} \tag{6.26}$$

$$\mathbf{k} = \begin{bmatrix} (k_1 + k_2) & (k_2 L_2 - k_1 L_1) \\ (k_2 L_2 - k_1 L_1) & (k_1 L_1^2 + k_2 L_2^2) \end{bmatrix} \tag{6.27}$$

and

$$\mathbf{u} = \begin{Bmatrix} y_G \\ \theta \end{Bmatrix} \tag{6.28}$$

It may be seen from Eqs. (6.23) and (6.24), or Eq. (6.25), or Eqs. (6.26) and (6.27) that the system is coupled through the stiffnesses for this choice of coordinates. Note that the equations completely decouple if the parameters of the system are such that $k_1 L_1 = k_2 L_2$.

Case 2: Translation of, and Rotation About, an Arbitrary Point

Consider the vertical translation of some arbitrary point "P" located a distance L_3 from the rear of the frame as shown in the corresponding kinetic diagram of Figure 6.7b, and the rotation about an axis through that point. Let L_4 locate the same point from the front end of the frame and L_P locate that point with respect to the center of mass, as indicated. In this case we choose the lateral displacement of point P, y_P, and the angular displacement, θ, to describe the motion of this two degree of freedom system. We next parallel the development of Case 1, but with the present set of coordinates.

We first note that the lateral displacement of the center of mass may be expressed in terms of the lateral displacement of point P and the rotational displacement. Hence,

$$y_G = y_P + L_P \sin \theta \tag{6.29}$$

and, after differentiating twice with respect to time,

$$\ddot{y}_G = \ddot{y}_P + L_P \left(\ddot{\theta} \cos \theta - \dot{\theta}^2 \sin \theta \right) \tag{6.30}$$

Newton's Second Law then gives the equation of lateral motion as

$$-k_1 (y_P - L_3 \sin \theta) - k_2 (y_P + L_4 \sin \theta) = m \left[\ddot{y}_P + L_P \left(\ddot{\theta} \cos \theta - \dot{\theta}^2 \sin \theta \right) \right]$$

where L_P is the distance between points P and G, as indicated in Figure 6.7b. For small angle motion, the above equation of translational motion simplifies to

$$m \ddot{y}_P + m L_P \ddot{\theta} + (k_1 + k_2) y_P + (k_2 L_4 - k_1 L_3) \theta = 0 \tag{6.31}$$

We next write the equation rotational motion about an axis through P. Applying Eq. (1.163) for the present system gives the equation of rotational motion of the frame as

$$k_2 (y_P + L_4 \sin \theta) L_4 \cos \theta - k_1 (y_P - L_3 \sin \theta) L_3 \cos \theta$$
$$= -I_G \ddot{\theta} - m \ddot{y}_G L_P \cos \theta \tag{6.32}$$

For small angle motion, the equation of rotational motion simplifies to

$$mL_P \ddot{y}_G + I_G \ddot{\theta} + (k_2 L_4 - k_1 L_3) y_P + (k_2 L_4^2 + k_1 L_3^2)\theta = 0 \qquad (6.33)$$

Equations (6.31) and (6.33) may be combined and expressed in matrix form as

$$\begin{bmatrix} m & mL_P \\ mL_P & I_G \end{bmatrix} \begin{Bmatrix} \ddot{y}_P \\ \ddot{\theta} \end{Bmatrix} + \begin{bmatrix} (k_1 + k_2) & (k_2 L_4 - k_1 L_3) \\ (k_2 L_4 - k_1 L_3) & (k_2 L_4^2 + k_1 L_3^2) \end{bmatrix} \begin{Bmatrix} y_P \\ \theta \end{Bmatrix} = \begin{Bmatrix} 0 \\ 0 \end{Bmatrix} \qquad (6.34)$$

It may be seen from Eq. (6.34) that the system is coupled both inertially and elastically. However, if the location of point P (and hence the ratio L_3/L_4) is chosen such that $L_3/L_4 = k_2/k_1$, then the stiffness matrix will be diagonal, and the system will only be coupled inertially. For this case the equation of motion reduces to the form

$$\mathbf{m\ddot{u} + ku = 0}$$

where

$$\mathbf{m} = \begin{bmatrix} m & mL_P \\ mL_P & I_G \end{bmatrix} \qquad (6.35)$$

and

$$\mathbf{k} = \begin{bmatrix} (k_1 + k_2) & (k_2 L_4 - k_1 L_3) \\ (k_2 L_4 - k_1 L_3) & (k_2 L_4^2 + k_1 L_3^2) \end{bmatrix} \rightarrow \begin{bmatrix} (k_1 + k_2) & 0 \\ 0 & (k_2 L_4^2 + k_1 L_3^2) \end{bmatrix} \qquad (6.36)$$

$$\mathbf{u} = \begin{Bmatrix} y_P \\ \theta \end{Bmatrix} \qquad (6.37)$$

Note that if point P corresponds to one of the edges of the frame, say the left edge, then $L_3 = L$ and $L_4 = 0$. It may be seen that, for this choice of coordinates, the equations that govern the system cannot be decoupled.

We thus see that, in general, the equations of motion of a system may be coupled through the mass matrix, the stiffness matrix, or through both. (For damped systems, coupling may occur through the damping matrix as well.) We also see that a proper choice of coordinates can simplify the equations of motion.

6.2 Lagrange's Equations

The equation of motion for the N-degree of freedom system comprised of a series of springs, dampers and masses in rectilinear motion considered in the previous section was easily derived by direct application of Newton's Laws of Motion. This, however, will not usually be the case when one considers complex systems that exhibit multi-dimensional motion. For such situations the vector approach is generally cumbersome and a scalar technique is often desirable. An approach from an area of mechanics known as Analytical Dynamics will help in this regard. This is the utilization of *Lagrange's Equations*. Though the development is somewhat abstract, the implementation is rather straightforward. Once it is mastered, the application of Lagrange's Equations for the derivation of the equations of motion for complex multi-degree of freedom systems can be accomplished with a fair amount of ease. In this section we develop Lagrange's Equations and demonstrate their utilization in deriving the equations of motion for multi-degree of freedom systems. We first introduce the concept of virtual work.

6.2.1 Virtual Work

To begin, let $\{q_1, q_2, ..., q_N\}$ represent the set of independent "generalized coordinates" used to describe the motion of some N-degree of freedom system. These may be, for example, the linear coordinates (u_1, u_2) used to describe the displacements of the two mass system connected in series, the angular coordinates $(\theta_1, \theta_2, \theta_3)$ employed to describe the motion of a triple pendulum, the combination of linear and angular coordinates (y, θ) adopted to describe the motion of the motorcycle frame discussed earlier, or some other type of coordinate description. Further, let

$$\vec{r}_l = \vec{r}_l(q_1, q_2, ..., q_N, t) \tag{6.38}$$

denote the position vector of mass m_l $(l = 1, 2, ..., N)$. Consider a set of *virtual* increments in position (*virtual displacements*) of the various mass elements of the system that are consistent with the constraints of the system, but are otherwise arbitrary. That is, consider the increments

$$\delta \vec{r}_l \quad (l = 1, 2, ..., N)$$

such that the constraints of the system are not violated. The operator δ is used to denote the differential increment rather than d since the increments are considered *virtual* (possible, but not actual). The virtual increments in position are thus referred to as *virtual displacements*. [As an example, suppose we were to consider the virtual motion of the ladder/rod that is constrained to move along the vertical wall and horizontal wall, as indicated in Figure 6.8. For this system, the virtual displacement (the increment in position) of mass A must be vertical and that of mass B must be horizontal in order to comply with the constraints imposed on system due the wall, the floor, and the rigid rod. Thus, virtual displacements through the walls or off the tracks are not permitted. Further, the possible displacements of the two masses/rod ends cannot be independent of one another but, rather, must be related through the geometry of the rigid rod. Thus, for the constrained rod,

$$s_A{}^2 + s_B{}^2 = L^2 \quad \rightarrow \quad \cancel{2}s_A \delta s_A + \cancel{2}s_B \delta s_B = 0$$

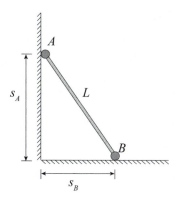

Figure 6.8 Example of a constrained system.

where s_A and s_B respectively locate the positions of mass A and mass B, as indicated.] Returning to the discussion of a general N-degree of freedom system, the *virtual work* done by the external force and by the resultant internal force acting on mass m_l, as the mass moves through the corresponding virtual displacement $\delta \vec{r}_l$, is then

$$\delta W_l = \left(\vec{F}_l + \vec{P}_l \right) \cdot \delta \vec{r}_l \tag{6.39}$$

where,

$$\vec{P}_l = \sum_{j=1}^{N} \vec{P}_{lj} \tag{6.40}$$

$\vec{P}_{lj}$ is the internal force exerted on mass m_l by mass m_j, and it is understood that $\vec{P}_{11} = \vec{P}_{22} = ... = \vec{P}_{NN} = \vec{0}$ (i.e., no mass exerts a force on itself). It follows from Newton's Third Law that $\vec{P}_{jl} = -\vec{P}_{lj}$ ($l, j = 1, 2, ..., N$), and hence that

$$\sum_{l=1}^{N} \vec{P}_l = \sum_{l=1}^{N} \sum_{j=1}^{N} \vec{P}_{lj} = (\vec{P}_{11} + \vec{P}_{12} + ... + \vec{P}_{1N}) + ... + (\vec{P}_{N1} + \vec{P}_{N2} + ... + \vec{P}_{NN}) = \vec{0} \tag{6.41}$$

Thus, adding the work done by the forces acting on each particle of the system as expressed by Eq. (6.39), and noting that the internal forces of the system sum to zero as expressed by Eq. (6.41), gives *the total virtual work done on the system* as

$$\delta W = \sum_{l=1}^{N} \delta W_l = \sum_{l=1}^{N} \vec{F}_l \cdot \delta \vec{r}_l \tag{6.42}$$

6.2.2 The Canonical Equations

Now that the concepts of generalized coordinates and virtual work have been established we proceed to develop the general statement of Lagrange's Equations from Newton's Second Law of Motion applied to the individual particles that comprise the system and converting them to an energy statement.

Writing Newton's Second Law for mass m_l, we have

$$\vec{F}_l + \vec{P}_l = m_l \frac{d^2 \vec{r}_l}{dt^2}$$

or

$$\vec{F}_l + \vec{P}_l - m_l \frac{d^2 \vec{r}_l}{dt^2} = 0 \quad (l = 1, 2, ..., N) \tag{6.43}$$

Taking the scalar dot product of Eq. (6.43) with its corresponding virtual displacement gives

$$\left[\vec{F}_l + \vec{P}_l - m_l \frac{d^2 \vec{r}_l}{dt^2} \right] \cdot \delta \vec{r}_l = 0 \quad (l = 1, 2, ..., N) \tag{6.44}$$

Summing the equations for all masses that comprise the system and noting Eq. (6.41) results in the statement

$$\sum_{l=1}^{N} \left[\left(\vec{F}_l - m_l \ddot{\vec{r}}_l \right) \cdot \delta \vec{r}_l \right] = 0 \tag{6.45}$$

which is seen to be a work type statement. We next wish to convert this statement to work-energy form.

It follows from Eq. (6.38) that

$$d\vec{r}_l = \frac{\partial \vec{r}_l}{\partial q_1} dq_1 + \frac{\partial \vec{r}_l}{\partial q_2} dq_2 + ... + \frac{\partial \vec{r}_l}{\partial q_N} dq_N = \sum_{j=1}^{N} \frac{\partial \vec{r}_l}{\partial q_j} dq_j \tag{6.46}$$

which, after dividing through by dt becomes

$$\dot{\vec{r}}_l \equiv \frac{d\vec{r}_l}{dt} = \frac{\partial \vec{r}_l}{\partial q_1} \dot{q}_1 + \frac{\partial \vec{r}_l}{\partial q_2} \dot{q}_2 + ... + \frac{\partial \vec{r}_l}{\partial q_N} \dot{q}_N = \sum_{j=1}^{N} \frac{\partial \vec{r}_l}{\partial q_j} \dot{q}_j \tag{6.47}$$

It follows from Eq. (6.47) that

$$\frac{\partial \dot{\vec{r}}_l}{\partial \dot{q}_j} = \frac{\partial \vec{r}_l}{\partial q_j} \quad (l, j = 1, 2, ..., N) \tag{6.48}$$

Since the virtual displacements correspond to possible displacements and therefore possess the functional properties of actual displacements, it follows that the virtual increment in position can be expressed in the form of Eq. (6.46). Hence,

$$\delta \vec{r}_l = \frac{\partial \vec{r}_l}{\partial q_1} \delta q_1 + \frac{\partial \vec{r}_l}{\partial q_2} \delta q_2 + ... + \frac{\partial \vec{r}_l}{\partial q_N} \delta q_N = \sum_{j=1}^{N} \frac{\partial \vec{r}_l}{\partial q_j} \delta q_j \tag{6.49}$$

With the identities stated by Eqs. (6.48) and (6.49) established, we now return to the evaluation of Eq. (6.45). We shall perform this evaluation term by term and in reverse order.

Let us take the second expression in Eq. (6.45), substitute Eq. (6.49), and interchange the order of summation. This gives

$$\sum_{l=1}^{N} m_l \ddot{\vec{r}}_l \cdot \delta \vec{r}_l = \sum_{l=1}^{N} m_l \ddot{\vec{r}}_l \cdot \sum_{j=1}^{N} \frac{\partial \vec{r}_l}{\partial q_j} \delta q_j = \sum_{j=1}^{N} \left(\sum_{l=1}^{N} m_l \ddot{\vec{r}}_l \cdot \frac{\partial \vec{r}_l}{\partial q_j} \right) \delta q_j$$

$$= \sum_{j=1}^{N} \sum_{l=1}^{N} \left[\frac{d}{dt} \left(m_l \dot{\vec{r}}_l \cdot \frac{\partial \vec{r}_l}{\partial q_j} \right) - m_l \dot{\vec{r}}_l \cdot \frac{d}{dt} \left(\frac{\partial \vec{r}_l}{\partial q_j} \right) \right] \delta q_j$$

If we next incorporate the identity specified by Eq. (6.48) in the first term of the last bracketed expression, and interchange the order of differentiation in the second term, the above equation takes the form

$$\sum_{l=1}^{N} m_l \ddot{\vec{r}}_l \cdot \delta \vec{r}_l = \sum_{j=1}^{N} \sum_{l=1}^{N} \left[\frac{d}{dt} \left(m_l \dot{\vec{r}}_l \cdot \frac{\partial \dot{\vec{r}}_l}{\partial \dot{q}_j} \right) - m_l \dot{\vec{r}}_l \cdot \frac{\partial \dot{\vec{r}}_l}{\partial q_j} \right] \delta q_j$$

$$= \sum_{j=1}^{N} \sum_{l=1}^{N} \left[\frac{d}{dt} \frac{\partial}{\partial \dot{q}_j} \left(\tfrac{1}{2} m_l \dot{\vec{r}}_l \cdot \dot{\vec{r}}_l \right) - \frac{\partial}{\partial q_j} \left(\tfrac{1}{2} m_l \dot{\vec{r}}_l \cdot \dot{\vec{r}}_l \right) \right] \delta q_j$$

Finally,

$$\sum_{l=1}^{N} m_l \ddot{\vec{r}}_l \cdot \delta \vec{r}_l = \sum_{j=1}^{N} \left(\frac{d}{dt} \frac{\partial T}{\partial \dot{q}_j} - \frac{\partial T}{\partial q_j} \right) \delta q_j \tag{6.50}$$

where

$$T = T(q_1, q_2, ..., q_N; \dot{q}_1, \dot{q}_2, ..., \dot{q}_N) = \sum_{l=1}^{N} \tfrac{1}{2} m_l \dot{\vec{r}}_l \cdot \dot{\vec{r}}_l \tag{6.51}$$

is the total kinetic energy of the system. To evaluate the first term in Eq. (6.45), let us first decompose the resultant force acting on a given mass into the sum of resultant conservative and nonconservative forces (see Sections 1.5.2 and 1.5.3). Thus, let

$$\vec{F}_l = \vec{F}_l^{(C)} + \vec{F}_l^{(NC)} \quad (l = 1, 2, ..., N) \tag{6.52}$$

where $\vec{F}_l^{(C)}$ is the resultant conservative force acting on mass m_l and $\vec{F}_l^{(NC)}$ is the corresponding resultant nonconservative force. The total work done on the system as it moves through the virtual displacements $\delta \vec{r}_l$ ($l = 1, 2, ..., N$) may then be similarly partitioned as

$$\delta W = \sum_{l=1}^{N} \vec{F}_l \bullet \delta \vec{r}_l = \sum_{l=1}^{N} \vec{F}_l^{(C)} \bullet \delta \vec{r}_l + \sum_{l=1}^{N} \vec{F}_l^{(NC)} \bullet \delta \vec{r}_l = \delta W^{(C)} + \delta W^{(NC)} \quad (6.53)$$

where $\delta W^{(C)}$ and $\delta W^{(NC)}$ represent the increment in total work done on the system by the conservative and by the nonconservative forces, respectively. As discussed in Section 1.5.2, the (change in) potential energy of a system is the negative of the work done by a conservative forces in moving the system between two configurations. The increment in potential energy then follows accordingly. Hence,

$$\delta W^{(C)} = -\delta U \quad (6.54)$$

where

$$U = U(q_1, q_2, ..., q_N) \quad (6.55)$$

is the total potential energy of the system. Upon, substitution of Eq. (6.54), Eq. (6.53) takes the form

$$\sum_{l=1}^{N} \vec{F}_l \bullet \delta \vec{r}_l = -\delta U + \delta W^{(NC)} \quad (6.56)$$

where it follows from Eq. (6.55) that

$$\delta U = \frac{\partial U}{\partial q_1} \delta q_1 + \frac{\partial U}{\partial q_2} \delta q_2 + ... + \frac{\partial U}{\partial q_N} \delta q_N = \sum_{j=1}^{N} \frac{\partial U}{\partial q_j} \delta q_j \quad (6.57)$$

Lastly, we wish to express the increment in total work of the nonconservative forces in terms of the generalized coordinates. This includes the work of any forces that are not included in the total potential energy. It follows from Eqs. (6.49) and (6.53) that

$$\delta W^{(NC)} = Q_1 \delta q_1 + Q_2 \delta q_2 + ... + Q_N \delta q_N = \sum_{j=1}^{N} Q_j \delta q_j \quad (6.58)$$

where

$$Q_j = \sum_{l=1}^{N} \vec{F}_l^{(NC)} \bullet \frac{\partial \vec{r}_l}{\partial q_j} \quad (j = 1, 2, ... N) \quad (6.59)$$

The set $\{Q_1, Q_2, ..., Q_N\}$ represents the set of "generalized" forces associated one to one with the elements of the set of generalized coordinates $\{q_1, q_2, ..., q_N\}$. Substitution of Eqs. (6.57) and (6.58) into Eq. (6.56) gives

$$\sum_{l=1}^{N} \vec{F}_l \cdot \delta \vec{r}_l = \sum_{j=1}^{N} \left(Q_j - \frac{\partial \mathcal{U}}{\partial q_j} \right) \delta q_j \tag{6.60}$$

Having evaluated both expressions of Eq. (6.45) in terms of the generalized coordinates, we may now substitute back to obtain an alternate form of that equation. Doing this results in

$$\sum_{j=1}^{N} \left\{ Q_j - \left[\frac{d}{dt} \left(\frac{\partial \mathcal{T}}{\partial \dot{q}_j} \right) - \frac{\partial \mathcal{T}}{\partial q_j} + \frac{\partial \mathcal{U}}{\partial q_j} \right] \right\} \delta q_j = 0 \tag{6.61}$$

Now, recall that the only restriction on the virtual displacements was that they must be compatible with the constraints imposed on the system. They are otherwise arbitrary. Since each δq_j ($j = 1, 2, ..., N$) appearing in Eq. (6.61) is arbitrary, that equation is identically satisfied only if the corresponding coefficients (the expressions within braces in that equation) vanish identically. This results in N equations of the form

$$\frac{d}{dt} \left(\frac{\partial L}{\partial \dot{q}_j} \right) - \frac{\partial L}{\partial q_j} = Q_j \quad (j = 1, 2, ..., N) \tag{6.62}$$

where

$$L \equiv \mathcal{T} - \mathcal{U} \tag{6.63}$$

is referred to as the *Lagrangian*. The equations defined by Eq. (6.62) are referred to as *Lagrange's Equations* and may be used to derive the equations of motion for multi-particle systems.

6.2.3 Implementation

The governing equations for any discrete system can be derived directly from Lagrange's Equations. A simple procedure for this purpose is delineated below.

1. Establish a set of independent generalized coordinates to describe the motion of the system.
2. Form the potential and kinetic energy functionals for the system and the virtual work of the applied forces.
3. Express the virtual work and potential and kinetic energy of the system in terms of the chosen set of independent coordinates.
4. Determine the generalized forces for each degree of freedom. The corresponding generalized forces may be identified as the coefficients of the variations (virtual increments) of the chosen set of generalized coordinates, as per Eq. (6.58).

5. Substitute the generalized forces and the pertinent potential and kinetic energy expressions into Eqs. (6.62) and perform the indicated operations for each degree of freedom. The result is the equations of motion for the system.
6. Express the equations of motion of the system in matrix form.

Several examples are presented next to elucidate this process.

Example 6.3

Use Lagrange's Equations to derive the equations of motion for the 3 mass, 3 spring system depicted in Figure E6.2-1.

Solution
The system is easily identified as having three degrees of freedom, and the displacements of the three masses, u_1, u_2 and u_3, respectively, are the obvious choice of generalized coordinates for this case. Hence, for the system under consideration, $N = 3$ and we have chosen $\{q_1, q_2, q_3\} \leftrightarrow \{u_1, u_2, u_3\}$. The virtual displacements are thus readily identified as δu_1, δu_2 and δu_3. The corresponding virtual work done by the applied forces as the masses move through these virtual displacements is then simply

$$\delta W^{(NC)} = F_1 \delta u_1 + F_2 \delta u_2 + F_3 \delta u_3 \tag{a}$$

where the generalized forces are identified as simply the applied forces. (What would the generalized forces be if we were to include damping? – See Example 6.4.) Thus, for the present case, we have the simple correspondence

$$\{Q_1, Q_2, Q_3\} \leftrightarrow \{F_1, F_2, F_3\} \tag{b}$$

Next, the total potential and kinetic energies of the system are easily expressed in terms of the chosen coordinates as

$$\mathcal{U} = \mathcal{U}(u_1, u_2, u_3) = \tfrac{1}{2} k_1 u_1^2 + \tfrac{1}{2} k_2 (u_2 - u_1)^2 + \tfrac{1}{2} k_3 (u_3 - u_2)^2 + \tfrac{1}{2} k_4 u_3^2 \tag{c}$$

$$\mathcal{T} = \mathcal{T}(\dot{u}_1, \dot{u}_2, \dot{u}_3) = \tfrac{1}{2} m_1 \dot{u}_1^2 + \tfrac{1}{2} m_2 \dot{u}_2^2 + \tfrac{1}{2} m_3 \dot{u}_1^2 \tag{d}$$

Recall that the Lagrangian is defined as the difference between the kinetic and potential energy of the system. Hence, $L = \mathcal{T} - \mathcal{U}$.

Now, substituting Eqs. (b)–(d) into Eq. (6.63) and evaluating Lagrange's Equations (6.62) gives

$j = 1$:

$$\frac{d}{dt}\left(\frac{\partial L}{\partial \dot{q}_1}\right) - \frac{\partial L}{\partial q_1} = Q_1 \quad \rightarrow \quad \frac{d}{dt}(m_1\dot{u}_1) - \left[-k_1 u_1 + k_2(u_2 - u_1)\right] = F_1$$

Regrouping terms results in the equation of motion for the first mass,

$$m_1\ddot{u}_1 + (k_1 + k_2)u_1 - k_2 u_2 = F_1 \qquad \text{(e)}$$

$j = 2$:

$$\frac{d}{dt}\left(\frac{\partial L}{\partial \dot{q}_2}\right) - \frac{\partial L}{\partial q_2} = Q_2 \quad \rightarrow \quad \frac{d}{dt}(m_2\dot{u}_2) + k_2(u_2 - u_1) + k_3(u_3 - u_2)(-1) = F_2$$

This results in the equation of motion for the second mass,

$$m_2\ddot{u}_2 - k_2 u_1 + (k_2 + k_3)u_2 - k_3 u_3 = F_2 \qquad \text{(f)}$$

$j = 3$:

$$\frac{d}{dt}\left(\frac{\partial L}{\partial \dot{q}_1}\right) - \frac{\partial L}{\partial q_1} = Q_1 \quad \rightarrow \quad \frac{d}{dt}(m_1\dot{u}_1) - \left[-k_1 u_1 + k_2(u_2 - u_1)\right] = F_1$$

which gives the equation of motion for the third mass,

$$m_3\ddot{u}_3 - k_3 u_2 + (k_3 + k_4)u_3 = F_3 \qquad \text{(g)}$$

Eqs. (e) – (f) may be combined in matrix form as

$$\begin{bmatrix} m_1 & 0 & 0 \\ 0 & m_2 & 0 \\ 0 & 0 & m_3 \end{bmatrix}\begin{Bmatrix} \ddot{u}_1 \\ \ddot{u}_2 \\ \ddot{u}_3 \end{Bmatrix} + \begin{bmatrix} (k_1 + k_2) & -k_2 & 0 \\ -k_2 & (k_2 + k_3) & -k_3 \\ 0 & -k_3 & (k_3 + k_4) \end{bmatrix}\begin{Bmatrix} u_1 \\ u_2 \\ u_3 \end{Bmatrix} = \begin{Bmatrix} F_1 \\ F_2 \\ F_3 \end{Bmatrix} \quad \triangleleft \text{(h)}$$

Equation (h) may be compared directly with Eqs. (6.2)–(6.6) for the case where $N = 3$, $u_0 \equiv 0$ and $\mathbf{c} = \mathbf{0}$.

Example 6.4

Use Lagrange's Equations to derive the equations of motion for the 3-mass 3-spring 3-damper system of Figure E6.4-1.

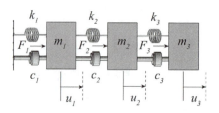

Figure E6.4-1

Solution

To derive the equations of motion for this system we must add the virtual work of the viscous damping forces to the virtual work of the applied forces when evaluating the total virtual work done by the nonconservative forces that act on the system. We may then proceed as we did for the undamped 3 mass system of Example 6.3. Thus, with the aid of the kinetic diagram depicted in Figure 6.4-2, the virtual work done by the nonconservative forces is given by

$$
\begin{aligned}
\delta \mathcal{W}^{(NC)} &= Q_1 \delta u_1 + Q_2 \delta u_2 + Q_3 \delta u_3 \\
&= \left[F_1 - c_1 \dot{u}_1 + c_2 (\dot{u}_2 - \dot{u}_1) \right] \delta u_1 + \left[F_2 - c_2 (\dot{u}_2 - \dot{u}_1) + c_3 (\dot{u}_3 - \dot{u}_2) \right] \delta u_2 \quad \text{(a)} \\
&\quad + \left[F_3 - c_3 (\dot{u}_3 - \dot{u}_2) - c_4 \dot{u}_3 \right] \delta u_3
\end{aligned}
$$

The generalized forces are then given by the coefficients of the virtual displacements in Eq. (a). Hence,

$$Q_1 = F_1 - (c_1 + c_2)\dot{u}_1 + c_2 \dot{u}_2 \tag{b}$$

$$Q_2 = F_2 + c_2 \dot{u}_1 - (c_2 + c_3)\dot{u}_2 + c_3 \dot{u}_3 \tag{c}$$

$$Q_3 = F_3 + c_3 \dot{u}_2 - (c_3 + c_4)\dot{u}_3 \tag{d}$$

The potential and kinetic energies for the damped system are the same as for the undamped system of the previous example. Therefore, incorporating the damping forces into the generalized forces in the development put forth in Example 6.3 modifies the right hand side of the resulting matrix equation of that example to the form

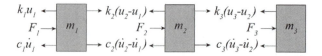

Figure E6.4-2 Kinetic diagram.

$$\begin{Bmatrix} F_1 \\ F_2 \\ F_3 \end{Bmatrix} \rightarrow \begin{Bmatrix} F_1 \\ F_2 \\ F_3 \end{Bmatrix} - \begin{bmatrix} (c_1 + c_2) & -c_2 & 0 \\ -c_2 & (c_2 + c_3) & -c_3 \\ 0 & -c_3 & (c_3 + c_4) \end{bmatrix} \begin{Bmatrix} \dot{u}_1 \\ \dot{u}_2 \\ \dot{u}_3 \end{Bmatrix} \tag{e}$$

Replacing Eq. (e) for the right-hand side of Eq. (h) of Example 6.3, then bringing the damping terms to the left-hand side of the resulting matrix equation, renders the governing equation for the damped three mass system to the form

$$\begin{bmatrix} m_1 & 0 & 0 \\ 0 & m_2 & 0 \\ 0 & 0 & m_3 \end{bmatrix} \begin{Bmatrix} \ddot{u}_1 \\ \ddot{u}_2 \\ \ddot{u}_3 \end{Bmatrix} + \begin{bmatrix} (c_1 + c_2) & -c_2 & 0 \\ -c_2 & (c_2 + c_3) & -c_3 \\ 0 & -c_3 & (c_3 + c_4) \end{bmatrix} \begin{Bmatrix} \dot{u}_1 \\ \dot{u}_2 \\ \dot{u}_3 \end{Bmatrix}$$
$$+ \begin{bmatrix} (k_1 + k_2) & -k_2 & 0 \\ -k_2 & (k_2 + k_3) & -k_3 \\ 0 & -k_3 & (k_3 + k_4) \end{bmatrix} \begin{Bmatrix} u_1 \\ u_2 \\ u_3 \end{Bmatrix} = \begin{Bmatrix} F_1 \\ F_2 \\ F_3 \end{Bmatrix} \tag{f}$$

Equation (f) may be compared directly with Eqs. (6.2)–(6.6) for the case where $N = 3$ and $u_0 \equiv 0$.

Example 6.5

Use Lagrange's Equations to derive the equations of motion for the double pendulum of Figure 6.4. Use the angular displacements θ_1 and θ_2 as the generalized coordinates for this two degree of freedom system.

Solution

For the current system $N = 2$ and $\{q_1, q_2\} \leftrightarrow \{\theta_1, \theta_2\}$. Before proceeding to form the energy functional and evaluate the virtual work of the applied forces it is convenient to relate the Cartesian and polar coordinates of each of the two masses of the system. It may seen from the figure that

$$x_1 = L_1 \sin\theta_1, \quad y_1 = L_1 \cos\theta_1 \tag{a-1,2}$$

and

$$x_2 = L_1 \sin\theta_1 + L_2 \sin\theta_2, \quad y_2 = L_1 \cos\theta_1 + L_2 \cos\theta_2 \tag{b-1,2}$$

These expressions will be helpful in implementing and interpreting what follows.

We next form the potential energy functional. Since the potential energy is defined in terms of its change (Section 1.5.2) we must first choose a datum. We shall choose the point corresponding to the rest configuration of the lower mass, m_2, as the common datum for both masses. The potential energy for the system is then given by

$$U = m_1 g h_1 + m_2 g h_2 = m_1 g(L_1 + L_2 - y_1) + m_2 g(L_1 + L_2 - y_2) \tag{c}$$

Substituting the coordinate transformation described by Eqs. (a) and (b) into Eq. (c) gives the potential energy in terms of the angular coordinates θ_1 and θ_2 as

$$U = m_1 g\left[L_1(1 - \cos\theta_1) + L_2\right] + m_2 g\left[L_1(1 - \cos\theta_1) + L_2(1 - \cos\theta_2)\right] \tag{d}$$

The kinetic energy of the system is easily written as

$$T = \tfrac{1}{2} m_1 v_1^2 + \tfrac{1}{2} m_2 v_2^2 = \tfrac{1}{2} m_1\left(\dot{x}_1^2 + \dot{y}_1^2\right) + \tfrac{1}{2} m_2\left(\dot{x}_2^2 + \dot{y}_2^2\right) \tag{e}$$

Substituting Eqs. (a) and (b) into Eq. (e) gives

$$
\begin{aligned}
T = \frac{1}{2} m_1 &\left[\left(L_1\dot{\theta}_1 \cos\theta_1\right)^2 + \left(-L_1\dot{\theta}_1 \sin\theta_1\right)^2\right] \\
+ \frac{1}{2} m_2 &\left[\left(L_1\dot{\theta}_1 \cos\theta_1 + L_2\dot{\theta}_2 \cos\theta_2\right)^2 + \left(-L_1\dot{\theta}_1 \sin\theta_1 - L_2\dot{\theta}_2 \sin\theta_2\right)^2\right]
\end{aligned} \tag{f}
$$

which, when expanded, takes the simple form

$$T = \tfrac{1}{2}(m_1 + m_2)L_1^2\dot{\theta}_1^2 + \tfrac{1}{2} m_2 L_2^2\dot{\theta}_2^2 + m_2 L_1 L_2 \dot{\theta}_1\dot{\theta}_2 \cos(\theta_1 - \theta_2) \tag{g}$$

It is instructive to note that Eq. (f) could be obtained directly by calculating the relative of velocity of m_2 with respect to m_1 and adding it to the velocity of the first mass using the reference frame depicted in Figure 6.4 and discussed in Section 6.1.2. Hence, expressing the velocity of each mass in terms of its components with respect to the path coordinates of the first mass at a given instant in time, we have

$$\vec{v}_1 = L_1\dot{\theta}_1\vec{e}_\xi \tag{h}$$

$$\vec{v}_2 = \vec{v}_1 + \vec{v}_{2/1} = L_1\dot{\theta}_1\vec{e}_\xi + L_2\dot{\theta}_2\left[\cos(\theta_2 - \theta_1)\vec{e}_\xi + \sin(\theta_2 - \theta_1)\vec{e}_\eta\right] \tag{i}$$

It follows directly that

$$v_1^2 = \vec{v}_1 \bullet \vec{v}_1 = L_1^2\dot{\theta}_1^2 \tag{j}$$

and

$$v_2^2 = \vec{v}_2 \bullet \vec{v}_2 = L_1^2\dot{\theta}_1^2 + 2L_1 L_2 \dot{\theta}_1^2\dot{\theta}_2^2 \cos(\theta_2 - \theta_1) + L_2^2\dot{\theta}_2^2 \tag{k}$$

Substitution of Eqs. (j) and (k) into Eq. (e) then gives the kinetic energy in the form of Eq. (f).

The last things we must determine before we can apply Lagrange's Equations are the generalized forces Q_1 and Q_2. To do this we must evaluate the virtual work of the applied forces F_1 and F_2. Since the indicated forces act along the horizontal, the virtual work of these forces is simply

$$\delta W^{(NC)} = F_1 \delta x_1 + F_2 \delta x_2 \tag{l}$$

However, we must express everything in terms of the chosen generalized coordinates, θ_1 and θ_2. Now, it follows from Eqs. (a) and (b) that

$$\delta x_1 = L_1 \cos \theta_1 \, \delta \theta_1 \tag{m-1}$$

and

$$\delta x_2 = L_1 \cos \theta_1 \, \delta \theta_1 + L_2 \cos \theta_2 \, \delta \theta_2 \tag{m-2}$$

Substitution of Eqs. (m-1) and (m-2) into Eq. (l) gives the virtual work of the applied forces in the desired form. Hence,

$$\delta W^{(NC)} = Q_1 \delta \theta_1 + Q_2 \delta \theta_2 = \left(F_1 + F_2\right) L_1 \cos \theta_1 \, \delta \theta_1 + F_2 L_2 \cos \theta_2 \, \delta \theta_2 \tag{n}$$

From Eq. (6.58), the generalized forces associated with the chosen coordinates are the coefficients of the virtual displacements $\delta \theta_1$ and $\delta \theta_2$. These may be read directly from Eq. (n) to give

$$Q_1 = \left(F_1 + F_2\right) L_1 \cos \theta_1 \tag{o-1}$$

$$Q_2 = F_2 L_2 \cos \theta_2 \tag{o-2}$$

[Note that the generalized forces may have also been computed by direct application of Eq. (6.59).] It may be seen that, for this case, the generalized forces are actually moments. This is, of course, as it should be since the virtual displacements are angular. Equations (d), (g) and (o) may now be substituted directly into Eqs. (6.62) and (6.63). Carrying through the required calculations, with $\{q_1, q_2\} \leftrightarrow \{\theta_1, \theta_2\}$, gives the equations of motion for the double pendulum. Hence,

$j = 1$:

$$\frac{d}{dt}\left(\frac{\partial L}{\partial \dot{q}_1}\right) - \frac{\partial L}{\partial q_1} = Q_1 \quad \Rightarrow$$

$$(m_1 + m_2)L_1^2\ddot{\theta}_1 + m_2 L_1 L_2 \ddot{\theta}_2 \cos(\theta_2 - \theta_1) - m_2 L_1 L_2 \dot{\theta}_2^2 \sin(\theta_2 - \theta_1)$$
$$+ (m_1 + m_2)gL_1 \sin\theta_1 = (F_1 + F_2)L_1 \cos\theta_1 \qquad (p)$$

$j = 2$:

$$\frac{d}{dt}\left(\frac{\partial L}{\partial \dot{q}_2}\right) - \frac{\partial L}{\partial q_2} = Q_2 \quad \Rightarrow$$

$$m_2 L_1 L_2 \left[\ddot{\theta}_1 \cos(\theta_2 - \theta_1) + \dot{\theta}_1^2 \sin(\theta_2 - \theta_1)\right]$$
$$+ m_2 L_2^2 \left[\ddot{\theta}_2 - \dot{\theta}_2^2 \sin(\theta_2 - \theta_1)\{1 - \cos(\theta_2 - \theta_1)\}\right] \qquad (q)$$
$$+ m_2 gL_2 \sin\theta_2 = F_2 L_2 \cos\theta_2$$

which are identical with Eqs. (6.14) and (6.16), respectively. Linearizing about $\theta_1 = \theta_2 = 0$ and putting the resulting expressions in matrix form gives the governing equation for small angle motion,

$$\begin{bmatrix} (m_1 + m_2)L_1^2 & m_2 L_1 L_2 \\ m_2 L_1 L_2 & m_2 L_2^2 \end{bmatrix}\begin{Bmatrix} \ddot{\theta}_1 \\ \ddot{\theta}_2 \end{Bmatrix} + \begin{bmatrix} (m_1 + m_2)gL_1 & 0 \\ 0 & m_2 gL_2 \end{bmatrix}\begin{Bmatrix} \theta_1 \\ \theta_2 \end{Bmatrix}$$
$$= \begin{Bmatrix} (F_1 + F_2)L_1 \\ F_2 L_2 \end{Bmatrix} \qquad \triangleleft (r)$$

which is, of course, identical to Eq. (6.18).

Example 6.6

A docked utility tram consists of a barrow of mass m_2 suspended from an overhead frame of mass m_1 by rods of length L as shown. The frame is latched to a rigid wall by an elastic coupler of effective stiffness k. A cable exerts a tension force F_1 on the frame and a controller exerts a torque M about the pivot point. Environmental forces are represented by the horizontal force F_2 acting though the attachment point of the barrow as indicated. If the stretch of track in the vicinity of the docking station is horizontal and the mass of the connecting rods and the spin of the barrow, as well as its moment of inertia about its own axis, are negligible derive the equations of motion for the system using Lagrange's Equations. Linearize the resulting equations by assuming small relative motion of the suspended car.

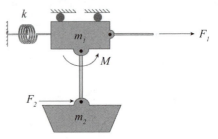

Figure E6.6-1 Docked utility tram.

Solution

The two body system evidently possesses two degrees of freedom. The horizontal coordinate u and the angular coordinate θ, shown in the adjacent figure, are the obvious choices for our generalized coordinates. The former measures the absolute displacement of the frame with respect to the fixed track and the latter measures the relative displacement of the barrow with respect to the frame.

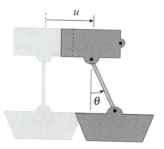

Figure E6.6-2 Coordinates.

Since we have both pure translation of the frame and combined translation and rotation of the barrow it is convenient to partition the motion of the latter in terms of its horizontal and vertical components. We shall take our origin at the equilibrium position of the frame. For consistency we shall set our datum to coincide with our datum at the level of the frame as well. Thus, let $(x_1 = u, 0)$ locate the frame and let (x_2, y_2) locate the car. It then follows that

$$x_2 = u + L\sin\theta, \quad y_2 = -L\cos\theta \tag{a-1,2}$$

The kinetic energy is then easily computed as

$$
\begin{aligned}
T &= \tfrac{1}{2}m_1\dot{u}^2 + \tfrac{1}{2}m_2\left[\dot{x}_2^{\,2} + \dot{y}_2^{\,2}\right] \\
&= \tfrac{1}{2}m_1\dot{u}^2 + \tfrac{1}{2}m_2\left\{\left[\dot{u} + L\dot{\theta}\cos\theta\right]^2 + \left[L\dot{\theta}\sin\theta\right]^2\right\} \\
&= \tfrac{1}{2}\left(m_1 + m_2\right)\dot{u}^2 + \tfrac{1}{2}m_2\left[L^2\dot{\theta}^2 + 2\dot{x}_1 L\dot{\theta}\cos\theta\right]
\end{aligned}
\tag{b}
$$

The potential energy of the system at some generic configuration is

$$\mathcal{U} = \tfrac{1}{2}ku^2 - mgL\cos\theta \tag{c}$$

Next, we determine the generalized forces associated with our chosen coordinates. To do this, we evaluate the virtual work done by the applied forces and moment and express the work in terms of the chosen coordinates u and θ. The generalized forces then correspond to the respective coefficients of the variations (virtual increments) of the generalized coordinates. Now, for a small virtual motion of the system it is clear that the virtual work done by the applied actions is

$$\delta\mathcal{W} = F_1\delta u + F_2\delta x_2 + M\delta\theta \tag{d}$$

We wish to express Eq. (d) in terms of the two independent coordinates x_1 and θ. This is accomplished by first evaluating the variation ("virtual differential") of Eq. (a-1), which gives the relation

$$\delta x_2 = \delta u + L \cos \theta \, \delta \theta \tag{e}$$

Substituting Eq. (e) into Eq. (d) and regrouping terms gives the virtual work in the desired form

$$\delta W = \left(F_1 + F_2 \right) \delta u + \left(F_2 L \cos \theta + M \right) \delta \theta \tag{f}$$

It follows form Eqs. (f) and (6.58) that

$$Q_1 = F_1 + F_2 \tag{g-1}$$

and

$$Q_2 = F_2 L \cos \theta + M \tag{g-2}$$

With the kinetic and potential energies expressed in terms of the chosen generalized coordinates, and the corresponding generalized forces evaluated, we can now substitute these expressions into Lagrange's Equations and determine the sought after equations of motion for the system. Thus, substituting Eqs. (b), (c), (g-1) and (g-2) into Eq. (6.62) and (6.63) and carrying through the indicated operations, with $\{q_1, q_2\} \leftrightarrow \{u, \theta\}$, gives the equations of motion for the system. Hence,

$j = 1$:

$$\frac{d}{dt}\left(\frac{\partial L}{\partial \dot{q}_1} \right) - \frac{\partial L}{\partial q_1} = Q_1 \quad \Rightarrow$$

$$\left(m_1 + m_2 \right) \ddot{u} + m_2 L \left[\ddot{\theta} \cos \theta - \dot{\theta}^2 \sin \theta \right] + ku = F_1 + F_2 \qquad \triangleleft \text{(h-1)}$$

$j = 2$:

$$\frac{d}{dt}\left(\frac{\partial L}{\partial \dot{q}_2} \right) - \frac{\partial L}{\partial q_2} = Q_2 \quad \Rightarrow$$

$$m_2 \left[L^2 \ddot{\theta} + L \left(\ddot{u} \cos \theta - \ddot{u}\dot{\theta} \sin \theta \right) \right]$$
$$+ m_2 \left[\ddot{u}L\dot{\theta} \sin \theta + gL \sin \theta \right] = F_2 L \cos \theta + M \qquad \triangleleft \text{(h-2)}$$

Linearizing Eqs. (h) and expressing the resulting equations in matrix form gives the equation of motion

$$
\begin{bmatrix} (m_1+m_2) & m_2L \\ m_2L & m_2L^2 \end{bmatrix}\begin{Bmatrix} \ddot{u} \\ \ddot{\theta} \end{Bmatrix} + \begin{bmatrix} k & 0 \\ 0 & m_2gL \end{bmatrix}\begin{Bmatrix} u \\ \theta \end{Bmatrix} = \begin{Bmatrix} (F_1+F_2) \\ (F_2L+M) \end{Bmatrix} \qquad \triangleleft \text{(i)}
$$

Example 6.7

Consider the motorcycle frame of Section 6.1.3. Suppose a rider of mass m_b sits rear of the center of mass a distance ℓ and that the padding and support for the seat is represented as a spring of stiffness k_b attached to the frame at that point, as shown. In addition, let the road conditions be such that the bases undergo the prescribed vertical deflections $y_{01}(t)$ and $y_{02}(t)$, respectively, and let the frame be subjected to dynamic loading represented by the vertical forces $F_1(t)$ and $F_2(t)$ applied at the indicated ends of the frame and directed downward and the force $F_b(t)$ applied to the rider. Use Lagrange's Equations to derive the equations of motion of the system. Linearize the equations for small rotations and express them in matrix form.

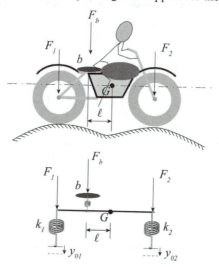

Figure E6.7-1 Motorcycle and rider. Equivalent system.

Solution

We shall choose the vertical deflection of the center of mass, y_G, the rotation of the frame, θ, and the vertical deflection of the rider, y_b, as our generalized coordinates for the present analysis. Thus, (q_1, q_2, q_3) $\leftrightarrow (y_G, \theta, y_b)$. (See Figure E6.7-2.) Further, we shall measure all deflections from their equilibrium positions. The deflections of the end points of the frame and of the point of the frame at which the seat is attached, y_1, y_2 and y_ℓ are expressed in terms of the chosen coordinates as follows:

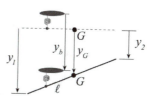

Figure E6.7-2 Coordinates.

$$y_1 = y_G - L_1 \sin\theta, \quad y_2 = y_G + L_2 \sin\theta, \quad y_\ell = y_G + \ell\sin\theta \qquad \text{(a-1, 2, 3)}$$

It follows that

$$\delta y_1 = \delta y_G - L_1 \cos\theta\,\delta\theta, \quad \delta y_2 = \delta y_G + L_2 \cos\theta\,\delta\theta \qquad \text{(b-1, 2)}$$

The virtual work of the applied forces is then

$$
\begin{aligned}
\delta W &= F_1\delta y_1 + F_2\delta y_2 + F_b\delta y_b \\
&= (F_1 + F_2)\delta y_G + (F_2 L_2 - F_1 L_1)\cos\theta\,\delta\theta + F_b\delta y_b
\end{aligned}
\qquad \text{(c)}
$$

from which we deduce the generalized forces

$$Q_1 = F_1 + F_2, \quad Q_2 = (F_2 L_2 - F_1 L_1)\cos\theta, \quad Q_3 = F_b \qquad \text{(d-1, 2, 3)}$$

The kinetic and potential energies for the system are, respectively,

$$T = \tfrac{1}{2}m\dot{y}_G^2 + \tfrac{1}{2}I_G\dot{\theta}^2 + \tfrac{1}{2}m_b\dot{y}_b^2 \qquad \text{(e)}$$

and

$$
\begin{aligned}
\mathcal{U} &= \tfrac{1}{2}k_1(y_G - L_1\sin\theta - y_{01})^2 \\
&\quad + \tfrac{1}{2}k_2(y_G + L_2\sin\theta - y_{02})^2 + \tfrac{1}{2}k_b(y_G + \ell\sin\theta - y_b)^2
\end{aligned}
\qquad \text{(f)}
$$

Substituting Eqs. (d), (e), and (f) into Eq. (6.62) and (6.63) and carrying through the indicated operations, with $(q_1, q_2, q_3) \leftrightarrow (y_G, \theta, y_b)$, gives the equations of motion for the system. Hence,

$j = 1$:

$$\frac{d}{dt}\left(\frac{\partial L}{\partial \dot{q}_1}\right) - \frac{\partial L}{\partial q_1} = Q_1 \;\Rightarrow$$

$$
\begin{aligned}
m\ddot{y}_G + (k_1 + k_2 + k_b)y_G + (k_2 L_2 - k_1 L_1 + k_b\ell)y_G \sin\theta - k_b y_b \\
= F_1 + F_2 + k_1 y_{01} + k_2 y_{02}
\end{aligned}
\qquad \triangleleft \text{(g-1)}
$$

$j = 2$:

$$\frac{d}{dt}\left(\frac{\partial L}{\partial \dot{q}_2}\right) - \frac{\partial L}{\partial q_2} = Q_2 \;\Rightarrow$$

$$I_G \ddot{\theta} + (k_2 L_2 - k_1 L_1 + k_b \ell) y_G \cos\theta$$
$$+ (k_2 L_2^2 + k_1 L_1^2 + k_b \ell^2) \sin\theta \cos\theta \qquad \triangleleft \text{(g-2)}$$
$$- k_b \ell y_b \cos\theta = F_2 L_2 - F_1 L_1 + (k_1 L_1 y_{01} - k_2 L_2 y_{02}) \cos\theta$$

$j = 3$:

$$\frac{d}{dt}\left(\frac{\partial L}{\partial \dot{q}_3}\right) - \frac{\partial L}{\partial q_3} = Q_3 \quad \Rightarrow$$

$$m_b \ddot{y}_b - k_b y_G + k_b \ell \sin\theta - k_b y_b = F_b \qquad \triangleleft \text{(g-3)}$$

Linearizing equations (g-1)–(g-3) for $\theta \ll 1$ and expressing the resulting equations in matrix form gives

$$\begin{bmatrix} m & 0 & 0 \\ 0 & I_G & 0 \\ 0 & 0 & m_b \end{bmatrix} \begin{Bmatrix} \ddot{y}_G \\ \ddot{\theta} \\ \ddot{y}_b \end{Bmatrix}$$
$$+ \begin{bmatrix} (k_1 + k_2 + k_b) & (k_2 L_2 - k_1 L_1 + k_b \ell) & -k_b \\ (k_2 L_2 - k_1 L_1 + k_b \ell) & (k_2 L_2^2 + k_1 L_1^2 + k_b \ell^2) & -k_b \ell \\ -k_b & -k_b \ell & k_b \end{bmatrix} \begin{Bmatrix} y_G \\ \theta \\ y_b \end{Bmatrix} \qquad \triangleleft \text{(h)}$$
$$= \begin{Bmatrix} (F_1 + F_2 + k_1 y_{01} + k_2 y_{02}) \\ (F_2 L_2 - F_1 L_1 + k_1 L_1 y_{01} - k_2 L_2 y_{02}) \\ F_b \end{Bmatrix}$$

It may be seen that if we let m_b, k_b and $F_b \to 0$ in Eqs. (g) and (h) we obtain the equations for the frame derived in Section 6.1.3 using Newton's Law's of Motion directly. If we let the excitations vanish identically, as well, Eq. (h) becomes identical with Eq. (6.25).

6.2.4 The Rayleigh Dissipation Function

In Example 6.4 we derived the equations of motion for a viscously damped system using Lagrange's Equations and included the effects of damping by calculating the virtual work of the damping forces to obtain the corresponding generalized forces. For systems with linear viscous damping it is possible to formulate a *dissipation function* which may be included on the left hand side of Lagrange's Equations in the sprit of the kinetic and potential energies to augment that formulation. This would replace the calculation of the virtual work and inclusion of the associated generalized forces due to damping. To develop this formulation we separate the forces due to damping from the remaining external forces in Eq. (6.45) and thereafter. Equation (6.45) then takes the form

$$\sum_{l=1}^{N}\left[\left(\vec{F}_l + \vec{F}_l^{(d)} - m_l\ddot{\vec{r}}_l\right)\cdot\delta\vec{r}_l\right] = 0 \tag{6.64}$$

where

$$\vec{F}_l^{(d)} = -\hat{c}_l\dot{\vec{r}}_l \tag{6.65}$$

and $\hat{c}_l$ is a linear combination of the coefficients of the dampers attached to mass l. In contrast to its interpretation in Eq. (6.45), $\vec{F}_l$ now represents all external forces that act on mass l except those due to damping. With this distinction, let us now proceed as in Section 6.2.2. All computations are the same, but now we shall evaluate the contributions of the damping forces separately. Toward this end, we determine the product

$$\sum_{l=1}^{N}\vec{F}_l^{(d)}\cdot\frac{\partial\vec{r}_l}{\partial q_j} = -\sum_{l=1}^{N}\hat{c}_l\dot{\vec{r}}_l\cdot\frac{\partial\dot{\vec{r}}_l}{\partial\dot{q}_j} = -\frac{\partial}{\partial\dot{q}_j}\sum_{l=1}^{N}\tfrac{1}{2}\hat{c}_l\dot{\vec{r}}_l\cdot\dot{\vec{r}}_l \tag{6.66}$$

Thus,

$$\sum_{l=1}^{N}\vec{F}_l^{(d)}\cdot\frac{\partial\vec{r}_l}{\partial q_j} = -\frac{\partial}{\partial\dot{q}_j}\mathcal{R} \tag{6.67}$$

where

$$\mathcal{R} = \mathcal{R}(\dot{q}_1,\dot{q}_2,...,\dot{q}_N) = \sum_{l=1}^{N}\tfrac{1}{2}\hat{c}_l\dot{\vec{r}}_l\cdot\dot{\vec{r}}_l \tag{6.68}$$

is called the *Rayleigh Dissipation Function*. Paralleling the development of Section 6.3.2, and incorporating Eq. (6.67), results in the alternate statement of Lagrange's Equations given by

$$\frac{d}{dt}\left(\frac{\partial L}{\partial\dot{q}_j}\right) - \frac{\partial L}{\partial q_j} + \frac{\partial\mathcal{R}}{\partial\dot{q}_j} = Q_j \quad (j = 1,2,...,N) \tag{6.69}$$

The equations of motion for a discrete system with viscous damping may be derived by formulating the Rayleigh dissipation function for that system, in addition to the potential and kinetic energies of the system and the virtual work of the applied forces, and substituting each into Eq. (6.69) and performing the indicated operations. Such a formulation may be advantageous when using certain numerical techniques for large scale systems. In this case we would construct the function from the damping forces. In the present context, the above approach provides an alternative to the basic formulation of Section 6.3.2 where the equations of motion for discrete systems are derived by computing the virtual work of the damping forces, identifying the corresponding generalized forces, formulating the kinetic and potential energies and utilizing the fundamental form of Lagrange's Equations given by Eq. (6.62), as was done in a very straightforward manner in Example 6.4.

Example 6.8

Formulate the Rayleigh dissipation function for the system of Example 6.4 and use it to derive the equations of motion for that system.

Solution
For this problem the generalized coordinates are simply the displacements of the masses from their equilibrium positions. Thus, $(q_1, q_2, q_3) \leftrightarrow (x_1, x_2, x_3)$. We seek a function whose derivatives with respect to the velocities of the masses will yield the corresponding damping forces. Toward this end let us consider the function

$$\mathcal{R} = \tfrac{1}{2}c_1\dot{x}_1^2 + \tfrac{1}{2}c_2\left(\dot{x}_2 - \dot{x}_1\right)^2 + \tfrac{1}{2}c_3\left(\dot{x}_3 - \dot{x}_2\right)^2 + \tfrac{1}{2}c_4\dot{x}_3^2 \tag{a}$$

Evaluating the derivatives of $\mathcal{R}$ with respect to each velocity gives the following:

$$\frac{\partial \mathcal{R}}{\partial \dot{x}_1} = (c_1 + c_2)\dot{x}_1 - c_2\dot{x}_2 \tag{b-1}$$

$$\frac{\partial \mathcal{R}}{\partial \dot{x}_2} = -c_2\dot{x}_1 + (c_2 + c_3)\dot{x}_3 - c_2\dot{x}_3 \tag{b-2}$$

$$\frac{\partial \mathcal{R}}{\partial \dot{x}_3} = -c_3\dot{x}_2 + (c_3 + c_4)\dot{x}_3 \tag{b-3}$$

The kinetic and potential energies are given by Eqs. (c) and (d) of Example 6.3. The virtual work done for the present case is simply

$$\delta W = F_1\delta x_1 + F_2\delta x_2 + F_3\delta x_3 \tag{c}$$

The generalized forces may be read directly from Eq. (c) and are seen to be simply the applied forces themselves. Substituting all of the above into Eq. (6.69) and carrying through the calculations gives the equations of motion stated in Eq. (f) of Example 6.4.

6.3 SYMMETRY OF THE SYSTEM MATRICES

The physical properties of linear mechanical systems were seen to be described by the mass, stiffness and damping matrices appearing in the general equation of motion

$$\mathbf{m\ddot{u}} + \mathbf{c\dot{u}} + \mathbf{ku} = \mathbf{F} \tag{6.70}$$

For each of the representative systems considered in Sections 6.2 and 6.3 these matrices were seen to be symmetric. In this section we show that the system matrices of all linear mechanical systems are symmetric when the corresponding motions are expressed in terms of a set of linearly independent coordinates.

6.3.1 The Stiffness Matrix

For the moment, let us restrict our attention to linear elastic systems subjected to quasi-static loading. Under these circumstances Eq. (6.70) reduces to the algebraic form

$$\mathbf{ku} = \mathbf{F} \tag{6.71}$$

Pre-multiplying Eq. (6.71) by $\mathbf{k}^{-1}$ gives the relation

$$\mathbf{u} = \mathbf{aF} \tag{6.72}$$

where

$$\mathbf{a} = \mathbf{k}^{-1} \tag{6.73}$$

is the *compliance matrix*, also known as the *flexibility matrix*. The symmetry of the stiffness matrix will be established by considering the work done by the applied loads.

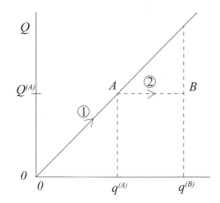

Figure 6.9 Load-deflection paths.

The constitutive (stress-strain) relations, and hence the load-deflection relations, for a linear elastic solid are, by definition, linear. Likewise, the load-deflection relations for a system comprised of linear elastic springs, Eq. (6.72), are linear as well. If we consider a single generalized force Q applied to a linear system, and the associated generalized displacement q of the system at the load point, the corresponding load-deflection path will plot as a straight line as shown in Figure 6.9. If the system is initially in the undeformed state when the force Q is applied, and Q is increased slowly from 0 to $Q^{(A)}$ the load-point deflection increases accordingly to the value $q^{(A)}$. The work done as the state of the system progresses along path OA is, by Eq. (1.85),

$$W^{(A)} = \int_{OA} \vec{F} \cdot d\vec{r} = \int_0^{q^{(A)}} Q \, dq = \tfrac{1}{2} Q^{(A)} q^{(A)} \tag{6.74}$$

That is, the work done by the force Q in bringing the system to state A corresponds to the area under the curve OA (the area of triangle $OAq^{(A)}$). Suppose now that the system is deformed further, say by some other force, so that the deflection under the first force increases to say $q^{(B)}$ while Q is maintained at the value $Q^{(A)}$ throughout the subsequent deformation process. During this process the system progresses along path AB in the Qq-plane (Figure 6.9). The additional work done by the constant force Q during this process is then

$$W^{(AB)} = \int_{AB} \vec{F} \cdot d\vec{r} = \int_{q^{(A)}}^{q^{(B)}} Q \, dq = Q^{(A)} q^{(BA)} \tag{6.75}$$

where

$$q^{(BA)} = q^{(B)} - q^{(A)} \tag{6.76}$$

(The integral is seen to correspond to area $q^{(A)}AB \, q^{(B)}$.) The total work done by the force applied at the point in question in bringing the system to state B is thus

$$W^{(OB)} = W^{(OA)} + W^{(AB)} = \tfrac{1}{2} Q^{(A)} q^{(A)} + Q^{(A)} q^{(BA)} \tag{6.77}$$

With the work of a given load during the segmented process established, let us next consider an arbitrary elastic system subjected to generalized forces at any two points l and j ($j, l = 1, 2, \ldots$) Let Q_l and Q_j represent these forces and let q_l and q_j represent the corresponding generalized displacements (generalized coordinates). Suppose that the force Q_j is first applied, increasing from zero and deforming the system from the reference (undeformed) state so that the point of application deflects to the value $q_j^{(j)}$ and point l deflects by the amount $q_l^{(j)}$. (See, for example, the frame in Figure 6.10a-1.) Next let a load be applied at point l. Let it increase slowly from zero deforming the system further so that the additional deflections at points j and l due to the load at point l are, respectively, $q_j^{(l)}$ and $q_l^{(l)}$. (See, for example, the frame in Figure 6.10a-2.)

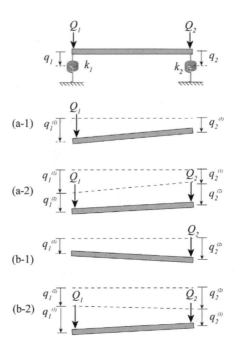

Figure 6.10 Physical example of load-deflection scenario.

With the aid of Eq. (6.77), the total work done by both forces in bringing the system to its current state is calculated as

$$\mathcal{W}^{(j,l)} = \tfrac{1}{2}Q_j q_j^{(j)} + Q_j q_j^{(l)} + \tfrac{1}{2}Q_l q_l^{(l)} \tag{6.78}$$

Now, form Eq. (6.72),

$$q_j^{(j)} = a_{jj}Q_j, \quad q_l^{(j)} = a_{lj}Q_j, \quad q_j^{(l)} = a_{jl}Q_l, \quad q_l^{(l)} = a_{ll}Q_l \tag{6.79}$$

Substituting Eqs. (6.79) into Eq. (6.78) gives the work to the present state as

$$\mathcal{W}^{(j,l)} = \tfrac{1}{2}a_{jj}Q_j^2 + a_{jl}Q_jQ_l + \tfrac{1}{2}a_{ll}Q_l^2 \tag{6.80}$$

Let us next reverse the order of application of the forces and compute the work done in bringing the system to its present state. Paralleling the development for the previous case, but with the order reversed as demonstrated for the example frame in Figures 6.10b-1 and 6.10b-2, the work done in bringing the system to the current state is

$$\mathcal{W}^{(l,j)} = \tfrac{1}{2}Q_l q_l^{(l)} + Q_l q_l^{(j)} + \tfrac{1}{2}Q_j q_j^{(j)} \tag{6.81}$$

This is also found by simply interchanging the indices in Eq. (6.78). Substituting Eqs. (6.79) into Eq. (6.81) gives the work for this load order as

$$W^{(l,j)} = \tfrac{1}{2}a_{ll}Q_l^2 + a_{lj}Q_lQ_j + \tfrac{1}{2}a_{jj}Q_j^2 \tag{6.82}$$

By definition of an elastic system all work is recoverable and the work done in bringing the system to its current configuration is independent of the loading path. That is, it is independent of the order in which the loads are applied. Thus,

$$W^{(l,j)} = W^{(j,l)} \tag{6.83}$$

Substituting Eqs. (6.80) and (6.82) into Eq. (6.83) gives the identity

$$a_{lj} = a_{jl} \tag{6.84}$$

Equation (6.84) is valid for any pair of points ($j,l = 1, 2, \ldots$) in the system and implies that the deflection at one point of an elastic system due the force at a second point of the system is equal to the deflection at the second point due to a force of equal magnitude applied at the first point. This is known as *Maxwell's Reciprocal Theorem* and is applicable to continuous as well as discrete systems. Since we are presently interested in discrete systems, we shall take $j,l = 1, 2, \ldots, N$. Hence, when written in matrix form, Eq. (6.84) implies that

$$\mathbf{a}^{\mathsf{T}} = \mathbf{a} \tag{6.85}$$

It then follows from Eq. (6.73) that

$$\mathbf{k}^{\mathsf{T}} = \mathbf{k} \tag{6.86}$$

The stiffness matrix is thus symmetric, which is what we set out to show.

6.3.2 The Mass Matrix

The inherent symmetry of the mass matrix, when written in terms of a set of linearly independent coordinates, may be shown by consideration of the kinetic energy of the system. Let the position vector of a generic mass of the system be described in terms of some set of generalized coordinates as described by Eq. (6.38). Substituting Eq. (6.47) into Eq. (6.51) gives the kinetic energy of the system as

$$T = \frac{1}{2}\sum_{n=1}^{N}\dot{\vec{r}}_n \cdot \dot{\vec{r}}_n = \frac{1}{2}\sum_{n=1}^{N}\left\{ \sum_{j=1}^{N}\frac{\partial \dot{\vec{r}}_n}{\partial q_j}\dot{q}_j \cdot \sum_{l=1}^{N}\frac{\partial \dot{\vec{r}}_n}{\partial q_l}\dot{q}_l \right\} \tag{6.87}$$

Changing the order of summation renders the kinetic energy to the form

$$T = \frac{1}{2}\sum_{j=1}^{N}\sum_{l=1}^{N}\left\{\sum_{n=1}^{N}\frac{\partial \dot{\vec{r}}_n}{\partial q_j}\cdot\frac{\partial \dot{\vec{r}}_n}{\partial q_l}\right\}\dot{q}_j\dot{q}_l$$

Hence, the kinetic energy of an N-degree of freedom system is

$$T = \sum_{j=1}^{N}\sum_{l=1}^{N}\tfrac{1}{2}m_{jl}\dot{q}_j\dot{q}_l = \tfrac{1}{2}\dot{\mathbf{q}}^{\mathsf{T}}\mathbf{m}\dot{\mathbf{q}} \tag{6.88}$$

where

$$m_{jl} = \sum_{n=1}^{N}\frac{\partial \dot{\vec{r}}_n}{\partial q_j}\cdot\frac{\partial \dot{\vec{r}}_n}{\partial q_l} \tag{6.89}$$

$$\mathbf{m} = \begin{bmatrix} m_{11} & \cdots & m_{1N} \\ \vdots & \ddots & \vdots \\ m_{N1} & \cdots & m_{NN} \end{bmatrix} \tag{6.90}$$

and

$$\dot{\mathbf{q}} = \begin{Bmatrix} \dot{q}_1 \\ \vdots \\ \dot{q}_N \end{Bmatrix} \tag{6.91}$$

It is evident from Eq. (6.89) that

$$m_{jl} = m_{lj} \quad (l, j = 1, 2, ..., N) \tag{6.92}$$

Thus,

$$\mathbf{m}^{\mathsf{T}} = \mathbf{m} \tag{6.93}$$

for any linear system whose motion is described by a set of generalized coordinates.

6.3.3 The Damping Matrix

To show the symmetry of the damping matrix for linear systems whose motions are expressed in terms of a set of generalized coordinates we shall proceed in an analogous manner as for the mass matrix, but with the Rayleigh dissipation function replacing the kinetic energy. Thus,

$$\mathcal{R} = \frac{1}{2}\sum_{l=1}^{N}\hat{c}_n\dot{\vec{r}}_n\bullet\dot{\vec{r}}_n = \frac{1}{2}\sum_{n=1}^{N}\hat{c}_n\left\{\sum_{j=1}^{N}\frac{\partial\dot{\vec{r}}_n}{\partial q_j}\dot{q}_j\bullet\sum_{l=1}^{N}\frac{\partial\dot{\vec{r}}_n}{\partial q_l}\dot{q}_l\right\}$$

$$= \frac{1}{2}\sum_{j=1}^{N}\sum_{l=1}^{N}\left\{\sum_{n=1}^{N}\hat{c}_n\frac{\partial\dot{\vec{r}}_n}{\partial q_j}\bullet\frac{\partial\dot{\vec{r}}_n}{\partial q_l}\right\}\dot{q}_j\dot{q}_l$$

(6.94)

Hence,

$$\mathcal{R} = \sum_{j=1}^{N}\sum_{l=1}^{N}\tfrac{1}{2}c_{jl}\dot{q}_j\dot{q}_l = \tfrac{1}{2}\dot{\mathbf{q}}^\mathsf{T}\mathbf{c}\dot{\mathbf{q}}$$

(6.95)

where

$$c_{jl} = \sum_{n=1}^{N}\hat{c}_n\frac{\partial\dot{\vec{r}}_n}{\partial q_j}\bullet\frac{\partial\dot{\vec{r}}_n}{\partial q_l}$$

(6.96)

and

$$\mathbf{c} = \begin{bmatrix} c_{11} & \cdots & c_{1N} \\ \vdots & \ddots & \vdots \\ c_{N1} & \cdots & c_{NN} \end{bmatrix}$$

(6.97)

It is evident, from Eq. (6.96), that

$$c_{jl} = c_{lj} \quad (l, j = 1, 2, ..., N)$$

(6.98)

and hence that

$$\mathbf{c}^\mathsf{T} = \mathbf{c}$$

(6.99)

The symmetry of the system matrices will be of great importance in our studies of both free and forced vibrations of multi-degree of freedom systems.

6.4 CONCLUDING REMARKS

In this chapter we considered the mathematical description of multi-degree of freedom systems. We first used the Newtonian approach to derive the equations of motion for certain representative systems. We then introduced the notions of generalized coordinates and virtual work and, with these, derived Lagrange's Equations — a general statement of the equations of motion expressed in terms of the potential and kinetic energies of the system and the virtual work of forces for which a potential function is not or cannot be written. Thus, once the generalized forces are identified for a given problem through the virtual work of the actual forces and moments, Lagrange's Equations basically provide a recipe for the derivation of the equations of motion for the specific system of interest. Derivations for selected systems were performed by direct application of Newton's Laws of Motion and by Lagrange's Equations and compared. It was seen that the derivation of the equations of motion for complex systems is simplified a great deal when implemented using Lagrange's Equations. Once the technique is mastered, Lagrange's Equations provide a convenient and relatively simple scalar procedure for deriving the equations of motion for complex mechanical systems. The chapter concluded with a proof of *Maxwell's Reciprocal Theorem* which, for discrete systems, implies that the stiffness matrix is symmetric for all linear systems. Finally, it was shown that the mass and damping matrices for all linear mechanical systems whose motions are expressed in terms of a set of linearly independent coordinates are symmetric as well.

BIBLIOGRAPHY

Baruh, H., *Analytical Dynamics*, McGraw-Hill, Boston, 1999.

Ginsberg, J.H., *Advanced Engineering Dynamics*, Cambridge, New York, 1998.

Greenwood, D.T., *Principles of Dynamics*, 2nd ed., Prentice-Hall, Englewood Cliffs, 1988.

Meirovitch, L., *Elements of Vibration Analysis*, 2nd ed., McGraw-Hill, New York, 1986.

Rayleigh, J.W.S., *The Theory of Sound*, Vol.1, Dover, New York, 1945.

Shabana, A.A., *Vibration of Discrete and Continuous Systems*, Springer, New York, 1997.

Whittaker, E.T., *A Treatise on the Analytical Dynamics of Particles and Rigid Bodies*, Cambridge, New York, 1970.

PROBLEMS

6.1   Use Lagrange's Equations to derive the equation of motion for a simple mass-spring-damper system.

6.2   Use Lagrange's Equations to derive the equation of motion of the simple pendulum.

6.3   Use Lagrange's Equations to derive the equation of motion for the timing device of Problem 2.19.

6.4   Use Lagrange's Equations to derive the equation of motion for the constrained rod shown in Figure P6.4.

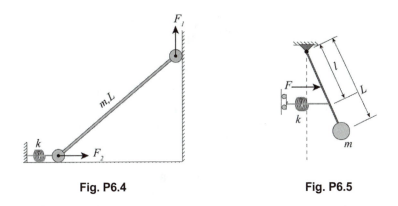

Fig. P6.4   **Fig. P6.5**

6.5 The pendulum with elastic constraint shown in Fig. P6.5 is subjected to a horizontally directed force applied to the mid-span of the rod as indicated. Use Lagrange's Equations to derive the equation of motion for the system. The spring is unstretched when $\theta = 0$. The masses of the rod, roller and spring are negligible.

6.6 Derive the equations of motion for the crankshaft system shown in Figure P6.6 using Lagrange's Equations. The spring is undeformed when the connecting pin A is directly above or below the hub of the wheel.

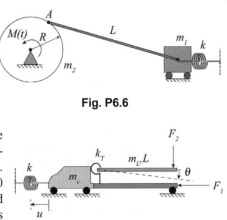

Fig. P6.6

6.7 Use Lagrange's Equations to derive the equations of motion for the constrained hook and ladder shown. The spring is untorqued when $\theta = 0$ and the tip of the ladder is subjected to a downward vertical force F_2 as indicated.

Fig. P6.7

6.8 A tram consists of a rigid frame of mass m_f from which a passenger compartment of mass m_c and radius of gyration r_G is pinned to the frame at the car's center of mass as shown in Figure P6.8. The wheels of the tram frame are of radius R and negligible mass m_w and roll without slip during motion. A motor applies a torque M to one of the wheels, thus driving the system, and the passenger compartment is subjected to a wind force whose resultant acts through a point that lies a distance a below the pin, as indicated. If an elastic guide cable of effective stiffness k is attached to the frame as shown, derive the equations of motion for the system using Lagrange's Equations. ($I_0 = m_c r_G^2$)

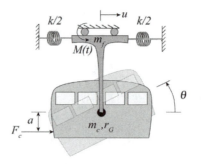

Fig. P6.8

6.9 Use Lagrange's Equations to derive the equations of motion for the coupled pendulum shown in Figure P6.9.

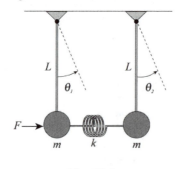

Fig. P6.9

6.10 A conveyor belt system consists of two flywheels of mass m_1 and radius R_1, and mass m_2 and radius R_2, respectively. The wheels are connected by an elastic belt as shown in Figure P6.10. If no slip occurs between the wheels and belt, the effective stiffness of each leg is k, as indicated. Finally, a motor applies a torque M_1 to the left wheel and a torque M_2 to the right wheel as shown. (**a**) Use Lagrange's Equations to derive the equations of motion for the system. (**b**) Check your answer using Newton's Laws of Motion.

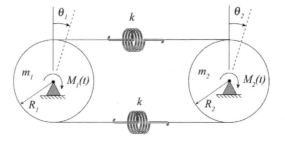

Fig. P6.10

6.11 A square raft of mass m and side L sits in water of specific gravity γ_w. A uniform vertical line force of intensity P acts downward at a distance a left of center of the span. (a) Use Lagrange's Equations to derive the 2-D equations of motion of the raft. (b) Check your answers using Newton's Laws of Motion.

6.12 Two identical bodies of mass m are connected by a spring of stiffness k and constrained to move in rectilinear motion as shown. Derive the equations of motion for the system.

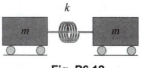

Fig. P6.12

6.13 Use Lagrange's Equations to derive the governing equations for two dimensional motion of the elastic dumbbell satellite shown in Figure P6.13.

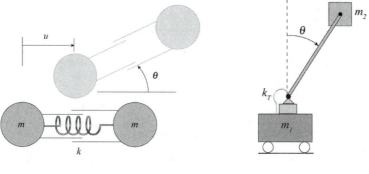

Fig. P6.13   **Fig. P6.14**

6.14 Use Lagrange's Equations to derive the equations of motion for the inverted elastic pendulum shown in Figure P6.14 if the massless rod is of length L.

6.15 Use Lagrange's Equations to derive the Equations of motion for the pulley system shown in Figure P6.15. The wheel has radius R and $I = mR^2$.

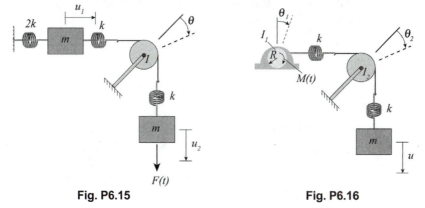

Fig. P6.15   **Fig. P6.16**

6.16 A motor supplies a torque M through a drive shaft whose moment of inertia is I_1. The shaft of radius R is connected to the pulley system shown in Figure P6.16. If the cable is elastic with effective stiffness k, as indicated, the moment of inertia of the pulley is I_2 and the cable supports a suspended mass m as indicated, derive the equations of motion of the system using Lagrange's Equations.

6.17 Use Lagrange's Equations to derive the equations of motion for the pulley system shown in Figure P6.17.

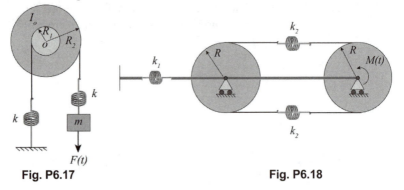

Fig. P6.17   **Fig. P6.18**

6.18 The elastic fan-belt system shown is restrained by an elastic rod of effective stiffness k_1 while a motor supplies a torque M to the right flywheel, as indicated. The mass of each wheel is m. Use Lagrange's Equations to derive the equations of motion for the system.

6.19 A rigid rod of length L, and mass m_a is connected to a rigid base of mass m_b through a torsional spring of stiffness k_T as shown. The base sits on an elastic support of stiffness k as indicated. Derive the equations of motion of the system using Lagrange's Equations.

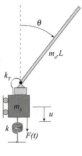

Fig. P6.19

6.20 A crane is attached to the geometric center of an offshore platform as shown. The total mass of the crane and platform is m and the mass of the boom is negligible. The square platform has side L and sits atop four identical floats of radius R. If the length of the horizontal projection of the boom is L_B, derive the equations that govern planar motion of the structure.

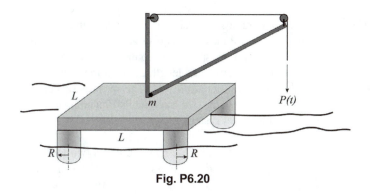

Fig. P6.20

6.21 A certain submarine is modeled as shown, for simple calculations of longitudinal motion. The mass of the hull and frame structure is $2m_s$ and that of the interior compartment is m_c. The hull and interior compartment are separated by springs of stiffness k, and the longitudinal stiffness of the hull is k_s as indicated. Derive the equations that govern longitudinal motion of the boat.

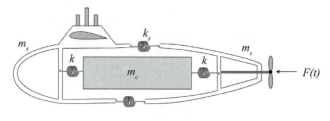

Fig. P6.21

6.22 Use Lagrange's Equations to derive the equations of motion for the linked system shown in Figure P6.22. The mass of the rigid connecting rod is negligible compared with the other masses of the system.

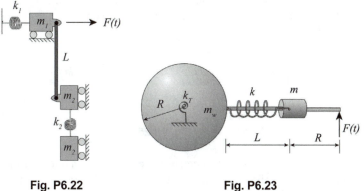

Fig. P6.22   **Fig. P6.23**

6.23 A rigid spoke of negligible mass extends radially from the periphery of a solid wheel of mass m_w and radius R, as shown. The hub of the wheel is attached to an elastic axle of negligible mass and equivalent torsional stiffness k_T. A sleeve of mass m is fitted around the spoke and connected to the wheel by an elastic spring of stiffness k and unstretched length L, and a transverse force $F(t)$ is applied to the end of the rod of length $L + R$. The spoke is sufficiently lubricated so that friction is not a concern. Use Lagrange's Equations to derive the equations of motion for the wheel system.

6.24 A rigid rod of length L and negligible mass connects two identical cylindrical floats, each possessing radius R and mass m_a. The system floats in a fluid of mass density ρ_f. A block of mass m_b is suspended from the center of the span by an elastic spring of stiffness k as shown, and a downward vertical force of magnitude F is applied to the suspended mass, as indicated. Derive the equations of motion of the system using Lagrange's Equations.

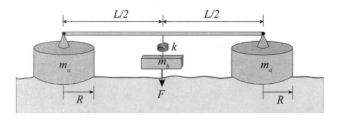

Fig. P6.24

6.25 Three identical rigid disks, each of mass m and radius R, are attached at their centers to an elastic shaft of area polar moment of inertia J and shear modulus G. The ends of the rod are embedded in rigid supports as shown. The spans between the disks and between the disks and the supports are each of length L. Derive the equations of angular motion for the system if the disks are subjected to the twisting moments M_1, M_2 and M_3, respectively.

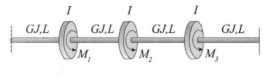

Fig. P6.25

6.26 Use Lagrange's Equations to derive the equations of motion for the triple pendulum whose bobs are subjected to horizontal forces F_1, F_2 and F_3, respectively.

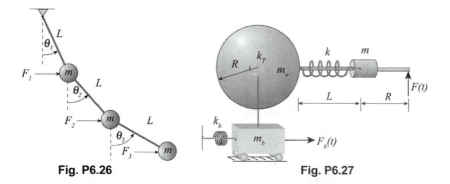

Fig. P6.26   **Fig. P6.27**

6.27 The assembly of Problem 6.23 is mounted on a movable base of mass m_b as shown. Use Lagrange's Equations to derive the equations of motion for the augmented system.

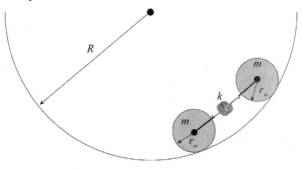

Fig. P6.28

6.28 Two wheels, each of mass m and radius r_w, are connected by an elastic coupler of effective stiffness k and undeformed length L. The system rolls without slip around a circular track of radius R, as shown. Derive the equations of motion of the wheel system.

6.29 Use Lagrange's Equations to derive the equations of motion for the coupled trio of pendulums shown.

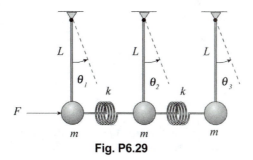

Fig. P6.29

6.30 A system consists of a rigid frame of mass m_a and length L that is supported at each end by elastic springs of stiffness k. Two identical blocks of mass m_b are mounted at either end of the frame atop mounts of stiffness k as shown. Use Lagrange's Equations to derive the equations of motion for the system.

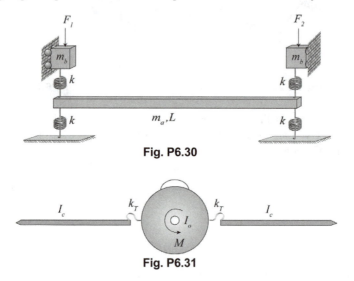

Fig. P6.30

Fig. P6.31

6.31 Consider an aircraft traveling at constant altitude and speed as it undergoes tight periodic rolling motion of the fuselage. Let the wings be modeled as equivalent rigid bodies with torsional springs of stiffness k_T at the fuselage wall. In addition, let each wing possess moment of inertia I_c about its respective connection point and let the fuselage of radius R have moment of inertia I_o about its axis. Derive the equations of rolling motion for the aircraft.

6.32 Derive the equation of motion for the system of Problem 6.19 if the support possesses damping of coefficient c.

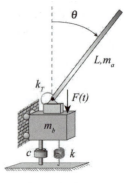

Fig. P6.32

6.33 Derive the equations of motion for the elastically coupled wheel system of Problem 6.28 if a damper of coefficient c is attached as shown in Figure P6.33.

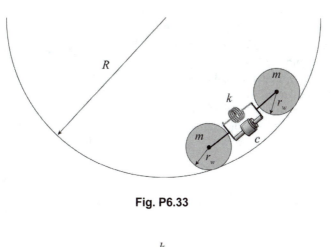

Fig. P6.33

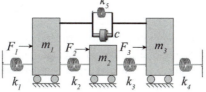

Fig. P6.34

6.34 Derive the equations of motion for the compound mass-spring-damper system shown in Figure P6.34.

6.35 Derive the equations of motion for the conveyor belt system of Problem 6.10 if the elastic belt is replaced by a viscoelastic belt of the same stiffness and damping coefficient c as shown.

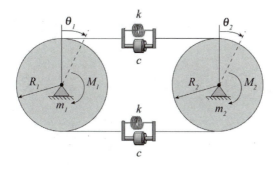

Fig. P6.35

6.36 Derive the equations of motion for the system of Problem 6.15 if a damper of coefficient $2c$ is inserted into the left support as shown, and the elastic cable is replaced by a viscoelastic cable of the same stiffness and damping coefficient c as indicated.

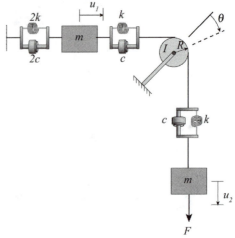

Fig. P6.36

6.37 Derive the equations of motion for longitudinal motion of the submarine of Problem 6.21 if damping of effective coefficient c_s is introduced to the hull and damping of coefficient c is introduced to the compartment mounts, as shown.

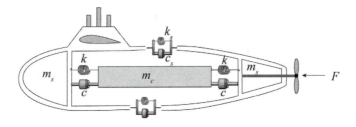

Fig. P6.37

7

Free Vibration of Multi-Degree of Freedom Systems

In this chapter we consider the behavior of discrete multi-degree of freedom systems that are free from externally applied dynamic forces. That is, we examine the response of such systems when each mass of the system is displaced and released in a manner that is consistent with the constraints imposed on it. We are thus interested in the behavior of the system when is left to move under its own volition. As for the case of single degree of freedom systems, it will be seen that the free vibration response yields fundamental information and parameters that define the inherent dynamical properties of the system.

7.1 THE GENERAL FREE VIBRATION PROBLEM AND ITS SOLUTION

It was seen in Chapter 6 that the equations that govern discrete multi-degree of freedom systems take the general matrix form of Eq. 6.2. We shall here consider the fundamental class of problems corresponding to undamped systems that are free from applied (external) forces. For this situation, Eq. 6.2 reduces to the form

$$\mathbf{m}\ddot{\mathbf{u}} + \mathbf{k}\mathbf{u} = \mathbf{0} \tag{7.1}$$

where, for an N degree of freedom system, $\mathbf{m}$ and $\mathbf{k}$ are the $N \times N$ mass and stiffness matrices of the system, respectively, and $\mathbf{u}$ is the corresponding $N \times 1$ displacement matrix. To solve Eq. (7.1), we parallel the approach taken for solving the corresponding scalar problem for single degree of freedom systems. We thus assume a solution of the form

$$\mathbf{u} = \mathbf{U}e^{i\omega t} \qquad (7.2)$$

where $\mathbf{U}$ is a column matrix with N, as yet, unknown constants, and ω is an, as yet, unknown constant as well. The column matrix $\mathbf{U}$ may be considered to be the spatial distribution of the response while the exponential function is the time dependence. Based on our experience with single degree of freedom systems, we anticipate that the time dependence may be harmonic. We therefore assume solutions of the form of Eq. (7.2). If we find harmonic forms that satisfy the governing equations then such forms are, by definition, solutions to those equations.

Substitution of Eq. (7.2) into Eq. (7.1), and factoring common terms, results in the form

$$\left[\mathbf{k} - \omega^2 \mathbf{m}\right]\mathbf{U}e^{i\omega t} = \mathbf{0}$$

Assuming that $\omega \geq 0$, we can divide through by the exponential term. This results in the equation

$$\left[\mathbf{k} - \omega^2 \mathbf{m}\right]\mathbf{U} = \mathbf{0} \qquad (7.3)$$

which may also be stated in the equivalent forms,

$$\mathbf{k}\mathbf{U} = \omega^2 \mathbf{m}\mathbf{U} \qquad (7.3')$$

and

$$\hat{\mathbf{k}}\mathbf{U} = \lambda\mathbf{U} \qquad (7.3'')$$

where

$$\hat{\mathbf{k}} = \mathbf{m}^{-1}\mathbf{k} \quad \text{and} \quad \lambda = \omega^2$$

Thus, solving Eq. (7.1) for $\mathbf{u}(t)$ is reduced to finding $(\omega^2, \mathbf{U})$ pairs that satisfy Eq. (7.3), or equivalently Eq. (7.3'). A problem of this type may be recognized as an *eigenvalue problem*, with the scalar parameter ω^2 identified as the *eigenvalue* and the column matrix $\mathbf{U}$ as the corresponding *eigenvector*. An N degree of freedom system will generally possess N eigenvalues and N eigenvectors. The solutions of the eigenvalue problem, when substituted into Eq. (7.2), will give the solution to Eq. (7.1) and hence the free vibration response of the system of interest.

Natural Frequencies

One obvious solution of Eq. (7.3) is *the trivial solution* $\mathbf{U} = \mathbf{0}$. This corresponds to the equilibrium configuration of the system. Though this is clearly a solution corresponding to a physically realizable configuration, it is evidently uninteresting as far as vibrations are concerned. We are thus interested in physical configurations associated with *nontrivial solutions*. From linear algebra we know that a matrix equation $\mathbf{Ax} = \mathbf{b}$ may be row reduced. If the rows or columns of the matrix $\mathbf{A}$ are linearly independent (that is, no row can be expressed as a linear combination of the other rows) then the corresponding matrix equation can be reduced to diagonal form, and the solution for **x** can be read directly. In matrix form,

$$
\mathbf{Ax=b} \rightarrow \bar{\mathbf{A}}\mathbf{x}=\bar{\mathbf{b}} : \quad
\begin{pmatrix} \bar{a}_1 & & 0 \\ & \ddots & \\ 0 & & \bar{a}_n \end{pmatrix}
\begin{Bmatrix} x_1 \\ \vdots \\ x_N \end{Bmatrix}
=
\begin{Bmatrix} \bar{b}_1 \\ \vdots \\ \bar{b}_N \end{Bmatrix}
\quad\Rightarrow\quad
\begin{Bmatrix} x_1 \\ \vdots \\ x_N \end{Bmatrix}
=
\begin{Bmatrix} \bar{b}_1/\bar{a}_1 \\ \vdots \\ \bar{b}_N/\bar{a}_N \end{Bmatrix}
$$

It is evident that $\det \bar{\mathbf{A}} \neq 0$. A matrix **A** whose rows are linearly independent is said to be *nonsingular*. Similarly, if Eq. (7.3) is row reduced to diagonal form then the matrix $\left[\mathbf{k}-\omega^2\mathbf{m}\right]$ is nonsingular and the solution, **U**, can be read directly. However, since the right-hand side of Eq. (7.3) is the null matrix, this will yield the trivial solution $\mathbf{U}=\mathbf{0}$. Therefore, *for nontrivial solutions we require that the matrix* $\left[\mathbf{k}-\omega^2\mathbf{m}\right]$ *be singular.* That is, we require that at least one of the rows (or columns) of the matrix can be expressed as a linear combination of the others. This means that the set of linear equations represented by Eq. (7.3) corresponds to, at most, $N-1$ equations for the N unknowns U_j ($j=1, 2, \ldots, N$). (The importance of this property will become apparent later in our analysis.) If the rows of the matrix $\left[\mathbf{k}-\omega^2\mathbf{m}\right]$ are linear dependent (i.e., at least one row can be written as a linear combination of the others) then when the corresponding system of equations is row reduced, at least one row will become all zeros. The determinant of such a matrix clearly vanishes identically. Thus, if **A** is a singular matrix, then

$$
\det \mathbf{A} =
\begin{vmatrix}
a_{11} & a_{12} & \cdots & a_{1N} \\
a_{21} & a_{22} & \cdots & a_{2N} \\
\vdots & \vdots & \ddots & \vdots \\
0 & 0 & \cdots & 0
\end{vmatrix}
= 0
$$

(This property is often taken as the definition of a singular matrix.) Therefore, for Eq. (7.3) to yield nontrivial solutions, we require that

$$
\mathcal{F}(\omega^2) \equiv \det\left[\mathbf{k}-\omega^2\mathbf{m}\right] = 0 \tag{7.4}
$$

If the elements of the mass and stiffness matrices, **k** and **m**, are specified then Eq. (7.4) results in an Nth order polynomial equation (the characteristic equation) in terms of the parameter ω^2 of the form

$$
\mathcal{F}(\omega^2) = \mu_N(\omega^2)^N + \mu_{N-1}(\omega^2)^{N-1} + \ldots + \mu_1(\omega^2) + \mu_0 = 0 \tag{7.5}
$$

where μ_j ($j=0, 1, 2, \ldots, N$) are functions of the mass and stiffness parameters for the particular system. The *characteristic equation* will yield N roots,

$$\omega^2 = \omega_1^{\ 2}, \omega_2^{\ 2}, ..., \omega_N^{\ 2} \tag{7.6}$$

(It is customary to label the frequencies in ascending order, according to their magnitude. That is, $\omega_1^{\ 2} < \omega_2^{\ 2} < ... < \omega_N^{\ 2}$.) Since the assumed form of the solution given by Eq. (7.2) is harmonic, the roots of Eq. (7.5) correspond to frequencies, and the characteristic equation is referred to as the *frequency equation* for the system. Recall that we are presently considering systems that are not subjected to applied dynamic forces. These frequencies are therefore referred to as the *natural frequencies of the system*. An N degree of freedom system is seen to possess N natural frequencies, each of which are dependent upon the parameters of the system through Eq. (7.4). The eigenvalues that satisfy Eq. (7.3) (the eigenvalues of $\mathbf{m}^{-1}\mathbf{k}$) thus correspond to the squares of the natural frequencies of the system. Recall that we seek the (ω^2, $\mathbf{U}$) pairs that satisfy Eq. (7.3). We must therefore determine the eigenvector associated with each eigenvalue (frequency). This is discussed following Example 7.1.

Example 7.1

The system shown consists of two identical masses of mass m connected by three identical elastic springs of stiffness k. If the outer springs are fixed at one end, as shown, determine the natural frequencies of the system.

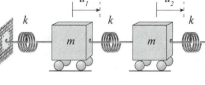

Figure E7.1

Solution
The equation of motion for this system is obtained from Eq. (b) of Example 6.1 by setting $m_1 = m_2 = m$ and $k_1 = k_2 = k_3 = k$, or equivalently from Eq. (6.2) by letting $N = 2$ and $\mathbf{C} = \mathbf{0}$ and inputting the given mass and stiffnesses into the resulting expression. Since we are interested in free vibrations, the force matrix is the null array $\mathbf{F} = \mathbf{0}$. This gives

$$\mathbf{m\ddot{u}} + \mathbf{ku} = \mathbf{0} \tag{a}$$

with

$$\mathbf{u} = \begin{Bmatrix} u_1 \\ u_2 \end{Bmatrix} \tag{b}$$

$$\mathbf{m} = \begin{bmatrix} m & 0 \\ 0 & m \end{bmatrix} = m \begin{bmatrix} 1 & 0 \\ 0 & 1 \end{bmatrix} \tag{c}$$

and

$$\mathbf{k} = \begin{bmatrix} 2k & -k \\ -k & 2k \end{bmatrix} = k \begin{bmatrix} 2 & -1 \\ -1 & 2 \end{bmatrix} \tag{d}$$

The frequency equation for the system is then obtained as follows:

$$\mathcal{F}(\omega^2) = \det\left[\mathbf{k} - \omega^2\mathbf{m}\right] = \begin{vmatrix} (2k - \omega^2 m) & -k \\ -k & (2k - \omega^2 m) \end{vmatrix} = 0 \qquad \text{(e)}$$

or

$$\mathcal{F}(\omega^2) = (2k - \omega^2 m)^2 - k^2 = 0$$

After rearranging, the above relation results in the algebraic equation

$$(\omega^2)^2 - 4\frac{k}{m}(\omega^2) + 3\left(\frac{k}{m}\right)^2 = 0 \qquad \text{(f)}$$

Equation (f) is readily solved for ω^2 to give the two roots (eigenvalues)

$$\omega^2 = \frac{k}{m}, \; 3\frac{k}{m} \qquad \text{(g)}$$

The natural frequencies of the system under consideration are then

$$\omega_1 = \sqrt{k/m}, \quad \omega_2 = \sqrt{3k/m} \qquad \triangleleft \text{(h)}$$

Natural Modes

Each natural frequency ω_j ($j = 1, 2, \ldots, N$) corresponds to a solution of Eq. (7.5). The associated eigenvector, the matrix $\mathbf{U}^{(j)}$, for the frequency ω_j is found by substituting that particular frequency into Eq. (7.3) and solving the resulting algebraic system of equations. Note that if the matrix $\mathbf{U}^{(j)}$ satisfies Eq. (7.3) then the matrix $\alpha\mathbf{U}^{(j)}$, where α is an arbitrary scalar, satisfies that equation as well. That is

$$\left[\mathbf{k} - \omega^2\mathbf{m}\right]\alpha\mathbf{U} = \mathbf{0}$$

The eigenvectors are thus, at most, unique to within a scalar multiplier. (Graphically this means that the length of the vector in N dimensional space is arbitrary. Hence, only the orientation of the vector in that space is determined.)

Recall that the matrix $\left[\mathbf{k} - \omega^2\mathbf{m}\right]$ is singular and therefore has, at most, $N - 1$ independent rows. Thus there are, at most, $N - 1$ independent algebraic equations to determine the N elements of the matrix $\mathbf{U}^{(j)}$. In matrix form,

$$
\begin{bmatrix}
(k_{11} - \omega_j^2 m_{11}) & k_{12} & \cdots & k_{1N} \\
k_{21} & (k_{22} - \omega_j^2 m_{22}) & \cdots & k_{2N} \\
\vdots & \vdots & \ddots & \vdots \\
k_{N1} & k_{N2} & \cdots & (k_{NN} - \omega_j^2 m_{NN})
\end{bmatrix}
\begin{Bmatrix}
U_1^{(j)} \\
U_2^{(j)} \\
\vdots \\
U_N^{(j)}
\end{Bmatrix}
=
\begin{Bmatrix}
0 \\
0 \\
\vdots \\
0
\end{Bmatrix}
\tag{7.7}
$$

It follows that, when expanded, the above matrix equation will yield, at most, $N - 1$ independent equations. This means that we can, at most, determine the relative magnitudes (the ratios) of the elements of $\mathbf{U}^{(j)}$. Therefore the direction, but not the length, of $\mathbf{U}^{(j)}$ is determined. Solving Eq. (7.7) for the elements of $\mathbf{U}^{(j)}$, gives

$$
\mathbf{U}^{(j)} =
\begin{Bmatrix}
U_1^{(j)} \\
U_2^{(j)} \\
\vdots \\
U_N^{(j)}
\end{Bmatrix}
\tag{7.8}
$$

to within a scalar multiplier. The lack of determinacy of an eigenvector is typically resolved by choosing the magnitude of one element, say the first, as unity. Alternatively the elements are *normalized* so as to render the length of the vector as unity. We shall discuss this process, in detail, later in this chapter. Since the eigenvectors that satisfy Eq. (7.3) define the relative motion of the various mass elements for a given frequency of the system they are referred to as the *modal vectors, modal matrices*, mode shapes or simply *the modes* of the system. When normalized (their lengths made unity) they are referred to as *normal modes*. It may be seen from Eq. (7.2) that the motion of the system associated with each $(\omega_j^2, \mathbf{U}^{(j)})$ pair corresponds to oscillations at the rate of that natural frequency, with the masses moving relative to one another in the proportions described by the modal matrix. It will be seen that any motion of the system consists of a linear combination of these fundamental motions.

Example 7.2

Determine the modal vectors for the two-mass three-spring system of Example 7.1.

Solution
Recall from Example 7.1 that, for the system under consideration,

$$
\mathbf{m} =
\begin{bmatrix}
m & 0 \\
0 & m
\end{bmatrix}
\tag{a}
$$

$$\mathbf{k} = \begin{bmatrix} 2k & -k \\ -k & 2k \end{bmatrix} \tag{b}$$

$$\omega_1 = \sqrt{k/m}, \quad \omega_2 = \sqrt{3k/m} \tag{c-1,2}$$

Substitution of the elements of the mass and stiffness matrices of Eqs. (a) and (b) into Eq. (7.7) with $N = 2$ gives

$$\begin{bmatrix} (2k - \omega_j^2) & -k \\ -k & (2k - \omega_j^2) \end{bmatrix} \begin{Bmatrix} U_1^{(j)} \\ U_2^{(j)} \end{Bmatrix} = \begin{Bmatrix} 0 \\ 0 \end{Bmatrix}, \quad (j = 1,2) \tag{d}$$

Each natural frequency – modal matrix pair $\left(\omega_j, \mathbf{U}^{(j)}\right)$ must satisfy Eq. (d).

$j = 1$:
The modal matrix associated with the frequency ω_1 is found by setting $j = 1$ in Eq. (d) and substituting Eq. (c-1). This gives

$$\begin{bmatrix} k & -k \\ -k & k \end{bmatrix} \begin{Bmatrix} U_1^{(1)} \\ U_2^{(1)} \end{Bmatrix} = \begin{Bmatrix} 0 \\ 0 \end{Bmatrix} \tag{e}$$

or, after expanding the matrix equation,

$$k U_1^{(1)} - k U_2^{(1)} = 0$$
$$-k U_1^{(1)} + k U_2^{(1)} = 0 \tag{f}$$

It follows that

$$U_2^{(1)} = U_1^{(1)} \tag{g}$$

Consider the square matrix, $\left[\mathbf{k} - \omega_1^2 \mathbf{m}\right]$, of Eq. (e). Note that the second row of that matrix is simply the negative of the first. Correspondingly, we observe that the second equation of Eq. (f) is simply the first equation multiplied by -1. This occurs because we rendered this matrix singular by setting its determinant to zero in order to obtain nontrivial solutions. Recall that when a matrix is singular, then at least one row can be expressed as a linear combination of the others. Since this singular matrix has only two rows then one is a scalar multiple of the other. We thus have only one distinct equation for the two unknowns $U_1^{(1)}$ and $U_2^{(1)}$. Therefore the modal matrix is unique to within a scalar multiple. Hence, from Eq. (g),

$$\mathbf{U}^{(1)} = \begin{Bmatrix} U_1^{(1)} \\ U_2^{(1)} \end{Bmatrix} = \alpha_1 \begin{Bmatrix} 1 \\ 1 \end{Bmatrix} \tag{h}$$

where α_1 is an arbitrary scalar constant. For definiteness, we <u>shall</u> choose $\alpha_1 = 1$. The modal vector associated with the frequency $\omega_1 = \sqrt{k/m}$ is then given by

$$\mathbf{U}^{(1)} = \begin{Bmatrix} 1 \\ 1 \end{Bmatrix} \tag{i}$$

Note that, as an alternative, we could have chosen the value of α_1 so that the corresponding vector has unit magnitude in some sense of normalization. If we do this in the conventional sense (we shall discuss normalization in Section 7.4.3), the modal matrix takes the form

$$\mathbf{U}^{(1)} = \frac{1}{\sqrt{2}} \begin{Bmatrix} 1 \\ 1 \end{Bmatrix} \tag{i'}$$

$j = 2$:

The modal matrix associated with the frequency ω_2 is found by setting $j = 2$ in Eq. (d) and substituting Eq. (c-2). This gives

$$\begin{bmatrix} -k & -k \\ -k & -k \end{bmatrix} \begin{Bmatrix} U_1^{(2)} \\ U_2^{(2)} \end{Bmatrix} = \begin{Bmatrix} 0 \\ 0 \end{Bmatrix} \tag{j}$$

Expanding Eq. (j) gives

$$\begin{aligned} -kU_1^{(2)} - kU_2^{(2)} &= 0 \\ -kU_1^{(2)} - kU_2^{(2)} &= 0 \end{aligned} \tag{k}$$

Hence,

$$U_2^{(2)} = -U_1^{(2)} \tag{l}$$

and

$$\mathbf{U}^{(2)} = \begin{Bmatrix} U_1^{(2)} \\ U_2^{(2)} \end{Bmatrix} = \alpha_2 \begin{Bmatrix} 1 \\ -1 \end{Bmatrix} \tag{m}$$

where α_2 is an arbitrary scalar constant. If we choose $\alpha_2 = 1$ for definiteness, the modal vector associated with the frequency $\omega_2 = \sqrt{3k/m}$ is given by

$$\mathbf{U}^{(2)} = \begin{Bmatrix} 1 \\ -1 \end{Bmatrix} \tag{n}$$

Alternatively, we could choose α_2 so that the associated vector has unit magnitude. Doing this gives

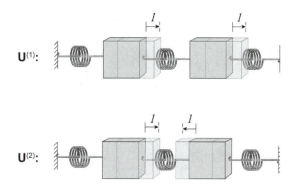

Figure E7.2-1 Natural modes of the system depicted in Figure E7.1.

$$\mathbf{U}^{(2)} = \frac{1}{\sqrt{2}}\begin{Bmatrix} 1 \\ -1 \end{Bmatrix} \tag{n'}$$

The depictions of the system for each of the modes are shown in Figure E7.2-1. It may be seen that when the system vibrates in the first mode, it does so at the frequency $\omega_1 = \sqrt{k/m}$, and with both masses moving in the same direction (left or right) and with the same magnitude. During this motion, the center spring is unstretched, and so the two masses behave as if they are a single mass of magnitude $2m$ as indicated in Figure E7.2-2. The combined stiffness of the outer spring during this mode is $2k$. Thus, for the first mode, the system behaves as a single degree of freedom system of mass $2m$ and stiffness $2k$ and hence ω_1 is given by Eq. (c-1).

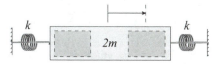

Figure E7.2-2 Effective behavior of system in first mode.

When the system vibrates in the second mode it does so at the frequency $\omega_2 = \sqrt{3k/m}$ and the motion of the second mass is the reflection of the motion of the first. When vibrating in this "accordion mode," the two masses move together and apart symmetrically about the center of the middle spring. The midpoint of the center spring remains stationary (and referred to as a *node*). Thus, the motion is equivalent to that of two independent masses of mass m attached to a fixed spring of half the actual length, and hence twice the stiffness $2k$ [see Eq. (1.8) of Section 1.2.1], as well as to an outer spring of stiffness k as shown in Figure 7.2-3. This, in turn, is equivalent two independent masses of mass m

attached to single fixed springs of stiffness $3k$, as indicated. The natural frequency of each of these single degree of freedom systems is then given by Eq. (c-2).

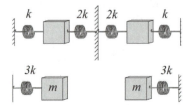

Figure E7.2-3 Effective behavior of system in second mode.

Free Vibration Response

Once the set of frequency-mode pairs $\{\omega_j^2, \mathbf{U}^{(j)} \mid j = 1, 2, ..., N\}$ is determined, then the set of solutions of the form of Eq. (7.2) is determined. Recall that the eigenvalues are the squares of the frequencies. Therefore, both $+\omega_j$ and $-\omega_j$, together with the modal vector $\mathbf{U}^{(j)}$, provide solutions of the desired form. Since the governing equation, Eq. (7.1), is linear the general solution consists of a linear combination of the individual modal solutions. Hence,

$$\mathbf{u}(t) = \sum_{j=1}^{N} \mathbf{u}^{(j)}(t) = \sum_{j=1}^{N} \mathbf{U}^{(j)}\left(C_1^{(j)} e^{i\omega_j t} + C_2^{(j)} e^{-i\omega_j t}\right) \tag{7.9}$$

where $C_1^{(j)}$ and $C_2^{(j)}$ $(j = 1, 2, ..., N)$ are constants of integration. Alternate forms of the response are found by substituting Euler's Formula, Eq. (1.61), into Eq. (7.9) and paralleling the development for the free vibration solution of single degree of freedom systems (see Section 2.1). This gives the response for the N degree of freedom system in the equivalent forms

$$\mathbf{u}(t) = \sum_{j=1}^{N} \mathbf{U}^{(j)}\left(A_1^{(j)} \cos \omega_j t + A_2^{(j)} \sin \omega_j t\right) \tag{7.10}$$

and

$$\mathbf{u}(t) = \sum_{j=1}^{N} \mathbf{U}^{(j)} A^{(j)} \cos\left(\omega_j t - \phi_j\right) \tag{7.11}$$

where the various integration constants are related as follows:

$$A^{(j)} = \sqrt{A_1^{(j)^2} + A_2^{(j)^2}} \tag{7.12}$$

$$\phi_j = \tan^{-1}\left(A_2^{(j)} / A_1^{(j)}\right) \tag{7.13}$$

$$A_1^{(j)} = C_1^{(j)} + C_2^{(j)} \tag{7.14}$$

$$A_2^{(j)} = i\left(C_1^{(j)} - C_2^{(j)}\right) \tag{7.15}$$

Consideration of Eq. (7.11) indicates that any free vibration response consists of the superposition of a harmonic vibration of each mode oscillating at its particular natural frequency. The constants $A^{(j)}$ ($j = 1, 2, ..., N$) are seen to correspond to the amplitudes of the modes and indicate the relative participation of each mode in the total response of the system. Similarly, the constants ϕ_j ($j = 1, 2, ..., N$) are seen to correspond to the phase angles of the individual modes during free vibration. The set of constants $\{A^{(j)}, \phi_j; (j = 1, 2, ..., N)\}$ are determined from the initial conditions

$$\mathbf{u}(0) = \begin{Bmatrix} u_1(0) \\ u_2(0) \\ \vdots \\ u_N(0) \end{Bmatrix}, \quad \dot{\mathbf{u}}(0) = \begin{Bmatrix} \dot{u}_1(0) \\ \dot{u}_2(0) \\ \vdots \\ \dot{u}_N(0) \end{Bmatrix} \tag{7.16}$$

imposed on the system. The initial displacements and initial velocities of each mass element of the system must be specified to determine the explicit form of the free vibration response.

Example 7.3

Determine the free vibration response of the two-mass three-spring system of Examples 7.1 and 7.2 if it is released from rest with the second mass held at its equilibrium position and the first mass held 1 unit of distance from its equilibrium position.

Solution

The specified initial conditions may be written in matrix form as

$$\mathbf{u}(0) = \begin{Bmatrix} 1 \\ 0 \end{Bmatrix}, \quad \dot{\mathbf{u}}(0) = \begin{Bmatrix} 0 \\ 0 \end{Bmatrix} \tag{a-1, 2}$$

Substituting the modal vectors given by Eqs. (i) and (o) of Example 7.2 into Eq. (7.11) with $N = 2$ gives the general free vibration response of the system as

$$\begin{Bmatrix} u_1(t) \\ u_2(t) \end{Bmatrix} = A^{(1)} \begin{Bmatrix} 1 \\ 1 \end{Bmatrix} \cos(\omega_1 t - \phi_1) + A^{(2)} \begin{Bmatrix} 1 \\ -1 \end{Bmatrix} \cos(\omega_2 t - \phi_2) \qquad (b)$$

where

$$\omega_1 = \sqrt{k/m}, \qquad \omega_2 = \sqrt{3k/m} \qquad (c\text{-}1, 2)$$

Imposing Eq. (a-1) on Eq. (b) gives

$$\begin{Bmatrix} 1 \\ 0 \end{Bmatrix} = \begin{Bmatrix} 1 \\ 1 \end{Bmatrix} A^{(1)} \cos(-\phi_1) + \begin{Bmatrix} 1 \\ -1 \end{Bmatrix} A^{(2)} \cos(-\phi_2) \qquad (d)$$

which, when expanded and noting that $\cos(-\phi) = \cos \phi$, takes the form

$$\begin{aligned} 1 &= A^{(1)} \cos \phi_1 + A^{(2)} \cos \phi_2 \\ 0 &= A^{(1)} \cos \phi_1 - A^{(2)} \cos \phi_2 \end{aligned} \qquad (e)$$

Adding and subtracting Eqs. (e) with one another gives the alternate pair of equations

$$\begin{aligned} 1 &= 2A^{(1)} \cos \phi_1 \\ 1 &= 2A^{(2)} \cos \phi_2 \end{aligned} \qquad (f)$$

If we next impose Eq. (a-2) on Eq. (b), and proceed in a similar manner, we obtain

$$\dot{\mathbf{u}}(0) = \begin{Bmatrix} 0 \\ 0 \end{Bmatrix} = -\begin{Bmatrix} 1 \\ 1 \end{Bmatrix} A^{(1)} \sqrt{\frac{k}{m}} \sin(-\phi_1) - \begin{Bmatrix} 1 \\ -1 \end{Bmatrix} A^{(2)} \sqrt{3} \sqrt{\frac{k}{m}} \sin(-\phi_2) \qquad (g)$$

which, when expanded and noting that $\sin(-\phi) = -\sin(\phi)$, takes the form

$$\begin{aligned} 0 &= A^{(1)} \sin \phi_1 + A^{(2)} \sqrt{3} \sin \phi_2 \\ 0 &= A^{(1)} \sin \phi_1 - A^{(2)} \sqrt{3} \sin \phi_2 \end{aligned} \qquad (h)$$

Adding and subtracting Eqs. (h) with one another gives the alternate pair of equations

$$\begin{aligned} 0 &= 2A^{(1)} \sin \phi_1 \\ 0 &= 2A^{(2)} \sin \phi_2 \end{aligned} \qquad (i)$$

The first of Eqs. (i) is satisfied if either $A^{(1)} = 0$ or $\sin\phi_1 = 0$. Since the first of Eqs. (f) cannot be satisfied if $A^{(1)} = 0$, we conclude that

$$\sin\phi_1 = 0$$

and hence that

$$\phi_1 = n\pi \quad (n = 0,1,2,...) \tag{j}$$

Similarly, the second of Eqs. (i) is satisfied if either $A^{(2)} = 0$ or $\sin\phi_2 = 0$. However, the second of Eqs. (f) cannot be satisfied if $A^{(2)} = 0$. We thus conclude that

$$\sin\phi_2 = 0$$

and hence that

$$\phi_2 = n\pi \quad (n = 0,1,2,...) \tag{k}$$

Substituting Eqs. (j) and (k) into Eqs. (f) we find that,

$$\begin{aligned}
A^{(1)} = A^{(2)} = +1/2 \quad (n = 0,2,4,...)\\
A^{(1)} = A^{(2)} = -1/2 \quad (n = 1,3,5,...)
\end{aligned} \tag{l}$$

Substitution of Eqs. (l) and (c) into Eq. (b) gives the free vibration response of the system resulting from the given initial conditions as

$$\mathbf{u}(t) = \begin{Bmatrix} u_1(t)\\ u_2(t) \end{Bmatrix} = \frac{1}{2}\begin{Bmatrix} 1\\ 1 \end{Bmatrix}\cos\sqrt{\tfrac{k}{m}}t + \frac{1}{2}\begin{Bmatrix} 1\\ -1 \end{Bmatrix}\cos\sqrt{\tfrac{3k}{m}}t \qquad \lhd \text{(m)}$$

Equation (m) can also be expressed in the equivalent form

$$\mathbf{u}(t) = \begin{Bmatrix} u_1(t)\\ u_2(t) \end{Bmatrix} = \frac{1}{\sqrt{2}}\begin{Bmatrix} 1/\sqrt{2}\\ 1/\sqrt{2} \end{Bmatrix}\cos\sqrt{\tfrac{k}{m}}t + \frac{1}{\sqrt{2}}\begin{Bmatrix} 1/\sqrt{2}\\ -1/\sqrt{2} \end{Bmatrix}\cos\sqrt{\tfrac{3k}{m}}t \qquad \lhd \text{(m')}$$

for which the modal vectors have unit magnitude [that is they have the alternate forms stated by Eqs. (i') and (o') of Example 7.2)]. It may be seen from either Eq. (m) or Eq. (m') that, for the given initial conditions, the amplitudes of both modes are the same. Thus, the degree of participation of each mode is the same.

Example 7.4

Consider the system of the previous example when each mass is initially displaced and held a unit distance from its equilibrium position and then both are released from rest simultaneously. That is, consider an initial displacement of the system in the form of the first modal vector and let the system be released from rest from this configuration.

Solution

The initial conditions of the two mass-three spring system for the present case are written in matrix form as

$$\mathbf{u}(0) = \begin{Bmatrix} 1 \\ 1 \end{Bmatrix}, \quad \dot{\mathbf{u}}(0) = \begin{Bmatrix} 0 \\ 0 \end{Bmatrix} \qquad \text{(a-1, 2)}$$

Recall from Eqs. (b) and (c) of Example 7.3 that the general form of the free vibration response of the system under consideration is

$$\begin{Bmatrix} u_1(t) \\ u_2(t) \end{Bmatrix} = A^{(1)} \begin{Bmatrix} 1 \\ 1 \end{Bmatrix} \cos(\omega_1 t - \phi_1) + A^{(2)} \begin{Bmatrix} 1 \\ -1 \end{Bmatrix} \cos(\omega_2 t - \phi_2) \qquad \text{(b)}$$

where

$$\omega_1 = \sqrt{k/m}, \quad \omega_2 = \sqrt{3k/m} \qquad \text{(c-1, 2)}$$

Imposing Eq. (a-1) on Eq. (b) results in the pair of algebraic equations

$$\begin{aligned} 1 &= A^{(1)} \cos\phi_1 + A^{(2)} \cos\phi_2 \\ 1 &= A^{(1)} \cos\phi_1 - A^{(2)} \cos\phi_2 \end{aligned} \qquad \text{(d)}$$

Adding and subtracting Eqs. (d) with one another gives the alternate pair of algebraic equations

$$\begin{aligned} 2 &= 2A^{(1)} \cos\phi_1 \\ 0 &= 2A^{(2)} \cos\phi_2 \end{aligned} \qquad \text{(e)}$$

Similarly, imposing Eq. (a-2) on Eq. (b) and adding and subtracting the resulting expressions gives the pair of algebraic equations

$$\begin{aligned} 0 &= 2A^{(1)} \sin\phi_1 \\ 0 &= 2A^{(2)} \sin\phi_2 \end{aligned} \qquad \text{(f)}$$

The first Eqs. (f) is satisfied if either $A^{(1)} = 0$ or $\sin\phi_1 = 0$. Since the first of Eqs. (e) cannot be satisfied if $A^{(1)} = 0$, we conclude that

$$\sin\phi_1 = 0$$

and hence that

$$\phi_1 = n\pi \quad (n = 0,1,2,...) \tag{g}$$

Substitution of Eq. (g) into the first of Eqs. (e) gives the amplitude of the first mode as

$$A^{(1)} = \frac{1}{\cos n\pi} = \begin{cases} +1 & (n = 0,2,4,...) \\ -1 & (n = 1,3,5,...) \end{cases} \tag{h}$$

Upon consideration of the second of Eqs. (e) and the second of Eqs. (f) it may be concluded that both equations are satisfied only if

$$A^{(2)} = 0 \tag{i}$$

Substitution of Eqs. (g), (h) and (i) into Eq. (b) gives the free vibration response for the given initial conditions as

$$\mathbf{u}(t) = \begin{Bmatrix} 1 \\ 1 \end{Bmatrix} \cos\sqrt{\tfrac{k}{m}}t = \sqrt{2} \begin{Bmatrix} 1/\sqrt{2} \\ 1/\sqrt{2} \end{Bmatrix} \cos\sqrt{\tfrac{k}{m}}t \qquad \lhd \text{(j)}$$

It is seen that when the initial displacement is in the form of the first mode, then the response of the system is solely comprised of that mode. The other mode does not participate. If the initial displacement was of the form of the second mode, the response of the system would be solely comprised of the second mode (see Problem 7.9).

Example 7.5

Consider the uniform double pendulum shown in Figure E7.5-1. (*a*) Determine the natural frequencies of the system. (*b*) Determine the corresponding natural modes, and sketch the mode shapes. (*c*) Determine the free vibration response of the pendulum.

Solution

The equations of motion for the double pendulum were derived in Chapter 6. (See Section 6.1.2 and Example 6.5.) The corresponding displacement matrix is

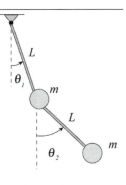

Figure E7.5-1

$$\mathbf{u}(t) = \left\{ \begin{matrix} \theta_1(t) \\ \theta_2(t) \end{matrix} \right\}$$

where θ_1 and θ_2 are measured from the vertical (rest) configuration, as indicated. For the particular system shown the masses are of equal magnitude, as are the lengths of the rods. Thus, $m_1 = m_2 = m$ and $L_1 = L_2 = L$. Substitution of these values into the mass and stiffness matrices of Eq. (6.25), or equivalently of Eq. (r) of Example 6.5, gives the mass and stiffness matrices for the present system as

$$\mathbf{m} = \begin{bmatrix} (m_1 + m_2)L_1^2 & m_2 L_1 L_2 \\ m_2 L_1 L_2 & m_2 L_2^2 \end{bmatrix} = mL^2 \begin{bmatrix} 2 & 1 \\ 1 & 1 \end{bmatrix} \tag{a}$$

$$\mathbf{k} = \begin{bmatrix} (m_1 + m_2)L_1 g & 0 \\ 0 & m_2 L_2 g \end{bmatrix} = mL^2 \begin{bmatrix} 2g/L & 0 \\ 0 & g/L \end{bmatrix} \tag{b}$$

(a)

We first establish the frequency equation for the system. Hence,

$$\det\left[\mathbf{k} - \omega^2 \mathbf{m}\right] = mL^2 \begin{vmatrix} 2(\frac{g}{L} - \omega^2) & -\omega^2 \\ -\omega^2 & (\frac{g}{L} - \omega^2) \end{vmatrix} = 0 \tag{c}$$

Expanding the determinant in Eq. (c) and dividing through by mL^2 gives the frequency equation

$$(\omega^2)^2 - 4\frac{g}{L}(\omega^2) + 2\left(\frac{g}{L}\right)^2 = 0 \tag{d}$$

The natural frequencies are next determined from this equation. Equation (d) is easily solved to give the two distinct roots

$$\omega^2 = \frac{g}{L}\left(2 \mp \sqrt{2}\right)$$

Thus,

$$\omega_1 = 0.765\sqrt{g/L}, \quad \omega_2 = 1.85\sqrt{g/L} \tag{e-1,2}$$

(b)

Substituting the mass and stiffness matrices given by Eqs. (a) and (b) into the equation

$$\left[\mathbf{k} - \omega_j^2 \mathbf{m}\right]\mathbf{U}^{(j)} = \mathbf{0} \quad (j = 1, 2)$$

gives the equations for the modal matrices as

$$mL^2 \begin{bmatrix} 2(\frac{g}{L} - \omega_j^2) & -\omega_j^2 \\ -\omega_j^2 & (\frac{g}{L} - \omega_j^2) \end{bmatrix} \begin{Bmatrix} \theta_1^{(j)} \\ \theta_2^{(j)} \end{Bmatrix} = \begin{Bmatrix} 0 \\ 0 \end{Bmatrix}, \quad (j = 1, 2) \tag{f}$$

$j = 1$:
Setting $j = 1$ in Eq. (f), substituting Eq. (e-1) and expanding the matrix equation gives two algebraic equations. The first is

$$2\left(\frac{g}{L} - (2 - \sqrt{2})\frac{g}{L}\right)U_1^{(1)} - (2 - \sqrt{2})\frac{g}{L}U_2^{(1)} = 0$$

which simplifies to the relation

$$U_2^{(1)} = \sqrt{2}\, U_1^{(1)} \tag{g}$$

(Note that the second equation corresponding to the second row of Eq. (f) reduces to Eq. (g) as well. Why?) The modal matrix associated with frequency ω_1 is then

$$\mathbf{U}^{(1)} = \alpha_1 \begin{Bmatrix} 1 \\ \sqrt{2} \end{Bmatrix} \tag{h}$$

where α_1 is arbitrary. For definiteness we shall let $\alpha_1 = 1$. Thus

$$\mathbf{U}^{(1)} = \begin{Bmatrix} 1 \\ \sqrt{2} \end{Bmatrix} \tag{h'}$$

$j = 2$:
Following the same procedure for $j = 2$ as we did for $j = 1$, we obtain the algebraic equation

$$2\left(\frac{g}{L} - (2 + \sqrt{2})\frac{g}{L}\right)U_1^{(2)} - (2 + \sqrt{2})\frac{g}{L}U_2^{(2)} = 0$$

which, upon performing the indicated additions, simplifies to the relation

$$U_2^{(2)} = -\sqrt{2}\, U_1^{(2)} \tag{i}$$

It follows that

$$\mathbf{U}^{(2)} = \alpha_2 \begin{Bmatrix} 1 \\ -\sqrt{2} \end{Bmatrix} \tag{j}$$

where α_2 is arbitrary. Setting $\alpha_2 = 1$ for definiteness gives the modal matrix associated with natural frequency ω_2 as

$$\mathbf{U}^{(2)} = \left\{ \begin{array}{c} 1 \\ -\sqrt{2} \end{array} \right\}$$

(j')

The two mode shapes are depicted in Figure E7.5-2.

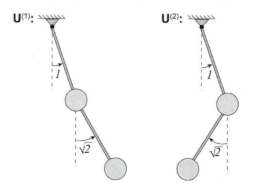

Figure E7.5-2 Natural modes for the double pendulum.

(c)

The general form of the free vibration response of the double pendulum is found by substituting the above natural frequencies and modal matrices into Eq. (7.11), with $N = 2$. This gives

$$\left\{ \begin{array}{c} \theta_1(t) \\ \theta_2(t) \end{array} \right\} = A^{(1)} \left\{ \begin{array}{c} 1 \\ \sqrt{2} \end{array} \right\} \cos\left(0.765\sqrt{\tfrac{g}{L}} - \phi_1\right) + A^{(2)} \left\{ \begin{array}{c} 1 \\ -\sqrt{2} \end{array} \right\} \cos\left(1.85\sqrt{\tfrac{g}{L}} - \phi_2\right) \quad (k)$$

Hence,

$$\theta_1(t) = A^{(1)} \cos\left(0.765\sqrt{\tfrac{g}{L}} - \phi_1\right) + A^{(2)} \cos\left(1.85\sqrt{\tfrac{g}{L}} - \phi_2\right)$$

$$\theta_2(t) = A^{(1)} \sqrt{2} \cos\left(0.765\sqrt{\tfrac{g}{L}} - \phi_1\right) - A^{(2)} \sqrt{2} \cos\left(1.85\sqrt{\tfrac{g}{L}} - \phi_2\right)$$

◁ (l)

The values of the amplitudes, $A^{(1)}$ and $A^{(2)}$, are found from the specific initial conditions imposed on the system.

Example 7.6

Determine the free vibration response associated with side-sway motion of the three-story building shown in Figure E7.6-1. The 3 floors are each of mass m, and the 12 identical elastic columns are each of length L and have bending stiffness EI.

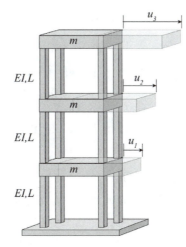

Figure E7.6-1 Three-story building.

Solution

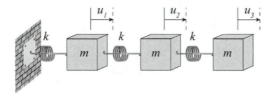

Figure E7.6-2 Equivalent system.

We first establish the equivalent discrete system shown in Figure E7.6-2. It is evident that the equivalent system has three degrees of freedom. The stiffnesses of the equivalent elastic springs can be obtained directly from Eq. (1.19) by setting $N = 3$, $E_1 I_1 = E_2 I_2 = E_3 I_3 = EI$ and $L_1 = L_2 = L_3 = L$. We thus have that

$$k_1 = k_2 = k_3 = k = 48EI/L^3 \tag{a}$$

The discrete representation for this structure therefore consists of three masses (the floors and roof), each of mass m, connected by three identical springs whose stiffnesses are given by Eq. (a).

The equation of motion for the equivalent system is easily determined by setting $m_1 = m_2 = m_3 = m$, $k_1 = k_2 = k_3 = k$, $k_4 = 0$ and $F_1 = F_2 = F_3 = 0$ in Eqs. (d)–(f) of Example 6.2 or Eq. (h) of Example 6.3. (Alternatively, the equation of motion could be derived directly using one of the methods discussed in Chapter 6.) Doing this we arrive at the governing equation given by

$$\mathbf{m\ddot{u}} + \mathbf{ku} = \mathbf{0} \tag{b}$$

where

$$\mathbf{m} = m \begin{bmatrix} 1 & 0 & 0 \\ 0 & 1 & 0 \\ 0 & 0 & 1 \end{bmatrix}, \quad \mathbf{k} = k \begin{bmatrix} 2 & -1 & 0 \\ -1 & 2 & -1 \\ 0 & -1 & 1 \end{bmatrix} \tag{c, d}$$

The eigenvalue problem, $\left[\mathbf{k} - \omega^2 \mathbf{m} \right] \mathbf{U} = \mathbf{0}$, for the system under consideration is then

$$\begin{bmatrix} (2k - \omega^2 m) & -k & 0 \\ -k & (2k - \omega^2 m) & -k \\ 0 & -k & (k - \omega^2 m) \end{bmatrix} \begin{Bmatrix} U_1 \\ U_2 \\ U_3 \end{Bmatrix} = \begin{Bmatrix} 0 \\ 0 \\ 0 \end{Bmatrix} \tag{e}$$

Since the numerical values of the parameters k and m are not specified it is convenient to express Eq. (e) in an alternate form by factoring out k. This gives

$$k \begin{bmatrix} (2 - \bar{\omega}^2) & -1 & 0 \\ -1 & (2 - \bar{\omega}^2) & -1 \\ 0 & -1 & (1 - \bar{\omega}^2) \end{bmatrix} \begin{Bmatrix} U_1 \\ U_2 \\ U_3 \end{Bmatrix} = \begin{Bmatrix} 0 \\ 0 \\ 0 \end{Bmatrix} \tag{f}$$

where

$$\bar{\omega}^2 = \frac{\omega^2}{k/m} \tag{g}$$

Once the eigenvalues $\bar{\omega}^2$ are determined, the corresponding natural frequencies ω^2 may be obtained using Eq. (g).

The characteristic equation (frequency equation) is found by requiring that the determinant of the square matrix of Eq. (f) vanish. Hence,

$$\mathcal{F}(\bar{\omega}^2) = \begin{vmatrix} (2 - \bar{\omega}^2) & -1 & 0 \\ -1 & (2 - \bar{\omega}^2) & -1 \\ 0 & -1 & (1 - \bar{\omega}^2) \end{vmatrix} = -(\bar{\omega}^2)^3 + 5(\bar{\omega}^2)^2 - 6(\bar{\omega}) + 1 = 0 \tag{h}$$

The roots of Eq. (h) may be found by classical means, or by using a root solving routine, or by using software that solves the complete eigenvalue problem defined by Eq. (f) directly, such as the MATLAB function "eig." We here use the MATLAB polynomial solver, "roots," to determine the zeros of the characteristic equation. This is done by constructing the matrix of coefficients of the polynomial appearing in Eq. (h), $P = [-1, 5, -6, 1]$, and invoking the MATLAB command roots(P). We thus obtain the zeroes of Eq. (h),

$$\bar{\omega}^2 = 0.1981, \ 1.555, \ 3.247 \tag{i}$$

Substitution of each eigenvalue of Eq. (i) into Eq. (g) gives the corresponding natural frequencies for the three-story structure,

$$\omega_1 = 0.4451\sqrt{k/m} = 3.08\sqrt{\frac{EI}{mL^3}} \tag{j$_1$}$$

$$\omega_2 = 1.247\sqrt{k/m} = 8.64\sqrt{\frac{EI}{mL^3}} \tag{j$_2$}$$

$$\omega_3 = 1.802\sqrt{k/m} = 12.5\sqrt{\frac{EI}{mL^3}} \tag{j$_3$}$$

To determine the associated modal vectors, we substitute a frequency from Eq. (j), or equivalently Eq. (i), into Eq. (f) and solve for the components of the corresponding modal vector. Since the matrix $\left[\mathbf{k} - \omega^2 \mathbf{m} \right] \mathbf{U} = \mathbf{0}$ was rendered singular by requiring that its determinant vanish, (at most) only two of the three scalar equations of Eq. (f) will be independent. We shall choose to solve the first and third equations since they each have a vanishing term. Hence,

$$
\begin{aligned}
(2 - \bar{\omega}_i^2)U_1^{(j)} - U_2^{(j)} &= 0 \\
-U_2^{(j)} + (1 - \bar{\omega}_i^2)U_3^{(j)} &= 0
\end{aligned}
\quad (j = 1, 2, 3) \tag{k}
$$

$\mathbf{U}^{(1)}$:
Substitution of the first value of Eq. (i) into Eqs. (k), with $j = 1$, gives

$$
\begin{aligned}
(2 - 0.1981)U_1^{(1)} - U_2^{(1)} &= 0 \quad \Rightarrow \quad U_2^{(1)} = 1.802U_1^{(1)} \\
-U_2^{(1)} + (1 - 0.1981)U_3^{(1)} &= 0 \quad \Rightarrow \quad U_3^{(1)} = 1.247U_2^{(1)} = 2.247U_1^{(1)}
\end{aligned}
$$

Hence,

$$\mathbf{U}^{(1)} = \alpha_1 \begin{Bmatrix} 1 \\ 1.802 \\ 2.247 \end{Bmatrix} \tag{l}$$

where α_1 is an arbitrary scalar. The scalar multiplier can be chosen as anything we like, such as unity as was done in prior examples. For this case, however, we shall choose α_1 so that $\mathbf{U}^{(1)}$ has unit magnitude in the conventional sense (i.e., we shall divide the above vector by its magnitude). Thus, let

$$\alpha_1 = \frac{1}{\sqrt{1^2 + 1.802^2 + 2.247^2}}$$

This gives the first natural mode as the normal mode

$$\mathbf{U}^{(1)} = \begin{Bmatrix} 0.328 \\ 0.591 \\ 0.737 \end{Bmatrix} \tag{l'}$$

$\mathbf{U}^{(2)}$:

Proceeding as for the first modal vector,

$$(2-1.555)U_1^{(2)} - U_2^{(2)} = 0 \quad \Rightarrow \quad U_2^{(2)} = 0.4450U_1^{(2)}$$
$$-U_2^{(2)} + (1-1.555)U_3^{(2)} = 0 \quad \Rightarrow \quad U_3^{(2)} = -1.802U_2^{(2)} = -0.8019U_1^{(2)}$$

Hence,

$$\mathbf{U}^{(2)} = \alpha_2 \begin{Bmatrix} 1 \\ 0.4450 \\ -0.8019 \end{Bmatrix} \quad \rightarrow \quad \mathbf{U}^{(2)} = \begin{Bmatrix} 0.737 \\ 0.328 \\ -0.591 \end{Bmatrix} \tag{m}$$

$\mathbf{U}^{(3)}$:

The third modal vector is obtained in the same manner as the first two. Thus,

$$(2-3.247)U_1^{(3)} - U_2^{(3)} = 0 \quad \Rightarrow \quad U_2^{(3)} = -1.247U_1^{(3)}$$
$$-U_2^{(3)} + (1-3.247)U_3^{(3)} = 0 \quad \Rightarrow \quad U_3^{(3)} = 0.4450U_2^{(3)} = 0.5550U_1^{(3)}$$

Solving the above system gives

$$\mathbf{U}^{(3)} = \alpha_3 \begin{Bmatrix} 1 \\ -1.247 \\ 0.5550 \end{Bmatrix} \quad \rightarrow \quad \mathbf{U}^{(3)} = \begin{Bmatrix} 0.591 \\ -0.737 \\ 0.328 \end{Bmatrix} \tag{n}$$

Depictions of the three-story structure in each of its three natural modes are displayed in Figures E7.6-3 (a)–(c). We are now ready to determine the general response of the structure.

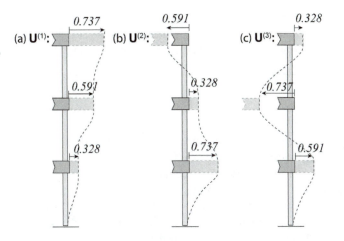

Figure E7.6-3 Natural modes of three-story building.

The general form of the free vibration response of a three degree of freedom system is given by Eq. (7.11) with $N = 3$. Hence,

$$\mathbf{u}(x,t) = \sum_{j=1}^{3} \mathbf{U}^{(j)} \cos(\omega_j t - \phi_j)$$

Substituting the calculated frequencies and modes into the above expression gives the free vibration response of the three story building as

$$\begin{Bmatrix} u_1(x,t) \\ u_2(x,t) \\ u_3(x,t) \end{Bmatrix} = \begin{Bmatrix} 0.328 \\ 0.591 \\ 0.737 \end{Bmatrix} A^{(1)} \cos\left(3.08\omega_0 t - \phi_1\right)$$

$$+ \begin{Bmatrix} 0.737 \\ 0.328 \\ -0.591 \end{Bmatrix} A^{(2)} \cos\left(8.64\omega_0 t - \phi_2\right) \tag{o}$$

$$+ \begin{Bmatrix} 0.591 \\ -0.737 \\ 0.328 \end{Bmatrix} A^{(3)} \cos\left(12.5\omega_0 t - \phi_3\right)$$

where

$$\omega_0 = \sqrt{\frac{EI}{mL^3}} \tag{p}$$

It is seen that, from a vibrations perspective, the particular system is defined by the single parameter ω_0. The amplitudes and phase angles, $A^{(j)}$ and ϕ_j, ($j = 1, 2, 3$) are determined from the specific initial conditions imposed on the structure.

Example 7.7

Consider free vibrations of the tram system of Example 6.6. (*a*) Determine the frequency equation for the general system. (*b*) Independent measurements of the natural frequency of the frame and spring alone, and of the barrow and support rod when attached to a fixed frame, are found to yield identical values. Determine the natural frequencies and modes of the coupled tram system if the mass of the barrow is twice that of the frame. (*c*) Determine the free vibration response for the system of part (*b*). Express your answers to (*b*) and (*c*) in terms of the common natural frequency measured for the detached subsystems.

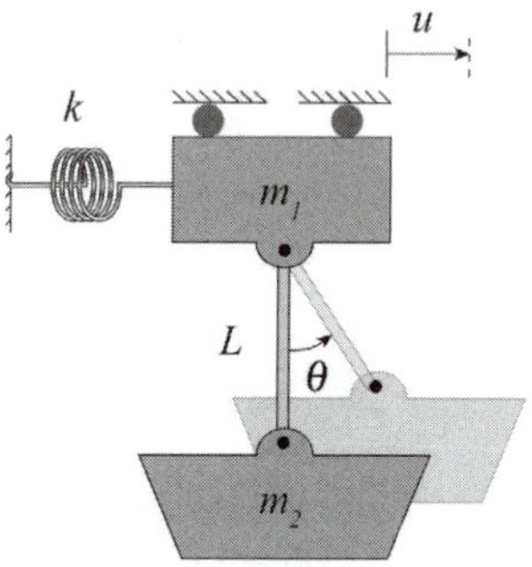

Figure E7.7-1

Solution

(*a*)

Setting $\mathbf{F} = \mathbf{0}$ in Eq. (i) of Example 6.6 gives the pertinent equation of motion,

$$\begin{bmatrix} (m_1 + m_2) & m_2 L \\ m_2 L & m_2 L^2 \end{bmatrix} \begin{Bmatrix} \ddot{u} \\ \ddot{\theta} \end{Bmatrix} + \begin{bmatrix} k & 0 \\ 0 & m_2 gL \end{bmatrix} \begin{Bmatrix} u \\ \theta \end{Bmatrix} = \begin{Bmatrix} 0 \\ 0 \end{Bmatrix} \tag{a}$$

The corresponding eigenvalue problem, $\left[\mathbf{k} - \omega^2 \mathbf{m} \right] \mathbf{U} = \mathbf{0}$, is then

$$\begin{bmatrix} \{k - \omega^2 (m_1 + m_2)\} & -\omega^2 m_2 L \\ -\omega^2 m_2 L & \{m_2 gL - \omega^2 m_2 L^2\} \end{bmatrix} \begin{Bmatrix} U_1 \\ U_2 \end{Bmatrix} = \begin{Bmatrix} 0 \\ 0 \end{Bmatrix} \tag{b}$$

Setting the determinant of the square matrix of Eq. (b) to zero and rearranging terms gives the frequency equation for the system as

$$\left(\omega^2 \right)^2 - \omega^2 \left[\left(1 + \frac{m_2}{m_1} \right) \frac{g}{L} + \frac{k}{m_1} \right] + \frac{k}{m_1} \frac{g}{L} = 0 \qquad \triangleleft \text{(c)}$$

(b)

Let ω_0 be the measured natural frequency of the detached subsystems. Hence,

$$\omega_0^2 = g/L = k/m_1 \tag{d}$$

Substituting Eq. (d) and the given mass ratio into Eq. (c) renders the frequency equation to the form

$$\left(\bar{\omega}^2\right)^2 - 4\bar{\omega}^2 + 1 = 0 \tag{e}$$

where

$$\bar{\omega} = \omega/\omega_0 \tag{f}$$

The roots of Eq. (e) are easily found to be

$$\bar{\omega}^2 = 2 \mp \sqrt{3} = 0.268, \; 3.73 \tag{g}$$

Substituting these roots into Eq. (e) gives the natural frequencies,

$$\omega_1 = 0.518\omega_0, \quad \omega_2 = 1.93\omega_0 \qquad \triangleleft \text{(h)}$$

To determine the natural modes let us expand Eq. (b) and divide the second equation by $m_2 L$. This gives the relation

$$-\bar{\omega}^2 U_1 + \left(1 - \bar{\omega}^2\right) L U_2 = 0 \tag{i}$$

Substituting the first root stated in Eq. (g) into Eq. (i) and solving for $U_2^{(1)}$ in terms of $U_1^{(1)}$ gives the first natural mode as

$$\mathbf{U}^{(1)} = \alpha_1 \begin{Bmatrix} 1 \\ 0.366/L \end{Bmatrix} \rightarrow \begin{Bmatrix} 1 \\ 0.366/L \end{Bmatrix} = \begin{Bmatrix} 1 \\ 0.366\,\omega_0^2/g \end{Bmatrix} \qquad \triangleleft \text{(j)}$$

where we have chosen the arbitrary scalar multiple α_1 to have unit value. Substituting the second root stated in Eq. (g) into Eq. (i) and proceeding in a similar manner gives the second natural mode as

$$\mathbf{U}^{(2)} = \alpha_2 \begin{Bmatrix} 1 \\ -1.37/L \end{Bmatrix} \rightarrow \begin{Bmatrix} 1 \\ -1.37/L \end{Bmatrix} = \begin{Bmatrix} 1 \\ -1.37\,\omega_0^2/g \end{Bmatrix} \qquad \triangleleft \text{(k)}$$

The natural modes for the system are depicted in Figure E7.7-2.

U(1) :

U(2) :

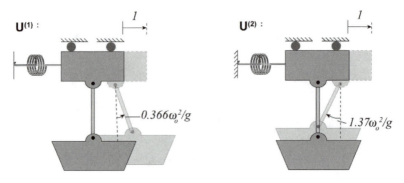

Figure E7.7-2 Natural modes of tram system.

(*c*)
Substitution of Eqs. (h), (j) and (k) into Eq. (7.11) with $N = 2$ results in the free vibration response of the system,

$$
\mathbf{u}(x,t) = \begin{Bmatrix} u(x,t) \\ \theta(x,t) \end{Bmatrix} = A^{(1)} \begin{Bmatrix} 1 \\ 0.366\,\omega_0^2/g \end{Bmatrix} \cos(0.518\omega_0 t - \phi_1)
$$
$$
+ A^{(2)} \begin{Bmatrix} 1 \\ -1.37\,\omega_0^2/g \end{Bmatrix} \cos(1.93\omega_0 t - \phi_2)
$$

◁ (l)

where the amplitudes and phase angles are found from the particular initial conditions imposed on the system.

Example 7.8

Determine the free vibration response for the motorcycle frame of Example 6.7 for the case where the stiffness supplied by each wheel is identical and the rider sits over the center of mass. The mass of the rider is 1/3 the mass of the bike, and the stiffness of the seat assemblage is 2/3 the effective stiffness of the tire and suspension system combination. For simplicity, assume that the frame can be treated as uniform.

Figure E7.8-1

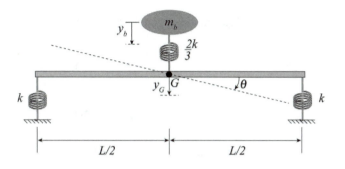

Figure E7.8-2 Equivalent system.

Solution

For the given frame and rider, $k_1 = k_2 = k$, $k_b = 2k/3$, $m_b = m/3$, $\ell = 0$ and $L_1 = L_2 = L/2$. The moment of inertia about an axis through the center of mass of the *uniform* frame (Figure E7.8-2) is then $I_G = mL^2/12$.

Recall that the vertical displacement of the center of mass of the frame, y_G, the rotational displacement, θ, and the vertical displacement of the rider, y_b, are chosen as the generalized coordinates. The corresponding displacement matrix is then

$$\mathbf{u}(t) = \begin{Bmatrix} y_G(t) \\ \theta(t) \\ y_b(t) \end{Bmatrix} \tag{a}$$

For the given parameters, the system matrices of Example 6.7 reduce to

$$\mathbf{k} \rightarrow \begin{bmatrix} 8k/3 & 0 & -2k/3 \\ 0 & kL^2/2 & 0 \\ -2k/3 & 0 & 2k/3 \end{bmatrix} \tag{b}$$

$$\mathbf{m} \rightarrow \begin{bmatrix} m & 0 & 0 \\ 0 & I_G & 0 \\ 0 & 0 & m/3 \end{bmatrix} \tag{c}$$

The corresponding eigenvalue problem is then

$$\begin{bmatrix} \left(8k/3 - \omega^2 m\right) & 0 & -2k/3 \\ 0 & \left(kL^2/2 - \omega^2 I_G\right) & 0 \\ -2k/3 & 0 & \left(2k - \omega^2 m\right)/3 \end{bmatrix} \begin{Bmatrix} U_1 \\ U_2 \\ U_3 \end{Bmatrix} = \begin{Bmatrix} 0 \\ 0 \\ 0 \end{Bmatrix} \tag{d}$$

We next determine the frequency equation for the system. Hence,

$$\begin{vmatrix} (8k/3 - \omega^2 m) & 0 & -2k/3 \\ 0 & (kL^2/2 - \omega^2 I_G) & 0 \\ -2k/3 & 0 & (2k - \omega^2 m)/3 \end{vmatrix} = 0 \qquad \text{(e)}$$

Expanding the above determinant gives the desired frequency equation,

$$\left(kL^2/2 - \omega^2 I_G\right)\left[12k^2 - 14k\,m(\omega^2) + 3m^2(\omega^2)^2\right] = 0 \qquad \text{(f)}$$

Equation (f) is easily factored to yield the natural frequencies,

$$\omega_1^2 = 1.132\,k/m, \quad \omega_2^2 = 3.535\,k/m, \quad \omega_3^2 = kL^2/2I_G = 6k/m \qquad \triangleleft \text{(g)}$$

We next determine the associated modal vectors. Substituting the first frequency into the first row of Eq. (d) gives

$$\left(\frac{8}{3}k - 1.132\frac{k}{m}\,m\right)U_1^{(1)} - \frac{2}{3}kU_3^{(1)} = 0 \quad \Rightarrow \quad U_3^{(1)} = 2.303 U_1^{(1)} \qquad \text{(h)}$$

[The third row of Eq. (d) gives the identical relation.] Substituting the first natural frequency into the second row of Eq. (d) gives

$$\left(2kL^2 - 1.132\frac{k}{m}\,m\right)U_2^{(1)} = 0 \quad \Rightarrow \quad U_2^{(1)} = 0 \qquad \text{(i)}$$

Combining Eqs. (h) and (i) in matrix form gives the modal vector associated with the first natural frequency,

$$\mathbf{U}^{(1)} = \left\{ \begin{array}{c} 1 \\ 0 \\ 2.303 \end{array} \right\} \qquad \triangleleft \text{(j)}$$

A physical depiction of the modal vector is shown in Figure E7.8-3a. Similar calculations give the modal vector associated with the second natural frequency. Hence,

$$\left(\frac{8}{3}k - 3.535\frac{k}{\cancel{m}}\cancel{m}\right)U_1^{(2)} - \frac{2}{3}kU_3^{(2)} = 0 \quad \Rightarrow \quad U_3^{(2)} = -1.303U_1^{(2)} \qquad \text{(k)}$$

and

$$\left(2\cancel{k}L^2 - 3.535\frac{\cancel{k}}{\cancel{m}}\cancel{m}\right)U_2^{(2)} = 0 \quad \Rightarrow \quad U_2^{(2)} = 0 \qquad \text{(l)}$$

which, when combining in matrix form gives the modal vector associated with the second natural frequency,

$$\mathbf{U}^{(2)} = \left\{\begin{array}{c} 1 \\ 0 \\ -1.303 \end{array}\right\} \qquad \lhd \text{(m)}$$

A physical depiction of the modal vector is shown in Figure E7.8-3b. To obtain the third modal vector we substitute the third natural frequency into the first and third rows of Eq. (d) to get

$$\left(\frac{8}{3}k - \frac{2kL^2}{mL^2/12}\right)U_1^{(3)} - \frac{2}{3}kU_3^{(3)} = 0 \qquad \text{(n)}$$

and

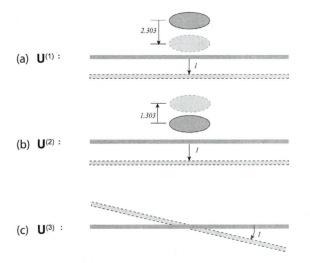

(a) $\mathbf{U}^{(1)}$:

2.303

(b) $\mathbf{U}^{(2)}$:

1.303

(c) $\mathbf{U}^{(3)}$:

Figure E7.8-3 Natural modes of motorcycle frame.

$$-\frac{2}{3}kU_1^{(3)} + \left(2 - \frac{2mL^2}{mL^2/12}\right)U_3^{(3)} = 0 \qquad (o)$$

These two equations may be solved to give

$$U_1^{(3)} = U_3^{(3)} = 0 \qquad (p)$$

Since Eqs. (n) and (o) yielded definite values, the second row of Eq. (d) will reflect the lack of determinacy. Substituting the third natural frequency into the second row of Eq. (d) gives the relation

$$\left(2kL^2 - \frac{2kL^2}{\cancel{V_G}}\cancel{V_G}\right)U_2^{(3)} = 0 \quad \rightarrow \quad 0 \cdot U_2^{(3)} = 0 \quad \Rightarrow \quad U_2^{(3)} = anything \qquad (q)$$

as we would expect since the square matrix of Eq. (d) must be singular. [For this frequency, all elements of the second row of the square matrix of Eq. (d) vanish.] The third modal vector is thus of the form

$$\mathbf{U}^{(3)} = \begin{Bmatrix} 0 \\ 1 \\ 0 \end{Bmatrix} \qquad \triangleleft (r)$$

As seen in the depiction of the mode shown in Figure E7.8-3c, the third mode corresponds to pure rotation. Finally, the free vibration response of the motorcycle frame is

$$\begin{Bmatrix} y_G(t) \\ \theta(t) \\ y_b(t) \end{Bmatrix} = A^{(1)} \begin{Bmatrix} 1 \\ 0 \\ 2.30 \end{Bmatrix} \cos(1.06\omega_0 t - \phi_1)$$

$$+ A^{(2)} \begin{Bmatrix} 1 \\ 0 \\ -1.30 \end{Bmatrix} \cos(1.88\omega_0 t - \phi_2) \qquad \triangleleft (s)$$

$$+ A^{(3)} \begin{Bmatrix} 0 \\ 1 \\ 0 \end{Bmatrix} \cos(2.45\omega_0 t - \phi_3)$$

where

$$\omega_0 = \sqrt{k/m} \qquad (t)$$

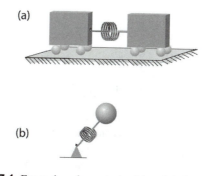

(a)

(b)

Figure 7.1 Examples of unrestrained (semi-definite) systems.

7.2 UNRESTRAINED SYSTEMS

Systems such as an aircraft in flight or a railroad train in transit are said to be *unrestrained* in the sense that they are free to translate and, in the case of the airplane, free to rotate. Even though the systems are not restrained in an overall sense, we know that such systems exhibit vibrations and we must be able to understand, predict and characterize their motions for engineering and performance purposes. Systems that are not fixed with respect to translation or rotation at one or more points are referred to as unrestrained, or *semi-definite*, systems. (Simple examples of such systems include the multiple mass-spring system of Figure 7.1a or the rotating elastic system depicted in Figure 7.1b.) With regard to vibrations, such systems have specific characteristics associated with them. The lack of constraint manifests itself as a set of "rigid body modes" for which there is no oscillatory behavior, together with the pure vibration modes that we have studied to this point. The occurrence of rigid body modes, their properties and implications, is best demonstrated by example.

Example 7.9

Consider rectilinear motion of the system comprised of two identical masses of mass m connected by a single linear spring of stiffness k, as shown in Figure E7.9-1. Determine (*a*) the natural frequencies, (*b*) the natural modes, and (*c*) the free vibration response of the system.

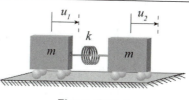

Figure E7.9-1

Solution

(*a*)

The equation of motion and mass and stiffness matrices may be found directly from Eqs. (b)–(e) of Example (6.1), with $m_1 = m_2 = m$, $k_2 = k$, $k_1 = k_3 = 0$ and $\mathbf{F} = \mathbf{0}$. This gives

$$\mathbf{m\ddot{u}} + \mathbf{ku} = 0 \tag{a}$$

where

$$\mathbf{m} = \begin{bmatrix} m & 0 \\ 0 & m \end{bmatrix} \quad \text{and} \quad \mathbf{k} = \begin{bmatrix} k & -k \\ -k & k \end{bmatrix} \tag{b, c}$$

Note that the second row of the stiffness matrix is simply the negative of the first row, and hence that the matrix **k** is singular. This is characteristic of unrestrained systems. Next, assuming a solution of the form

$$\mathbf{u} = \mathbf{U}e^{i\omega t} \tag{d}$$

and substituting into Eq. (a) results in the eigenvalue problem

$$\left[\mathbf{k} - \omega^2\mathbf{m}\right]\mathbf{U} = \begin{bmatrix} (k - \omega^2 m) & -k \\ -k & (k - \omega^2 m) \end{bmatrix}\begin{Bmatrix} U_1 \\ U_2 \end{Bmatrix} = \begin{Bmatrix} 0 \\ 0 \end{Bmatrix} \tag{e}$$

The frequency equation is then

$$\mathcal{F}(\omega^2) = \begin{vmatrix} (k - \omega^2 m) & -k \\ -k & (k - \omega^2 m) \end{vmatrix} = (k - \omega^2 m)^2 - (-k)^2 = 0 \tag{f}$$

which, when expanded, takes the form

$$\omega^2 m\left[\omega^2 m - 2k\right] = 0 \tag{g}$$

The roots of Eq. (g) may be read directly as

$$\omega^2 = 0, 2k/m \tag{h}$$

from which we obtain the natural frequencies

$$\omega_1 = 0, \quad \omega_2 = \sqrt{\frac{2k}{m}} \tag{i-1, 2}$$

We see that the first natural frequency is zero, indicating no oscillation for that mode. This is due to the lack of constraint of the system. We shall see its implications in what follows. Let us first determine the associated natural modes.

(b)

$\mathbf{U}^{(1)}$:

Substitution of Eq. (i-1) into Eq. (e) gives the algebraic relation

$$(k - 0^2 m)U_1^{(1)} - kU_2^{(1)} = 0 \quad \Rightarrow \quad U_1^{(1)} = U_2^{(1)} \tag{j}$$

Hence,

$$\mathbf{U}^{(1)} = \alpha_1 \begin{Bmatrix} 1 \\ 1 \end{Bmatrix} \rightarrow \begin{Bmatrix} 1 \\ 1 \end{Bmatrix} \tag{k}$$

where α_1 is any scalar and we have chose its value to be unity. Note that this is the "rigid body" mode seen in Example 7.2. However, in this case there is no oscillation ($\omega_1 = 0$) for the rigid body mode due to the lack of constraint. Thus, the first mode simply corresponds to a rigid body translation of the entire system. (See Figure E7.9-2a.)

$\mathbf{U}^{(2)}$:

Substitution of Eq. (i-2) into Eq. (e) gives the algebraic relation

$$\left(k - \frac{2k}{m}m\right)U_1^{(2)} - kU_2^{(2)} = 0 \quad \Rightarrow \quad U_2^{(2)} = -U_1^{(2)} \tag{l}$$

Hence,

$$\mathbf{U}^{(2)} = \alpha_2 \begin{Bmatrix} 1 \\ -1 \end{Bmatrix} \rightarrow \begin{Bmatrix} 1 \\ -1 \end{Bmatrix} \tag{m}$$

where α_2 is arbitrary and we have chosen it to have unit value. This is the accordion mode discussed in Example 7.2, but with the oscillations now at the frequency given by Eq. (i-2). (See Figure E7.9-2b.)

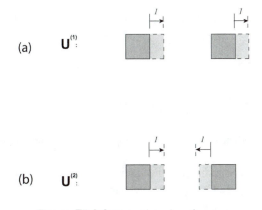

(a) $\mathbf{U}^{(1)}$:

(b) $\mathbf{U}^{(2)}$:

Figure E7.9-2 Natural modes of system.

(*c*)
Substitution of the roots $\pm\omega_{1,2}$ given by Eqs.(i) and the associated modal matrices given by Eqs. (k) and (m) into Eq. (d) and superposing these solutions gives the solution of Eq. (a) for the unstrained system. Note that both $\omega = +0$ and $\omega = -0$ give the same solution, a constant. In addition, note that multiplying a constant by t also yields a solution corresponding to vanishing ω. The free vibration response of the unrestrained two mass – one spring system is thus

$$\mathbf{u}(t) = \begin{Bmatrix} u_1(t) \\ u_2(t) \end{Bmatrix} = \left(A_1^{(1)} + A_2^{(1)} t \right) \begin{Bmatrix} 1 \\ 1 \end{Bmatrix} + A^{(2)} \cos\left(\sqrt{\tfrac{2k}{m}} t - \phi_2 \right) \begin{Bmatrix} 1 \\ -1 \end{Bmatrix} \tag{n}$$

Consideration of Eq. (n) shows that the free vibration response of the unrestrained system consists of a rigid body displacement of the entire system, with the corresponding rigid body displacement increasing linearly with time, together with the two masses moving harmonically relative to the center of the translating spring and 180 degrees out of phase with one another (i.e., moving toward and away from one another at the same rate). The rigid body translation is a result of the lack of restraint. Note that since the system is constrained to translate in one dimension there is only one rigid body mode. Thus, imagine that the mass spring system is sitting on a frictionless surface or track when it is suddenly tossed, or struck with a baseball bat or a golf club. Equation (n) tells us that, after being released (or struck), the system will move off in the direction of the initial velocity and will vibrate relative to the position of the center of the spring (the center of mass of the system) with frequency $\omega_2 = \sqrt{2k/m}$. (See Figure E7.9-3.)

Figure E7.9-3 Motion of system.

7.3 PROPERTIES OF MODAL VECTORS

In this section we shall study the general properties of the modal vectors of discrete multi-degree of freedom systems. These properties will be central to our study of forced, as well as, free vibrations of mechanical systems. We first introduce the concept of the scalar product of two modal vectors.

7.3.1 The Scalar Product

Fundamental to the characterization of the natural modes of a system is the scalar product of two modal vectors. We introduce this concept in the present section. Though we are interested in general N-degree of freedom systems, and hence in N-dimensional space, it is instructive to first discuss vectors in three-dimensional space since such vectors are familiar to us and easier to visualize.

Consider the Cartesian reference frame with axes (x_1, x_2, x_3) in three-dimensional space, and let $\vec{e}_1, \vec{e}_2, \vec{e}_3$ be corresponding unit vectors directed along these axes, as shown in Figure 7.2. Further, let $\vec{u}$ and $\vec{v}$ be two vectors in that space as indicated. The two vectors may be expressed in terms of their components with respect to the particular coordinate system chosen as follows;

$$\vec{u} = u_1\vec{e}_1 + u_2\vec{e}_2 + u_3\vec{e}_3$$
$$\vec{v} = v_1\vec{e}_1 + v_2\vec{e}_2 + v_3\vec{e}_3 \tag{7.17}$$

Taking the scalar dot product of $\vec{u}$ and $\vec{v}$ results in the familiar relation

$$\vec{u} \bullet \vec{v} = u_1 v_1 + u_2 v_2 + u_3 v_3 = \vec{v} \bullet \vec{u} \tag{7.18}$$

Let us next construct two column matrices, $\mathbf{u}$ and $\mathbf{v}$, whose elements correspond to the components of $\vec{u}$ and $\vec{v}$, respectively. Hence, let

$$\mathbf{u} = \begin{Bmatrix} u_1 \\ u_2 \\ u_3 \end{Bmatrix}, \quad \mathbf{v} = \begin{Bmatrix} v_1 \\ v_2 \\ v_3 \end{Bmatrix} \tag{7.19}$$

We shall define the scalar product of $\mathbf{u}$ and $\mathbf{v}$ as follows;

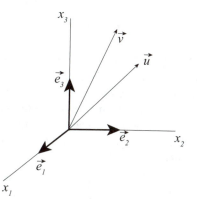

Figure 7.2 Vectors and coordinate system.

$$\langle \mathbf{u}, \mathbf{v} \rangle \equiv \mathbf{u}^\mathsf{T} \mathbf{v} = \begin{bmatrix} u_1 & u_2 & u_3 \end{bmatrix} \begin{Bmatrix} v_1 \\ v_2 \\ v_3 \end{Bmatrix} = u_1 v_1 + u_2 v_2 + u_3 v_3 = \mathbf{v}^\mathsf{T} \mathbf{u} = \langle \mathbf{v}, \mathbf{u} \rangle \quad (7.20)$$

Comparison of Eqs. (7.20) and (7.18) shows the two to be equivalent statements.

Orthogonality

The basic definition and geometric interpretation of the scalar dot product between two vectors is given by the relation

$$\vec{u} \cdot \vec{v} = \| \vec{u} \| \, \| \vec{v} \| \cos \theta \qquad (7.21)$$

where θ is the angle between the vectors. It follows that

$$\text{if} \quad \vec{u} \cdot \vec{v} = 0 \quad \text{then} \quad \vec{u} \perp \vec{v} \qquad (7.22)$$

Likewise, we say that,

$$\text{if} \quad \langle \mathbf{u}, \mathbf{v} \rangle = 0 \quad \text{then} \quad \mathbf{u} \perp \mathbf{v}.$$

The Weighted Scalar Product

It is useful for our study of vibrations, as well as for fundamental purposes, to extend the concept of the scalar product beyond the elementary definition and operation discussed above. Therefore, let us consider some square symmetric matrix $\mathbf{A} = \mathbf{A}^\mathsf{T}$ and two column matrices $\mathbf{u}$ and $\mathbf{v}$. The weighted scalar product of $\mathbf{u}$ and $\mathbf{v}$ with respect to the weight matrix $\mathbf{A}$ is defined as

$$\langle \mathbf{u}, \mathbf{v} \rangle_{\mathbf{A}} \equiv \mathbf{u}^\mathsf{T} \mathbf{A} \mathbf{v} \qquad (7.23)$$

The operation described by Eq. (7.23) may be interpreted algebraically as taking a scalar product of the two vectors $\mathbf{u}$ and $\mathbf{v}$, but weighting the contributions of the various products of the elements (components) differently according to the *weight matrix* $\mathbf{A}$. The operation may be interpreted geometrically as first stretching and rotating the vector $\mathbf{v}$ to obtain a new vector $\hat{\mathbf{v}} = \mathbf{A}\mathbf{v}$, then taking the conventional dot product between the vectors $\mathbf{u}$ and $\hat{\mathbf{v}}$. Orthogonality of two vectors with respect to the weight matrix is defined accordingly.

Orthogonality

If the weighted scalar product of **u** and **v** vanishes, then **u** and **v** are said to be orthogonal (with respect to **A**). Stated symbolically,

$$\text{if } \langle \mathbf{u}, \mathbf{v} \rangle_{\mathbf{A}} = 0 \text{ then } \mathbf{u} \underset{\mathbf{A}}{\perp} \mathbf{v}$$

All of the concepts and operations discussed above for three-dimensional vectors are applicable to vectors of any dimension, say N. Hence,

$$\langle \mathbf{u}, \mathbf{v} \rangle \equiv \mathbf{u}^{\mathsf{T}} \mathbf{v} = \sum_{j=1}^{N} u_j v_j = \mathbf{v}^{\mathsf{T}} \mathbf{u} \tag{7.24}$$

and

$$\langle \mathbf{u}, \mathbf{v} \rangle_{\mathbf{A}} \equiv \mathbf{u}^{\mathsf{T}} \mathbf{A} \mathbf{v} = \sum_{l=1}^{N} \sum_{j=1}^{N} u_l a_{lj} v_j = \sum_{l=1}^{N} \sum_{j=1}^{N} u_l a_{jl} v_j = \mathbf{v}^{\mathsf{T}} \mathbf{A} \mathbf{u} \tag{7.25}$$

where

$$\mathbf{u} = \begin{Bmatrix} u_1 \\ u_2 \\ \vdots \\ u_N \end{Bmatrix}, \quad \mathbf{v} = \begin{Bmatrix} v_1 \\ v_2 \\ \vdots \\ v_N \end{Bmatrix}, \quad \mathbf{A} = \begin{bmatrix} a_{11} & a_{12} & \cdots & a_{1N} \\ a_{21} & a_{22} & \cdots & a_{2N} \\ \vdots & \vdots & \ddots & \vdots \\ a_{N1} & a_{N2} & \cdots & a_{NN} \end{bmatrix} = \mathbf{A}^{\mathsf{T}}$$

It may be seen that the conventional scalar product corresponds to the weighted scalar product with $\mathbf{A} = \mathbf{I}$ (the $N \times N$ identity matrix). With the scalar product and orthogonality defined, we are now ready to establish the properties and characteristics of the normal modes.

7.3.2 Orthogonality

The mutual orthogonality of the modal vectors for multi-degree of freedom systems is an important property that is central to the understanding and solution of vibration problems. In this section we establish the mutual orthogonality of the modes associated with distinct roots of the frequency equation, and examine the characteristics of the modes associated with repeated roots of the frequency equation and the manner in which they can be rendered mutually orthogonal.

Consider a set of frequency-mode pairs, $\{\omega_j^2, \mathbf{U}^{(j)} \mid j = 1, 2, ..., N\}$ for a given N-degree of freedom system. Let us focus our attention on two generic frequency-mode pairs, say the lth and j^{th}, and recall that each pair satisfies Eq. (7.3). Hence,

$$\mathbf{k} \mathbf{U}^{(l)} = \omega_l^2 \, \mathbf{m} \mathbf{U}^{(l)}$$
$$\mathbf{k} \mathbf{U}^{(j)} = \omega_j^2 \, \mathbf{m} \mathbf{U}^{(j)} \tag{7.26}$$

Pre-multiplying the first equation by $\mathbf{U}^{(j)\mathsf{T}}$ results in the equality

$$\mathbf{U}^{(j)\mathsf{T}}\mathbf{k}\mathbf{U}^{(l)} = \omega_l^{\,2}\mathbf{U}^{(j)\mathsf{T}}\mathbf{m}\mathbf{U}^{(l)} \tag{7.27}$$

Taking the transpose of the second of Eqs. (7.26) gives

$$\mathbf{U}^{(j)\mathsf{T}}\mathbf{k}^{\mathsf{T}} = \omega_j^{\,2}\mathbf{U}^{(j)\mathsf{T}}\mathbf{m}^{\mathsf{T}}$$

Now, post-multiplying the above expression by $\mathbf{U}^{(l)}$ and using the fact that the mass and stiffness matrices are symmetric ($\mathbf{m} = \mathbf{m}^{\mathsf{T}}$, $\mathbf{k} = \mathbf{k}^{\mathsf{T}}$) results in the equality

$$\mathbf{U}^{(j)\mathsf{T}}\mathbf{k}\mathbf{U}^{(l)} = \omega_j^{\,2}\mathbf{U}^{(j)\mathsf{T}}\mathbf{m}\mathbf{U}^{(l)} \tag{7.28}$$

Distinct Frequencies

Subtracting Eq. (7.28) from Eq. (7.27) results in the relation

$$0 = (\omega_l^{\,2} - \omega_j^{\,2})\left\langle \mathbf{U}^{(j)}, \mathbf{U}^{(l)} \right\rangle_{\mathbf{m}} \tag{7.29}$$

where

$$\left\langle \mathbf{U}^{(l)}, \mathbf{U}^{(j)} \right\rangle_{\mathbf{m}} = \mathbf{U}^{(l)\mathsf{T}}\mathbf{m}\mathbf{U}^{(j)} \tag{7.30}$$

is the weighted scalar product of the lth and j^{th} modal vectors with respect to the mass matrix $\mathbf{m}$. It is seen from Eq. (7.29) that that the modal vectors are mutually orthogonal with respect to the mass matrix provided that the corresponding frequencies are distinct (i.e., they are not equal). Thus,

$$\mathbf{U}^{(l)} \underset{\mathbf{m}}{\perp} \mathbf{U}^{(j)} \text{ provided that } \omega_l^{\,2} \neq \omega_j^{\,2} \tag{7.31}$$

It may be seen from Eq. (7.27) that if the scalar product of the two modal vectors with respect to $\mathbf{m}$ vanishes, then it also vanishes with respect to $\mathbf{k}$. Thus,

$$\mathbf{U}^{(l)} \underset{\mathbf{k}}{\perp} \mathbf{U}^{(j)} \text{ provided that } \omega_l^{\,2} \neq \omega_j^{\,2} \tag{7.32}$$

We have thus proven the following theorem: *modal vectors associated with distinct natural frequencies are mutually orthogonal with respect to both the mass matrix and the stiffness matrix.*

Example 7.10

Verify that the natural modes for the double pendulum of Example 7.5 are mutually orthogonal.

Solution

From the solution of Example 7.5, the natural frequencies and associated modal matrices for the double pendulum are

$$\omega_1^2 = \frac{g}{L}\left(2 - \sqrt{2}\right), \quad \mathbf{U}^{(1)} = \left\{\begin{matrix} 1 \\ \sqrt{2} \end{matrix}\right\} \qquad \text{(a-1, 2)}$$

and

$$\omega_2^2 = \frac{g}{L}\left(2 + \sqrt{2}\right), \quad \mathbf{U}^{(2)} = \left\{\begin{matrix} 1 \\ -\sqrt{2} \end{matrix}\right\} \qquad \text{(b-1, 2)}$$

Clearly $\omega_1^2 \neq \omega_2^2$ and so, from the theorem of this section, we know that $\mathbf{U}^{(1)}$ and $\mathbf{U}^{(2)}$ are mutually orthogonal with respect to both the mass matrix of the system and the system matrix for the system. This may be verified by direct substitution of the modal matrices and mass and stiffness matrices into the corresponding weighted scalar products. Upon carrying through the calculations, we see that

$$\left\langle \mathbf{U}^{(1)}, \mathbf{U}^{(2)} \right\rangle_m = \mathbf{U}^{(1)\mathsf{T}} \mathbf{m} \mathbf{U}^{(2)} = \begin{bmatrix} 1 & \sqrt{2} \end{bmatrix} mL^2 \begin{bmatrix} 2 & 1 \\ 1 & 1 \end{bmatrix} \left\{\begin{matrix} 1 \\ -\sqrt{2} \end{matrix}\right\} = 0 \qquad \text{(c)}$$

therefore,

$$\mathbf{U}^{(1)} \underset{m}{\perp} \mathbf{U}^{(2)} \qquad \triangleleft$$

Similarly,

$$\left\langle \mathbf{U}^{(1)}, \mathbf{U}^{(2)} \right\rangle_k = \mathbf{U}^{(1)\mathsf{T}} \mathbf{k} \mathbf{U}^{(2)} = \begin{bmatrix} 1 & \sqrt{2} \end{bmatrix} mL^2 \begin{bmatrix} 2g/L & 0 \\ 0 & g/L \end{bmatrix} \left\{\begin{matrix} 1 \\ -\sqrt{2} \end{matrix}\right\} = 0 \qquad \text{(d)}$$

which implies that

$$\mathbf{U}^{(1)} \underset{k}{\perp} \mathbf{U}^{(2)} \qquad \triangleleft$$

Equations (c) and (d) verify that the modal vectors are mutually orthogonal with respect to both the mass and the stiffness matrices.

Repeated Frequencies

Suppose now that two roots of the frequency equation are repeated. For example, for a three degree of freedom system the frequency equation would take the form

$$\mathcal{F}(\omega^2) = (\omega^2 - \alpha^2)^2(\omega^2 - \beta^2) = 0 \Rightarrow \omega^2 = \alpha^2, \alpha^2, \beta^2$$

For the purpose of this discussion, and ease of visualization, let us consider such a case for a three degree of freedom system. The corresponding results and interpretations can then be abstracted to systems with any number of degrees of freedom.

Consider a three degree of freedom system with two repeated frequencies, as discussed above. Thus, let

$$\omega_1^2 = \omega_2^2 = \alpha^2, \text{ and } \omega_3^2 = \beta^2$$

Then, from the theorem of the previous section, we know that $\mathbf{U}^{(3)}$ is orthogonal (with respect to both $\mathbf{m}$ and $\mathbf{k}$) to the modal vectors associated with the repeated frequency. We therefore know that the latter vectors lie in the plane whose normal is parallel to $\mathbf{U}^{(3)}$, as depicted in Figure 7.3). However, the aforementioned theorem gives us no further information about the modal vectors associated with the repeated frequencies. Nevertheless, we do know that the repeated frequencies and associated modal vectors must satisfy Eq. (7.3). Hence,

$$\mathbf{kU}^{(1)} = \alpha^2 \mathbf{mU}^{(1)}$$
$$\mathbf{kU}^{(2)} = \alpha^2 \mathbf{mU}^{(2)} \tag{7.33}$$

where $\mathbf{U}^{(1)}$ and $\mathbf{U}^{(2)}$ are two such vectors. Let us multiply the first equation by a scalar constant a, and the second equation by a scalar constant b, where a and b are otherwise arbitrary, and add the resulting equations. We then have that

$$\mathbf{k\tilde{U}} = \alpha^2 \mathbf{m\tilde{U}} \tag{7.34}$$

where

$$\tilde{\mathbf{U}} = a\mathbf{U}^{(1)} + b\mathbf{U}^{(2)} \tag{7.35}$$

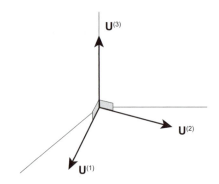

Figure 7.3 Vectors associated with repeated roots (lying in horizontal plane) and vector associated with differing root (perpendicular to plane).

It is seen that if $\mathbf{U}^{(1)}$ and $\mathbf{U}^{(2)}$ are modal vectors associated with the repeated frequency α then, since a and b are arbitrary, any linear combination of them is also a modal vector. Therefore, any vector that is orthogonal to $\mathbf{U}^{(3)}$, and hence lies in the plane whose normal is parallel to $\mathbf{U}^{(3)}$, is a modal vector associated with the repeated frequency. This lack of determinacy is explained algebraically by recalling the nature of the matrix $\left[\mathbf{k} - \omega^2 \mathbf{m}\right]\mathbf{U} = \mathbf{0}$. Recall that this matrix is singular – that is, at least one row of the matrix can be expressed as a linear combination of the other rows. However, when two roots are repeated it implies that two rows are linear dependent on the other rows. This adds an additional degree of indeterminacy. For a three degree of freedom system there are thus at most two independent rows of $\left[\mathbf{k} - \omega^2 \mathbf{m}\right]\mathbf{U} = \mathbf{0}$, but this is reduced to one when two frequencies are repeated. This lack of determinacy is reflected in the result that *any* vector orthogonal to the third modal vector is a modal vector associated with the repeated frequency. The above discussion is readily extended to N degree of freedom systems. Therefore, for systems possessing N degrees of freedom, any vector lying in the hyperplane that is orthogonal to the remaining modal vectors is a modal vector corresponding to the repeated frequency. For the purpose of analysis, it is convenient to choose two mutually orthogonal vectors lying in that hyperplane as the modal vectors for the repeated frequency.

Example 7.11

A floating platform is comprised of a board of length L that sits atop two identical floats, each of mass $m/2$ and cross-sectional area A, as indicated. The mass of the board is negligible compared with the mass of the floats and the mass density of the fluid is ρ_f. Determine the natural frequencies and natural modes for the floating platform. Assume small rotations of the platform.

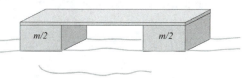

Figure E7.11-1 Floating platform.

Solution

The effects of buoyancy may be accounted for through equivalent springs of stiffness $k = \rho_f A g$, as shown in Figure E7.11-2, where g is the gravitational acceleration (see Section 1.3). The equations of motion for the equivalent system can then be derived by direct application of Newton's Laws, or by using Lagrange's Equations. As the present system is a simple one, the former approach is easily implemented. In either case, let us choose the centerspan vertical deflection, y_G (positive downward), and the rotational displacement, θ (positive clockwise), as the generalized coordinates to describe the motion of this system. Newton's Second Law and the corresponding angular momentum principle are then, respectively, expressed for the current system as

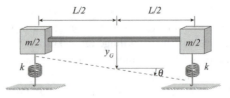

Figure E7.11-2 Equivalent system.

$$\sum F = m\ddot{y}_G, \quad \sum M_G = I_G\ddot{\theta} \qquad \text{(a-1, 2)}$$

The kinetic diagram for the deflected system is shown in Figure E7.11-3. Upon implementing Eq. (a-1), we have

$$-k\left(y_G + \frac{L}{2}\sin\theta\right) - k\left(y_G - \frac{L}{2}\sin\theta\right) = m\ddot{y}_G$$

which simplifies to the standard form

$$\ddot{y}_G + \frac{2k}{m}y_G = 0 \qquad \text{(b)}$$

Similarly, taking about point G and implementing Eq. (a-2) gives the relation

$$k\left(y_G + \frac{L}{2}\sin\theta\right)\frac{L}{2}\cos\theta - k\left(y_G - \frac{L}{2}\sin\theta\right)\frac{L}{2}\cos\theta = -2\frac{m}{2}\left(\frac{L}{2}\right)^2\ddot{\theta}$$

which, after regrouping terms, reduces to the form

$$m\ddot{\theta} + 2k\sin\theta\cos\theta = 0 \qquad \text{(c)}$$

For small angle motions of the platform, we may linearize Eq. (c). The equation for rotational motion then simplifies to the standard form

$$\ddot{\theta} + \frac{2k}{m}\theta = 0 \qquad \text{(d)}$$

Equations (b) and (d) may be expressed in matrix form as

$$\begin{bmatrix} m & 0 \\ 0 & m \end{bmatrix}\begin{Bmatrix} \ddot{y}_G \\ \ddot{\theta} \end{Bmatrix} + \begin{bmatrix} 2k & 0 \\ 0 & 2k \end{bmatrix}\begin{Bmatrix} y_G \\ \theta \end{Bmatrix} = \begin{Bmatrix} 0 \\ 0 \end{Bmatrix} \qquad \text{(e)}$$

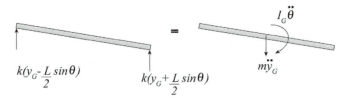

Figure E7.11-3 Kinetic diagram.

A comparison of the equations for translational and rotational motion, Eqs. (b) and (d), reveals two interesting properties of the system. First, we may note that the two equations are uncoupled for the chosen generalized coordinates. (Coordinates which have this property are referred to as principal, or modal, coordinates and will be discussed formally in the next chapter.) Second, it is readily seen that

$$\omega_1 = \omega_2 = \sqrt{\frac{2k}{m}} \tag{f}$$

That is, the natural frequencies for the two motions are the same. Formally, the eigenvalue problem for this system takes the form

$$\begin{bmatrix} (2k - \omega^2 m) & 0 \\ 0 & (2k - \omega^2 m) \end{bmatrix} \begin{Bmatrix} Y \\ \Theta \end{Bmatrix} = \begin{Bmatrix} 0 \\ 0 \end{Bmatrix} \tag{g}$$

The corresponding frequency equation is then

$$\mathcal{F}(\omega^2) = \det\left[\mathbf{k} - \omega^2 \mathbf{m}\right] = (2k - \omega^2 m)^2 = 0 \tag{h}$$

which yields the roots

$$\omega^2 = \frac{2k}{m}, \frac{2k}{m} \tag{i}$$

The roots of the frequency equation are clearly *repeated* (not distinct) and, of course, yield the identical frequencies stated in Eq. (f).

To determine the modal matrices we substitute the frequencies into Eq. (g) and solve for $\mathbf{U}^{(j)} = [Y^{(j)}\ \Theta^{(j)}]^T$ ($j = 1, 2$). Since the roots are repeated (the two natural frequencies are the same) this gives, for both modes,

$$\left(2k - \frac{2k}{m} m\right) Y^{(1,2)} + 0 \cdot \Theta^{(1,2)} = 0$$

$$0 \cdot Y^{(1,2)} + \left(2k - \frac{2k}{m} m\right) \Theta^{(1,2)} = 0$$

(j)

Each of which reduces to

$$0 \cdot Y^{(1,2)} + 0 \cdot \Theta^{(1,2)} = 0 \tag{k}$$

It is seen from Eq. (k) that both of the components of the modal vectors can take on any value. Thus, any 2 row column matrix will satisfy Eq. (g) and, therefore, correspond to a modal matrix. Two convenient pairs are given by

$$\mathbf{U}^{(1)} = \begin{Bmatrix} 1 \\ 0 \end{Bmatrix}, \quad \mathbf{U}^{(2)} = \begin{Bmatrix} 0 \\ 1 \end{Bmatrix}$$

or

$$\mathbf{U}^{(1)} = \frac{1}{\sqrt{2}} \begin{Bmatrix} 1 \\ 1 \end{Bmatrix}, \quad \mathbf{U}^{(2)} = \frac{1}{\sqrt{2}} \begin{Bmatrix} 1 \\ -1 \end{Bmatrix}$$

Physical depictions of each of the two pairs of modal vectors are sketched in Figure E7.11-4.

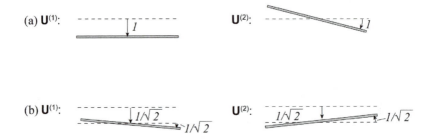

(a) $\mathbf{U}^{(1)}$: $\mathbf{U}^{(2)}$:

(b) $\mathbf{U}^{(1)}$: $\mathbf{U}^{(2)}$:

Figure E7.11-4 Natural modes for floating platform.

7.3.3 Normalization

It was shown in Section 7.1 that the modal matrices are determined to within, at most, a scalar multiple. One way to introduce definiteness to the modal matrix is to set an element of the matrix, typically the first element, to one. Alternatively, we can normalize the corresponding vector so that its magnitude is unity. That is, we can render the modal vectors unit vectors. When this is done to a set of modal vectors, the modes are said to be normalized and the resulting vectors are referred to as the *normal modes* of the system. In the present section, we shall discuss several options for constructing a set of normal modes.

A unit vector is a vector whose scalar product with itself is one. In this regard, the scale or metric may be defined in several ways. We can set the conventional scalar product of a modal vector with itself to unity, or we can set a weighted scalar product of the modal vector with itself to unity. For the latter case we have two obvious candidates for the weight matrix; the mass matrix or the stiffness matrix. The problem may be stated mathematically as follows; since a modal matrix is determined to within a scalar multiplier, a typical modal vector for an N-degree of freedom system will be of the form

$$\mathbf{U} = \alpha \begin{Bmatrix} U_1 \\ U_2 \\ \vdots \\ U_N \end{Bmatrix} \tag{7.36}$$

where α is arbitrary. To construct the corresponding *normal mode*, we take the scalar product of this vector with itself, set the resulting expression to one and solve for α. Then, substitute the calculated value of α back into Eq. (7.36) to obtain the corresponding normal mode.

Conventional Scalar Product

To normalize the modal vector in terms of the conventional scalar product, we take the product as

$$\langle \mathbf{U}, \mathbf{U} \rangle = \mathbf{U}^{\mathrm{T}} \mathbf{U} = 1 \tag{7.37}$$

or

$$\alpha \begin{bmatrix} U_1 & U_2 & \cdots & U_N \end{bmatrix} \alpha \begin{Bmatrix} U_1 \\ U_2 \\ \vdots \\ U_N \end{Bmatrix} = 1$$

Solving for α and substituting this value back into Eq. (7.37) gives the corresponding normal mode as

$$\bar{\mathbf{U}} = \frac{1}{\sqrt{U_1^2 + U_2^2 + \ldots + U_N^2}} \begin{Bmatrix} U_1 \\ U_2 \\ \vdots \\ U_N \end{Bmatrix} \tag{7.38}$$

It may be seen that this result would be obtained by simply dividing the original vector by its magnitude. (See Example 7.6.) Equivalently, the arbitrary multiplier, α, is eliminated if we simply divide $\mathbf{U}$ by its magnitude (the square root of scalar product of the modal vector with itself). Thus,

$$\bar{\mathbf{U}} = \frac{\mathbf{U}}{\|\mathbf{U}\|} = \frac{1}{\sqrt{\langle \mathbf{U}, \mathbf{U} \rangle}} \mathbf{U} \tag{7.39}$$

Weighted Scalar Product

The modal vectors may also be normalized in terms of a weighted scalar product. The arbitrary scalar multiplier is eliminated for a given modal vector if we normalize with respect to some weight matrix, say the mass matrix or the stiffness matrix. Hence,

$$\bar{\mathbf{U}} = \frac{\mathbf{U}}{\|\mathbf{U}\|_m} = \frac{1}{\sqrt{\langle \mathbf{U}, \mathbf{U} \rangle_m}} \mathbf{U} = \frac{1}{\sqrt{\mathbf{U}^{\mathsf{T}} \mathbf{m} \mathbf{U}}} \mathbf{U} \tag{7.40}$$

or

$$\bar{\mathbf{U}} = \frac{\mathbf{U}}{\|\mathbf{U}\|_k} = \frac{1}{\sqrt{\langle \mathbf{U}, \mathbf{U} \rangle_k}} \mathbf{U} = \frac{1}{\sqrt{\mathbf{U}^{\mathsf{T}} \mathbf{k} \mathbf{U}}} \mathbf{U} \tag{7.41}$$

Normalizing the modal vectors in a consistent manner allows for the evaluation of the relative contribution of each mode in a given response. For the analysis of forced vibrations, it is often convenient to normalize with respect to the mass matrix as described by Eq. (7.40).

Example 7.12

Determine the normal modes for the two-mass three-spring system of Examples 7.1 and 7.2. Use the mass matrix as the weighting measure.

Solution

The modes for the system of interest were determined in Example 2.2 to be

$$\mathbf{U}^{(1)} = \begin{Bmatrix} 1 \\ 1 \end{Bmatrix}, \quad \mathbf{U}^{(2)} = \begin{Bmatrix} 1 \\ -1 \end{Bmatrix} \tag{a-1, 2}$$

The corresponding scalar products with respect to the mass are then

$$\langle \mathbf{U}^{(1)}, \mathbf{U}^{(1)} \rangle_m = \begin{bmatrix} 1 & 1 \end{bmatrix} \begin{bmatrix} m & 0 \\ 0 & m \end{bmatrix} \begin{Bmatrix} 1 \\ 1 \end{Bmatrix} = 2m \tag{b}$$

$$\langle \mathbf{U}^{(2)}, \mathbf{U}^{(2)} \rangle_{\mathbf{m}} = \begin{bmatrix} 1 & -1 \end{bmatrix} \begin{bmatrix} m & 0 \\ 0 & m \end{bmatrix} \begin{Bmatrix} 1 \\ -1 \end{Bmatrix} = 2m \tag{c}$$

Substituting Eqs. (a-1) and (b), and Eqs. (a-2) and (c), into Eq. (7.40) gives the desired normal modes,

$$\bar{\mathbf{U}}^{(1)} = \frac{1}{\sqrt{2m}} \begin{Bmatrix} 1 \\ 1 \end{Bmatrix}, \quad \bar{\mathbf{U}}^{(2)} = \frac{1}{\sqrt{2m}} \begin{Bmatrix} 1 \\ -1 \end{Bmatrix} \qquad \triangleleft (d\text{-}1,2)$$

7.4 SYSTEMS WITH VISCOUS DAMPING

To this point, we have considered the fundamental problem of free vibrations of multi-degree of freedom systems without damping. The problem is an important one, both as a basis for analysis and because we know from our discussions of single degree of freedom systems that damped oscillations eventually die out. Nevertheless, there are situations in which the effects of damping are important and/or the understanding of these effects is germane. We now consider free vibrations of multi-degree of freedom systems with viscous damping.

Recall from Chapter 6 that the equation of motion for an N-degree of freedom system with viscous damping is of the general form

$$\mathbf{m}\ddot{\mathbf{u}} + \mathbf{c}\dot{\mathbf{u}} + \mathbf{k}\mathbf{u} = \mathbf{0} \tag{7.42}$$

We shall first obtain the free vibration response of an arbitrary N-degree of freedom system by solving Eq. (7.42) directly. We will then gain further insight by considering the system response in state space.

7.4.1 System Response

We shall here approach the problem for damped systems in a manner similar to that for undamped systems. We thus assume a solution of the form

$$\mathbf{u}(t) = \mathbf{U}e^{\alpha t} \tag{7.43}$$

where α and the elements of $\mathbf{U}$ are constants to be determined. Substituting Eq. (7.43) into Eq. (7.42) results in the characteristic value problem

$$\left[\alpha^2 \mathbf{m} + \alpha \mathbf{c} + \mathbf{k} \right] \mathbf{U} = \mathbf{0} \tag{7.44}$$

The free vibration problem is thus reduced to finding $(\alpha, \mathbf{U})$ pairs that satisfy Eq. (7.44). For nontrivial solutions we require that

$$\det\left[\alpha^2 \mathbf{m} + \alpha \mathbf{c} + \mathbf{k}\right] = \mathcal{F}(\alpha) = 0 \tag{7.45}$$

which, when expanded, results in a characteristic equation for the unknown exponent α in the form of a polynomial of order $2N$. For dissipative systems, the roots of the characteristic equation will be complex with negative real parts or, for large damping, they will be real and negative. That is

$$\alpha_j = -\mu_j \pm i\omega_j \quad (j = 1, 2, ..., N) \tag{7.46}$$

or, for large damping,

$$\alpha_j = -\mu_j \quad (j = 1, 2, ..., 2N) \tag{7.47}$$

where $\mu_j > 0$. The characteristic equation could also yield roots of both of the aforementioned types. We shall first consider the case of complex roots of the characteristic equation.

For each complex α_j there corresponds a $\mathbf{U}^{(j)}$ which will, in general, be complex. That is,

$$\mathbf{U}^{(j)} = \mathbf{U}_R^{(j)} \pm i\mathbf{U}_I^{(j)} \quad (j = 1, 2, ..., N) \tag{7.48}$$

[For the conjugate root, the matrix operator in Eq. (7.44) is the complex conjugate of the original operator. The corresponding vector is then the complex conjugate of the original vector.] Substitution of Eqs. (7.46) and (7.48) into Eq. (7.43) gives the corresponding solution

$$\mathbf{u}^{(j)}(t) = \left\{\mathbf{U}_R^{(j)} + i\mathbf{U}_I^{(j)}\right\} \hat{A}^{(j)} e^{(-\mu_j + i\omega_j)t} + \left\{\mathbf{U}_R^{(j)} - i\mathbf{U}_I^{(j)}\right\} \hat{B}^{(j)} e^{(-\mu_j - i\omega_j)t} \tag{7.49}$$

which, after using Euler's Formula, takes the alternate form

$$\mathbf{u}^{(j)}(t) = e^{-\mu_j t} \mathbf{U}_R^{(j)} \left[\bar{A}^{(j)} \cos\omega_j t + \bar{B}^{(j)} \sin\omega_j t\right]$$
$$+ e^{-\mu_j t} \mathbf{U}_I^{(j)} \left[\bar{B}^{(j)} \cos\omega_j t - \bar{A}^{(j)} \sin\omega_j t\right] \tag{7.50}$$

where

$$\bar{A}^{(j)} = \hat{A}^{(j)} + \hat{B}^{(j)}, \quad \bar{B}^{(j)} = i\left(\hat{A}^{(j)} - \hat{B}^{(j)}\right)$$

Proceeding as in Section 2.1 renders the solution to the form

$$\mathbf{u}^{(j)}(t) = A^{(j)} e^{-\mu_j t} \left\{ \mathbf{U}_R^{(j)} \cos(\omega_j t - \phi_j) + \mathbf{U}_I^{(j)} \cos(\omega_j t - \widehat{\phi}_j) \right\} \qquad (7.51)$$

where

$$A^{(j)} = \sqrt{\widehat{A}^{(j)2} + \widehat{B}^{(j)2}}, \quad \phi_j = \mathrm{Tan}^{-1}\left(\widehat{B}^{(j)}/\widehat{A}^{(j)}\right)$$

$$\widehat{\phi}_j = \tan^{-1}\left(-\widehat{A}^{(j)}/\widehat{B}^{(j)}\right) = \phi_j - \pi/2$$

The general free vibration response is comprised of a linear combination of all solutions of the above form. Hence,

$$\mathbf{u}(t) = \sum_{j=1}^{N} \mathbf{u}^{(j)}(t)$$

$$= \sum_{j=1}^{N} A^{(j)} e^{-\mu_j t} \left\{ \mathbf{U}_R^{(j)} \cos(\omega_j t - \phi_j) + \mathbf{U}_I^{(j)} \sin(\omega_j t - \phi_j) \right\} \qquad (7.52)$$

where the constants $A^{(j)}$ and ϕ_j are determined from the initial conditions. It may be seen that the motion of the system at a given frequency consists of two motions that are out of phase with one another. It is also seen that a separate rate of decay is associated with each (damped) natural frequency.

For large damping the characteristic values are all real and negative and the corresponding response is of the form

$$\mathbf{u}(t) = \sum_{j=1}^{2N} A^{(j)} e^{-\mu_j t} \mathbf{U}^{(j)} \qquad (7.53)$$

More generally, the response of a damped system may be comprised of some combination of the elemental solutions stated in Eqs. (7.52) and (7.53).

The analysis and behavior of viscously damped systems that are free from external forces is demonstrated by the following example.

Example 7.13

Consider the uniform frame of mass m and length L supported at its ends as shown. (*a*) Determine the free vibration response of the frame if $k_1 = k_2 = k$, and no damping exists, and (*b*) when the stiffnesses and dampers are such that $k_1 = k_2 = k$, $c_2 = 2c_1$ and $c_1/\sqrt{2km} = 0.1$. (*c*) Repeat part (*b*) if $c_2 = 2c_1$ and $c_1/\sqrt{2km} = 1.0$. (*d*) Obtain the characteristic values and the general form of the free vibration response if $c_2 = 2c_1$ and $c_1/\sqrt{2km} = 0.1$.

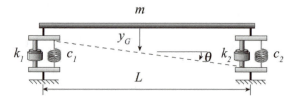

Figure E7.13 Uniform frame with viscoelastic supports.

Solution
The equation of motion for the system is obtained by including the damping forces in the development of Section 6.1.3. This gives, for the system under consideration,

$$
\begin{bmatrix} m & 0 \\ 0 & I_G \end{bmatrix} \begin{Bmatrix} \ddot{y}_G \\ \ddot{\theta} \end{Bmatrix} + \begin{bmatrix} (c_1 + c_2) & (c_2 - c_1)L/2 \\ (c_2 - c_1)L/2 & (c_1 + c_2)L^2/4 \end{bmatrix} \begin{Bmatrix} \dot{y}_G \\ \dot{\theta} \end{Bmatrix}
$$
$$
+ \begin{bmatrix} 2k & 0 \\ 0 & kL^2/2 \end{bmatrix} \begin{Bmatrix} y_G \\ \theta \end{Bmatrix} = \begin{Bmatrix} 0 \\ 0 \end{Bmatrix} \tag{a}
$$

where, for a uniform frame, $I_G = mL^2/12$. The system is seen to be coupled through the damping matrix.

(a)
For vanishing the damping ($c_1 = c_2 = 0$) the natural frequencies and modal vectors are easily seen to be

$$
\omega_1 = \sqrt{2k/L}, \quad \mathbf{U}^{(1)} = \begin{Bmatrix} 1 \\ 0 \end{Bmatrix} \tag{b-1, 2}
$$

$$
\omega_2 = \sqrt{2I_G/L^2} = \sqrt{6k/m}, \quad \mathbf{U}^{(2)} = \begin{Bmatrix} 0 \\ 1 \end{Bmatrix} \tag{b-3, 4}
$$

Note that the second natural frequency and mode shape correspond to the third mode of Example 7.12 since in that example the rider was over the center of mass of the frame and so did not influence the rotation of the system. The response of the system is then

$$
\begin{Bmatrix} y_G \\ \theta \end{Bmatrix} = A^{(1)} \begin{Bmatrix} 1 \\ 0 \end{Bmatrix} \cos\left(\sqrt{\tfrac{2k}{m}}\, t - \phi_1\right) + A^{(2)} \begin{Bmatrix} 1 \\ 0 \end{Bmatrix} \cos\left(\sqrt{\tfrac{6k}{m}}\, t - \phi_2\right) \tag{c}
$$

(b)

The characteristic value problem for the damped system takes the form

$$\begin{bmatrix} \{\alpha^2 m + \alpha(c_1 + c_2) + 2k\} & \alpha(c_2 - c_1)L/2 \\ \alpha(c_2 - c_1)L/2 & \{\alpha^2 I_G + \alpha(c_1 + c_2)L^2/4 + kL^2/2\} \end{bmatrix} \begin{Bmatrix} U_1 \\ U_2 \end{Bmatrix} = \begin{Bmatrix} 0 \\ 0 \end{Bmatrix} \quad \text{(d)}$$

Setting the determinant of the square matrix of Eq. (d) to zero (and dividing the resulting equation by $mI_G\omega_0^4$) gives the characteristic equation

$$\mathcal{F}(\hat{\alpha}) = \hat{\alpha}^4 + \hat{\alpha}^3 \left[4(\eta_1 + \eta_2)\right] + \hat{\alpha}^2 \left[4(1 + 3\eta_1\eta_2)\right] + \hat{\alpha}\left[6(\eta_1 + \eta_2)\right] + 3 = 0 \quad \text{(e)}$$

where

$$\hat{\alpha} = \alpha/\omega_0 \quad \text{(f)}$$

$$\omega_0 = \sqrt{2k/m} \quad \text{(g)}$$

$$\eta_1 = c_1/\sqrt{2mk}, \quad \eta_2 = c_2/\sqrt{2mk} \quad \text{(h-1, 2)}$$

For $\eta_2 = 2\eta_1$ and $\eta_1 = 0.1$ we find, using the MATLAB solver "roots," that

$$\hat{\alpha}_{a,b}^{(1)} = -0.1499 \pm 0.9964i$$
$$\hat{\alpha}_{a,b}^{(2)} = -0.4501 \pm 1.6590i \quad \text{(i)}$$

The first row of Eq. (d) gives, for $c_2 \ne c_1$,

$$U_2^{(j)} = -2\frac{\alpha^{(j)2} m + \alpha^{(j)}(c_1 + c_2) + 2k}{\alpha^{(j)}(c_2 - c_1)}\frac{U_1^{(j)}}{L} = -\left(P + iQ\right)U_1^{(j)}/L \quad \text{(j)}$$

where

$$P^{(j)} = 2\frac{\left\| \hat{\alpha}^{(j)} \right\|^2 \left[-\hat{\mu}_j + (\eta_1 + \eta_2)\right] - \hat{\mu}_j}{\left\| \hat{\alpha}^{(j)} \right\|^2 (\eta_2 - \eta_1)} \quad \text{(k)}$$

$$Q^{(j)} = \frac{2\hat{\omega}_j \left[\left\| \hat{\alpha}^{(j)} \right\|^2 - 1\right]}{\left\| \hat{\alpha}^{(j)} \right\|^2 (\eta_2 - \eta_1)} \quad \text{(l)}$$

$$\left\| \hat{\alpha}^{(j)} \right\|^2 = \hat{\mu}_j^2 + \hat{\omega}_j^2 \quad \text{(m)}$$

Substituting the given parameters and the first of Eq. (i) into Eqs. (j)–(l) gives, for $j = 1$,

$$U_2^{(1)} = -\left(0.0491 + 0.300i\right)U_1^{(1)}/L \tag{n}$$

Hence,

$$\mathbf{U}^{(1)} = \mathbf{U}_R^{(1)} + i\mathbf{U}_I^{(1)} = \left\{\begin{matrix} L \\ -0.0491 \end{matrix}\right\} + i\left\{\begin{matrix} 0 \\ -0.300 \end{matrix}\right\} \tag{o}$$

Similarly, for $j = 2$,

$$U_2^{(2)} = \left(6.05 - 22.0i\right)U_1^{(2)}/L \tag{p}$$

and hence

$$\mathbf{U}^{(2)} = \mathbf{U}_R^{(2)} + i\mathbf{U}_I^{(2)} = \left\{\begin{matrix} L \\ 6.05 \end{matrix}\right\} + i\left\{\begin{matrix} 0 \\ -22.0 \end{matrix}\right\} \tag{q}$$

Substituting Eqs. (o) and (q) into Eq. (7.52) gives the free vibration response of the system,

$$\left\{\begin{matrix} y_G(t)/L \\ \theta(t) \end{matrix}\right\} =$$

$$A^{(1)}e^{-0.150\omega_0 t}\left[\left\{\begin{matrix} 1 \\ -0.0491 \end{matrix}\right\}\cos(0.996\omega_0 t - \phi_1) + \left\{\begin{matrix} 0 \\ -0.300 \end{matrix}\right\}\sin(0.996\omega_0 t - \phi_1)\right]$$

$$+ A^{(2)}e^{-0.450\omega_0 t}\left[\left\{\begin{matrix} 1 \\ 6.05 \end{matrix}\right\}\cos(1.66\omega_0 t - \phi_2) + \left\{\begin{matrix} 1 \\ -22.0 \end{matrix}\right\}\sin(1.66\omega_0 t - \phi_2)\right]$$

$$\lhd \text{ (r)}$$

where ω_0 is defined by Eq. (g). The system is seen to move with damped harmonic vibrations that decay exponentially with time.

(c)
Substituting $\eta_2 = 2.0$ and $\eta_1 = 1.0$ into Eq. (e) and finding the corresponding zeros of the characteristic equation using the MATLAB solver "roots," we obtain

$$\hat{\alpha} = -0.2611, \, -0.6470, \, -1.9406, \, -9.1514 \tag{s}$$

In this case the characteristic values are all real. To obtain the associated vectors for $c_2 \neq c_1$ we have, from the first row of Eq. (d),

$$U_2^{(j)} = -\frac{\hat{\alpha}^{(j)2} + \hat{\alpha}^{(j)}(\eta_1 + \eta_2) + 1}{\hat{\alpha}^{(j)}(\eta_2 - \eta_1)L/2} U_1^{(j)} \quad (j = 1-4) \tag{t}$$

Substitution of each root listed in Eq. (s) into Eq. (t) gives the corresponding vectors

$$\mathbf{U}^{(1)} = \begin{Bmatrix} L \\ 2.182 \end{Bmatrix}, \ \mathbf{U}^{(2)} = \begin{Bmatrix} L \\ -1.615 \end{Bmatrix}, \ \mathbf{U}^{(3)} = \begin{Bmatrix} L \\ -1.088 \end{Bmatrix}, \ \mathbf{U}^{(4)} = \begin{Bmatrix} L \\ -12.52 \end{Bmatrix} \tag{u}$$

Finally, we substitute each $(\alpha, \mathbf{U})$ pair into Eq. (7.53) to obtain the free vibration response

$$\begin{Bmatrix} y_G/L \\ \theta \end{Bmatrix} = A^{(1)} e^{-0.261\omega_0 t} \begin{Bmatrix} 1 \\ 2.18 \end{Bmatrix} + A^{(2)} e^{-0.647\omega_0 t} \begin{Bmatrix} 1 \\ -1.61 \end{Bmatrix}$$

$$+ A^{(3)} e^{-1.94\omega_0 t} \begin{Bmatrix} 1 \\ -1.09 \end{Bmatrix} + A^{(4)} e^{-9.15\omega_0 t} \begin{Bmatrix} 1 \\ -12.6 \end{Bmatrix} \tag{v}$$ ◁

For this case we see that the system response is a purely decaying motion.

(d)
Substituting $\eta_2 = 1.0$ and $\eta_1 = 0.5$ into Eq. (e) and finding the corresponding zeros of the characteristic equation using the MATLAB solver "roots," we obtain

$$\hat{\alpha} = -0.6129, \ -4.0251, \ -0.6810 \pm 0.8673 \tag{w}$$

We see that for this case we have both real and complex roots. The free vibration response is then

$$\begin{Bmatrix} y_G/L \\ \theta \end{Bmatrix} = A^{(1)} e^{-0.613\omega_0 t} \begin{Bmatrix} 1 \\ 2.98 \end{Bmatrix} + A^{(2)} e^{-4.03\omega_0 t} \begin{Bmatrix} 1 \\ 11.1 \end{Bmatrix}$$

$$+ A^{(3)} e^{-0.681\omega_0 t} \left[\begin{Bmatrix} 1 \\ -1.04 \end{Bmatrix} \cos(0.867\omega_0 t - \phi_3) + \begin{Bmatrix} 0 \\ -0.616 \end{Bmatrix} \sin(0.867\omega_0 t - \phi_3) \right]$$ ◁

$$\tag{x}$$

where the corresponding vectors are obtained by substituting each of the two real roots listed in Eq. (w) into Eq. (t) and substituting the real and imaginary parts of the complex roots into Eq. (j). For this last case we see that the system response is comprised of a purely decaying motion together with decaying harmonic oscillations.

To offset the lack of determinacy of the characteristic vectors, an element of the vector can be set to unity. However, as for undamped systems, it is often desirable to introduce a common scale and thus to normalize the vectors in some way. One approach is to extend the procedures employed for undamped systems. In this regard, complex characteristic vectors can be normalized by setting the *Hermitian scalar product* of the vector with itself to unity. That is we set the scalar product a vector with its complex conjugate to unity. Stated mathematically, to normalize $\mathbf{U}^{(j)}$ we may set

$$\left\langle \mathbf{U}^{(j)\mathbf{C}}, \mathbf{U}^{(j)} \right\rangle = \left[\mathbf{U}_R^{(j)} - i\mathbf{U}_I^{(j)} \right]^T \left\{ \mathbf{U}_R^{(j)} + i\mathbf{U}_I^{(j)} \right\} = \mathbf{U}_R^{(j)\mathrm{T}}\mathbf{U}_R^{(j)} + \mathbf{U}_I^{(j)\mathrm{T}}\mathbf{U}_I^{(j)} = 1 \quad (7.54)$$

where $\mathbf{U}^\mathbf{C}$ represents the complex conjugate of $\mathbf{U}$. Equivalently, to obtain the normalized vector, the characteristic vector may be divided by the square root of the aforementioned product (its *Hermitian length*). Hence,

$$\bar{\mathbf{U}}^{(j)} = \frac{\mathbf{U}^{(j)}}{\sqrt{\left\langle \mathbf{U}^{(j)\mathbf{C}}, \mathbf{U}^{(j)} \right\rangle}} \quad (7.55)$$

The characteristic vectors can be similarly scaled by dividing each by the square root of the corresponding weighted scalar product of the vector and its conjugate, taken with respect to a real symmetric system matrix such as $\mathbf{m}$ or $\mathbf{k}$. Hence,

$$\bar{\mathbf{U}}^{(j)} = \frac{\mathbf{U}^{(j)}}{\sqrt{\left\langle \mathbf{U}^{(j)\mathbf{C}}, \mathbf{U}^{(j)} \right\rangle_\mathbf{m}}} \quad (7.56)$$

Characteristic vectors associated with real characteristic values will be real and therefore can be normalized as discussed in Section 7.3.3. Other, means of normalization for damped systems are suggested in the next section. Finally, in contrast to the modal vectors of systems with no damping, the characteristic vectors associated with damped systems are generally not mutually orthogonal in the conventional sense. To understand their relation we next examine the corresponding problem in the context of its state space representation. This will yield more general orthogonality relations and also suggest procedures of normalizing vectors, both of which will be pertinent to forced vibration damped systems.

7.4.2 State Space Representation

As an alternative to the approach of Section 7.4.1 we may consider the vibration problem in the context of its state space (the space of the generalized displacements and velocities). It will be seen that the two approaches lead to the same results and that additional insight into the nature of the characteristic vectors is gained from the latter formulation.

Formulation and Solution

To formulate the problem in its state space let us first write the velocity matrix as

$$\dot{u} = I\dot{u} \tag{7.57}$$

where I represents the $N \times N$ identity matrix. Next, let us pre-multiply Eq. (7.42) by m^{-1} and solve for $\ddot{u}$. This gives

$$\ddot{u} = -m^{-1}ku - m^{-1}c\dot{u} \tag{7.58}$$

Equations (7.57) and (7.58) may be combined in matrix form as

$$\dot{z} = Sz \tag{7.59}$$

where

$$z = \begin{Bmatrix} u \\ \dot{u} \end{Bmatrix} \tag{7.60}$$

is the $2N \times 1$ *state vector*, and

$$S = \begin{bmatrix} 0 & I \\ -m^{-1}k & -m^{-1}c \end{bmatrix} \tag{7.61}$$

is the $2N \times 2N$ system matrix. (Note that, in general, S is not symmetric.) The free vibration problem is now recast in terms of the state vector z. We wish to solve Eq. (7.59).

To solve for the state vector as a function of time let us seek solutions of the form

$$z = \hat{U}e^{\alpha t} \tag{7.62}$$

where

$$\hat{U} = \begin{Bmatrix} U \\ V \end{Bmatrix} \tag{7.63}$$

is a $2N \times 1$ array of, as yet, unknown (complex) constants that we have partitioned into two $N \times 1$ vectors U and V for convenience. It is evident that $V = \alpha U$. We next substitute Eq. (7.62) into Eq. (7.59) and arrive at the (complex) eigenvalue problem

$$\left[S - \alpha \hat{I} \right] \hat{U} = 0 \tag{7.64}$$

where

$$\hat{\mathbf{I}} = \begin{bmatrix} \mathbf{I} & 0 \\ 0 & \mathbf{I} \end{bmatrix}$$

is the $2N \times 2N$ identity matrix. The characteristic equation for the above eigenvalue problem is then

$$\det\left[\mathbf{S} - \alpha\,\hat{\mathbf{I}}\right] = \mathcal{F}(\alpha) = 0 \tag{7.65}$$

which yields $2N$ roots, $\alpha = \alpha_1, \alpha_2, \dots \alpha_{2N}$. For each eigenvalue α_j ($j = 1, 2, \dots, 2N$) there is an associated eigenvector $\hat{\mathbf{U}}^{(j)}$. More precisely, $\hat{\mathbf{U}}^{(j)}$ is said to be the *right eigenvector* of the nonsymmetric matrix $\mathbf{S}$. The solution of Eq. (7.59) is then comprised of a linear combination of all such solutions. Hence,

$$\mathbf{z}(t) = \sum_{j=1}^{2N} A^{(j)} e^{\alpha_j t}\, \hat{\mathbf{U}}^{(j)} \tag{7.66}$$

It follows from Eqs. (7.60), (7.63) and (7.66) that

$$\mathbf{u}(t) = \sum_{j=1}^{2N} A^{(j)} e^{\alpha_j t}\, \mathbf{U}^{(j)} \tag{7.67}$$

The response corresponding to complex roots with negative real parts or to negative real roots follows directly from the development of the preceding section beginning with Eqs. (7.46) and (7.47), and leading to Eqs. (7.52) and (7.53). It is pertinent to note that the eigenvalues, α_j ($j = 1, 2, \dots, 2N$), and the associated subeigenvectors, $\mathbf{U}^{(j)}$, of $\mathbf{S}$ correspond directly with the roots and vectors of the solution described in Section 7.4.1. This may be seen by utilizing the identity

$$\begin{vmatrix} \mathbf{A}_{11} & \mathbf{A}_{12} \\ \mathbf{A}_{21} & \mathbf{A}_{22} \end{vmatrix} = \left|\mathbf{A}_{11}\right|\left|\mathbf{A}_{22} - \mathbf{A}_{21}\,\mathbf{A}_{11}^{-1}\,\mathbf{A}_{12}\right| \tag{7.68}$$

for the determinant of a partitioned matrix in Eq. (7.65) and noting that $\left[\alpha\mathbf{I}\right]^{-1} = \alpha^{-1}\mathbf{I}$. Doing this results in a characteristic equation that is identical to Eq. (7.45). The roots of Eq. (7.45) therefore correspond to the eigenvalues of $\mathbf{S}$. It follows that the vector comprised of the first N rows of the eigenvector of the system matrix $\mathbf{S}$, the subvector $\mathbf{U}$ appearing in Eq. (7.63), corresponds to the characteristic vector $\mathbf{U}$ of Eq. (7.44) for the same value of α. The various forms of Eq. (7.67) for complex and real eigenvalues therefore correspond directly to the forms derived in the previous section by way of direct solution of the equation of motion, Eq. (7.42).

Example 7.14

Determine the response of the system of Example 7.13(d) using the approach of this section.

Solution

It is convenient to nondimensionalize the equations of motion before beginning. (We effectively did this for Example 7.13 when expressed the characteristic equation in terms of ratios of the system parameters and divided the translation of the center of mass by the length of the frame.) Formally, let us introduce the normalized displacement

$$\bar{y}_G = y_G/L \tag{a}$$

and the normalized timescale

$$\tau = \omega_0 t = \sqrt{\frac{2k}{m}}\, t \tag{b}$$

It follows that for any function f,

$$\frac{df}{dt} = \frac{df}{d\tau}\frac{d\tau}{dt} = \omega_0 \frac{df}{d\tau} \tag{c}$$

Introducing Eqs. (a)–(c) into Eq. (a) of Example 7.13 and dividing the first row by m and the second row by $I_G = mL^2/12$ renders the equation of motion to the nondimensional form

$$\begin{bmatrix} 1 & 0 \\ 0 & 1 \end{bmatrix} \begin{Bmatrix} d^2\bar{y}_G/d\tau^2 \\ d^2\theta/d\tau^2 \end{Bmatrix}$$
$$+ \begin{bmatrix} (\eta_1 + \eta_2) & (\eta_2 - \eta_1)/2 \\ 6(\eta_2 - \eta_1) & 3(\eta_1 + \eta_2) \end{bmatrix} \begin{Bmatrix} d\bar{y}_G/d\tau \\ d\theta/d\tau \end{Bmatrix} + \begin{bmatrix} 1 & 0 \\ 0 & 3 \end{bmatrix} \begin{Bmatrix} \bar{y}_G \\ \theta \end{Bmatrix} = \begin{Bmatrix} 0 \\ 0 \end{Bmatrix} \tag{d}$$

We next construct the pertinent system matrix **S** by substituting the nondimensional mass, damping and stiffness matrices into Eq. (7.61). Hence,

$$\mathbf{S} = \begin{bmatrix} 0 & 0 & 1 & 0 \\ 0 & 0 & 0 & 1 \\ -1 & 0 & -(\eta_1 + \eta_2) & -(\eta_2 - \eta_1)/2 \\ 0 & -3 & -6(\eta_2 - \eta_1) & -3(\eta_1 + \eta_2) \end{bmatrix} \tag{e}$$

The eigenvalue problem for the free vibration problem is then

$$
\begin{bmatrix}
-\hat{\alpha} & 0 & 1 & 0 \\
0 & -\hat{\alpha} & 0 & 1 \\
-1 & 0 & -(\eta_1 + \eta_2 + \hat{\alpha}) & -(\eta_2 - \eta_1)/2 \\
0 & -3 & -6(\eta_2 - \eta_1) & -3(\eta_1 + \eta_2) - \hat{\alpha}
\end{bmatrix}
\begin{Bmatrix} U_1 \\ U_2 \\ V_1 \\ V_2 \end{Bmatrix}
=
\begin{Bmatrix} 0 \\ 0 \\ 0 \\ 0 \end{Bmatrix}
\tag{f}
$$

We next require that the determinant of the square matrix of Eq. (f) vanishes. This results in the characteristic equation

$$
\mathcal{F}(\hat{\alpha}) = \hat{\alpha}^4 + \hat{\alpha}^3 \left[4(\eta_1 + \eta_2) \right] + \hat{\alpha}^2 \left[4(1 + 3\eta_1\eta_2) \right] + \hat{\alpha} \left[6(\eta_1 + \eta_2) \right] + 3 = 0
\tag{g}
$$

which is seen to be identical to the characteristic equation of Example 7.13, as it should be. The roots of the characteristic equation, for $\eta_2 = 1.0$ and $\eta_1 = 0.5$, are then

$$
\hat{\alpha} = -0.6129, \ -4.0251, \ -0.6810 \pm 0.8673
\tag{h}
$$

Substituting the first row of Eq. (f) into the third row and solving the resulting expression for U_2 in terms of U_1 gives, for each respective characteristic value,

$$
\mathbf{U}^{(1)} = \begin{Bmatrix} 1 \\ 2.98 \end{Bmatrix}, \quad
\mathbf{U}^{(2)} = \begin{Bmatrix} 1 \\ 11.1 \end{Bmatrix}, \quad
\mathbf{U}^{(3,4)} = \begin{Bmatrix} 1 \\ -1.04 \mp 0.616i \end{Bmatrix}
\tag{i}
$$

The response of the system is then

$$
\begin{Bmatrix} y_G/L \\ \theta \end{Bmatrix} = A^{(1)} e^{-0.613\omega_0 t} \begin{Bmatrix} 1 \\ 2.98 \end{Bmatrix} + A^{(2)} e^{-4.03\omega_0 t} \begin{Bmatrix} 1 \\ 11.1 \end{Bmatrix}
$$

$$
+ A^{(3)} e^{-0.681\omega_0 t} \left[\begin{Bmatrix} 1 \\ -1.04 \end{Bmatrix} \cos(0.867\omega_0 t - \phi_3) + \begin{Bmatrix} 0 \\ -0.616 \end{Bmatrix} \sin(0.867\omega_0 t - \phi_3) \right]
$$

◁

$$
\tag{j}
$$

Orthogonality

The (right) eigenvectors $\hat{\mathbf{U}}^{(j)}$ ($j = 1, 2, \ldots, 2N$) of the system matrix $\mathbf{S}$ are not, in general, mutually orthogonal in the conventional sense. However, a broader view of the problem reveals an orthogonality relation between the right eigenvectors and the corresponding members of a related set of vectors, the left eigenvectors of $\mathbf{S}$. We establish this relation in the following development. Toward this end, let us consider the eigenvalue problem

$$\hat{\mathbf{W}}^{\mathrm{T}}\mathbf{S} = \alpha\hat{\mathbf{W}}^{\mathrm{T}} \tag{7.69}$$

where $\mathbf{S}$ is the system matrix defined by Eq. (7.61) and we wish to determine $(\alpha, \hat{\mathbf{W}})$ pairs that satisfy this equation. The $2N \times 1$ (complex) vector $\hat{\mathbf{W}}$ is referred as the *left eigenvector* of $\mathbf{S}$, due to its positioning in Eq. (7.69). Taking the transpose of Eq. (7.69) gives the equivalent relation

$$\mathbf{S}^{\mathrm{T}}\hat{\mathbf{W}} = \alpha\hat{\mathbf{W}}$$

which may also be written in the form

$$\left[\mathbf{S}^{\mathrm{T}} - \alpha\hat{\mathbf{i}}\right]\hat{\mathbf{W}} = \mathbf{0} \tag{7.70}$$

The problem is therefore equivalent to the determination of the eigenvalues and eigenvectors of $\mathbf{S}^{\mathrm{T}}$. Since the determinant of a matrix is equal to the determinant of its transpose, we have that

$$\left|\mathbf{S}^{\mathrm{T}} - \alpha\hat{\mathbf{i}}\right| = \left|\mathbf{S} - \alpha\hat{\mathbf{i}}\right| = \mathcal{F}(\alpha) = 0 \tag{7.71}$$

Thus, the eigenvalues of $\mathbf{S}^{\mathrm{T}}$ are the same as those of $\mathbf{S}$ ($\alpha = \alpha_1, \alpha_2, \dots \alpha_{2N}$). Substitution of each eigenvalue into Eq. (7.64) and solving for the corresponding vector components generates the associated right eigenvectors $\hat{\mathbf{U}}^{(j)}$ ($j = 1, 2, \dots, 2N$). Likewise, substitution of each eigenvalue into Eq. (7.70) and solving for the corresponding vector components generates the associated left eigenvectors $\hat{\mathbf{W}}^{(j)}$ ($j = 1, 2, \dots, 2N$). It follows that the $\hat{\mathbf{U}}^{(j)}$ and $\hat{\mathbf{W}}^{(l)}$ respectively satisfy the equations

$$\mathbf{S}\hat{\mathbf{U}}^{(j)} = \alpha_j\,\hat{\mathbf{U}}^{(j)} \tag{7.72}$$

$$\hat{\mathbf{W}}^{(l)\mathrm{T}}\mathbf{S} = \alpha_l\hat{\mathbf{W}}^{(l)\mathrm{T}} \tag{7.73}$$

Let us multiply Eq. (7.72) on the left by $\hat{\mathbf{W}}^{(l)\mathrm{T}}$ and multiply Eq. (7.73) on the right by $\hat{\mathbf{U}}^{(j)}$, and then subtract the latter from the former. Doing this gives the relation

$$0 = (\alpha_j - \alpha_l)\hat{\mathbf{W}}^{(l)\mathrm{T}}\,\hat{\mathbf{U}}^{(j)} \tag{7.74}$$

It follows that if $\alpha_j \neq \alpha_l$ then

$$0 = \hat{\mathbf{W}}^{(l)\mathrm{T}}\,\hat{\mathbf{U}}^{(j)} \tag{7.75}$$

Hence, for distinct eigenvalues, the left eigenvectors of **S** are mutually orthogonal to the right eigenvectors and vice versa. The modal vectors are thus orthogonal in this sense. In addition, it follows from Eq. (7.72) or (7.73) that

$$\hat{\mathbf{W}}^{(l)\mathrm{T}}\mathbf{S}\,\hat{\mathbf{U}}^{(j)} = 0 \qquad (7.76)$$

That is, for distinct eigenvalues, the weighted scalar product with respect to **S** of the left eigenvectors with the right eigenvectors vanishes. Thus, for distinct eigenvalues, the left and right eigenvectors are orthogonal to one another in this sense as well. The above orthogonality relations will prove useful when considering forced vibration of damped systems.

Normalization

The othogonality relations of Eqs. (7.75) and (7.76) suggest normalization of the eigenvectors by setting

$$\hat{\mathbf{W}}^{(j)\mathrm{T}}\hat{\mathbf{U}}^{(j)} = 1 \qquad (7.77)$$

or by setting

$$\hat{\mathbf{W}}^{(j)\mathrm{T}}\mathbf{S}\,\hat{\mathbf{U}}^{(j)} = 1 \qquad (7.78)$$

The state space formulation and the associated orthogonality conditions and normalization procedure will prove useful when considering forced vibration of damped multi-degree of freedom systems.

7.5 EVALUATION OF AMPLITUDES AND PHASE ANGLES

It was shown in this chapter that the free vibration response of discrete systems consists of a linear combination of the modal vectors with harmonic time signatures (undamped systems), or exponentially decaying time signatures – harmonic or otherwise (damped systems), the amplitudes and phase angles of which are evaluated by imposing the initial conditions. For an N-degree of freedom system this results in a system of $2N$ equations in the $2N$ unknowns $A^{(j)}$, ϕ_j ($j = 1, 2, …, N$), which may be solved by conventional algebraic means as demonstrated in Example 7.3 and 7.4. As an alternative to solving simultaneous algebraic equations, whether for numerical reasons or for fundamental purposes, the amplitudes and phase angles can be evaluated explicitly by exploiting the mutual orthogonality of the modal vectors. This procedure is discussed in the present section. We will first discuss this for undamped systems.

7.5.1 Undamped Systems

Consider an N-degree of freedom system which is in the initial configuration

$$\mathbf{u}(0) = \mathbf{u}_0, \quad \dot{\mathbf{u}}(0) = \mathbf{v}_0$$

at the instant it is released. Let us recall the general form of the free vibration response for N-degree of freedom systems without damping, Eq. (7.11):

$$\mathbf{u}(t) = \sum_{j=1}^{N} \mathbf{U}^{(j)} \left[A_1^{(j)} \cos \omega_j t + A_2^{(j)} \sin \omega_j t \right] = \sum_{j=1}^{N} \mathbf{U}^{(j)} A^{(j)} \cos(\omega_j t - \phi_j)$$

where

$$A^{(j)} = \sqrt{A_1^{(j)2} + A_2^{(j)2}}$$

$$\phi_j = \tan^{-1}\left(A_2^{(j)} / A_1^{(j)} \right)$$

and ω_j and $\mathbf{U}^{(j)}$ correspond to the jth frequency-mode pair of the undamped system. Imposing the initial conditions on the above modal expansion results in the identities

$$\mathbf{u}_0 = \sum_{j=1}^{N} \mathbf{U}^{(j)} A_1^{(j)}$$

$$\mathbf{v}_0 = \sum_{j=1}^{N} \mathbf{U}^{(j)} \omega_j A_2^{(j)}$$

When expanded, the above equalities represent $2N$ equations in $2N$ unknowns that may be solved for the indicated amplitudes and phase angles. Let us next multiply the above relations by $\mathbf{U}^{(l)\mathsf{T}}\mathbf{m}$. Hence,

$$\mathbf{U}^{(l)\mathsf{T}}\mathbf{m}\mathbf{u}_0 = \sum_{j=1}^{N} \mathbf{U}^{(l)\mathsf{T}}\mathbf{m}\mathbf{U}^{(j)} A_1^{(j)}$$

$$\mathbf{U}^{(l)\mathsf{T}}\mathbf{m}\mathbf{v}_0 = \sum_{j=1}^{N} \mathbf{U}^{(l)\mathsf{T}}\mathbf{m}\mathbf{U}^{(j)} \omega_j A_2^{(j)}$$

Exploiting the mutual orthogonality of the modal vectors, Eq. (7.29), renders the above expressions to the forms

$$A_1^{(j)} = \mathcal{X}^{(j)} \tag{7.79}$$

and

$$A_2^{(j)} = \mathcal{Y}^{(j)} \tag{7.80}$$

where

$$\mathcal{X}^{(j)} = \frac{\mathbf{U}^{(j)\mathsf{T}}\mathbf{m}\mathbf{u}_0}{\left\|\mathbf{U}^{(j)}\right\|_{\mathbf{m}}^2} \tag{7.81}$$

$$\mathcal{Y}^{(j)} = \frac{\mathbf{U}^{(j)\mathsf{T}}\mathbf{m}\mathbf{v}_0}{\omega_j\left\|\mathbf{U}^{(j)}\right\|_{\mathbf{m}}^2} \tag{7.82}$$

In doing this we have decoupled the modes, and have thus reduced the problem to evaluating the above expressions. It should be noted that if the above development is paralleled with the mass matrix replaced by the stiffness matrix, we again arrive at Eqs. (7.79) and (7.80) but with the equivalent statements

$$\mathcal{X}^{(j)} = \frac{\mathbf{U}^{(j)\mathsf{T}}\mathbf{k}\mathbf{u}_0}{\left\|\mathbf{U}^{(j)}\right\|_{\mathbf{k}}^2} \tag{7.83}$$

$$\mathcal{Y}^{(j)} = \frac{\mathbf{U}^{(j)\mathsf{T}}\mathbf{k}\mathbf{v}_0}{\omega_j\left\|\mathbf{U}^{(j)}\right\|_{\mathbf{k}}^2} \tag{7.84}$$

Either form, Eqs. (7.81) and (7.82) or Eqs. (7.83) and (7.84) may be used. The former form would evidently be advantageous when the particular system is coupled through the stiffness matrix while the latter form would be advantageous when the system is coupled through the mass matrix. The amplitudes and phase angles are then respectively evaluated as

$$A^{(j)} = \sqrt{\mathcal{X}^{(j)2} + \mathcal{Y}^{(j)2}} \tag{7.85}$$

and

$$\phi_j = \tan^{-1}\left(\mathcal{Y}^{(j)}/\mathcal{X}^{(j)}\right) \tag{7.86}$$

Finally, *for a system that is released from rest,* $\mathbf{v}_0 = 0$. It follows that $\phi_j = 0$ ($j = 1, 2, \dots, N$) and

$$A^{(j)} = \mathcal{X}^{(j)} = \frac{\mathbf{U}^{(j)\mathsf{T}}\mathbf{m}\mathbf{u}_0}{\left\|\mathbf{U}^{(j)}\right\|_{\mathbf{m}}^2} = \frac{\mathbf{U}^{(j)\mathsf{T}}\mathbf{k}\mathbf{u}_0}{\left\|\mathbf{U}^{(j)}\right\|_{\mathbf{k}}^2} \quad (j = 1, 2, \dots, N) \tag{7.87}$$

The expressions established in this section allow for the direct evaluation of the amplitudes and phase angles for any and all modes of a given N-degree of freedom system in free vibration.

7.5.2 Systems with General Viscous Damping

The evaluation of the amplitudes and phase angles for systems with general (linear) viscous damping is achieved in an analogous manner to that for undamped systems. It is convenient to consider the evaluation in state space. Toward this end, consider a damped N-degree of freedom system that is in the initial configuration

$$\mathbf{z}(0) = \mathbf{z}_0$$

where

$$\mathbf{z}_0 = \begin{Bmatrix} \mathbf{u}_0 \\ \mathbf{v}_0 \end{Bmatrix}$$

is the state vector at $t = 0$. Let us next recall the general form of the response in state space given by Eq. (7.67). Hence,

$$\mathbf{z}(t) = \sum_{j=1}^{2N} A^{(j)} e^{\alpha_j t} \hat{\mathbf{U}}^{(j)}$$

where α_j and $\hat{\mathbf{U}}^{(j)}$ correspond to the jth (complex) eigenvalue and corresponding (complex) right eigenvector of the system matrix $\mathbf{S}$ and $A^{(j)}$ is a complex amplitude. Imposing the initial conditions on the above expansion yields the identity

$$\mathbf{z}_0 = \sum_{j=1}^{2N} A^{(j)} \hat{\mathbf{U}}^{(j)}$$

When expanded, the above equality represents $2N$ equations in $2N$ unknowns that may be solved for the indicated (complex) amplitudes. Let us next multiply the above expression on the left by the product $\hat{\mathbf{W}}^{(l)\mathsf{T}} \mathbf{S}$, where $\hat{\mathbf{W}}^{(l)}$ is the lth right eigenvector of $\mathbf{S}$. This results in the identity

$$\hat{\mathbf{W}}^{(l)\mathsf{T}} \mathbf{S} \mathbf{z}_0 = \sum_{j=1}^{2N} A^{(j)} \, \hat{\mathbf{W}}^{(l)\mathsf{T}} \mathbf{S} \hat{\mathbf{U}}^{(j)}$$

Exploiting the statement of orthogonality of the left eigenvectors of $\mathbf{S}$ with the corresponding right eigenvectors of $\mathbf{S}$, Eq. (7.76), and solving for $A^{(j)}$ gives the relation

$$A^{(j)} = \frac{\hat{\mathbf{W}}^{(j)\mathsf{T}}\mathbf{S}\mathbf{z}_0}{\hat{\mathbf{W}}^{(j)\mathsf{T}}\mathbf{S}\hat{\mathbf{W}}^{(j)}} \quad (j = 1, 2, ..., 2N) \tag{7.88}$$

The expressions established in this section allow for the direct evaluation of the complex amplitudes, and hence of the amplitudes and phase angles associated with any and all modes of a given viscously damped N-degree of freedom system in free vibration.

7.6 CONLUDING REMARKS

The vibration of multi-degree of freedom systems is germane to a variety of applications and engineering systems. In this chapter we laid the groundwork for the study of such systems by considering the motion of systems when they are free from externally applied dynamic forces and moments. It was seen that the free vibration problem reduces to the solution of an eigenvalue problem, the eigenvalues of which correspond to the squares of the natural frequencies for an undamped system and the eigenvectors of which correspond to the associated natural modes (modal matrices/vectors). The characteristic equation for such problems is therefore referred to as the frequency equation for the system. Through this analysis it was seen that a discrete system will possess the same number of natural frequencies and modes as degrees of freedom, with their values and form dependent upon the values of the physical parameters that describe the system. Each mode represents a natural motion of the system that oscillates at the corresponding natural frequency, and the individual elements of the modal vector physically represent the relative amplitudes of the motions of the individual members that comprise the system when vibrating in that mode. The free vibration response of a discrete system is comprised of a linear combination of the natural modes undergoing harmonic vibration. The degree of participation of the various modes is measured by their amplitudes and associated phase angles, which depend on the specific initial conditions imposed on the system. For systems with viscous damping the characteristic values and associated vectors are generally complex, with the imaginary part of each eigenvalue corresponding to a (damped) natural frequency and the (negative) real part of the eigenvalue being an associated damping factor. A discussion of the general properties of modal vectors, including normalization and orthogonality, was presented. These properties will be central to the understanding of vibrations of discrete systems subjected to applied dynamic loading, which is studied in the next chapter.

BIBLIOGRAPHY

Ginsberg, J.H., *Mechanical and Structural Vibrations: Theory and Applications*, Wiley, New York, 2001.

Hildebrand, F.B., *Methods of Applied Mathematics*, 2nd ed., Prentice-Hall, Englewood Cliffs, 1965.

Meirovitch, L., *Elements of Vibration Analysis*, 2nd ed., McGraw-Hill, New York, 1986.

Meirovitch, L., *Fundamentals of Vibrations*, McGraw-Hill, New York, 2001.

Rayleigh, J.W.S., *The Theory of Sound*, Vol.1, Dover, New York, 1945.

Thomson, W.T., *Theory of Vibration with Applications*, 4th ed., Prentice-Hall, Englewood Cliffs, 1993.

Weaver, W. Jr., Timoshenko, S.P. and Young, D.H., *Vibration Problems in Engineering*, 5th ed., Wiley-Interscience, 1990.

Whittaker, E.T., *A Treatise on the Analytical Dynamics of Particles and Rigid Bodies*, Cambridge, New York, 1970.

Zwillinger, D., *CRC Standard Mathematical Tables and Formulae*, 31st ed., Chapman and Hall/CRC, Boca Raton, 2003.

PROBLEMS

7.1 Consider the constrained hook and ladder system of Problem 6.7 when $k_T = kL^2$ and $m_1 = 10m_2 = 10m$. (a) Determine the natural frequencies and modal vectors for the system. (b) Sketch and label the physical configuration of the system for each mode. (c) Establish the free vibration response of the system.

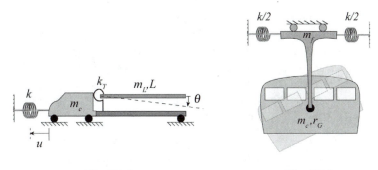

Fig. P7.1   **Fig. P7.2**

7.2 Consider the tram of Problem 6.8 when $m_C = 5m_F$. (a) Determine the natural frequencies and modal vectors for the system. (b) Sketch and label the physical configuration of the system for each mode. (c) Establish the free vibration response of the system.

7.3 Consider the coupled pendulums of Problem 6.9. (a) Determine the natural frequencies and modal vectors for the system. (b) Sketch and label the physical configuration of the system for each mode. (c) Establish the free vibration response of the system.

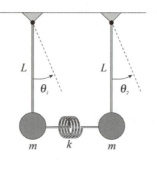

Fig. P7.3

7.4 Consider the special case where the pendulum system of Problem 7.3 has the property that $k/m \ll g/L$. If the pendulums are released from rest when in the configuration $\theta_1(0) = \theta_0$ and $\theta_2(0) = 0$, show that the response is of the form

$$\theta_1(t) \cong A_1(t)\cos\omega_a t = \theta_0 \cos\omega_b t \cos\omega_a t$$
$$\theta_2(t) \cong A_2(t)\sin\omega_a t = \theta_0 \sin\omega_b t \sin\omega_a t$$

where

$$\omega_a = (\omega_2 + \omega_1)/2, \quad \omega_b = (\omega_2 - \omega_1)/2$$

Plot the response. What type of behavior does the pendulum system exhibit?

7.5 Consider the system of Problem 6.15. (a) Determine the natural frequencies and modal vectors for the system if the wheel is of radius R and $I = 2mR^2$. (b) Sketch and label the physical configuration of the system for each mode. (c) Establish the free vibration response of the system.

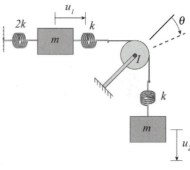

Fig. P7.5

7.6 Suppose each mass of the system of Problem 7.5 is displaced a distance u_0 and held in that position. If the masses are subsequently released from rest, what is the response of the system?

7.7 The brakes of the truck of Problem 7.1 are engaged and hold it stationary while a fireman slowly mounts the ladder, bringing it to equilibrium at an angle of deflection θ_0. At a certain instant, the fireman jumps off the ladder and the brakes are simultaneously released by the driver. Determine the response of the system.

7.8 The coupled pendulums of Problem 7.3 are at rest when the right bob is struck, giving it a velocity v_0. Determine the response of the system.

7.9 Consider the two-mass three-spring system of Example 7.2. Suppose a rigid brace is slowly inserted between the two masses so that one mass is displaced a distance u_0 to the left and the other a distance u_0 the right. Determine the response of the system if the brace is suddenly removed.

7.10 Consider the inverted pendulum of Problem 6.14. Determine the natural frequencies and modal vectors for the system if $m_1 = 5m_2$ and the length of the massless rod is L. (b) Sketch and label the physical configuration of the system for each mode. (c) Establish the free vibration response of the system.

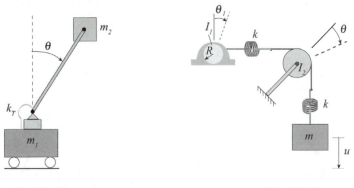

Fig. P7.10   **Fig. P7.11**

7.11 Consider the system of Problem 6.16. Determine the natural frequencies and modal vectors for the system if $I_2 = 2I_1 = 2mR^2$. (b) Sketch and label the physical configuration of the system for each mode. (c) Establish the free vibration response of the system.

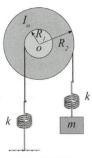

7.12 Consider the system of Problem 6.17 ($R_2 = 2R_1 = 2R$, $I_0 = mR^2$). Determine the natural frequencies and modal vectors for the system. (b) Sketch and label the physical configuration of the system for each mode. (c) Establish the free vibration response of the system.

Fig. P7.12

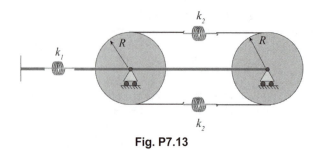

Fig. P7.13

7.13 Consider the elastically restrained fan belt of Problem 6.18, with $k_2 = 2k_1 = 2k$, $m_2 = m_1 = m$. Determine the natural frequencies and modal vectors for the system. (b) Sketch and label the physical configuration of the system for each mode. (c) Establish the free vibration response of the system.

7.14 Consider the system of Problem 6.19 when $m_b = 2m_a = 2m$ and $k_T = kL^2$. (a) Determine the natural frequencies and modal vectors for the system. (b) Sketch and label the physical configuration of the system for each mode. (c) Establish the free vibration response of the system.

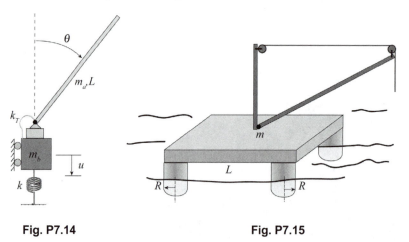

Fig. P7.14   **Fig. P7.15**

7.15 Consider the offshore platform of Problem 6.20. (a) Determine the natural frequencies and modal vectors for the platform when $L = 10R$. (b) Sketch and label the physical configuration of the system for each mode. (c) Establish the free vibration response of the platform.

7.16 Consider the linked system of Problem 6.22 when $k_1 = k_2 = k$ and $2m_2 = m_1 = m$. (a) Determine the natural frequencies and modal vectors for the system. (b) Sketch and label the physical configuration of the system for each mode. (c) Establish the free vibration response of the system.

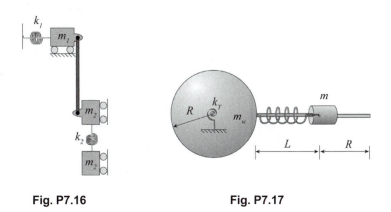

| Fig. P7.16 | Fig. P7.17 |

7.17 Consider the mechanism of Problem 6.23 when $m_w = 3m$ and $k_T = kR^2$ and $R = L$ (the unstretched spring length). Determine the natural frequencies and modal vectors for the system. (b) Sketch and label the physical configuration of the system for each mode. (c) Establish the free vibration response of the system.

7.18 Consider the floating system of Problem 6.24 when $m_b = m_a/2$ and $k = \rho_f g R^2 /2$, Determine the natural frequencies and modal vectors for the system. (b) Sketch and label the physical configuration of the system for each mode. (c) Establish the free vibration response of the system.

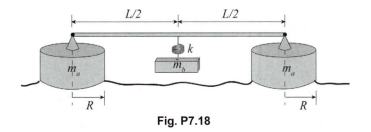

Fig. P7.18

7.19 Consider the shaft system of Problem 6.25. Determine the natural frequencies and modal vectors for the system. (b) Sketch and label the physical configuration of the system for each mode. (c) Establish the free vibration response of the system.

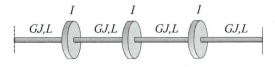

Fig. P7.19

7.20 Consider the triple pendulum of Problem 6.26. Determine the natural frequencies and modal vectors for the system. (b) Sketch and label the physical configuration of the system for each mode. (c) Establish the free vibration response of the system.

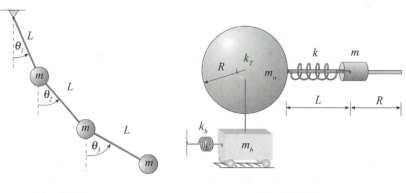

Fig. P7.20   Fig. P7.21

7.21 Consider the system of Problem 6.27, when $m_b = m_w = 3m$ and $k_T = kR^2$ and $R = L$. Determine the natural frequencies and modal vectors for the system. (b) Sketch and label the physical configuration of the system for each mode. (c) Establish the free vibration response of the system.

7.22 Consider the coupled pendulums of Problem 6.29. Determine the natural frequencies and modal vectors for the system. (b) Sketch and label the physical configuration of the system for each mode. (c) Establish the free vibration response of the system.

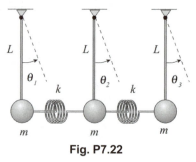

Fig. P7.22

7.23 Consider the frame system of Problem 6.30 with $m_b/m_a = 0.25$. Determine the natural frequencies and modal vectors for the system. (b) Sketch and label the physical configuration of the system for each mode. (c) Establish the free vibration response of the system.

7.24 Consider the conveyor belt system of Problem 6.10 with $m_2 = m_1 = m$ and $R_1 = R_2 = R$. Determine the natural frequencies and modal vectors for the system. (b) Sketch and label the physical configuration of the system for each mode. (c) Establish the free vibration response of the system.

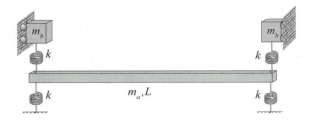

Fig. P7.23

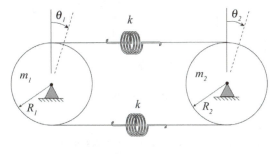

Fig. P7.24

7.25 Consider the dumbbell satellite of Problem 6.13, where the undeformed length of the access tube is L. Determine the natural frequencies and modal vectors for two dimensional motion of the system. (b) Sketch and label the physical configuration of the system for each mode. (c) Establish the free vibration response of the system.

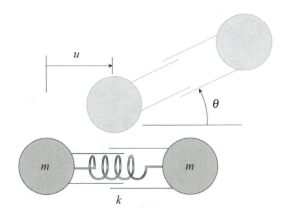

Fig. P7.25

7.26 The submarine of Problem 6.21 is modeled as shown, for simple calculations of longitudinal motion. The mass of the hull and frame structure is $2m_s$ and that of the interior compartment is $m_c = 0.5m_s$. The hull and interior compartment are separated by springs of stiffness k, and the longitudinal stiffness of the hull is k_s ($= 2k$) as indicated. (a) Determine the natural frequencies and modes for the boat. (b) Sketch and label the physical configuration of the structure for each mode. (c) Establish the free (longitudinal) vibration response of the submarine.

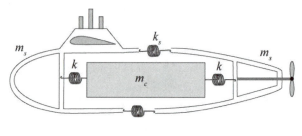

Fig. P7.26

7.27 Consider the elastically coupled two wheel system of Problem 6.28. The vehicle is comprised of two wheels, each of mass m and radius r_w, that are connected by an elastic coupler of effective stiffness k and undeformed length L. The system rolls without slip around a circular track of radius R, as shown. (a) Determine the natural frequencies and modal vectors for the specific vehicle where $L = 4r_w$ and $R = 20r_w$. (b) Determine the small angle motion of the vehicle if it is released from rest when the front wheel is in the position θ_{01} and the rear wheel is in the position θ_{02}, where the angles are measured from the bottom of the valley.

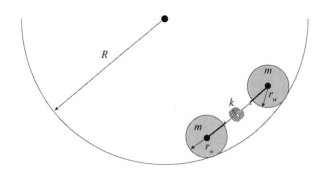

Fig. P7.27

7.28 Consider the aircraft of Problem 6.31 where the wings are modeled as equivalent rigid bodies with torsional springs of stiffness k_T at the fuselage wall, each wing possesses moment of inertia I_c about its respective connection point and the fuselage has moment of inertia $I_o = I_c$ about its axis. Determine the natural frequencies and modal vectors for pure rolling motion of the fuselage of radius R.

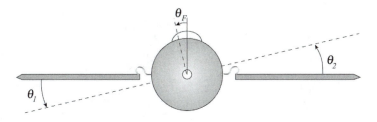

Fig. P7.28

7.29 Normalize the modes for the hook and ladder system of Problem 7.1, (a) in the conventional sense and (b) with respect to the mass matrix.

7.30 Normalize the modes for the tram system of Problem 7.2, (a) in the conventional sense and (b) with respect to the mass matrix.

7.31 Normalize the modes for the conveyor belt system of Problem 7.24, (a) in the conventional sense and (b) with respect to the mass matrix.

7.32 Normalize the modes for the shaft system of Problem 7.19, (a) in the conventional sense and (b) with respect to the mass matrix.

7.33 Verify that the modes computed in Problem 7.1 are mutually orthogonal with respect to both **m** and **k**.

7.34 Verify that the modes computed in Problem 7.2 are mutually orthogonal with respect to both **m** and **k**.

7.35 Verify that the modes computed in Problem 7.24 are mutually orthogonal with respect to both **m** and **k**.

7.36 Verify that the modes computed in Problem 7.19 are mutually orthogonal with respect to both **m** and **k**.

7.37 Determine the general free vibration response of a two-mass three-spring three-damper system where $m_1 = m_2 = m$, $k_1 = k_2 = k_3 = k$, $c_1 = c_2 = c_3 = c$ and $c^2/km = 0.04$.

7.38 Determine the general free vibration response of the system of Problem 6.32 when $m_b = 2m_a = 2m$, $k_T = kL^2$ and $c^2/km = 0.04$.

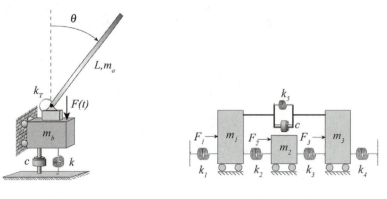

Fig. P7.38   **Fig. P7.39**

7.39 Determine the general free vibration response of the system of Problem 6.34 when $m_1 = m_2 = m_3 = m$, $k_1 = k_2 = k_3 = k_4 = 0.5\ k_5 = k$ and $c^2/km = 0.04$.

7.40 Determine the general free vibration response of the system of Problem 6.36 for $c^2/km = 0.04$ and $I = 2mR^2$.

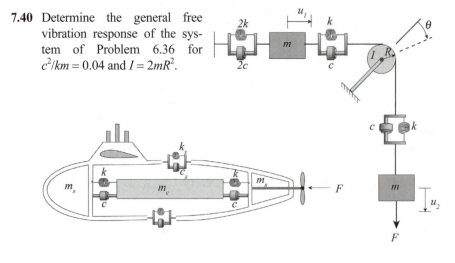

Fig. P7.41   **Fig. P7.40**

7.41 Determine the general free vibration response of the submarine of Problem 6.37 when $k_s = 2k$, $c_s = 2c$, $m_s = 2m_c$ and $c^2/km_c = 0.01$.

8

Forced Vibration of Multi-Degree of Freedom Systems

Mechanical systems are generally subjected to a variety of forces and force types during the course of their operation. Such forces may cause desirable or undesirable motions of the system, with the former required for its effective operation and the latter having detrimental, if not catastrophic, consequences. In any event, it is clearly of interest to understand and predict the effects that time dependent forces have on mechanical and structural systems. In the previous chapter we discussed the motion of discrete multi-degree of freedom systems when they are free to move under their own volition. It was seen therein that each system has fundamental motions associated with it called modes, and that each mode oscillates at its own natural frequency. It was also shown that any free vibration of the system is comprised of some combination of these fundamental motions. In this chapter we will examine the behavior of discrete multi-degree of freedom systems that are subjected to external forces. It will be seen that, as for free vibrations, the response of such systems to time dependent forcing is described by a combination of the basic motions, or modes, as well. In this light, the general approach that we shall take to study the behavior of multi-degree of freedom systems to forces of all types will be via the fundamental method known as *Modal Analysis*. This approach not only provides a vehicle for solving forced vibration problems for any type of excitation but it also unveils important physical characteristics of the excited system and the basic mechanisms involved. After an extensive discussion for undamped systems, the procedure is extended to systems possessing a restricted type of viscous damping (Rayleigh Damping). The chapter finishes with an abstraction of modal analysis to state space for multi-degree of freedom systems with general linear viscous damping. To prepare for our study, we introduce the concept of *modal coordinates* in Section 8.2. We begin, however, with a simple solution for the steady state response of undamped systems subjected to harmonic excitation, and a discussion of the vibration absorber for elementary systems.

8.1 INTRODUCTION

In this section we present a simple solution for the steady state response of undamped multi-degree of freedom systems subjected to external excitations possessing a (synchronous) harmonic time signature. The same approach is then employed to examine a simple application — that of a vibration absorber appended to a single degree of freedom system.

8.1.1 Steady State Response to Harmonic Excitation

Consider an N-degree of freedom system subjected to external forces, all of which possess the same harmonic time signature. The equation of motion for such a system will be of the general form

$$\mathbf{m}\ddot{\mathbf{u}}(t) + \mathbf{k}\mathbf{u}(t) = \mathbf{F_0}e^{i\Omega t} \tag{8.1}$$

where Ω is the excitation frequency. We wish to obtain the particular solution to Eq. (8.1). Toward this end, let us assume a solution of the form

$$\mathbf{u_p}(t) = \mathbf{H}e^{i\Omega t} \tag{8.2}$$

Substituting Eq. (8.2) into Eq. (8.1) and solving for $\mathbf{H}$ gives

$$\mathbf{H} = \left[\mathbf{k} - \Omega^2\mathbf{m}\right]^{-1}\mathbf{F_0} \tag{8.3}$$

We now substitute Eq. (8.3) into Eq. (8.2) to obtain the steady state response,

$$\mathbf{u_p}(t) = \left[\mathbf{k} - \Omega^2\mathbf{m}\right]^{-1}\mathbf{F_0}e^{i\Omega t} \tag{8.4}$$

Employing Kramer's Rule results in the equivalent form

$$\mathbf{u_p}(t) = \frac{\text{adj}\left[\mathbf{k} - \Omega^2\mathbf{m}\right]}{\det\left[\mathbf{k} - \Omega^2\mathbf{m}\right]}\mathbf{F_0}e^{i\Omega t} \tag{8.5}$$

(A similar solution is found for damped systems in Section 8.8.1.) We know from our discussions of free vibrations (Chapter 7) that $\det\left[\mathbf{k} - \omega^2\mathbf{m}\right] = 0$ is the characteristic equation that yields the natural frequency of the system. We thus see that when $\Omega = \omega$ we have a resonance condition and the above solution is no longer valid. (Calculation of the resonance solution will be demonstrated in Example 8.6b-ii of Section 8.5.) The solution defined by Eq. (8.4) or Eq. (8.5) is mathematically equivalent to that which would be obtained using modal analysis (Sections 8.2–8.5), though it does not reveal the fundamental characteristics of the response to the extent that modal analysis does. It is simple to apply in principle, but requires computation of the inverse of an $N \times N$ matrix. This can prove cumbersome for large scale systems. The response is, however, readily obtained for simple systems as seen in the following example.

Example 8.1

Consider the two-mass three-spring system of Example 6.1. (*a*) Determine the steady state response of the system when each mass is subjected to a force that varies harmonically in time with frequency Ω and the corresponding magnitudes are F_1^0 and F_2^0, respectively. (*b*) Use the results from Part (a) to evaluate the response of a system for which $k_1 = k_2 = k_3 = k$ and $m_1 = m_2 = m$, when the excitation is such that $F_2 = 0$ and $F_1(t) = F_0 \sin \Omega t$.

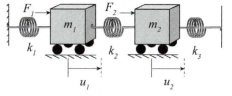

Solution

(*a*)

Substituting the specific force system under consideration into Eq. (b) of Example 6.1, gives the equation of motion for the system as

$$\begin{bmatrix} m_1 & 0 \\ 0 & m_2 \end{bmatrix}\begin{Bmatrix} \ddot{u}_1 \\ \ddot{u}_2 \end{Bmatrix} + \begin{bmatrix} k_1 + k_2 & -k_2 \\ -k_2 & k_2 + k_3 \end{bmatrix}\begin{Bmatrix} u_1 \\ u_2 \end{Bmatrix} = \begin{Bmatrix} F_1^0 \\ F_2^0 \end{Bmatrix} e^{i\Omega t} \tag{a}$$

Now, the determinant and the adjoint for the simple two degree of freedom system under consideration are easily computed as

$$\det\left[\mathbf{k} - \Omega^2 \mathbf{m}\right] = (k_1 + k_2 - \Omega^2 m_1)(k_2 + k_3 - \Omega^2 m_2) - k_2^2 \tag{b}$$

and

$$\text{adj}\left[\mathbf{k} - \Omega^2 \mathbf{m}\right] = \begin{bmatrix} (k_2 + k_3 - \Omega^2 m_2) & k_2 \\ k_2 & (k_1 + k_2 - \Omega^2 m_1) \end{bmatrix} \tag{c}$$

Substitution of Eqs. (b) and (c) into Eq. (8.5) gives the steady state response

$$u_1(t) = \frac{(k_2 + k_3 - \Omega^2 m_2)F_1^0 + k_2 F_2^0}{(k_1 + k_2 - \Omega^2 m_1)(k_2 + k_3 - \Omega^2 m_2) - k_2^2} e^{i\Omega t} \qquad \lhd \text{(d-1)}$$

$$u_2(t) = \frac{k_2 F_1^0 + (k_1 + k_2 - \Omega^2 m_1)F_2^0}{(k_1 + k_2 - \Omega^2 m_1)(k_2 + k_3 - \Omega^2 m_2) - k_2^2} e^{i\Omega t} \qquad \lhd \text{(d-2)}$$

(b)
For the case where $k_1 = k_2 = k_3 = k$ and $m_1 = m_2 = m$ and $F_2 = 0$ the response of the system simplifies to the form

$$u_1(t) = \frac{F_1^0}{k} \sin \Omega t \, \frac{\left[2 - (\Omega/\omega_1)^2 \right]}{\left[1 - (\Omega/\omega_1)^2 \right]\left[1 - (\Omega/\omega_2)^2 \right]}$$

$$\qquad \lhd \text{(e)}$$

$$u_2(t) = \frac{F_1^0}{3k} \sin \Omega t \, \frac{1}{\left[1 - (\Omega/\omega_1)^2 \right]\left[1 - (\Omega/\omega_2)^2 \right]}$$

where

$$\omega_1 = \sqrt{k/m}, \qquad \omega_2 = \sqrt{3k/m} \qquad \text{(f)}$$

are the natural frequencies of the system (see Example 7.1).

8.1.2 The Simple Vibration Absorber

It was seen in Sections 3.3.2 and 8.1.1 that when an undamped system is excited harmonically at one of its natural frequencies a resonance condition occurs whereby large amplitude vibrations occur. It was also seen that if the excitation frequency of a single degree of freedom system is sufficiently above (or below) the natural frequency then the oscillations are well behaved. In fact, when operating at excitation frequencies sufficiently above the resonance frequency the amplitude of the steady state oscillations are lower then the deflection that would be induced by a static load of the same magnitude. If vibrations are to be avoided then the system can be designed to operate in this range. (It then becomes a practical issue as to how to ramp up an initially quiescent system past the resonance frequency to the desired operating range.) Suppose, however, that the normal operating range is at, or near, the resonance frequency. In addition, suppose that the machine or device cannot be redesigned. That is, we wish to use the system as is. How may we resolve this issue? We know from our studies of free vibrations in Chapter 7 and related discussions of the present chapter that a multi-degree of freedom system will have the same number of natural frequencies, and hence the same number of resonance conditions, as the number of degrees of freedom. Therefore, one practical solution is to add an extra degree of freedom to the system. This will shift the natural frequencies and hence the conditions for resonance. Alternatively, we may wish to induce large motions of the added mass, leaving the original mass undergoing relatively small motions, in effect *isolating* it from vibration. We examine this situation for an originally single degree of freedom system in the present section.

Consider a single degree of freedom system represented as the mass-spring system shown in Figure 8.1a. Suppose the system operates in an environment where it is subjected to a harmonic force $F(t)$, as indicated. Let us further suppose that the operating frequency is close to the natural frequency of the system. One approach to remedy the situation is to change the mass or stiffness. However, suppose that this is not an option, say for reasons of functionality, practicality, economics or aesthetics. As an alternative, let us attach a spring of stiffness k_{ab} and mass m_{ab} to the original system of mass m and stiffness k as shown in Figure 8.1b. Let us examine the response of the augmented system compared with that of the original system.

The equation of motion for the augmented system is easily derived, or is obtained directly from Eq. (b) of Example 6.1 by setting $m_1 = m$, $m_2 = m_{ab}$, $k_1 = k$, $k_2 = k_{ab}$, $k_3 = 0$, $F_1 = F(t)$ and $F_2 = 0$. Hence,

$$\begin{bmatrix} m & 0 \\ 0 & m_{ab} \end{bmatrix} \begin{Bmatrix} \ddot{u}_1(t) \\ \ddot{u}_2(t) \end{Bmatrix} + \begin{bmatrix} (k+k_{ab}) & -k_{ab} \\ -k_{ab} & k_{ab} \end{bmatrix} \begin{Bmatrix} u_1(t) \\ u_2(t) \end{Bmatrix} = \begin{Bmatrix} F(t) \\ 0 \end{Bmatrix} \qquad (8.6)$$

We seek the steady state response (i.e., the particular solution) of the augmented, now two degree of freedom, system for excitations of the form

$$F(t) = F_0 e^{i\Omega t} \qquad (8.7)$$

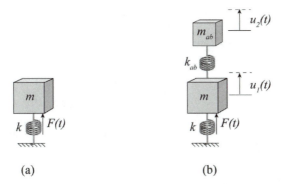

(a)   (b)

Figure 8.1 Single degree of freedom system and vibration absorber: (a) the system alone, (b) system with absorber.

For the purposes of the present discussion we bypass modal analysis for this very simple system and loading, and seek a particular solution to Eq. (8.6) of the form

$$\begin{Bmatrix} u_1(t) \\ u_2(t) \end{Bmatrix}_p = \begin{Bmatrix} H_1 \\ H_2 \end{Bmatrix} e^{i\Omega t} \tag{8.8}$$

where H_1 and H_2 are to be determined. Substituting Eqs. (8.7) and (8.8) into Eq. (8.6) results in the pair of equations

$$\begin{aligned} \left[(k + k_{ab}) - \Omega^2 m \right] H_1 - k_{ab} H_2 &= F_0 \\ -k_{ab} H_1 + \left[k_{ab} - \Omega^2 m_{ab} \right] H_2 &= 0 \end{aligned} \tag{8.9}$$

which are easily solved for the unknown amplitudes H_1 and H_2. Hence,

$$H_1 = \frac{f_0 \left(\bar{\omega}^2 - \bar{\Omega}^2 \right)}{\left(1 - \bar{\Omega}^2 \right) \left(\bar{\omega}^2 - \bar{\Omega}^2 \right) - \bar{k} \bar{\Omega}^2} \tag{8.10}$$

and

$$H_2 = \frac{f_0 \bar{\omega}^2}{\left(1 - \bar{\Omega}^2 \right) \left(\bar{\omega}^2 - \bar{\Omega}^2 \right) - \bar{k} \bar{\Omega}^2} \tag{8.11}$$

where

$$f_0 = F_0/k \tag{8.12}$$

$$\bar{\Omega} = \Omega/\omega \tag{8.13}$$

$$\omega^2 = k/m \tag{8.14}$$

$$\bar{\omega}^2 = \bar{k}/\bar{m} \tag{8.15}$$

$$\bar{k} = k_{ab}/k \tag{8.16}$$

$$\bar{m} = m_{ab}/m \tag{8.17}$$

Substituting Eqs. (8.10) and (8.11) into Eq. (8.8) gives the steady state response of the system as

$$\begin{Bmatrix} u_1(t) \\ u_2(t) \end{Bmatrix} = \frac{f_0 e^{i\Omega t}}{\left(1 - \bar{\Omega}^2 \right) \left(\bar{\omega}^2 - \bar{\Omega}^2 \right) - \bar{k} \bar{\Omega}^2} \begin{Bmatrix} \bar{\omega}^2 - \bar{\Omega}^2 \\ \bar{\omega}^2 \end{Bmatrix} \tag{8.18}$$

Now, we are interested in the effect of the added mass and spring on the steady state response of the original (base) mass. The augmented magnification factor for the motion of the base mass is then

$$\hat{\Gamma} = \frac{\|H_1\|}{f_0} = \left\| \frac{\left(\bar{\omega}^2 - \bar{\Omega}^2\right)}{\left(1 - \bar{\Omega}^2\right)\left(\bar{\omega}^2 - \bar{\Omega}^2\right) - \bar{k}\bar{\Omega}^2} \right\| \tag{8.19}$$

When the augmented magnification factor is expressed in the above form it is seen that the added mass and spring modify the original magnification factor, and hence influence the steady state response of the system, through the ratios of the added mass to the original mass and the added stiffness to the original stiffness. (It is readily seen that the magnification factor for the original single degree of freedom system is recovered when $\bar{\omega} = \bar{k} = 0$.)

The augmented magnification factor is displayed along with the original magnification factor in Figure 8.2 for the case where $\bar{k} = 0.2$ and $\bar{\omega} = 1$. It is seen from the figure that the resonance condition at $\bar{\Omega} = 1$ has been removed and that there is no motion of the base mass at this excitation frequency. Furthermore, there is a finite range of excitation frequencies near $\bar{\Omega} = 1$ for which the amplitude of the displacement of the base mass is less than that which would be induced by a static load of the same magnitude. The drawback is, of course, that since the augmented system has two degrees of freedom there are now two resonance conditions as indicated by the two peaks appearing in the figure. The practical problem of passing through the lower resonance frequency to get to the operating range when starting up an initially quiescent system would still have to be addressed.

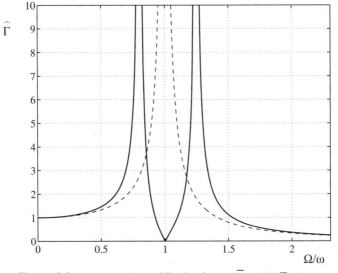

Figure 8.2 Augmented magnification factor $(\bar{k} = 0.2, \bar{\omega} = 1)$.

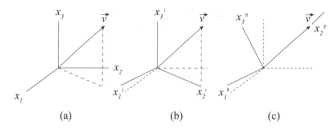

Figure 8.3 Vector shown with various coordinate systems: (a) arbitrary, (b) vector lies in a coordinate plane, (c) vector parallel to a coordinate axis.

8.2 MODAL COORDINATES

To study the forced vibration of multi-degree of freedom systems it is often convenient, as well as informative, to express a given problem in terms of its most fundamental set of coordinates known as modal coordinates. We introduce the definition and nature of these coordinates in the present section.

Let us consider a vector $\vec{v}$ in some three-dimensional space. If we express the vector in terms of its components with respect to some coordinate system then the vector will generally have three nonvanishing components (Figure 8.3a). If the coordinate system is chosen so that the same vector lies in a coordinate plane, then the vector will generally have only two nonvanishing components (Figure 8.3b). Lastly, if the coordinate system is chosen so that one of the axes is aligned with the vector, then the vector will have only one nonvanishing component (Figure 8.3c). This property will also be true for a vector in a space of any number of dimensions, say N. Suppose now that we have a set of mutually orthogonal vectors of a given dimension (say N) in some space. If we choose a coordinate system whose axes are parallel to the vectors of the given set then each of the mutually orthogonal vectors will possess only one nonvanishing component, the one corresponding to the coordinate axis aligned with the vector. Such coordinates are referred to as *principal coordinates* and find application in many fields. From our study of free vibrations, we know that the modal vectors for an undamped system form a mutually orthogonal set of vectors. The corresponding principle coordinates for this case are referred to as *modal coordinates*. These concepts may be generalized for the case of damped systems. In the present section we introduce modal coordinates and examine their important implications.

8.2.1 Principal Coordinates

The following discussion is presented for vectors in three dimensions for ease of visualization. However, the concepts and results are readily extended to N-dimensional space and are thus applicable to systems with any number of degrees of freedom. With this in mind, let us consider a Cartesian reference frame with axes (x_1,

x_2, x_3) in three-dimensional space and let $\vec{e}^{(1)}$, $\vec{e}^{(2)}$ and $\vec{e}^{(3)}$ represent the corresponding unit vectors directed along these axes as shown in Figure 8.4a. Consider, also, three mutually orthogonal, but otherwise arbitrary, vectors $\vec{u}, \vec{v}$ and $\vec{w}$, as indicated. The three vectors may be expressed in terms of their components with respect to the given set of axes in the form

$$\vec{u} = u_1 \vec{e}^{(1)} + u_2 \vec{e}^{(2)} + u_3 \vec{e}^{(3)}$$
$$\vec{v} = v_1 \vec{e}^{(1)} + v_2 \vec{e}^{(2)} + v_3 \vec{e}^{(3)}$$
$$\vec{w} = w_1 \vec{e}^{(1)} + w_2 \vec{e}^{(2)} + w_3 \vec{e}^{(3)}$$

The corresponding matrices of the components of these vectors with respect to the given coordinate system are then

$$\mathbf{u} = \begin{Bmatrix} u_1 \\ u_2 \\ u_3 \end{Bmatrix}, \quad \mathbf{v} = \begin{Bmatrix} v_1 \\ v_2 \\ v_3 \end{Bmatrix}, \quad \mathbf{w} = \begin{Bmatrix} w_1 \\ w_2 \\ w_3 \end{Bmatrix}$$

Consider next a second set of coordinate axes $(\tilde{x}_1, \tilde{x}_2, \tilde{x}_3)$ that are obtained from the first set of axes (x_1, x_2, x_3) by a rotation about some axis through the origin, and that are aligned with the three orthogonal vectors as shown in Figure 8.4b. Consider also, the same mutually orthogonal vectors expressed in terms of their components with respect this second set of axes. Hence,

$$\vec{u} = u\,\vec{n}^{(1)}$$
$$\vec{v} = v\,\vec{n}^{(2)}$$
$$\vec{w} = w\,\vec{n}^{(3)}$$

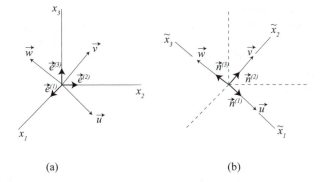

(a) (b)

Figure 8.4 Three mutually orthogonal vectors: (a) shown with arbitrary coordinate system, (b) shown with principal coordinate system.

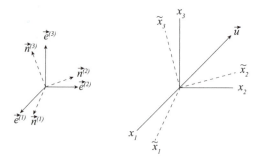

Figure 8.5 Vector displayed with two coordinate systems. (Unit basis vectors for each system shown at left.)

where $\vec{n}^{(1)}$, $\vec{n}^{(2)}$ and $\vec{n}^{(3)}$ represent the unit vectors along the second set of axes, as shown. The corresponding matrices of components of these vectors expressed with respect to the coordinates $(\tilde{x}_1, \tilde{x}_2, \tilde{x}_3)$ are then

$$
\tilde{\mathbf{u}} = \begin{Bmatrix} u \\ 0 \\ 0 \end{Bmatrix}, \quad \tilde{\mathbf{v}} = \begin{Bmatrix} 0 \\ v \\ 0 \end{Bmatrix}, \quad \tilde{\mathbf{w}} = \begin{Bmatrix} 0 \\ 0 \\ w \end{Bmatrix}
$$

In this context the coordinates $(\tilde{x}_1, \tilde{x}_2, \tilde{x}_3)$ are referred to as the principal coordinates for the given set of vectors. The obvious advantage of choosing to express the vectors in terms of the principle coordinates is that the associated matrices have only one nonzero element. The matrices are thus decoupled in this sense.

8.2.2 Coordinate Transformations

The components of a vector expressed in terms of one set of coordinates can be related to the components of that same vector expressed in terms of another set of coordinates by a system of linear equations whose coefficients are dependent on the angles between the two sets of axes. This is readily seen when we consider a three-dimensional vector $\vec{u}$ and the two coordinate systems (x_1, x_2, x_3) and $(\tilde{x}_1, \tilde{x}_2, \tilde{x}_3)$, as depicted in Figure 8.5. Hence,

$$
\vec{u} = \begin{cases} u_1 \vec{e}^{(1)} + u_2 \vec{e}^{(2)} + u_3 \vec{e}^{(3)} \\ \tilde{u}_1 \vec{n}^{(1)} + \tilde{u}_2 \vec{n}^{(2)} + \tilde{u}_3 \vec{n}^{(3)} \end{cases}
$$

Taking the successive scalar products of the unit vectors of the second coordinate system with the first equation gives the set of linear equations

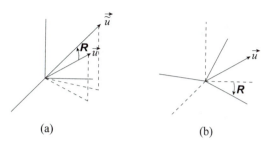

Figure 8.6 Vector under coordinate transformation.

$$\tilde{u}_1 = r_{11}u_1 + r_{12}u_2 + r_{13}u_3$$
$$\tilde{u}_2 = r_{21}u_1 + r_{22}u_2 + r_{23}u_3$$
$$\tilde{u}_3 = r_{31}u_1 + r_{32}u_2 + r_{33}u_3$$

where $r_{ij} = \vec{n}^{(l)} \cdot \vec{e}^{(j)} = \cos(\tilde{x}_l, x_j)$ $(l, j = 1-3)$ is the cosine of the angle between the $\tilde{x}_l$ axis and the x_j axis. This system of linear equations can be expressed in matrix form as

$$\tilde{\mathbf{u}} = \mathbf{R}\mathbf{u} \qquad (8.20)$$

where $\tilde{\mathbf{u}} = \{\tilde{u}_l\}$ $\mathbf{R} = \left[r_{ij} \right]$ and $\mathbf{u} = \{u_j\}$. In the matrix equation, the column matrix $\tilde{\mathbf{u}}$ corresponds to the matrix of components of vector $\tilde{u}$ obtained by rotating and stretching the original vector, as shown in Figure 8.6a. (If the columns/rows of $\mathbf{R}$ are orthonormal, then the length of the vector is preserved during the operation and $\mathbf{R}$ represents a pure rotation.) Alternatively, the operation defined by Eq. (8.20) represents a rotation (and stretch) of the axes, as shown in Figure 8.6b, and $\mathbf{u}$ corresponds to the matrix of components of the same vector expressed with respect to the set of coordinates obtained by rotation (and stretch) of the original set of coordinates as discussed earlier. The above discussion can be extended to vectors of any dimension, say N.

Example 8.2

The vector shown in Figure E8.2-1 is of magnitude A, lies in the $x_2 x_3$-plane and makes equal angles between these two axes.

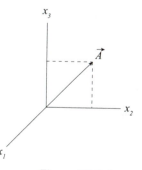

Figure E8.2-1

Determine the components of the vector $\vec{A}$ with respect to the given coordinate system, and also with respect to a coordinate system related to the first by the transformation

$$\mathbf{R} = \begin{bmatrix} 1 & 0 & 0 \\ 0 & \sqrt{2}/2 & \sqrt{2}/2 \\ 0 & -\sqrt{2}/2 & \sqrt{2}/2 \end{bmatrix}$$

How is the second coordinate system aligned?

Solution

The given vector may be expressed in matrix form as

$$\mathbf{v} = A \begin{Bmatrix} 0 \\ 1/\sqrt{2} \\ 1/\sqrt{2} \end{Bmatrix} \tag{a}$$

Thus, $v_1 = 0$, $v_2 = v_3 = A/\sqrt{2}$. To determine the components of the vector with respect to the second system we employ the coordinate transformation defined by Eq. (8.20). Hence,

$$\tilde{\mathbf{v}} = \mathbf{R}\mathbf{v} = \begin{bmatrix} 1 & 0 & 0 \\ 0 & \sqrt{2}/2 & \sqrt{2}/2 \\ 0 & -\sqrt{2}/2 & \sqrt{2}/2 \end{bmatrix} \begin{Bmatrix} 0 \\ A/\sqrt{2} \\ A/\sqrt{2} \end{Bmatrix} = A \begin{Bmatrix} 0 \\ 1 \\ 0 \end{Bmatrix} \tag{b}$$

The components of the same vector with respect to the new coordinates are thus $\tilde{v}_1 = 0$, $\tilde{v}_2 = A$, and $\tilde{v}_3 = 0$. The new coordinates for this particular case are evidently principle coordinates and correspond to a set of axes oriented with respect to the original coordinate axes as shown in Figure E8.2-2. It is seen that the given coordinate transformation, $\mathbf{R}$, corresponds to a 45° rotation of the x_2 and x_3 axes about the x_1 axis.

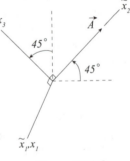

Figure E8.2-2

8.2.3 Modal Coordinates

In the previous section it was seen that the components of a vector expressed in two different coordinate systems are related by a simple matrix operation. We next consider a particular transformation based on the modal vectors of an undamped multi-degree of freedom system, and the coordinates associated with this transformation.

Consider the set of modal vectors of an N-degree of freedom system, and let us extend the interpretation of Eqs. (8.20) to N-dimensional space. Further, let us construct a particular matrix $\mathbf{B}$ such that the elements of each column correspond to the components of a modal vector for the system of interest. Hence, let

$$\mathbf{B} \equiv \begin{bmatrix} \mathbf{U}^{(1)} & \mathbf{U}^{(2)} & \cdots & \mathbf{U}^{(N)} \end{bmatrix} = \begin{bmatrix} U_1^{(1)} & U_1^{(2)} & \cdots & U_1^{(N)} \\ U_2^{(1)} & U_2^{(2)} & \cdots & U_2^{(N)} \\ \vdots & \vdots & \ddots & \vdots \\ U_N^{(1)} & U_N^{(2)} & \cdots & U_N^{(N)} \end{bmatrix} \qquad (8.21)$$

The matrix $\mathbf{B}$ is called the *modal matrix* of the given system. It follows that the rows of the transpose of the modal matrix correspond to the transposes of the modal vectors. Hence,

$$\mathbf{B}^{\mathsf{T}} = \begin{bmatrix} \mathbf{U}^{(1)\mathsf{T}} \\ \vdots \\ \mathbf{U}^{(N)\mathsf{T}} \end{bmatrix} \qquad (8.22)$$

Equivalently,

$$b_{ij} = U_i^{(j)}, \quad b_{ij}^{\mathsf{T}} = U_j^{(l)} \quad (l, j = 1, 2, ..., N) \qquad (8.23)$$

Let us now consider the motion of an N-degree of freedom system and the transformation

$$\mathbf{u}(t) = \mathbf{B}\,\boldsymbol{\eta}(t) \qquad (8.24)$$

where $\mathbf{u}(t)$ is some matrix that characterizes the motion, say the displacement matrix, and $\mathbf{B}$ is the modal matrix for the system as defined by Eq. (8.21). The inverse transformation follows directly as

$$\boldsymbol{\eta}(t) = \mathbf{B}^{-1}\mathbf{u}(t)$$

It is seen, upon comparison with Eq. (8.20), that Eq. (8.24) represents the relation between the displacements of the system expressed in two different coordinate systems. Since the elements of the matrix $\mathbf{u}(t)$ correspond to the physical displacements of the system expressed in terms of the (generalized) physical coordinates chosen to describe the motion, then the elements of the matrix $\boldsymbol{\eta}(t)$ correspond to the displacements expressed in terms of a different set of coordinates. These latter coordinates are referred to as the *modal coordinates* for the system. It will be seen that modal coordinates are, in fact, principal coordinates for a multi-degree of freedom system and that the implications of transforming to modal coordinates are profound.

Example 8.3 – Free Vibration of an *N*-Degree of Freedom System

Express the general form of the free vibration response of a discrete *N*-degree of freedom system in the form of Eq. (8.24) and identify the modal coordinates for this case.

Solution
The general form of the free vibration response for the system under consideration is given by Eq. (7.11) as

$$\mathbf{u}(t) = \sum_{j=1}^{N} \mathbf{U}^{(j)} A^{(j)} \cos(\omega_j t - \phi_j) \tag{a}$$

The summation appearing in Eq. (a) may be written as the product of a square matrix and a column vector as follows:

$$\begin{Bmatrix} u_1(t) \\ u_2(t) \\ \vdots \\ u_N(t) \end{Bmatrix} = \begin{bmatrix} \mathbf{U}^{(1)} & \mathbf{U}^{(2)} & \cdots & \mathbf{U}^{(N)} \end{bmatrix} \begin{Bmatrix} A^{(1)} \cos(\omega_1 t - \phi_1) \\ A^{(2)} \cos(\omega_2 t - \phi_2) \\ \vdots \\ A^{(N)} \cos(\omega_N t - \phi_N) \end{Bmatrix} \tag{b}$$

If we compare Eq. (b) with Eq. (8.24) we see that the square matrix is the modal matrix for the system and

$$\boldsymbol{\eta}(t) = \begin{Bmatrix} \eta_1(t) \\ \eta_2(t) \\ \vdots \\ \eta_N(t) \end{Bmatrix} = \begin{Bmatrix} A_1 \cos(\omega_1 t - \phi_1) \\ A_2 \cos(\omega_2 t - \phi_2) \\ \vdots \\ A_N \cos(\omega_N t - \phi_N) \end{Bmatrix} \tag{c}$$

It follows from Eq. (c) that the time dependent coefficients

$$\eta_j(t) = A^{(j)} \cos(\omega_j t - \phi_j) \quad (j = 1, 2, ..., N) \tag{d}$$

correspond to the *modal displacements for the free vibration problem.*

Example 8.4

Consider the floating platform of Example 7.11. Suppose the platform is impacted in such a way that a downward directed impulse of magnitude $\mathcal{I}$ is imparted to the left edge of the platform. Determine the motion of the platform.

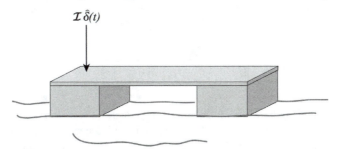

$$\mathcal{I}\,\hat{\delta}(t)$$

Solution

For this case, we chose the vertical deflection of the center, $y_G(t)$ (positive downward), and the rotation of the float, $\theta(t)$, as our generalized coordinates to describe the motion of the system. If we include the force $F_1(t)$ (positive downward) applied at the left edge of the float in the derivation of the governing equations in Example 7.11 we obtain the equations of motion

$$m\,\ddot{y}_G + 2k\,y_G = F_1(t)$$
$$mL^2\ddot{\theta} + 2k\,L^2\theta = L\,F_1(t) \tag{a}$$

which may be stated in matrix form as

$$\begin{bmatrix} m & 0 \\ 0 & mL^2 \end{bmatrix}\begin{Bmatrix} \ddot{y}_G \\ \ddot{\theta} \end{Bmatrix} + \begin{bmatrix} 2k & 0 \\ 0 & 2kL^2 \end{bmatrix}\begin{Bmatrix} y_G \\ \theta \end{Bmatrix} = \begin{Bmatrix} F_1(t) \\ LF_1(t) \end{Bmatrix} \tag{a'}$$

where, for the given loading,

$$F_1(t) = \mathcal{I}\,\delta(t) \tag{b}$$

It is easily shown from the above equations (as was done in Example 7.11) that the natural frequencies and modes for the system are

$$\omega_1 = \omega_2 = \sqrt{2k/m} \tag{c}$$

and

$$\mathbf{U}^{(1)} = \begin{Bmatrix} 1 \\ 0 \end{Bmatrix}, \quad \mathbf{U}^{(2)} = \begin{Bmatrix} 0 \\ 1 \end{Bmatrix} \tag{d}$$

Now, it is seen that the governing equations (a) or (a') are uncoupled. This allows us to solve for each coordinate separately. In this regard, the solution of each equation follows directly from Eq. (4.22). The response of the impacted platform is thus

$$y_G(t) = \frac{I}{\sqrt{2km}} \sin\left(\sqrt{\tfrac{2k}{m}}\, t\right) \mathcal{H}(t)$$

$$\theta(t) = \frac{I}{L\sqrt{2km}} \sin\left(\sqrt{\tfrac{2k}{m}}\, t\right) \mathcal{H}(t)$$

(e)

For this particular problem the governing equations were very easy to solve because they were uncoupled. They were uncoupled because of the particular set of coordinates that we chose to describe the motion of the platform. Let's examine this a bit more closely.

Let us form the modal matrix **B** from the modal vectors stated in Eq. (d). Hence,

$$\mathbf{B} = \begin{bmatrix} 1 & 0 \\ 0 & 1 \end{bmatrix}$$

(f)

That is, the modal matrix is simply the identity matrix. Let us next apply the transformation defined by Eq. (8.24) to the chosen coordinates. Doing this we see that

$$\begin{Bmatrix} y_G(t) \\ \theta(t) \end{Bmatrix} = \begin{bmatrix} 1 & 0 \\ 0 & 1 \end{bmatrix} \begin{Bmatrix} \eta_1(t) \\ \eta_2(t) \end{Bmatrix} \quad \Rightarrow \quad \begin{matrix} y_G(t) = \eta_1(t) \\ \theta(t) = \eta_2(t) \end{matrix}$$

(g)

It is seen that the chosen set of coordinates $y_G(t)$ and $\theta(t)$ are, in fact, the modal (principal) coordinates for the system. We could have deduced this at the outset since the governing equations are completely decoupled. In this example we formulated the problem in terms of modal coordinates by "chance." A natural choice of displacement measures happened to correspond to modal coordinates. This is not usually the case. In general, the equations of motion for multi-degree of freedom systems are coupled and it is generally desirable to transform to the modal coordinates for a given system. The modal coordinates are not typically obvious or necessarily physical displacements. A rational procedure for transforming a given problem to modal coordinates is presented in the next section.

In general, a mapping of the form of Eq. (8.24) corresponds to a transformation from modal coordinates to physical coordinates. The transformation to modal coordinates and its inverse is central to the solution and fundamental understanding of forced vibration problems. We discuss this process and its ramifications in detail in the next section.

8.3 GENERAL MOTION IN TERMS OF THE NATURAL MODES

When the N vectors that comprise a set in some N-dimensional space are linearly independent and mutually orthogonal then any vector in that space can be expressed as a linear combination of the vectors of that set. This is, in fact, what is done when we express a three-dimensional vector in terms of the unit vectors aligned with the coordinate axes. From a vibrations perspective, this means that if the modal vectors for a given system are linearly independent then the response of the system under any loading (the solution for any problem) can be expressed as a linear combination of the modal vectors for that system. This property of the modal vectors will be central to the analysis of forced vibration problems. In this section we show that the modal vectors for an undamped system are indeed linearly independent and, therefore, that any displacement array is comprised of a linear combination of the natural modes of the system.

8.3.1 Linear Independence of the Set of Modal Vectors

The vectors of a given set are said to be *linearly independent* if no vector of that set can be expressed as a linear combination of the other vectors of that set. Equivalently, the vectors of a given set, say the set of modal vectors $\{\mathbf{U}^{(1)}, \mathbf{U}^{(2)}, \ldots, \mathbf{U}^{(N)}\}$ for an undamped system, are linearly independent if the equation

$$\sum_{l=1}^{N} a_l \mathbf{U}^{(l)} = \mathbf{0} \tag{8.25}$$

can only be satisfied if all of the scalar coefficients a_l $(l = 1, 2, \ldots, N)$ vanish. That is, if $\forall a_l = 0$ $(l = 1, 2, \ldots, N)$. We must show that this is the case for a set of modal vectors. To do this, let us first premultiply each term of Eq. (8.25) by $\mathbf{U}^{(j)\mathsf{T}}\mathbf{m}$. This gives

$$\sum_{l=1}^{N} a_l \mathbf{U}^{(j)\mathsf{T}}\mathbf{m}\mathbf{U}^{(l)} = 0 \tag{8.26}$$

Recall from Section 7.3.2 that the modal vectors corresponding to distinct natural frequencies of an undamped system are mutually orthogonal with respect to the mass matrix. Hence,

$$\mathbf{U}^{(j)\mathsf{T}}\mathbf{m}\mathbf{U}^{(l)} = 0 \quad \text{for all } l \neq j \tag{8.27}$$

Substitution of Eq. (8.27) into Eq. (8.26) results in the simple statement

$$a_j \mathbf{U}^{(j)\mathrm{T}} \mathbf{m} \mathbf{U}^{(j)} = 0 \tag{8.28}$$

Now,

$$\mathbf{U}^{(j)\mathrm{T}} \mathbf{m} \mathbf{U}^{(j)} = \left\langle \mathbf{U}^{(j)}, \mathbf{U}^{(j)} \right\rangle_{\mathbf{m}} = \left\| \mathbf{U}^{(j)} \right\|_{\mathbf{m}}^2 \neq 0$$

It follows from Eq. (8.28) that $a_j = 0$. Since Eqs. (8.26) – (8.28) hold for each and every mode $\mathbf{U}^{(j)}$ ($j = 1, 2, \ldots, N$) then

$$a_j = 0 \text{ for all } j = 1, 2, \ldots, N \tag{8.29}$$

The modal vectors are therefore linearly independent.

8.3.2 Modal Expansion

It was shown in Section 8.3.1 that the mutually orthogonal modal vectors of an un-damped system are linearly independent. This suggests that the modal vectors form a basis in their vector space in much the same way as the standard unit vectors along a set of coordinate axes do in three-dimensional physical space. It follows that, just as a physical vector can be expressed as a linear combination of the basis vectors in physi-cal 3-D space, the $N \times 1$ displacement vector (matrix of displacements) for a given system may be expressed as a linear combination of its modal vectors in the corre-sponding N-dimensional space. The response of an undamped system may therefore be expressed in the general form

$$\mathbf{u}(t) = \sum_{j=1}^{N} \mathbf{U}^{(j)} \eta_j(t) \tag{8.30}$$

where $\mathbf{u}$ is the array of displacements, $\mathbf{U}^{(j)}$ is the jth modal vector and η_j is a corre-sponding time dependent coefficient. Equation (8.30) basically tells us that the mo-tion of an undamped system, whether free or forced, is comprised of some combina-tion of the natural modes. If we expand Eq. (8.30) and regroup terms we see that the series representation can be written in matrix form as

$$\mathbf{u}(t) = \mathbf{U}^{(1)} \eta_1(t) + \mathbf{U}^{(2)} \eta_2(t) + \ldots + \mathbf{U}^{(N)} \eta_N(y) = \begin{bmatrix} \mathbf{U}^{(1)} & \mathbf{U}^{(2)} & \cdots & \mathbf{U}^{(N)} \end{bmatrix} \begin{Bmatrix} \eta_1(t) \\ \eta_2(t) \\ \vdots \\ \eta_N(t) \end{Bmatrix}$$

The square matrix in the above representation is seen to be the modal matrix defined by Eq. (8.21). The modal expansion, Eq. (8.30), can thus be written in the equivalent form

$$\mathbf{u}(t) = \mathbf{B}\,\boldsymbol{\eta}(t) \tag{8.31}$$

where

$$\boldsymbol{\eta}(t) = \begin{Bmatrix} \eta_1(t) \\ \eta_2(t) \\ \vdots \\ \eta_N(t) \end{Bmatrix} \tag{8.32}$$

It may be seen that Eq. (8.31) is identical to Eq. (8.24). It follows that the coefficients of the modal expansion correspond to the modal coordinates of the system. The elements of $\boldsymbol{\eta}$ are then the *modal displacements*. With Eq. (8.30), and equivalently Eq. (8.31), established we now proceed to the problem of forced vibrations of discrete systems.

8.4 DECOMPOSITION OF THE FORCED VIBRATION PROBLEM

In this section, the forced vibration problem for an a general undamped N-degree of freedom system expressed in terms of the physical displacements of the mass elements is transformed to its statement in terms of modal coordinates. We will see that, when this is done, the governing system of equations decouples into a system of N uncoupled equivalent single degree of freedom systems.

Consider an undamped N-degree of freedom system. It was seen in Chapter 6 that the equation of motion for any system of this class is of the general form

$$\mathbf{m}\ddot{\mathbf{u}} + \mathbf{k}\mathbf{u} = \mathbf{F} \tag{8.33}$$

where

$$\mathbf{u} = \mathbf{u}(t) = \begin{Bmatrix} u_1(t) \\ u_2(t) \\ \vdots \\ u_N(t) \end{Bmatrix} \tag{8.34}$$

is the displacement matrix,

$$\mathbf{F} = \mathbf{F}(t) = \begin{Bmatrix} F_1(t) \\ F_2(t) \\ \vdots \\ F_N(t) \end{Bmatrix} \tag{8.35}$$

is the force matrix,

$$
\mathbf{m} =
\begin{bmatrix}
m_{11} & m_{12} & \cdots & m_{1N} \\
m_{21} & m_{22} & \cdots & m_{2N} \\
\vdots & \vdots & \ddots & \vdots \\
m_{N1} & m_{N2} & \cdots & m_{NN}
\end{bmatrix}
= \mathbf{m}^{\mathsf{T}}
\tag{8.36}
$$

is the mass matrix, and

$$
\mathbf{k} =
\begin{bmatrix}
k_{11} & k_{12} & \cdots & k_{1N} \\
k_{21} & k_{22} & \cdots & k_{2N} \\
\vdots & \vdots & \ddots & \vdots \\
k_{N1} & k_{N2} & \cdots & k_{NN}
\end{bmatrix}
= \mathbf{k}^{\mathsf{T}}
\tag{8.37}
$$

is the stiffness matrix. In general, we wish to determine the response, $\mathbf{u}(t)$, due to a given system of generalized forces $\mathbf{F}(t)$. In order to do this we shall take advantage of the properties of the modes associated with the system as discussed in Sections 7.3, 8.2 and 8.3. Using these properties, we will transform the governing equation so that it is expressed in terms of the modal (principal) coordinates. In doing so, we shall decouple the individual equations from one another, effectively isolating a set of un-coupled single degree of freedom systems each of which corresponds to an individual mode. The response of these effective 1 d.o.f. systems will be seen to correspond to the response of the individual modes. Once the responses of these effective 1 d.o.f. systems are determined, we can transform back to the original coordinates and obtain the displacements of the system as a function of time.

Recall from our discussions in Section 8.1.3 that the modal expansion de-scribed by Eq. (8.31) is equivalent to a transformation between the displacements described by the physical coordinates of the system and the displacements expressed in terms of the modal coordinates of the system. It will be seen that writing the forced vibration problem in terms of the modal coordinates greatly simplifies the analysis and offers insight into the nature of the response. With this in mind, let us substitute Eq. (8.31) into Eq. (8.33), and then multiply the resulting equation on the left by $\mathbf{B}^{\mathsf{T}}$. This gives the equation

$$
\mathbf{B}^{\mathsf{T}}\mathbf{m}\mathbf{B}\ddot{\boldsymbol{\eta}} + \mathbf{B}^{\mathsf{T}}\mathbf{k}\mathbf{B}\boldsymbol{\eta} = \mathbf{B}^{\mathsf{T}}\mathbf{F}
$$

which may be written in the familiar form

$$
\tilde{\mathbf{m}}\ddot{\boldsymbol{\eta}} + \tilde{\mathbf{k}}\boldsymbol{\eta} = \tilde{\mathbf{F}}
\tag{8.38}
$$

where

$$\tilde{\mathbf{m}} \equiv \mathbf{B}^{\mathrm{T}} \mathbf{m} \mathbf{B} \tag{8.39}$$

$$\tilde{\mathbf{k}} \equiv \mathbf{B}^{\mathrm{T}} \mathbf{k} \mathbf{B} \tag{8.40}$$

and

$$\tilde{\mathbf{F}} \equiv \mathbf{B}^{\mathrm{T}} \mathbf{F} \tag{8.41}$$

Equation (8.38) is the equation of motion expressed in terms of modal coordinates. The matrices $\tilde{\mathbf{m}}$ and $\tilde{\mathbf{k}}$ shall be referred to as the associated (modal) mass and (modal) stiffness matrices, respectively, and $\tilde{\mathbf{F}}$ as the corresponding (modal) force matrix. We next examine their form and their implications.

Consider the transformed mass matrix defined by Eq. (8.39) and let us express it in expanded form by substituting Eqs. (8.21), (8.22) and (8.36). Hence,

$$\tilde{\mathbf{m}} = \begin{bmatrix} \mathbf{U}^{(1)\mathrm{T}} \\ \vdots \\ \mathbf{U}^{(N)\mathrm{T}} \end{bmatrix} \begin{bmatrix} m_{11} & \cdots & m_{1N} \\ \vdots & \ddots & \vdots \\ m_{N1} & \cdots & m_{NN} \end{bmatrix} \begin{bmatrix} \mathbf{U}^{(1)} & \cdots & \mathbf{U}^{(N)} \end{bmatrix}$$

which, after multiplying through, gives

$$\tilde{\mathbf{m}} = \begin{bmatrix} \mathbf{U}^{(1)\mathrm{T}} \mathbf{m} \mathbf{U}^{(1)} & \mathbf{U}^{(1)\mathrm{T}} \mathbf{m} \mathbf{U}^{(2)} & \cdots & \mathbf{U}^{(1)\mathrm{T}} \mathbf{m} \mathbf{U}^{(N)} \\ \mathbf{U}^{(2)\mathrm{T}} \mathbf{m} \mathbf{U}^{(1)} & \mathbf{U}^{(2)\mathrm{T}} \mathbf{m} \mathbf{U}^{(2)} & \cdots & \mathbf{U}^{(2)\mathrm{T}} \mathbf{m} \mathbf{U}^{(N)} \\ \vdots & \vdots & \ddots & \vdots \\ \mathbf{U}^{(N)\mathrm{T}} \mathbf{m} \mathbf{U}^{(1)} & \mathbf{U}^{(N)\mathrm{T}} \mathbf{m} \mathbf{U}^{(2)} & \cdots & \mathbf{U}^{(N)\mathrm{T}} \mathbf{m} \mathbf{U}^{(N)} \end{bmatrix} \tag{8.42}$$

Let us next recall that the modal vectors are mutually orthogonal with respect to the mass and stiffness matrices. (See Section 7.4.2.) That is,

$$\mathbf{U}^{(l)\mathrm{T}} \mathbf{m} \mathbf{U}^{(j)} = 0$$

$$\omega_l^2 \neq \omega_j^2 \quad (l, j = 1, 2, ..., N)$$

$$\mathbf{U}^{(l)\mathrm{T}} \mathbf{k} \mathbf{U}^{(j)} = 0$$

Upon incorporating these properties into Eq. (8.42), it may be seen that the transformed mass matrix takes the diagonal form

$$\tilde{\mathbf{m}} = \begin{bmatrix} \tilde{m}_1 & 0 & \cdots & 0 \\ 0 & \tilde{m}_2 & \cdots & 0 \\ \vdots & \vdots & \ddots & \vdots \\ 0 & 0 & \cdots & \tilde{m}_N \end{bmatrix} \tag{8.43}$$

where

$$\tilde{m}_j = \mathbf{U}^{(j)\mathrm{T}}\mathbf{m}\mathbf{U}^{(j)} \quad (j = 1, 2, ..., N) \tag{8.44}$$

The modal mass matrix is seen to be diagonal due to the mutual orthogonality of the modal vectors as discussed in Section 7.3. It is useful to note that if we choose to normalize the modal vectors with respect to $\mathbf{m}$, as discussed in Section 7.3.3, then all of the nonvanishing elements of the transformed mass matrix are unity. That is, if we normalize the modal vectors such that

$$\left\langle \mathbf{U}^{(j)}, \mathbf{U}^{(j)} \right\rangle_{\mathbf{m}} \equiv \mathbf{U}^{(j)\mathrm{T}}\mathbf{m}\mathbf{U}^{(j)} = 1 \quad (j = 1, 2, ...N)$$

then

$$\tilde{\mathbf{m}} = \begin{bmatrix} 1 & \cdots & 0 \\ \vdots & \ddots & \vdots \\ 0 & \cdots & 1 \end{bmatrix} = \mathbf{I}_{N \times N}$$

Consider next the transformed stiffness matrix defined by Eq. (8.40). Paralleling the discussion for the mass matrix, we find that

$$\tilde{\mathbf{k}} = \begin{bmatrix} \mathbf{U}^{(1)\mathrm{T}} \\ \vdots \\ \mathbf{U}^{(N)\mathrm{T}} \end{bmatrix} \begin{bmatrix} k_{11} & \cdots & k_{1N} \\ \vdots & \ddots & \vdots \\ k_{N1} & \cdots & k_{NN} \end{bmatrix} \begin{bmatrix} \mathbf{U}^{(1)} & \cdots & \mathbf{U}^{(N)} \end{bmatrix} \tag{8.45}$$

Performing the indicated multiplications and incorporating the mutual orthogonality of the modal vectors gives the modal stiffness matrix as

$$\tilde{\mathbf{k}} = \begin{bmatrix} \tilde{k}_1 & \cdots & 0 \\ \vdots & \ddots & \vdots \\ 0 & \cdots & \tilde{k}_N \end{bmatrix} \tag{8.46}$$

where

$$\tilde{k}_j = \mathbf{U}^{(j)\mathrm{T}}\mathbf{k}\mathbf{U}^{(j)} \quad (j = 1, 2, ..., N) \tag{8.47}$$

The transformed stiffness matrix is thus seen to be diagonal as well. Further, if we choose to normalize the modal vectors such that

$$\left\langle \mathbf{U}^{(i)}, \mathbf{U}^{(i)} \right\rangle_{\mathbf{k}} \equiv \mathbf{U}^{(i)\mathrm{T}}\mathbf{k}\mathbf{U}^{(i)} = 1 \quad (i = 1, 2, ..., N)$$

(i.e., with respect to $\mathbf{k}$ rather than with respect to $\mathbf{m}$), then

$$\mathbf{\bar{k}} = \begin{bmatrix} 1 & \cdots & 0 \\ \vdots & \ddots & \vdots \\ 0 & \cdots & 1 \end{bmatrix} = \mathbf{I}_{N \times N}$$

The choice of normalization is at the discretion of the analyst. Each form may offer certain advantages when performing modal analysis for a particular system.

Now that we have established that the modal mass and modal stiffness matrices are of diagonal form we return to our discussion of the transformed equation of motion, Eq. (8.38). Substituting Eqs. (8.42) and (8.45) into Eq. (8.38), gives

$$\begin{bmatrix} \tilde{m}_1 & \cdots & 0 \\ \vdots & \ddots & \vdots \\ 0 & \cdots & \tilde{m}_N \end{bmatrix} \begin{Bmatrix} \ddot{\eta}_1 \\ \vdots \\ \ddot{\eta}_N \end{Bmatrix} + \begin{bmatrix} \tilde{k}_1 & \cdots & 0 \\ \vdots & \ddots & \vdots \\ 0 & \cdots & \tilde{k}_N \end{bmatrix} \begin{Bmatrix} \eta_1 \\ \vdots \\ \eta_N \end{Bmatrix} = \begin{Bmatrix} \tilde{F}_1 \\ \vdots \\ \tilde{F}_N \end{Bmatrix}$$

which, when expanded, results in the (uncoupled) system of equations of the form

$$\begin{aligned} \tilde{m}_1 \ddot{\eta}_1 + \tilde{k}_1 \eta_1 &= \tilde{F}_1(t) \\ \tilde{m}_2 \ddot{\eta}_2 + \tilde{k}_2 \eta_2 &= \tilde{F}_2(t) \\ &\vdots \\ \tilde{m}_N \ddot{\eta}_N + \tilde{k}_N \eta_N &= \tilde{F}_N(t) \end{aligned}$$

(8.48)

Recall that each frequency-mode pair satisfies Eq. (7.3). Therefore, for the jth mode,

$$\mathbf{k}\mathbf{U}^{(j)} = \omega_j^2 \mathbf{m}\mathbf{U}^{(j)}$$

It follows that

$$\mathbf{U}^{(j)\mathsf{T}}\mathbf{k}\mathbf{U}^{(j)} = \omega_j^2 \mathbf{U}^{(j)\mathsf{T}}\mathbf{m}\mathbf{U}^{(j)}$$

Incorporating the definitions of Eqs. (8.44) and (8.47) into the above identity gives the relation,

$$\tilde{k}_j = \omega_j^2 \tilde{m}_j \quad (j = 1, 2, ..., N)$$

(8.49)

Hence,

$$\mathbf{\tilde{k}} = \mathbf{\Lambda}\mathbf{\tilde{m}}$$

(8.50)

where

$$\mathbf{\Lambda} = \begin{bmatrix} \omega_1^2 & \cdots & 0 \\ \vdots & \ddots & \vdots \\ 0 & \cdots & \omega_N^2 \end{bmatrix}$$

(8.51)

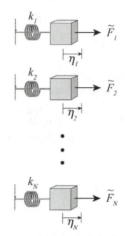

Figure 8.7 Equivalent system of uncoupled single degree of freedom systems.

With the aid of Eq. (8.49), the transformed system of equations (8.48) may be expressed in the "standard form"

$$\ddot{\eta}_1 + \omega_1^2 \eta_1 = \omega_1^2 \tilde{f}_1$$
$$\ddot{\eta}_2 + \omega_2^2 \eta_2 = \omega_2^2 \tilde{f}_2$$
$$\vdots \qquad \vdots \qquad \vdots$$
$$\ddot{\eta}_N + \omega_N^2 \eta_1 = \omega_N^2 \tilde{f}_N$$

(8.52)

where

$$\tilde{f}_j(t) = \frac{\tilde{F}_j(t)}{\tilde{k}_j} = \frac{\tilde{F}_j(t)}{\omega_j^2 \tilde{m}_j} \quad (j = 1, 2, ..., N)$$

(8.53)

The above system may also be written in the matrix form

$$\ddot{\boldsymbol{\eta}} + \boldsymbol{\Lambda}\boldsymbol{\eta} = \boldsymbol{\Lambda}\tilde{\mathbf{f}}$$

(8.54)

where

$$\tilde{\mathbf{f}} = \tilde{\mathbf{k}}^{-1}\tilde{\mathbf{F}} = \left[\boldsymbol{\Lambda}\tilde{\mathbf{m}}\right]^{-1}\tilde{\mathbf{F}} = \begin{Bmatrix} \tilde{f}_1(t) \\ \tilde{f}_2(t) \\ \vdots \\ \tilde{f}_N(t) \end{Bmatrix}$$

(8.55)

It may be seen that Eqs. (8.48), and equivalently Eqs. (8.52), are completely decoupled. Each individual equation therefore corresponds to the response of each individual mode acting as an independent single degree of freedom system of mass $\tilde{m}_j$ and stiffness $\tilde{k}_j$ subjected to the force $\tilde{F}_j(t)$. A physical representation of this equivalence is depicted in Figure 8.8. Since the quantities η_j $(j = 1, 2, ..., N)$ are seen to correspond to the displacements of the modal masses, the corresponding forces may be interpreted as the portion of the external forces that is distributed to the particular mode indicated. This occurs since each modal vector is aligned with its corresponding coordinate (as in Figure 8.5b) when the problem is expressed in terms of the modal coordinates. Each of Eqs. (8.48) may be solved using the techniques established in Chapters 3, 4 and 5 for single degree of freedom systems. Once the response for each mode is determined, these values comprise the matrix $\eta(t)$, which may be substituted back into Eq. (8.31) to give the desired physical response.

Example 8.5

Consider the double pendulum of Example 7.5. (*a*) Compute the modal mass matrix and modal stiffness matrix for the system. (*b*) Use Eq. (8.49) to compute the corresponding natural frequencies and compare them with those computed in Example 7.5.

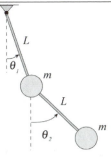

Solution

From Examples 6.5 and 7.5, the mass, stiffness and displacement matrices for the double pendulum are

$$\mathbf{m} = mL^2 \begin{bmatrix} 2 & 1 \\ 1 & 1 \end{bmatrix}, \quad \mathbf{k} = mL^2 \frac{g}{L} \begin{bmatrix} 2 & 0 \\ 0 & 1 \end{bmatrix}, \quad \mathbf{u}(t) = \begin{Bmatrix} \theta_1(t) \\ \theta_2(t) \end{Bmatrix} \tag{a}$$

The corresponding modal vectors were calculated to be

$$\mathbf{U}^{(1)} = \begin{Bmatrix} 1 \\ \sqrt{2} \end{Bmatrix}, \quad \mathbf{U}^{(2)} = \begin{Bmatrix} 1 \\ -\sqrt{2} \end{Bmatrix} \tag{b}$$

(*a*)
Substituting the modal vectors into Eq. (8.21), with $N = 2$, gives the modal matrix for the pendulum as

$$\mathbf{B} = \begin{bmatrix} 1 & 1 \\ \sqrt{2} & -\sqrt{2} \end{bmatrix} \tag{c}$$

We next compute the transformed mass and stiffness and matrices for the double pendulum by incorporating the pertinent system matrices stated in Eq. (a) and the modal matrix (c) into Eqs. (8.39) and (8.40), respectively. We thus find the modal mass and stiffness matrices

$$\tilde{\mathbf{m}} = \begin{bmatrix} 1 & \sqrt{2} \\ 1 & -\sqrt{2} \end{bmatrix} mL^2 \begin{bmatrix} 2 & 1 \\ 1 & 1 \end{bmatrix} \begin{bmatrix} 1 & 1 \\ \sqrt{2} & -\sqrt{2} \end{bmatrix} = 2mL^2 \begin{bmatrix} 2+\sqrt{2} & 0 \\ 0 & 2-\sqrt{2} \end{bmatrix} \quad \text{(d)}$$

$$\tilde{\mathbf{k}} = \begin{bmatrix} 1 & \sqrt{2} \\ 1 & -\sqrt{2} \end{bmatrix} mL^2 \frac{g}{L} \begin{bmatrix} 2 & 0 \\ 0 & 1 \end{bmatrix} \begin{bmatrix} 1 & 1 \\ \sqrt{2} & -\sqrt{2} \end{bmatrix} = 4mL^2 \frac{g}{L} \begin{bmatrix} 1 & 0 \\ 0 & 1 \end{bmatrix} \quad \text{(e)}$$

(b)
We see that both $\tilde{\mathbf{m}}$ and $\tilde{\mathbf{k}}$ are diagonal as they should be. The natural frequencies of the uncoupled equivalent single degree of freedom systems are then, from Eq. (8.49),

$$\omega_1 = \sqrt{\frac{\tilde{k}_1}{\tilde{m}_1}} = \sqrt{\frac{4mL^2 g/L}{2mL^2 \left(2+\sqrt{2}\right)}} = \sqrt{\frac{2}{2+\sqrt{2}} \frac{g}{L}} = 0.765 \sqrt{\frac{g}{L}} \;\; \sqrt{} \quad \text{(f-1)}$$

$$\omega_2 = \sqrt{\frac{\tilde{k}_2}{\tilde{m}_2}} = \sqrt{\frac{4mL^2 g/L}{2mL^2 \left(2-\sqrt{2}\right)}} = \sqrt{\frac{2}{2-\sqrt{2}} \frac{g}{L}} = 1.85 \sqrt{\frac{g}{L}} \;\; \sqrt{} \quad \text{(f-2)}$$

It may be seen that the frequencies computed above are identical with those computed in Example 7.5, as they should be.

8.5 SOLUTION OF FORCED VIBRATION PROBLEMS

It was shown in Section 8.4 that the forced vibration problem may be mapped to modal coordinates and stated in the form of Eqs. (8.48), or equivalently Eqs. (8.52) or (8.54). We next establish an algorithm for the solution of such problems based on this transformation. The resulting procedure is referred to as modal analysis. In this regard, we first identify the generalized coordinates to describe the motion of the particular system of interest and then derive the corresponding equations of motion. We next solve the free vibration problem for the system and determine the natural frequencies and the associated natural modes. Once the set of modal vectors is determined we then form the modal matrix and compute its transpose. The transformed mass, stiffness and force matrices may then be determined using Eqs. (8.39)–(8.41). It should be noted that Eq. (8.49) may be substituted for either Eq. (8.39) or Eq. (8.40) to render the computation more efficient. At this point the problem is expressed as a system of uncoupled single degree of freedom systems as defined by Eqs. (8.48), or equivalently by Eqs. (8.52) or (8.54). The associated modal forces correspond to the portions of the external forces distributed to the individual modes.

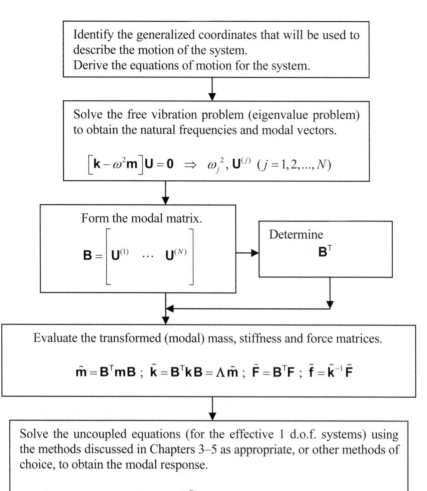

Figure 8.8 Recipe for modal analysis of discrete systems subjected to external forcing.

These equations can be solved for the modal displacements directly, using the methods of Chapters Three and Four, as appropriate. Once the modal displacements have been determined, we may substitute them into Eq. (8.31) to obtain the physical response of the system. The recipe for this procedure is summarized in the flow chart displayed in Figure 8.8.

Example 8.6

Consider the system comprised of two identical masses and three identical springs of Examples 7.1 and 7.2. (*a*) Determine the steady state response of the system if the left mass is subjected to a harmonic force $F_1(t) = F_a \sin \Omega_a t$, as indicated. (*b*) Evaluate the motion of the system (*i*) when $\Omega_a = \sqrt{4k/m}$, (*ii*) when $\Omega_a = \sqrt{3k/m}$ and (*iii*) when $\Omega_a = \sqrt{2k/m}$.

Solution

(*a*)

For the given two degree of freedom system, we have from Example 7.1,

$$\mathbf{m} = \begin{bmatrix} m & 0 \\ 0 & m \end{bmatrix}, \quad \mathbf{k} = \begin{bmatrix} 2k & -k \\ -k & 2k \end{bmatrix}, \quad \mathbf{u}(t) = \begin{Bmatrix} u_1(t) \\ u_2(t) \end{Bmatrix} \tag{a}$$

For the present problem the force array is of the form

$$\mathbf{F}(t) = \begin{Bmatrix} F_1(t) \\ F_2(t) \end{Bmatrix} = \begin{Bmatrix} F_a \sin \Omega_a t \\ 0 \end{Bmatrix} \tag{b}$$

To solve the problem using modal analysis we must first determine the natural frequencies and modal vectors for the system in question. Fortunately we have already determined these quantities in Examples 7.1 and 7.2, respectively. Thus,

$$\mathbf{U}^{(1)} = \begin{Bmatrix} 1 \\ 1 \end{Bmatrix}, \quad \omega_1 = \sqrt{\frac{k}{m}} \tag{c-1, 2}$$

and

$$\mathbf{U}^{(2)} = \begin{Bmatrix} 1 \\ -1 \end{Bmatrix}, \quad \omega_2 = \sqrt{\frac{3k}{m}} \tag{d-1, 2}$$

The modal matrix for the current system is obtained by substituting the above modes into Eq. (8.21) (with $N = 2$). This gives the modal matrix as

$$\mathbf{B} = \begin{bmatrix} 1 & 1 \\ 1 & -1 \end{bmatrix} \tag{e}$$

Note that $\mathbf{B}^{\mathrm{T}} = \mathbf{B}$ for the present case. We next compute the modal force matrix, modal mass and modal stiffness matrices. The modal force matrix is found by substituting Eqs. (b) and (e) into Eq. (8.41). This gives

$$\tilde{\mathbf{F}} = \mathbf{B}^{\mathrm{T}}\mathbf{F} = \begin{bmatrix} 1 & 1 \\ 1 & -1 \end{bmatrix} \begin{Bmatrix} F_a \sin\Omega_a t \\ 0 \end{Bmatrix} = \begin{Bmatrix} F_a \sin\Omega_a t \\ F_a \sin\Omega_a t \end{Bmatrix} \tag{f}$$

It is seen from Eq. (f) that the applied force is distributed equally between the two modes. We next evaluate the modal mass and modal stiffness matrices by substituting Eq. (e) together with the pertinent matrices of Eq. (a) into Eqs. (8.39) and (8.40), respectively. We then get

$$\tilde{\mathbf{m}} = \mathbf{B}^{\mathrm{T}}\mathbf{m}\mathbf{B} = \begin{bmatrix} 1 & 1 \\ 1 & -1 \end{bmatrix} \begin{bmatrix} m & 0 \\ 0 & m \end{bmatrix} \begin{bmatrix} 1 & 1 \\ 1 & -1 \end{bmatrix} = \begin{bmatrix} 2m & 0 \\ 0 & 2m \end{bmatrix} \tag{g}$$

and

$$\tilde{\mathbf{k}} = \mathbf{B}^{\mathrm{T}}\mathbf{k}\mathbf{B} = \begin{bmatrix} 1 & 1 \\ 1 & -1 \end{bmatrix} \begin{bmatrix} 2k & -k \\ -k & 2k \end{bmatrix} \begin{bmatrix} 1 & 1 \\ 1 & -1 \end{bmatrix} = \begin{bmatrix} 2k & 0 \\ 0 & 6k \end{bmatrix} \tag{h}$$

Alternatively, we can compute the modal stiffness using Eqs. (8.50) and (8.51). Hence,

$$\tilde{\mathbf{k}} = \mathbf{\Lambda}\tilde{\mathbf{m}} = \begin{bmatrix} k/m & 0 \\ 0 & 3k/m \end{bmatrix} \begin{bmatrix} 2m & 0 \\ 0 & 2m \end{bmatrix} = \begin{bmatrix} 2k & 0 \\ 0 & 6k \end{bmatrix} \tag{h'}$$

It is seen that the latter computation of $\tilde{\mathbf{k}}$ is simpler since, for the present system, $\mathbf{m}$ is diagonal. Note that both $\tilde{\mathbf{m}}$ and $\tilde{\mathbf{k}}$ are diagonal, as they should be according to Eqs. (8.43) and (8.46). The modal equations are then obtained by substituting the elements of the matrices of Eqs. (f), (g) and (h) into Eqs. (8.52) and (8.53). Doing this results in the uncoupled pair of equations

$$\ddot{\eta}_1 + \omega_1^2\eta_1 = \omega_1^2 \tilde{f}_{a1} \sin\Omega_a t$$
$$\ddot{\eta}_2 + \omega_2^2\eta_2 = \omega_2^2 \tilde{f}_{a2} \sin\Omega_a t \tag{i-1, 2}$$

where ω_1 and ω_2 are given by Eqs. (c-2) and (d-2), and

$$\tilde{f}_{a1} = \frac{\tilde{F}_1}{k_1} = \frac{F_a}{2k} \quad \text{and} \quad \tilde{f}_{a2} = \frac{\tilde{F}_2}{k_2} = \frac{F_a}{6k} \tag{j-1, 2}$$

It may be seen that Eq. (i-1) corresponds to an effective single degree of freedom system whose mass is $2m$ and whose spring stiffness is $2k$, while Eq. (i-2) represents a single degree of freedom system whose mass is 2m and whose spring stiffness is 6k. In each case the mass is subjected to a force $F_a \sin \Omega_a t$. (See Figure E8.6-2.) We now proceed to obtain the modal responses.

The steady state response of each effective single degree of freedom (modal) system may be obtained by direct application of the solution for harmonic excitation given by Eq. (3.28). We thus have that

$$
\eta_1(t) = \frac{\tilde{f}_{a1}}{1 - \left(\Omega_a/\omega_1\right)^2} \sin \Omega_a t = \frac{F_a/2k}{1 - \left(\Omega_a^2 m/k\right)} \sin \Omega_a t
$$

$$
\eta_2(t) = \frac{\tilde{f}_{a2}}{1 - \left(\Omega_a/\omega_2\right)^2} \sin \Omega_a t = \frac{F_a/6k}{1 - \left(\Omega_a^2 m/3k\right)} \sin \Omega_a t
$$

(k)

The matrix of modal displacements is then

$$
\boldsymbol{\eta}(t) = \frac{F_a}{2k} \sin \Omega_a t \left\{ \begin{array}{c} \dfrac{1}{1 - \left(\Omega_a^2 m/k\right)} \\[3mm] \dfrac{1/3}{1 - \left(\Omega_a^2 m/3k\right)} \end{array} \right\}
$$

(l)

Finally, the steady state response of the two degree of freedom system of interest is obtained by transforming back to physical coordinates using Eq. (8.31). Substituting the modal matrix, Eq. (e), into Eq. (8.31) and carrying through the indicated multiplication gives the relation between the physical displacements and the modal coordinates for the given system as

$$
\left\{ \begin{array}{c} u_1(t) \\ u_2(t) \end{array} \right\} = \begin{bmatrix} 1 & 1 \\ 1 & -1 \end{bmatrix} \left\{ \begin{array}{c} \eta_1(t) \\ \eta_2(t) \end{array} \right\} = \left\{ \begin{array}{c} \eta_1(t) + \eta_2(t) \\ \eta_1(t) - \eta_2(t) \end{array} \right\}
$$

(m)

Inserting the specific values of the modal displacements from Eqs. (k) or (l) gives the explicit form of the response,

$$
u_1(t) = \frac{F_a}{2k} \sin \Omega_a t \left[\frac{1}{1 - \left(\Omega_a^2 m/k\right)} + \frac{1/3}{1 - \left(\Omega_a^2 m/3k\right)} \right]
$$

$$
u_2(t) = \frac{F_a}{2k} \sin \Omega_a t \left[\frac{1}{1 - \left(\Omega_a^2 m/k\right)} - \frac{1/3}{1 - \left(\Omega_a^2 m/3k\right)} \right]
$$

◁ (n)

It is easily verified that the responses given by Eq. (e) of Example 8.1 and by Eq. (n) of the present example are identical as, of course, they should be.

(b-i)

Substituting $\Omega_a = \sqrt{4k/m}$ into Eq. (n) gives the motion of the system for this excitation frequency as

$$u_1(t) = -\frac{2}{3}\frac{F_a}{k}\sin\left(\sqrt{\tfrac{4k}{m}}t\right) = \frac{2}{3}\frac{F_a}{k}\sin\left(\sqrt{\tfrac{4k}{m}}t - \pi\right)$$

$$u_2(t) = \frac{F_a}{3k}\sin\left(\sqrt{\tfrac{4k}{m}}t\right)$$

◁ (o)

The motion of the first mass is seen to be 180° out of phase with the applied force, while the motion of the second mass is seen to be in phase with the excitation.

(b-ii)

For this case $\Omega_a = 3k/m = \omega_2$. This evidently corresponds to a resonance condition and the solution for the second modal displacement given by Eq. (k) is not valid. For this excitation frequency Eq. (i-2) takes the form

$$\ddot{\eta}_2 + \omega_2^2\eta_2 = \omega_2^2\tilde{f}_{a2}\sin\omega_2 t \tag{p}$$

The resonance solution for this case follows directly from Eq. (3.34). Hence,

$$\eta_2(t) = \frac{1}{2}\tilde{f}_{a2}\omega_2 t\sin(\omega_2 t - \pi/2) = -\frac{F_a}{12k}\sqrt{\frac{3k}{m}}\,t\cos\left(\sqrt{\tfrac{3k}{m}}t\right) \tag{q}$$

Substituting Eqs. (k-1) and (q) into Eq. (m) gives the motion of the system for this excitation frequency as

$$u_1(t) = -\frac{F_a}{4k}\left[\sqrt{\frac{k}{3m}}\,t\cos\left(\sqrt{\tfrac{3k}{m}}t\right) + \sin\left(\sqrt{\tfrac{3k}{m}}t\right)\right]$$

$$u_2(t) = \frac{F_a}{4k}\left[\sqrt{\frac{k}{3m}}\,t\cos\left(\sqrt{\tfrac{3k}{m}}t\right) - \sin\left(\sqrt{\tfrac{3k}{m}}t\right)\right]$$

◁ (r)

(b-iii)

Substituting $\Omega_a = \sqrt{2k/m}$ into Eq. (n) gives the motion of the system for this excitation frequency as

$$u_1(t) = 0$$

$$u_2(t) = -\frac{F_a}{k}\sin\left(\sqrt{\tfrac{2k}{m}}t\right) = \frac{F_a}{k}\sin\left(\sqrt{\tfrac{2k}{m}}t - \pi\right)$$

◁ (s)

It is seen that for this excitation frequency the first mass remains stationary and is, in effect, "isolated" from the influence of the applied force. In addition, the motion of the second mass is 180° out of phase with the applied force.

Example 8.7

Consider the two degree of freedom system of Example 8.6 when it subjected to the two forces $F_1(t)$ and $F_2(t)$, as shown. Determine the steady state response of the system if (*a*) $F_1(t) = 0$ and $F_2(t) = F_b \sin \Omega_b t$ and (*b*) $F_1(t) = F_a \sin \Omega_a t$ and $F_2(t) = F_b \sin \Omega_b t$.

Solution
(*a*) *External force applied to right mass.*
Since the system under consideration is the same as the system of Example 8.2, the natural frequencies and modal vectors, and the mass and stiffness matrices are those identified in that example. It also follows that the modal matrix, and hence the modal mass and modal stiffness matrices are those computed in Example 8.2. The problem differs only in the force matrix. For the present problem,

$$\mathbf{F}(t) = \left\{ \begin{array}{c} 0 \\ F_b \sin \Omega t \end{array} \right\} \tag{a}$$

The modal force matrix is then

$$\tilde{\mathbf{F}}(t) = \begin{bmatrix} 1 & 1 \\ 1 & -1 \end{bmatrix} \left\{ \begin{array}{c} 0 \\ F_b \sin \Omega_b t \end{array} \right\} = \left\{ \begin{array}{c} F_b \sin \Omega_b t \\ -F_b \sin \Omega_b t \end{array} \right\} \tag{b}$$

and hence,

$$\tilde{\mathbf{f}}(t) = \left\{ \begin{array}{c} \tilde{F}_1/\tilde{k}_1 \\ \tilde{F}_2/\tilde{k}_2 \end{array} \right\} = \frac{F_b \sin \Omega_b t}{2k} \left\{ \begin{array}{c} 1 \\ 1/3 \end{array} \right\} \tag{c}$$

The equations that govern the modal displacements are then

$$\ddot{\eta}_1 + \omega_1^2 \eta_1 = \omega_1^2 \tilde{f}_{b1} \sin \Omega_b t$$
$$\ddot{\eta}_2 + \omega_2^2 \eta_2 = -\omega_2^2 \tilde{f}_{b2} \sin \Omega_b t \tag{d}$$

where

$$\tilde{f}_{b1} = \frac{\tilde{F}_1}{\tilde{k}_1} = \frac{F_b}{2k} \quad \text{and} \quad \tilde{f}_{b2} = \frac{\tilde{F}_2}{\tilde{k}_2} = \frac{F_b}{6k} \tag{e}$$

The solutions to Eqs. (d) are obtained by direct application of Eq. (3.28). This gives the modal response as

$$\boldsymbol{\eta}(t) = \begin{Bmatrix} \eta_1(t) \\ \eta_2(t) \end{Bmatrix} = \frac{F_b}{2k} \sin \Omega_b t \begin{Bmatrix} \dfrac{1}{1-\left(\Omega_b^2 m/k\right)} \\ \dfrac{-1/3}{1-\left(\Omega_b^2 m/3k\right)} \end{Bmatrix}$$ (f)

The physical displacements are next found by substituting Eq. (f) into Eq. (o) of Example (8.2). Doing this, we obtain the displacements

$$u_1(t) = \frac{F_b}{2k} \sin \Omega_b t \left[\frac{1}{1-\left(\Omega_b^2 \, m/k\right)} - \frac{1/3}{1-\left(\Omega_b^2 \, m/3k\right)} \right]$$

◁ (g)

$$u_2(t) = \frac{F_b}{2k} \sin \Omega_b t \left[\frac{1}{1-\left(\Omega_b^2 \, m/k\right)} + \frac{1/3}{1-\left(\Omega_b^2 \, m/3k\right)} \right]$$

(b) *External forces applied to both masses.*
Since the governing equations are linear, the response to the combined loading may be superposed. Thus, adding Eq. (p) of Example 8.2 and Eq. (g) of the present example gives the response to the combined loading as

$$u_1(t) = \frac{F_a}{2k} \sin \Omega_a t \left[\frac{1}{1-\left(\Omega_a^2 \, m/k\right)} + \frac{1/3}{1-\left(\Omega_a^2 \, m/3k\right)} \right]$$

◁ (h-1)

$$+ \frac{F_b}{2k} \sin \Omega_b t \left[\frac{1}{1-\left(\Omega_b^2 \, m/k\right)} - \frac{1/3}{1-\left(\Omega_b^2 \, m/3k\right)} \right]$$

$$u_2(t) = \frac{F_a}{2k} \sin \Omega_a t \left[\frac{1}{1-\left(\Omega_a^2 \, m/k\right)} - \frac{1/3}{1-\left(\Omega_a^2 \, m/3k\right)} \right]$$

◁ (h-2)

$$+ \frac{F_b}{2k} \sin \Omega_b t \left[\frac{1}{1-\left(\Omega_b^2 \, m/k\right)} + \frac{1/3}{1-\left(\Omega_b^2 \, m/3k\right)} \right]$$

Example 8.8

Consider the three-story building comprised of 12 identical columns, each of length L and bending stiffness EI shown in the figure. The three floors supported by the columns are each of mass m as indicated. Determine the steady state response of the structure if the base is excited harmonically in the form $u_0(t) = h_0 \sin \Omega t$.

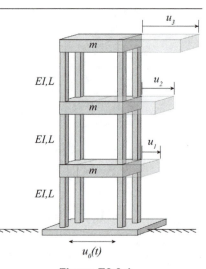

Figure E8.8-1

Solution

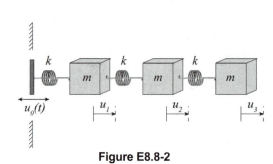

Figure E8.8-2

This is the structure of Example 7.6 subjected to base excitation. As such, the three story building under consideration may be represented as the equivalent three-mass three-spring system shown in Figure 8.8-2, where the stiffness of each equivalent spring is obtained directly from Eq. (1.19) as

$$k = 48EI/L^3 \tag{a}$$

The equivalent discrete system corresponds to the system considered in Example 6.2. Hence, the equations of motion that govern the equivalent system are, from Eq. (c) of Example 6.2,

$$\begin{bmatrix} m & 0 & 0 \\ 0 & m & 0 \\ 0 & 0 & m \end{bmatrix} \begin{Bmatrix} \ddot{u}_1 \\ \ddot{u}_2 \\ \ddot{u}_3 \end{Bmatrix} + \begin{bmatrix} 2k & -k & 0 \\ -k & 2k & -k \\ 0 & -k & k \end{bmatrix} \begin{Bmatrix} u_1 \\ u_2 \\ u_3 \end{Bmatrix} = \begin{Bmatrix} k u_0(t) \\ 0 \\ 0 \end{Bmatrix} = \begin{Bmatrix} k h_0 \sin \Omega t \\ 0 \\ 0 \end{Bmatrix} \tag{b}$$

The governing equations could have also been obtained by direct application of Eqs. (6.2)–(6.7) for $N = 3$, with $m_1 = m_2 = m_3 = m$, $k_1 = k_2 = k_3 = k$ and $k_4 = 0$.

The natural frequencies and corresponding modes for the system under consideration are, from Example 7.6,

$$\omega_1 = 0.445\sqrt{k/m} = 3.08\omega_0 \, , \quad \mathbf{U}^{(1)} = \begin{Bmatrix} 0.328 \\ 0.591 \\ 0.737 \end{Bmatrix} \tag{c}$$

$$\omega_2 = 1.25\sqrt{k/m} = 8.64\omega_0 \, , \quad \mathbf{U}^{(2)} = \begin{Bmatrix} 0.737 \\ 0.328 \\ -0.591 \end{Bmatrix} \tag{d}$$

$$\omega_3 = 1.80\sqrt{k/m} = 12.5\omega_0 \, , \quad \mathbf{U}^{(3)} = \begin{Bmatrix} 0.591 \\ -0.737 \\ 0.328 \end{Bmatrix} \tag{e}$$

where

$$\omega_0 = \sqrt{\frac{EI}{mL^3}} \tag{f}$$

The modal matrix is then

$$\mathbf{B} = \begin{bmatrix} 0.328 & 0.737 & 0.591 \\ 0.591 & 0.328 & -0.737 \\ 0.737 & -0.591 & 0.328 \end{bmatrix} \tag{g}$$

and the modal mass, stiffness and force matrices are computed as follows. The modal mass matrix is computed directly as

$$\tilde{\mathbf{m}} = \mathbf{B}^{\mathsf{T}}\mathbf{m}\mathbf{B} = m \begin{bmatrix} 1 & 0 & 0 \\ 0 & 1 & 0 \\ 0 & 0 & 1 \end{bmatrix} \tag{h}$$

Since the mass matrix $\mathbf{m}$ is diagonal, and the stiffness matrix $\mathbf{k}$ is full (stiffness coupling), the computation of the modal stiffness matrix for this system is simplified somewhat by using Eq. (8.51). Hence,

$$\tilde{\mathbf{k}} = \mathbf{\Lambda}\tilde{\mathbf{m}}$$

$$= \frac{k}{m} \begin{bmatrix} 0.198 & 0 & 0 \\ 0 & 1.56 & 0 \\ 0 & 0 & 3.25 \end{bmatrix} m \begin{bmatrix} 1 & 0 & 0 \\ 0 & 1 & 0 \\ 0 & 0 & 1 \end{bmatrix} = k \begin{bmatrix} 0.198 & 0 & 0 \\ 0 & 1.56 & 0 \\ 0 & 0 & 3.25 \end{bmatrix} \tag{i}$$

The modal force matrix is easily computed as

$$\tilde{\mathbf{F}} = \mathbf{B}^{\mathsf{T}}\mathbf{F} = k h_0 \sin \Omega t \begin{Bmatrix} 0.328 \\ 0.737 \\ 0.591 \end{Bmatrix} \tag{j}$$

Having computed the modal mass, modal stiffness and modal force matrices, the modal equations for the three-story building are then

$$m \ddot{\eta}_1 + 0.198 k \eta = 0.328 k h_0 \sin \Omega t \tag{k-1}$$

$$m \ddot{\eta}_2 + 1.56 k \eta_2 = 0.737 k h_0 \sin \Omega t \tag{k-2}$$

$$m \ddot{\eta}_3 + 3.25 k \eta_3 = 0.591 k h_0 \sin \Omega t \tag{k-3}$$

or, in standard form,

$$\ddot{\eta}_1 + \omega_1^2 \eta_1 = \omega_1^2 \tilde{f}_0^{(1)} \sin \Omega t \; ; \quad \tilde{f}_0^{(1)} = \frac{0.328 k h_0}{0.198 k} = 1.66 h_0 \tag{l-1}$$

$$\ddot{\eta}_2 + \omega_2^2 \eta_2 = \omega_2^2 \tilde{f}_0^{(2)} \sin \Omega t \; ; \quad \tilde{f}_0^{(2)} = \frac{0.737 k h_0}{1.56 k} = 0.472 h_0 \tag{l-2}$$

$$\ddot{\eta}_3 + \omega_3^2 \eta_3 = \omega_3^2 \tilde{f}_0^{(1)} \sin \Omega t \; ; \quad \tilde{f}_0^{(3)} = \frac{0.591 k h_0}{3.25 k} = 0.182 h_0 \tag{l-3}$$

The steady state response for each mode [i.e., the solutions to Eqs. (l-1)–(l-3)] can be written directly from Eq. (3.28). Hence, letting $\bar{\Omega}_j \equiv \Omega / \omega_j$ ($j = 1$, 2, 3), we obtain

$$\eta_1(t) = \frac{\tilde{f}_0^{(1)}}{1 - \bar{\Omega}_1^{\,2}} \sin \Omega t = \frac{1.66 h_0}{1 - 0.105 \left(\Omega^2 / \omega_0^2 \right)} \sin \Omega t \tag{m-1}$$

$$\eta_2(t) = \frac{\tilde{f}_0^{(2)}}{1 - \bar{\Omega}_2^{\,2}} \sin \Omega t = \frac{0.472 h_0}{1 - 0.0134 \left(\Omega^2 / \omega_0^2 \right)} \sin \Omega t \tag{m-2}$$

$$\eta_3(t) = \frac{\tilde{f}_0^{(3)}}{1 - \bar{\Omega}_3^{\,2}} \sin \Omega t = \frac{0.182 h_0}{1 - 0.00640 \left(\Omega^2 / \omega_0^2 \right)} \sin \Omega t \tag{m-3}$$

Finally, the response of the structure is obtained by transforming back to physical coordinates using Eq. (8.30). Hence,

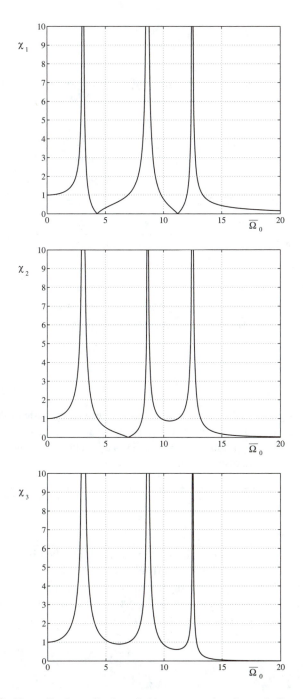

Figure E8.8-3 Normalized amplitudes of side-sway motion for each floor of three-story building.

$$\mathbf{u}(t) = \sum_{j=1}^{3} \mathbf{U}^{(j)} \eta_j(t)$$

which gives the steady state response of the three story building as

$$u_1(t) = \chi_1\left(\bar{\Omega}_0\right) h_0 \sin \Omega t$$

$$= \left(\frac{0.544}{1 - 0.105\bar{\Omega}_0^{\ 2}} + \frac{0.348}{1 - 0.0134\bar{\Omega}_0^{\ 2}} + \frac{0.106}{1 - 0.00640\bar{\Omega}_0^{\ 2}} \right) h_0 \sin \Omega t \qquad \lhd \text{(n-1)}$$

$$u_2(t) = \chi_2\left(\bar{\Omega}_0\right) h_0 \sin \Omega t$$

$$= \left(\frac{0.981}{1 - 0.105\bar{\Omega}_0^{\ 2}} + \frac{0.155}{1 - 0.0134\bar{\Omega}_0^{\ 2}} - \frac{0.134}{1 - 0.00640\bar{\Omega}_0^{\ 2}} \right) h_0 \sin \Omega t \qquad \lhd \text{(n-2)}$$

$$u_3(t) = \chi_3\left(\bar{\Omega}_0\right) h_0 \sin \Omega t$$

$$= \left(\frac{1.22}{1 - 0.105\bar{\Omega}_0^{\ 2}} - \frac{0.279}{1 - 0.0134\bar{\Omega}_0^{\ 2}} + \frac{0.0597}{1 - 0.00640\bar{\Omega}_0^{\ 2}} \right) h_0 \sin \Omega t \qquad \lhd \text{(n-3)}$$

where

$$\bar{\Omega}_0 = \Omega / \omega_0 \qquad \qquad \text{(o)}$$

and ω_0 is given by Eq. (f). It may be seen that the amplitude of the response of each mass is dependent upon the normalized forcing frequency, $\bar{\Omega}_0$. Plots of the normalized amplitudes χ_j $(j = 1, 2, 3)$ as a function of the normalized excitation frequency are displayed in Figure E8.8-3.

Example 8.9

Two identical railroad cars, each of mass m, are attached by an elastic coupler of effective stiffness k, as shown. The cars are initially at rest when a third car collides with the system and imparts an impact of magnitude $\mathcal{I}$ to the left car, as indicated. Determine the motion of the coupled cars following impact.

Solution

The equation of motion for the system is easily derived, or may be found directly from Eq. (b) of Example 6.1 by setting $m_1 = m_2 = m$, $k_2 = k$, $k_1 = k_3 = 0$, $F_2 = 0$ and $F_1(t) = \mathcal{I}\,\hat{\delta}(t)$. The resulting equation is

$$\begin{bmatrix} m & 0 \\ 0 & m \end{bmatrix} \begin{Bmatrix} \ddot{u}_1(t) \\ \ddot{u}_2(t) \end{Bmatrix} + \begin{bmatrix} k & -k \\ -k & k \end{bmatrix} \begin{Bmatrix} u_1(t) \\ u_2(t) \end{Bmatrix} = \begin{Bmatrix} \mathcal{I}\,\hat{\delta}(t) \\ 0 \end{Bmatrix} \tag{a}$$

The free vibration problem for the system corresponds to that for the unrestrained system of Example 7.9. The natural frequencies and modal vectors were computed therein as

$$\omega_1 = 0, \quad \mathbf{U}^{(1)} = \begin{Bmatrix} 1 \\ 1 \end{Bmatrix} \tag{b}$$

$$\omega_2 = \sqrt{2k/m}, \quad \mathbf{U}^{(2)} = \begin{Bmatrix} 1 \\ -1 \end{Bmatrix} \tag{c}$$

Recall that the first mode corresponds to a rigid body mode. The modal matrix follows directly as

$$\mathbf{B} = \begin{bmatrix} 1 & 1 \\ 1 & -1 \end{bmatrix} = \mathbf{B}^{\mathsf{T}} \tag{d}$$

We next compute the modal mass, modal stiffness and modal forces. Hence,

$$\tilde{\mathbf{m}} = \begin{bmatrix} 1 & 1 \\ 1 & -1 \end{bmatrix} \begin{bmatrix} m & 0 \\ 0 & m \end{bmatrix} \begin{bmatrix} 1 & 1 \\ 1 & -1 \end{bmatrix} = \begin{bmatrix} 2m & 0 \\ 0 & 2m \end{bmatrix} \tag{e}$$

$$\tilde{\mathbf{k}} = \begin{bmatrix} 1 & 1 \\ 1 & -1 \end{bmatrix} \begin{bmatrix} k & -k \\ -k & k \end{bmatrix} \begin{bmatrix} 1 & 1 \\ 1 & -1 \end{bmatrix} = \begin{bmatrix} 0 & 0 \\ 0 & 4k \end{bmatrix} \tag{f}$$

and

$$\tilde{\mathbf{F}}(t) = \begin{bmatrix} 1 & 1 \\ 1 & -1 \end{bmatrix} \begin{Bmatrix} \hat{\mathcal{I}}\,\hat{\delta}(t) \\ 0 \end{Bmatrix} = \hat{\mathcal{I}}\,\hat{\delta}(t) \begin{Bmatrix} 1 \\ 1 \end{Bmatrix} \tag{g}$$

It is seen from Eq. (g) that the applied force excites both modes. Note also, from Eq. (f), that the equivalent 1 d.o.f. system associated with the first mode (i.e., the rigid body mode) possesses vanishing stiffness. The corresponding modal equations are

$$2m\ddot{\eta}_1(t) = \mathcal{I}\,\hat{\delta}(t)$$

$$2m\ddot{\eta}_2(t) + 4k\eta_2(t) = \mathcal{I}\,\hat{\delta}(t) \tag{h}$$

Integrating the first of Eqs. (h) and incorporating Eq. (4.21) gives

$$\eta_1(t) = \frac{\mathcal{I}}{2m}t\,\mathcal{H}(t) \tag{i-1}$$

It is seen that the first modal displacement corresponds to a rectilinear motion at constant velocity. The solution to the second of Eqs. (h) is found by direct application of Eq. (4.21). We thus obtain the second modal displacement as

$$\eta_2(t) = \frac{\mathcal{I}}{\sqrt{8km}}\sin\sqrt{\tfrac{2k}{m}}t\,\,\mathcal{H}(t) \tag{i-2}$$

Mapping back to physical space gives the response of the system as

$$\begin{Bmatrix} u_1(t) \\ u_2(t) \end{Bmatrix} = \begin{bmatrix} 1 & 1 \\ 1 & -1 \end{bmatrix}\begin{Bmatrix} \eta_1(t) \\ \eta_2(t) \end{Bmatrix} = \begin{Bmatrix} \eta_1(t) + \eta_2(t) \\ \eta_1(t) - \eta_2(t) \end{Bmatrix} \tag{j}$$

Substituting the Eqs. (i-1) and (i-2) into Eq. (j) gives the motion of the railroad cars as

$$u_1(t) = \frac{\mathcal{I}}{2m}\left(t + \frac{1}{\sqrt{2k/m}}\sin\sqrt{\tfrac{2k}{m}}t \right)\mathcal{H}(t)$$

$$u_2(t) = \frac{\mathcal{I}}{2m}\left(t - \frac{1}{\sqrt{2k/m}}\sin\sqrt{\tfrac{2k}{m}}t \right)\mathcal{H}(t) \tag{k}$$

The motion of the system is seen to be comprised of a rigid body translation of both cars traveling together at constant speed $v_0 = \mathcal{I}/2m$ combined with the two cars vibrating relative to one another in an "accordion mode" at frequency $\omega = \sqrt{2k/m}$.

Example 8.10

Determine the response of the double pendulum of Examples 6.5 and 7.5 if the bottom mass is subjected to the horizontally directed triangular pulse indicated.

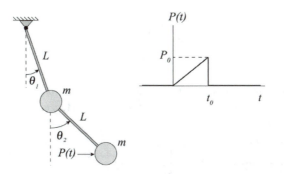

Figure E8.10-1 Double pendulum subjected to triangular pulse.

Solution
From the development in Section 6.1, and from Examples 6.5 and 7.5, the governing equations for the pendulum are

$$\mathbf{m\ddot{u}}+\mathbf{ku}=\mathbf{F}$$

where

$$\mathbf{u}=\mathbf{u}(t)=\begin{Bmatrix} \theta_1(t) \\ \theta_2(t) \end{Bmatrix}$$

and the system matrices for the uniform pendulum are

$$\mathbf{m}=mL^2\begin{bmatrix} 2 & 1 \\ 1 & 1 \end{bmatrix}, \quad \mathbf{k}=mL^2\frac{g}{L}\begin{bmatrix} 2 & 0 \\ 0 & 1 \end{bmatrix} \qquad \text{(a, b)}$$

Further, the natural frequencies and associated modal vectors are, from Example 7.5,

$$\omega_1=0.765\sqrt{g/L}, \quad \mathbf{U}^{(1)}=\begin{Bmatrix} 1 \\ \sqrt{2} \end{Bmatrix} \qquad \text{(c-1, 2)}$$

$$\omega_2=1.85\sqrt{g/L}, \quad \mathbf{U}^{(2)}=\begin{Bmatrix} 1 \\ -\sqrt{2} \end{Bmatrix} \qquad \text{(c-3, 4)}$$

For the present problem, the force matrix is

$$\mathbf{F}(t)=\begin{Bmatrix} (F_1+F_2)L_1 \\ F_2L_2 \end{Bmatrix} \rightarrow \begin{Bmatrix} P(t)L \\ P(t)L \end{Bmatrix}=P(t)L\begin{Bmatrix} 1 \\ 1 \end{Bmatrix} \qquad \text{(d)}$$

It follows from Eqs. (c-2) and (c-4) that the modal matrix for the system is

$$\mathbf{B} = \begin{bmatrix} 1 & 1 \\ \sqrt{2} & -\sqrt{2} \end{bmatrix} \tag{e}$$

The modal mass, modal stiffness and modal force matrices are next obtained using Eqs. (8.39), (8.40) and (8.41), respectively. Hence,

$$\tilde{\mathbf{m}} = \begin{bmatrix} 1 & \sqrt{2} \\ 1 & -\sqrt{2} \end{bmatrix} mL^2 \begin{bmatrix} 2 & 1 \\ 1 & 1 \end{bmatrix} \begin{bmatrix} 1 & 1 \\ \sqrt{2} & -\sqrt{2} \end{bmatrix} = 2mL^2 \begin{bmatrix} 2+\sqrt{2} & 0 \\ 0 & 2-\sqrt{2} \end{bmatrix} \tag{f}$$

$$\tilde{\mathbf{k}} = \begin{bmatrix} 1 & \sqrt{2} \\ 1 & -\sqrt{2} \end{bmatrix} mL^2 \frac{g}{L} \begin{bmatrix} 2 & 0 \\ 0 & 1 \end{bmatrix} \begin{bmatrix} 1 & 1 \\ \sqrt{2} & -\sqrt{2} \end{bmatrix} = 4mL^2 \frac{g}{L} \begin{bmatrix} 1 & 0 \\ 0 & 1 \end{bmatrix} \tag{g}$$

Since the present system has only two degrees of freedom, is was simple enough to compute the modal mass matrix directly. It may be seen, however, that the stiffness matrix for this system is diagonal while the mass matrix is full (inertia coupling). The computation of the modal mass matrix for systems of this type is generally simplified by using Eq. (8.50). Thus, we may also compute the modal mass as follows:

$$\tilde{\mathbf{m}} = \mathbf{\Lambda}^{-1}\tilde{\mathbf{k}} = \frac{L}{g}\begin{bmatrix} 1/(2-\sqrt{2}) & 0 \\ 0 & 1/(2+\sqrt{2}) \end{bmatrix} 4mL^2 \frac{g}{L}\begin{bmatrix} 1 & 0 \\ 0 & 1 \end{bmatrix}$$

$$= 2mL^2 \begin{bmatrix} 2+\sqrt{2} & 0 \\ 0 & 2-\sqrt{2} \end{bmatrix}$$

Finally, the modal force matrix is computed as

$$\tilde{\mathbf{F}}(t) = \begin{bmatrix} 1 & \sqrt{2} \\ 1 & -\sqrt{2} \end{bmatrix} \begin{Bmatrix} P(t)L \\ P(t)L \end{Bmatrix} = P(t)\begin{Bmatrix} \tilde{L}_1 \\ \tilde{L}_2 \end{Bmatrix} \tag{h}$$

where

$$\tilde{L}_1 = L\left(1+\sqrt{2}\right), \quad \tilde{L}_2 = L\left(1-\sqrt{2}\right) \tag{i}$$

With the modal mass, modal stiffness and modal force matrices for the system established, the transformed (modal) equations for the double pendulum take the specific form

$$\tilde{m}_1\ddot{\eta}_1 + \tilde{k}_1\eta_1 = P(t)\tilde{L}_1$$
$$\tilde{m}_2\ddot{\eta}_2 + \tilde{k}_2\eta_2 = P(t)\tilde{L}_2$$

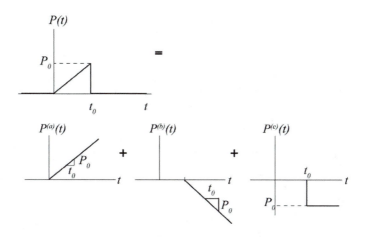

Figure E8.10-2 Decomposition of triangular pulse into the sum of three pulses.

or

$$\ddot{\eta}_1 + \omega_1^2 \eta_1 = \omega_1^2 \tilde{f}_1(t)$$
$$\ddot{\eta}_2 + \omega_2^2 \eta_2 = \omega_2^2 \tilde{f}_2(t)$$

(j)

where

$$\tilde{f}_1(t) = \omega_1^2 P(t) \tilde{L}_1 / \tilde{k}_1$$
$$\tilde{f}_2(t) = \omega_2^2 P(t) \tilde{L}_2 / \tilde{k}_2$$

(k)

Equations (j) may be solved for the modal displacements once we establish an analytical form for the pulse $P(t)$. Toward this end, the given triangular pulse may be constructed as the sum of two ramp loads and a step load as shown in Figure E8.10-2. Mathematically,

$$P(t) = P^{(a)}(t) + P^{(b)}(t) + P^{(c)}(t)$$

(l)

where

$$P^{(a)}(t) = \frac{P_0}{t_0} t \, \mathcal{H}(t)$$

(m-1)

$$P^{(b)}(t) = P_0 \left[1 - (t/t_0) \right] \mathcal{H}(t - t_0)$$

(m-2)

$$P^{(c)}(t) = -P_0 \, \mathcal{H}(t - t_0)$$

(m-3)

The response of each 1 d.o.f. system [the solutions to Eqs. (j)] is then the sum of the responses to the individual pulses (m-1), (m-2) and (m-3). From the discussions of Sections 4.4, 4.5 and 4.6 we thus have that

$$\eta_1(t) = \eta_1^{(a)}(t) + \eta_1^{(b)}(t) + \eta_1^{(c)}(t)$$

$$= \frac{P_0 \tilde{L}_1 / \tilde{k}_1}{t_0} \left[\mathcal{R}_1(t) - \mathcal{R}_1(t - t_0) \right] - \frac{P_0 \tilde{L}_1}{\tilde{k}_1} S_1(t - t_0) \tag{n-1}$$

and

$$\eta_2(t) = \eta_2^{(a)}(t) + \eta_2^{(b)}(t) + \eta_2^{(c)}(t)$$

$$= \frac{P_0 \tilde{L}_2 / \tilde{k}_2}{t_0} \left[\mathcal{R}_2(t) - \mathcal{R}_2(t - t_0) \right] - \frac{P_0 \tilde{L}_2}{\tilde{k}_2} S_2(t - t_0) \tag{n-2}$$

where, from Eqs. (4.32) and (4.39),

$$\mathcal{R}_j(t) = \left[t - \frac{\sin \omega_j t}{\omega_j} \right] \mathcal{H}(t) \quad (j = 1, 2) \tag{o}$$

$$S_j(t - t_0) = \left[1 - \cos \omega_j (t - t_0) \right] \mathcal{H}(t - t_0) \quad (j = 1, 2) \tag{p}$$

The response of the double pendulum to the triangular pulse imparted on the bottom mass is then found by substituting Eqs. (e) and (n) into Eq. (8.31). Doing this, we find that

$$\begin{Bmatrix} \theta_1(t) \\ \theta_2(t) \end{Bmatrix} = \begin{bmatrix} 1 & 1 \\ \sqrt{2} & -\sqrt{2} \end{bmatrix} \begin{Bmatrix} \eta_1(t) \\ \eta_2(t) \end{Bmatrix} = \begin{Bmatrix} (\eta_1 + \eta_2) \\ \sqrt{2}(\eta_1 - \eta_2) \end{Bmatrix} \tag{q}$$

and hence that

$$\theta_1(t) = \frac{P_0}{t_0} \left[t \left(\frac{\tilde{L}_1}{\tilde{k}_1} + \frac{\tilde{L}_2}{\tilde{k}_2} \right) - \frac{\tilde{L}_1}{\tilde{k}_1} \frac{\sin \omega_1 t}{\omega_1} - \frac{\tilde{L}_2}{\tilde{k}_2} \frac{\sin \omega_2 t}{\omega_2} \right] \mathcal{H}(t)$$

$$- \frac{P_0}{t_0} \left[(t - t_0) \left(\frac{\tilde{L}_1}{\tilde{k}_1} + \frac{\tilde{L}_2}{\tilde{k}_2} \right) - \frac{\tilde{L}_1}{\tilde{k}_1} \frac{\sin \omega_1 (t - t_0)}{\omega_1} - \frac{\tilde{L}_2}{\tilde{k}_2} \frac{\sin \omega_2 (t - t_0)}{\omega_2} \right] \mathcal{H}(t - t_0)$$

$$- P_0 \left[\frac{\tilde{L}_1}{\tilde{k}_1} + \frac{\tilde{L}_2}{\tilde{k}_2} - \frac{\tilde{L}_1}{\tilde{k}_1} \cos \omega_1 (t - t_0) - \frac{\tilde{L}_2}{\tilde{k}_2} \cos \omega_2 (t - t_0) \right] \mathcal{H}(t - t_0) \tag{r-1}$$

$$\theta_2(t) = \frac{P_0}{t_0}\left[t\left(\frac{\tilde{L}_1}{\tilde{k}_1} - \frac{\tilde{L}_2}{\tilde{k}_2}\right) - \frac{\tilde{L}_1}{\tilde{k}_1}\frac{\sin\omega_1 t}{\omega_1} + \frac{\tilde{L}_2}{\tilde{k}_2}\frac{\sin\omega_2 t}{\omega_2}\right]\mathcal{H}(t)$$

$$-\frac{P_0}{t_0}\left[(t-t_0)\left(\frac{\tilde{L}_1}{\tilde{k}_1} - \frac{\tilde{L}_2}{\tilde{k}_2}\right) - \frac{\tilde{L}_1}{\tilde{k}_1}\frac{\sin\omega_1(t-t_0)}{\omega_1} + \frac{\tilde{L}_2}{\tilde{k}_2}\frac{\sin\omega_2(t-t_0)}{\omega_2}\right]\mathcal{H}(t-t_0)$$

$$-P_0\left[\frac{\tilde{L}_1}{\tilde{k}_1} - \frac{\tilde{L}_2}{\tilde{k}_2} - \frac{\tilde{L}_1}{\tilde{k}_1}\cos\omega_1(t-t_0) + \frac{\tilde{L}_2}{\tilde{k}_2}\cos\omega_2(t-t_0)\right]\mathcal{H}(t-t_0)$$

$$\text{(r-2)}$$

Finally, upon substituting Eqs. (g) and (i) into Eqs. (r-1) and (r-2), we have

$$\theta_1(t) = \frac{P_0}{2mg}\left[\frac{t}{t_0} - \frac{(1+\sqrt{2})}{2\omega_1 t_0}\sin\omega_1 t - \frac{(1-\sqrt{2})}{2\omega_2 t_0}\sin\omega_2 t\right]\mathcal{H}(t)$$

$$-\frac{P_0}{2mg}\left[\frac{t}{t_0} - \frac{(1+\sqrt{2})}{2}\left\{\frac{\sin\omega_1(t-t_0)}{\omega_1 t_0} + \cos\omega_1(t-t_0)\right\}\right.$$

$$\left.-\frac{(1-\sqrt{2})}{2}\left\{\frac{\sin\omega_2(t-t_0)}{\omega_2 t_0} + \cos\omega_2(t-t_0)\right\}\right]\mathcal{H}(t-t_0)$$

$$\triangleleft\text{(s-1)}$$

$$\theta_2(t) = \frac{P_0}{2mg}\left[\frac{t}{t_0} - \frac{(1+\sqrt{2})}{2\omega_1 t_0}\sin\omega_1 t + \frac{(1-\sqrt{2})}{2\omega_2 t_0}\sin\omega_2 t\right]\mathcal{H}(t)$$

$$-\frac{P_0}{2mg}\left[\frac{t}{t_0} - \frac{(1+\sqrt{2})}{2}\left\{\frac{\sin\omega_1(t-t_0)}{\omega_1 t_0} + \cos\omega_1(t-t_0)\right\}\right.$$

$$\left.+\frac{(1-\sqrt{2})}{2}\left\{\frac{\sin\omega_2(t-t_0)}{\omega_2 t_0} + \cos\omega_2(t-t_0)\right\}\right]\mathcal{H}(t-t_0)$$

$$\triangleleft\text{(s-2)}$$

where ω_1 and ω_2 are given by Eqs. (c-1) and (c-3) respectively.

It may be seen from Eqs. (s-1) and (s-2) that once the pulse subsides (i.e., when $t > t_0$) the linear time dependence in the response cancels. It may be noted that this also occurs for the sine terms in the expression for either θ_1 or θ_2 if the duration of the pulse, t_0, is a multiple of the corresponding natural period. That is, Eq. (s-1) or Eq. (s-2) will simplify to the indicated cosine terms if $t_0 = 2n\pi/\omega_1$ or if $t_0 = 2n\pi/\omega_2$, respectively, where n is any integer.

Example 8.11

The tram of Examples 6.6 and 7.7 is shut down and at rest and, as a consequence, the cable is slack and the controller is off when a wind gust strikes the
structure. Determine the motion of the tram if the effect of the wind gust may
be modeled as a rectangular pulse of magnitude F_0 and duration τ.

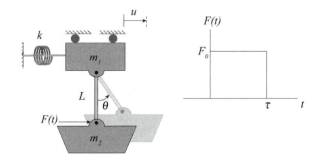

Figure E8.11 Tram subjected to wind load.

Solution
For the specific system in question, the mass of the barrow is twice the mass of
the frame. That is, $m_2 = 2m_1 = 2m$. Further, for the present conditions, $F_1 = 0$, $M = 0$ and

$$F_2 = F(t) = F_0\left[\mathcal{H}(t) - \mathcal{H}(t-\tau)\right] \tag{a}$$

The equation of motion of the system is then, from Example 6.6,

$$\begin{bmatrix} 3m & 2mL \\ 2mL & 2mL^2 \end{bmatrix}\begin{Bmatrix} \ddot{u} \\ \ddot{\theta} \end{Bmatrix} + \begin{bmatrix} k & 0 \\ 0 & 2mgL \end{bmatrix}\begin{Bmatrix} u \\ \theta \end{Bmatrix} = \begin{Bmatrix} F(t) \\ LF(t) \end{Bmatrix} \tag{b}$$

The natural frequencies and modes for this system were computed in Example
7.7 as

$$\omega_1 = 0.518\omega_0, \quad \mathbf{U}^{(1)} = \begin{Bmatrix} 1 \\ 0.366\,\omega_0^2/g \end{Bmatrix} \tag{c-1, 2}$$

$$\omega_2 = 1.93\omega_0, \quad \mathbf{U}^{(2)} = \begin{Bmatrix} 1 \\ -1.37\,\omega_0^2/g \end{Bmatrix} \tag{d-1, 2}$$

where the properties of the particular system under consideration are such that

$$\omega_0^2 = k/m = g/L \tag{e}$$

The corresponding modal matrix is then

$$\mathbf{B} = \begin{bmatrix} 1 & 1 \\ 0.366\,\omega_0^2/g & -1.37\,\omega_0^2/g \end{bmatrix} \tag{f}$$

With the modal matrix established, we next compute the modal stiffness and modal force matrices. Hence,

$$\tilde{\mathbf{k}} = \begin{bmatrix} 1 & 0.366\,\omega_0^2/g \\ 1 & -1.37\,\omega_0^2/g \end{bmatrix} \begin{bmatrix} k & 0 \\ 0 & 2mgL \end{bmatrix} \begin{bmatrix} 1 & 1 \\ 0.366\,\omega_0^2/g & -1.37\,\omega_0^2/g \end{bmatrix}$$

$$\Rightarrow \quad \tilde{\mathbf{k}} = \begin{bmatrix} 1.27k & 0 \\ 0 & 4.76k \end{bmatrix} \tag{g}$$

and

$$\tilde{\mathbf{F}} = \begin{bmatrix} 1 & 0.366\,\omega_0^2/g \\ 1 & -1.37\,\omega_0^2/g \end{bmatrix} \begin{Bmatrix} F(t) \\ L\,F(t) \end{Bmatrix} = \begin{Bmatrix} 1.37F(t) \\ -0.366F(t) \end{Bmatrix} \tag{h}$$

The modal equations are then

$$\ddot{\eta}_1(t) + \omega_1^2 \eta_1(t) = \omega_1^2 \tilde{f}_0^{(1)} \left[\mathcal{H}(t) - \mathcal{H}(t-\tau) \right]$$

$$\ddot{\eta}_2(t) + \omega_2^2 \eta_1(t) = \omega_2^2 \tilde{f}_0^{(2)} \left[\mathcal{H}(t) - \mathcal{H}(t-\tau) \right] \tag{i}$$

where

$$\tilde{f}_0^{(1)} = \tilde{F}_0^{(1)}/\tilde{k}_1 = 1.08\,F_0/k$$

$$\tilde{f}_0^{(2)} = \tilde{F}_0^{(2)}/\tilde{k}_2 = -0.0769\,F_0/k \tag{j}$$

The solutions to equations (i) follow directly from Eq. (4.43). The modal displacements are therefore

$$\eta_1(t) = \tilde{f}_0^{(1)} \left\{ \left[1 - \cos(\omega_1 t - \phi_1) \right] \mathcal{H}(t) - \left[1 - \cos\{\omega_1(t-\tau) - \phi_1\} \right] \mathcal{H}(t-\tau) \right\}$$

$$\eta_2(t) = \tilde{f}_0^{(2)} \left\{ \left[1 - \cos(\omega_2 t - \phi_2) \right] \mathcal{H}(t) - \left[1 - \cos\{\omega_2(t-\tau) - \phi_2\} \right] \mathcal{H}(t-\tau) \right\} \tag{k}$$

Transforming back to physical space gives the desired response. Hence,

$$\begin{Bmatrix} u(t) \\ \theta(t) \end{Bmatrix} = \begin{bmatrix} 1 & 1 \\ 0.366\,\omega_0^2/g & -1.37\,\omega_0^2/g \end{bmatrix} \begin{Bmatrix} \eta_1(t) \\ \eta_2(t) \end{Bmatrix}$$

$$= \begin{Bmatrix} \eta_1(t) + \eta_2(t) \\ (0.366\eta_1 - 1.37\eta_2)\omega_0^2/g \end{Bmatrix} \tag{l}$$

Finally,

$$u(t)/L = \frac{F_0}{mg}\left[\mathcal{H}(t) - \mathcal{H}(t - \tau)\right]$$

$$-\frac{F_0}{mg}\left[1.08\cos(0.518\omega_0 t - \phi_1) - 0.0769\cos(1.93\omega_0 t - \phi_1)\right]\mathcal{H}(t)$$

$$+\frac{F_0}{mg}\left[1.08\cos\{0.518\omega_0(t - \tau) - \phi_1\} - 0.0769\cos\{1.93\omega_0(t - \tau) - \phi_1\}\right]\mathcal{H}(t - \tau)$$

◁ (m-1)

$$\theta(t) = \frac{F_0}{2mg}\left[\mathcal{H}(t) - \mathcal{H}(t - \tau)\right]$$

$$-\frac{F_0}{mg}\left[0.395\cos(0.518\omega_0 t - \phi_1) + 0.105\cos(1.93\omega_0 t - \phi_1)\right]\mathcal{H}(t)$$

$$+\frac{F_0}{mg}\left[0.395\cos\{0.518\omega_0(t - \tau) - \phi_1\} + 0.105\cos\{1.93\omega_0(t - \tau) - \phi_1\}\right]\mathcal{H}(t - \tau)$$

◁ (m-2)

Example 8.12

The motorcycle of Examples 6.7 and 7.8 is traveling at constant speed v_0 on a horizontal road when it encounters a small depression in the road. If the depression is described by the function $y(\xi) = h_0\left[1 - \cos(2\pi\xi/\lambda)\right]$, determine the response of the motorcycle that results from riding through the dip.

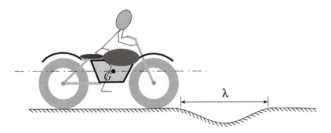

Figure E8.12-1 Motorcycle approaching depression in road.

Solution

The problem is similar to Example 4.7, but now for a three-degree of freedom system with two wheels (Figure E8.12-2). The spatial equation that describes the geometry of the depression is given as

$$y(\xi) = h_0\left[1 - \cos(2\pi\xi/\lambda)\right] \tag{a}$$

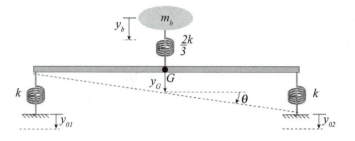

Figure E8.12-2 Equivalent system.

where h_0 is the depth of the depression and λ is the corresponding length as shown. Since the motorcycle moves at constant speed v_0 the problem may be viewed as if the vehicle is fixed and each base undergoes a prescribed vertical motion, where the motion of the second base lags the first by the time increment

$$t_{lag} = v_0/L \tag{b}$$

If the first wheel encounters the depression at $t = 0$ then, while passing through the dip, the horizontal motion of the first wheel is

$$\xi = v_0 t \tag{c}$$

Similarly, while passing through the depression, the horizontal motion of the second wheel is

$$\xi = v_0 t' = v_0(t - t_{lag}) \tag{d}$$

Substituting Eq. (c) into Eq. (a) gives the vertical motion of the front wheel as,

$$y_{01}(t) = h_0\left[1 - \cos\Omega t\right]\left[\mathcal{H}(t) - \mathcal{H}(t - t_\lambda)\right] \tag{e}$$

where

$$\Omega = 2\pi/t_\lambda \tag{f}$$

and

$$t_\lambda = \lambda/v_0 \tag{g}$$

is the time it takes for the wheel to traverse the depression. Likewise, the vertical motion of the second wheel is given by

$$y_{02}(t) = h_0\left[1 - \cos\Omega(t - t_{lag})\right]\left[\mathcal{H}(t - t_{lag}) - \mathcal{H}(t - t_{lag} - t_\lambda)\right] \tag{h}$$

Now, the equations of motion for the particular system under consideration are found by substituting the given system parameters into Eq. (h) of Example 6.7. We then have

$$
\begin{bmatrix} m & 0 & 0 \\ 0 & I_G & 0 \\ 0 & 0 & m/3 \end{bmatrix} \begin{Bmatrix} \ddot{y}_G \\ \ddot{\theta} \\ \ddot{y}_b \end{Bmatrix} + \begin{bmatrix} 8k/3 & 0 & -2k/3 \\ 0 & 2kL^2 & 0 \\ -2k/3 & 0 & 2k/3 \end{bmatrix} \begin{Bmatrix} y_G \\ \theta \\ y_b \end{Bmatrix} = \begin{Bmatrix} k(y_{01} + y_{02}) \\ kL(y_{01} - y_{02})/2 \\ 0 \end{Bmatrix} \quad \text{(i)}
$$

where

$$
I_G = mL^2/12 \tag{j}
$$

We wish to solve Eq. (8) subject to the base motions defined by Eqs. (e) and (g). To do this we first recall the natural frequencies and modal vectors that were computed in Example 7.8. Thus,

$$
\omega_1^2 = 1.132\,k/m, \quad \mathbf{U}^{(1)} = \begin{Bmatrix} 1 \\ 0 \\ 2.303 \end{Bmatrix} \tag{k}
$$

$$
\omega_2^2 = 3.535\,k/m, \quad \mathbf{U}^{(1)} = \begin{Bmatrix} 1 \\ 0 \\ -1.303 \end{Bmatrix} \tag{l}
$$

$$
\omega_3^2 = 6\,k/m, \quad \mathbf{U}^{(1)} = \begin{Bmatrix} 0 \\ 1 \\ 0 \end{Bmatrix} \tag{m}
$$

The corresponding modal matrix is then

$$
\mathbf{B} = \begin{bmatrix} 1 & 1 & 0 \\ 0 & 0 & 1 \\ 2.303 & -1.303 & 0 \end{bmatrix} \tag{n}
$$

We next compute the modal mass, modal stiffness and modal force matrices. Hence,

$$\tilde{m} = \begin{bmatrix} 1 & 0 & 2.303 \\ 1 & 0 & -1.303 \\ 0 & 1 & 0 \end{bmatrix} \begin{bmatrix} m & 0 & 0 \\ 0 & I_G & 0 \\ 0 & 0 & m/3 \end{bmatrix} \begin{bmatrix} 1 & 1 & 0 \\ 0 & 0 & 1 \\ 2.303 & -1.303 & 0 \end{bmatrix}$$

$$= \begin{bmatrix} 2.768m & 0 & 0 \\ 0 & 1.566m & 0 \\ 0 & 0 & mL^2/12 \end{bmatrix} \tag{o}$$

$$\tilde{k} = \Lambda \tilde{m} = \frac{k}{m} \begin{bmatrix} 1.132 & 0 & 0 \\ 0 & 3.535 & 0 \\ 0 & 0 & 6 \end{bmatrix} \begin{bmatrix} 2.768m & 0 & 0 \\ 0 & 1.566m & 0 \\ 0 & 0 & I_G \end{bmatrix}$$

$$= k \begin{bmatrix} 3.133 & 0 & 0 \\ 0 & 5.536 & 0 \\ 0 & 0 & L^2/2 \end{bmatrix} \tag{p}$$

and

$$\tilde{F} = \begin{bmatrix} 1 & 0 & 2.303 \\ 1 & 0 & -1.303 \\ 0 & 1 & 0 \end{bmatrix} \begin{Bmatrix} k(y_{01} + y_{02}) \\ kL(y_{01} - y_{02})/2 \\ 0 \end{Bmatrix} = \begin{Bmatrix} k(y_{01} + y_{02}) \\ k(y_{01} + y_{02}) \\ kL(y_{01} - y_{02})/2 \end{Bmatrix} \tag{q}$$

It is seen from Eq. (q) that all three modes are excited. Substituting the above values into Eq. (8.52) gives the corresponding equations for the modal displacements,

$$\ddot{\eta}_1(t) + \omega_1^2 \eta_1(t) = \omega_1^2 \tilde{f}_1(t) = \omega_1^2 \left[0.3192 \{ y_{01}(t) + y_{02}(t) \} \right] \tag{r-1}$$

$$\ddot{\eta}_2(t) + \omega_2^2 \eta_2(t) = \omega_2^2 \tilde{f}_2(t) = \omega_2^2 \left[0.1806 \{ y_{01}(t) + y_{02}(t) \} \right] \tag{r-2}$$

$$\ddot{\eta}_3(t) + \omega_3^2 \eta_3(t) = \omega_3^2 \tilde{f}_3(t) = \omega_3^2 \left[\{ y_{01}(t) - y_{02}(t) \}/L \right] \tag{r-3}$$

where $y_{01}(t)$ and $y_{02}(t)$ are given by Eqs. (e) and (h), respectively. Equations (r-1), (r-2) and (r-3) can be solved using the methods of Chapter 4. For this particular case, we can simplify our analysis by taking advantage of calculations already performed in a previous example and using superposition. In this regard, the solution to Example 4.7 can be utilized by incorporating the current values of the parameters. Doing this we obtain

$$\eta_1(t) = h^{(1)}\left[1+\left(\Omega/\omega_1\right)^2 \beta_1 \cos\omega_1 t - \beta_1 \cos\Omega t\right]\mathcal{H}(t)$$

$$-h^{(1)}\left[1+\left(\Omega/\omega_1\right)^2 \beta_1 \cos\omega_1(t-t_\lambda) - \beta_1 \cos\Omega(t-t_\lambda)\right]\mathcal{H}(t-t_\lambda)$$

$$+h^{(1)}\left[1+\left(\Omega/\omega_1\right)^2 \beta_1 \cos\omega_1(t-t_{lag}) - \beta_1 \cos\Omega(t-t_{lag})\right]\mathcal{H}(t-t_{lag})$$

$$-h^{(1)}\left[1+\left(\Omega/\omega_1\right)^2 \beta_1 \cos\omega_1(t-t_{lag}-t_\lambda) - \beta_1 \cos\Omega(t-t_{lag}-t_\lambda)\right]\mathcal{H}(t-t_{lag}-t_\lambda)$$

(s-1)

$$\eta_2(t) = h^{(2)}\left[1+\left(\Omega/\omega_2\right)^2 \beta_2 \cos\omega_2 t - \beta_2 \cos\Omega t\right]\mathcal{H}(t)$$

$$-h^{(2)}\left[1+\left(\Omega/\omega_2\right)^2 \beta_2 \cos\omega_2(t-t_\lambda) - \beta_2 \cos\Omega(t-t_\lambda)\right]\mathcal{H}(t-t_\lambda)$$

$$+h^{(2)}\left[1+\left(\Omega/\omega_2\right)^2 \beta_2 \cos\omega_2(t-t_{lag}) - \beta_2 \cos\Omega(t-t_{lag})\right]\mathcal{H}(t-t_{lag})$$

$$-h^{(2)}\left[1+\left(\Omega/\omega_2\right)^2 \beta_2 \cos\omega_2(t-t_{lag}-t_\lambda) - \beta_2 \cos\Omega(t-t_{lag}-t_\lambda)\right]\mathcal{H}(t-t_{lag}-t_\lambda)$$

(s-2)

$$\eta_3(t) = h^{(3)}\left[1+\left(\Omega/\omega_3\right)^2 \beta_3 \cos\omega_3 t - \beta_3 \cos\Omega t\right]\mathcal{H}(t)$$

$$-h^{(3)}\left[1+\left(\Omega/\omega_3\right)^2 \beta_3 \cos\omega_3(t-t_\lambda) - \beta_3 \cos\Omega(t-t_\lambda)\right]\mathcal{H}(t-t_\lambda)$$

$$+h^{(3)}\left[1+\left(\Omega/\omega_3\right)^2 \beta_3 \cos\omega_3(t-t_{lag}) - \beta_3 \cos\Omega(t-t_{lag})\right]\mathcal{H}(t-t_{lag})$$

$$-h^{(3)}\left[1+\left(\Omega/\omega_3\right)^2 \beta_3 \cos\omega_3(t-t_{lag}-t_\lambda) - \beta_3 \cos\Omega(t-t_{lag}-t_\lambda)\right]\mathcal{H}(t-t_{lag}-t_\lambda)$$

(s-3)

where

$$\beta_j = \frac{1}{1-\left(\Omega/\omega_j\right)^2} \quad (j=1,2,3) \tag{t}$$

$$h^{(1)} = 0.3192h_0, \quad h^{(2)} = 0.1806h_0, \quad h^{(3)} = h_0/L \tag{u-1, 2, 3}$$

The physical displacements are finally obtained by transforming back to the original coordinates using Eq. (8.31). Hence,

$$\begin{Bmatrix} y_G(t) \\ \theta(t) \\ y_b(t) \end{Bmatrix} = \begin{bmatrix} 1 & 1 & 0 \\ 0 & 0 & 1 \\ 2.303 & -1.303 & 0 \end{bmatrix} \begin{Bmatrix} \eta_1(t) \\ \eta_2(t) \\ \eta_3(t) \end{Bmatrix} \tag{v}$$

which when expanded gives

$$y_G(t) = \eta_1(t) + \eta_2(t) \qquad\qquad \triangleleft \text{(w-1)}$$

$$\theta(t) = \eta_3(t) \qquad\qquad \triangleleft \text{(w-2)}$$

$$y_p(t) = 2.303\eta_1(t) - 1.303\eta_2(t) \qquad\qquad \triangleleft \text{(w-3)}$$

where $\eta_1(t)$, $\eta_2(t)$ and $\eta_3(t)$ are given by Eqs. (s-1), (s-2) and (s-3), respectively. It is seen that the rotational coordinate is a modal coordinate. Let's examine the motion of the rider in detail.

The detailed response of the rider is found by substituting Eqs. (s-1) and (s-2) into Eq. (w-3) and evaluating the resulting expression when the front wheel is rolling through the depression ($0 < t < t_\lambda$), after it passes the dip but before the second wheel encounters the depression ($t_\lambda < t < t_{lag}$), when the second wheel rolls through the depression ($t_{lag} < t < t_{lag} + t_\lambda$) and after the second wheel passes the dip ($t > t_{lag}$). The explicit forms are detailed below.

$0 < t < t_\lambda$:
During the interval when the first wheel is going through the depression, the motion of the rider is

$$\frac{y_b(t)}{h_0} = \frac{1}{2} + \left[\frac{0.7351}{\left(1 - 34.87\bar{v}_0^{\,2}\right)} - \frac{0.2353}{\left(1 - 11.17\bar{v}_0^{\,2}\right)} \right] \cos\left(2\pi\bar{v}_0\omega_0 t\right)$$

$$+ \frac{25.63\bar{v}_0^{\,2}}{\left(1 - 34.87\bar{v}_0^{\,2}\right)} \cos(1.064\omega_0 t) - \frac{2.628\bar{v}_0^{\,2}}{\left(1 - 11.17\bar{v}_0^{\,2}\right)} \cos(1.880\omega_0 t)$$

$$\triangleleft \text{(x-1)}$$

where

$$\bar{v}_0 = \frac{v_0}{\omega_0 \lambda} \qquad\qquad \text{(y)}$$

and

$$\omega_0 = \sqrt{k/m} \qquad\qquad \text{(z)}$$

$t_\lambda < t < t_{lag}$:
After the first wheel has gone through the depression, but before the second wheel begins its transversal, the motion of the rider is

$$\frac{y_b(t)}{h_0} = \frac{25.63\bar{v}_0^{\,2}}{\left(1 - 34.87\bar{v}_0^{\,2}\right)} \left[\cos(1.064\omega_0 t) - \cos\{1.064\omega_0(t - t_\lambda)\}\right]$$

$$\triangleleft \text{(x-2)}$$

$$- \frac{2.628\bar{v}_0^{\,2}}{\left(1 - 11.17\bar{v}_0^{\,2}\right)} \left[\cos(1.880\omega_0 t) - \cos\{1.880\omega_0(t - t_\lambda)\}\right]$$

$t_{lag} < t < t_{lag} + t_\lambda$:

As the second wheel rolls through the depression the motion of the rider is given by

$$\frac{y_b(t)}{h_0} = \frac{1}{2} + \left[\frac{0.7351}{\left(1-34.87\overline{v}_0^{\,2}\right)} - \frac{0.2353}{\left(1-11.17\overline{v}_0^{\,2}\right)}\right]\cos\{2\pi\overline{v}_0\omega_0(t-t_{lag})\}$$

$$+\frac{25.63\overline{v}_0^{\,2}}{\left(1-34.87\overline{v}_0^{\,2}\right)}\left[\cos(1.064\omega_0 t)-\cos\{1.064\omega_0(t-t_\lambda)\}+\cos\{1.064\omega_0(t-t_{lag})\}\right]$$

$$-\frac{2.628\overline{v}_0^{\,2}}{\left(1-11.17\overline{v}_0^{\,2}\right)}\left[\cos(1.880\omega_0 t)-\cos\{1.880\omega_0(t-t_\lambda)\}+\cos\{1.880\omega_0(t-t_{lag})\}\right]$$

◁ (x-3)

$t > t_{lag}$:

After the second wheel has passed the depression the motion of the rider is

$$\frac{y_b(t)}{h_0} = \frac{25.63\overline{v}_0^{\,2}}{\left(1-34.87\overline{v}_0^{\,2}\right)}\left\{\left[\cos(1.064\omega_0 t)-\cos\{1.064\omega_0(t-t_\lambda)\}\right]\right.$$

$$\left.+\left[\cos\{1.064\omega_0(t-t_{lag})\}-\cos\{1.064\omega_0(t-t_{lag}-t_\lambda)\}\right]\right\}$$

$$-\frac{2.628\overline{v}_0^{\,2}}{\left(1-11.17\overline{v}_0^{\,2}\right)}\left\{\left[\cos(1.880\omega_0 t)-\cos\{1.880\omega_0(t-t_\lambda)\}\right]\right.$$

$$\left.+\left[\cos\{1.880\omega_0(t-t_{lag})\}-\cos\{1.880\omega_0(t-t_{lag}-t_\lambda)\}\right]\right\}$$

◁ (x-4)

8.6 MODE ISOLATION

In certain situations it may be desirable to excite one particular mode of a system, but none of the other modes. In the present section we examine how this may be accomplished.

Recall from Section 8.2.2 that when a vector **v** is operated on by a linear transformation **R**, it results in a vector $\tilde{\mathbf{v}}$, such that

$$\tilde{\mathbf{v}} = \mathbf{R}\,\mathbf{v}$$

As discussed earlier, the matrix $\mathbf{v}$ may be considered to be the matrix of components of a vector with respect to a certain set of coordinates, and the matrix $\tilde{\mathbf{v}}$ may be thought of as the matrix of components of that same vector expressed in terms of a different set of coordinates. With this in mind let us consider an N-degree of freedom system with mass matrix $\mathbf{m}$ and stiffness matrix $\mathbf{k}$, and let us focus on a particular modal matrix, say $\mathbf{U}^{(j)}$. Further, let us consider the transformed modal vector $\tilde{\mathbf{U}}^{(j)}$, where either

$$\tilde{\mathbf{U}}^{(j)} = \mathbf{m}\mathbf{U}^{(j)} \tag{8.56}$$

or

$$\tilde{\mathbf{U}}^{(j)} = \mathbf{k}\mathbf{U}^{(j)} \tag{8.57}$$

Next, let us consider the specific class of force systems whose matrices are proportional to the transformed modal vector. That is, let us consider force matrices of the form

$$\mathbf{F}(t) = \lambda(t)\tilde{\mathbf{U}}^{(j)} \tag{8.58}$$

where $\lambda(t)$ is a scalar function with appropriate units. The corresponding matrix of modal forces is then obtained by substituting Eq. (8.58) into Eq. (8.41) to obtain

$$\tilde{\mathbf{F}}(t) = \lambda(t)\mathbf{B}^{\mathsf{T}}\tilde{\mathbf{U}}^{(j)} \tag{8.59}$$

Recalling the definition of the modal matrix $\mathbf{B}$, Eq. (8.21), and expressing this in Eq. (8.59), we see that

$$\tilde{\mathbf{F}}(t) = \lambda(t)\begin{bmatrix} \mathbf{U}^{(1)\mathsf{T}} \\ \vdots \\ \mathbf{U}^{(N)\mathsf{T}} \end{bmatrix}\tilde{\mathbf{U}}^{(j)} = \lambda(t)\begin{Bmatrix} \mathbf{U}^{(1)\mathsf{T}}\tilde{\mathbf{U}}^{(j)} \\ \vdots \\ \mathbf{U}^{(N)\mathsf{T}}\tilde{\mathbf{U}}^{(j)} \end{Bmatrix} \tag{8.60}$$

If the transformation defined by Eq. (8.56) is used in Eq. (8.60), and if we exploit the mutual orthogonality of the modal vectors and also incorporate Eq. (8.44), then

$$\tilde{\mathbf{F}} = \lambda(t)\begin{Bmatrix} \mathbf{U}^{(1)\mathsf{T}}\mathbf{m}\mathbf{U}^{(j)} \\ \vdots \\ \mathbf{U}^{(j)\mathsf{T}}\mathbf{m}\mathbf{U}^{(j)} \\ \vdots \\ \mathbf{U}^{(N)\mathsf{T}}\mathbf{m}\mathbf{U}^{(j)} \end{Bmatrix} = \lambda(t)\begin{Bmatrix} 0 \\ \vdots \\ 0 \\ \tilde{m}_j \\ 0 \\ \vdots \\ 0 \end{Bmatrix} = \lambda(t)\tilde{m}_j\begin{Bmatrix} 0 \\ \vdots \\ 0 \\ 1 \\ 0 \\ \vdots \\ 0 \end{Bmatrix} \tag{8.61}$$

Similarly, if the transformation defined by Eq. (8.57) is used and we incorporate Eq. (8.47), then

$$
\tilde{\mathbf{F}} = \lambda(t) \left\{ \begin{matrix} \mathbf{U}^{(1)\mathrm{T}} \mathbf{k} \mathbf{U}^{(j)} \\ \vdots \\ \mathbf{U}^{(j)\mathrm{T}} \mathbf{k} \mathbf{U}^{(j)} \\ \vdots \\ \mathbf{U}^{(N)\mathrm{T}} \mathbf{k} \mathbf{U}^{(j)} \end{matrix} \right\} = \lambda(t) \left\{ \begin{matrix} 0 \\ \vdots \\ 0 \\ \tilde{k}_j \\ 0 \\ \vdots \\ 0 \end{matrix} \right\} = \lambda(t) \tilde{k}_j \left\{ \begin{matrix} 0 \\ \vdots \\ 0 \\ 1 \\ 0 \\ \vdots \\ 0 \end{matrix} \right\} \tag{8.62}
$$

In each case, it may be seen that the matrix of modal forces has only one nonvanishing component; that corresponding to the jth mode. It is thus seen, from (8.48), that when the matrix of applied forces is of the form of Eq. (8.58), with $\mathbf{U}^{(j)}$ given by either Eq. (8.56) or Eq. (8.57), then only the jth mode is excited by the given set of forces. Finally, it may be seen that for those systems for which the mass matrix or the stiffness matrix is a scalar matrix (a matrix that is proportional to the identity matrix) then a single mode will be excited if the force matrix is directly proportional to that mode.

Example 8.13

Consider a system that may be modeled as the double pendulum of Examples 7.5 and 8.10. Suppose that we wish to examine the response of the system to impulses applied to the masses of the system, as indicated. In addition, suppose that it is desired to observe the motion of each mode individually, and that we have a mechanism that will pulse each mass simultaneously. (*a*) How should

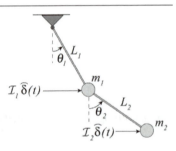

we choose the relative magnitudes of the pulses so that the system responds in the first mode only? (*b*) Demonstrate that the forcing chosen does, in fact, activate the desired mode in each case. (*c*) determine the response of the system to the chosen forcing function.

Solution

This is evidently a mode isolation problem. We would therefore configure the forcing mechanism to apply the pulses so that, in each case, the matrix of applied forces is proportional to either Eq. (8.56) or Eq. (8.57).

Let us recall from Example (7.5) that the two modes for this system are

$$\mathbf{U}^{(1)} = \begin{Bmatrix} 1 \\ \sqrt{2} \end{Bmatrix}, \quad \mathbf{U}^{(2)} = \begin{Bmatrix} 1 \\ -\sqrt{2} \end{Bmatrix} \tag{a-1, 2}$$

and hence that the modal matrix is

$$\mathbf{B} = \begin{bmatrix} 1 & 1 \\ \sqrt{2} & -\sqrt{2} \end{bmatrix} \tag{b}$$

Let us also recall that the mass and stiffness matrices for the system are, respectively,

$$\mathbf{m} = mL^2 \begin{bmatrix} 2 & 1 \\ 1 & 1 \end{bmatrix}, \quad \mathbf{k} = mgL \begin{bmatrix} 2 & 0 \\ 0 & 1 \end{bmatrix} \tag{c-1, 2}$$

For the loading under consideration, the time dependence of the forces corresponds to an impulse. Hence,

$$\lambda(t) \sim \hat{\mathcal{I}}_0 \hat{\delta}(t) \tag{d}$$

(*a*)
Since, for this particular system, the stiffness matrix is diagonal, let's design our forcing using Eq. (8.57). Hence,

$$\mathbf{F}(t) = \lambda(t)\tilde{\mathbf{U}}^{(1)} = \lambda(t)\mathbf{k}\mathbf{U}^{(1)}$$
$$= \lambda(t)mgL \begin{bmatrix} 2 & 0 \\ 0 & 1 \end{bmatrix} \begin{Bmatrix} 1 \\ \sqrt{2} \end{Bmatrix} = \hat{\mathcal{I}}_0 \hat{\delta}(t) \begin{Bmatrix} 2 \\ \sqrt{2} \end{Bmatrix} \tag{e}$$

where, from Eq. (d), we have taken $\lambda(t) = \hat{\mathcal{I}}_0 \hat{\delta}(t)/mgL$.

(*b*)
Now, to demonstrate that the pair of pulses given by Eq. (e) will excite only the first mode let us compute the corresponding modal forces. Thus,

$$\tilde{\mathbf{F}}(t) = \mathbf{B}^{\mathsf{T}}\mathbf{F}(t) = \begin{bmatrix} 1 & \sqrt{2} \\ 1 & -\sqrt{2} \end{bmatrix} \hat{\mathcal{I}}_0 \hat{\delta}(t) \begin{Bmatrix} 2 \\ \sqrt{2} \end{Bmatrix} = 4\hat{\mathcal{I}}_0 \hat{\delta}(t) \begin{Bmatrix} 1 \\ 0 \end{Bmatrix} \tag{f}$$

It is seen from Eq. (f) that only the first element of the modal force matrix, that corresponding to the first mode, is nonzero. Therefore, only the first mode is excited.

(c)
Paralleling the analysis of Example 8.6 for the present loading, the modal equations for this system are

$$\tilde{m}_1\ddot{\eta}_1 + \tilde{k}_1\eta_1 = 4\hat{\mathcal{I}}_0\delta(t)$$
$$\tilde{m}_2\ddot{\eta}_2 + \tilde{k}_2\eta_2 = 0 \Rightarrow \eta_2(t) = 0$$ (g)

where the modal mass and modal stiffness matrices for the double pendulum are given by Eqs. (f) and (g) of Example 8.5. Equation (g-2) clearly yields the trivial solution since the second mode is not forced as concluded in Part (b). The solution to Eq. (g-1) is obtained directly from Eqs. (4.22) and (4.23). Identifying $\mathcal{I} = 4\hat{\mathcal{I}}_0$, $\zeta = 0$, and the mass and stiffness as the modal mass and modal stiffness for the mode in question gives

$$\eta_1(t) = \frac{4\hat{\mathcal{I}}_0}{\tilde{m}_1\omega_1}\sin\omega_1 t\, \mathcal{H}(t)$$ (h)

Transforming to physical coordinates gives the response of the pendulum as

$$\begin{Bmatrix} \theta_1(t) \\ \theta_2(t) \end{Bmatrix} = \begin{bmatrix} 1 & 1 \\ \sqrt{2} & -\sqrt{2} \end{bmatrix} \begin{Bmatrix} \eta_1(t) \\ 0 \end{Bmatrix} = \eta_1(t) \begin{Bmatrix} 1 \\ \sqrt{2} \end{Bmatrix} = \frac{4\hat{\mathcal{I}}_0}{\tilde{m}_1\omega_1}\sin\omega_1 t\, \mathcal{H}(t) \begin{Bmatrix} 1 \\ \sqrt{2} \end{Bmatrix}$$ (i)

and finally,

$$\begin{Bmatrix} \theta_1(t) \\ \theta_2(t) \end{Bmatrix} = \frac{5.23\hat{\mathcal{I}}_0}{m\sqrt{gL^3}}\sin(0.765\sqrt{g/L}\, t)\mathcal{H}(t) \begin{Bmatrix} 1 \\ \sqrt{2} \end{Bmatrix}$$ ◁ (j)

Example 8.14

Consider the two degree of freedom mass-spring system of Example 8.7(b) for the case where $\Omega_a = \Omega_b = \Omega$. (a) Determine the relative amplitudes of the applied forces, F_a and F_b, if we wish to excite (i) the first mode alone or (ii) the second mode alone. (b) Determine the corresponding response for each case.

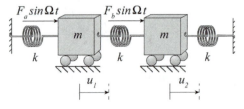

Solution

(a)
For the loading under consideration, the force system is of the form

$$\mathbf{F}(t) = \left\{ \begin{matrix} F_a \\ F_b \end{matrix} \right\} \sin \Omega t \qquad (a)$$

Hence

$$\lambda(t) = \lambda_0 \sin \Omega t \qquad (b)$$

where the amplitude λ_0 may be any scalar. Since, for this particular system, the mass matrix is diagonal we shall choose to construct the force matrix using Eq. (8.56) for ease of computation.

(i) *Response in the form of Mode 1:*
For this case,

$$\mathbf{F}(t) = \lambda(t)\mathbf{m}\mathbf{U}^{(1)} = \lambda(t)\begin{bmatrix} 1 & 0 \\ 0 & 1 \end{bmatrix}\left\{ \begin{matrix} 1 \\ 1 \end{matrix} \right\} = 2\lambda(t)\left\{ \begin{matrix} 1 \\ 1 \end{matrix} \right\} \qquad (c)$$

We thus see that only the first mode will be excited if we apply a force system that is proportional to the first mode. For the particular case of harmonic loading described by Eq. (a), we see from Eqs. (b) and (c) that

$$F_a = F_b = 2\lambda_0 \qquad (d)$$

and any force system such that

$$F_b / F_a = 1$$

will accomplish our objective.

(ii) *Response in the form of Mode 2:*
For this case,

$$\mathbf{F}(t) = \lambda(t)\mathbf{m}\mathbf{U}^{(2)} = \lambda(t)\begin{bmatrix} 1 & 0 \\ 0 & 1 \end{bmatrix}\left\{ \begin{matrix} 1 \\ -1 \end{matrix} \right\} = 2\lambda(t)\left\{ \begin{matrix} 1 \\ -1 \end{matrix} \right\} \qquad (e)$$

It is seen that, as for the previous case, the second mode alone will be excited if the force matrix is constructed to be proportional to the second mode. For the particular loading under consideration is seen that

$$F_b = -F_a = 2\lambda_0 \qquad (f)$$

and any force system such that

$$F_b/F_a = -1$$

will achieve the desired response. The results for this system should have been anticipated since the mass matrix for the system is a *scalar matrix*. That is, it is proportional to the identity matrix.

(b)
The response to the harmonic loadings described by Eqs. (c) and (d), and by Eqs. (e) and (f), may be found directly from the solution of Example 8.5 by setting $\Omega_a = \Omega_b = \Omega$, and also setting $F_a = F_b = 2\lambda_0$ or $F_b = -F_a = 2\lambda_0$ accordingly. Thus, for the first case,

$$\begin{Bmatrix} u_1(t) \\ u_2(t) \end{Bmatrix} = \frac{2\lambda_0/k}{1-(\Omega/\omega_1)^2} \sin\Omega t \begin{Bmatrix} 1 \\ 1 \end{Bmatrix} \tag{g}$$

where we recall from Example 8.5 that $\omega_1 = \sqrt{k/m}$. For the second case,

$$\begin{Bmatrix} u_1(t) \\ u_2(t) \end{Bmatrix} = \frac{2\lambda_0/k}{1-(\Omega/\omega_2)^2} \sin\Omega t \begin{Bmatrix} 1 \\ -1 \end{Bmatrix} \tag{h}$$

where $\omega_2 = \sqrt{3k/m}$.

8.7 RAYLEIGH DAMPING

To this point we have restricted our attention to the ideal case of undamped systems subjected to external dynamic loading. In this and the next section we relax this restriction and consider the behavior of systems that possess viscous damping. We defer our discussion of the general case to Section 8.8 and presently consider a specific class of systems with viscous damping, those for which the damping matrix is proportional to the mass and/or stiffness matrices. Damping of this type is known as *Rayleigh Damping*, and is also referred to as *proportional damping*. It will be seen that damping of this type allows decoupling of the equations of motion through modal analysis of forced systems, based on the natural modes of the corresponding undamped systems. We first recall the general equations of motion for damped systems.

Systems with linear damping are governed by equations of the form of Eq. (6.10). We repeat them here for clarity of the present discussion. Hence,

$$\mathbf{m\ddot{u} + c\dot{u} + ku = F} \tag{8.63}$$

We next introduce *proportional damping*, that is damping where the matrix of damping coefficients (the damping matrix), $\mathbf{c}$, is linearly proportional to the mass and stiffness matrices. We thus consider systems for which

$$\mathbf{c} = \alpha\mathbf{m} + \beta\mathbf{k} \tag{8.64}$$

where α and β are scalar material constants. We shall seek a solution in the form of an expansion in terms of the modal vectors of the undamped system. That is we assume a solution in the form of Eq. (8.31). Substituting Eq. (8.31) into Eq. (8.63) and pre-multiplying the resulting expression by $\mathbf{B}^{\mathrm{T}}$ results in the transformed equation of motion

$$\left[\mathbf{B}^{\mathrm{T}}\mathbf{m}\mathbf{B}\right]\ddot{\boldsymbol{\eta}} + \left[\mathbf{B}^{\mathrm{T}}\mathbf{c}\mathbf{B}\right]\dot{\boldsymbol{\eta}} + \left[\mathbf{B}^{\mathrm{T}}\mathbf{m}\mathbf{B}\right]\boldsymbol{\eta} = \mathbf{B}^{\mathrm{T}}\mathbf{F}$$

which may be written as

$$\tilde{\mathbf{m}}\ddot{\boldsymbol{\eta}} + \tilde{\mathbf{c}}\dot{\boldsymbol{\eta}} + \tilde{\mathbf{k}}\boldsymbol{\eta} = \tilde{\mathbf{F}} \tag{8.65}$$

where, from the development in Section 8.4,

$$\tilde{\mathbf{m}} = \mathbf{B}^{\mathrm{T}}\mathbf{m}\mathbf{B} = \begin{bmatrix} \tilde{m}_1 & 0 & \cdots & 0 \\ 0 & \tilde{m}_2 & \cdots & 0 \\ \vdots & \vdots & \ddots & \vdots \\ 0 & 0 & \cdots & \tilde{m}_N \end{bmatrix}$$

is the modal mass matrix,

$$\tilde{\mathbf{k}} = \mathbf{B}^{\mathrm{T}}\mathbf{k}\mathbf{B} = \begin{bmatrix} \tilde{k}_1 & 0 & \cdots & 0 \\ 0 & \tilde{k}_2 & \cdots & 0 \\ \vdots & \vdots & \ddots & \vdots \\ 0 & 0 & \cdots & \tilde{k}_N \end{bmatrix}$$

is the modal stiffness matrix and

$$\tilde{\mathbf{F}} = \mathbf{B}^{\mathrm{T}}\mathbf{F} = \begin{Bmatrix} \tilde{F}_1 \\ \tilde{F}_2 \\ \vdots \\ \tilde{F}_N \end{Bmatrix}$$

is the modal force matrix. Further, incorporating Eq. (8.64) into the transformation of the damping matrix gives

$$\tilde{\mathbf{c}} = \mathbf{B}^\mathsf{T}\mathbf{c}\mathbf{B} = \alpha\mathbf{B}^\mathsf{T}\mathbf{m}\mathbf{B} + \beta\mathbf{B}^\mathsf{T}\mathbf{k}\mathbf{B} = \alpha\tilde{\mathbf{m}} + \beta\tilde{\mathbf{k}} \qquad (8.66)$$

Thus, the modal damping matrix for a system with proportional damping takes the diagonal form

$$\tilde{\mathbf{c}} = \begin{bmatrix} \tilde{c}_1 & 0 & \cdots & 0 \\ 0 & \tilde{c}_2 & \cdots & 0 \\ \vdots & \vdots & \ddots & \vdots \\ 0 & 0 & \cdots & \tilde{c}_N \end{bmatrix} \qquad (8.67)$$

where

$$\tilde{c}_j = \alpha\tilde{m}_j + \beta\tilde{k}_j \quad (j = 1, 2, ..., N) \qquad (8.68)$$

Since the damping matrix is a linear combination of the mass and stiffness matrices of the system, the transformed damping matrix is diagonal like the modal mass and modal stiffness matrices. Therefore, the transformed equations of motion for systems with Rayleigh Damping are decoupled in the same way as those for undamped systems. Expanding Eq. (8.65) gives the system of N uncoupled differential equations for the modal displacements η_j ($j = 1, 2, ..., N$),

$$\begin{aligned} \tilde{m}_1\ddot{\eta}_1 + \tilde{c}_1\dot{\eta}_1 + \tilde{k}_1\eta_1 &= \tilde{F}_1 \\ \tilde{m}_2\ddot{\eta}_2 + \tilde{c}_1\dot{\eta}_2 + \tilde{k}_1\eta_2 &= \tilde{F}_2 \\ &\vdots \\ \tilde{m}_N\ddot{\eta}_N + \tilde{c}_N\dot{\eta}_N + \tilde{k}_N\eta_N &= \tilde{F}_N \end{aligned} \qquad (8.69)$$

which may be thought of as corresponding to the governing equations of the N equivalent single degree of freedom systems shown in Figure 8.7. These equations may be put in the standard form,

$$\begin{aligned} \ddot{\eta}_1 + 2\omega_1\tilde{\zeta}_1\dot{\eta}_1 + \omega_1^2\eta_1 &= \omega_1^2\tilde{f}_1 \\ \ddot{\eta}_2 + 2\omega_2\tilde{\zeta}_2\dot{\eta}_2 + \omega_2^2\eta_2 &= \omega_2^2\tilde{f}_2 \\ &\vdots \\ \ddot{\eta}_N + 2\omega_N\tilde{\zeta}_N\dot{\eta}_N + \omega_N^2\eta_N &= \omega_N^2\tilde{f}_N \end{aligned} \qquad (8.70)$$

where

$$\tilde{\zeta}_i = \tilde{c}_i/2\omega_i\tilde{m}_i \qquad (8.71)$$

and

$$\tilde{f}_i(t) = \tilde{F}_i(t)/\tilde{k}_i \qquad (8.72)$$

For given forcing, each of Eqs. (8.70) may be solved using the techniques already discussed for single degree of freedom systems in Chapters 3–5. Once the modal displacements (the solutions to the above equations) have been obtained, they may be mapped back to physical coordinates using Eq. (8.31). This gives the forced response of the proportionally damped system.

Example 8.15

Consider the system of Example 8.6 but let it now possess three identical viscous dampers with coefficient $c = 0.2\sqrt{mk}$ as shown. Determine the steady state response of the system if the left mass is subjected to the harmonic force $F_1(t) = F_a \sin\Omega_a t$.

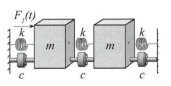

Solution

The equation of motion for the system is easily derived directly, or by using Eq. 6. with $N = 2$, $k_1 = k_2 = k_3 = k$, $c_1 = c_2 = c_3 = c$ and $m_1 = m_2 = m$. Hence,

$$\mathbf{m} = \begin{bmatrix} m & 0 \\ 0 & m \end{bmatrix}, \quad \mathbf{k} = \begin{bmatrix} 2k & -k \\ -k & 2k \end{bmatrix}, \quad \mathbf{u} = \begin{Bmatrix} u_1(t) \\ u_2(t) \end{Bmatrix}, \quad \mathbf{F} = \begin{Bmatrix} F_1(t) \\ F_2(t) \end{Bmatrix} = \begin{Bmatrix} F_a \sin\Omega_a t \\ 0 \end{Bmatrix} \quad (a)$$

and

$$\mathbf{c} = \begin{bmatrix} 2c & -c \\ -c & 2c \end{bmatrix} = \frac{c}{k}\begin{bmatrix} 2k & -k \\ -k & 2k \end{bmatrix} = \beta\mathbf{k} \quad (b)$$

where

$$\alpha = 0, \quad \beta = \frac{c}{k} = \frac{0.2}{\sqrt{k/m}} \quad (c)$$

This is clearly a system with Rayleigh (Proportional) Damping.

The analysis for the present system is identical to that for the undamped system of Example 8.4 up to Eq. (j) of that example. Thus, the modal mass, modal stiffness and modal force matrices remain the same. However, the modal equations must now be modified to include the effects of proportional damping. For the present system we have, after substituting the given and calculated material parameters into Eq. (8.67),

$$\tilde{\mathbf{c}} = \alpha\tilde{\mathbf{m}} + \beta\tilde{\mathbf{k}} = 0 + \frac{0.2}{\sqrt{k/m}}\begin{bmatrix} 2k & 0 \\ 0 & 6k \end{bmatrix} = \begin{bmatrix} 0.4\sqrt{mk} & 0 \\ 0 & 1.2\sqrt{mk} \end{bmatrix} \quad (d)$$

Hence,

$$\tilde{\zeta}_1 = \frac{\tilde{c}_1}{2\omega_1\tilde{m}_1} = \frac{0.4\sqrt{mk}}{2\sqrt{k/m}\,(2m)} = 0.1 \tag{e-1}$$

$$\tilde{\zeta}_2 = \frac{\tilde{c}_2}{2\omega_2\tilde{m}_2} = \frac{1.2\sqrt{mk}}{2\sqrt{3k/m}\,(2m)} = 0.1732 \tag{e-2}$$

Applying Eqs. (8.70) to the present problem gives the equations governing the modal displacements,

$$\ddot{\eta}_1 + 2\omega_1\tilde{\zeta}_1\dot{\eta}_1 + \omega_1^2\eta_1 = \omega_1^2\tilde{f}_{a1}\sin\Omega_a t = \omega_1^2\frac{F_a}{2k}\sin\Omega_a t$$
$$\ddot{\eta}_2 + 2\omega_2\tilde{\zeta}_2\dot{\eta}_2 + \omega_2^2\eta_2 = \omega_2^2\tilde{f}_{a2}\sin\Omega_a t = \omega_2^2\frac{F_a}{6k}\sin\Omega_a t \tag{f}$$

where $\tilde{\zeta}_1$, $\tilde{\zeta}_2$ and ω_1, ω_2 are respectively given by Eqs. (e-1) and (e-2) above, and Eqs. (c-2) and (d-2) of Example 8.5, respectively. The solutions of Eqs. (f) are obtained by direct application of Eq. (3.54), (3.50) and (3.51) for each case. We thus obtain

$$\eta_1(t) = \frac{F_a}{2k}\Gamma(\bar{\Omega}_1,\tilde{\zeta}_1)\sin(\Omega_a t - \Phi_1)$$
$$\eta_2(t) = \frac{F_a}{6k}\Gamma(\bar{\Omega}_2,\tilde{\zeta}_2)\sin(\Omega_a t - \Phi_2) \tag{g}$$

where

$$\Gamma(\bar{\Omega}_j,\tilde{\zeta}_j) = \frac{1}{\sqrt{\left[1-\bar{\Omega}_j^2\right]^2 + \left[2\tilde{\zeta}_j\bar{\Omega}_j\right]^2}} \quad (j=1,2) \tag{h}$$

$$\Phi_j = \tan^{-1}\left(\frac{2\tilde{\zeta}_j\bar{\Omega}_j}{1-\bar{\Omega}_j^2}\right) \quad (j=1,2) \tag{i}$$

and

$$\bar{\Omega}_j = \Omega_a/\omega_j \tag{j}$$

Mapping back to physical coordinates gives the desired steady state response. Hence,

$$\mathbf{u} = \mathbf{B}\boldsymbol{\eta} = \mathbf{U}^{(1)}\eta_1(t) + \mathbf{U}^{(2)}\eta_2(t)$$

which, after substituting Eqs. (g), gives

$$\begin{Bmatrix} u_1(t) \\ u_2(t) \end{Bmatrix} = \begin{Bmatrix} 1 \\ 1 \end{Bmatrix} \frac{F_a}{2k} \Gamma(\bar{\Omega}_1, \tilde{\zeta}_1) \sin(\Omega_a t - \Phi_1)$$

$$+ \begin{Bmatrix} 1 \\ -1 \end{Bmatrix} \frac{F_a}{6k} \Gamma(\bar{\Omega}_2, \tilde{\zeta}_2) \sin(\Omega_a t - \Phi_2) \tag{k}$$

Expanding Eq. (k) and substituting Eqs. (e) and (h) gives the explicit form of the displacements as functions of time,

$$u_1(t) = \frac{F_a}{2k} \frac{1}{\sqrt{\left[1 - (\Omega_a m/k)^2\right]^2 + 0.0400(\Omega_a m/k)^2}} \sin(\Omega_a t - \Phi_1)$$

$$+ \frac{F_a}{6k} \frac{1}{\sqrt{\left[1 - (\Omega_a m/3k)^2\right]^2 + 0.1200(\Omega_a m/3k)^2}} \sin(\Omega_a t - \Phi_2) \tag{l-1}$$

$$u_2(t) = \frac{F_a}{2k} \frac{1}{\sqrt{\left[1 - (\Omega_a m/k)^2\right]^2 + 0.0400(\Omega_a m/k)^2}} \sin(\Omega_a t - \Phi_1)$$

$$- \frac{F_a}{6k} \frac{1}{\sqrt{\left[1 - (\Omega_a m/3k)^2\right]^2 + 0.1200(\Omega_a m/3k)^2}} \sin(\Omega_a t - \Phi_2) \tag{l-2}$$

8.8 SYSTEMS WITH GENERAL VISCOUS DAMPING

To this point we have considered forced vibration of undamped systems or the special case of systems with proportional damping. It was seen that the modal vectors were linearly independent and mutually orthogonal and therefore that any response is comprised of a linear combination of the modal vectors. This property formed the basis of modal analysis, which we proceeded to apply to problems of forced vibration. Because of the special property of the damping matrix for systems with Rayleigh Damping, we were able to apply the same approach to these systems as well. That is, we were able to determine a solution in terms of the modal vectors for the undamped system. It was seen in Section 7.1 that the characteristic vectors for a system with viscous damping, are not, in general, mutually orthogonal in the conventional sense. However, when the motion of such systems is represented in state space then the right eigenvectors of the $2N \times 2N$ system matrix **S** are mutually orthogonal with respect to their counterparts, the left eigenvectors of **S**. It is this property that will allow us to establish a generalization of modal analysis for damped systems. We first show that the response of any N-degree of freedom system can be expressed as a linear combination of the right eigenvectors of the system matrix $\mathbf{S}$. Before proceeding to the general problem in state space, we first determine a simple solution for damped multi-degree of freedom systems subjected to harmonic excitation.

8.8.1 Steady State Response to Harmonic Excitation

In this section we determine a simple solution for damped multi-degree of freedom systems. To accomplish this we shall parallel the related development for undamped systems discussed in Section 8.1.1. Toward this end, let us consider an N-degree of freedom system with general linear viscous damping subjected to external forces, all of which possess the same harmonic time signature. The equation of motion for such a system is of the general form

$$\mathbf{m\ddot{u}}(t) + \mathbf{c\dot{u}}(t) + \mathbf{ku}(t) = \mathbf{F}_0 e^{i\Omega t} \tag{8.73}$$

where Ω is the excitation frequency. We wish to obtain the particular solution to Eq. (8.73). Let us therefore assume a solution of the form

$$\mathbf{u}_p(t) = \mathbf{H} e^{i\Omega t}$$

and substitute it into Eq. (8.73). Solving the resulting expression for $\mathbf{H}$ gives the particular solution, and hence the steady state response,

$$\mathbf{u}_p(t) = \left[\mathbf{k} - \Omega^2 \mathbf{m} + i\Omega \mathbf{c}\right]^{-1} \mathbf{F}_0 e^{i\Omega t} \tag{8.74}$$

or, equivalently,

$$\mathbf{u}_p(t) = \frac{\mathrm{adj}\left[\mathbf{k} - \Omega^2 \mathbf{m} + i\Omega \mathbf{c}\right]}{\det\left[\mathbf{k} - \Omega^2 \mathbf{m} + i\Omega \mathbf{c}\right]} \mathbf{F}_0 e^{i\Omega t} \tag{8.75}$$

The use of the above solution is demonstrated in by the following example. We then proceed to a more general technique, applicable to any type of excitation.

Example 8.16

Consider the elastically supported frame of mass m and length L of Examples 7.13d and 7.14 , where the stiffnesses of the two identical springs at each end of the frame have the magnitude $k = mg/L$, and the damping coefficients have the values $c_2 = 2c_1$ and $c_1/\sqrt{2km} = 0.1$. Let a force $F(t) = F_0 \sin\Omega t$ (positive downward) be applied to the left edge of the frame, as shown. Determine the steady state motion of the frame when $F_0 = mg$ and $\Omega = 1.5g/L$.

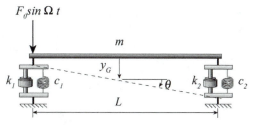

Solution

We shall describe the motion of the system in terms of the rotation of the frame, $\theta(t)$ (positive clockwise), and the transverse deflection of the center of mass, $y_G(t)$ (positive downward). To solve the problem, we shall normalize the deflection and the time, and hence the system parameters and equations of motion, as we did in Example 7.14. We thus introduce the normalized deflection of center of mass,

$$\bar{y}_G = y_G/L \tag{a}$$

and the normalized timescale

$$\tau = \omega_0 t = \sqrt{\frac{2k}{m}}\, t \tag{b}$$

where,

$$\omega_0^2 = k/m = g/L \tag{c}$$

Hence, for any function f,

$$\frac{df}{dt} = \frac{df}{d\tau}\frac{d\tau}{dt} = \omega_0 \frac{df}{d\tau} \tag{d}$$

Introducing Eqs. (a)–(d) into Eq. (a) of Example 7.13 and dividing the first row by m and the second row by $I_G = mL^2/12$ renders the equation of motion to the nondimensional form

$$
\begin{bmatrix} 1 & 0 \\ 0 & 1 \end{bmatrix}
\begin{Bmatrix} d^2\bar{y}_G/d\tau^2 \\ d^2\theta/d\tau^2 \end{Bmatrix}
$$
$$
+ \begin{bmatrix} (\eta_1 + \eta_2) & (\eta_2 - \eta_1)/2 \\ 6(\eta_2 - \eta_1) & 3(\eta_1 + \eta_2) \end{bmatrix}
\begin{Bmatrix} d\bar{y}_G/d\tau \\ d\theta/d\tau \end{Bmatrix}
+ \begin{bmatrix} 1 & 0 \\ 0 & 3 \end{bmatrix}
\begin{Bmatrix} \bar{y}_G \\ \theta \end{Bmatrix}
= \begin{Bmatrix} \bar{F} \\ \bar{M}_G \end{Bmatrix} \tag{e}
$$

where

$$\bar{F}(\tau) = F(\tau)/\omega_0^2 mL = \sin(1.5\tau) \tag{f-1}$$

and

$$\bar{M}_G(\tau) = M_G(\tau)\big/\omega_0^2 I_G = 12\, M_G(\tau)\big/\omega_0^2 mL^2 = 6\sin(1.5\tau) \qquad \text{(f-2)}$$

are the normalized resultant applied force and moment about an axis through G. For the particular system in question, $\eta_1 = 0.5$ and $\eta_2 = 1.0$. It then follows that, for the system in question, the complex form of the normalized force vector is

$$\mathbf{F}(t) = \begin{Bmatrix} 1 \\ 6 \end{Bmatrix} e^{i1.5\tau} \qquad \text{(g)}$$

Further, for the given parameters,

$$\left[\mathbf{k} - \Omega^2 \mathbf{m} + i\Omega \mathbf{c} \right] = \begin{bmatrix} (-1.25 + 2.25i) & 0.375i \\ 4.5i & (0.75 + 6.75i) \end{bmatrix} \qquad \text{(h)}$$

We next compute the inverse of the above matrix using the MATLAB "inv" facility. Hence,

$$\left[\mathbf{k} - \Omega^2 \mathbf{m} + i\Omega \mathbf{c} \right]^{-1} = \begin{bmatrix} -(0.2220 + 0.3637i) & (0.0100 + 0.0213i) \\ (0.1196 + 0.2558i) & (0.0113 - 0.1611i) \end{bmatrix} \qquad \text{(i)}$$

The complex form of the response is then found by substituting Eqs. (g) and (i) into Eq. (8.74). Doing this we obtain

$$\begin{Bmatrix} \bar{y}_G(t)/L \\ \theta(t) \end{Bmatrix} = \begin{bmatrix} -(0.2220 + 0.3637i) & (0.0100 + 0.0213i) \\ (0.1196 + 0.2558i) & (0.0113 - 0.1611i) \end{bmatrix} \begin{Bmatrix} 1 \\ 6 \end{Bmatrix} e^{i1.5\tau}$$

$$= \begin{Bmatrix} -(0.1622 + 0.2358i) \\ (0.1871 - 0.7109i) \end{Bmatrix} e^{i1.5\tau} \qquad \text{(j)}$$

Restating the exponential using Euler's Formula and carrying through the indicated multiplications gives

$$\bar{y}_G(t)/L = [-0.1622\cos(1.5\tau) + 0.2358\sin(1.5\tau)]$$
$$- i[0.2358\cos(1.5\tau) + 0.1622\sin(1.5\tau)] \qquad \text{(k-1)}$$

$$\theta(t) = [0.1871\cos(1.5\tau) + 0.7109\sin(1.5\tau)]$$
$$+ i[-0.7109\cos(1.5\tau) + 0.1871\sin(1.5\tau)] \qquad \text{(k-2)}$$

Since the excitation is a sine function, the imaginary part of the above solution corresponds to the response of the system. The steady state response of the system is then

$$\bar{y}_G(t)/L = -[0.2358\cos(1.5\omega_0 t) + 0.1622\sin(1.5\omega_0 t)] \qquad \triangleleft \text{(k-1)}$$

$$\theta(t) = -0.7109\cos(1.5\omega_0 t) + 0.1871\sin(1.5\omega_0 t) \qquad \triangleleft \text{(k-2)}$$

8.8.2 Eigenvector Expansion

The solution presented in Section 8.8.1 is restricted to harmonic excitation. In this section we develop a procedure for damped systems that is a generalization of modal analysis, and is applicable to any type of loading. To accomplish this we must represent the forced vibration problem in its corresponding state space, as we did for free vibrations of damped systems in Section 7.4.2. We then seek to express the response of the forced system in terms of the (right) eigenvectors of the state space system matrix computed for the associated free vibration problem.

In order to express the state vector in terms of the right eigenvectors of the system matrix $\mathbf{S}$ defined by Eq. (7.61) we must first show that the vectors are linearly independent. This is done by generalizing the development of Section 8.2.1 in the context of the state space representation. Toward this end, let us consider the sets of $2N \times 1$ (complex) right and left eigenvectors of the (nonsymmetric) system matrix $\mathbf{S}$,

$$\left\{ \hat{\mathbf{U}}^{(j)} \,\middle|\, j = 1, 2, \ldots, 2N \right\} \text{ and } \left\{ \hat{\mathbf{W}}^{(j)} \,\middle|\, j = 1, 2, \ldots, 2N \right\}$$

respectively, where $\mathbf{S}$ is defined by Eq. (7.61) and $\hat{\mathbf{U}}$ and $\hat{\mathbf{W}}$ are defined by Eqs. (7.63), (7.64) and (7.69). If the right eigenvectors are linearly independent then the equation

$$\sum_{j=1}^{2N} a_j \hat{\mathbf{U}}^{(j)} = 0 \qquad (8.76)$$

is satisfied only if all $a_j = 0$ ($j = 1, 2, \ldots, N$). We proceed to show that this is the case. Let us next multiply Eq. (8.76) by $\hat{\mathbf{W}}^{(l)\mathsf{T}}$. This gives

$$\sum_{j=1}^{2N} a_j \hat{\mathbf{W}}^{(l)\mathsf{T}} \hat{\mathbf{U}}^{(j)} = 0 \qquad (8.77)$$

We next incorporate the orthogonality relation stated by Eq. (7.75) into Eq. (8.77). This results in the statement

$$a_l \hat{\mathbf{W}}^{(l)\mathsf{T}} \hat{\mathbf{U}}^{(l)} = 0 \qquad (8.78)$$

from which we conclude that $a_l = 0$ for all $l = 1, 2, \ldots, 2N$. The right eigenvectors are therefore linearly independent. The linear independence of the right eigenvectors and their orthogonality with the left eigenvectors suggests that a $2N \times 1$ complex vector $\mathbf{z}$ defined in the same vector space may be expressed as a linear combination of the eigenvectors of $\mathbf{S}$. That is,

$$\mathbf{z}(t) = \sum_{j=1}^{2N} \hat{\mathbf{U}}^{(j)} \hat{\eta}_j(t) \tag{8.79}$$

where the time dependent coefficients of the eigenvectors are, in general, complex. We shall use such an expansion in the following development.

8.8.3 Decomposition of the Forced Vibration Problem

Let us consider the forced vibration problem

$$\mathbf{m}\ddot{\mathbf{u}} + \mathbf{c}\dot{\mathbf{u}} + \mathbf{k}\mathbf{u} = \mathbf{F} \tag{8.80}$$

for a damped N-degree of freedom system, where $\mathbf{m}$, $\mathbf{c}$ and $\mathbf{k}$ are the $N \times N$ mass, damping and stiffness matrices of the system and $\mathbf{u}$ and $\mathbf{F}$ are the $N \times 1$ displacement and forces matrices, respectively. We shall approach the problem by consideration of the corresponding representation in state space. Hence, paralleling the development of Section 7.4.2, we first write

$$\dot{\mathbf{u}} = \mathbf{I}\mathbf{u} \tag{8.81}$$

where $\mathbf{I}$ is the $N \times N$ identity matrix. Multiplying Eq. (8.80) by $\mathbf{m}^{-1}$ and solving for $\ddot{\mathbf{u}}$ gives

$$\ddot{\mathbf{u}} = -\mathbf{m}^{-1}\mathbf{k}\mathbf{u} - \mathbf{m}^{-1}\mathbf{c}\dot{\mathbf{u}} + \mathbf{m}^{-1}\mathbf{F} \tag{8.82}$$

We next combine Eqs. (8.81) and (8.82) in matrix form giving the state space representation of the forced vibration problem as

$$\dot{\mathbf{z}} - \mathbf{S}\mathbf{z} = \hat{\mathbf{F}} \tag{8.83}$$

where

$$\mathbf{z}(t) = \begin{Bmatrix} \mathbf{u}(t) \\ \dot{\mathbf{u}}(t) \end{Bmatrix} \tag{8.84}$$

is the $2N \times 1$ state vector,

$$\hat{\mathbf{F}} = \begin{bmatrix} \mathbf{0} \\ \mathbf{m}^{-1} \end{bmatrix} \mathbf{F} \tag{8.85}$$

is the corresponding $2N \times 1$ force vector and

$$\mathbf{S} = \begin{bmatrix} \mathbf{0} & \mathbf{I} \\ -\mathbf{m}^{-1}\mathbf{k} & -\mathbf{m}^{-1}\mathbf{c} \end{bmatrix} \tag{8.86}$$

is the $2N \times 2N$ system matrix. In the remainder of this development we shall normalize the eigenvectors according to Eq. (7.77). That is, we shall eliminate a degree of indeterminacy of the eigenvectors by enforcing the relation

$$\hat{\mathbf{W}}^{(j)\mathrm{T}}\hat{\mathbf{U}}^{(j)} = 1 \quad (j = 1, 2, ..., 2N) \tag{8.87}$$

It then follows, from either Eq. (7.72) or Eq. (7.73), that

$$\hat{\mathbf{W}}^{(j)\mathrm{T}}\mathbf{S}\hat{\mathbf{U}}^{(j)} = \alpha_j \quad (j = 1, 2, ..., 2N) \tag{8.88}$$

where α_j is the jth eigenvalue of $\mathbf{S}$. Let us next introduce the $2N \times 2N$ right and left "modal" matrices

$$\hat{\mathbf{B}} = \begin{bmatrix} \hat{\mathbf{U}}^{(1)} & \cdots & \hat{\mathbf{U}}^{(2N)} \end{bmatrix} \quad \text{and} \quad \hat{\mathbf{A}} = \begin{bmatrix} \hat{\mathbf{W}}^{(1)} & \cdots & \hat{\mathbf{W}}^{(2N)} \end{bmatrix} \tag{8.89}$$

respectively. With the aid of these matrices and the orthogonality relations, Eqs. (7.75) and (7.76), Eqs. (8.87) and (8.88) can be expressed in the equivalent matrix forms

$$\hat{\mathbf{A}}^{\mathrm{T}}\hat{\mathbf{B}} = \hat{\mathbf{I}} \tag{8.90}$$

and

$$\hat{\mathbf{A}}^{\mathrm{T}}\mathbf{S}\hat{\mathbf{B}} = \boldsymbol{\Lambda} \tag{8.91}$$

where

$$\boldsymbol{\Lambda} = \begin{bmatrix} \alpha_1 & & 0 \\ & \ddots & \\ 0 & & \alpha_{2N} \end{bmatrix} \tag{8.92}$$

is the $2N \times 2N$ diagonal matrix of eigenvalues and

$$\hat{\mathbf{I}} = \begin{bmatrix} \mathbf{I} & \mathbf{0} \\ \mathbf{0} & \mathbf{I} \end{bmatrix}$$

is the $2N \times 2N$ identity matrix. It follows from Eq. (8.90) that, for the adopted normalization,

$$\hat{\mathbf{A}}^{\mathsf{T}} = \hat{\mathbf{B}}^{-1} \tag{8.93}$$

With the above established, we shall seek a solution to Eq. (8.83) as an expansion of the right eigenvectors of $\mathbf{S}$. Hence, let

$$\mathbf{z}(t) = \sum_{j=1}^{2N} \hat{\mathbf{U}}^{(j)} \hat{\eta}_j(t) = \hat{\mathbf{B}} \hat{\boldsymbol{\eta}}(t) \tag{8.94}$$

where

$$\hat{\boldsymbol{\eta}}(t) = \begin{Bmatrix} \hat{\eta}_1(t) \\ \hat{\eta}_2(t) \\ \vdots \\ \hat{\eta}_{2N}(t) \end{Bmatrix}$$

is the $2N \times 1$ matrix of the, as yet unknown, time dependent coefficients of the eigenvectors. Let us next substitute Eq. (8.94) into Eq. (8.83) and multiply the resulting expression on the left by $\hat{\mathbf{A}}^{\mathsf{T}}$. This results in the relation

$$\hat{\mathbf{A}}^{\mathsf{T}} \hat{\mathbf{B}} \dot{\hat{\boldsymbol{\eta}}}(t) - \hat{\mathbf{A}}^{\mathsf{T}} \mathbf{S} \, \hat{\mathbf{B}} \, \dot{\hat{\boldsymbol{\eta}}}(t) = \hat{\mathbf{A}}^{\mathsf{T}} \hat{\mathbf{F}}(t)$$

Substituting Eqs. (8.90) and (8.91) into the above identity reduces the transformed equation of motion to the form

$$\dot{\hat{\boldsymbol{\eta}}}(t) - \boldsymbol{\Lambda} \, \hat{\boldsymbol{\eta}}(t) = \hat{\mathbf{f}}(t) \tag{8.95}$$

where

$$\hat{\mathbf{f}}(t) = \hat{\mathbf{A}}^{\mathsf{T}} \hat{\mathbf{F}}(t) = \hat{\mathbf{B}}^{-1} \hat{\mathbf{F}}(t) \tag{8.96}$$

Recall that $\boldsymbol{\Lambda}$ is diagonal. We thus see that the transformed ("modal") equations of motion are decoupled. If we expand Eq. (8.95) we have the $2N$ uncoupled first order ordinary differential equations

$$\begin{aligned} \dot{\hat{\eta}}_1(t) - \alpha_1 \hat{\eta}_1(t) &= \hat{f}_1(t) \\ \dot{\hat{\eta}}_2(t) - \alpha_2 \hat{\eta}_2(t) &= \hat{f}_2(t) \\ &\vdots \\ \dot{\hat{\eta}}_{2N}(t) - \alpha_{2N} \hat{\eta}_{2N}(t) &= \hat{f}_{2N}(t) \end{aligned} \tag{8.97}$$

where $\hat{f}_j(t)$ is the jth element of $\hat{\mathbf{f}}(t)$. Each equation of the above system can be solved directly for the corresponding transformed ("modal") displacement $\hat{\eta}_j(t)$. Once these have been found for a given system with specified forcing, their values can be substituted back into Eq. (8.94) to obtain the state vector as a function of time. The physical displacements of the system are then the first N elements of $\mathbf{z}(t)$. The last N rows of $\mathbf{z}(t)$ are the corresponding velocities.

8.8.4 Solution of Forced Vibration Problems

The decomposition developed in the previous section allows for solution of forced vibration problems of damped N-degree of freedom systems. The general procedure is summarized below.

1. Identify a system of generalized coordinates to describe the motion of the system. Derive the equations of motion in terms of these coordinates.
2. Construct the system matrix $\mathbf{S}$ from the mass, damping and stiffness matrices. Solve the (right) eigenvalue problem to obtain the associated eigenvalues and corresponding (right) eigenvectors of $\mathbf{S}$.
3. Formulate the (right) "modal matrix" $\hat{\mathbf{B}}$ from the eigenvectors and compute its inverse. The latter corresponds to the transpose of the left "modal matrix" of $\mathbf{S}$. That is $\hat{\mathbf{B}}^{-1} = \hat{\mathbf{A}}^T$.
4. Evaluate the state space force vector $\hat{\mathbf{F}}(t)$ from the physical force array and then determine the transformed forces $\hat{f}_j(t)$ ($j = 1, 2, ..., 2N$). Solve the uncoupled transformed equations of motion to obtain the corresponding transformed state variables $\hat{\eta}_j(t)$.
5. Map back to the state space to obtain the state vector $\mathbf{z}(t)$. The first N rows of $\mathbf{z}(t)$ correspond to the physical displacements $\mathbf{u}(t)$.

The complementary solutions of Eq. (8.97) correspond to the free vibration solution discussed in Section 7.4. The particular solutions to Eq. (8.97) depend, of course, on the specific form of the transformed force. The above procedure is applicable to any type of dynamic loading. We consider impulse loading and general harmonic forcing below.

Impulse Loading

Let us take the Laplace Transform of the jth equation of Eq. (8.97). Doing this, using Eq. (5.18) and solving for $L\{\hat{\eta}_j(t)\}$ gives

$$L\{\hat{\eta}_j(t)\} = \frac{L\{\hat{f}_j(t)\} + \hat{\eta}_j(0)}{s - \alpha_j} \tag{8.98}$$

For a system that is initially at rest, $\hat{\eta}_j(0) = 0$. Now, let us suppose that the system is subjected to impulse loading. We therefore consider forces of the form

$$\hat{f}_j(t) = \hat{f}_j^0 \hat{\delta}(t) \tag{8.99}$$

where $\hat{\delta}(t)$ is the Dirac Delta Function. The Laplace transform of this force is, from Eq. (5.3),

$$L\{\hat{f}_j(t)\} = L\{\hat{f}_j^0 \hat{\delta}(t)\} = \hat{f}_j^0 \tag{8.100}$$

Substituting this specific force into Eq. (8.98) gives the Laplace transform of the "modal" displacement as

$$L\{\hat{\eta}_j(t)\} = \frac{\hat{f}_j^0 + \hat{\eta}_j(0)}{s - \alpha_j} \tag{8.101}$$

Equation (8.101) is inverted by direct application of Eq. (5.8) to give

$$\hat{\eta}_j(t) = \left[\hat{f}_j^0 + \hat{\eta}_j(0)\right] e^{\alpha_j t} \mathcal{H}(t) = \left[\hat{f}_j^0 + \hat{\eta}_j(0)\right] e^{(-\mu_j \pm \omega_j)t} \mathcal{H}(t) \tag{8.102}$$

where $\mathcal{H}(t)$ is the Heaviside Step Function. Mapping back to the state space using Eq. (8.94) gives the general form of the state vector as

$$z_l(t) = \sum_{j=1}^{2N} \hat{b}_{lj} \left[\hat{f}_j^0 + \hat{\eta}_j(0)\right] e^{\alpha_j t} \mathcal{H}(t) \quad (l = 1, 2, ..., 2N) \tag{8.103}$$

The physical displacements of the system correspond to the first N elements of the state vector. That is, $u_j(t) = z_j(t)$ $(j = 1, 2, ..., N)$.

Example 8.17

Consider the elastically supported frame of mass m and length L of Examples 7.13d, 7.14 and 8.16, where the stiffnesses of the two identical springs at each end of the frame have the magnitude $k = mg/L$ and the damping coefficients have the values $c_2 = 2c_1$ and $c_1/\sqrt{2km} = 0.1$. Let the frame be struck above the left support in such a way that an impulse of magnitude $\mathcal{I} = mg/\omega_0$ is imparted to the upper surface, as shown. Determine the response of the frame.

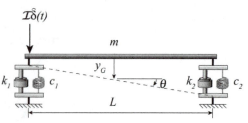

Solution

In keeping with our prior analyses of this system we choose the rotation of the frame, $\theta(t)$ (positive clockwise), and the transverse deflection of the center of mass, $y_G(t)$ (positive downward) as our generalized coordinates and normalize them accordingly. The equations of motion are then

$$\begin{bmatrix} 1 & 0 \\ 0 & 1 \end{bmatrix} \begin{Bmatrix} d^2\bar{y}_G/d\tau^2 \\ d^2\theta/d\tau^2 \end{Bmatrix}$$

$$+ \begin{bmatrix} (\eta_1+\eta_2) & (\eta_2-\eta_1)/2 \\ 6(\eta_2-\eta_1) & 3(\eta_1+\eta_2) \end{bmatrix} \begin{Bmatrix} d\bar{y}_G/d\tau \\ d\theta/d\tau \end{Bmatrix} + \begin{bmatrix} 1 & 0 \\ 0 & 3 \end{bmatrix} \begin{Bmatrix} \bar{y}_G \\ \theta \end{Bmatrix} = \begin{Bmatrix} \bar{F} \\ \bar{M}_G \end{Bmatrix} \qquad (a)$$

where

$$\bar{y}_G = y_G/L \qquad (b)$$

$$\tau = \omega_0 t = \sqrt{\frac{2k}{m}}\, t \qquad (c)$$

$$\omega_0^2 = k/m = g/L \qquad (d)$$

For the present loading condition,

$$\bar{F}(\tau) = F(\tau)/\omega_0^2 mL = \hat{\delta}(\tau)/\omega_0 \qquad (e\text{-}1)$$

and

$$\bar{M}_G(\tau) = M_G(\tau)/\omega_0^2 I_G = 12\, M_G(\tau)/\omega_0^2 mL^2 = 6\hat{\delta}(\tau)/\omega_0 \qquad (e\text{-}2)$$

are the normalized resultant applied force at and moment about an axis through point G. We next construct the pertinent system matrix **S** by substituting the nondimensional mass, damping and stiffness matrices into Eq. (7.61). For the particular system in question, $\eta_1 = 0.5$ and $\eta_2 = 1.0$. Hence,

$$\mathbf{S} = \begin{bmatrix} 0 & 0 & 1 & 0 \\ 0 & 0 & 0 & 1 \\ -1 & 0 & -(\eta_1+\eta_2) & -(\eta_2-\eta_1)/2 \\ 0 & -3 & -6(\eta_2-\eta_1) & -3(\eta_1+\eta_2) \end{bmatrix} = \begin{bmatrix} 0 & 0 & 1 & 0 \\ 0 & 0 & 0 & 1 \\ -1 & 0 & -1.5 & -0.25 \\ 0 & -3 & -3.0 & -4.5 \end{bmatrix} \qquad (f)$$

The state space force vector is then

$$\hat{\mathbf{F}}(\tau) = \begin{bmatrix} 0 & 0 \\ 0 & 0 \\ 1 & 0 \\ 0 & 1 \end{bmatrix} \begin{Bmatrix} \overline{F}(\tau) \\ \overline{M}_G(\tau) \end{Bmatrix} = \begin{Bmatrix} 0 \\ 0 \\ \overline{F}(\tau) \\ \overline{M}_G(\tau) \end{Bmatrix} = \begin{Bmatrix} 0 \\ 0 \\ 1 \\ 6 \end{Bmatrix} \frac{\hat{\delta}(\tau)}{\omega_0} \tag{g}$$

With the problem formulated, we next proceed to calculate the response. Using the MATLAB "eig" routine to solve the eigenvalue problem gives the eigenvalues of **S**,

$$\hat{\alpha} = -4.035, -(0.6810 - 0.8673i), -(0.6810 + 0.8673i), -0.6129 \tag{h}$$

and the corresponding matrix of (right) eigenvectors

$$\hat{\mathbf{B}} = \begin{bmatrix} 0.0216 & (0.4001+0.1545i) & (0.4001-0.1545i) & 0.2714 \\ 0.2401 & -(0.3193+0.4066i) & -(0.3193-0.4066i) & 0.8082 \\ -0.0871 & -(0.4065-0.2418i) & -(0.4065+0.2418i) & -0.1664 \\ -0.9666 & 0.5701 & 0.5701 & -0.4954 \end{bmatrix} \tag{i}$$

[Note that MATLAB normalizes the eigenvectors in the conventional sense, as per Eq. (7.39).] We next compute the inverse of the "modal matrix" using the MATLAB "inv" routine. Hence,

$$\hat{\mathbf{B}}^{-1} =$$
$$\begin{bmatrix} -0.3068 & -0.8508 & -1.235 & -1.142 \\ (0.3644-1.052i) & (-0.2564+0.2162i) & -(0.6640+1.032i) & (0.0043+0.1232i) \\ (0.3644+1.052i) & -(0.2564+0.2162i) & (-0.6640+1.032i) & (0.0043-0.1232i) \\ 1.437 & 1.070 & 0.8809 & 0.2186 \end{bmatrix} \tag{j}$$

With the above established, we now compute the "modal" forces, using Eq. (8.96). Hence,

$$\hat{\mathbf{f}}(t) =$$
$$\begin{bmatrix} -0.3068 & -0.8508 & -1.235 & -1.142 \\ (0.3644-1.052i) & (-0.2564+0.2162i) & -(0.6640+1.032i) & (0.0043+0.1232i) \\ (0.3644+1.052i) & -(0.2564+0.2162i) & (-0.6640+1.032i) & (0.0043-0.1232i) \\ 1.437 & 1.070 & 0.8809 & 0.2186 \end{bmatrix} \begin{Bmatrix} 0 \\ 0 \\ 1 \\ 6 \end{Bmatrix} \frac{\hat{\delta}(\tau)}{\omega_0}$$

which, after carrying through the indicated multiplications gives

$$\hat{\mathbf{f}}(t) = \begin{Bmatrix} -8.0839 \\ -0.6381 - 0.2929i \\ -0.6381 + 0.2929i \\ 2.1925 \end{Bmatrix} \dfrac{\delta(\tau)}{\omega_0}$$

(k)

The "modal" displacements are next found by substituting Eq. (k) into Eq. (8.113). Hence,

$$\omega_0 \hat{\eta}_1(\tau) = -8.084e^{-4.035\tau} \mathcal{H}(\tau)$$
$$\omega_0 \hat{\eta}_2(\tau) = -(0.6381 + 0.2929i)e^{-(0.6381+0.2929i)\tau} \mathcal{H}(\tau)$$
$$\omega_0 \hat{\eta}_3(\tau) = -(0.6381 - 0.2929i)e^{-(0.6381-0.2929i)\tau} \mathcal{H}(\tau)$$
$$\omega_0 \hat{\eta}_4(\tau) = 2.193e^{-0.6129\tau} \mathcal{H}(\tau)$$

(l)

To obtain the physical displacements, we map back to the state space as follows:

$$\omega_0 \bar{y}_G(\tau) = \omega_0 \sum_{j=1}^{4} \hat{b}_{1j} \hat{\eta}_j \, \mathcal{H}(\tau)$$

$$= 0.02160(-8.084)e^{-4.035\tau} \mathcal{H}(\tau)$$
$$- (0.4001 + 0.1545i)(0.6381 + 0.2929i)e^{-(0.6381+0.2929i)\tau} \mathcal{H}(\tau)$$
$$+ (0.4001 - 0.1545i)(0.6381 - 0.2929i)e^{-(0.6381-0.2929i)\tau} \mathcal{H}(\tau)$$
$$+ 0.2714(2.193)e^{-0.6129\tau} \mathcal{H}(\tau)$$

$$\omega_0 \theta(\tau) = \omega_0 \sum_{j=1}^{4} \hat{b}_{2j} \hat{\eta}_j(\tau) \mathcal{H}(\tau)$$

$$= 0.2401(-8.084)e^{-4.035\tau} \mathcal{H}(\tau)$$
$$- (0.3193 + 0.4066i)(-0.6381 - 0.2929i)e^{-(0.6381+0.2929i)\tau} \mathcal{H}(\tau)$$
$$- (0.3193 - 0.4066i)(-0.6381 + 0.2929i)e^{-(0.6381-0.2929i)\tau} \mathcal{H}(\tau)$$
$$+ 0.8082(2.193)e^{-0.6129\tau} \mathcal{H}(\tau)$$

Carrying through the indicated multiplication gives the response of the frame as

$$y_G(t)\omega_0/L = -0.1746e^{-4.035\omega_0 t} \mathcal{H}(t) + 0.5952e^{-0.6129\omega_0 t} \mathcal{H}(t)$$
$$- e^{-0.6810\omega_0 t} [0.4200\cos(0.8673\omega_0 t) - 0.4316\sin(0.8673\omega_0 t)] \mathcal{H}(t)$$

◁ (m-1)

$$\omega_0 \theta(t) = -1.941 e^{-4.035\omega_0 t} \mathcal{H}(t) + 1.772 e^{-0.6129\omega_0 t} \mathcal{H}(t)$$

$$+ e^{-0.6810\omega_0 t} \left[0.1612 \cos(0.8673\omega_0 t) - 0.7060 \sin(0.8673\omega_0 t) \right] \mathcal{H}(t) \qquad \triangleleft \text{(m-2)}$$

Arbitrary Loading

The response of a system to arbitrary loading is found by paralleling the arguments of Section 4.3 and incorporating the impulse response defined above to obtain the corresponding "modal displacements." Doing this we find that, for a system initially at rest,

$$\hat{\eta}_j(t) = \int_0^t \hat{f}_j(\tau) e^{\alpha_j(t-\tau)} d\tau \, \mathcal{H}(t) \qquad (8.104)$$

The above expression may be evaluated for any given $\hat{f}_j(\tau)$. Once the "modal displacements" are determined for a given set of forces, they may be mapped back to the state space using Eq. (8.94) to obtain the state vector $\mathbf{z}(t)$. The physical displacements of the system then correspond to the first N elements of this vector. Equation (8.94) is evaluated for step loading and ramp loading below.

Step Loading
Consider forces of the form

$$\hat{f}_j(t) = \hat{f}_j^0 \mathcal{H}(t) \qquad (8.105)$$

where $\hat{f}_j^0 = $ (complex) constant. Substituting the above step load function into Eq. (8.104) and evaluating the resulting integral gives the corresponding displacements

$$\hat{\eta}_j(t) = \frac{\hat{f}_j^0}{\alpha_j} \left[e^{\alpha_j t} - 1 \right] \mathcal{H}(t) \qquad (8.106)$$

Ramp Loading
Consider forces of the form

$$\hat{f}_j(t) = \hat{f}_j^0 \tau \mathcal{H}(t) \qquad (8.107)$$

where $\hat{f}_j^0 = $ (complex) constant. Substituting the above ramp load function into Eq. (8.104) and evaluating the resulting integral gives the corresponding displacements

$$\hat{\eta}_j(t) = \frac{\hat{f}_j^0}{\alpha_j^2}\left[e^{\alpha_j t} - (1 + \alpha_j t)\right]\mathcal{H}(t)$$

(8.108)

Harmonic Excitation

We close out the section by considering the state space representation of damped systems subjected to harmonic excitation. Since such an analysis involves inversion of the matrix of (right) eigenvectors of the system matrix it does not have the computational advantage over the procedure of Section 8.8.1 for large systems as does modal analysis over the approach of Section 8.1.1 for undamped systems in this regard. Further, since the eigenvectors are complex, the procedure does not offer the physical interpretation of its counterpart for undamped systems either. It is nevertheless instructive to apply this approach to harmonic loading of damped systems for the purposes of continuity and comparison, as well as for utility.

Suppose a multi-degree of freedom system is subjected to a system of generalized forces that vary harmonically in time with the same excitation frequency. Let the "modal" forces be of the form

$$\hat{f}_j(t) = \hat{f}_j^0 e^{i\Omega t}$$

(8.109)

where $\hat{f}_j^0$ = complex constant. The corresponding differential equation is then, from Eq. (8.97),

$$\dot{\hat{\eta}}_j(t) - \alpha_j \hat{\eta}_j(t) = \hat{f}_j^0 e^{i\Omega t}$$

(8.110)

To obtain the particular solution, let us assume a solution of the form

$$\hat{\eta}_j(t) = H_j^0 e^{i\Omega t}$$

(8.111)

Substituting the above form into Eq. (8.110) and solving for H_j^0 gives

$$H_j^0 = -\frac{\hat{f}_j^0}{\alpha_j - i\Omega}$$

(8.112)

Thus,

$$\hat{\eta}_j(t) = -\frac{\hat{f}_j^0}{\alpha_j - i\Omega} e^{i\Omega t}$$

(8.113)

As discussed in Section 7.4, for dissipative systems, the eigenvalues of **S** will be either negative and real or complex with negative real parts. We thus consider eigenvalues of the general form

$$\alpha_j = -\mu_j \pm i\omega_j \tag{8.114}$$

where ω_j is a damped natural frequency. Further, the magnitude of the "modal" force will generally be complex. Let us therefore write the complex magnitude of the transformed force in terms of its real and imaginary parts as

$$\hat{f}_j^0 = \hat{f}_j^{0R} + i\hat{f}_j^{0I} \tag{8.115}$$

Substituting Eqs. (8.114) and (8.115) into Eq. (8.113) gives the solution

$$\hat{\eta}_j(t) = \left\| H_j^0 \right\| e^{i(\Omega t + \Psi)} \tag{8.116}$$

where

$$\left\| H_j^0 \right\|^2 = \left(H_j^R \right)^2 + \left(H_j^I \right)^2 \tag{8.117}$$

$$\Psi = \tan^{-1}\left(H_j^I / H_j^R \right) \tag{8.118}$$

$$H_j^R = \frac{\mu_j \hat{f}_j^{0R} - (\Omega \pm \omega_j)\hat{f}_j^{0I}}{\mu_j^2 + (\Omega \pm \omega_j)^2} \tag{8.119}$$

$$H_j^I = \frac{\mu_j \hat{f}_j^{0I} + (\Omega \pm \omega_j)\hat{f}_j^{0R}}{\mu_j^2 + (\Omega \pm \omega_j)^2} \tag{8.120}$$

We may then transform back to the state space using Eq. (8.94). The first N rows of the resulting state vector correspond to the physical displacements $u_j(t)$ ($j = 1, 2, \ldots, N$). Euler's Formula can then be used to express the displacements in terms of harmonic functions. Then, as per the related discussions of Section 3.3.3, if the time dependence of the applied force $\mathbf{F}(t)$ is a cosine function, then the displacements due to this force correspond to the real part of the solution. That is, $u_j(t) = \operatorname{Re} z_j(t)$ ($j = 1, 2, \ldots, N$). Likewise, if the same applied force is a sine function, then the displacements due to that force correspond to the imaginary part of the solution. That is, $u_j(t) = \operatorname{Im} z_j(t)$ ($j = 1, 2, \ldots, N$). When considering force systems with multiple time dependencies, the results for each may be superposed.

Example 8.18

Consider once again the elastically supported frame of Examples 8.16 and 8.17. The frame of mass m and length L is supported at each edge by identical springs and disparate dampers. The springs each possess stiffness $k = mg/L$ and the dampers have the properties $c_2 = 2c_1$ and $c_1/\sqrt{2km} = 0.1$. If a force $F(t) = F_0 \sin \Omega t$ (positive downward) is applied to the left edge of the frame, as shown in Figure E8.16, determine the steady state motion of the frame when $F_0 = mg$ and $\Omega = 1.5g/L$.

Solution

The formulation of the problem is identical to that for Example 8.17 except for the time dependence of the excitation. Hence,

$$\begin{bmatrix} 1 & 0 \\ 0 & 1 \end{bmatrix} \begin{Bmatrix} d^2\bar{y}_G/d\tau^2 \\ d^2\theta/d\tau^2 \end{Bmatrix}$$

$$+ \begin{bmatrix} (\eta_1 + \eta_2) & (\eta_2 - \eta_1)/2 \\ 6(\eta_2 - \eta_1) & 3(\eta_1 + \eta_2) \end{bmatrix} \begin{Bmatrix} d\bar{y}_G/d\tau \\ d\theta/d\tau \end{Bmatrix} + \begin{bmatrix} 1 & 0 \\ 0 & 3 \end{bmatrix} \begin{Bmatrix} \bar{y}_G \\ \theta \end{Bmatrix} = \begin{Bmatrix} \bar{F} \\ \bar{M}_G \end{Bmatrix} \qquad \text{(a)}$$

where

$$\bar{y}_G = y_G/L \qquad \text{(b)}$$

$$\tau = \omega_0 t = \sqrt{\frac{2k}{m}}\, t \qquad \text{(c)}$$

$$\omega_0^2 = k/m = g/L \qquad \text{(d)}$$

For the present loading condition,

$$\bar{F}(\tau) = F(\tau)/\omega_0^2 mL = \sin(1.5\tau) \qquad \text{(f-1)}$$

and

$$\bar{M}_G(\tau) = M_G(\tau)/\omega_0^2 I_G = 12\, M_G(\tau)/\omega_0^2 mL^2 = 6\sin(1.5\tau) \qquad \text{(f-2)}$$

are the normalized resultant applied force at and moment about an axis through G. The state space force vector is then

$$\hat{\mathbf{F}}(\tau) = \begin{bmatrix} 0 & 0 \\ 0 & 0 \\ 1 & 0 \\ 0 & 1 \end{bmatrix} \begin{Bmatrix} \bar{F}(\tau) \\ \bar{M}_G(\tau) \end{Bmatrix} = \begin{Bmatrix} 0 \\ 0 \\ \bar{F}(\tau) \\ \bar{M}_G(\tau) \end{Bmatrix} = \begin{Bmatrix} 0 \\ 0 \\ 1 \\ 6 \end{Bmatrix} e^{i1.5\tau} \qquad \text{(g)}$$

where we have adopted the exponential form for the harmonic time dependence to facilitate the calculation. We will extract the appropriate portion of the solution at the end of our computations. For the particular system in question, $\eta_1 = 0.5$ and $\eta_2 = 1.0$. Hence the system matrix is

$$
\mathbf{S} =
\begin{bmatrix}
0 & 0 & 1 & 0 \\
0 & 0 & 0 & 1 \\
-1 & 0 & -(\eta_1 + \eta_2) & -(\eta_2 - \eta_1)/2 \\
0 & -3 & -6(\eta_2 - \eta_1) & -3(\eta_1 + \eta_2)
\end{bmatrix}
=
\begin{bmatrix}
0 & 0 & 1 & 0 \\
0 & 0 & 0 & 1 \\
-1 & 0 & -1.5 & -0.25 \\
0 & -3 & -3.0 & -4.5
\end{bmatrix}
\tag{h}
$$

as computed in Example 8.17. Now that the problem has been formulated, we proceed to calculate the response.

The eigenvalues of $\mathbf{S}$, the matrix of the associated right eigenvectors $\hat{\mathbf{B}}$ and its inverse were computed in Example 8.17 as

$$
\hat{\alpha} = -4.035, -(0.6810 - 0.8673i), -(0.6810 + 0.8673i), -0.6129
\tag{i}
$$

$$
\hat{\mathbf{B}} =
\begin{bmatrix}
0.0216 & (0.4001 + 0.1545i) & (0.4001 - 0.1545i) & 0.2714 \\
0.2401 & -(0.3193 + 0.4066i) & -(0.3193 - 0.4066i) & 0.8082 \\
-0.0871 & -(0.4065 - 0.2418i) & -(0.4065 + 0.2418i) & -0.1664 \\
-0.9666 & 0.5701 & 0.5701 & -0.4954
\end{bmatrix}
\tag{j}
$$

and

$$
\hat{\mathbf{B}}^{-1} =
$$

$$
\begin{bmatrix}
-0.3068 & -0.8508 & -1.235 & -1.142 \\
(0.3644 - 1.052i) & (-0.2564 + 0.2162i) & -(0.6640 + 1.032i) & (0.0043 + 0.1232i) \\
(0.3644 + 1.052i) & -(0.2564 + 0.2162i) & (-0.6640 + 1.032i) & (0.0043 - 0.1232i) \\
1.437 & 1.070 & 0.8809 & 0.2186
\end{bmatrix}
\tag{k}
$$

respectively. The "modal" forces for the present loading are computed, using Eq. (8.96),

$\hat{\mathbf{f}}(t) =$

$$
\begin{bmatrix}
-0.3068 & -0.8508 & -1.235 & -1.142 \\
(0.3644-1.052i) & (-0.2564+0.2162i) & -(0.6640+1.032i) & (0.0043+0.1232i) \\
(0.3644+1.052i) & -(0.2564+0.2162i) & (-0.6640+1.032i) & (0.0043-0.1232i) \\
1.437 & 1.070 & 0.8809 & 0.2186
\end{bmatrix}
\begin{bmatrix} 0 \\ 0 \\ 1 \\ 6 \end{bmatrix} e^{i1.5\tau}
$$

which, after carrying through the indicated multiplications gives

$$
\hat{\mathbf{f}}(t) =
\begin{Bmatrix}
-8.0839 \\
-0.6381-0.2929i \\
-0.6381+0.2929i \\
2.1925
\end{Bmatrix} e^{i1.5\tau}
\tag{l}
$$

Substituting Eq. (l) into Eq. (8.113) gives the matrix of modal displacements

$$
\boldsymbol{\eta}(t) =
\begin{Bmatrix}
-1.7635-0.6572i \\
-0.7174-0.2364i \\
0.0427-0.2818i \\
0.5118+1.2525i
\end{Bmatrix} e^{i1.5\tau}
\tag{m}
$$

We next map back to the state space to obtain the state vector as a function of time. Substituting Eqs. (j) and (m) into Eq. (8.94) gives

$$
\mathbf{z}(t) =
\begin{bmatrix}
0.0216 & (0.4001+0.1545i) & (0.4001-0.1545i) & 0.2714 \\
0.2401 & -(0.3193+0.4066i) & -(0.3193-0.4066i) & 0.8082 \\
-0.0871 & -(0.4065-0.2418i) & -(0.4065+0.2418i) & -0.1664 \\
-0.9666 & 0.5701 & 0.5701 & -0.4954
\end{bmatrix}
$$

$$
\bullet \begin{Bmatrix}
-1.7635-0.6572i \\
-0.7174-0.2364i \\
0.0427-0.2818i \\
0.5118+1.2525i
\end{Bmatrix} e^{i1.5\tau}
$$

Carrying the indicated matrix multiplication results in the state vector

$$
\mathbf{z}(t) =
\begin{Bmatrix}
-0.1622+0.2358i \\
0.1871-0.7109i \\
0.3538-0.2433i \\
1.0663+0.2807i
\end{Bmatrix} e^{i1.5\tau}
\tag{n}
$$

We now extract the first two rows of the state vector to obtain the displacements

$$\begin{Bmatrix} \overline{y}_G(t) \\ \theta(t) \end{Bmatrix} = \begin{Bmatrix} -0.1622 + 0.2358i \\ 0.1871 - 0.7109i \end{Bmatrix} e^{i1.5\tau} \tag{o}$$

Expanding Eq. (o) and using Euler's Formula gives the complex response of the system,

$$\overline{y}_G(\tau) = \left[-0.1622\cos(1.5\tau) + 0.2358\sin(1.5\tau) \right]$$
$$-i\left[0.2358\cos(1.5\tau) + 0.1622\sin(1.5\tau) \right] \tag{p-1}$$

and

$$\theta(\tau) = \left[0.1871\cos(1.5\tau) + 0.7109\sin(1.5\tau) \right]$$
$$+i\left[-0.7109\cos(1.5\tau) + 0.1871\sin(1.5\tau) \right] \tag{p-2}$$

Finally, since the excitation is a sine function, the physical response corresponds to the imaginary part of the computed displacements. The steady state motion of the frame is thus

$$y_G(t)/L = -\left[0.2358\cos(1.5\omega_0 t) + 0.1622\sin(1.5\omega_0 t) \right] \qquad \triangleleft \text{(q-1)}$$

$$\theta(t) = -0.7109\cos(1.5\omega_0 t) + 0.1871\sin(1.5\omega_0 t) \qquad \triangleleft \text{(q-2)}$$

The response calculated above compares exactly with the response calculated in Example 8.16 as, of course, it should.

8.9 CONCLUDING REMARKS

In this chapter we have considered the vibration of systems possessing multiple degrees of freedom when they are subjected to a variety of dynamic load conditions. We studied the behavior of systems with no damping, systems with a special type of viscous damping known as Rayleigh Damping and systems with general viscous damping. For the undamped case we began by developing a simple solution for systems subjected to harmonic loading. The advantage of such an approach was in its conceptual simplicity. Its implementation, however, requires the inversion of an $N \times N$ matrix, which may be computationally intensive for large systems. In addition, while this approach provides a convenient mathematical solution, little is gained in the way of detailing the fundamental mechanisms involved when a system is forced.

The concepts of coordinate transformations, principal coordinates and specifically modal coordinates were introduced. A fundamental approach called modal analysis was developed based on these ideas. In this approach, the response of the system is expressed as a linear combination of the corresponding modal vectors. Such an approach may be applied to study systems subjected to any type of forcing, including but not limited to harmonic excitation. The governing equations are transformed so that they are expressed in terms of modal coordinates (the time dependent coefficients of the modal vectors). When this is done, the governing equations map to a system of uncoupled ordinary differential equations of the individual modal coordinates and time. In the process, the manner in which the applied load is distributed to the individual modes is revealed and the modal coordinates are seen to correspond to the displacements of equivalent single degree of freedom systems associated with each mode. These equations may be solved by established methods and then mapped back to the physical space yielding the time response of the system. In this form the contribution of each individual mode to the overall response is unveiled. It is also seen how particular modes may be excited individually. In addition to the obvious physical significance of this procedure, the computations do not involve matrix inversion but rather involve simple matrix multiplication. Modal analysis was extended to systems with Rayleigh damping. In this case the damping matrix is proportional to the mass and stiffness matrices of the system, the constants of proportionality being properties of the system. Because of this special form, the response of such systems to forcing may be expressed in terms of the modal vectors for the corresponding undamped system. As a consequence of this, the analysis and results maintain all the benefits seen for the undamped systems. The chapter finished with a discussion of forced multi-degree of freedom systems possessing general viscous damping. For these systems a solution for harmonic excitation was first presented. This approach was seen to possess the simplicity and convenience, as well as the drawbacks, of the corresponding analysis for undamped systems considered in the introduction. For general loading, the concept of modal analysis was generalized and applied to the case of damped systems. To do this, the equations of motion were represented in the corresponding state space. Though analogous to that for undamped systems, the analysis for damped systems is much more cumbersome. The system matrix for the N-degree of freedom in this case is not symmetric and the associated (right) eigenvectors of the associated $2N \times 2N$ system matrix are not mutually orthogonal. They are, however, orthogonal to their counterparts — the left eigenvectors of the system matrix. Further, the procedure involves the inversion of a $2N \times 2N$ matrix of eigenvectors.

To this point we have studied discrete systems. That is, systems with a discrete distribution of mass. In the remaining chapters we abstract the present discussion to infinite degree of freedom systems possessing a continuous distribution of mass.

BIBLIOGRAPHY

Ginsberg, J.H., *Mechanical and Structural Vibrations: Theory and Applications*, Wiley, New York, 2001.

Hildebrand, F.B., *Methods of Applied Mathematics*, 2nd ed., Prentice-Hall, Englewood Cliffs, 1965.

Meirovitch, L., *Elements of Vibration Analysis*, 2nd ed., McGraw-Hill, New York, 1986.

Meirovitch, L., *Fundamentals of Vibrations*, McGraw-Hill, Boston, 2001.

Rayleigh, J.W.S., *The Theory of Sound*, Vol.1, Dover, New York, 1945.

Thomson, W.T., *Theory of Vibration with Applications*, 4th ed., Prentice-Hall, Englewood Cliffs, 1993.

PROBLEMS

8.1   Consider the system of Problems 6.15 and 7.5. Use the method of Section 8.1.1 to determine the steady state response of the system if $F(t) = F_0 \cos \Omega t$.

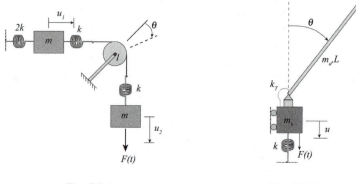

Fig. P8.1   **Fig. P8.2**

8.2   Consider the system of Problems 6.19 and 7.14 when $m_b = 2m_a = 2m$ and $k_T = kL^2$. Use the method of Section 8.1.1 to determine the steady state response of the system for small angle motion of the rod if a force $F(t) = F_0 \sin \Omega t$ is applied to the base mass.

8.3   Design a vibration absorber for the beam and mass structure of Example 3.4.

8.4   A machine is placed on the pontoon of Problem 3.4. At its operating frequency, the machine excites the pontoon near its resonance frequency. Design a vibration absorber for the pontoon. The mass of the machine is small compared with the mass of the float.

8.5   Show that the results of Examples 8.1b and 8.6a are the same.

8.6   Consider the constrained hook and ladder system of Problems 6.7 and 7.1 when $kL^2 = k_T$ and $m_1 = 10m_2 = 10m$. Determine the response of the system when a fireman of weight W suddenly mounts the end of the ladder.

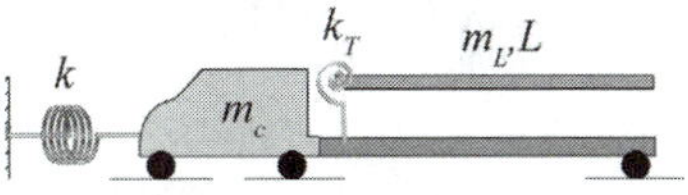

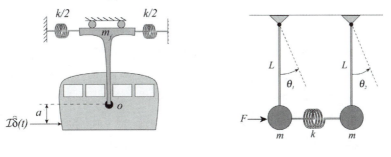

Fig. P8.6

8.7   Consider the tram of Problems 6.8 and 7.2 when $m_C = 5m_F$. Determine the response of the system if, when at rest, a horizontal impulse of magnitude $\mathcal{I}$ strikes the bottom of the car as shown.

Fig. P8.7   **Fig. P8.8**

8.8   Consider the coupled pendulums of Problems 6.9, 7.3. Determine the steady state response of the system when the left mass is subjected to the horizontal force $F(t) = F_0 \sin \Omega t$.

8.9   Consider the system of Problem 6.15, 7.5 and 8.1. Use modal analysis to determine the steady state response of the system when $F(t) = F_0 \cos \Omega t$.

8.10 An automobile is traveling at constant speed v_0 over a buckled road. Determine the motion of the car if the buckle is described as $y(x) = h_0 \sin(2\pi x/\lambda)$ where λ is the wavelength of the buckle and h_0 is its rise. Assume that the frame of the car may be modeled as a uniform rod of mass m and length L and the combined stiffness of the tires and suspension in both the front and the back is k.

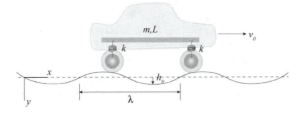

Fig. P8.10

8.11 An automobile is traveling at constant speed v_0 along a flat road when it encounters as speed bump described by $y(\xi) = -h_0\left[1 - \cos\left(2\pi\xi/\lambda\right)\right]$ where λ is the wavelength of the bump and h_0 is the height. Determine the motion of the car as it traverses the bump. Assume that the frame of the car may be modeled as a uniform rod of mass m and length L and the combined stiffness of the tires and suspension in both the front and the back is k.

Fig. P8.11

8.12 The conveyor belt system of Problems 6.10 and 7.24 ($m_1 = m_2 = m$, $R_1 = R_2 = R$) is at rest when a constant torque M_0 is suddenly applied to the left flywheel. Determine the response of the system.

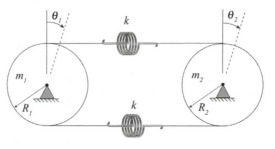

Fig. P8.12/P8.13

8.13 The conveyor belt system of Problems 6.10 and 7.24 ($m_1 = m_2 = m$, $R_1 = R_2 = R$) is at rest when it is loaded by a torque applied to the right flywheel. Determine the motion of the system if the magnitude of the applied torque is increased at the constant rate $\dot{F}$ to the level F_0 where it remains thereafter.

8.14 Consider the offshore platform of Problems 6.20 and 7.15. Each side of the square platform of mass m has length $L = 10R$. The cable of a small crane is fixed to the center of the platform and is attached to a diving bell of mass m_b that floats on the calm surface of the ocean. If the hoist suddenly engages and lifts the bell at constant speed, determine the motion of the platform. The boom of the crane is aligned so that its horizontal projection is of length L and coincides with the central axis of the platform. The dominant motion of the platform may, therefore, be considered planar. Assume that the effects of small angle motion of the bell on the tension in the cable can be neglected and that the

mass of the boom and cable is negligible compared with the mass of the platform and crane base.

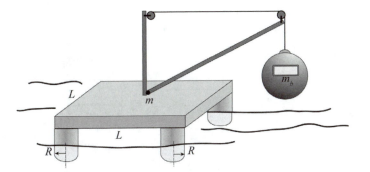

Fig. P8.14

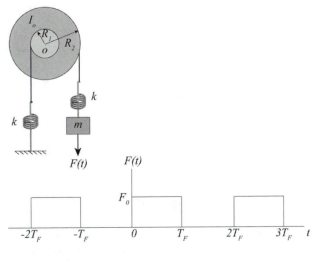

Fig. P8.15

8.15 Consider the pulley system of Problems 6.17 and 7.12 when $R_2 = 2R_1 = 2R$ and
$I_0 = mR^2$. Determine the steady state response of the system if the block is subjected to the periodic rectangular pulse of magnitude F_0 and duration T_F shown.

8.16 Consider the simple model of the submarine of Problems 6.21 and 7.26. Let the
mass of the hull and frame structure be $2m_s$ and the mass of the internal compartment be $m_c = m_s/2$. In addition, let the stiffness of each of the elastic mounts
between the internal compartment and the frame be k, and let the longitudinal
stiffness of the hull be $k_s = 2k$. The propeller exerts a force F_0 that is sufficient
to overcome the drag force and maintain the constant speed v_0. An imperfection

in the propeller shaft induces a small harmonic perturbation in the thrust of frequency Ω. The actual thrust applied is then of the form $F(t) = F_0[1 + \varepsilon_0 \sin \Omega t]$ where $\varepsilon_0 \ll 1$. If we neglect the corresponding perturbation of the drag force, determine the response of the submarine.

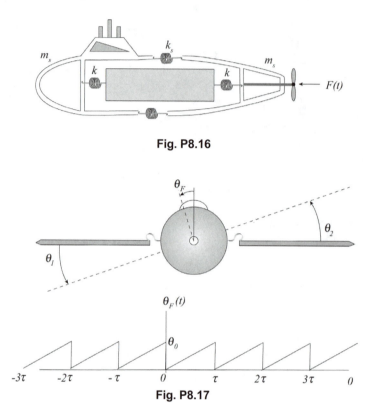

Fig. P8.16

Fig. P8.17

8.17 Consider the aircraft of Problems 6.31 and 7.28 where the wings are modeled as equivalent rigid bodies with torsional springs of stiffness k_T at the fuselage wall, each wing possesses moment of inertia I_c about its respective connection point and the fuselage of radius R has moment of inertia $I_o = I_c$ about its axis. Let the plane be traveling at constant altitude and speed as it undergoes a maneuver inducing a tight periodic rolling motion of the fuselage with the sawtooth time history shown. Determine the (perturbed) steady state motion of the aircraft under these conditions.

8.18 The dumbbell satellite of Problems 6.13 and 7.25 spins about its axis at the constant rate ω_0 as it travels in a fixed orbit at constant speed v_0, as shown. The orbit is maintained such that spin axis is always perpendicular to the surface of the earth during the orbit. The two compartments are each of mass m and an elastic access tube of undeformed length L, effective stiffness k and negligible mass connects them. To prepare for a docking maneuver thrusters are suddenly activated applying a couple of magnitude M_0 for time duration τ and are then suddenly shut off. Determine the motion of the satellite during this maneuver. Assume that deformations of the access tube do not significantly alter the moment inertia of the system about the spin axis.

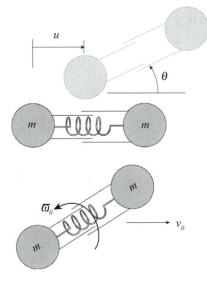

Fig. P8.18

8.19 Consider the elastically restrained conveyor belt of Problems 6.18 and 7.13 when $k_2 = 2k$ and $m_1 = m_2 = m$. Determine the steady state response if a torque $M(t) = M_0 \sin \Omega t$ is applied to the right flywheel as shown.

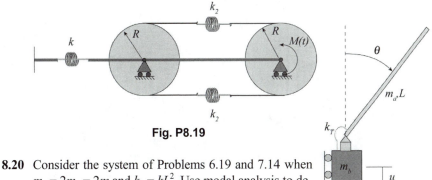

Fig. P8.19

8.20 Consider the system of Problems 6.19 and 7.14 when $m_b = 2m_a = 2m$ and $k_T = kL^2$. Use modal analysis to determine the steady state response of the system if a force $F(t) = F_0 \sin \Omega t$ is applied to the base mass.

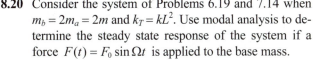

Fig. P8.20

8.21 Consider the linked system of Problems 6.22 and 7.16 when $k_1 = k_2 = k$ and $2m_2$ $= m_1 = m$. Determine the response of the system if the upper link is subjected to a (temporally) symmetric triangular pulse of magnitude F_0 and duration $\tau = m/k$.

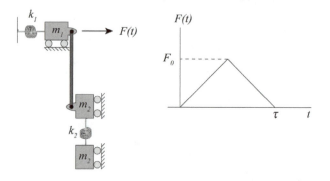

Fig. P8.21

8.22 Consider the mechanism of Problems 6.23 and 7.17 when $m_w = 3m$, $k_T = kR^2$ and the undeformed length of the coil equals the radius of the disk (i.e., $L = R$). Determine the steady state response of the system if the end of the arm is excited by the transverse harmonic load $F(t) = F_0 \sin \Omega t$.

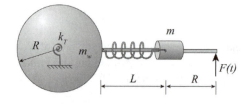

Fig. P8.22

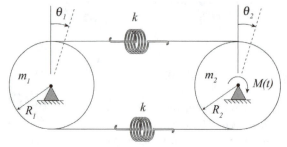

Fig. P8.23

8.23 The conveyor belt system of Problems 6.10 and 7.24 ($m_2 = m_1 = m$ and $R_2 = R_1 = R$) is at rest when it is loaded by a torque applied to the right flywheel. Determine the motion of the system if the magnitude of the applied torque is increased at the constant rate $\dot{M}$ to the level M_0 where it remains thereafter.

8.24 Consider the floating platform of Problems 6.24 and 7.18 when $m_b = m_a/2$ and $k = \rho_f g R^2 /2$. Determine the response of the system if $F(t) = F_0 \sin \Omega t$.

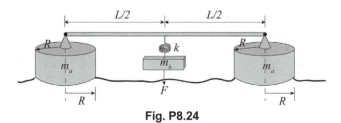

Fig. P8.24

8.25 Consider the shaft system of Problems 6.25 and 7.19. Determine the response of the system if the center disk is initially at rest when it is suddenly twisted by a constant torque of magnitude M_0.

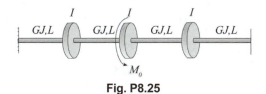

Fig. P8.25

8.26 Consider the triple pendulum of Problems 6.26 and 7.20. Determine the response of the system if it is initially at rest when the bottom most bob is impacted horizontally by an impulse of magnitude $\mathcal{I}$.

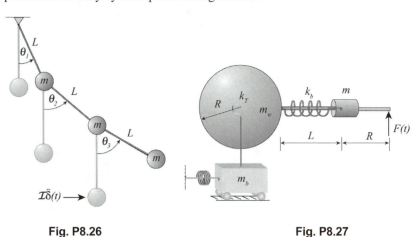

Fig. P8.26   **Fig. P8.27**

8.27 Consider the mechanism of Problems 6.27 and 7.21 (Figure P8.27) when $m_w = m_b = 3m$, $k_T R^2 = k_b = k$ and the undeformed length of the coil equals the radius of the disk (i.e., $L = R$). Determine the steady state response of the system if the end of the arm is excited by the transverse harmonic load $F(t) = F_0 \sin \Omega t$.

8.28 Consider the coupled pendulums of Problems 6.29 and 7.22. Determine the response of the system when the leftmost bob is subjected to a horizontally directed rectangular pulse of magnitude F_0 and duration $L/2g$.

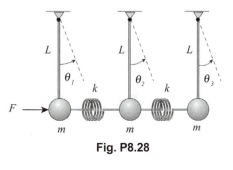

Fig. P8.28

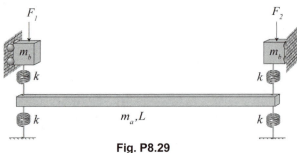

Fig. P8.29

8.29 Consider the frame system of Problems 6.30 and 7.23, where $m_b/m_a = 0.25$. Determine the motion of the system when $F_1(t) = F_{01} \sin \Omega_1 t$ (and $F_2 = 0$).

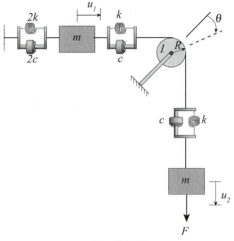

Fig. P8.30

8.30 Consider the system of Problems 6.36 and 7.40 ($c^2/km = 0.04$, $I = 2mR^2$). Determine the steady state response of the system if $F(t) = F_0 \cos \Omega t$.

8.31 Determine the response of the two-mass three-spring three-damper system of Example 7.32 when the left mass is subjected to the pulse $F(t) = \mathcal{I}\hat{\delta}(t)$.

8.32 Consider the system of Problem 6.32 and 7.38 when $m_b = 2m_a = 2m$, $k_T = kL^2$ and $c^2/km = 0.04$. Determine the steady state response of the system if a force $F(t) = F_0 \sin \Omega t$ is applied to the base mass.

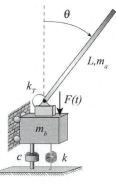

Fig. P8.32

8.33 Repeat Problem 8.16, this time taking into account the effect of the perturbation on the drag force where the corresponding coefficient is $c_d = c$.

8.34 Consider the simple model of the submarine of Problem 6.37, where the mass of the internal compartment is m_c, the mass of the hull and frame structure is $m_s = 2m_c$, the stiffness of each of the elastic mounts between the internal compartment and the hull frame is k, the damping factor for each of the mounts is $c = 0.1\sqrt{km_c}$, the longitudinal stiffness of the hull is $k_s = 2k$ and the effective damping factor of the hull is $c_s = 2c$. The propeller exerts a force F_0 that is sufficient to overcome the drag force and maintain the constant speed v_0. An imperfection in the propeller shaft induces a small harmonic perturbation in the thrust of frequency Ω. The actual thrust applied is then of the form $F(t) = F_0[1 + \varepsilon_0 \sin \Omega t]$ where $\varepsilon_0 \ll 1$. If we neglect the corresponding perturbation of the drag force, determine the response of the submarine.

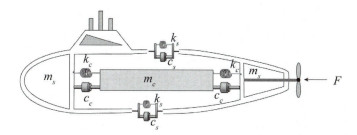

Fig. P8.34

8.35 Determine the response of the frame of Example 8.17 if it is struck at the left support by a rectangular pulse of amplitude F_0 and duration t_p.

8.36 The conveyor belt system of Problem 6.35 has the properties $m_1 = m_2 = m$, $R_1 = R_2 = R$ and $c^2/km = 0.16$. It is at rest when a constant torque M_0 is suddenly applied to the left flywheel. Determine the response of the system.

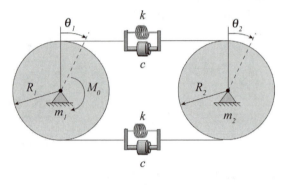

Fig. P8.36

9

Dynamics of One-Dimensional Continua

A material continuum is a medium or body that possesses a continuous, not necessarily uniform, distribution of mass and other material properties over its domain of definition. At the macroscopic scale many engineering systems lie in this category. One-dimensional continua are the simplest types of continua and correspond to bodies whose deformations are determined in terms of one spatial variable and time. Such bodies include elastic rods, strings and cables, and elastic beams and columns. To this point we have restricted our attention to systems for which an elastic rod or beam was a part, and for which the mass of the rod or beam was negligible compared with other mass measures of the total system. In this and subsequent chapters we shall study the motion of the elastic body itself. It will be seen that much, if not all, of the capabilities that were developed for discrete systems can be abstracted and generalized for continuous systems with the introduction of a few new concepts. As such, the study of vibrations of continuous systems will be seen to be completely analogous to that of discrete systems. In this chapter we derive the equations of motion of the systems of interest using elementary means. We begin by discussing the correlations between the mathematical representations of discrete systems and mathematically one-dimensional continua.

9.1 MATHEMATICAL DESCRIPTION OF 1-D CONTINUA

In the previous three chapters it was seen that the properties, external forces and motion of discrete systems are described by matrices. In contrast, the properties, external forces and behavior of continuous systems are naturally described by continuous functions. The purpose of this section is to provide a smooth and logical segue from discrete to continuous systems. In this section we view discrete and continuous systems from a unified perspective, with one being viewed as a limiting case of the other. This correlation will allow for a smooth extension of previously established concepts that will aid in the transition to, and in our interpretation of the behavior of, continuous systems.

9.1.1 Correspondence Between Discrete and Continuous Systems

Consider the discrete system of masses and springs aligned in series as shown in Figure 9.1a, and let the coordinate x_j ($j = 1, 2, ..., N$) correspond to the equilibrium position of mass m_j ($j = 1, 2, ..., N$) as indicated. The coordinate, x, originates at the left boundary of the system, and the system is defined on the domain $0 \leq x \leq L$. Let u_j ($j = 1, 2, ..., N$) correspond to the displacement of the indicated mass and let F_j ($j = 1, 2, ..., N$) be the corresponding external force acting on that mass as defined in the previous three chapters. With the introduction of the coordinate system x, the displacements and external forces may be thought of as values of the functions $u(x,t)$ and $F(x,t)$ defined at the discrete equilibrium coordinates as follows:

$$u(x_j,t) = u_j(t) , \quad F(x_j,t) = F_j(t) \quad (j = 1,2,...,N) \tag{9.1}$$

The forms displayed in Eq. (9.1) are thus alternative forms of the matrix representations

$$\mathbf{u}(t) = \begin{Bmatrix} u_1(t) \\ u_2(t) \\ \vdots \\ u_N(t) \end{Bmatrix}, \quad \mathbf{F}(t) = \begin{Bmatrix} F_1(t) \\ F_2(t) \\ \vdots \\ F_N(t) \end{Bmatrix} \tag{9.2}$$

Let us next consider the system to be comprised of a progressively increasing number of masses distributed over the fixed domain defined earlier. It follows that the distance between adjacent masses,

$$\Delta x_j \equiv x_{j+1} - x_j \quad (j = 0,1,2,...,N; \ x_0 \equiv 0) \tag{9.3}$$

decreases accordingly. Let ℓ denote the average axial length of the masses. It then follows that, in the limit as $N \to \infty$ and $\Delta x_j \to dx$ ($j = 0, 1, 2, ..., N$),

$$\mathbf{u}(t) \to u(x,t) \quad \text{and} \quad \mathbf{F}(t)/\ell \to p(x,t) \tag{9.4}$$

where $p(x,t)$ is the applied force per unit length of the continuous body.

Let us next consider the stiffness matrix for the system of Figure 9.1a and the product $\mathbf{ku}$ as appears in the corresponding equation of motion, Eq. (8.32). In the spirit of Eq. (9.1), let us introduce the notation

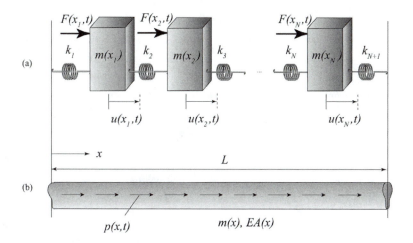

Figure 9.1 Discrete and continuous systems exhibiting longitudinal motion: (a) N-degree of freedom system, (b) continuous rod.

$$k(x_j) = \ell k_j \quad (j = 1, 2, ..., N) \tag{9.5}$$

and

$$\Delta k_j \equiv k_{j+1} - k_j \quad (j = 1, 2, ..., N) \tag{9.6}$$

Now, let us consider the jth row of the matrix product **ku**. Noting that

$$k(x_{j+1}) = k(x_j + \Delta x_j) = k(x_j) + \Delta k(x_j) \tag{9.7}$$

and substituting Eqs. (9.5)–(9.7) into the jth row of **ku** gives

$$\ell \{\mathbf{ku}\}_j = -k(x_j)\left[u(x_j) - \Delta u(x_{j-1})\right] + \left[k(x_j) + k(x_{j+1})\right]u(x_j)$$
$$- \left[k(x_j) + \Delta k(x_j)\right]\left[u(x_j) + \Delta u(x_j)\right]$$

which, after rearranging terms, takes the form

$$\ell \{\mathbf{ku}\}_j = -k(x_j)\overbrace{\left[\Delta u(x_j) - \Delta u(x_{j-1})\right]}^{\Delta\Delta u(x_j)} + \overbrace{\left[k(x_{j+1}) - k(x_j)\right]}^{\Delta k(x_j)}u(x_j)$$
$$- \Delta k(x_j)u(x_j) - \Delta k(x_j)\Delta u(x_j)$$

Hence,

$$\ell \{\mathbf{ku}\}_j = -k(x_j)\Delta\Delta u(x_j) - \Delta k(x_j)\Delta u(x_j) \tag{9.8}$$

Dividing Eq. (9.8) by ℓ^2, and multiplying and dividing by $(\Delta x_j)^2$, gives

$$\frac{1}{\ell}\{\mathbf{ku}\}_j = -\left(\frac{\Delta x_j}{\ell}\right)^2 \left\{ k(x_j)\frac{\Delta\Delta u(x_j)}{(\Delta x_j)^2} + \frac{\Delta k(x_j)}{\Delta x_j}\frac{\Delta u(x_j)}{\Delta x_j} \right\}$$

Letting $N \to \infty$ and $\Delta x_j \to 0$ simultaneously, $\Delta x_j/\ell \to 1$ and we find that

$$\frac{1}{\ell}\mathbf{ku} \;\to\; \Bbbk u(x,t) = -\frac{\partial}{\partial x}k(x)\frac{\partial}{\partial x}u(x,t) \tag{9.9}$$

where $\Bbbk u(x,t)$ is read "$\Bbbk$ operating on $u(x,t)$." It follows that

$$\frac{1}{\ell}\mathbf{k} \to \Bbbk = -\frac{\partial}{\partial x}k(x)\frac{\partial}{\partial x} \tag{9.10}$$

It is thus seen that, in the limit, the stiffness matrix tends to the differential *stiffness operator* $\Bbbk$. With the limit of the stiffness matrix established, we next evaluate the limit of the mass matrix.

It was seen in Section 6.2 that the mass matrix for the system of Figure 9.1a is of diagonal form. If we write the masses of the system as

$$m_j = \ell\, m(x_j) \quad (j = 1, 2, ..., N) \tag{9.11}$$

then the corresponding mass matrix takes the form

$$\mathbf{m} = \begin{bmatrix} m_1 & 0 & \cdots & 0 \\ 0 & m_2 & & 0 \\ \vdots & & \ddots & \vdots \\ 0 & 0 & \cdots & m_N \end{bmatrix} = \ell \begin{bmatrix} m(x_1) & 0 & \cdots & 0 \\ 0 & m(x_2) & & 0 \\ \vdots & & \ddots & \vdots \\ 0 & 0 & \cdots & m(x_N) \end{bmatrix} \tag{9.12}$$

It is seen that the elements of **m** are such that

$$m_{lj} = \ell m(x_j)\hat{\delta}_{lj} \quad (l, j = 1, 2, ..., N) \tag{9.13}$$

where $\hat{\delta}_{lj}$ is known as Knonecker's Delta and has the property that

$$\hat{\delta}_{lj} = \begin{cases} 1 & (\text{when } l = j) \\ 0 & (\text{when } l \neq j) \end{cases} \tag{9.14}$$

In the limit,

$$\frac{1}{\ell}\mathbf{m\ddot{u}} = \left\{ \begin{array}{c} m(x_1)\ddot{u}_1(x_1,t) \\ m(x_2)\ddot{u}_2(x_2,t) \\ \vdots \\ m(x_N)\ddot{u}_N(x_N,t) \end{array} \right\} \rightarrow \mathbb{m}\ddot{u}(x,t) = m(x)\ddot{u}(x,t) \qquad (9.15)$$

Hence, for the present system,

$$\frac{1}{\ell}\mathbf{m} \rightarrow \mathbb{m} = m(x) \qquad (9.16)$$

where $m(x)$ is interpreted as the mass per unit length of the structure. In general, the mass matrix for a discrete system will not be diagonal. For such systems, the limit of the mass matrix will be a differential operator, $\mathbb{m}$, in the spirit of the stiffness operator.

With the above limits established, the limit of the matrix equation that governs the system of Figure 9.1a, Eq. (8.32), takes the form

$$\mathbb{m}\ddot{u}(x,t) + \mathbb{k}u(x,t) = p(x,t) \qquad (9.17)$$

We thus see that, in the present context, a function is simply a limiting case of a column matrix or vector. We also see that the limiting case of a matrix operator (square matrix) is a differential operator. The limiting case of the discrete system of Figure 9.1a corresponds to the representation that describes longitudinal motion of an elastic rod (Figure 9.1b) and will be considered in detail in Section 9.3. The correlation between certain other discrete systems and their continuous counterparts discussed in later sections is described below.

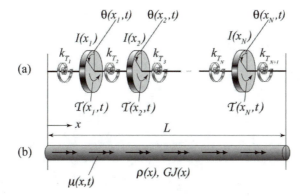

Figure 9.2 Discrete and continuous systems exhibiting torsional motion: (a) *N*-degree of freedom system, (b) continuous rod.

Suppose now that the masses of the discrete system of Figure 9.1a are replaced by rigid disks of known mass moment of inertia that are attached to massless rigid axles that are connected in series by torsional springs that rotate about the axis of the collinear axles as shown in Figure 9.2a. Correspondingly, let the linear displacements, $u_j(t)$, of the original system be replaced by angular displacements, $\theta_j(t)$, and let the external forces $F_j(t)$ be replaced by externally applied torques, $T_j(t)$. If we then proceed as we did for the original system, we arrive at a representation for the torsion of elastic rods in terms of the angular displacement field $\theta(x,t)$ and the applied distributed torque $\mu(x,t)$ shown in Figure 9.2b. This system will be considered in detail in Section 9.4.

Finally, consider a system of collinear rigid rods of known length and mass density, whose centers are located at the coordinates x_j, and let the rods be connected in series by torsional springs, as shown in Figure 9.3a. For this system, the torsional springs rotate about axes that are perpendicular to the axes of the rods as indicated. Let the rods be subjected to externally applied transverse forces, $Q_j(t)$, (forces that act perpendicular to the axis of the rods) and let us use the transverse displacement $w_j(t)$ = $\Delta x_j \sin \psi_j$, as the measure of displacement, where $\psi_j(t)$ is the angle of rotation of the rod, and Δx_j is defined by Eq. (9.3). If we proceed as we did for the two systems discussed previously we arrive at the representation for flexure of elastic beams in terms of the transverse displacement field $w(x,t)$ and the distributed transverse load $q(x,t)$ (Figure 9.3b). This system will be considered in detail in Sections 9.6–9.9.

To conclude, it is seen that, in the limit, column matrices tend to functions and the corresponding matrix operators tend to differential operators. The detailed derivation of the mathematical models for the systems discussed above and others, including the associated mass and stiffness operators, will be considered in later sections of this chapter.

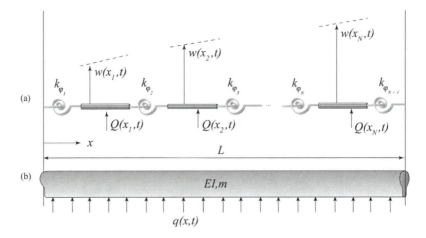

Figure 9.3 Discrete and continuous systems exhibiting flexural motion: (a) *N*-degree of freedom system, (b) continuous beam.

9.1.2 The Scalar Product and Orthogonality

It was seen in Chapter 8 that the orthogonality of the modal vectors is central to the analysis and comprehension of forced vibration of discrete systems. Since continua and their motions are described by continuous functions, and since it was shown in the previous section that functions are generalizations of column matrices and vectors, it may be anticipated that the orthogonality of functions will be important in our study of continuous systems and that the definitions and operations pertaining to scalar products and orthogonality, etc., can be abstracted/extended to functions. We next establish the corresponding definitions and operations. Other properties, such as normalization, will be established in later chapters as needed.

The Conventional Scalar Product

Consider two vectors **u** and **v** expressed in the forms of Eqs. (9.1) and (9.2). It follows from Eq. (7.24) that the scalar product of **u** and **v** is given by

$$\langle \mathbf{u}, \mathbf{v} \rangle = \mathbf{u}^{\mathsf{T}} \mathbf{v} = \sum_{j=1}^{N} u_j(t) v_j(t) = \sum_{j=1}^{N} u(x_j, t) v(x_j, t) \tag{9.18}$$

We next expand Eq. (9.18), multiply by ℓ, note that $u(x_{j+1}) = u(x_j + \Delta x_j)$ and take the limit as $N \to \infty$ and $\Delta x_j \to dx$. Recognizing that in the limit summation becomes integration, we see that the scalar product of two functions defined on the domain $0 \le x \le L$ is given by

$$\langle u, v \rangle = \int_0^L u(x,t) v(x,t) \, dx \tag{9.19}$$

As for the case of discrete vectors, two functions are said to be mutually orthogonal if their scalar product vanishes. Stated mathematically,

$$\text{if } \langle u, v \rangle = 0 \text{ then } u(x,t) \perp v(x,t) \tag{9.20}$$

Example 9.1

Determine the scalar product of the functions $f(x) = \sin(2n\pi x/L)$ and $g(x) = \sin(2p\pi x/L)$ on the domain $0 \le x \le L$, where $n \ne p$ are positive integers greater than zero.

Solution
From Eq. (9.19),

$$\langle f,g\rangle = \int_0^L f(x)g(x)dx = \int_0^L \sin\left(2n\pi x/L\right)\sin\left(2p\pi x/L\right)dx$$

$$= \frac{\sin\{2\pi(n-p)\}}{2\pi(n-p)/L} - \frac{\sin\{2\pi(n+p)\}}{2\pi(n+p)/L} = 0-0 = 0$$

◁

Thus, $f \perp g$ on $[0, L]$.

Note that the given functions would be orthogonal on the domain $[-L, L]$ as well. Compare the above functions with the functions used as the basis for conventional Fourier Series (Section 3.6). It may be seen that a Fourier Series is an expansion in terms of a set of mutually orthogonal functions.

The Weighted Scalar Product

It follows from Eq. (7.25), that the weighted scalar product of two vectors **u** and **v** with respect to a matrix **m** is given by

$$\langle\mathbf{u},\mathbf{v}\rangle_m = \mathbf{u}^\mathsf{T}\mathbf{m}\mathbf{v} = \sum_{l=1}^N\sum_{j=1}^N u_l(t)m_{lj}\,v_j(t) = \sum_{l=1}^N\sum_{j=1}^N u(x_l,t)m_{lj}\,v(x_j,t) \quad (9.21)$$

It is seen, upon incorporating Eq. (9.13) into Eq. (9.21), that the weighted scalar product with respect to a *diagonal* mass matrix simplifies to the form

$$\langle\mathbf{u},\mathbf{v}\rangle_m = \ell\sum_{j=1}^N u(x_j,t)m(x_j)v(x_j,t) \quad (9.22)$$

Taking the limit as $N \to \infty$ and $\ell, \Delta x_j \to dx$ gives the scalar product of the two functions $u(x,t)$ and $v(x,t)$ with respect to the *weight function* $m(x)$. Hence,

$$\langle u,v\rangle_m = \int_0^L u(x,t)m(x)v(x,t)dx \quad (9.23)$$

If the weighted scalar product vanishes, the functions $u(x,t)$ and $v(x,t)$ are said to be orthogonal with respect to the weight function $m(x)$. Stated mathematically,

$$\text{if } \langle u,v\rangle_m = 0 \quad \text{then} \quad u(x,t)\underset{m}{\perp}v(x,t) \quad (9.24)$$

The weighted scalar product of two vectors, **u** and **v**, with respect to the stiffness matrix **k** is defined as discussed in Section 7.3.1. Hence,

$$\langle \mathbf{u}, \mathbf{v} \rangle_{\mathbf{k}} = \mathbf{u}^{\mathsf{T}} \mathbf{k} \mathbf{v} = \sum_{l=1}^{N} \sum_{j=1}^{N} u_l(t) k_{lj} v_j(t) = \sum_{l=1}^{N} \sum_{j=1}^{N} u(x_l, t) k_{lj} v(x_j, t) \quad (9.25)$$

It follows from prior discussions that taking the limit as $N \to \infty$ and $\Delta x_j \to dx$ gives the weighted scalar product of the two functions $u(x,t)$ and $v(x,t)$ *with respect to the stiffness operator* $\mathbb{k}$. Hence,

$$\langle u, v \rangle_{\mathbb{k}} \equiv \int_0^L u(x,t) \mathbb{k} v(x,t) \, dx \quad (9.26)$$

If the weighted scalar product vanishes, the functions $u(x,t)$ and $v(x,t)$ may be said to be orthogonal with respect to the stiffness operator $\mathbb{k}$. Stated mathematically,

$$\text{if } \langle u, v \rangle_{\mathbb{k}} = 0 \quad \text{then} \quad u(x,t) \underset{\mathbb{k}}{\perp} v(x,t) \quad (9.27)$$

The general case of the scalar product and corresponding statement of orthogonality, with respect to a differential mass operator follows as for the stiffness, with the mass function being the operator as a special case as discussed below.

In general, the scalar product of two functions $u(x,t)$ and $v(x,t)$ with respect to some linear differential operator $\mathbb{d}$ is defined in an analogous fashion to that for the case above. The mutual orthogonality of the two functions with respect to the differential operator follows directly. Thus, in general, the weighted scalar product of two functions, $u(x,t)$ and $v(x,t)$, with respect to a differential operator

$$\mathbb{d} = \frac{\partial^n}{\partial x^n} S(x) \frac{\partial^p}{\partial x^p} \quad (9.28)$$

where $S(x)$ represents some system property, is defined as

$$\langle u, v \rangle_{\mathbb{d}} \equiv \int_0^L u(x,t) \mathbb{d} v(x,t) \, dx \quad (9.29)$$

The corresponding statement of orthogonality follows directly. Hence,

$$\text{if } \langle u, v \rangle_{\mathbb{d}} = 0 \quad \text{then} \quad u(x,t) \underset{\mathbb{d}}{\perp} v(x,t) \quad (9.30)$$

It is seen that the scalar product and orthogonality with respect to weight functions follows directly from Eqs. (9.28)–(9.30) for operators where the order of the spatial derivatives is reduced to zero (i.e., when $n = p = 0$). Finally, it is evident that $\langle u, v \rangle_{\mathbb{d}} = \langle u, \mathbb{d} v \rangle$.

Vector Functions and Differential Matrix Operators

For more complex systems the above definitions may be generalized. For certain types of structures the displacement field may be described by more than one scalar function. In this case the displacement is described by a vector (matrix) function of the form

$$\mathbf{u}(x,t) = \begin{Bmatrix} u_1(x,t) \\ u_2(x,t) \\ \vdots \\ u_N(x,t) \end{Bmatrix}$$

and the corresponding mass and stiffness operators are of the general form

$$\mathbf{d} = \begin{bmatrix} d_{11} & d_{12} & \cdots & d_{1N} \\ d_{21} & d_{22} & \cdots & d_{2N} \\ \vdots & \vdots & \ddots & \vdots \\ d_{N1} & d_{N1} & \cdots & d_{NN} \end{bmatrix}$$

where d_{lj} $(l,j = 1, 2, \ldots, N)$ are differential operators. For such systems we extend the definition of scalar product between two vector functions. In this case we define the scalar product between two vector functions $\mathbf{u}(x,t)$ and $\mathbf{v}(x,t)$ as

$$\langle \mathbf{u}, \mathbf{v} \rangle_{\mathbf{d}} \equiv \int_0^L \mathbf{u}^{\mathsf{T}} \mathbf{d} \mathbf{v}\, dx \tag{9.31}$$

Correspondingly the vector functions are said to be orthogonal with respect to the differential operator if the scalar product vanishes. That is

$$\text{if} \quad \langle \mathbf{u}, \mathbf{v} \rangle_{\mathbf{d}} \equiv \int_0^L \mathbf{u}^{\mathsf{T}} \mathbf{d} \mathbf{v}\, dx = 0 \quad \text{then} \quad \mathbf{u} \underset{\mathbf{d}}{\perp} \mathbf{v} \tag{9.32}$$

Finally, it is evident that $\langle \mathbf{u}, \mathbf{v} \rangle_{\mathbf{d}} = \langle \mathbf{u}, \mathbf{d}\mathbf{v} \rangle$. The above definitions can be extended to include multi-dimensional domains as well.

9.2 CHARACTERIZATION OF LOCAL DEFORMATION

To study the motion of continua, in particular the vibrations of continua, we are interested in the relative motion of adjacent material particles. Such relative motion is generally termed deformation of a material body. In the present and the next two chapters we shall be concerned with one-dimensional continua. Such systems typically correspond to continua that are geometrically long and thin. More precisely, we shall consider structures for which one dimension, the axial, is much larger than the others. We shall herein be interested in two types of deformation, stretching and distortion, and shall limit our discussion to those measures pertinent to one-dimensional continua.

9.2.1 Relative Extension of a Material Line Element

One mode of deformation can be characterized by examination of the relative extension/contraction of a line element in a continuous medium or body. This measure is often referred to as the normal strain. When the strain is measured in a direction parallel to the major axis of a long thin body, the normal strain is also referred to as the axial strain. In all of our discussions we shall consider infinitesimal strains. That is, we shall consider strains whose magnitudes are small compared with unity.

Pure Translation

We first consider an element of a material line that is originally aligned parallel to a coordinate axis, say the x-axis, and remains oriented parallel to that axis throughout its motion. Thus, let us consider a line element of initial length dx emanating from coordinate x, as shown in Figure 9.4. During the course of its motion, the element translates and stretches so that the left end of the element is currently located at coordinate $\bar{x}$ and the current length of the element is $d\bar{x}$ as indicated. Let $u(x,t)$ correspond to the displacement of the left end of the element. For continuous displacements of the material line, the right end then displaces an amount $u + (\partial u/\partial x)dx$, as shown. It then follows, as may be seen from Figure 9.4, that

$$d\bar{x} = dx + \frac{\partial u}{\partial x}dx$$

The relative extension, ε_{xx}, is then

$$\varepsilon_{xx} = \frac{d\bar{x} - dx}{dx} = \frac{\partial u}{\partial x} \tag{9.33}$$

and is typically referred to as the *normal strain*.

Translation and Local Rotation

Let us now consider the same material line initially emanating from coordinate (x,z) and of initial length dx as shown in Figure 9.5. However, we now remove the restriction of motions of pure translation and allow small rotations in a plane, say the xz-plane, as well as extension and contraction. Let u and w respectively correspond to the displacements in the x and z directions of the left end of the element as shown.

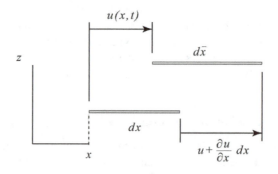

Figure 9.4 Deformation of material line element in pure translation.

The right ends then displace correspondingly, as indicated. Since the element is no longer parallel to the *x*-axis when in the current configuration, let the current length of the element be denoted as *ds*, and the projections of *ds* onto the coordinate axes be denoted as $d\bar{x}$ and $d\bar{z}$, respectively. We wish to evaluate the relative extension of the line element in terms of the displacements and their spatial gradients. It may be seen from Figure 9.5 that

$$ds = \sqrt{d\bar{x}^2 + d\bar{z}^2} = \sqrt{\left(dx + \frac{\partial u}{\partial x}dx\right)^2 + \left(\frac{\partial w}{\partial x}dx\right)^2} = dx\sqrt{1 + 2\frac{\partial u}{\partial x} + \left(\frac{\partial u}{\partial x}\right)^2 + \left(\frac{\partial w}{\partial x}\right)^2}$$

which, after expressing the radical by its series expansion gives the relation

$$ds = dx\left[1 + \frac{\partial u}{\partial x} + \frac{1}{2}\left(\frac{\partial u}{\partial x}\right)^2 + \frac{1}{2}\left(\frac{\partial w}{\partial x}\right)^2 + \dots\right] \qquad (9.34)$$

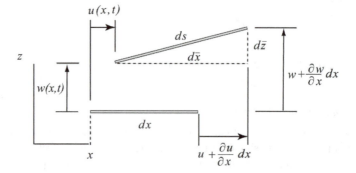

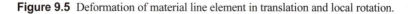

Figure 9.5 Deformation of material line element in translation and local rotation.

The relative extension of the material line element is then, to second order,

$$\frac{ds - dx}{dx} = \frac{\partial u}{\partial x} + \frac{1}{2}\left(\frac{\partial u}{\partial x}\right)^2 + \frac{1}{2}\left(\frac{\partial w}{\partial x}\right)^2 \cong \frac{\partial u}{\partial x} + \frac{1}{2}\left(\frac{\partial w}{\partial x}\right)^2 \tag{9.35}$$

where, for small displacement gradients, the square of a quantity is neglected when compared with that quantity to the first power. We shall consider two classifications of infinitesimal strain: small strain with "moderate" rotations, and small strain with "small" rotations.

Small Strain, Moderate Rotations

We here consider infinitesimal extension ratios. We, therefore, consider deformations where the magnitude of each term in Eq. (9.35) is small compared with unity. If

$$\left(\frac{\partial w}{\partial x}\right)^2 \sim O\left(\frac{\partial u}{\partial x}\right)$$

then the nonlinear term must be retained and the infinitesimal strain ε_{xx} is given by

$$\varepsilon_{xx} = \frac{\partial u}{\partial x} + \frac{1}{2}\left(\frac{\partial w}{\partial x}\right)^2 \tag{9.36}$$

The nonlinear term in the above expression may be identified as the angle of rotation of the line element in the xz-plane since, for small angles, the tangent is approximated by the angle itself and vice-versa. It is seen that the axial motion and transverse motion are coupled through this term. If we considered the element to move out of the plane as well as in it, and we paralleled the above development, an analogous nonlinear term corresponding to out of plane rotation would be added to the right hand side of Eq. (9.36).

Small Strain, Small Rotations

When the nonlinear term in Eq. (9.36) is small compared to the first, the rotations are said to be "small." In this case the nonlinear term is often neglected, resulting in a common definition of infinitesimal strain. Stated mathematically, if

$$\frac{\partial w}{\partial x} \sim O\left(\frac{\partial u}{\partial x}\right)$$

then the infinitesimal strain is often taken as

$$\varepsilon_{xx} \approx \frac{\partial u}{\partial x} \tag{9.37}$$

This is the common form of infinitesimal normal strain used in linear problems. It may be seen that, for this case, the current length, ds, is approximated by its projection, dx (Figure 9.5). When this approximation is made, all coupling between axial and transverse motion is ignored. The modeler should be aware of this limitation as such omissions can have significant ramifications in certain settings, even for small rotations.

9.2.2 Distortion

Consider a differential element whose faces are initially parallel to the coordinate planes. In particular, let us focus on the face parallel to the xz-plane, as shown in Figure 9.6. During motion, the element will generally translate rotate, stretch and distort. We shall here be concerned with the distortion of the face shown. Thus, let the edges of the element, originally perpendicular to one another, be currently oriented at angles ϕ_{xz} and ψ_{xz} with respect to the coordinate planes as indicated. We shall label the subtended angle as θ_{xz} as shown. For small angle changes, $\phi_{xz} \ll 1$ and $\psi_{xz} \ll 1$, it follows that

$$\phi_{xz} \approx \mathrm{Tan}\,\phi_{xz} = \frac{\partial w}{\partial x} \quad \text{and} \quad \psi_{xz} \approx \mathrm{Tan}\,\psi_{xz} = \frac{\partial u}{\partial z} \tag{9.38}$$

The change in angle parallel to the xz-plane is then

$$\gamma_{xz} = \frac{\pi}{2} - \theta_{xz} \approx \frac{\partial w}{\partial x} + \frac{\partial u}{\partial z} \tag{9.39}$$

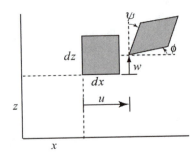

Figure 9.6 Distortion of differential material element.

The corresponding *shear strain* is formally defined as half the angle change, hence

$$\varepsilon_{xz} = \frac{1}{2}\left(\frac{\partial w}{\partial x} + \frac{\partial u}{\partial z}\right) \tag{9.40}$$

Analogous measures are defined for distortion on other planes. With the relation between discrete and continuous systems understood, and the measures of characterizing deformation established, we now examine several models for studying various motions of one-dimensional continua.

9.3 LONGITUDINAL MOTION OF ELASTIC RODS

In this section we derive the equation of motion, and the general boundary and initial conditions that govern the longitudinal motion of elastic rods. That is, we shall here be interested in motions of material particles whose directions parallel that of the axis of the rod. For small deformations of thin rods, rods whose axial deformations as well as lateral dimensions are small compared with the overall length of the rod, the axial motion and associated stress may be considered to be approximately uniform over a given cross section. Similarly, the Poisson effect (the lateral contraction/extension that accompanies axial stretching/compression) may also be neglected, rendering the problem mathematically one-dimensional in nature. In what follows, we shall consider a moderately general case, where the material properties of the rod may vary smoothly along its length but are invariant over a given cross section. We thus consider an orthotropic rod in this sense. Similarly, the shape of the rod is arbitrary and the cross-sectional area may vary in the axial direction. The case of a uniform isotropic rod is thus recovered as a special case.

Consider an elastic rod of length L and cross-sectional area $A(x)$, where the coordinate x parallels the axis of the rod and originates at its left end as shown in Figure 9.7a. Let the rod be comprised of material of mass density $\rho(x)$ and elastic modulus $E(x)$. The mass per unit length of the rod is then $m(x) = \rho(x)A(x)$. In addition, let the rod be subjected to the externally applied distributed axial force $p(x,t)$ as depicted in Figure 9.1b. As discussed in the preceding paragraph, we assume that the stress and deformation is uniform over a cross section and, therefore, that they each are a function of the axial coordinate x and time t only. Since we neglect the Poisson effect as well, the axial stress $\sigma(x,t)$ is solely dependent on the axial strain $\varepsilon(x,t)$ through the elementary form of Hooke's Law. Hence,

$$\sigma(x,t) = E(x)\varepsilon(x,t) = E(x)\frac{\partial u}{\partial x} \tag{9.41}$$

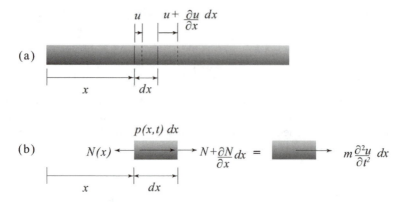

Figure 9.7 Longitudinal motion of rod: (a) kinematical description, (b) kinetic diagram of differential rod element.

where $u(x,t)$ corresponds to the axial displacement of the cross section located at coordinate x when in the rest configuration. Since the stress is assumed to be uniform over a cross section, the resultant membrane force, $N(x,t)$, acting on a cross section is then

$$N(x,t) = k_a(x)\frac{\partial u(x,t)}{\partial x} \tag{9.42}$$

where

$$k_a(x) = E(x)A(x) \tag{9.43}$$

is identified as the axial stiffness (per unit length) of the rod.

With the internal force measure for the rod established, we now proceed to derive the equation of motion. Toward this end, we first examine the kinetic diagram for a generic differential element of the rod shown in Figure 9.7b. We next apply Newton's Second Law to the element, which takes the form

$$p(x,t)dx + \left[N(x,t) + \frac{\partial N(x,t)}{\partial x}dx\right] - N(x,t) = m(x)dx\frac{\partial^2 u}{\partial t^2}$$

Rearranging terms and incorporating Eq. (9.42) gives the local form of the equation of motion for the rod as

$$m(x)\frac{\partial^2 u(x,t)}{\partial t^2} - \frac{\partial}{\partial x}k_a(x)\frac{\partial}{\partial x}u(x,t) = p(x,t) \tag{9.44}$$

Equation (9.44) is seen to be a second order partial differential equation for the axial displacement and may be identified as the one-dimensional wave equation. As such, the solution of this equation requires the specification of two boundary conditions and two initial conditions. We consider the boundary conditions first.

To obtain a solution to the equation of motion, one term (but not both terms) of the work $W = Nu$ must be prescribed at two points on the structure, typically at the boundaries. (Note that if one term is prescribed then its conjugate represents the corresponding response or reaction and therefore cannot be specified independently.) The general forms of the boundary conditions for the elastic rod are stated below. At the left edge of the rod, we must specify either

$$N(0,t) = k_a \left. \frac{\partial u}{\partial x} \right|_{x=0} = \mathcal{N}_0(t) \quad \text{or} \quad u(0,t) = h_0(t) \tag{9.45}$$

where $\mathcal{N}_0(t)$ is a prescribed edge load or $h_0(t)$ is a prescribed edge displacement. Similarly, at the right edge of the rod, we must specify either

$$N(L,t) = k_a \left. \frac{\partial u}{\partial x} \right|_{x=L} = \mathcal{N}_L(t) \quad \text{or} \quad u(L,t) = h_L(t) \tag{9.46}$$

where $\mathcal{N}_L(t)$ is a prescribed edge load or $h_L(t)$ is a prescribed edge displacement.

In addition to the boundary conditions, the initial displacement and initial velocity of each material point in the rod must be specified as well. The solution of the equation of motion therefore requires the initial conditions

$$u(x,0) = u_0(x) \tag{9.47}$$

and

$$\left. \frac{\partial u}{\partial t} \right|_{t=0} = v_0(x) \tag{9.48}$$

where $u_0(x)$ and $v_0(x)$ are prescribed functions that describe the initial state of the rod. The boundary value problem corresponding to longitudinal motion of an elastic rod is defined by Eqs. (9.44)–(9.48).

Before leaving our discussion of elastic rods it is useful to rewrite the equation of motion, Eq. (9.44), in operator form. This will aid in the comparison of various systems as well as with the interpretation and solution in this and subsequent chapters. Hence, let us define the *stiffness operator* for the rod as

$$\mathbb{k} = -\frac{\partial}{\partial x} k_a(x) \frac{\partial}{\partial x} \tag{9.49}$$

Comparison of Eq. (9.49) with Eq. (9.10) shows the stiffness operator for the rod to be of the identical form as that via the limiting process of Section 9.1. For completeness, let us similarly define the *mass operator* $\mathbb{m}$ for the present system as

$$\mathbb{m} = m(x) \tag{9.50}$$

Incorporating Eqs. (9.49) and (9.50) into Eq. (9.44) gives the equation of motion for the rod in operator form. The alternate form of the equation of motion governing longitudinal motion of elastic rods is then

$$\mathbb{m}\frac{\partial^2 u}{\partial t^2} + \mathbb{k}u = p(x,t) \tag{9.51}$$

where $\mathbb{k}$ and $\mathbb{m}$ for the rod are defined above. Comparison of Eq. (9.51) with Eq. (8.32) shows the mass and stiffness operators to be completely analogous to the mass and stiffness matrices for discrete systems. In fact, the operators for longitudinal motion of a rod were shown to be limiting cases of these matrices in Section 9.1.

Example 9.2

State the boundary conditions for longitudinal motion of an elastic rod for the two cases shown.

Solution
In both case (i) and case (ii) the left end of the rod is fixed. Thus, for both cases,

$$u(0,t) = 0 \tag{a}$$

For case (i) the right end of the rod is subjected to the force $F(t) = F_0 \sin \Omega t$. Hence,

$$N(L,t) = F_0 \sin \Omega t \quad \Rightarrow \quad EA\frac{\partial u}{\partial x}\bigg|_{x=L} = F_0 \sin \Omega t \tag{b}$$

For case (ii) the right end of the rod is free. That is the applied load is zero on this edge. Since the right edge is stress free, the boundary condition is

$$EA\frac{\partial u}{\partial x}\bigg|_{x=L} = 0 \tag{c}$$

Example 9.3

Determine the boundary conditions for an elastic rod contained between two elastic walls of stiffness k_w.

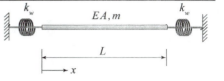

Solution

To establish the boundary conditions for this system, consider the displacements at the edges of the rod to be positive. Then, the spring at the left edge of the rod is extended and the spring at the right end of the rod is compressed. It follows that a tensile load is applied at $x = 0$ while a compressive load is applied to the rod at $x = L$. The boundary conditions are then

$$N(0,t) = k_w u(0,t) \quad \Rightarrow \quad EA\frac{\partial u}{\partial x}\bigg|_{x=0} = k_w u(0,t) \tag{a}$$

and

$$N(L,t) = -k_w u(L,t) \quad \Rightarrow \quad EA\frac{\partial u}{\partial x}\bigg|_{x=L} = -k_w u(L,t) \tag{b}$$

Example 9.4

Determine the boundary conditions for the elastic rod shown, if the support undergoes the horizontal motion $A_0 \sin\Omega t$ and a rigid block of mass m is attached to the right end.

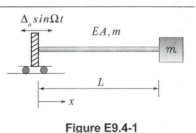

Figure E9.4-1

Solution

The left edge of the rod is affixed to the support. Therefore, the displacement at the left edge is prescribed. The corresponding boundary condition is then

$$u(0,t) = A_0 \sin\Omega t \tag{a}$$

To establish the boundary condition at the right edge of the rod, it is helpful to consider the kinetic diagram for the mass and rod (Fig. E9.4-2). From Newton's Third Law, the force acting on the rigid mass equal and opposite to the resultant membrane force acting on the edge of the rod, as indicated in the figure.

Figure E9.4-2 Kinetic diagram of end mass.

Applying Newton's Second Law to the rigid mass gives

$$-N(L,T) = m \frac{\partial^2 u}{\partial t^2}\bigg|_{x=L}$$

Hence,

$$-EA \frac{\partial u}{\partial x}\bigg|_{x=L} = m \frac{\partial^2 u}{\partial t^2}\bigg|_{x=L} \qquad (b)$$

9.4 TORSIONAL MOTION OF ELASTIC RODS

In this section we derive the equation of motion and general boundary and initial conditions that govern the torsional motion of solid elastic rods of circular cross section. The small strain theory of torsion of elastic rods attributed to St. Venant predicts that no warping of cross sections or axial extension accompanies twisting of rods with circular cross sections. Though we restrict our attention to rods of circular cross section we shall, however, allow the radius, mass density and shear modulus of the rod to vary in the axial direction.

Consider the circular elastic rod of length L, radius $R(x)$, shear modulus $G(x)$, and mass density $\rho(x)$, where the axial coordinate x originates at the left end of the rod as shown in Figure 9.8a. In addition, let the rod be subjected to the distributed twisting moment (torque per unit length) $\mu(x,t)$ as indicated. The St. Venant assumption holds that a cross section rotates uniformly about its axis. The angular displacement of a cross section is then simply a function of the axial coordinate and time, while the axial and radial displacements vanish identically. It follows that the linear displacement, u_θ, of a material particle in the plane of the cross section varies linearly with the radial coordinate r measured from the axis of the rod (Figure 9.8b). Hence,

$$u_\phi(x,r,t) = r\,\theta(x,t) \qquad (9.52)$$

$$u_x(x,r,t) \equiv 0, \quad u_r(x,r,t) \equiv 0 \qquad (9.53)$$

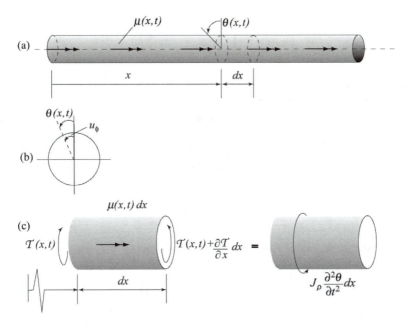

Figure 9.8 Torsional motion of circular rod: (a) deformation and loading, (b) rotation of cross section, (c) kinetic diagram of differential rod element.

where $\theta(x,t)$ is the angular displacement of the cross section and r is the radial coordinate measured from the axis of the rod.

With the linear displacement defined in terms of the rotational displacement, the shear stress, $\tau(x,t)$, acting on the cross section is readily obtained in terms of the rotation of the cross-section using Hooke's law. Hence,

$$\tau(x,r,t) = 2G\varepsilon_{r\theta} = G\left(\frac{\partial u_\phi}{\partial x} + \frac{1}{r}\frac{\partial u_x}{\partial \phi}\right) = Gr\frac{\partial \theta(x,t)}{\partial x} \tag{9.54}$$

Multiplying the shear stress by rdA, where dA is the differential area on the cross section, substituting Eq. (9.54), and integrating the resulting expression over the area of the cross-section gives the resultant torque, $T(x,t)$ (positive counterclockwise), that acts on the cross section. We thus obtain the constitutive relation

$$T(x,t) = k_T(x)\frac{\partial \theta}{\partial x} \tag{9.55}$$

where

$$k_T(x) = G(x)J(x) \tag{9.56}$$

is the *torsional stiffness* (per unit length) of the circular rod and

$$J(x) \equiv \int_{A(x)} r^2 \, dA = \tfrac{1}{2} \pi R^4 (x) \qquad (9.57)$$

is the geometric polar moment of inertia of the cross section. In this sense, the rod is defined in terms of the parameters k_T, ρ and L. With the constitutive relation for the rod established, we now proceed to derive the equation of torsional motion for elastic rods of circular cross section.

We first consider the kinetic diagram of a differential element of the rod shown in Figure 9.8c. Noting that the mass polar moment of inertia of the uniform circular cross-section is simply

$$J_\rho (x) = \rho(x) J(x) \qquad (9.58)$$

We now express Eq. (1.162) for the element and obtain

$$\mu(x,t) dx + \left[T(x,t) + \frac{\partial T}{\partial x} dx \right] - T(x,t) = J_\rho(x) \frac{\partial^2 \theta}{\partial t^2}$$

Rearranging terms and incorporating Eq. (9.55) gives the local equation of motion for the rod as

$$J_\rho(x) \frac{\partial^2 \theta(x,t)}{\partial t^2} - \frac{\partial}{\partial x} k_T(x) \frac{\partial}{\partial x} \theta(x,t) = \mu(x,t) \qquad (9.59)$$

which is seen to correspond to the one-dimensional wave equation in terms of the rotational displacement. As for the case of longitudinal motion, the solution requires the specification of two boundary conditions and two initial conditions.

To obtain a solution to the equation of motion we must specify one term of the torsional work $W = T\theta$ at two points along the rod, typically at the boundaries. Thus, at the left edge, we must specify either

$$T(0,t) = k_T \left. \frac{\partial \theta}{\partial x} \right|_{x=0} = T_0(t) \quad \text{or} \quad \theta(0,t) = h_0(t) \qquad (9.60)$$

where $T_0(t)$ is a prescribed torque or $h_0(t)$ is a prescribed rotation. Similarly, at the right edge of the rod, we must specify either

$$T(L,t) = k_T \left. \frac{\partial \theta}{\partial x} \right|_{x=L} = T_L(t) \quad \text{or} \quad \theta(L,t) = h_L(t) \qquad (9.61)$$

where $T_L(t)$ is a prescribed torque or $h_L(t)$ is a prescribed rotation.

The solution of the equation of motion also requires the specification of the initial angular displacement and initial angular velocity of each cross section of the rod. The initial conditions for torsional motion of the rod are thus of the form

$$\theta(x,0) = \theta_0(x) \tag{9.62}$$

and

$$\left.\frac{\partial \theta}{\partial t}\right|_{\theta=0} = \chi_0(x) \tag{9.63}$$

where $\theta_0(x)$ and $\chi_0(x)$ are prescribed functions that describe the initial state of the rod. It may be see that the problem of torsional motion defined above is directly analogous to the problem of longitudinal motion defined in Section 9.2, with the axial displacement replaced by the rotation $\theta(x,t)$, the axial force replaced by the resultant torque $T(x,t)$ and the distributed axial load replaced by the distributed twisting moment $\mu(x,t)$.

To complete our discussion, we rewrite the equation of motion in operator form. We thus identify the stiffness operator and mass operators, $\Bbbk$ and $\Bbbm$, for the present system as

$$\Bbbk = -\frac{\partial}{\partial x} k_T(x) \frac{\partial}{\partial x} \tag{9.64}$$

and

$$\Bbbm = J_\rho(x) \tag{9.65}$$

Incorporation of the above relations into Eq. (9.59) gives the alternate form of the equation of torsional motion,

$$\Bbbm \frac{\partial^2 \theta}{\partial t^2} + \Bbbk \theta = \mu(x,t) \tag{9.66}$$

The correspondence between Eq. (9.66) and its associated operators with the equation of motion for discrete systems and its associated matrices is evident.

Example 9.5

Determine the boundary conditions for the rod supported rod by identical torsional springs of stiffness k_θ at each edge.

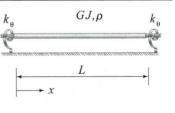

Solution

Recall that the torques are taken as positive in the counterclockwise sense. Thus, if the angle of rotation is positive at the left end of the rod, then the restoring torque applied by the spring is positive. In contrast, if the angle of rotation at the right end of the rod is positive, then the restoring torque produced by the spring is negative. (Compare with the extension/compression of the linear springs in Example 9.3.) It follows that the boundary conditions for this system are then

$$
k_T(x)\frac{\partial \theta}{\partial x}\bigg|_{x=0} = k_\theta \theta(0,t) \tag{a}
$$

and

$$
k_T(x)\frac{\partial \theta}{\partial x}\bigg|_{x=L} = -k_\theta \theta(L,t) \tag{b}
$$

9.5 TRANSVERSE MOTION OF STRINGS AND CABLES

Strings and cables are employed in musical instruments and many engineering systems where flexible tension carrying members are needed. Strings and cables are long thin continua possessing negligible resistance to bending compared with their resistance to axial deformation. As such, they are idealized mathematically as structures with vanishing bending stiffness and hence with no global resistance to axial compression. Correspondingly, the stress distribution over a cross section is characterized by the resultant axial force. From a vibrations perspective, we are primarily interested in the dominant transverse motion of these systems. Since strings and cables are extremely flexible, their transverse motion and axial effects are coupled, primarily through the axial tension. We must therefore examine the equation of axial motion as well as the equation of transverse motion for these systems. As for the elastic rods discussed earlier, we shall consider the properties of the string to be uniform through a cross section, but allow them to vary along the axis of the structure. We shall neglect the Poisson effect as well.

Consider a string of length L, cross-sectional area $\mathcal{A}(x)$ and mass density $\rho(x)$, where the coordinate x originates at the left end of the structure as shown in Figure 9.9a. Let the string be subjected to the external distributed transverse force (force per unit length) $q(x,t)$ and external distributed axial force $p(x,t)$. In addition, let $w(x,t)$ represent the transverse displacement of a material particle originally located at coordinate x when the string is undeformed, as indicated, and let $N(x,t) \geq 0$ represent the internal tension acting on the corresponding cross-section. The kinetic diagram for a representative differential element of the string is depicted in Figure 9.9b. We shall apply Newton's Second Law to the generic string element shown, first in the transverse direction and then in the axial direction. In what follows, we restrict our attention to smooth motions of the string with "moderate rotations" of the cross section. That is, we restrict our attention to motions for which the angle of rotation, φ (Figure 9.9b), at any cross section is such that

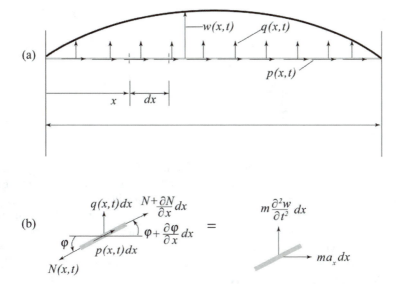

(a)

(b)

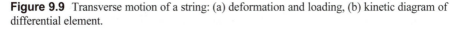

Figure 9.9 Transverse motion of a string: (a) deformation and loading, (b) kinetic diagram of differential element.

$$\sin \varphi \approx \tan \varphi = \frac{\partial w}{\partial x} \ll 1 \tag{9.67}$$

Applying Newton's Second Law of Motion in the transverse direction then gives

$$q(x,t)dx + p(x,t)dx \frac{\partial w}{\partial x} - N \frac{\partial w}{\partial x}$$
$$+ \left(N + \frac{\partial N}{\partial x} dx \right) \left[\frac{\partial w}{\partial x} + \frac{\partial}{\partial x} \left(\frac{\partial w}{\partial x} \right) dx \right] = m(x) dx \frac{\partial^2 w}{\partial t^2} \tag{9.68}$$

where

$$m(x) = \rho(x)A(x) \tag{9.69}$$

is the mass per unit length of the string. Expanding Eq. (9.68), neglecting terms of order $(dx)^2$ compared with dx and rearranging the resulting expression gives the equation of transverse motion for the string as

$$m \frac{\partial^2 w}{\partial t^2} - N \frac{\partial^2 w}{\partial x^2} - \left[\frac{\partial N}{\partial x} + p(x,t) \right] \frac{\partial w}{\partial x} = q(x,t) \tag{9.70}$$

Insight into the internal tension is obtained by considering the equation of motion for the axial direction. Newton's Second Law gives

$$pdx + \left(N + \frac{\partial N}{\partial x} dx \right) - N = mdx\,a_x \qquad (9.71)$$

which, after rearranging terms, takes the form

$$ma_x - \frac{\partial N}{\partial x} = p(x,t) \qquad (9.72)$$

where a_x represents the axial acceleration of a material particle originally located at coordinate x. Since the string has negligible bending stiffness, it may be anticipated that the transverse motion of a given point on the string will be much greater than the axial motion of that point. If, for example, we consider the motion of a guitar string during a cycle, the distance traversed in the transverse direction by a material particle will be much larger than the distance traveled in the axial direction by that same particle. The particle will necessarily travel much more rapidly in the transverse direction. It follows that the kinetic energy of transverse motion is much greater than the kinetic energy of axial motion. Likewise, the axial component of the acceleration will be negligible compared with the corresponding transverse component. If we neglect the kinetic energy of axial motion and, equivalently, the axial component of the acceleration in our formulation, then Eq. (9.72) reduces to the statement

$$\frac{\partial N}{\partial x} \cong -p \qquad (9.73)$$

Incorporation of Eq. (9.73) into Eq. (9.70) reduces the equation of transverse motion of the string to the form

$$m\frac{\partial^2 w}{\partial t^2} - N\frac{\partial^2 w}{\partial x^2} = q(x,t) \qquad (9.74)$$

It is seen from Eq. (9.73) that the tension in the string is effectively a spatial integral of the axial body force $p(x,t)$. It is further seen that if the distributed axial load is independent of time then the tension in the string is effectively independent of time as well. That is,

$$\text{if} \quad p = p(x) \quad \text{then} \quad N = N(x)$$

It follows, as a special case, that if the axial body force is uniform, as well as independent of time then the internal tension is effectively constant.

The equation of transverse motion for strings and cables, Eq. (9.74), is seen to be of the same form as the equations that govern longitudinal motion and torsional motion of uniform elastic rods, the one-dimensional wave equation. It is also seen that the internal tension supplies the stiffness per unit length of the string. Regardless, we must specify two boundary conditions and two initial conditions to complete our formulation. For the highly flexible string under consideration, the internal force acting in the transverse direction is the projection of the corresponding internal tension in that direction (see Figure 9.9b). We must therefore specify one term of the transverse work, $\mathcal{W} = (N\,\partial w/\partial x)w$, at two points of the string. The boundary conditions for the string are thus

$$N\frac{\partial w}{\partial x}\bigg|_{x=0} = Q_0(t) \quad \text{or} \quad w(0,t) = h_0(t) \tag{9.75}$$

and

$$N\frac{\partial w}{\partial x}\bigg|_{x=L} = Q_L(t) \quad \text{or} \quad w(L,t) = h_L(t) \tag{9.76}$$

where $Q_0(t)$ and $Q_L(t)$ are prescribed transverse edge loads, and $h_0(t)$ and $h_L(t)$ are prescribed edge displacements. To finish our formulation for transverse motion we must specify the initial transverse displacement and the initial transverse velocity for each point on the string. The corresponding initial conditions thus take the form

$$w(x,0) = w_0(x) \tag{9.77}$$

and

$$\frac{\partial w}{\partial t}\bigg|_{t=0} = v_0(x) \tag{9.78}$$

where $w_0(x)$ and $v_0(x)$ are prescribed functions that describe the initial state of the string.

We complete our discussion by writing the equation of motion in operator form. We thus introduce the stiffness and mass operators

$$\Bbbk = -N\frac{\partial^2}{\partial x^2} \tag{9.79}$$

and

$$\Bbbm = m(x) \tag{9.80}$$

respectively, into Eq. (9.74). This results in the familiar form

$$\Bbbm\frac{\partial^2 w}{\partial t^2} + \Bbbk w = q(x,t) \tag{9.81}$$

Example 9.6

A sign of mass m hangs from a uniform chain of length L and mass per unit length $m \ll m/L$. Determine the explicit form of the transverse equation of (small) motion for the cable. Also determine the associated boundary conditions if (i) the lateral motion of the sign is restricted as shown, and (ii) if the sign hangs freely (Figure E9.6-1). Assume that the weight of the sign is much greater than the weight of the chain.

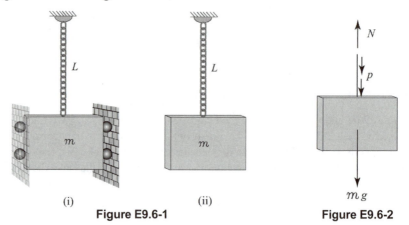

(i)   (ii)

Figure E9.6-1   **Figure E9.6-2**

Solution

The tension in the cable is easily obtained by considering the kinetic diagram of a section of the cable and sign (Figure E9.6-2). In keeping with earlier discussions, we assume that the axial acceleration of the system is negligible. Furthermore, $m(L-x) \ll m$ and hence $p = mg$ is negligible. We thus have that

$$N = mg \tag{a}$$

Substitution of Eq. (a) into Eq. (9.79) gives the stiffness operator as

$$\Bbbk = -mg \frac{\partial^2}{\partial x^2} \tag{b}$$

Equation (9.74), or equivalently Eq. (9.81), gives the explicit form of the equation of motion as

$$m \frac{\partial^2 w}{\partial t^2} - mg \frac{\partial^2 w}{\partial x^2} = 0 \tag{c}$$

For both case (i) and case (ii), the boundary condition at the fixed support at $x = 0$, is

$$w(0,t) = 0 \tag{d}$$

The boundary condition at the end of the chain that is attached to the sign depends on the restraints on the sign.

Case (i):
For this case the sign cannot move laterally. Hence, the boundary condition for the string is

$$w(L,t) = 0 \tag{e-i}$$

Case (ii):
In this case, the sign is free to move laterally. The corresponding boundary condition is thus

$$mg \left. \frac{\partial w}{\partial x} \right|_{x=L} = 0 \implies \left. \frac{\partial w}{\partial x} \right|_{x=L} = 0 \tag{e-ii}$$

Therefore, for small motions the sign may swing laterally, but any rotation of the sign is neglected.

9.6 TRANSVERSE MOTION OF ELASTIC BEAMS

Long thin structural elements that are primarily excited by end moments and/or transverse loading are encountered in many practical situations. Unlike strings and cables, such bodies possess substantial resistance to bending and are utilized for this purpose. These objects are referred to as beams and their mathematically one-dimensional representations are referred to as beam theories. In this section we develop and discuss several fundamental beam theories often employed in linear vibration analysis. The section also includes discussions of geometrically nonlinear beam theory and translating beams. We begin with an account of the basic geometrical assumptions and material relations.

9.6.1 Kinematical and Constitutive Relations

A beam theory is a mathematically one-dimensional representation of a long and thin three-dimensional body that undergoes flexure. Such a theory is developed by exploiting the fact that the thickness and width of the beam are small compared with the overall length of the beam. Standard theories incorporate a linear variation of the strain and stress through the thickness that is strictly true for pure bending (i.e., when the structure is subjected to stress distributions applied at its edges that are statically equivalent to only a moment). The beam theory similarly incorporates the assumption that the variations of the stress, strain and displacement through the width of the beam are negligible. Such theories have been quite successful in predicting the behavior of these structures for many other loading conditions, when the stresses acting on surfaces with normals perpendicular to the axis of the beam are small compared with the stresses acting on the cross sections of the beam. With the variation through the thickness and width assumed apriori, the problem is reduced to finding the deflection of the neutral axis of the beam and is thus transformed into a mathematically one-dimensional boundary value problem as discussed below.

Consider an elastic beam, and let the x-axis coincide with the centroid of the beam in the rest configuration, as shown in Figure 9.10₁. In addition, let the z-axis originate at the x-axis, and run perpendicular to it in the thickness direction. Let $u_x(x,z,t)$ and $u_z(x,z,t)$ respectively correspond to the axial and transverse displacements of the material particle originally located at the indicated coordinates. Further, let $u(x,t)$ and $w(x,t)$ represent the corresponding displacements of the material particles on the neutral surface $z = 0$. We next assume the kinematical relations attributed to Kirchoff,

$$u_x(x,z,t) = u(x,t) - z\varphi(x,t) \tag{9.82}$$

$$u_z(x,z,t) \cong w(x,t) \tag{9.83}$$

where $\varphi(x,t)$ represents the in-plane rotation of the cross section of the beam originally located at coordinate x (Figure 9.10₂). An analogous relation between the infinitesimal axial strain $\varepsilon_{xx}(x,z,t)$, where

$$\varepsilon_{xx}(x,z,t) = \frac{\partial u_x}{\partial x} \tag{9.84}$$

and its counterpart at the neutral surface

$$\varepsilon(x,t) = \frac{\partial u}{\partial x} \tag{9.85}$$

is found by substituting Eq. (9.82) into Eq. (9.84). This gives the strain distribution

$$\varepsilon_{xx}(x,z,t) = \varepsilon(x,t) - z\kappa(x,t) \tag{9.86}$$

where

$$\kappa(x,t) = \frac{\partial \varphi}{\partial x} \tag{9.87}$$

is the curvature of the neutral axis of the beam at the point originally located at coordinate x. (In future we shall simply say the displacement, strain, curvature, etc., "at x," with the interpretation being taken to mean at the point originally located at coordinate x.)

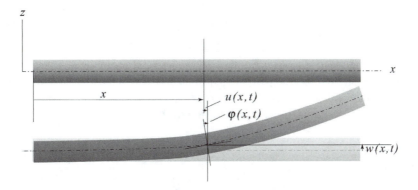

Figure 9.10 Kinematical measures of the motion of a beam.

With the basic kinematical assumptions established we next introduce alternate measures of the internal stress distribution. Based on the Kirchoff assumptions, we implicitly neglect the Poisson effect in the beam. The axial stress, $\sigma_{xx} = \sigma(x,z,t)$, is therefore related to the strain by the one-dimensional statement of Hooke's Law,

$$\sigma(x,z,t) = E(x)\varepsilon_{xx}(x,z,t) \qquad (9.88)$$

Since we wish to construct a mathematically one-dimensional theory, we shall express the stress distribution acting on a cross section by statically equivalent forces and moments. In this regard, the normal stress distribution acting on a cross section is statically equivalent to a resultant normal force and moment, while the associated shear stress distribution is statically equivalent to a resultant transverse shear force. The resultant membrane (axial) force, $N(x,t)$, acting on the cross section at x is found by integrating the axial stress over the area of the cross section. Hence,

$$N(x,t) = \int_A \sigma(x,z,t)\,dA \qquad (9.89)$$

Substituting Eq. (9.86) and (9.88) into Eq. (9.89) gives

$$N(x,t) = \int_A E(x)\varepsilon_{xx}(x,z,t)\,dA = E(x)\varepsilon(x,t)\int_A dA - E(x)\kappa(x,t)\int_A z\,dA$$

Recall that z originates at the x-axis, the area centroid. Therefore, by definition of the area centroid,

$$\int_{A(x)} z\,dA = 0 \qquad (9.90)$$

The resultant membrane force is thus given by the constitutive relation

$$N(x,t) = EA\varepsilon(x,t) = EA\frac{\partial u}{\partial x} \qquad (9.91)$$

The resultant moment produced by the axial stress field about an axis that passes through the centroidal axis and is perpendicular to it is obtained by taking the moment of each individual differential force, $\sigma\,dA$, and summing all such moments. The resultant moment, referred to as the bending moment, acting on cross section x is then

$$M(x,t) = \int_A \sigma(x,z,t)\,z\,dA \qquad (9.92)$$

Substituting Eq. (9.86) and (9.88) into Eq. (9.92) gives the relation

$$M(x,t) = E(x)\varepsilon(x,t)\int_A z\,dA - E(x)\kappa(x,t)\int_A z^2\,dA$$

Substituting Eq. (9.90) into the above expression gives the bending moment about the axis perpendicular to the neutral axis as

$$M(x,t) = -EI\kappa(x,t) \qquad (9.93)$$

where

$$I = I(x) \equiv \int_{A(x)} z^2\,dA \qquad (9.94)$$

The resultant shear force, $Q(x,t)$, on cross-section x is simply

$$Q(x,t) = \int_{A(x)} \tau(x,z,t)\,dA \qquad (9.95)$$

where $\tau(x,z,t)$ is the transverse shear stress acting on that cross section. With the kinematical assumptions and force measures introduced in this section, the description of the beam is given in terms of the displacements and strains of the centroidal surface and the resultant forces and moments at a cross section, all of which are functions of one spatial variable and time. We now derive the equations of motion in terms of these variables.

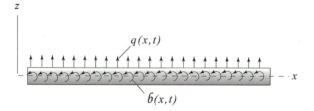

Figure 9.11 Beam with distributed transverse force and body couples.

9.6.2 Kinetics

Consider a beam that is loaded by both normal and shear stresses over its upper and lower surfaces. In keeping with the resultant internal forces and moments discussed in the previous section, the external forces may be expressed as distributed transverse loads and distributed (body) couples, the latter taken about axes that go through the neutral surface and are perpendicular to the xz-plane (i.e., that are parallel to the y-axis). The distributed transverse forces, $q(x,t)$, may be considered to be the sum of any body forces acting in the transverse direction and the difference (jump) in the normal stresses applied on the outer surfaces, while the body couples, $b(x,t)$, may be considered as the moments about the y-axis of the shear stresses acting on the outer surfaces of the beam plus any intrinsic body couples (see Figure 9.11). To derive the equations of motion for the beam let us consider a generic differential element of length dx. The corresponding kinetic diagram is expressed in terms of the displacement and force parameters defined in the previous section and is depicted in Figure 9.12. We consider both translation in the transverse direction and rotation of the element. Since the bending stiffness for beams is finite, as is the resistance to shear, the corresponding rotations and distortions are relatively small. Because of this, the projection of the membrane force in the transverse direction is generally much smaller than the resultant transverse shear and is neglected in elementary (linear) beam theory. Recall that the reverse was true for strings and cables. The reader should be aware, however, that such terms must be retained in situations where the coupling of axial and transverse motions is important, such as for predicting dynamic, as well as static, buckling. For the present theories, the statement of Newton's Second Law in the transverse direction may be written directly from the kinetic diagram as

$$q(x,t)dx + \left[Q(x,t) + \frac{\partial Q}{\partial x} dx \right] - Q(x,t) = m(x)dx \frac{\partial^2 w}{\partial t^2}$$

which reduces to

$$q(x,t) + \frac{\partial Q}{\partial x} = m(x) \frac{\partial^2 w}{\partial t^2} \tag{9.96}$$

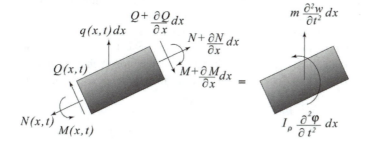

Figure 9.12 Kinetic diagram for beam element.

The pertinent statement of Eq. (1.162) about an axis through the center of the element follows from the kinetic diagram as well. Hence, assuming the right hand rule,

$$6(x,t)dx - \left[M(x,t) + \frac{\partial M}{\partial x} dx \right] + M(x,t)$$

$$+ \left[Q(x,t) + \frac{\partial Q}{\partial x} dx \right] \frac{dx}{2} + Q(x,t) \frac{dx}{2} = I_\rho(x) dx \frac{\partial^2 \varphi}{\partial t^2}$$

which, neglecting terms of $O(dx)^2$, reduces to

$$Q(x,t) = \frac{\partial M}{\partial x} + I_\rho(x) \frac{\partial^2 \varphi}{\partial t^2} - 6(x,t) \tag{9.97}$$

where

$$I_\rho(x) = \rho(x)I(x) \tag{9.98}$$

is typically referred to as the *rotatory inertia* of the beam and corresponds to the mass moment of inertia of the element per unit length about an axis perpendicular to the neutral axis. Substitution of Eq. (9.97) into Eq. (9.96), and incorporating Eqs. (9.87) and (9.93), eliminates the internal forces and moments and results in the single equation

$$q(x,t) - \frac{\partial 6}{\partial x} - \frac{\partial^2}{\partial x^2} EI \frac{\partial \varphi}{\partial x} = m(x) \frac{\partial^2 w}{\partial t^2} - \frac{\partial}{\partial x} I_\rho \frac{\partial^2 \varphi}{\partial t^2} \tag{9.99}$$

expressed in terms of the transverse displacement w and in-plane rotation φ. The particular form of this equation depends upon further kinematical assumptions.

9.6.3 Euler-Bernoulli Beam Theory

The simplest and most common beam theory is that attributed to Euler and Bernoulli. For this model the assumptions and developments of Sections 9.6.1 and 9.6.2 are incorporated, with the exception that the rotatory inertia is not taken into account. That is, the effects of the rotatory inertia are neglected compared with those of the linear inertia (the mass per unit length). This omission, in effect, treats the mass distribution as if it is concentrated along the neutral axis of the beam. The deformations associated with transverse shear are not included as well. These assumptions are reasonable provided that the beam is thin and, for vibratory behavior, that the effective wave lengths of the individual modes are sufficiently large compared with the thickness.

Kinematics and the Equation of Motion

For small and smooth deflections, the angle of rotation (the tangent angle) of the neutral axis may be approximated by the tangent itself (the small angle approximation). We thus have that

$$\varphi(x,t) \cong \frac{\partial w}{\partial x} \tag{9.100}$$

from which it follows that

$$\kappa(x,t) = \frac{\partial \varphi}{\partial x} \cong \frac{\partial^2 w}{\partial x^2} \tag{9.101}$$

$$M(x,t) = -EI\kappa(x,t) \cong -EI\frac{\partial^2 w}{\partial x^2} \tag{9.102}$$

and

$$\frac{\partial^2 \varphi}{\partial t^2} \cong \frac{\partial^2}{\partial t^2}\frac{\partial w}{\partial x} = \frac{\partial}{\partial x}\frac{\partial^2 w}{\partial t^2} \tag{9.103}$$

Substituting Eqs. (9.100)–(9.103) into Eq. (9.99), and setting $I_\rho = 0$ gives the equation of motion for Euler-Bernoulli Beams,

$$m(x)\frac{\partial^2 w}{\partial t^2} + \frac{\partial^2}{\partial x^2}k_b(x)\frac{\partial^2}{\partial x^2}w = q(x,t) - \frac{\partial b}{\partial x} \tag{9.104}$$

where

$$k_b(x) = E(x)I(x) \tag{9.105}$$

is the *bending stiffness* of the beam. It may be seen that the governing equation is expressed solely in terms of the transverse displacement, the external distributed load and the properties of the beam. Equation (9.104) is known as the *Euler-Bernoulli Beam Equation.*

Boundary and Initial Conditions

The equation of motion is seen to be a fourth order partial differential equation in space and second order in time. A solution will therefore require the specification of four boundary conditions and two initial conditions.

For the boundary conditions, we must specify one term from the work of transverse translational motion and one term from the work of rotational motion at two points on the beam. That is, we must specify one term from each of the work functionals $\mathcal{W}_Q = Qw$ and $\mathcal{W}_M = M\varphi$. The *boundary conditions* at the left end of a beam of length L are thus

$$Q(0,t) = -\frac{\partial}{\partial x}EI\frac{\partial^2 w}{\partial x^2}\bigg|_{x=0} = \mathcal{Q}_0(t) \quad \text{or} \quad w(0,t) = h_0(t) \tag{9.106}$$

$$M(0,t) = -EI\frac{\partial^2 w}{\partial x^2}\bigg|_{x=0} = \mathcal{M}_0(t) \quad \text{or} \quad \frac{\partial w}{\partial x}\bigg|_{x=0} = \mathcal{R}_0(t) \tag{9.107}$$

where $\mathcal{Q}_0(t)$ or $h_0(t)$, and $\mathcal{M}_0(t)$ or $f_0(t)$ are prescribed functions. Similarly, the boundary conditions at the right end of the beam are

$$Q(L,t) = -\frac{\partial}{\partial x}EI\frac{\partial^2 w}{\partial x^2}\bigg|_{x=L} = \mathcal{Q}_L(t) \quad \text{or} \quad w(L,t) = h_L(t) \tag{9.108}$$

$$M(L,t) = -EI\frac{\partial^2 w}{\partial x^2}\bigg|_{x=L} = \mathcal{M}_L(t) \quad \text{or} \quad \frac{\partial w}{\partial x}\bigg|_{x=L} = f_L(t) \tag{9.109}$$

where $\mathcal{Q}_L(t)$ or $h_L(t)$, and $\mathcal{M}_L(t)$ or $f_L(t)$ are prescribed functions.

In addition to the boundary conditions, the initial transverse displacement and transverse velocity of each point in the beam must be specified. The initial conditions therefore take the form

$$w(x,0) = w_0(x) \tag{9.110}$$

and

$$\frac{\partial w}{\partial t}\bigg|_{t=0} = v_0(x) \tag{9.111}$$

where $w_0(x)$ and $v_0(x)$ are prescribed functions that describe the initial state of the beam. Note that when the initial transverse displacement, $w_0(x)$, is specified for each point on the beam then the initial rotation, $w_0'(x)$, of each point is specified as well. Similarly, when the initial transverse velocity, $v_0(x)$, is specified for each point on the beam than the initial angular velocity, $v_0'(x)$, of each point is also specified.

Mass and Stiffness Operators

The equation of transverse motion for Euler-Bernoulli beams can be written in compact form by introducing the stiffness and mass operators, $\Bbbk$ and $\Bbbm$ respectively, as follows

$$\Bbbk = \frac{\partial^2}{\partial x^2}k_b(x)\frac{\partial^2}{\partial x^2} \tag{9.112}$$

$$\Bbbm = m(x) \tag{9.113}$$

Incorporating the above operators into Eq. (9.104) renders the equation of transverse motion for elementary beams to the familiar form

$$m\frac{\partial^2 w}{\partial t^2} + kw = q(x,t) \tag{9.114}$$

Example 9.7

State the boundary conditions for a cantilever beam that is fixed at its left end, as shown, if the beam is modeled using Euler-Bernoulli Theory.

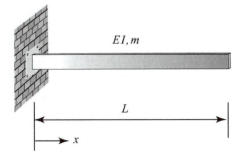

Solution

The left end of the beam is fixed with regard to both translation and rotation. Therefore, the corresponding boundary conditions at the support are

$$w(0,t) = 0 \quad \text{and} \quad \frac{\partial w}{\partial x}\bigg|_{x=0} = 0 \tag{a-1,2}$$

Since the right end of the beam is unsupported and no external load acts on that edge, the exposed cross section is stress free. It follows that the bending moment and resultant transverse shear force both vanish on this surface. The boundary conditions at the right end of the bend are therefore

$$-EI\frac{\partial^2 w}{\partial x^2}\bigg|_{x=L} = 0 \quad \text{and} \quad -\frac{\partial}{\partial x}EI\frac{\partial^2 w}{\partial x^2}\bigg|_{x=0} = 0 \tag{b-1,2}$$

Example 9.8

The beam shown is pinned at its left end and is embedded in an elastic wall of rotational stiffness k_φ and translational stiffness k_w. Deduce the boundary conditions for the beam if it is modeled using Euler-Bernoulli Beam Theory.

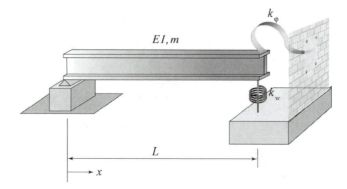

Solution

The support at the left edge is such that the beam is free to rotate about an axis through that point, but it cannot translate in the transverse direction. The corresponding boundary conditions are thus

$$-EI\frac{\partial^2 w}{\partial x^2}\bigg|_{x=0} = 0 \quad \text{and} \quad w(0,t) = 0 \tag{a-1,2}$$

The elastic wall exerts a restoring moment and restoring force on the right end of the beam. The latter is in the form of a transverse shear force. Therefore, the corresponding boundary conditions are

$$-EI\frac{\partial^2 w}{\partial x^2}\bigg|_{x=L} = -k_\varphi \frac{\partial w}{\partial x}(L,t) \tag{b-1}$$

and

$$-\frac{\partial}{\partial x}EI\frac{\partial^2 w}{\partial x^2}\bigg|_{x=L} = k_w w(L,t) \tag{b-2}$$

Example 9.9

State the boundary conditions for the problem of flexural motion of a uniform beam supporting a small block of mass m at its edge if the support undergoes the prescribed lateral motion indicated. The dimensions of the block and the weight of the beam may be neglected.

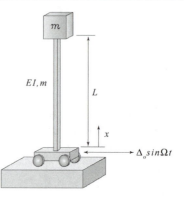

Figure E9.9-1

Solution

The boundary conditions at the support follow directly as

$$w(0,t) = \Delta_0 \sin \Omega t \quad \text{and} \quad \left. \frac{\partial w}{\partial x} \right|_{x=0} = 0 \qquad \text{(a-1, 2)}$$

Since the dimensions of the block may be neglected, the moment of inertia of the block is neglected as well. The edge of the beam is thus free to rotate. Hence,

$$-EI \left. \frac{\partial^2 w}{\partial x^2} \right|_{x=L} = 0 \qquad \text{(b-1)}$$

The last boundary condition follows directly from the kinetic diagram of the supported mass (Figure E9.9-2) and gives

$$-Q(L,t) = EI \left. \frac{\partial^3 w}{\partial x^3} \right|_{x=L} = m \left. \frac{\partial^2 w}{\partial t^2} \right|_{x=L} \qquad \text{(b-2)}$$

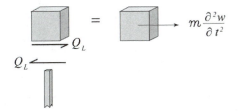

Figure E9.9-2 Kinetic diagram of supported mass.

9.6.4 Rayleigh Beam Theory

We next study the beam theory attributed to J. W. S. Rayleigh. This model incorporates the rotatory inertia into the model already established for the Euler-Bernoulli Beam Theory.

Equation of Motion

To derive the equation of motion for Rayleigh Beams, we parallel the development of the previous section but now include the effects of rotatory inertia. Substituting Eqs.

(9.100)–(9.103) into Eq. (9.99) and retaining the rotatory inertia gives the equation of motion for Rayleigh Beams as

$$m(x)\frac{\partial^2 w}{\partial t^2} - \frac{\partial}{\partial x} I_\rho(x)\frac{\partial}{\partial x}\frac{\partial^2 w}{\partial t^2} + \frac{\partial^2}{\partial x^2}k_b(x)\frac{\partial^2}{\partial x^2}w = q(x,t) - \frac{\partial b}{\partial x} \quad (9.115)$$

Boundary and Initial Conditions

The equation of transverse motion for Rayleigh Beams is a fourth order partial differential equation in space and second order in time. Therefore, as for Euler-Bernoulli Beams, we must specify one term from the work of transverse translational motion and one term from the work of rotational motion at two points on the beam. That is, we must specify one term of each of the work functionals $\mathcal{W}_Q = Qw$ and $\mathcal{W}_M = M\varphi$. For Rayleigh Beams, however, we must be careful to account for the rotatory inertia in these conditions. Specifically, the rotatory inertia enters the condition for transverse shear through Eq. (9.97). The remaining conditions are the same as for Euler-Bernoulli Beams. Hence, the *boundary conditions* for a Rayleigh Beam of length L are thus, at the left end of the beam,

$$Q(0,t) = -\left[\frac{\partial}{\partial x}EI\frac{\partial^2 w}{\partial x^2} - I_\rho\frac{\partial^3 w}{\partial t^2 \partial x}\right]_{x=0} = \mathcal{Q}_0(t) \quad \text{or} \quad w(0,t) = h_0(t) \quad (9.116)$$

$$M(0,t) = -EI\frac{\partial^2 w}{\partial x^2}\bigg|_{x=0} = \mathcal{M}_0(t) \quad \text{or} \quad \frac{\partial w}{\partial x}\bigg|_{x=0} = f_0(t) \quad (9.117)$$

and, at the right end of the beam,

$$Q(L,t) = -\left[\frac{\partial}{\partial x}EI\frac{\partial^2 w}{\partial x^2} - I_\rho\frac{\partial^3 w}{\partial t^2 \partial x}\right]_{x=L} = \mathcal{Q}_L(t) \quad \text{or} \quad w(L,t) = h_L(t) \quad (9.118)$$

$$M(L,t) = -EI\frac{\partial^2 w}{\partial x^2}\bigg|_{x=L} = \mathcal{M}_L(t) \quad \text{or} \quad \frac{\partial w}{\partial x}\bigg|_{x=L} = f_L(t) \quad (9.119)$$

where $\mathcal{Q}_0(t)$ or $h_0(t)$, $\mathcal{M}_0(t)$ or $f_0(t)$, $\mathcal{Q}_L(t)$ or $h_L(t)$, and $\mathcal{M}_L(t)$ or $f_L(t)$ are prescribed functions.

The initial conditions are of the form

$$w(x,0) = w_0(x) \quad (9.120)$$

and

$$\frac{\partial w}{\partial t}\bigg|_{t=0} = v_0(x) \qquad (9.121)$$

where $w_0(x)$ and $v_0(x)$ are prescribed functions that describe the initial state of the beam. As for Euler-Bernoulli Beams, when these two functions are specified, the initial rotations and rotation rates are specified as well.

Mass and Stiffness Operators

The equations of transverse motion for Rayleigh beams can be written in compact form by introducing the stiffness and mass operators, $\Bbbk$ and $\Bbb{m}$ respectively, as follows

$$\Bbbk = \frac{\partial^2}{\partial x^2} k_b(x) \frac{\partial^2}{\partial x^2} \qquad (9.122)$$

$$\Bbb{m} = m(x) - \frac{\partial}{\partial x} I_\rho(x) \frac{\partial}{\partial x} \qquad (9.123)$$

Incorporating the above operators in Eq. (9.115) renders the equation of transverse motion for elementary beams to the familiar form

$$\Bbb{m} \frac{\partial^2 w}{\partial t^2} + \Bbbk w = q(x,t) \qquad (9.124)$$

Example 9.10

Deduce the boundary conditions for the beam of Example 9.9 if it has local mass moment of inertia I_ρ and is modeled using Rayleigh Beam Theory.

Solution

The boundary conditions for displacement and moment are the same as for Euler-Bernoulli Theory. The first three conditions, Eqs. (a), (b) and (c), established in Example 9.8 hold for the present model as well. The condition for transverse shear does, however, differ. The last boundary condition of Example 9.8, Eq. (d), is replaced by the condition

$$-Q(L,t) = \left[EI \frac{\partial^3 w}{\partial x^3} - I_\rho \frac{\partial^3 w}{\partial t^2 \partial x} \right]_{x=L} = m \frac{\partial^2 w}{\partial t^2}\bigg|_{x=L} \qquad \lhd \text{(d)}$$

The beam theories presented herein are predicated on the assumption that the thickness deformations are negligible. The transverse deflections therefore arise from the rotations of the cross sections. In the next section we consider modifications of the current beam theory to include a measure of deformation due to transverse shear.

9.6.5 Timoshenko Beam Theory

The beam theories discussed in Sections 9.6.3 and 9.6.4 are very successful at predicting the flexural behavior of structures whose thicknesses are very small compared with their lengths. From a vibrations perspective, such beam theories yield satisfactory results for situations where the wave lengths of the deformation are relatively large compared with the thickness of the beam. For shorter beams, or for situations where we are interested in vibrations whose wavelengths are not so restricted, a modified beam theory is needed. Alternatively, or if the behavior of interest was such that the thickness vibrations and associated behavior is pertinent, we would investigate the problem from the much more complex two or three-dimensional elastodynamics point of view. We here present the former approach, and include the effects of transverse shear as a correction to classical beam theory consistent with the desired mathematically one-dimensional representations discussed to this point.

Both the Euler-Bernoulli and the Rayleigh beam theories discussed in Sections 9.6.3 and 9.6.4, respectively, each neglect the contribution of shear deformation. This may be seen by substituting Eqs.(9.82) and (9.83) into Eq. (9.39), which yields vanishing angle of distortion, identically. This omission is satisfactory provided the ratio of the flexural wave length to the thickness of the vibrating structure is sufficiently large. (For static problems the critical ratio is less acute and corresponds to the ratio of the overall length of the beam to its thickness.) Typically, the shear stresses and transverse normal stresses in the structures of interest are much smaller than the axial normal stress, which justifies the basic assumptions of the elementary beam theories. However, a static analysis of a cantilever beam of rectangular cross section shows that the maximum value of the average shear stress acting on a cross section becomes the same order of magnitude as the corresponding maximum normal stress when the length to thickness ratio is less than about three. On this basis, we anticipate that shear deformation will become important in vibration problems for which the wave length is sufficiently small as well. The following theory adds a correction for shear deformation to the basic beam theories discussed to this point.

Correction for Transverse Shear

Let us consider the transverse shear stress, $\sigma_{xz} = \tau(x,z,t)$, acting on a cross section, and the associated shear strain, $\varepsilon_{xz}(x,z,t)$, where

$$\varepsilon_{xz}(x,z,t) = \gamma_{xz}(x,z,t)/2 \qquad (9.125)$$

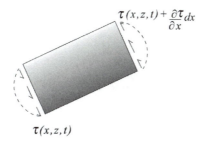

$\tau(x,z,t)+\dfrac{\partial\tau}{\partial x}dx$

$\tau(x,z,t)$

Figure 9.13 Transverse shear stress distribution on cross sections of beam element.

and $\gamma_{xz}(x,z,t)$ represents the corresponding angle change (shear distortion) as discussed in Section 9.2.2. Hooke's Law for shear then gives the relation

$$\tau(x,z,t) = 2G(x)\varepsilon_{xz}(x,z,t) = G(x)\gamma_{xz}(x,z,t) \qquad (9.126)$$

where $G(x)$ is the shear modulus of the elastic material that comprises the beam. The shear stress acting on a cross section of a beam will generally vary through the thickness, as suggested by the kinetic diagram for the partial beam element shown in Figure 9.13. It follows that the associated shear distortion will be nonuniform over a cross section as well. Nevertheless, since a beam theory is a mathematically one-dimensional representation of a three-dimensional body, we shall represent the effects of shear distortion by a single "shear angle" associated with transverse shear for a given cross section. We thus define the *shear angle* for a beam as

$$\gamma(x,t) = \frac{1}{k\,A(x)} \int_{A(x)} \gamma_{xz}(x,z,t)\,dA \qquad (9.127)$$

where k is a "shape factor" that depends on the geometry of the cross section and is sometimes referred to as the Timoshenko Shear Coefficient. (Specific values of the shape factor may be determined, for example, by matching results predicted by the above beam theory with results predicted by an "exact" three-dimensional elastodynamics analysis, deduced from a static elasticity solution, or measured from experiments. The value $k = 2/3$ is often associated with rectangular cross sections, while the value $k = 3/4$ is similarly associated with circular cross sections.) The shear angle defined above is seen to correspond to a weighted average of the shear strain over the cross-section.

To incorporate the above shear description into a mathematically one-dimensional theory we express the shear stress τ distribution acting on the cross section in terms of the corresponding resultant transverse shear force Q and the shape factor k. Substitution of Eqs. (9.126) and (9.127) into Eq. (9.95) results in the constitutive relation

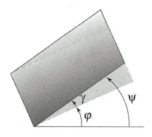

Figure 9.14 Rotation of element due to transverse shear and bending.

$$Q(x,t) = k_s(x)\gamma(x,t) \tag{9.128}$$

where

$$k_s(x) = \hat{k} A(x)G(x) \tag{9.129}$$

is the *shear stiffness* of the beam. With the introduction of the shear deformation described above, the total angle of rotation, ψ, of the centroidal axis is now comprised of that due to bending, φ, and that due to shear, γ (Figure 9.14). Hence,

$$\varphi(x,t) + \gamma(x,t) = \psi(x,t) \cong \frac{\partial w}{\partial x} \tag{9.130}$$

Substitution of Eq. (9.130) into Eq. (9.128) gives the useful identity

$$\gamma(x,t) = \frac{\partial w}{\partial x} - \varphi(x,t) = \frac{Q(x,t)}{k_s(x)} \tag{9.131}$$

Substitution of Eqs. (9.130) and (9.131) into Eq. (9.93) gives the constitutive relation

$$M(x,t) = -EI\frac{\partial}{\partial x}\left[\frac{\partial w}{\partial x} - \gamma(x,t)\right] = -EI\left[\frac{\partial^2 w}{\partial x^2} - \frac{\partial}{\partial x}\frac{Q}{k_s}\right] \tag{9.132}$$

The constitutive relation for the membrane force for the present case remains the same as for the beam theories discussed previously, and is given by Eq. (9.91). With the kinematical and constitutive relations established, we may now derive the equations of motion.

Governing Equations, Boundary Conditions and Initial Conditions

Substitution of Eqs. (9.128), (9.131) and (9.132) into Eqs. (9.96) and (9.97) gives the equations of motion

$$m(x)\frac{\partial^2 w}{\partial t^2} - \frac{\partial}{\partial x}\left[k_s(x)\left\{\frac{\partial w}{\partial x} - \varphi(x,t)\right\}\right] = q(x,t) \tag{9.133}$$

$$I_\rho \frac{\partial^2 \varphi}{\partial t^2} - k_s(x)\left\{\frac{\partial w}{\partial x} - \varphi(x,t)\right\} - \frac{\partial}{\partial x} EI \frac{\partial \varphi}{\partial x} = b(x,t) \tag{9.134}$$

The above equations can be written in operator/matrix form as follows

$$\mathbf{m}\frac{\partial^2 \mathbf{u}}{\partial t^2} + \mathbf{k}\mathbf{u} = \mathbf{F} \tag{9.135}$$

where

$$\mathbf{m} = \begin{bmatrix} m & 0 \\ 0 & I_\rho \end{bmatrix} \tag{9.136}$$

$$\mathbf{k} = \begin{bmatrix} -\dfrac{\partial}{\partial x} k_s \dfrac{\partial}{\partial x} & k_s \dfrac{\partial}{\partial x} \\ -k_s \dfrac{\partial}{\partial x} & k_s - \dfrac{\partial}{\partial x} EI \dfrac{\partial}{\partial x} \end{bmatrix} \tag{9.137}$$

$$\mathbf{u} = \begin{Bmatrix} w(x,t) \\ \varphi(x,t) \end{Bmatrix} \tag{9.138}$$

$$\mathbf{F} = \begin{Bmatrix} q(x,t) \\ b(x,t) \end{Bmatrix} \tag{9.139}$$

Beams whose descriptions include both the shear correction and rotatory inertia are referred to as *Timoshenko Beams*. Beams whose descriptions include the shear correction but neglect the rotatory inertia are generally referred to as *shear beams*.

By incorporating the shear angle into the mathematical model, we have added an additional degree of freedom at each cross section. Consequently, the governing equations, Eqs. (9.133) and (9.134), are seen to be coupled second order partial differential equations of two dependent variables, the transverse displacement w and the bending rotation (of the cross section) φ. The associated boundary conditions follow from the work of rotation and the work of transverse translation as discussed for the elementary beam theories. For the present beam theory, the *boundary conditions* take the general forms

$$M(0,t) = -EI\frac{\partial \varphi}{\partial x}\bigg|_{x=0} = M_0(t) \quad \text{or} \quad \varphi(0,t) = f_0(t) \tag{9.140}$$

$$Q(0,t) = k_s \left[\frac{\partial w}{\partial x} - \varphi \right]_{x=0} = \mathcal{Q}_0(t) \quad \text{or} \quad w(0,t) = h_0(t) \tag{9.141}$$

$$M(L,t) = -EI \frac{\partial \varphi}{\partial x}\bigg|_{x=L} = \mathcal{M}_L(t) \quad \text{or} \quad \varphi(L,t) = f_L(t) \tag{9.142}$$

$$Q(L,t) = k_s \left[\frac{\partial w}{\partial x} - \varphi \right]_{x=L} = \mathcal{Q}_L(t) \quad \text{or} \quad w(L,t) = h_L(t) \tag{9.143}$$

where $\mathcal{M}_0(t)$ or $f_0(t)$, $\mathcal{Q}_0(t)$ or $h_0(t)$, $\mathcal{M}_L(t)$ or $f_L(t)$, and $\mathcal{Q}_L(t)$ or $h_L(t)$ are prescribed functions.

Since the governing equations are both second order in time, and since we now have two dependent variables, we must independently specify two initial conditions for each variable. In particular, we must specify the initial transverse displacement and velocity for each particle on the centroidal surface and the initial rotation and rate of rotation of each cross section. The *initial conditions* thus take the general forms

$$w(x,0) = w_0(x) , \quad \frac{\partial w}{\partial t}\bigg|_{t=0} = v_0(x) \tag{9.144}$$

$$\varphi(x,0) = \varphi_0(x) , \quad \frac{\partial \varphi}{\partial t}\bigg|_{t=0} = \chi_0(x) \tag{9.145}$$

where $w_0(x)$, $v_0(x)$, $\varphi_0(x)$ and $\chi_0(x)$ are prescribed functions.

Example 9.11

Deduce the boundary conditions for the beam of Examples 9.8 and 9.10 if it has local mass moment of inertia I_ρ, shape factor k_s, and is modeled using Timoshenko Beam Theory.

Solution

To determine the appropriate boundary conditions, we parallel the discussion of the beam of Example 9.8, but now apply Eqs. (9.140)–(9.143). Doing this gives the boundary conditions at the left end of the beam as

$$-EI \frac{\partial \varphi}{\partial x}\bigg|_{x=0} = 0 \quad \text{and} \quad w(0,t)=0 \tag{a-1,2}$$

Similarly, the boundary conditions at the elastic wall are then

$$-EI\left.\frac{\partial\varphi}{\partial x}\right|_{x=L} = -k_\varphi\,\varphi(L,t) \qquad\qquad\text{(b-1)}$$

and

$$k_sAG\left[\frac{\partial w}{\partial x} - \varphi\right]_{x=L} = k_w w(L,t) \qquad\qquad\text{(b-2)}$$

Uniform Beams

The coupled equations of motion, Eqs. (9.133) and (9.134), or the equivalent matrix form, Eq. (9.135), correspond to the fundamental description for Timoshenko Beams. It is readily solved for uniform beams, and it is this form that will be employed in the remainder of this text. The governing equations can, however, be simplified to a single equation. We therefore present the following development for completeness.

For beams whose material properties are constant, the equations of motion, Eqs. (9.133) and (9.134), can be consolidated and simplified to some degree. Toward this end, let us first substitute Eq. (9.134) into Eq. (9.133) to get

$$m(x)\frac{\partial^2 w}{\partial t^2} - \frac{\partial}{\partial x}\left(I_\rho\frac{\partial^2\varphi}{\partial t^2}\right) + \frac{\partial^2}{\partial x^2}\left(EI\frac{\partial\varphi}{\partial x}\right) = q(x,t) \qquad\qquad\text{(9.146)}$$

which may replace either equation in the general formulation. Next, for uniform beams (beams whose material properties are independent of x), Eq. (9.146) may be rewritten in the form

$$m\frac{\partial^2 w}{\partial t^2} + \left(EI\frac{\partial^2}{\partial x^2} - I_\rho\frac{\partial^2}{\partial t^2}\right)\frac{\partial\varphi}{\partial x} = q(x,t) \qquad\qquad\text{(9.147)}$$

Solving Eq. (9.133) for $\partial\varphi/\partial x$ gives

$$\frac{\partial\varphi}{\partial x} = \frac{q(x,t)}{k_s} + \frac{\partial^2 w}{\partial x^2} - \frac{m}{k_s}\frac{\partial^2 w}{\partial t^2} \qquad\qquad\text{(9.148)}$$

Substituting Eq. (9.148) into Eq. (9.147) gives a single equation of motion in terms of the transverse deflection. Hence,

$$
m\frac{\partial^2 w}{\partial t^2} + \frac{I_\rho m}{k_s}\frac{\partial^2}{\partial t^2}\frac{\partial^2 w}{\partial t^2} - \left(I_\rho + \frac{mEI}{k_s}\right)\frac{\partial^2}{\partial x^2}\frac{\partial^2 w}{\partial t^2} + EI\frac{\partial^2}{\partial x^2}\frac{\partial^2 w}{\partial x^2}
$$
$$
= q(x,t) + \frac{I_\rho}{k_s}\frac{\partial^2 q}{\partial t^2} - \frac{EI}{k_s}\frac{\partial^2 q}{\partial x^2}
$$

(9.149)

Equation (9.149) is often referred to as the *Timoshenko Beam Equation*. Once this equation is solved and $w(x,t)$ is determined, the resulting function can be substituted into Eq. (9.148) and integrated to give the corresponding rotations due to bending. The bending moments and shear forces can then be calculated using previously defined formulae.

The Timoshenko Beam Equation, Eq. (9.149), is written in operator form as

$$
\mathbb{m}\frac{\partial^2 w}{\partial t^2} + \mathbb{k}w = F(x,t)
$$

(9.150)

where

$$
\mathbb{m} = m + \left(I_\rho + \frac{mEI}{k_s}\right)\frac{\partial^2}{\partial x^2} + \frac{mI_\rho}{k_s}\frac{\partial^2}{\partial t^2}
$$

(9.151)

$$
\mathbb{k} = EI\frac{\partial^4}{\partial x^4}
$$

(9.152)

and

$$
F(x,t) = q(x,t) + \frac{I_\rho}{k_s}\frac{\partial^2 q}{\partial t^2} + \frac{EI}{k_s}\frac{\partial^2 q}{\partial x^2}
$$

(9.153)

As discussed earlier, the coupled equations presented in matrix form correspond to the fundamental description for Timoshenko Beams and is readily solved for uniform systems. The reduced form for uniform systems discussed above allows an alternate approach and is presented here for completeness.

9.7 GEOMETRICALLY NONLINEAR BEAM THEORY

In many situations, the coupling between the transverse motion and axial motion of a beam is important. This may occur in problems of static or dynamic buckling, or simply when the rotations of the beam's axis are sufficiently large as described in Section 9.2. Thin straight structures that carry compressive axial loads as well as transverse loads are referred to as beam-columns. For problems of this nature, the linear beam theories discussed in Sections 9.6.3–9.6.5 are inadequate. In fact, linear beam theory cannot predict buckling behavior at all. We must therefore construct, or extend the beam theories discussed in prior sections to include the effects of geometric nonlinearities. For simplicity, compactness and utility, we shall neglect the effects of shear deformation and rotatory inertia. Those effects may be added on directly if desired.

To derive the geometrically nonlinear beam theory we must incorporate the strain-displacement relation for moderate rotations, Eq. (9.36) into our development. We next parallel the development presented in Section 9.6.1, proceeding exactly as we did for linear beam theory. Everything then follows identically as in Section 9.6.1, with the exception that the constitutive relation for the membrane force given by Eq. (9.91) is now replaced by the constitutive relation

$$N(x,t) = EA\,\varepsilon(x,t) = EA\left[\frac{\partial u}{\partial x} + \frac{1}{2}\left(\frac{\partial w}{\partial x}\right)^2\right] \tag{9.154}$$

We next derive the equations of motion for the beam based on the kinetic diagram for a generic beam element shown in Figure 9.11. To derive the equation of transverse motion, we parallel the corresponding development in Sections 9.6.2 and 9.6.3. However, for the present case, we include the projections of the membrane force in the transverse direction as was done for strings and cables in Section 9.5. This is necessarily consistent with Eq. (9.154). Adding the contribution of the membrane force to the equation of transverse of motion, Eq. (9.96), and to the equation of rotational motion, Eq. (9.97), respectively gives the relations

$$q(x,t) + p(x,t)\frac{\partial w}{\partial x} + \frac{\partial Q}{\partial x} = m(x)\frac{\partial^2 w}{\partial t^2} \tag{9.155}$$

$$Q(x,t) = \frac{\partial M}{\partial x} + N\frac{\partial w}{\partial x} \tag{9.156}$$

Substituting Eqs. (9.102) and (9.156) into Eq. (9.155) and paralleling the rest of the development of Section 9.6.2 gives the *equation of transverse motion* for the structure as

$$m(x)\frac{\partial^2 w}{\partial t^2} + \frac{\partial^2}{\partial x^2}k_b(x)\frac{\partial^2 w}{\partial x^2} - N(x,t)\frac{\partial^2 w}{\partial x^2}$$
$$- \left[\frac{\partial N}{\partial x} + p(x,t)\right]\frac{\partial w}{\partial x} = q(x,t) \tag{9.157}$$

where $N(x,t)$ is given by Eq. (9.154). Application of Newton's Second Law in the axial direction gives the *equation of longitudinal motion*

$$m(x)\frac{\partial^2 u}{\partial t^2} - \frac{\partial N}{\partial x} = p(x,t) \tag{9.158}$$

which is of the same form as for strings and cables (Section 9.5). Paralleling the arguments made for strings and cables, in many situations, the longitudinal component of the acceleration may be neglected compared with transverse component. In this case, Eq. (9.158) simplifies to the form

$$\frac{\partial N}{\partial x} \approx -p(x,t) \tag{9.159}$$

Substitution of Eq. (9.159) into Eq. (9.157) renders the equation of transverse motion to the form

$$m(x)\frac{\partial^2 w}{\partial t^2} + \frac{\partial^2}{\partial x^2}k_b(x)\frac{\partial^2}{\partial x^2}w - N(x,t)\frac{\partial^2 w}{\partial x^2} = q(x,t) \tag{9.160}$$

The boundary conditions follow as

$$Q(0,t) = -\frac{\partial}{\partial x}EI\frac{\partial^2 w}{\partial x^2}\bigg|_{x=0} + N\frac{\partial w}{\partial x}\bigg|_{x=0} = \mathcal{Q}_0(t) \quad \text{or} \quad w(0,t) = h_0(t) \tag{9.161}$$

$$M(0,t) = -EI\frac{\partial^2 w}{\partial x^2}\bigg|_{x=0} = \mathcal{M}_0(t) \quad \text{or} \quad \frac{\partial w}{\partial x}\bigg|_{x=0} = f_0(t) \tag{9.162}$$

$$Q(L,t) = -\frac{\partial}{\partial x}EI\frac{\partial^2 w}{\partial x^2}\bigg|_{x=L} + N\frac{\partial w}{\partial x}\bigg|_{x=L} = \mathcal{Q}_L(t) \quad \text{or} \quad w(L,t) = h_L(t) \tag{9.163}$$

$$M(L,t) = -EI\frac{\partial^2 w}{\partial x^2}\bigg|_{x=L} = \mathcal{M}_L(t) \quad \text{or} \quad \frac{\partial w}{\partial x}\bigg|_{x=L} = f_L(t) \tag{9.164}$$

where $\mathcal{Q}_0(t)$ or $h_0(t)$, $\mathcal{M}_0(t)$ or $f_0(t)$, $\mathcal{Q}_L(t)$ or $h_L(t)$, and $\mathcal{M}_L(t)$ or $f_L(t)$ are prescribed functions. As for Euler-Bernoulli Beams, the initial conditions are of the form

$$w(x,0) = w_0(x) \tag{9.165}$$

and

$$\frac{\partial w}{\partial t}\bigg|_{t=0} = v_0(x) \tag{9.166}$$

If no distributed axial load (axial body force) acts on the structure ($p = 0$) then, from Eq. (9.159),

$$N = N_0 = \text{constant} \tag{9.167}$$

For this case, Eq. (9.160) may be written in the form of Eq. (9.124) with $\mathbb{m} = m(x)$ and

$$\mathbb{k} = \frac{\partial^2}{\partial x^2} EI \frac{\partial^2}{\partial x^2} - N_0 \frac{\partial^2}{\partial x^2} \tag{9.168}$$

Thus, if N_0 is prescribed, the motion of the geometrically nonlinear beam is seen to be governed by a linear differential equation of standard operator form.

Example 9.12

Determine the equation of motion and associated boundary conditions for the uniform beam-column subjected to a compressive edge load, $P(t)$, as shown.

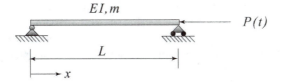

Solution
For the problem at hand, $q = p = 0$. The membrane force may be determined by consideration of the kinetic diagram for the beam element at the loaded edge, as shown. If we neglect the axial component of the acceleration, the membrane force in the beam is simply

$$N = -P(t) \tag{a}$$

Inserting Eq. (a) into Eq. (9.160) gives the explicit form of the equation of transverse motion as

$$EI \frac{\partial^4 w}{\partial x^4} + P(t) \frac{\partial^2 w}{\partial x^2} + m \frac{\partial^2 w}{\partial t^2} = 0 \qquad \triangleleft \text{(b)}$$

It is seen that in this problem, the forcing function enters the equation as a pre-scribed time dependent coefficient of the second term of the governing differential equation. The boundary conditions follow directly from Eqs. (9.116)–(9.119) as

$$w(0,t) = 0 , \quad -EI \frac{\partial^2 w}{\partial x^2}\bigg|_{x=0} = 0 \qquad \triangleleft \text{(c-1,2)}$$

$$w(L,t) = 0 , \quad -EI\frac{\partial^2 w}{\partial x^2}\bigg|_{x=L} = 0 \qquad \lhd (d\text{-}1,2)$$

9.8 TRANSLATING 1-D CONTINUA

In many situations, a structure or device is in overall motion and that motion, as well as other sources, induces vibrations of the system. Such situations include vehicular structures, support excited structures and parts of mechanisms, to name but a few. When the base motion of a structure is in the transverse direction alone, the base motion may be introduced as a boundary condition as discussed earlier. However, when the support motion includes motion in the axial direction, the reference frame for measuring deformation in the conventional sense is accelerating, or is at least translating with constant velocity. This is so since it is natural to choose a coordinate system that travels with the moving base so that the deformation measures defined in Section 9.2 may be employed directly. When this is done, the motion of the reference frame must be accounted for when evaluating time rates of change. This alters the form of the kinematical measures that describe the motion of a material particle and, ultimately, the equations of motion. In this section we examine the motion of a geometrically nonlinear beam-column that is translating in a given plane. The case of a moving string or cable is obtained as a special case by letting the bending stiffness vanish, and so is treated in this context. We begin by establishing a description of velocity and with respect to a translating reference frame, for material particles that comprise a translating beam-column.

9.8.1 Kinematics of a Material Particle

Consider the translating elastic beam-column of initial length L, shown in Figure 9.15. Let a point on the structure, say at its base, be moving with the prescribed motion in the xz-plane described by $\chi(t) = (\chi_x, \chi_z)$, where $\chi_x(t)$ and $\chi_z(t)$ respectively correspond to the longitudinal and transverse components of the displacement, as indicated. Let each particle along the axis of the structure be labeled by its coordinates in the rest configuration, $\mathbf{X} = (X,0)$. To describe the motion of the individual particles of the structure, let us follow a generic material particle $\mathbf{X}$ as the structure translates and deforms. Let the particle $\mathbf{X}$ move to its present position $\boldsymbol{\xi} = (\xi_x, \xi_z)$ at time t. Hence,

$$\mathbf{X} = (X,0) \rightarrow \boldsymbol{\xi} = (\xi_x, \xi_z)$$

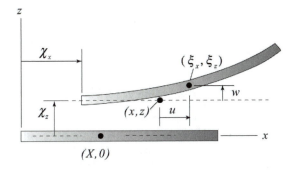

Figure 9.15 Deformation of translating beam-column.

To measure the deformation, let us introduce the (fictitious) intermediate configuration corresponding to the rigid body translation of the entire undeformed structure. [Since the structure is elastic, its deformation is independent of the deformation history and thus independent of the path traversed by a material particle as it moves from the point $(X,0)$ to the point (ξ_x, ξ_z). Deformations can then be measured relative to this convenient configuration.] Let the point $\mathbf{x} = (x,z)$ correspond to the location of the image of particle $\mathbf{X}$ in the intermediate configuration at time t. The coordinates $\mathbf{x}$ thus correspond to the projection of ξ onto the rigidly displaced image of the rod at time t. Hence,

$$\mathbf{x} = \mathbf{X} + \chi \tag{9.169}$$

or, in component form,

$$x = X + \chi_x(t)$$
$$z = \chi_z(t) \tag{9.170}$$

The relative deflection, the displacement of particle $\mathbf{X}$ at time t measured relative to the intermediate configuration, is then

$$\mathbf{u} = (u, w) = \xi - \mathbf{x} \tag{9.171}$$

The velocity of particle $\mathbf{X}$ is then

$$\mathbf{v} \equiv \left.\frac{\partial \xi}{\partial t}\right|_{\mathbf{X}} = \left.\frac{\partial \mathbf{x}}{\partial t}\right|_{\mathbf{X}} + \left.\frac{\partial \mathbf{u}}{\partial t}\right|_{\mathbf{X}} \tag{9.172}$$

Now, from Eq. (9.169),

$$\frac{\partial \mathbf{x}}{\partial t}\bigg|_\mathbf{x} = \dot{\boldsymbol{\chi}}(t) \tag{9.173}$$

In component form,

$$\frac{\partial x}{\partial t} = \dot{\chi}_x(t) \,, \qquad \frac{\partial z}{\partial t} = \dot{\chi}_z(t) \tag{9.174}$$

Further, applying the chain rule to the second term on the right-hand side of Eq. (9.172) and incorporating Eq. (9.173), we have that

$$\frac{\partial \mathbf{u}}{\partial t}\bigg|_\mathbf{x} = \frac{\partial \mathbf{u}}{\partial t}\bigg|_\mathbf{x} + \dot{\chi}_x \frac{\partial \mathbf{u}}{\partial x} \tag{9.175}$$

In Eq. (9.175), $\partial \mathbf{u}/\partial t$ corresponds to the time rate of change of $\mathbf{u}$ for any material particle as seen by an observer fixed at $\mathbf{x}$, while

$$\frac{\partial \mathbf{u}}{\partial t} + \dot{\chi}_x \frac{\partial \mathbf{u}}{\partial x}$$

corresponds to the velocity of a material particle as seen by an observer translating with the support. The convective term, $\dot{\chi}_x \, \partial \mathbf{u}/\partial x$ accounts for the fact that the particle at point $\mathbf{x}$ is changing, and we are following the particular particle $\mathbf{X}$.

Substituting Eqs. (9.173) and (9.175) into Eq. (9.172) gives the velocity in terms of the intermediate coordinates as

$$\mathbf{v}(x,t) = \dot{\boldsymbol{\chi}}(t) + \frac{\partial \mathbf{u}}{\partial t}\bigg|_\mathbf{x} + \dot{\chi}_x \frac{\partial \mathbf{u}}{\partial x} \tag{9.176}$$

In component form,

$$v_z(x,t) = \dot{\chi}_z + \frac{\partial w}{\partial t} + \dot{\chi}_x \frac{\partial w}{\partial x} \tag{9.177}$$

$$v_x(x,t) = \dot{\chi}_x + \frac{\partial u}{\partial t} + \dot{\chi}_x \frac{\partial u}{\partial x} \cong \dot{\chi}_x + \frac{\partial u}{\partial t} \tag{9.178}$$

If the kinetic energy of relative axial motion is small compared with the kinetic energy of transverse motion, or if it is small compared with the bulk kinetic energy of axial motion, then we may make the approximation that

$$v_x \approx \dot{\chi}_x(t) \tag{9.179}$$

which is consistent with similar approximations discussed for strings and cables in Section 9.5 and for geometrically nonlinear beams in Section 9.7.

To evaluate the acceleration, we proceed in an analogous fashion to the evaluation of the velocity. Thus, taking the time derivative of the velocity of a material particle, holding the particle constant and employing the chain rule, we obtain the acceleration of particle **X** as

$$\mathbf{a} \equiv \left.\frac{\partial \mathbf{v}}{\partial t}\right|_{\mathbf{X}} = \left.\frac{\partial \mathbf{v}}{\partial t}\right|_{\mathbf{x}} + \dot{\chi}_x \frac{\partial \mathbf{v}}{\partial x} \tag{9.180}$$

The transverse and axial components of the acceleration are then, respectively,

$$a_z = \ddot{\chi}_z + \frac{\partial^2 w}{\partial t^2} + \ddot{\chi}_x \frac{\partial w}{\partial x} + 2\dot{\chi}_x \frac{\partial^2 w}{\partial x \partial t} + \dot{\chi}_x{}^2 \frac{\partial^2 w}{\partial x^2} \tag{9.181}$$

$$a_x = \ddot{\chi}_x\left(1 + \frac{\partial u}{\partial x}\right) + \frac{\partial^2 u}{\partial t^2} + 2\dot{\chi}_x \frac{\partial^2 u}{\partial x \partial t} + \dot{\chi}_x{}^2 \frac{\partial^2 u}{\partial x^2}$$

$$\cong \ddot{\chi}_x + \frac{\partial^2 u}{\partial t^2} + 2\dot{\chi}_x \frac{\partial^2 u}{\partial x \partial t} + \dot{\chi}_x{}^2 \frac{\partial^2 u}{\partial x^2} \tag{9.182}$$

If the axial acceleration relative to the support is neglected, in keeping with approximations made in Sections 9.5 and 9.7, then the axial component of the acceleration simplifies to

$$a_x \approx \ddot{\chi}_x \tag{9.183}$$

Now that the expressions for velocity and acceleration of a material particle have been established, we may proceed to derive the equations of motion of the translating structure.

9.8.2 Kinetics

To derive the equations of motion, we consider the kinetic diagram of a generic element at the current time t and parallel the development of Section 9.7 (Figure 9.12). Application of Newton's Second Law in the coordinate directions gives the equations

$$q(x,t) - \frac{\partial^2}{\partial x^2} k_b(x) \frac{\partial^2 w}{\partial x^2} + N(x,t) \frac{\partial^2 w}{\partial x^2}$$

$$+ \left[\frac{\partial N}{\partial x} + p(x,t)\right]\frac{\partial w}{\partial x} = m(x)a_z(x,t) \tag{9.184}$$

$$p(x,t) + \frac{\partial N}{\partial x} = m(x)a_x(x,t) \tag{9.185}$$

where N is given by Eq. (9.154). Substituting Eqs. (9.181) and (9.182) into Eqs. (9.184) and (9.185) and rearranging terms gives the equations of transverse and longitudinal motion, respectively, as

$$
\frac{\partial^2}{\partial x^2} EI \frac{\partial^2}{\partial x^2} w - \left(N - m\dot{\chi}_x^{\,2} \right) \frac{\partial^2 w}{\partial x^2}
$$
$$
- \left[\frac{\partial N}{\partial x} + p(x,t) - m\ddot{\chi}_x \right] \frac{\partial w}{\partial x} + m \left[\frac{\partial^2 w}{\partial t^2} + 2\dot{\chi}_x \frac{\partial^2 w}{\partial x \partial t} \right] = q(x,t) - m\ddot{\chi}_z(t) \tag{9.186}
$$

and

$$
-\frac{\partial N}{\partial x} + m \left[\ddot{\chi}_x + \frac{\partial^2 u}{\partial t^2} + 2\dot{\chi}_x \frac{\partial^2 u}{\partial x \partial t} + \dot{\chi}_x^{\,2} \frac{\partial^2 u}{\partial x^2} \right] = p(x,t) \tag{9.187}
$$

If the longitudinal motion relative to the support is neglected, then the longitudinal equation of motion simplifies to the form

$$
\frac{\partial N}{\partial x} - m\ddot{\chi}_x \approx -p(x,t) \tag{9.188}
$$

Incorporating Eq. (9.188) into Eq. (9.186) renders the equation of transverse motion to the form

$$
\frac{\partial^2}{\partial x^2} EI \frac{\partial^2 w}{\partial x^2} - \left(N - m\dot{\chi}_x^{\,2} \right) \frac{\partial^2 w}{\partial x^2}
$$
$$
+ m \left(\frac{\partial^2 w}{\partial t^2} + 2\dot{\chi}_x \frac{\partial^2 w}{\partial x \partial t} \right) = q(x,t) - m\ddot{\chi}_z(t) \tag{9.189}
$$

If we consider an observer moving with a beam element, then the above equation may be thought of as representing an equivalent beam with stationary supports subjected to the effective distributed transverse load

$$
\hat{q}(x,t) = q(x,t) - m\ddot{\chi}_x(t)
$$

and effective membrane force

$$
\hat{N}(x,t) = N(x,t) - m\dot{\chi}_x^{\,2}(t)
$$

The quantity $\hat{N}$ may be viewed as an analog of the "stagnation pressure" in fluid mechanics, with $m\dot{\chi}_x^{\,2}$ the analog of the "dynamic pressure" and N the analog of the "static pressure." For an observer moving with the beam element, and thus rotating with it as the beam bends, the second term in the effective acceleration,

$$2\dot{\chi}_x \frac{\partial^2 w}{\partial x \partial t} = 2\dot{\chi}_x \frac{\partial \varphi}{\partial t}$$

may be interpreted as a Coriolis-like acceleration.

Boundary Conditions

The boundary conditions follow from the interpretation of the effective membrane force. (A derivation using the Calculus of Variations and Hamilton's Principle yields the above equations of motion and the boundary conditions below, and thus gives credence to this interpretation.) The associated boundary conditions are thus

$$\left[-\frac{\partial}{\partial x}\left(EI\frac{\partial^2 w}{\partial x^2} \right) + \left(N - m\dot{\chi}_x^{\;2} \right)\frac{\partial w}{\partial x} \right]_{x=\chi_x} = Q_0(t) \quad \text{or} \quad w\big|_{x=\chi_x} = 0 \quad (9.190)$$

$$-EI\frac{\partial^2 w}{\partial x^2}\bigg|_{x=\chi_x} = M_0(t) \quad \text{or} \quad \frac{\partial w}{\partial x}\bigg|_{x=\chi_x} = 0 \qquad (9.191)$$

$$\left[-\frac{\partial}{\partial x}\left(EI\frac{\partial^2 w}{\partial x^2} \right) + \left(N - m\dot{\chi}_x^{\;2} \right)\frac{\partial w}{\partial x} \right]_{x=\chi_x+L} = Q_L(t) \quad \text{or} \quad w\big|_{x=\chi_x+L} = 0 \quad (9.192)$$

$$-EI\frac{\partial^2 w}{\partial x^2}\bigg|_{x=\chi_x+L} = M_L(t) \quad \text{or} \quad \frac{\partial w}{\partial x}\bigg|_{x=\chi_x+L} = 0 \qquad (9.193)$$

where $Q_0(t)$, $M_0(t)$, $Q_L(t)$ and $M_L(t)$ are prescribed functions. For the special case of strings and cables, $EI \to 0$ in Eq. (9.189) and the boundary conditions for shear, Eqs. (9.190) and (9.192), while the boundary conditions defined by Eqs. (9.191) and (9.193) are omitted.

Example 9.13

Consider a fan belt operating at steady state as shown in the figure. If the constant speed of the belt is v_0 and the constant tension in the belt is N_0, establish the governing equation and boundary conditions for the upper straight segment of the belt between the rollers.

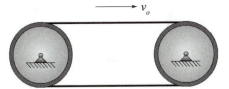

Solution

For the present problem, $EI = 0$, $\chi_z \equiv 0$ and $\dot{\chi}_x = v_0$. Substituting these values into Eq. (9.189) gives the equation of motion of the belt as

$$m\left(\frac{\partial^2 w}{\partial t^2} + 2v_0 \frac{\partial^2 w}{\partial x \partial t}\right) - \left(N_0 - mv_0^2\right)\frac{\partial^2 w}{\partial x^2} = 0 \qquad\qquad \triangleleft \text{(a)}$$

The boundary conditions are simply

$$w(0,t) = 0 \quad \text{and} \quad w(L,t) = 0 \qquad\qquad \triangleleft \text{(b-1,2)}$$

Example 9.14

Determine the equation of motion and associated boundary conditions for a uniform cantilevered beam-column supporting a large point mass $m \gg mL$ if the support undergoes the elliptical motion described parametrically by the equations $u_0(t) = \alpha\Delta_0(1 - \cos\Omega t)$ and $w_0 = \Delta_0 \sin\Omega t$, where $\Delta_0/L \ll 1$ and $\alpha < 1$.

Solution

For the problem at hand, $p = q = 0$. The kinetic diagram and associated equation of motion for the point mass show that

$$N\big|_{\chi_x + L} \cong m\left[g + \ddot{\chi}_x\right] \qquad\qquad \text{(a)}$$

Integrating the equation of longitudinal motion, Eq. (9.188), and incorporating Eq. (a) gives the membrane force as

$$\begin{aligned}
N(x,t) &= N\big|_{\chi_x + L} + \int_{\chi_x}^{\chi_x + L} m\left[g + a_x\right]dx \\
&\cong m\left[g + \ddot{\chi}_x\right] + m\left[g + \ddot{\chi}_x\right]\left(\chi_x + L - x\right) \approx m\left[g + \ddot{\chi}_x\right] \\
&= m\left[g + \Omega^2\alpha\Delta_0 \cos\Omega t\right]
\end{aligned} \qquad \text{(b)}$$

or

$$N(x,t) = N_0 + n_0 \cos\Omega t \qquad\qquad \text{(c)}$$

where

$$N_0 = mg \qquad\qquad \text{(d)}$$

and

$$n_0 = \alpha\Delta_0\Omega^2 \qquad\qquad \text{(e)}$$

Next,

$$N - m\dot{\chi}_x^2 = m\left(g + \alpha\Delta_0\Omega^2\cos\Omega t\right) - m\alpha^2\Delta_0^2\Omega^2\sin^2\Omega t$$

$$= mL\left[\left(g/L\right) + \Omega^2\left(\alpha\Delta_0/L\right)\cos\Omega t\right] - mL^2\left(\alpha\Delta_0/L\right)^2\Omega^2\sin^2\Omega t \quad \text{(f)}$$

$$\approx m\left(g + \alpha\Delta_0\Omega^2\sin\Omega t\right)$$

Substituting Eq. (f) into Eq. (9.189) gives the desired equation of motion,

$$EI\frac{\partial^4 w}{\partial x^4} + m\left(g + \alpha\Delta_0\Omega^2\cos\Omega t\right)\frac{\partial^2 w}{\partial x^2}$$

$$+ m\left(\frac{\partial^2 w}{\partial t^2} + 2\alpha\Delta_0\Omega\sin\Omega t\frac{\partial^2 w}{\partial t\partial x}\right) = m\Delta_0\Omega^2\cos\Omega t \quad \text{(d)}$$

9.9 CONCLUDING REMARKS

In this chapter we defined measures to characterize the local behavior of continua and developed a number of representations that describe the motion of mathematically one-dimensional continua. These mathematical models pertain to long, thin bodies. That is, they represent structures where one characteristic length is much larger than the others. We studied both linear and geometrically nonlinear structures. These included longitudinal motion of elastic rods, torsional motion of rods of circular cross section, the motion of strings and cables, and several representations for flexural motion of elastic structures possessing various degrees and types of complexity, including the coupling of flexural and axial motions. The linear systems of Sections 9.3–9.6 were seen to each be described by equations of motion of the general form

$$\mathbb{m}\frac{\partial^2 u}{\partial t^2} + \mathbb{k}u = F(x,t) \quad (9.194)$$

where $\mathbb{m}$ and $\mathbb{k}$ are differential (or scalar) operators, $u(x,t)$ is a displacement function that characterizes the motion of the body and $F(x,t)$ corresponds to an appropriate distributed external force. For Timoshenko Beam Theory, a matrix equation of the same general form governs the motion of the structure. For this case, the mass and stiffness operators are 2×2 matrices of differential operators, the motion is characterized by a 2×1 matrix of displacement functions and the applied force is characterized by a 2×1 matrix of force distributions. The mathematical representation of each linear system is summarized in Table 9.1. It was seen that the parameters and operators that describe the motion of continuous systems are abstractions of those that describe discrete systems, and that both classes of system lie within the same general framework.

Table 9.1 Parameters for Various 1-D Continua

$u(x,t)$	$\mathbb{k}$	$\mathbb{m}$	$F(x,t)$
Longitudinal Motion of Rods: $u(x,t)$	$\mathbb{k} = -\dfrac{\partial}{\partial x}EA\dfrac{\partial}{\partial x}$	$m(x) = \rho(x)A(x)$	$p(x,t)$
Torsional Motion of Rods: $\theta(x,t)$	$\mathbb{k} = -\dfrac{\partial}{\partial x}GJ\dfrac{\partial}{\partial x}$	$J_\rho(x) = \rho(x)J(x)$	$T(x,t)$
Transverse Motion of Strings and Cables: $w(x,t)$	$\mathbb{k} = -N\dfrac{\partial^2}{\partial x^2}$	$m(x) = \rho(x)A(x)$	$q(x,t)$
Euler-Bernoulli Beams: $w(x,t)$	$\mathbb{k} = \dfrac{\partial^2}{\partial x^2}EI\dfrac{\partial^2}{\partial x^2}$	$m(x) = \rho(x)A(x)$	$q(x,t)$ $-\partial b/\partial x$
Nonlinear E.-B. Beams with const. axial force: $w(x,t)$	$\mathbb{k} = \dfrac{\partial^2}{\partial x^2}EI\dfrac{\partial^2}{\partial x^2} - N_0\dfrac{\partial^2}{\partial x^2}$	$m(x) = \rho(x)A(x)$	$q(x,t)$ $(p=0)$
Rayleigh Beams: $w(x,t)$	$\mathbb{k} = \dfrac{\partial^2}{\partial x^2}EI\dfrac{\partial^2}{\partial x^2}$	$\mathbb{m} = m(x) - \dfrac{\partial}{\partial x}I_\rho(x)\dfrac{\partial}{\partial x}$	$q(x,t)$ $-\partial b/\partial x$
Timoshenko Beams: $\mathbf{u} = \begin{Bmatrix} w(x,t) \\ \varphi(x,t) \end{Bmatrix}$	$\mathbf{k} = \begin{bmatrix} \mathbb{k}_{ij} \end{bmatrix}$ where $\mathbb{k}_{11} = -\partial\mathbb{k}_{12}/\partial x$ $\mathbb{k}_{12} = -\mathbb{k}_{21} = k_s\,\partial/\partial x$ $\mathbb{k}_{22} = k_s - \dfrac{\partial}{\partial x}EI\dfrac{\partial}{\partial x}$	$\mathbf{m} = \begin{bmatrix} m(x) & 0 \\ 0 & I_\rho(x) \end{bmatrix}$	$\mathbf{F} = \begin{Bmatrix} q \\ b \end{Bmatrix}$

BIBLIOGRAPHY

Archibold, F.R. and Elmslie, A.G., "The Vibration of a String Having a Uniform Motion Along Its Length," *ASME Journal of Applied Mechanics*, 25, 1958, 347–348.

Bottega, W.J., "Dynamics and Stability of Support Excited Beam-Columns with End Mass," *Dynamics and Stability of Systems*, 1, 1986, 201–215.

Fung, Y.C., *Foundations of Solid Mechanics*, Prentice-Hall, Englewood Cliffs, 1965.

Graff, K.F., *Wave Motion in Elastic Solids*, Dover, New York, 1991.

Rayleigh, J.W.S., *Theory of Sound*, Vol. 1, Dover, New York, 1945.

Weaver, W. Jr., Timoshenko, S.P. and Young, D.P., *Vibration Problems in Engineering*, 5th ed., Wiley-Interscience, 1990.

PROBLEMS

9.1 Determine the scalar product of the functions $f(x) = \cos(2n\pi x/L)$ and $g(x) = \cos(2p\pi x/L)$ on the domain $0 \leq x \leq L$, where $n \neq p$ are positive integers greater than zero.

9.2 Determine the scalar product of the functions $f(x) = \sin(2n\pi x/L)$ and $g(x) = \cos(2p\pi x/L)$ on the domain $0 \leq x \leq L$, where $n \neq p$ are positive integers greater than zero.

9.3 State the equation of motion and deduce the boundary conditions for longitudinal motion of the rod shown in Figure P9.3.

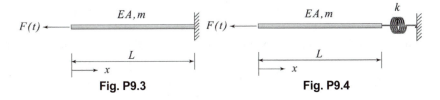

Fig. P9.3   Fig. P9.4

9.4 State the equation of motion and deduce the boundary conditions for longitudinal motion of the rod shown in Figure P9.4.

9.5 State the equation of motion and deduce the boundary conditions for longitudinal motion of the rod shown in Figure P9.5.

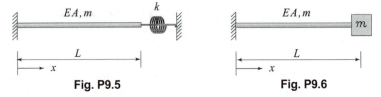

Fig. P9.5   Fig. P9.6

9.6 State the equation of motion and deduce the boundary conditions for longitudinal motion of the rod shown in Figure P9.6.

9.7 State the equation of motion and deduce the boundary conditions for torsional motion of the rod shown in Figure P9.7.

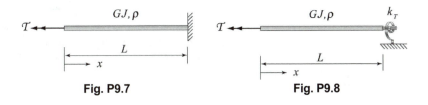

Fig. P9.7   Fig. P9.8

9.8 State the equation of motion and deduce the boundary conditions for torsional motion of the rod shown in Figure P9.8.

9.9 State the equation of motion and deduce the boundary conditions for torsional motion of the rod shown in Figure P9.9.

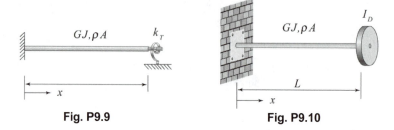

Fig. P9.9   **Fig. P9.10**

9.10 State the equation of motion and boundary conditions for torsional motion of an elastic rod that is fixed at on end and is attached to a rigid disk of mass moment of inertia I_D at its free end.

9.11 State the governing equation and boundary conditions for transverse motion of the active segment of a guitar string of mass density ρ and cross-sectional area A that is under static tension N, when a musician presses on the string with his finger at a fret located a distance L from the bridge. The stiffness of the musician's finger is k.

9.12 State the governing equation and boundary conditions for transverse motion of a cable of mass density ρ and cross-sectional area A that is under static tension N and is adhered to elastic mounts of stiffness k at each end.

9.13 State the equation of motion and deduce the boundary conditions for flexural motion of the simply supported beam shown when it is modeled mathematically using Euler-Bernoulli Beam Theory.

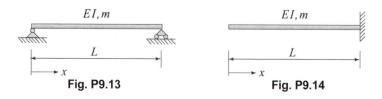

Fig. P9.13   **Fig. P9.14**

9.14 State the equation of motion and deduce the boundary conditions for flexural motion of the cantilevered beam shown when it is modeled mathematically using Euler-Bernoulli Beam Theory.

9.15 State the equation of motion and deduce the boundary conditions for flexural motion of the beam supported by elastic hinges at each end, as shown, when it is modeled mathematically using Euler-Bernoulli Beam Theory.

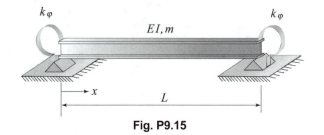

Fig. P9.15

9.16 State the equation of motion and deduce the boundary conditions for flexural motion of the elastically clamped beam shown when it is modeled mathematically using Euler-Bernoulli Beam Theory.

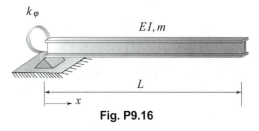

Fig. P9.16

9.17 State the equation of motion and deduce the boundary conditions for flexural motion of the beam supported at one end by an elastic clamp and at the other by an elastic mount, as shown, when the structure is modeled mathematically using Euler-Bernoulli Beam Theory.

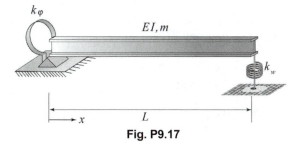

Fig. P9.17

9.18 State the equation of motion and deduce the boundary conditions for flexural motion of the beam supported by elastic mounts at each end, as shown, when it is modeled mathematically using Euler-Bernoulli Beam Theory.

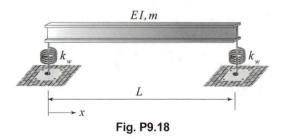

Fig. P9.18

9.19 State the equation of motion and deduce the boundary conditions for flexural motion of the an elastic beam that is embedded in a rigid wall at one and sits on an elastic foundation at the other when the structure is modeled mathematically using Euler-Bernoulli Beam Theory.

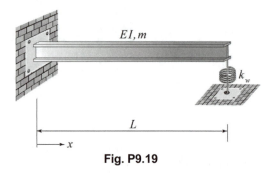

Fig. P9.19

9.20 Repeat Problem 9.13 for a beam of local mass moment of inertia I_ρ if the structure is modeled using Rayleigh Beam Theory.

9.21 Repeat Problem 9.14 for a beam of local mass moment of inertia I_ρ if the structure is modeled using Rayleigh Beam Theory.

9.22 Repeat Problem 9.15 for a beam of local mass moment of inertia I_ρ if the structure is modeled using Rayleigh Beam Theory.

9.23 Repeat Problem 9.16 for a beam of local mass moment of inertia I_ρ if the structure is modeled using Rayleigh Beam Theory.

9.24 Repeat Problem 9.17 for a beam of local mass moment of inertia I_ρ if the structure is modeled using Rayleigh Beam Theory.

9.25 Repeat Problem 9.18 for a beam of local mass moment of inertia I_ρ if the structure is modeled using Rayleigh Beam Theory.

9.26 Repeat Problem 9.19 for a beam of local mass moment of inertia I_ρ if the structure is modeled using Timoshenko Beam Theory.

9.27 Repeat Problem 9.13 for a beam of local mass moment of inertia I_ρ if the structure is modeled using Timoshenko Beam Theory.

9.28 Repeat Problem 9.14 for a beam of local mass moment of inertia I_ρ if the structure is modeled using Timoshenko Beam Theory.

9.29 Repeat Problem 9.15 for a beam of local mass moment of inertia I_ρ if the structure is modeled using Timoshenko Beam Theory.

9.30 Repeat Problem 9.16 for a beam of local mass moment of inertia I_ρ if the structure is modeled using Timoshenko Beam Theory.

9.31 Repeat Problem 9.17 for a beam of local mass moment of inertia I_ρ if the structure is modeled using Timoshenko Beam Theory.

9.32 Repeat Problem 9.18 for a beam of local mass moment of inertia I_ρ if the structure is modeled using Timoshenko Beam Theory.

9.33 Repeat Problem 9.19 for a beam of local mass moment of inertia I_ρ if the structure is modeled using Timoshenko Beam Theory.

9.34 State the equation of transverse motion and the associated boundary conditions for the elastic beam-column shown in Figure P9.34. The mass of the movable support is negligible.

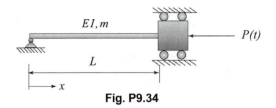

Fig. P9.34

9.35 State the equation of transverse motion and the associated boundary conditions for the beam-column with elastic clamp shown in Figure P9.35.

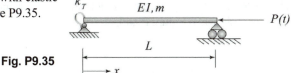

Fig. P9.35

9.36 Determine the equation of lateral motion and establish the corresponding boundary conditions for the inner segments of cable of the pulley system shown, if the cable has mass per unit length m and the mass of the pulley wheels is negligible compared with that of the suspended weight.

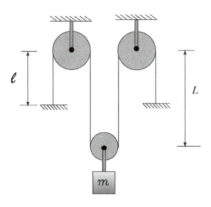

Fig. P9.36

9.37 Determine the equation of lateral motion for the inner segments of cable of the pulley system of Problem 9.36 after the left most segment of the cable is cut and before its free end passes through the first pulley wheel.

9.38 Show that, for the case of lateral motion of the support of a beam, the linearized version of Eq. (9.189) converges to Eq. (9.104) for proper change of variables. (Hint: Note that in Example 9.9 w corresponds to "absolute displacement.")

9.39 Determine the equation of motion and boundary conditions for the translating beam shown in Figure P9.39.

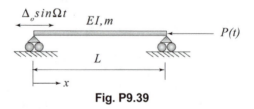

Fig. P9.39

9.40 Determine the equation of motion and boundary conditions for the translating beam shown in Figure P9.40.

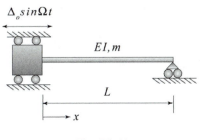

Fig. P9.40

9.41 Determine the equation of motion and boundary conditions for the translating beam shown in Figure P9.41.

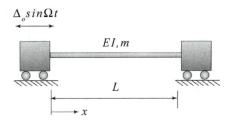

Fig. P9.41

10

Free Vibration of One-Dimensional Continua

One-dimensional continua are three-dimensional bodies or media whose deformations and motions are described mathematically as functions of a single spatial variable. In this chapter we consider the motion of mathematically one-dimensional continua when they are free from externally applied dynamic forces. That is, we examine free vibrations of one-dimensional continua. The study of free vibrations reveals fundamental characteristics of the system, as well as behavior that is of interest in its own right. The presentation and discussion will parallel earlier discussions pertaining to discrete systems. We specifically consider the behavior of linear continua, including longitudinal motion of elastic rods, torsional motion of elastic rods, transverse motion of strings and cables, and the flexural motion of elastic beams based on Euler-Bernoulli, Rayleigh and Timoshenko theories. We also study the free vibrations of geometrically nonlinear beams and beam-columns with constant axial load. We begin with a discussion of the general free vibration problem for mathematically one-dimensional continua.

10.1 THE GENERAL FREE VIBRATION PROBLEM

The equations of motion for linear one-dimensional continua are described by the general form of Eq. (9.194). For free vibrations, the distributed external load vanishes and the equations of motion take the general form

$$m \frac{\partial^2 u}{\partial t^2} + \mathbb{k} u = 0 \qquad (10.1)$$

where $u(x,t)$ represents the appropriate displacement field, $\mathbb{m}$ is a scalar or differential mass operator and $\mathbb{k}$ is a differential stiffness operator. The motion of Timoshenko Beams is governed by a matrix equation of the form of Eq. (10.1), where the displacement measure consists of a column matrix whose two elements are functions that correspond to the transverse displacement and the rotation of the cross-section, respectively. In addition, the mass and stiffness operators for Timoshenko Beams are each 2 by 2 matrices whose elements are scalar and differential operators, respectively.

Equation (10.1), together with an appropriate set of boundary and initial conditions, defines the free vibration problem. To solve this problem, we proceed in an analogous fashion to that for discrete systems. We thus seek a response of the form

$$u(x,t) = U(x)e^{i \omega t} \qquad (10.2)$$

Note that the spatial function $U(x)$ is completely analogous to the corresponding column vector for discrete systems introduced in Eq. (7.2). Substituting the assumed form, Eq. (10.2), into the general equation of motion, Eq. (10.1), and recalling that $\mathbb{m}$ and $\mathbb{k}$ operate only on the spatial variable x, gives

$$-\omega^2 \mathbb{m} U(x) e^{i \omega t} + \mathbb{k} U(x) e^{i \omega t} = 0$$

which results in the eigenvalue problem

$$\left[\mathbb{k} - \omega^2 \mathbb{m} \right] U(x) = 0 \qquad (10.3)$$

The free vibration problem is thus reduced to the determination of $\{\omega^2, U(x)\}$ pairs that satisfy Eq. (10.3), where the parameter ω^2 is identified as the *eigenvalue* and the function $U(x)$ is identified as the corresponding *eigenfunction*. Since deformable continua possess an infinite number of degrees of freedom, the eigenvalue problem yields an infinity of frequency-modal function pairs. It may be seen from Eq. (10.3) that the modal functions are unique to within, at most, a constant multiplier. The value of this constant is arbitrary and is typically chosen as unity, or it is chosen so as to render the magnitude of the modal function unity. The latter option is discussed in Section 10.7 and the mutual orthogonality of the modal functions is discussed in Section 10.8.

The solution of the eigenvalue problem depends on the specific stiffness and mass operators, as well as the boundary conditions, for a particular system under consideration. Each frequency-mode pair, $\{\omega_j^2, U^{(j)}(x) \mid (j = 1, 2, ...)\}$, obtained in this way corresponds to a solution of Eq. (10.1) in the form of Eq. (10.2). The general solution then consists of a linear combination of all such solutions. Thus, the free vibration response takes the general form

$$u(x,t) = \sum_{j=1}^{\infty} C^{(j)} U^{(j)}(x) e^{i\omega_j t} = \sum_{j=1}^{\infty} U^{(j)}(x) A^{(j)} \cos(\omega_j t - \phi_j) \qquad (10.4)$$

where the modal amplitudes and associated phase angles, $A^{(j)}$ and ϕ_j ($j = 1, 2, \ldots$) respectively, are determined from the initial conditions. In the next five sections we consider various second order and fourth order systems and analyze the corresponding types of motion.

10.2 FREE VIBRATION OF UNIFORM SECOND ORDER SYSTEMS

We next consider the free vibration problem for systems with a second order stiffness operator. In particular, we consider longitudinal motion of uniform rods, torsional motion of uniform rods with circular cross sections and the transverse motion of uniform strings and cables under constant tension. As all of these systems are governed by the same differential equation, the one-dimensional wave equation, the general solution is the same. We therefore consider the common problem first and then present specific results for each case.

10.2.1 The General Free Vibration Problem and Its Solution

For systems with uniform properties the local mass and stiffness measures, m and k, are constants. The stiffness operator then takes the form

$$\mathbb{k} = -k \frac{\partial^2}{\partial x^2} \qquad (10.5)$$

and the equation of motion is given by the one-dimensional wave equation,

$$\frac{\partial^2 u}{\partial t^2} - c^2 \frac{\partial^2 u}{\partial x^2} = 0 \qquad (10.6)$$

where

$$c^2 = k/m \qquad (10.7)$$

is a characteristic wave speed. Substitution of Eq. (10.2) into Eq. (10.6), or equivalently substituting Eq. (10.5) into Eq. (10.3), results in the eigenvalue problem defined by the ordinary differential equation

$$U''(x) + \beta^2 U(x) = 0 \qquad (10.8)$$

where

$$\beta^2 = \frac{\omega^2}{c^2} = \frac{\omega^2}{k/m} \tag{10.9}$$

and superposed primes denote total differentiation with respect to x. Equation (10.8) is seen to correspond to the harmonic equation in space. It therefore yields the solution

$$U(x) = A_1 \cos \beta x + A_2 \sin \beta x \tag{10.10}$$

which corresponds to *the general form of the modal functions for uniform second order systems*. The parameter β is referred to as the *wave number* and is interpreted as a spatial frequency. To complement this parameter we introduce the corresponding *wave length*

$$\lambda = 2\pi/\beta \tag{10.11}$$

which is interpreted as the corresponding spatial period. The parameter β, and hence ω, as well as the integration constants A_1 and A_2 depend on the specific boundary conditions for the particular system under consideration. In general there will be an infinite number of values of β, and hence of ω. Once these values have been determined, each may be substituted into Eq. (10.10) to give the associated modal function. The resulting expression may then be substituted back into Eq. (10.2) to give the corresponding solution. The sum of all such solutions, Eq. (10.4), then corresponds to the free vibration response of the given system. In the remainder of this section we shall apply the above results to study three types of vibratory motion of second order one-dimensional continua.

10.2.2 Longitudinal Vibration of Elastic Rods

Consider a uniform elastic rod of length L, mass per unit length $m = \rho A$ and axial stiffness $k_a = EA$. For longitudinal motion (Section 9.3) the displacement measure is the axial displacement $u(x,t)$ (Figure 10.1) and, from Eq. (9.44) or (9.51), the equation of motion for free vibrations is

$$m\frac{\partial^2 u}{\partial t^2} - EA\frac{\partial^2 u}{\partial x^2} = 0 \tag{10.12}$$

Letting $u(x,t) \to u(x,t)$, $U(x) \to U(x)$, $k \to k_a = EA$ and $m = \rho A$ (Section 9.3) in the general analysis of Sections 10.1 and 10.2.1 gives the free vibration response for longitudinal motion of rods as

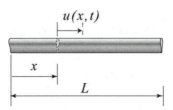

Figure 10.1 Longitudinal motion of a thin rod.

$$u(x,t) = \sum_{j=1}^{\infty} U^{(j)}(x) A^{(j)} \cos(\omega_j t - \phi_j) \qquad (10.13)$$

where

$$U^{(j)}(x) = A_1^{(j)} \cos \beta_j x + A_2^{(j)} \sin \beta_j x \qquad (10.14)$$

is the jth modal function and

$$\beta_j^{\,2} = \frac{\omega_j^{\,2}}{k_a/m} = \frac{\omega_j^{\,2}}{c_a^{\,2}} \qquad (10.15)$$

and

$$c_a = \sqrt{\frac{EA}{m}} = \sqrt{\frac{E}{\rho}} \qquad (10.16)$$

are, respectively the jth *wave number and wave speed for axial motion*. (The axial wave speed is the speed of a longitudinal wave in an elastic medium with vanishing Poisson's ratio.) The specific wave numbers, and hence the specific natural frequencies and corresponding modal functions, are determined by the particular boundary conditions imposed on a given structure as demonstrated in the following examples.

Example 10.1

Consider a uniform elastic rod of length L, membrane stiffness EA and mass per unit length m. (*a*) Determine the natural frequencies and modal functions for when the structure is fixed at its left end as shown. (*b*) Plot and label the first three modal functions. (*c*) Determine the general form of the free vibration response of the rod.

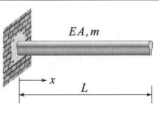

Figure E10.1-1

Solution

(*a*)

The physical boundary conditions for this rod are, from Example 9.2-ii,

$$u(0,t) = 0 , \quad EA\frac{\partial u}{\partial x}\bigg|_{x=L} = 0 \quad\quad\quad (\text{a-1,2})$$

To obtain the boundary conditions for the modal functions we substitute the assumed harmonic response,

$$u(x,t) = U(x)e^{i\omega t}$$

as per Eq. (10.2), into Eqs. (a-1) and (a-2) to get

$$U(0)e^{i\omega t} = 0 \;\Rightarrow\; U(0) = 0 \quad\quad\quad (\text{b-1})$$

$$EA\,U'(L)e^{i\omega t} = 0 \;\Rightarrow\; U'(L) = 0 \quad\quad\quad (\text{b-2})$$

To determine the natural frequencies and modal functions, we next impose conditions (b-1) and (b-2) on the general form of the modal function given by Eq. (10.14) to get

$$U(0) = 0 = A_1 \cos(0) + A_2 \sin(0) = A_1 \quad\quad\quad (\text{c-1})$$

and

$$U'(L) = 0 = -\beta A_1 \sin \beta L + \beta A_2 \cos \beta L \quad\quad\quad (\text{c-2})$$

Substituting Eq. (c-1) into Eq. (c-2) results in the identity

$$\beta A_2 \cos \beta L = 0 \quad\quad\quad (\text{d})$$

It is seen from Eq. (d) that if either $\beta = 0$ or $A_2 = 0$ then Eq. (10.14) yields the trivial solution. Hence, for vibratory motion, we must have that $\beta A_2 \neq 0$. Equation (d) then reduces to the *frequency equation* for the given rod,

$$\cos \beta L = 0 \qu\quad\quad\quad (\text{e})$$

It is seen that any value of β such that

$$\beta L = (2j-1)\pi/2 \quad (j = 1,2,...) \quad\quad\quad (\text{f})$$

satisfies Eq. (e). We thus have an infinite number of wave numbers,

$$\beta_j = (2j-1)\frac{\pi}{2L} \quad (j=1,2,...) \tag{g}$$

The corresponding frequencies are found from Eq. (10.15) and Eq. (g) as

$$\omega_j = (2j-1)\frac{\pi}{2}\omega_0 \quad (j=1,2,...) \tag{◁ (h)}$$

where

$$\omega_0 \equiv \sqrt{\frac{EA}{mL^2}} = \frac{c_a}{L} \tag{i}$$

Substituting Eqs. (c-1) and (g) into Eq. (10.14) gives the corresponding modal functions as

$$U^{(j)}(x) = A_2^{(j)} \sin\beta_j x = A_2^{(j)} \sin\{(2j-1)\pi x/2L\} \quad (j=1,2,...)$$

The value of the integration constant $A_2^{(j)}$ is arbitrary so we shall set it equal to one. (Alternatively, we can normalize the modal functions so that their magnitudes are one.) The modal functions for the rod under consideration are then

$$U^{(j)}(x) = \sin\beta_j x = \sin\{(2j-1)\pi x/2L\} \quad (j=1,2,...) \tag{◁ (j)}$$

The first three modal functions are displayed in Figure E10.1-2 and are discussed below.

(b)
The graphs depicted in Figure E10.1-2 are plots of the axial deformation as a function of x. Since the wave number, β_j, for a given mode is the spatial frequency for that mode, the corresponding wave length, λ_j, (the spatial period) is found by substituting Eq. (k) into Eq. (10.11) giving

$$\lambda_j \equiv \frac{2\pi}{\beta_j} = \frac{4L}{2j-1} \quad (j=1,2,...) \tag{k}$$

To interpret these results we note, from Eq. (9.37), that the slopes of these curves at any point correspond to the axial modal strain at that point. Let us consider the first mode, $U^{(1)}$, displayed in Figure E10.1-2. Since the wavelength for the first mode is greater than the length of the rod ($\lambda_1 = 4L$), the entire rod will be either in tension or compression when oscillating in that mode. It is seen that the maximum deformation occurs at the origin and the strain vanishes at the free end of the rod, $x = L$. It follows from Eq. (h), that the rod oscillates at the rate $\omega_1 = \pi\omega_0/2$ in this mode. Let us next consider the second mode. For this case the wavelength is calculated to be $\lambda_2 = 4L/3$, which is slightly larger than the length of the rod. It is seen that a "node" occurs ($U^{(2)} = 0$) at $x = 2L/3$. It is observed from Figure E10.1-2 that the slope, and hence the axial strain for the second mode, is positive (negative) for $0 < x < L/3$ and therefore that the rod is in tension (compression) in this region, with the maximum extension (contraction) occurring as we approach the origin. Conversely, the rod is seen to be in compression (extension) on $L/3 < x < L$ with the maximum deformation occurring as we approach the node. (Note that an inflection point of the U vs. x curve occurs at the node.) The magnitude of the deformation is seen to decrease monotonically as $x \to L/3$ and as $x \to L$. It follows from Eq. (h) that, in the second mode, the rod oscillates at the frequency $\omega_2 = 3\pi c_0/2$ with the regions of the structure alternating between tension and compression as indicated. For the third mode displayed in Figure E10.1-2, $U^{(3)}$, the wave length is calculated as $\lambda_3 = 4L/5$ and is seen to be less than the total length of the rod. For this case we observe nodes at two points, $x = 2L/5$ and $x = 4L/5$, and inflection points at both locations. In this mode, the structure is seen to be in tension (compression) on $0 < x < L/5$, in compression (tension) on $L/5 < x < 3L/5$, and in tension (compression) on $3L/5 < x < L$. It is also seen that the maximum deformation occurs as x approaches the origin and as x approaches the nodes, and that the minimum deformation occurs at the relative maxima and minima at $x = L/5$, $3L/5$ and L. In this mode the structure oscillates at the frequency $\omega_3 = 5\pi c_0/2$, with the regions defined earlier alternating between tension and compression as indicated.

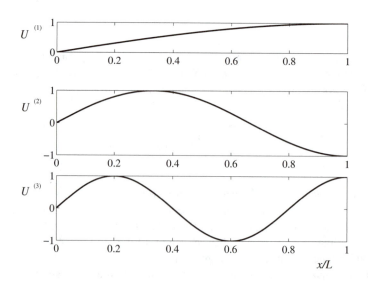

Figure E10.1-2 The first three longitudinal modes of an elastic rod with one edge fixed.

(*c*)

Substituting each mode defined by Eq. (j) into Eq. (10.13) gives the free vibration response of the rod as

$$u(x,t) = \sum_{j=1}^{\infty} A^{(j)} \sin\{(2j-1)\pi x/2L\}\cos(\omega_j t - \phi_j) \qquad \triangleleft (l)$$

where ω_j is given by Eq. (h) and the amplitudes and phase angles, $A^{(j)}$ and ϕ_j ($j = 1, 2, \ldots$), are determined from the initial conditions. The calculation of the amplitudes and phase is facilitated by the mutual orthogonality of the modal functions, which is discussed in Section 10.8. We therefore defer discussion and implementation of this calculation to that section.

Example 10.2

Consider the elastic rod of length L, axial stiffness EA and mass per unit length m that is fixed at its left end and is attached to a block of mass $m = \alpha mL$ on its right end as shown, where α is a dimensionless number. (*a*) Determine the frequency equation and the general form of the corresponding mode shapes for the structure. (*b*) Evaluate the first three frequencies for a structure with mass ratio $\alpha = 2$. (*c*) Evaluate the general free vibration response of the rod for the structure of case (b).

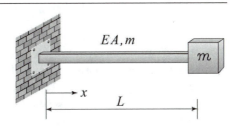

Solution

(*a*)

The left end of the rod is fixed. Hence,

$$u(0,t) = 0 = U(0)e^{i\omega t} \quad \Rightarrow \quad U(0) = 0 \qquad (a)$$

The boundary condition at the right end follows directly from Eq. (b) of Example 9.4. Substitution of the assumed form of the response, $u(x,t) = U(x)e^{i\omega t}$, into that boundary condition gives the modal boundary condition. Hence,

$$-EA\frac{\partial u}{\partial x}\bigg|_{x=L} = m\frac{\partial^2 u}{\partial t^2}\bigg|_{x=L} \quad \Rightarrow \quad -EAU'(L)e^{i\omega t} = -m\omega^2 U(L)e^{i\omega t}$$

or

$$U'(L) - \frac{m}{m} \beta^2 U(L) = 0 \qquad \text{(b)}$$

Now that the boundary conditions for the modal functions have been established, the frequencies and modal functions can be determined by imposing them on the general form of the modal function given by Eq. (10.14). Imposing condition (a) gives

$$U(0) = 0 = A_1 \qquad \text{(c)}$$

After substituting back into the general form of Eq. (10.14) we have that

$$U(x) = A_2 \sin \beta x \qquad \text{(d)}$$

Imposing condition (b) on Eq. (d) results in the statement

$$\beta A_2 \left[\cos \beta L - \beta \frac{m}{m} \sin \beta L \right] = 0$$

It is evident from Eq. (d) that both $\beta = 0$ and $A_2 = 0$ result in the trivial solution. Hence, for nontrivial solutions $\beta A_2 \neq 0$, which gives the frequency equation as

$$\cos \beta L - \alpha \, \beta L \sin \beta L = 0 \qquad \lhd \text{(e)}$$

The roots of Eq. (e) can be determined numerically for a given value of the mass ratio α. The corresponding mode shapes are then of the form

$$U^{(j)}(x) = A_2^{(j)} \sin \beta_j x \quad (j = 1, 2, \ldots)$$

where β_j corresponds to the jth root of Eq. (e). Since $A_2^{(j)}$ is arbitrary we shall set it equal to one. Rendering the modal function for the rod to the form

$$U^{(j)}(x) = \sin \beta_j x \quad (j = 1, 2, \ldots) \qquad \lhd \text{(f)}$$

(b)
The first three roots of Eq. (e) for $\alpha = 2$ are obtained using the MATLAB routine "fzero." Hence,
$$\beta L = 0.6533, \, 3.292, \, 6.362, \, \ldots \qquad \text{(g)}$$

The natural frequencies then follow from Eq. (10.15) as

$$\omega_j = \beta_j L \sqrt{EA/mL^2} \qquad \text{(h)}$$

from which we obtain

$$\omega_1 = 0.6533\omega_0, \quad \omega_2 = 3.292\omega_0, \quad \omega_3 = 6.362\omega_0, \quad ... \qquad \lhd (i)$$

where

$$\omega_0 = \sqrt{EA/mL^2} \qquad (j)$$

(c)
The general free vibration response is found by substituting Eq. (f) into Eq. (10.13). This gives the response of the rod as

$$
\begin{aligned}
u(x,t) &= \sum_{j=1}^{\infty} A^{(j)} \sin \beta_j x \, \cos(\omega_j t - \phi_j) \\
&= A^{(1)} \sin(0.6533\omega_0 x/L) \cos(0.6533\omega_0 t - \phi_1) \\
&\quad + A^{(2)} \sin(3.292\omega_0 x/L) \cos(3.292\omega_0 t - \phi_2) \qquad \lhd (k) \\
&\quad + A^{(3)} \sin(6.362\omega_0 x/L) \cos(6.362\omega_0 t - \phi_3) \\
&\quad + ...
\end{aligned}
$$

where $A^{(j)}$ and ϕ_j ($j = 1, 2, ...$) are determined from the specific form of the initial conditions.

Example 10.3
Evaluate the first three natural frequencies for the rod structure of Example 10.2 for the mass ratios $\alpha = 1, 2, 5, 10, 20$ and 50. Use these results to assess the validity of the equivalent single degree of freedom system of described in Section 1.2.1 to model structures of this type.

Solution
The stiffnesses of the "massless" springs of the equivalent single degree of freedom systems introduced in Chapter 1 were based on the assumption that the mass of the elastic body, in this case a uniform rod, is small compared with the mass of the attached body. They were then calculated using a static analysis (i.e., neglecting the inertia of the structure). It was argued, on physical grounds, that the behavior predicted by such simple mathematical models should approximate the fundamental mode of the elastic structure. We here assess the applicability of such a simplified model by comparing the natural frequencies computed using the 1 d.o.f. model with those computed by using the rod solution of Example 10.2. We first interpret the parameters of the equivalent single

degree of freedom system of Section 1.2.1 in the context of the present structure.

From Eq. (1.8), the stiffness of the equivalent single degree of freedom
system is $k_{eq} = EA/L$. It follows that the natural frequency, ω^*, of the equivalent mass-spring system is

$$\omega^* = \sqrt{\frac{EA}{mL}} = \frac{\omega_0}{\sqrt{\alpha}} \qquad (a)$$

where

$$\omega_0 = \sqrt{\frac{EA}{mL^2}} \qquad (b)$$

With the above established, we next compute the values of the natural frequencies, for different mass ratios, for the simple model and for the rod model of
Example 10.2. A comparison of the two will allow us to assess to what extent
the simple model represents the rod model.

The first three natural frequencies are computed for the given values of
the mass ratio α by calculating the roots of Eq. (e) of Example 10.2 using the
MATLAB routine "fzero." These values are displayed in the Table E10.3 along
with the corresponding values of the approximate single degree of freedom system. It is seen that the frequency predicted by the simple single degree of freedom model approximates that of the first mode of the rod system and converges
to within two significant figures when the attached mass is twenty times larger
than the mass of the beam. It is further seen that the frequency predicted by the
single degree of freedom system model converges to within three significant
figures of the beam model when the attached mass is fifty times the mass of the
column.

Table E10.3

The first three natural frequencies of an elastic rod with a concentrated end mass for
various values of the mass ratio together with the natural frequencies of the corresponding "equivalent" single degree of freedom systems

$\alpha = m/mL$	ω^*/ω_0	ω_1/ω_0	ω_2/ω_0	ω_3/ω_0
1	1.000	0.8603	3.426	6.437
2	0.7071	0.6533	3.292	6.362
5	0.4472	0.4328	3.204	6.315
10	0.3162	0.3111	3.173	6.299
20	0.2236	0.2218	3.157	6.291
50	0.1414	0.1410	3.148	6.286

*equivalent 1 d.o.f. system

10.2.3 Torsional Vibration of Elastic Rods

Consider a uniform elastic rod of circular cross section, length L, polar moment of inertia $J_\rho = \rho J$ and torsional stiffness $k_T = GJ$. For torsional motion of rods (Section 9.4) the displacement measure is the cross-sectional rotation $\theta(x,t)$ (Figure 10.2) and, from (Eq. 9.59), the equation of motion for free vibrations is given by

$$J_\rho \frac{\partial^2 \theta}{\partial t^2} - GJ \frac{\partial^2 \theta}{\partial x^2} = 0 \tag{10.17}$$

Letting $u(x,t) \to \theta(x,t)$, $U(x) \to \Theta(x)$ and $m \to J_\rho = \rho J$ in the general analysis of Sections 10.1 and 10.2.1 gives the free vibration response for torsional motion of rods as

$$\theta(x,t) = \sum_{j=1}^{\infty} \Theta^{(j)}(x) A^{(j)} \cos(\omega_j t - \phi_j) \tag{10.18}$$

where

$$\Theta^{(j)}(x) = A_1^{(j)} \cos \beta_j x + A_2^{(j)} \sin \beta_j x \tag{10.19}$$

is the jth modal function. Furthermore,

$$\beta_j^2 = \frac{\omega_j^2}{k_T / J_\rho} = \frac{\omega_j^2}{c_T^2} \tag{10.20}$$

and

$$c_T = \sqrt{\frac{GJ}{J_\rho}} = \sqrt{\frac{G}{\rho}} \tag{10.21}$$

are, respectively, the jth *wave number and wave speed for torsional motion*. (It may be noted that the torsional wave speed corresponds to the speed of shear waves in an elastic body.) The specific wave numbers, and hence the specific natural frequencies and corresponding modal functions, are determined by the particular boundary conditions imposed on a given structure as demonstrated in the following example.

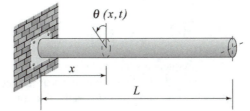

Figure 10.2 Torsional motion of a thin circular rod.

Example 10.4

Consider torsional motion of a uniform elastic rod of circular cross-section, stiffness GJ, length L and mass density ρ, when it is fixed at its right end and is supported at its left end by a torsional spring of stiffness $k_\theta = \alpha GJ/L$, where α is a dimensionless number, as shown. (*a*) Derive the frequency equation and general form of the mode shapes for the rod, and establish the general free vibration response of the structure. (*b*) Determine the first three natural frequencies for the case where $\alpha = 1$ and plot the corresponding modes. (*c*) Evaluate the general free vibration response of the rod of part (*b*).

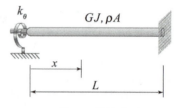

Figure E10.4-1

Solution

(*a*)

Substituting the value of the spring stiffness into Eq. (a) of Example 9.5 gives the boundary condition at the left end of the rod as

$$L\frac{\partial\theta}{\partial x}\bigg|_{x=0} = \alpha\theta(0,t) \tag{a-1}$$

Since the right end of the rod is fixed, the corresponding boundary condition is then

$$\theta(L,t) = 0 \tag{a-2}$$

Substitution of the assumed form of the solution for free torsional vibrations,

$$\theta(x,t) = \Theta(x)e^{i\omega t}$$

as per Eq. (10.2), into Eqs. (a-1) and (a-2) gives the boundary conditions for the modal functions as

$$L\Theta'(0)e^{i\omega t} = \alpha\,\Theta(0)e^{i\omega t} \quad\Rightarrow\quad L\Theta'(0) = \alpha\,\Theta(0) \tag{b-1}$$

and

$$\Theta(L)e^{i\omega t} = 0 \quad\Rightarrow\quad \Theta(L) = 0 \tag{b-2}$$

To obtain the natural frequencies and corresponding modal functions we impose the boundary conditions on the general form of the modal function given by Eq. (10.19). Imposing condition (b-1) gives the relation

$$\beta L\,A_2 = \alpha A_1 \tag{c}$$

Imposing condition (b-2) on the general form of the modal function and substituting Eq. (c) on the resulting expression gives

$$\frac{A_2}{\alpha}\left[\beta L\cos\beta L+\alpha\sin\beta L\right]=0$$

It may be seen from Eq. (10.19) and Eq. (c) that $A_2=A_1=0$ yields the trivial solution, as does $\beta=0$. Hence, for nontrivial solutions, the expression within the brackets must vanish. We thus obtain the frequency equation for the rod in question,

$$\beta L\cos\beta L+\alpha\sin\beta L=0 \hspace{3cm} \triangleleft\text{(d)}$$

Equation (d) may be solved numerically to obtain the natural frequencies of the rod. Substituting Eq. (c) into Eq. (10.19) gives the modal function associated with the jth natural frequency as

$$\Theta^{(j)}(x)=\frac{A_2^{(j)}}{\alpha}\left[\beta_j L\cos\beta_j x+\alpha\sin\beta_j x\right]$$

Since the value of $A_2^{(j)}$ is arbitrary we shall set $A_2^{(j)}=\alpha$. The jth modal function is then

$$\Theta^{(j)}(x)=\beta_j L\cos\beta_j x+\alpha\sin\beta_j x \hspace{3cm} \triangleleft\text{(e)}$$

With the frequencies determined from Eq. (d), substituting Eq. (e) into Eq. (10.19) gives the free vibration response of the rod as

$$\theta(x,t)=\sum_{j=1}^{\infty}A^{(j)}\left[\beta_j L\cos\beta_j x+\alpha\sin\beta_j x\right]\cos(\omega_j t-\phi_j) \hspace{1.5cm} \triangleleft\text{(f)}$$

(b)
The first three roots of Eq. (d) are determined numerically using the MATLAB routine "fzero," which gives, for systems with $\alpha=1$,

$$\beta_1 L=2.029\ ,\hspace{0.6cm}\beta_2 L=4.913\ ,\hspace{0.6cm}\beta_3 L=7.979,\hspace{0.3cm}\ldots \hspace{2cm}\text{(g)}$$

The associated natural frequencies are obtained by substituting the above roots into Eq. (10.20). Hence,

$$\omega_1=2.029\omega_0\ ,\hspace{0.5cm}\omega_2=4.913\omega_0\ ,\hspace{0.5cm}\omega_3=7.979\omega_0\ ,\hspace{0.3cm}\ldots \hspace{1.5cm}\triangleleft\text{(h)}$$

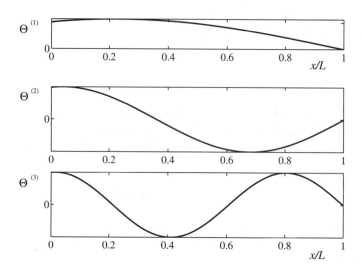

Figure E10.4-2 The first three torsional modes of elastic rod with one elastic support and one fixed support.

where

$$\omega_0 = c_T / L = \sqrt{\frac{G}{\rho L^2}} \tag{i}$$

The first three modes are displayed in Figure E10.4-2. The values shown correspond to the rotational displacement as a function of x and are interpreted in this context. The wave lengths for the first three modes are calculated as

$$\lambda_1 = \frac{2\pi L}{2.029} = 3.097L, \quad \lambda_2 = \frac{2\pi L}{4.913} = 1.279L, \quad \lambda_3 = \frac{2\pi L}{7.979} = 0.7875L \tag{j}$$

The displacement measure is the rotation of the cross section located at coordinate x, measured positive counterclockwise. The first mode then corresponds to deformations of the rod where all rotations are in the same sense. The sense of the motions alternate between positive and negative as the structure oscillates at the first natural frequency, but maintains the same proportion of the relative rotations of the cross sections throughout the motion. The second mode is seen to correspond to deformations where part of the rod, say that which contains the support, rotates in the positive (negative) sense while the remaining segment of the structure rotates in the negative (positive) sense. The deformations alternate in direction as the rod oscillates at the second natural frequency, with the proportions of the relative rotation of the cross sections maintained throughout the motion. Lastly, the third mode consists of three regions, with the middle segment deformed in the opposite sense of the outer segments. The sense the rotation of each segment alternates between positive and negative as the structure oscillates at the third natural frequency and maintains the same relative proportions throughout the motion.

(*c*)

The free vibration response of the structure with stiffness ratio $\alpha = 1$ is then determined by substituting the wave numbers and natural frequencies stated in Eqs. (g) and (h) into Eq. (f). We thus have that

$$
\begin{aligned}
\theta(x,t) = A^{(1)} & \left[2.029\cos\left(2.029x/L\right)+\sin\left(2.029x/L\right)\right]\cos\left(2.029\omega_0 t - \phi_1\right) \\
+ A^{(2)} & \left[4.913\cos\left(4.913x/L\right)+\sin\left(4.913x/L\right)\right]\cos\left(4.913\omega_0 t - \phi_2\right) \\
+ A^{(3)} & \left[7.979\cos\left(7.979x/L\right)+\sin\left(7.979x/L\right)\right]\cos\left(7.979\omega_0 t - \phi_2\right) \\
+ & \ldots
\end{aligned}
\quad \triangleleft \text{(k)}
$$

10.2.4 Transverse Vibration of Strings and Cables

Consider a string or cable of length L and mass per unit length $m = \rho A$. In addition, let the string be under uniform tension N_0. For these systems (Section 9.5) the displacement measure is the transverse displacement $w(x, t)$ (Figure 10.3) and, from Eq. (9.74), the equation of motion for free vibrations is

$$
m\frac{\partial^2 w}{\partial t^2} - N_0 \frac{\partial^2 w}{\partial x^2} = 0 \tag{10.22}
$$

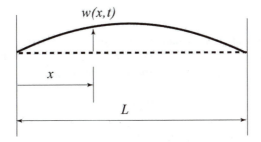

Figure 10.3 Transverse motion of a string.

Letting $u(x,t) \rightarrow w(x,t)$, $U(x) \rightarrow W(x)$, and $k \rightarrow N_0$ in the analysis of Sections 10.1 and 10.2.1 gives the free vibration response of the string or cable as

$$w(x,t) = \sum_{j=1}^{\infty} W^{(j)}(x) A^{(j)} \cos(\omega_j t - \phi_j) \tag{10.23}$$

where

$$W^{(j)}(x) = A_1^{(j)} \cos \beta_j x + A_2^{(j)} \sin \beta_j x \tag{10.24}$$

is the jth modal function. In addition,

$$\beta_j^{\,2} = \frac{\omega_j^{\,2}}{k/m} = \frac{\omega_j^{\,2}}{c_0^{\,2}} \tag{10.25}$$

and

$$c_0^{\,2} = N_0/m \tag{10.26}$$

are, respectively, the jth wave number and the wave speed of a transverse disturbance propagating along the length of the string. The specific wave numbers, and hence the specific natural frequencies and corresponding modal functions, are determined by the particular boundary conditions imposed on a given structure as demonstrated in the following example.

Example 10.5

Consider a string that is under uniform tension and is fixed at both ends, as shown. (*a*) Determine the natural frequencies and corresponding modes. (*b*) Plot the first three modes. (*c*) Establish the general free vibration response of the string. (*d*) Apply the results of part (c) to the chain of Example 9.6-i.

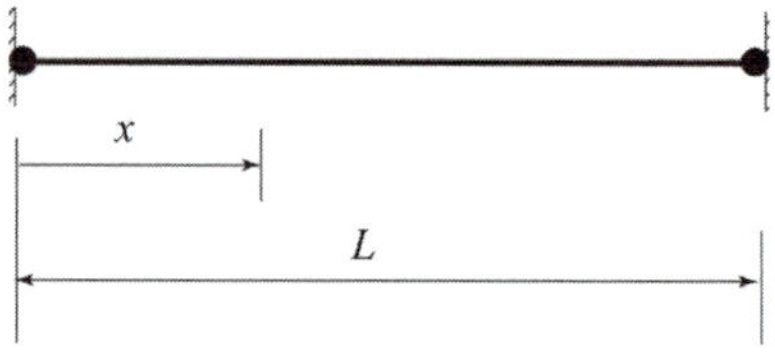

Figure E10.5-1

Solution

(*a*)

We first determine the boundary conditions for the modal functions from the physical boundary conditions. Since the string is fixed at both ends, the boundary conditions for the transverse displacement are simply

$$w(0,t) = 0, \quad w(L,T) = 0 \tag{a-1, 2}$$

Substitution of the assumed form of the solution,

$$w(x,t) = W(x)e^{i\omega t}$$

as per Eq.(10.2), into Eqs. (a-1) and (a-2) gives the modal boundary conditions as follows,

$$W(0)e^{i\omega t} = 0 \quad \Rightarrow \quad W(0) = 0 \tag{b-1}$$

$$W(L)e^{i\omega t} = 0 \quad \Rightarrow \quad W(L) = 0 \tag{b-2}$$

To determine the natural frequencies and the corresponding modal functions we impose the boundary conditions on the general form of the modal function given by Eq. (10.24). Imposing Eq. (b-1) gives

$$A_1 = 0 \tag{c}$$

Imposing Eq. (b-2) yields the relation

$$A_1 \cos \beta L + A_2 \sin \beta L = 0$$

which, upon incorporating Eq. (c), becomes

$$A_2 \sin \beta L = 0 \tag{d}$$

It may be seen from Eq. (10.24) and Eq. (d) that $A_2 = 0$ yields the trivial solution. Thus, for nontrivial solutions, we have require that

$$\sin \beta L = 0 \tag{e}$$

This is the frequency equation for the string which has the roots

$$\beta_j L = j\pi \quad (j = 1, 2, ...) \tag{f}$$

Substitution of Eq. (f) into Eq. (10.25) gives the natural frequencies for the string as

$$\omega_j = j\pi \sqrt{\frac{N_0}{mL^2}} \quad (j = 1, 2, ...) \qquad \triangleleft \text{(g)}$$

Substitution of Eq. (f) into Eq. (10.24) gives the corresponding modal functions

$$W^{(j)}(x) = A_2^{(j)} \sin \beta_j x \quad (j = 1, 2, ...)$$

Since the value of $A_2^{(j)}$ is arbitrary we shall set it equal to one. The modal functions then take the form

$$W^{(j)}(x) = \sin(j\pi x/L) \quad (j = 1, 2, ...) \qquad \triangleleft \text{(h)}$$

(b)
The first three mode shapes are displayed in Fig. E10.5-2. The corresponding
wave lengths are calculated from Eq. (f) and Eq. (10.11) to be, $\lambda_1 = 2L$, $\lambda_2 = L$
and $\lambda_3 = 2L/3$, respectively. Since the deflections are in the transverse direction
the plots displayed in the figure are, in fact, physical depictions of the first three
modes. Each mode oscillates at the associated frequency, $\omega_1 = \pi\omega_0$, $\omega_2 = 2\pi\omega_0$
and $\omega_3 = 3\pi\omega_0$, with the deflections alternating over the corresponding period
$T_1 = 2/\omega_0$, $T_2 = 1/\omega_0$ and $T_3 = 2/3\omega_0$.

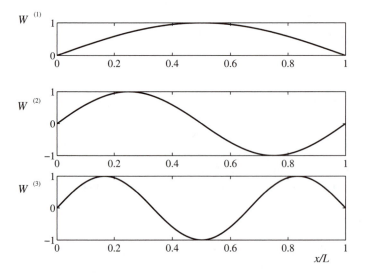

Figure E10.5-2 The first three modes for a string that is fixed at both ends.

(c)
Substitution of Eq. (h) into Eq. (10.23) gives the free vibration response

$$w(x,t) = \sum_{j=1}^{\infty} A^{(j)} \sin(j\pi x/L) \cos\left(j\pi\sqrt{N_0/mL^2}\,t - \phi_j\right) \qquad \triangleleft \text{(i)}$$

(d)
For the system of Example 9.5 the tension in the cable is due to the weight of
the hanging sign. Hence, $N_0 = mg$, where $m \gg mL$ is the mass of the sign. It
then follows from Eq. (h) that

$$\omega_j = j\pi\,\omega_0 \tag{j}$$

where

$$\omega_0 = \sqrt{\alpha g / L} \tag{k}$$

and

$$\alpha = m/mL \gg 1 \tag{l}$$

The free vibration response of the chain is then, from Eq. (i),

$$w(x,t) = \sum_{j=1}^{\infty} A^{(j)} \sin(j\pi x/L)\cos\!\left(j\pi\sqrt{\alpha g/L}\,t - \phi_j \right) \tag*{◁ (m)}$$

10.3 FREE VIBRATION OF EULER-BERNOULLI BEAMS

Consider a uniform Euler-Bernoulli Beam (Section 9.6.3) of length L, bending stiffness EI and mass per unit length $m = \rho A$. For this structure the displacement measure is the transverse displacement $w(x,t)$ (Figure 10.4) and the stiffness operator is a fourth order differential operator. For uniform beams the stiffness operator takes the form

$$\Bbbk = EI\,\frac{\partial^4}{\partial x^4}$$

The equation of motion for an unforced beam follows from Eq. (9.104) as

$$m\frac{\partial^2 w}{\partial t^2} + EI\frac{\partial^4 w}{\partial x^4} = 0 \tag{10.27}$$

Further, for these structures, $u(x,t) \rightarrow w(x,t)$ in the analysis of Section 10.1. We thus seek solutions of the form

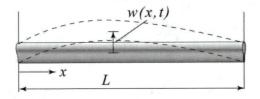

Figure 10.4 Transverse motion of a beam.

$$w(x,t) = W(x)e^{i\omega t} \qquad (10.28)$$

as per Eq. (10.2). Substituting Eq. (10.28) into Eq. (10.27) results in the specific form of the eigenvalue problem for Euler-Bernoulli Beams defined by the differential equation

$$W''''(x) - \beta^4 W(x) = 0 \qquad (10.29)$$

where

$$\beta^4 = \omega^2 \, m/EI \qquad (10.30)$$

To solve Eq. (10.29) we assume a solution of the form

$$W(x) = Ae^{sx} \qquad (10.31)$$

where the constants A and s are to be determined. Substituting Eq. (10.31) into Eq. (10.29) and solving the resulting equation for s gives the values

$$s = \pm\beta, \ \pm i\beta \qquad (10.32)$$

where β is defined by Eq. (10.30). Each value of s corresponds to a solution of the form of Eq. (10.31). The general solution of Eq. (10.29) is then comprised of a linear combination of all such solutions. Summing these solutions and using Eqs. (1.61) and (1.63) gives the general form of the *modal functions for uniform Euler-Bernoulli Beams*,

$$W(x) = A_1 \cosh\beta x + A_2 \sinh\beta x + A_3 \cos\beta x + A_4 \sin\beta x \qquad (10.33)$$

The integration constants and the parameter β, and hence the frequency ω, are determined from the specific boundary conditions for a given system. As for the second order systems discussed in Section 10.2, the modal functions are seen to be unique to within a constant multiplier. This constant is typically set to one, or chosen so as to render the magnitude of the modal function unity (Section 10.6). Once the specific frequencies are determined they may be substituted into Eq. (10.33) to evaluate the corresponding modal functions, with each frequency-mode pair yielding a solution of the form of Eq. (10.28). The sum of these solutions, as per Eq. (10.4), corresponds to the free vibration response of the beam given by

$$w(x,t) = \sum_{j=1}^{\infty} W^{(j)}(x) A^{(j)} \cos(\omega_j t - \phi_j) \qquad (10.34)$$

Example 10.6

Consider the simply supported uniform beam of Figure E10.6-1 and its representation using Euler-Bernoulli Theory. Determine the frequency equation, the natural frequencies, the natural modes and the general free vibration response.

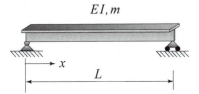

EI, m

x

L

Figure E10.6-1 Simply supported beam.

Solution

The boundary conditions for the modal functions are determined by substituting the assumed form of the solution, Eq. (10.28), into the physical boundary conditions. Hence,

$$w(0,t) = 0 = W(0)e^{i\omega t} \quad \Rightarrow \quad W(0) = 0 \tag{a-1}$$

$$M(L,t) = -EI\frac{\partial^2 w}{\partial x^2}\bigg|_{x=0} = 0 = -EIW''(0)e^{i\omega t} \quad \Rightarrow \quad W''(0) = 0 \tag{a-2}$$

$$w(L,t) = 0 = W(L)e^{i\omega t} \quad \Rightarrow \quad W(L) = 0 \tag{a-3}$$

$$M(L,t) = -EI\frac{\partial^2 w}{\partial x^2}\bigg|_{x=L} = 0 = -EIW''(L)e^{i\omega t} \quad \Rightarrow \quad W''(L) = 0 \tag{a-4}$$

Imposing the above boundary conditions on the general form of the modal function, Eq. (10.33), gives the following system of algebraic equations for the integration constants:

$$A_1 + A_3 = 0 \tag{b-1}$$

$$\beta^2\left(A_1 - A_3\right) = 0 \tag{b-2}$$

$$A_1 \cosh \beta L + A_2 \sinh \beta L + A_3 \cos \beta L + A_4 \sin \beta L = 0 \tag{b-3}$$

$$\beta^2\left(A_1 \cosh \beta L + A_2 \sinh \beta L - A_3 \cos \beta L - A_4 \sin \beta L\right) = 0 \tag{b-4}$$

It is seen from Eq. (10.33) that $\beta = 0$ corresponds to the trivial solution. It follows from Eqs. (b-1) and (b-2) that, for nontrivial solutions ($\beta \neq 0$),

$$A_1 = A_3 = 0 \qquad\qquad\qquad (\text{c-1, 2})$$

Substituting Eqs. (c-1) and (c-2) into Eqs. (b-3) and (b-4) and adding and subtracting the resulting expressions gives the relations

$$2A_2 \sinh \beta L = 0 \qquad\qquad\qquad (\text{d-1})$$

$$2A_4 \sin \beta L = 0 \qquad\qquad\qquad (\text{d-2})$$

It follows from Eqs. (d-1) and (d-2) that, for nontrivial solutions ($\beta \neq 0$),

$$A_2 = 0 \qquad\qquad\qquad (\text{e})$$

and

$$\sin \beta L = 0 \qquad\qquad\qquad \triangleleft (\text{f})$$

Equation (f) is the frequency equation for the simply supported Euler-Bernoulli beam and yields the roots

$$\beta_j L = j\pi \quad (j = 1, 2, \ldots) \qquad\qquad\qquad (\text{g})$$

It follows from Eq. (10.30) that

$$\omega_j = (j\pi)^2 \omega_0 \qquad\qquad\qquad (\text{h})$$

where

$$\omega_0 = \sqrt{\frac{EI}{mL^4}} \qquad\qquad\qquad (\text{i})$$

Substitution of Eqs. (c-1), (c-2), (e) and (g) into Eq. (10.33) gives

$$W^{(j)}(x) = A_4^{(j)} \sin (j\pi x / L) \qquad\qquad\qquad (\text{j})$$

Since $A_4^{(j)}$ is arbitrary we shall set it equal to one. Alternatively, we could choose $A_4^{(j)}$ so that the modal function has unit magnitude (Section 10.6). The modal functions for the simply supported Euler-Bernoulli Beam are seen to be of identical form with those of the string of Example 10.5 (though the natural frequencies of the beam and string, of course, differ). The plots of the first three modes of the simply supported beam are depicted in Figure 10.6-2 for the benefit of the reader. Since the deflections are transverse to the axis of the beam, the plots displayed in the figure correspond to physical descriptions of the first three mode shapes. The free vibration response is obtained by substituting each of the modes and frequencies into Eq. (10.34). Doing this gives the response as

$$w(x,t) = \sum_{j=1}^{\infty} A^{(j)} \sin(j\pi x/L) \cos\left\{(j\pi)^2 \omega_0 t - \phi_j\right\}$$ (k)

where ω_0 is given by Eq. (i) and the amplitudes and phase angles $A^{(j)}$ and ϕ_j ($j = 1, 2, \ldots$) are determined from the initial conditions.

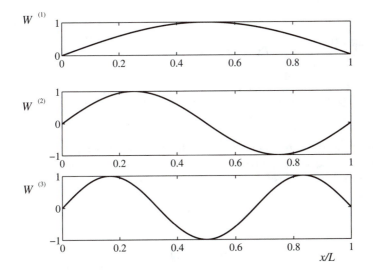

Figure E10.6-2 The first three modes for a simply supported Euler-Bernoulli Beam.

Example 10.7

Consider the uniform cantilever beam of Figure E10.7-1 and its representation using Euler-Bernoulli Theory. Determine the frequency equation, the first three natural frequencies, the natural modes and the general free vibration response for the structure. Plot and label the first three modes of the beam.

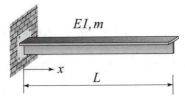

Figure E10.7-1 Cantilevered beam.

Solution

The boundary conditions for the modal functions are determined by substituting the assumed form of the solution, Eq. (10.28), into the physical boundary conditions established in Example 9.7. Hence,

$$w(0,t) = 0 = W(0)e^{i\omega t} \quad \Rightarrow \quad W(0) = 0 \tag{a-1}$$

$$\left.\frac{\partial w}{\partial x}\right|_{x=0} = 0 = W'(0)e^{i\omega t} \quad \Rightarrow \quad W'(0) = 0 \tag{a-2}$$

$$M(L,t) = -EI\left.\frac{\partial^2 w}{\partial x^2}\right|_{x=L} = 0 = -EIW''(L)e^{i\omega t} \quad \Rightarrow \quad W''(L) = 0 \tag{a-3}$$

$$Q(L,t) = -EI\left.\frac{\partial^3 w}{\partial x^3}\right|_{x=L} = 0 = -EIW'''(L)e^{i\omega t} \quad \Rightarrow \quad W'''(L) = 0 \tag{a-4}$$

Imposing the above boundary conditions on the general form of the modal function, Eq. (10.33), gives the following system of algebraic equations for the integration constants:

$$A_1 + A_3 = 0 \tag{b-1}$$

$$\beta\left(A_2 + A_4\right) = 0 \tag{b-2}$$

$$\beta^2\left(A_1 \cosh \beta L + A_2 \sinh \beta L - A_3 \cos \beta L - A_4 \sin \beta L\right) = 0 \tag{b-3}$$

$$\beta^3\left(A_1 \sinh \beta L + A_2 \cosh \beta L + A_3 \sin \beta L - A_4 \cos \beta L\right) = 0 \tag{b-4}$$

Upon noting Eq. (b-1), it is seen from Eq. (10.33) that $\beta = 0$ corresponds to the trivial solution. We are therefore interested in nonvanishing values of β and can divide Eqs. (b-2), (b-3) and (b-4) by β, β^2 and β^3, respectively. The remainder of the analysis is simplified if we eliminate A_3 and A_4 for the above system. Substituting Eqs. (b-1) and (b-2) into Eqs. (b-3) and (b-4) results in the pair of algebraic equations

$$A_1(\cosh \beta L + \cos \beta L) + A_2(\sinh \beta L + \sin \beta L) = 0 \tag{c-1}$$

$$A_1(\sinh \beta L - \sin \beta L) + A_2(\cosh \beta L + \cos \beta L) = 0 \tag{c-2}$$

or, in matrix form,

$$\begin{bmatrix} (\cosh \beta L + \cos \beta L) & (\sinh \beta L + \sin \beta L) \\ (\sinh \beta L - \sin \beta L) & (\cosh \beta L + \cos \beta L) \end{bmatrix} \begin{Bmatrix} A_1 \\ A_2 \end{Bmatrix} = \begin{Bmatrix} 0 \\ 0 \end{Bmatrix} \qquad \text{(d)}$$

For nontrivial solutions, we require that the determinant of the square matrix in Eq. (d) vanish. This gives the transcendental equation

$$(\cosh \beta L + \cos \beta L)^2 - (\sinh^2 \beta L - \sin^2 \beta L) = 0$$

Expanding the above relation and utilizing simple trigonometric identities results in the frequency equation for the uniform cantilever beam,

$$\cosh \beta L \, \cos \beta L + 1 = 0 \qquad\qquad\qquad \lhd \text{(e)}$$

which may be solved numerically for βL. The values of the first three roots are found using MATLAB's "fzero" routine to be

$$\beta_1 L = 1.875, \quad \beta_2 L = 4.694, \quad \beta_3 L = 7.855, \quad ... \qquad \text{(f)}$$

Substituting each root into Eq. (10.30) and solving for ω gives the first three natural frequencies,

$$\omega_1 = 3.516\sqrt{EI/mL^4}, \quad \omega_2 = 22.03\sqrt{EI/mL^4}, \quad \omega_3 = 61.70\sqrt{EI/mL^4}, \quad ... \quad \lhd \text{(g)}$$

Solving Eq. (c-1) for A_2 and substituting the result, along with Eqs. (b-1), (b-2) and (e), into Eq. (10.33) gives the corresponding modal functions

$$W^{(j)}(x) = \hat{A}_1^{(j)} \Big[\big(\sinh \beta_j L + \sin \beta_j L \big) \big\{ \cosh \beta_j x - \cos \beta_j x \big\}$$
$$- \big(\cosh \beta_j L + \cos \beta_j L \big) \big\{ \sinh \beta_j x - \sin \beta_j x \big\} \Big] \quad (j = 1, 2, ...)$$

where

$$\hat{A}_1^{(j)} = \frac{A_1}{\sinh \beta L + \sin \beta L}$$

Since the value of A_1 is arbitrary it may be set equal to one. When this is done, the jth modal function for the uniform cantilevered beam takes the form

$$W^{(j)}(x) = \Big[\big(\sinh \beta_j L + \sin \beta_j L \big) \big\{ \cosh \beta_j x - \cos \beta_j x \big\}$$
$$- \big(\cosh \beta_j L + \cos \beta_j L \big) \big\{ \sinh \beta_j x - \sin \beta_j x \big\} \Big] \Big/ \big(\sinh \beta_j L + \sin \beta_j L \big)$$
$$\lhd \text{(h)}$$

The first three modal functions are displayed in Figure E10.7-2. Since the deflections are transverse to the axis of the beam the plots displayed in the figure are, in fact, a physical depiction of the first three mode shapes.

The free vibration response is obtained by substituting each of the modes and frequencies into Eq. (10.34). This gives

$$w(x,t) = \sum_{j=1}^{\infty}\Big[\big(\sinh \beta_j L + \sin \beta_j L\big)\big\{\cosh \beta_j x - \cos \beta_j x\big\}$$

$$-\big(\cosh \beta_j L + \cos \beta_j L\big)\big\{\sinh \beta_j x - \sin \beta_j x\big\}\Big]\frac{A^{(j)}\cos(\omega_j t - \phi_j)}{\big(\sinh \beta_j L + \sin \beta_j L\big)}$$

◁ (i)

where β_j and ω_j $(j = 1, 2, \ldots)$ are given by the roots of Eq. (e) together with Eq. (10.30).

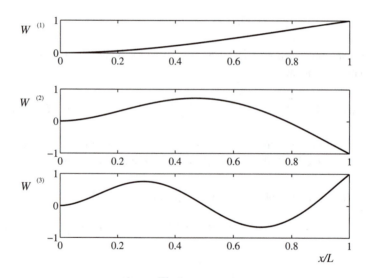

Figure E10.7-2 The first three flexural modes for a cantilevered Euler-Bernoulli Beam (shown scaled by their maximum values).

Example 10.8

Consider a uniform cantilever beam of length L, bending stiffness EI and mass per unit length m that supports a concentrated mass $m = \alpha mL$ at its free end, as shown, using Euler-Bernoulli Theory. (*a*) Determine the frequency equation and the general form of the mode shapes for the structure. (**b**) Evaluate the first three natural frequencies for a structure with mass ratio $\alpha = 2$. (**c**) Establish the general free vibration response of the beam for $\alpha = 2$.

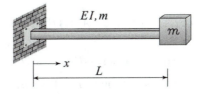

Solution

(**a**)

The boundary conditions for the modal functions are determined by substituting the assumed form of the solution, Eq. (10.28), into the physical boundary conditions established in Examples 9.9 and 10.7. The two boundary conditions at the support follow from Example 10.7. If the physical dimensions of the concentrated mass are negligible, the boundary condition for moment at the far end of the beam is also the same as for Example 10.7. The fourth boundary condition, the condition for shear at the far end of the beam, is obtained by inspection of the kinetic diagram of the concentrated mass (see Example 9.9) and proceeding as for the others. Hence,

$$W(0) = 0, \quad W'(0) = 0, \quad W''(L) = 0 \qquad \text{(a-1, 2, 3)}$$

From Eq. (b-2) of Example 9.9, the boundary condition for transverse shear in the rod is

$$EI \left.\frac{\partial^3 w}{\partial x^3}\right|_{x=L} = m \left.\frac{\partial^2 w}{\partial t^2}\right|_{x=L} \quad \Rightarrow \quad EI W'''(L)e^{i\omega t} = -\omega^2 m W(L)e^{i\omega t}$$

The frequency appears explicitly in the above condition due to the presence of the inertia of the concentrated mass in the boundary condition. However, the frequency and wave number are directly related through Eq. (10.30). Incorporating this relation into the right hand side of the last boundary condition renders the boundary condition for transverse shear to the form

$$W'''(L) + \beta^4 \alpha L W(L) = 0 \qquad \text{(a-4)}$$

Imposing the above boundary conditions on the general form of the modal function, Eq. (10.33), gives the following system of algebraic equations for the integration constants:

$$A_1 + A_3 = 0 \tag{b-1}$$

$$\beta\left(A_2 + A_4\right) = 0 \tag{b-2}$$

$$\beta^2\left(A_1 \cosh \beta L + A_2 \sinh \beta L - A_3 \cos \beta L - A_4 \sin \beta L\right) = 0 \tag{b-3}$$

$$\begin{aligned}
&\beta^3\left(A_1 \sinh \beta L + A_2 \cosh \beta L + A_3 \sin \beta L - A_4 \cos \beta L\right)\\
&- \beta^3 \alpha \beta L\left(A_1 \cosh \beta L + A_2 \sinh \beta L + A_3 \cos \beta L + A_4 \sin \beta L\right) = 0
\end{aligned} \tag{b-4}$$

As for the previous example, it is seen from Eqs. (b-1) and (10.33) that $\beta = 0$ corresponds to the trivial solution. We can therefore divide Eqs. (b-2), (b-3) and (b-4) by β, β^2 and β^3, respectively. Eliminating A_3 and A_4 in the above system by substituting Eqs. (b-1) and (b-2) into Eqs. (b-3) and (b-4) results in the pair of algebraic equations

$$A_1(\cosh \beta L + \cos \beta L) + A_2(\sinh \beta L + \sin \beta L) = 0 \tag{c-1}$$

$$\begin{aligned}
&A_1\left[(\sinh \beta L - \sin \beta L) + \alpha \beta L(\cosh \beta L - \cos \beta L)\right]\\
&+ A_2\left[(\cosh \beta L + \cos \beta L) + \beta L(\sinh \beta L - \sin \beta L)\right] = 0
\end{aligned} \tag{c-2}$$

Arranging Eqs. (c-1) and (c-2) in matrix form and, since we are interested in nontrivial solutions, setting the determinant of the coefficients of the integration constants to zero results in the frequency equation

$$\left(1 + \cosh \beta L \cos \beta L\right) + \alpha \beta L\left(\sinh \beta L \cos \beta L - \cosh \beta L \sin \beta L\right) = 0 \quad \lhd \text{(d)}$$

The roots of the above equation yield the wave numbers, and hence the natural frequencies, for the given structure.

Equation (c-1) gives A_3 in terms of A_1. Substituting this expression into Eqs. (b-1) and (b-2) gives the remaining constants, A_3 and A_4, in terms of A_1. Substituting the resulting expressions into Eq. (10.33) then gives modal functions for the structure. We thus obtain,

$$\begin{aligned}
W^{(j)}(x) = \widehat{A}_1^{(j)}\Big[&(\sinh \beta_j L + \sin \beta_j L)\{\cosh \beta_j x - \cos \beta_j x\}\\
&- (\cosh \beta_j L + \cos \beta_j L)\{\sinh \beta_j x - \sin \beta_j x\}\Big]\\
&\hspace{6cm}(j = 1, 2, ...)
\end{aligned}$$

where

$$\widehat{A}_1^{(j)} = A_1^{(j)}\big/(\sinh \beta_j L + \sin \beta_j L)$$

Since $A_1^{(j)}$ is arbitrary we shall choose it equal to one. The jth modal function for the structure is then

$$W^{(j)}(x) = \Big[(\sinh \beta_j L + \sin \beta_j L)\{\cosh \beta_j x - \cos \beta_j x\}$$
$$-(\cosh \beta_j L + \cos \beta_j L)\{\sinh \beta_j x - \sin \beta_j x\}\Big]\big/(\sinh \beta_j L + \sin \beta_j L)$$
◁ (e)

The general form of the response of the Euler-Bernoulli Beam with tip mass then follows as

$$w(x,t) = \sum_{j=1}^{\infty} A^{(j)}\Big[(\sinh \beta_j L + \sin \beta_j L)\{\cosh \beta_j x - \cos \beta_j x\}$$
◁ (f)
$$-(\cosh \beta_j L + \cos \beta_j L)\{\sinh \beta_j x - \sin \beta_j x\}\Big]\frac{A^{(j)}\cos(\omega_j t - \phi_j)}{(\sinh \beta_j L + \sin \beta_j L)}$$

It is seen that the form of the modal functions for the present case is the same as for the cantilevered beam without the tip mass. This is because the two boundary conditions at the support are the same for both structures. The difference in the two structures, that is the effect of the tip mass, is manifested through Eq. (d) and hence through the natural frequencies and wave numbers which we calculate next.

(b)
The first three roots of Eq. (d) are obtained for a structure with $\alpha = 2$ using the MATLAB routine "fzero" as

$$\beta L = 1.076, \; 3.983, \; 7.103, \; ... \tag{g}$$

From Eq. (10.30),
$$\omega_j = (\beta_j L)^2 \omega_0 \quad (j=1,2,...) \tag{h}$$
where
$$\omega_0 = \sqrt{EI/mL^4} \tag{i}$$

The natural frequencies corresponding to the roots stated in Eq. (g) are then

$$\omega = 1.158\,\omega_0, \; 15.86\,\omega_0, \; 50.45\,\omega_0, \; ... \tag{j}$$

The first three modal functions follow from Eq. (e) as

$$W^{(1)}(x) = \{\cosh(1.076x/L) - \cos(1.076x/L)\}$$
$$-0.971\{\sinh(1.076x/L) - \sin(1.076x/L)\} \qquad \text{(k-1)}$$

$$W^{(2)}(x) = \{\cosh(3.983x/L) - \cos(3.983x/L)\}$$
$$-\{\sinh(3.983x/L) - \sin(3.983x/L)\} \qquad \text{(k-2)}$$

$$W^{(3)}(x) = \{\cosh(7.103x/L) - \cos(7.103x/L)\}$$
$$\{\sinh(7.103x/L) - \sin(7.103x/L)\} \qquad \text{(k-3)}$$

(*c*)
The response of the structure with $\alpha = 2$ is found by substituting the computed frequencies and modes into Eq. (f). Thus,

$$w(x,t) = A^{(1)}\Big[\{\cosh(1.076x/L) - \cos(1.076x/L)\}$$
$$-0.971\{\sinh(1.076x/L) - \sin(1.076x/L)\}\Big]\cos(1.158\omega_0 t - \phi_1)$$
$$+ A^{(2)}\Big[\{\cosh(3.983x/L) - \cos(3.983x/L)\}$$
$$-\{\sinh(3.983x/L) - \sin(3.983x/L)\}\Big]\cos(15.86\omega_0 t - \phi_2)$$
$$+ A^{(3)}\Big[\{\cosh(7.103x/L) - \cos(7.103x/L)\}$$
$$-\{\sinh(7.103x/L) - \sin(7.103x/L)\}\Big]\cos(50.45\omega_0 t - \phi_3)$$
$$+ \dots$$
$$\lhd \text{(l)}$$

Example 10.9
Evaluate the first three natural frequencies for the beam structure of Example 10.8 for the mass ratios $\alpha = 1, 2, 5, 10, 20, 50$ and 100. Use these results to assess the validity of the equivalent single degree of freedom system described in Section 1.2.2 to model beam structures of this type.

Solution
The stiffnesses of the "massless" springs of the equivalent single degree of freedom systems introduced in Chapter 1 were based on the assumption that the mass of the elastic body, in this case a uniform rod, is small compared with the

mass of the attached body. They were then calculated using a static analysis (i.e., neglecting the inertia of the structure). It was argued, on physical grounds, that the behavior predicted by such simple mathematical models should approximate the fundamental mode of the elastic structure. We here assess the applicability of such a simplified model by comparing the natural frequencies computed using the 1 d.o.f. model with those computed by using the cantilever beam solution of Example 10.8. We first interpret the parameters of the equivalent single degree of freedom system of Section 1.2.2 in the context of the present structure.

From Eq. (1.148), the stiffness of the equivalent single degree of freedom system is $k_{eq} = 3EI/L^3$. It follows that the natural frequency, $\omega*$, of the equivalent mass-spring system is

$$\omega* = \sqrt{\frac{3EI}{mL^3}} = \omega_0 \sqrt{\frac{3}{\alpha}} \qquad (a)$$

where

$$\omega_0 = \sqrt{\frac{EI}{mL^4}} \qquad (b)$$

Hence,

$$\omega*/\omega_0 = \sqrt{3/\alpha} \qquad (c)$$

With the above established, we next compute the values of the natural frequencies, for different mass ratios, for the simple model and for the beam model of Example 10.8. A comparison of the two will allow us to assess to what extent the simple model represents the rod model.

The first three natural frequencies are computed for the given values of the mass ratio α by calculating the roots of Eq. (d) of Example 10.8 using the MATLAB routine "fzero." These values are displayed in the Table E10.9 along with the corresponding values of the approximate single degree of freedom system. It is seen that the frequency predicted by the simple single degree of freedom model approximates that of the first mode of the beam system and converges to within two significant figures when the attached mass is ten times larger than the mass of the beam. It is further seen that the frequency predicted by the single degree of freedom system model converges to within three significant figures of the beam model when the attached mass is one hundred times the mass of the column.

Table E10.9

The first three natural frequencies of a uniform cantilevered beam with a concentrated end mass for various values of the mass ratio, and the natural frequency predicted for the "equivalent" 1 d.o.f. system

$\alpha = m/mL$	ω^*/ω_0	ω_1/ω_0	ω_2/ω_0	ω_3/ω_0
1	1.732	1.558	16.25	50.89
2	1.225	1.158	15.86	50.45
5	0.7746	0.7569	15.60	50.16
10	0.5477	0.5414	15.52	50.07
20	0.3873	0.3850	15.46	50.01
50	0.2449	0.2443	15.43	49.98
100	0.1732	0.1730	15.43	49.97

*equivalent 1 d.o.f. system

Example 10.10

Consider the (transversely) free-free beam shown in Figure E10.10-1. (*a*) Establish the modal boundary conditions, determine the frequency equation and the general form of the modal functions, and the general free vibration response. (*b*) Determine the first three natural frequencies and corresponding modal functions. Plot the first three modes.

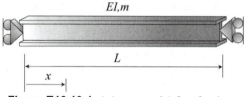

EI,m

L

x

Figure E10.10-1 A (transversely) free-free beam.

Solution

The structure is evidently an unrestrained system in that it can move freely in the vertical direction. From our discussion of unrestrained discrete systems (Section 7.2) we anticipate the existence of a "rigid body mode" along with modes that describe the deformation of the beam.

(*a*)

The beam is clearly free to rotate and free to translate in the vertical direction at the end points $x = 0$ and $x = L$. The support therefore exerts no moment or transverse force at these points. Therefore, the bending moment and transverse shear in the beam must vanish at the boundaries. Hence,

$$-EI \frac{\partial^2 w}{\partial x^2}\bigg|_{x=0} = 0, \quad -EI \frac{\partial^3 w}{\partial x^3}\bigg|_{x=0} = 0 \qquad \text{(a-1, 2)}$$

and

$$-EI \frac{\partial^2 w}{\partial x^2}\bigg|_{x=L} = 0, \quad -EI \frac{\partial^3 w}{\partial x^3}\bigg|_{x=L} = 0 \qquad \text{(a-3, 4)}$$

Substitution of the harmonic form

$$w(x,t) = W(x)e^{i\omega t}$$

into Eqs. (a-1)–(a-4) and dividing through by $e^{i\omega t}$ gives the modal boundary conditions

$$W''(0) = 0, \quad W'''(0) = 0 \qquad \text{(b-1, 2)}$$

and

$$W''(L) = 0, \quad W'''(L) = 0 \qquad \text{(b-3, 4)}$$

Imposing Eqs. (b-1)–(b-4) on the general form of the modal function, Eq. (10.33), results in the system of equations

$$\beta^2 \left(A_1 - A_3 \right) = 0 \qquad \text{(c-1)}$$

$$\beta^3 \left(A_2 - A_4 \right) = 0 \qquad \text{(c-2)}$$

$$\beta^2 \left(A_1 \cosh \beta L + A_2 \sinh \beta L - A_3 \cos \beta L - A_4 \sin \beta L \right) = 0 \qquad \text{(c-3)}$$

$$\beta^3 \left(A_1 \sinh \beta L + A_2 \cosh \beta L + A_3 \sin \beta L - A_4 \cos \beta L \right) = 0 \qquad \text{(c-4)}$$

In each of Eqs. (c-1)–(c-4), either $\beta = 0$ or the terms in parentheses vanish. Let's consider each case.

$\beta = 0$:
For this case, Eqs. (10.30) and (10.33) give the corresponding natural frequency and associated mode as

$$\omega = \omega_{RB} = 0 \qquad \text{(d-1)}$$

$$W(x) = W^{(RB)} = A_1 + A_3 = \text{constant} \qquad \text{(d-2)}$$

This is evidently the rigid body mode and corresponds to a rigid body displacement with no oscillations.

$\beta \neq 0$:

For this case, Eqs. (c-1) and (c-2) yield

$$A_3 = A_1, \quad A_4 = A_2 \tag{e-1, 2}$$

Substituting Eqs. (e-1) and (e-2) into Eqs. (c-3) and (c-4) and arranging the resulting expressions in matrix form gives

$$\begin{bmatrix} (\cosh \beta L - \cos \beta L) & (\sinh \beta L - \sin \beta L) \\ (\sinh \beta L + \sin \beta L) & (\cosh \beta L - \cos \beta L) \end{bmatrix} \begin{Bmatrix} A_1 \\ A_2 \end{Bmatrix} = \begin{Bmatrix} 0 \\ 0 \end{Bmatrix} \tag{f}$$

For nontrivial solutions, we require that the determinant of the square matrix of Eq. (f) vanish. Imposing this condition yields the frequency equation,

$$\cosh \beta L \cos \beta L = 1 \tag{g}$$

Equation (g) yields an infinite number of roots β_j ($j = 1, 2, \ldots$) and hence, via Eq. (10.30), an infinite number of frequencies

$$\omega_j = \omega_0 \left(\beta_j L \right)^2 \tag{h}$$

where

$$\omega_0 = \sqrt{EI/mL^4} \tag{i}$$

Substituting Eqs. (e-1), (e-2) and the first row of Eq. (f) into Eq. (10.33) gives the corresponding modal functions for the beam as

$$W^{(j)}(x) =$$

$$\hat{A}^{(j)} \left[(\cosh \beta x + \cos \beta x) - \frac{(\sinh \beta L + \sin \beta L)}{(\cosh \beta L - \cos \beta L)} (\sinh \beta x + \sin \beta x) \right] \tag{j}$$

$$(j = 1, 2, \ldots)$$

The value of $\hat{A}^{(j)}$ is arbitrary. We shall set it equal to one. The free vibration response is then given by

$$w(x,t) = A_0 t + B_0 + \sum_{j=1}^{\infty} W^{(j)}(x) A^{(j)} \cos(\omega_j t - \phi_j) \tag{k}$$

where the constants A_0, B_0, $A^{(j)}$ and ϕ_j ($j = 1, 2, \ldots$) are determine from the initial conditions (Section 10.8).

(**b**)

Since the first mode is the rigid body mode described earlier, the second and third modes will correspond to the first two roots of Eq. (g). These are found using the MATLAB routine "fzero" and give

$$\beta_1 L = 4.730, \quad \beta_2 L = 7.853 \tag{1}$$

The corresponding natural frequencies are then computed using Eq. (h) to give

$$\omega_1 = 22.37\omega_0, \quad \omega_2 = 61.67\omega_0 \tag{m}$$

where ω_0 is defined by Eq. (i). The first three modes — the rigid body mode, $W^{(RB)}$, and the first two deformation modes, $W^{(1)}$ and $W^{(2)}$ — are displayed in Figure E10.10-2.

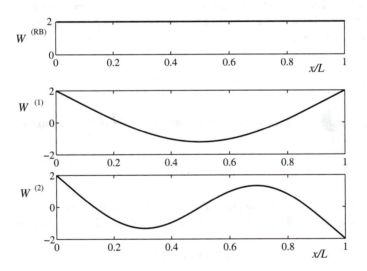

Figure E10.10-2 The first three natural modes of the free-free beam.

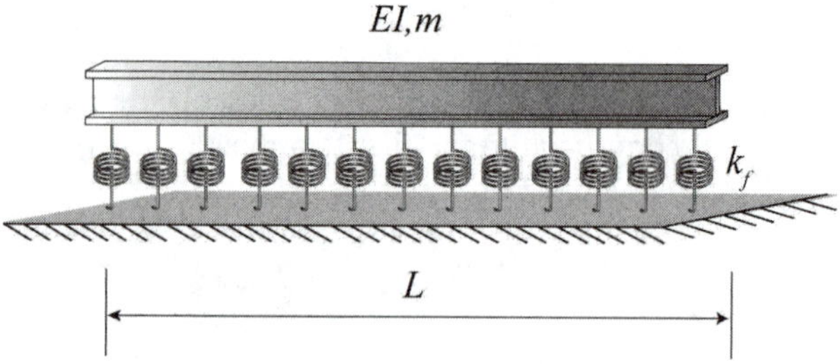

Figure 10.5 Beam on an elastic foundation.

Beam on Elastic Foundation

In certain situations a beam rests on a support throughout its span, such as when a railroad track sits on soil or other deformable media. The simplest mathematical model that accounts for the elastic compliance of the supporting media is the representation of the foundation as a continuous distribution of elastic springs, as shown in Figure 10.5. For such a model, let the stiffness per unit surface area of the *elastic foundation* be denoted as k_f. If we include the restoring force imparted on a deflected beam in the development of Sections 9.6.1–9.6.3, the corresponding equation of motion for free vibration of a uniform Euler-Bernoulli Beam on a uniform elastic foundation takes the form

$$m\frac{\partial^2 w}{\partial t^2} + k_f w + EI\frac{\partial^4 w}{\partial x^4} = 0 \tag{10.35}$$

To solve the free vibration problem we proceed as for prior structures and assume a solution in the form

$$w(x,t) = W(x)e^{i\omega t}$$

and substitute it into the equation of motion. This results in the identical eigenvalue problem as that for all other Euler-Bernoulli Beams considered to this point,

$$W''''(x) - \beta^4 W(x) = 0$$

but now,

$$\beta^4 = \omega^2 \frac{m}{EI} - \frac{k_f}{EI} \tag{10.36}$$

The corresponding modal functions are therefore of the general form given by Eq. (10.33). Thus, for a given beam and set of boundary conditions the roots β_j $(j = 1, 2, ...)$ of the characteristic equation, and the corresponding modal functions $W^{(j)}(x)$, will be identical to their counterparts for beams without foundation. Similarly, the free vibration response of the structure is given by Eq. (10.34). However, from Eq. (10.36), the natural frequencies associated with each mode for the present type of structure are given by

$$\omega_j = \omega_0 \sqrt{\left(\beta_j L\right)^4 + \left(k_f L^4 / EI\right)} \quad (j = 1, 2, ...) \tag{10.37}$$

where

$$\omega_0 = \sqrt{EI/mL^4} \tag{10.38}$$

Comparing Eq. (10.37) with Eq. (10.30) we see that the effect of the presence of the elastic foundation is to raise the natural frequencies of the beam. In addition, if we consider a beam with free-free edges, such as that of Example 10.10, we see the that the added constraints imposed by the foundation render the rigid body mode ($\beta = 0$) bounded and oscillatory ($\omega_{RB} \neq 0$), as may be anticipated on physical grounds. Thus, a beam on an elastic foundation, that possesses the same material properties and the same boundary conditions as its counterpart without the foundation, will freely vibrate in the same forms as its less constrained counterpart, but at higher natural frequencies.

10.4 FREE VIBRATION OF EULER-BERNOULLI BEAM-COLUMNS

In this section we consider the free vibrations of uniform Euler-Bernoulli Beam-Columns. These are beams for which the geometrically nonlinear term in the strain-displacement relation, Eq. (9.34), is retained. This causes the membrane force to appear in the equation of motion as well, thus accounting for bending-stretching coupling effects. As discussed at the end of Section 9.7, the membrane force may be treated as constant when no body force acts on the structure in the axial direction. If we consider the case where the membrane force is compressive, $N_0 = -P_0$, then the structure is referred to as a beam-column and, from Eq. (9.160), the equation of motion is

$$m\frac{\partial^2 w}{\partial t^2} + EI\frac{\partial^4 w}{\partial x^4} + P_0 \frac{\partial^2 w}{\partial x^2} = 0 \tag{10.39}$$

As for the linear beam of the previous section, we seek a solution of the form of Eq. (10.28) and substitute that expression into Eq. (10.39). Doing this results in the eigenvalue problem defined by the differential equation

$$EI\,W'''' + P_0 W'' - \omega^2 mW = 0 \tag{10.40}$$

To solve the above equation we proceed as for the linear Euler-Bernoulli Beam problem and assume a solution of the form of Eq. (10.31). We then substitute that expression into Eq. (10.40) and solve the resulting equation for s. Summing solutions for each root gives *the general form of the modal function for Euler Beam-Columns* as

$$W(x) = A_1 \cosh \alpha x + A_2 \sinh \alpha x + A_3 \cos \beta x + A_3 \sin \beta x \tag{10.41}$$

where

$$\alpha = \sqrt{\tfrac{1}{2}\left[a - \left(P_0/EI\right)\right]} \tag{10.42}$$

$$\beta = \sqrt{\tfrac{1}{2}\left[a + \left(P_0/EI\right)\right]} \tag{10.43}$$

and

$$a = \sqrt{\left(P_0/EI\right)^2 + 4\omega^2 \left(m/EI\right)} \tag{10.44}$$

The integration constants and the natural frequencies are found by imposing the boundary conditions for the particular structure in question on Eq. (10.41). Once the specific frequencies are determined they may be substituted into Eq. (10.41) to evaluate the corresponding modal functions. Each frequency-modal function pair evaluated in this way yields a solution of the form of Eq. (10.28). The sum of these solutions, as per Eq. (10.4), is the free vibration response of the beam-column given by

$$w(x,t) = \sum_{j=1}^{\infty} W^{(j)}(x) A^{(j)} \cos(\omega_j t - \phi_j) \tag{10.45}$$

Example 10.11

Consider the cantilevered beam-column subjected to a constant compressive load of magnitude P_0 at its free end, as shown, where P_0 is less than the critical static buckling load. (**a**) Determine the frequency equation, general form of the mode shapes and general free vibration response of the structure. (**b**) Assess the effects of the axial load on the natural frequencies of the structure. (**c**) Plot the first three modes for the case where the applied load is half the static buckling load.

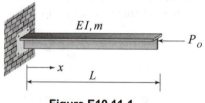

Figure E10.11-1

Solution

(**a**)

The boundary conditions for the beam-column follow from Eqs. (9.161)–(9.164) as

$$w(0,t) = 0, \quad \left.\frac{\partial w}{\partial x}\right|_{x=0} = 0 \qquad \text{(a-1, 2)}$$

$$-EI\left.\frac{\partial^2 w}{\partial x^2}\right|_{x=L} = 0, \quad -\left[EI\frac{\partial^3 w}{\partial x^3} + P_0\frac{\partial w}{\partial x}\right]_{x=L} = 0 \qquad \text{(a-3, 4)}$$

Substituting Eq. (10.28) into the above conditions gives the corresponding boundary conditions for the modal functions as

$$W(0) = 0, \quad W'(0) = 0 \qquad \text{(c-1, 2)}$$

$$W''(L) = 0, \quad EIW'''(L) + P_0 W'(L) = 0 \qquad \text{(c-3, 4)}$$

We next impose the modal boundary conditions, Eqs. (c-1) – (c-4), on Eq. (10.41). This gives the following systems of algebraic equations for the integration constants:

$$A_1 + A_3 = 0 \qquad \text{(d-1)}$$

$$\alpha A_2 + \beta A_4 = 0 \qquad \text{(d-2)}$$

$$A_1\alpha^2 \cosh \alpha L + A_2\alpha^2 \sinh \alpha L - A_3\beta^2 \cos \beta L - A_4\beta^2 \sin \beta L = 0 \qquad \text{(d-3)}$$

$$\alpha\beta^2 \left[A_1 \sinh \alpha L + A_2 \cosh \beta L \right] + \alpha^2\beta \left[A_3 \sin \beta L - A_4 \cos \beta L \right] = 0 \qquad \text{(d-4)}$$

The computations can be simplified somewhat by substituting Eqs. (d-1) and (d-2) into Eqs. (d-3) and (d-4) and writing the resulting equations in matrix form as

$$\begin{bmatrix} (\alpha^2 \cosh \alpha L + \beta^2 \cos \beta L) & (\alpha^2 \sinh \alpha L + \alpha\beta \sin \beta L) \\ (\alpha\beta^2 \sinh \alpha L - \alpha^2\beta \sin \beta L) & (\alpha\beta^2 \cosh \alpha L + \alpha^3 \cos \beta L) \end{bmatrix} \begin{Bmatrix} A_1 \\ A_2 \end{Bmatrix} = \begin{Bmatrix} 0 \\ 0 \end{Bmatrix} \qquad \text{(e)}$$

For nontrivial solutions we require that the determinant of the matrix of coefficients in Eq. (e) vanish. This results in the frequency equation for the beam-column,

$$2\alpha\omega^2 \left(m/EI \right) + \alpha\left(\alpha^4 + \beta^4 \right)\cosh \alpha L \cos \beta L$$
$$- \alpha^2\beta\left(P_0/EI \right)\sinh \alpha L \sin \beta L = 0 \qquad \triangleleft \text{(f)}$$

which may be solved numerically to obtain the associated set of natural frequencies.

Solving the first of Eqs. (e) for A_2 in terms of A_1 and substituting the resulting expression, along with equations (d-1) and (d-2), into Eq. (10.41) gives the modal functions for the beam-column as

$$W^{(j)}(x) = \hat{A}_1^{(j)}\left[\left(\alpha_j^2 \sinh \alpha_j L + \alpha_j \beta_j \sin \beta_j L\right)\left\{\cosh \alpha_j x - \cos \beta_j x\right\}\right.$$
$$\left. - \left(\alpha_j^2 \cosh \alpha_j L + \beta_j^2 \cos \beta_j L\right)\left\{\sinh \alpha_j x - \left(\alpha_j/\beta_j\right)\sin \beta_j x\right\}\right]$$

where

$$\hat{A}_1^{(j)} = A_1^{(j)}\left/\left(\alpha_j^2 \sinh \alpha_j L + \alpha_j \beta_j \sin \beta_j L\right)\right.$$

and α_j and β_j correspond to the values of α and β, Eqs. (10.42) and (10.43), evaluated at the jth natural frequency ω_j. Since its value is arbitrary we shall set $A_1^{(j)} = 1$. The modal function for the beam-column is then

$$W^{(j)}(x) = \left\{\cosh \alpha_j x - \cos \beta_j x\right\}$$
$$- \frac{\left(\alpha_j^2 \cosh \alpha_j L + \beta_j^2 \cos \beta_j L\right)}{\left(\alpha_j^2 \sinh \alpha_j L + \alpha_j \beta_j \sin \beta_j L\right)}\left\{\sinh \alpha_j x - \left(\alpha_j/\beta_j\right)\sin \beta_j x\right\} \qquad \lhd \text{(g)}$$

Substitution of Eq. (g) into Eq. (10.45) gives the free vibration response of the beam-column as

$$w(x,t) = \sum_{j=1}^{\infty} W^{(j)}(x)A^{(j)}\cos(\omega_j t - \phi_j) \qquad \lhd \text{(i)}$$

where $W^{(j)}$ is given by Eq. (g) and ω_j is the jth root of Eq. (f). The modal amplitudes, $A^{(j)}$, and corresponding phase angles, ϕ_j, depend on the specific initial conditions imposed on the structure.

(b)
For a cantilevered beam-column, the critical load is

$$P_{cr} = \frac{\pi^2}{4L^2}EI$$

We thus consider applied loads, P_0, for which

$$\frac{P_0}{EI} = \lambda \frac{\pi^2}{4L^2} \quad (\lambda < 1) \qquad \text{(j)}$$

where $\lambda\ (< 1)$ is a constant.

Table E10.11

The first three natural frequencies of a cantilevered beam-column for various values of the axial loading parameter

λ	ω_1/ω_0	ω_2/ω_0	ω_3/ω_0
0	3.516	22.03	61.70
0.005	3.508	22.03	61.69
0.01	3.500	22.02	61.68
0.05	3.433	21.94	61.62
0.10	3.348	21.85	61.54
0.20	3.168	21.67	61.39
0.50	2.535	21.11	60.92
0.99	0.3662	20.15	60.15

The frequency equation, Eq. (f), can now be rewritten in the form

$$2\alpha L (\omega/\omega_0)^2 + \alpha L \left[(\alpha L)^4 + (\beta L)^4 \right] \cosh \alpha L \cos \beta L$$
$$- (\alpha L)^2 (\beta L)(\lambda \pi^2/4) \sinh \alpha L \sin \beta L = 0 \tag{k}$$

with

$$\alpha L = \sqrt{\tfrac{1}{2}\left[aL^2 - (\lambda \pi^2/4) \right]} \tag{l-1}$$

$$\beta L = \sqrt{\tfrac{1}{2}\left[aL^2 + (\lambda \pi^2/4) \right]} \tag{l-2}$$

and

$$aL^2 = \sqrt{(\lambda \pi^2/4)^2 + 4(\omega/\omega_0)^2} \tag{l-3}$$

The first three roots of Eq. (k) are evaluated for a range of values of λ using the MATLAB routine "fzero." The results are summarized in Table E10.11. It is seen that the compressive load makes the structure more flexible. (Conversely, a tensile load makes the structure stiffer – that is, it makes it more resistant to bending.) We also remark that for compressive loads greater than or equal to the static buckling load $(\lambda \geq 1)$, the branch corresponding to the first natural frequency disappears.

(c)

The first three natural modes of an elastic beam-column are displayed in Figure E10.11-2, for the case where the applied load is half the static buckling load.

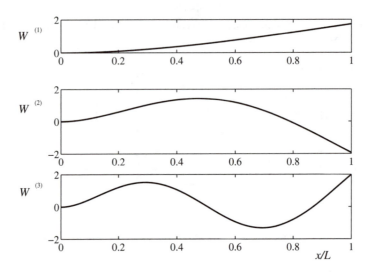

Figure E10.11-2 The first three modes for a cantilevered beam-column with $\lambda = 0.5$.

10.5 FREE VIBRATION OF RAYLEIGH BEAMS

Rayleigh Beam Theory adds a correction to Euler-Bernoulli Beam Theory to account for the effects of rotatory inertia on the motion of a beam. The governing equation for free vibration of Rayleigh Beams (Section 9.6.4) is obtained by letting the right-hand side of Eq. (9.115) vanish. In this section we consider the motion of uniform Rayleigh Beams (beams whose material and geometric properties are invariant throughout the length of the beam). The corresponding equation of motion is then

$$m\frac{\partial^2 w}{\partial t^2} - I_\rho \frac{\partial^4 w}{\partial x^2 \partial t^2} + EI\frac{\partial^4 w}{\partial x^4} = 0 \qquad (10.46)$$

where $I_\rho = \rho I$ is the mass moment of inertia of the cross-section. All other parameters are the same as those for Euler-Bernoulli Beams. We remark that, for the general analysis of Section 10.1, $u(x,t) \rightarrow w(x,t)$, $U(x) \rightarrow W(x)$ and

$$\mathbb{m} = m - I_\rho \frac{\partial^2}{\partial x^2}$$

For free vibrations, we seek solutions of the form

$$w(x,t) = W(x)e^{i\varpi t} \tag{10.47}$$

Substituting the assumed form of the response, Eq. (10.47), into the equation of motion, Eq. (10.46), results in the eigenvalue problem defined by the differential equation

$$W'''' + \varpi^2 W'' - \left(\varpi^2/r_G^2\right)W = 0 \tag{10.48}$$

where

$$\varpi^2 \equiv \omega^2/c_a^2 \tag{10.49}$$

$$c_a^2 = E/\rho \tag{10.50}$$

from Eq. (10.16),

$$r_G^2 = I/A \tag{10.51}$$

is the radius of gyration of the cross-section and ρ is the mass density. A comparison of Eq. (10.48) with Eq. (10.40) shows that the eigenvalue problem for Rayleigh Beams is governed by the same equation as that for Euler Beam-Columns, with suitable changes in parameters. We could therefore write the solution directly from Eq. (10.41). However, to keep the section self contained, it is useful to establish the solution once again. Hence, we assume a solution of the form

$$W(x) = Ae^{sx} \tag{10.52}$$

and substitute the above form into Eq. (10.48). This results in the characteristic equation

$$s^4 + \varpi^2 s^2 - \left(\varpi^2/r_G^2\right) = 0 \tag{10.53}$$

which yields the roots

$$s = \pm\alpha, \ \pm i\beta \tag{10.54}$$

where

$$\alpha = \varpi\sqrt{(R-1)/2} \tag{10.55}$$

$$\beta = \varpi\sqrt{(R+1)/2} \tag{10.56}$$

and

$$R = \sqrt{1 + \left(4/\varpi^2 r_G^2\right)} \tag{10.57}$$

Each root corresponds to a solution to Eq. (10.48) of the form of Eq. (10.58). Summing all four solutions and employing the trigonometric identities introduced in Section 1.3 gives the general form of *the modal functions for uniform Rayleigh Beams* as

$$W(x) = A_1\cosh\alpha x + A_2\sinh\alpha x + A_3\cos\beta x + A_4\sin\beta x \tag{10.58}$$

Note the similarity with the modal functions for Euler-Bernoulli Beams and Beam-Columns. The integration constants and the natural frequencies are found by imposing the boundary conditions for the particular structure in question on Eq. (10.58). Once the specific frequencies are determined they may be substituted into Eq. (10.58) to evaluate the corresponding modal functions. Each frequency-mode pair evaluated in this way yields a solution of the form of Eq. (10.47). The sum of these solutions, as per Eq. (10.4), corresponds to the free vibration response of the beam-column given by

$$w(x,t) = \sum_{j=1}^{\infty} W^{(j)}(x) A^{(j)} \cos(\omega_j t - \phi_j) \tag{10.59}$$

Example 10.12

Consider a uniform cantilevered elastic beam and its representation using Rayleigh Beam Theory. (*a*) Determine the frequency equation for a uniform beam that is supported at its left end. (*b*) Evaluate the first five values of the natural frequency of the beam for various values of the rotatory inertia (expressed in terms of its radius of gyration) and compare the results with those computed in Example 10.7 using Euler-Bernoulli Beam Theory.

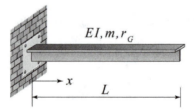

EI, m, r_G

L

Solution
(*a*)
The boundary conditions for a beam that is fixed at the origin and is completely free at $x = L$ follow from Eqs. (9.116)–(9.119) as

$$w(0,t) = 0, \qquad \left. \frac{\partial w}{\partial x} \right|_{x=0} = 0 \tag{a-1, 2}$$

$$-EI \left. \frac{\partial^2 w}{\partial x^2} \right|_{x=L} = 0, \qquad -\left[EI \frac{\partial^3 w}{\partial x^3} - I_\rho \frac{\partial^3 w}{\partial t^2 \partial x} \right]_{x=L} = 0 \tag{a-3, 4}$$

Substituting Eq. (10.47) into the above conditions gives the corresponding boundary conditions for the modal functions,

$$W(0) = 0, \quad W'(0) = 0 \tag{b-1, 2}$$

$$W''(L) = 0, \quad W'''(L) + \varpi^2 W'(L) = 0 \tag{b-3, 4}$$

Note that since the rotatory inertia is included in the present model, the frequency appears in the boundary condition for shear, Eq. (b-4), through the parameter ϖ. Imposing conditions (b-1)–(b-4) on Eq. (10.58) results in the system of algebraic equations for the integration constants given by

$$A_1 + A_3 = 0 \tag{c-1}$$

$$\alpha A_2 + \beta A_4 = 0 \tag{c-2}$$

$$A_1 \alpha^2 \cosh \alpha L + A_2 \alpha^2 \sinh \alpha L - A_3 \beta^2 \cos \beta L - A_4 \beta^2 \sin \beta L = 0 \tag{c-3}$$

$$\alpha \beta^2 \left[A_1 \sinh \alpha L + A_2 \cosh \beta L \right] + \alpha^2 \beta \left[A_3 \sin \beta L - A_4 \cos \beta L \right] = 0 \tag{c-4}$$

The ensuing computations are simplified somewhat by substituting Eqs. (c-1) and (c-2) into Eqs. (c-3) and (c-4) and writing the resulting equations in matrix form as

$$\begin{bmatrix} (\alpha^2 \cosh \alpha L + \beta^2 \cos \beta L) & (\alpha^2 \sinh \alpha L + \alpha \beta \sin \beta L) \\ (\alpha \beta^2 \sinh \alpha L - \alpha^2 \beta \sin \beta L) & (\alpha \beta^2 \cosh \alpha L + \alpha^3 \cos \beta L) \end{bmatrix} \begin{Bmatrix} A_1 \\ A_2 \end{Bmatrix} = \begin{Bmatrix} 0 \\ 0 \end{Bmatrix} \tag{d}$$

For nontrivial solutions we require that the determinant of the matrix of coefficients in Eq. (d) vanish. This results in the frequency equation for the Rayleigh Beam,

$$2\alpha \frac{\varpi^2}{r_G^2} + \alpha \left(\alpha^4 + \beta^4 \right) \cosh \alpha L \cos \beta L - \alpha^2 \beta \varpi^2 \sinh \alpha L \sin \beta L = 0 \quad \triangleleft (e)$$

Equation (e) may be solved numerically to obtain the set of natural frequencies for the Rayleigh Beam. The modal function for the beam is then

$$W^{(j)}(x) = \{\cosh \alpha_j x - \cos \beta_j x\}$$

$$-\frac{\left(\alpha_j^2 \cosh \alpha_j L + \beta_j^2 \cos \beta_j L\right)}{\left(\alpha_j^2 \sinh \alpha_j L + \alpha_j \beta_j \sin \beta_j L\right)}\left\{\sinh \alpha_j x - \left(\alpha_j/\beta_j\right)\sin \beta_j x\right\} \qquad \triangleleft (f)$$

where and α_j and β_j correspond to the values of α and β, Eqs. (10.55) and (10.56), evaluated at the jth natural frequency ω_j. Substitution of Eq. (f) into Eq. (10.59) gives the free vibration response of the cantilevered Rayleigh Beam as

$$w(x,t) = \sum_{j=1}^{\infty} W^{(j)}(x) A^{(j)} \cos(\omega_j t - \phi_j) \qquad \triangleleft (g)$$

where $W^{(j)}$ is given by Eq. (g) and ω_j is the jth root of Eq. (f). The modal amplitudes, $A^{(j)}$, and corresponding phase angles, ϕ_j, depend on the specific initial conditions imposed on the beam.

(b)
The frequency equation for the cantilevered beam, Eq. (e), can be rewritten in a form convenient for computation by multiplying it by L^5 and grouping terms accordingly. Doing this renders the frequency equation to the form

$$2\alpha L \bar{\omega}^2 + \alpha L \left[(\alpha L)^4 + (\beta L)^4\right]\cosh \alpha L \cos \beta L$$
$$-(\alpha L)^2 \beta L \bar{\omega}^2 \bar{r}_G^2 \sinh \alpha L \sin \beta L = 0 \qquad (h)$$

where

$$\bar{\omega} = \omega/\omega_0 \qquad (i\text{-}1)$$

$$\bar{r}_G = r_G/L \qquad (i\text{-}2)$$

$$\alpha L = \bar{\omega}\bar{r}_G \sqrt{(R-1)/2} \qquad (j\text{-}1)$$

$$\beta L = \bar{\omega}\bar{r}_G \sqrt{(R+1)/2} \qquad (j\text{-}2)$$

$$R = \sqrt{1 + 4\bar{\omega}^{-2}\bar{r}_G^{-4}} \qquad (j\text{-}3)$$

Table E10.12

The first five natural frequencies of a uniform cantilevered Rayleigh Beam for various values of rotatory inertia (radius of gyration)

$\overline{r}_G = r_G/L$	ω_1/ω_0	ω_2/ω_0	ω_3/ω_0	ω_4/ω_0	ω_5/ω_0
0.000*	3.516*	22.03*	61.70*	120.8*	199.7*
0.010	3.515	22.00	61.46	120.0	197.6
0.050	3.496	21.19	56.48	103.8	159.7
0.100	3.437	19.14	46.49	78.21	111.8
0.150	3.344	16.75	37.81	60.20	82.92
0.200	3.226	14.58	31.38	48.27	65.15

*Euler-Bernoulli Theory

The first five natural frequencies of the cantilevered Rayleigh Beam, the first five roots of Eq. (h), are evaluated using the MATLAB "fzero" routine for various values of the normalized radius of gyration. The results are tabulated in Table E10.12 and are compared with the first five natural frequencies computed using Euler-Bernoulli Beam Theory ($\overline{r}_G = 0$) as per Example 10.7. It is seen that the rotatory inertia influences the higher modes significantly, within the range of inertias considered.

10.6 FREE VIBRATION OF TIMOSHENKO BEAMS

In this section we discuss the vibration of Timoshenko Beams when moving under their own volition. Recall from Section 9.6.5 that Timoshenko Beams include a correction for transverse shear deformation in an average sense through the thickness and also include the effects of rotatory inertia as for Rayleigh Beams. Beams that include the correction for transverse shear but not the effects of rotatory inertia are referred to as shear beams.

Timoshenko Beams and Shear Beams differ from Euler-Bernoulli Beams and Rayleigh Beams in that there is an additional set of degrees of freedom corresponding to the shear deformations. Two of the three displacement measures (transverse displacement, rotation due to bending and distortion due to shear deformation) are independent. We shall use the transverse displacement, $w(x,t)$, and the cross-sectional rotation, $\varphi(x,t)$, to describe the motion of the beam. For free vibrations, the governing equations for Timoshenko Beams, Eq. (9.135), reduce to the form

$$\mathbf{m}\frac{\partial^2 \mathbf{u}}{\partial t^2} + \mathbf{k}\mathbf{u} = \mathbf{0} \qquad (10.60)$$

where

$$\mathbf{m} = \begin{bmatrix} m & 0 \\ 0 & I_\rho \end{bmatrix} \tag{10.61}$$

$$\mathbf{k} = \begin{bmatrix} -\dfrac{\partial}{\partial x} k_s \dfrac{\partial}{\partial x} & \dfrac{\partial}{\partial x} k_s \\[2ex] -k_s \dfrac{\partial}{\partial x} & k_s - \dfrac{\partial}{\partial x} EI \dfrac{\partial}{\partial x} \end{bmatrix} \tag{10.62}$$

$$\mathbf{u} = \begin{Bmatrix} w(x,t) \\ \varphi(x,t) \end{Bmatrix} \tag{10.63}$$

and, recalling Eq. (9.129),

$$k_s = k_s AG \tag{10.64}$$

where k_s is a shape factor (see Section 9.6.5). For free vibrations we seek solutions to Eq. (10.60) of the form

$$\mathbf{u}(x,t) = \mathbf{U}(x)e^{i\omega t} \tag{10.65}$$

where

$$\mathbf{U}(x) = \begin{Bmatrix} W(x) \\ \vartheta(x) \end{Bmatrix} \tag{10.66}$$

Substitution of Eq. (10.65) into Eq. (10.60) gives the corresponding eigenvalue problem

$$\left[\mathbf{k} - \omega^2 \mathbf{m} \right] \mathbf{U} = 0$$

or, in explicit form,

$$\begin{bmatrix} -\left(k_s \dfrac{d^2}{dx^2} + \omega^2 m \right) & k_s \dfrac{d}{dx} \\[2ex] -k_s \dfrac{d}{dx} & \left(k_s - EI \dfrac{d^2}{dx^2} - \omega^2 I_\rho \right) \end{bmatrix} \begin{Bmatrix} W(x) \\ \vartheta(x) \end{Bmatrix} = \begin{Bmatrix} 0 \\ 0 \end{Bmatrix} \tag{10.67}$$

which, after expanding, yields the coupled equations

$$-k_s W'' - \omega^2 m W + k_s \vartheta' = 0$$
$$-k_s W' - EI\, \vartheta'' + (k_s - \omega^2 I_\rho)\vartheta = 0 \tag{10.68}$$

To solve the above system let us assume a solution of the form

$$\mathbf{U}(x) = \mathbf{A}e^{sx} = \begin{Bmatrix} A \\ B \end{Bmatrix} e^{sx} \tag{10.69}$$

Substitution of (10.69) into Eq. (10.67) gives the algebraic equation

$$\begin{bmatrix} -(k_s s^2 + \omega^2 m) & k_s s \\ -k_s s & (k_s - EI s^2 - \omega^2 I_\rho) \end{bmatrix} \begin{Bmatrix} A \\ B \end{Bmatrix} e^{sx} = \begin{Bmatrix} 0 \\ 0 \end{Bmatrix} \tag{10.70}$$

For nontrivial solutions we require that the determinant of the square matrix in the above equation vanish. This results in the characteristic equation for s,

$$s^4 + \varpi^2 \left(1 + \overline{E}\right)s^2 + \varpi^2 \left(\varpi^2 \overline{E} - r_G^{-2}\right) = 0 \tag{10.71}$$

where

$$\overline{E} = \frac{E}{k_s G} \tag{10.72}$$

and ϖ and r_G are as defined in Eqs. (10.49) and (10.51) respectively. Solving Eq. (10.71) for s gives the roots

$$s = \pm\alpha, \pm i\beta \tag{10.73}$$

where

$$\alpha = \varpi \left[\tfrac{1}{2} \{ R - (1 + \overline{E}) \} \right]^{\frac{1}{2}} \tag{10.74}$$

$$\beta = \varpi \left[\tfrac{1}{2} \{ R + (1 + \overline{E}) \} \right]^{\frac{1}{2}} \tag{10.75}$$

$$R = \sqrt{\left(\overline{E} - 1\right)^2 + \left(4/\varpi^2 r_G^2\right)} \tag{10.76}$$

Note that for infinite shear modulus $\overline{E} \to 0$, and the characteristic equation and the parameters α, β and R reduce to the corresponding expressions for Rayleigh Beams, as they should. Substituting each root back into Eq. (10.69) and summing all such solutions gives the general form of the modal functions as

$$W(x) = A_1^* e^{\alpha x} + A_2^* e^{-\alpha x} + A_3^* e^{i\beta x} + A_4^* e^{-i\beta x} \tag{10.77}$$

$$\vartheta(x) = B_1^* e^{\alpha x} + B_2^* e^{-\alpha x} + B_3^* e^{i\beta x} + B_4^* e^{-i\beta x} \tag{10.78}$$

The constants of integration of the modal rotation, $\Theta(x)$, are related to the constants of integration of the associated modal deflection, $W(x)$, through Eq. (10.70). Since the matrix of coefficients was rendered singular, both rows of that matrix equation yield the same information. From the first row we find that

$$B^* = \left(s + \frac{\varpi^2 \overline{E}}{s} \right) A^* \tag{10.79}$$

Evaluating Eq. (10.79) for each root of the characteristic equation for s gives the corresponding relations

$$B^*_{1,2} = \pm \left(\alpha + \frac{\varpi^2 E}{\alpha} \right) A^*_{1,2} \tag{10.80}$$

and

$$B^*_{3,4} = \pm i \left(\beta - \frac{\varpi^2 E}{\beta} \right) A^*_{3,4} = \left(\beta - \frac{\varpi^2 E}{\beta} \right) e^{\pm i \pi / 2} A^*_{3,4} \tag{10.81}$$

Substituting Eqs. (10.80) and (10.81) into Eq. (10.78) gives the modal rotation as

$$\Theta(x) = \frac{1}{\alpha} \left(\alpha^2 + \varpi^2 \overline{E} \right) \left[A^*_1 e^{\alpha x} - A^*_2 e^{-\alpha x} \right]$$
$$+ \frac{1}{\beta} \left(\beta^2 - \varpi^2 \overline{E} \right) \left[A^*_3 e^{i(\beta x + \pi / 2)} + A^*_4 e^{-i(\beta x + \pi / 2)} \right] \tag{10.82}$$

Using Eqs. (1.61) and (1.63) in Eqs. (10.77) and (10.82) renders the modal functions to the forms

$$W(x) = A_1 \cosh \alpha x + A_2 \sinh \alpha x + A_3 \cos \beta x + A_4 \sin \beta x \tag{10.83}$$

and

$$\Theta(x) = g_\alpha \left[A_1 \sinh \alpha x + A_2 \cosh \alpha x \right] - g_\beta \left[A_3 \sin \beta x - A_4 \cos \beta x \right] \tag{10.84}$$

where

$$g_\alpha = \frac{1}{\alpha} \left(\alpha^2 + \varpi^2 \overline{E} \right), \quad g_\beta = \frac{1}{\beta} \left(\beta^2 - \varpi^2 \overline{E} \right) \tag{10.85}$$

and

$$A_1 = A^*_1 + A^*_2, \quad A_2 = A^*_1 - A^*_2, \quad A_3 = A^*_3 + A^*_4, \quad A_4 = i \left(A^*_3 - A^*_4 \right) \tag{10.86}$$

The remainder of the analysis proceeds as for the free vibration analysis of Euler-Bernoulli Beams, etc. and depends upon the particular support conditions imposed on the structure. Once the natural frequencies are obtained they are substituted into Eqs. (10.83) and (10.84) for each mode. The general response is then found by substituting the modes and corresponding frequencies into Eq. (10.65) and summing over all such solutions. The free vibration response of the Timoshenko beam is then

$$\mathbf{u}(x,t) = \begin{Bmatrix} w(x,t) \\ \varphi(x,t) \end{Bmatrix} = \sum_{j=1}^{\infty} \begin{Bmatrix} W^{(j)}(x) \\ \vartheta^{(j)}(x) \end{Bmatrix} A^{(j)} \cos(\omega_j t - \phi_j) \tag{10.87}$$

Example 10.13

Consider a uniform cantilevered elastic beam and its representation using Timoshenko Beam Theory. (**a**) Determine the frequency equation for a uniform beam that is supported at its left end. (**b**) Evaluate the first five values of the natural frequency of a beam with $r_G/L = 0.1$, for various values of the modulus ratio $E/kG = 0.1$. Compare the results with those computed in Example 10.12 using Rayleigh Beam Theory.

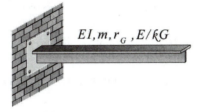

$$EI, m, r_G, E/kG$$

Solution

(**a**)

The boundary conditions follow from Eqs. (9.141)–(9.144) as

$$w(0,t) = 0, \quad \varphi(0,t) = 0 \tag{a-1, 2}$$

$$-EI \left. \frac{\partial \varphi}{\partial x} \right|_{x=L} = 0, \quad k_s \left[\frac{\partial w}{\partial x} - \varphi \right]_{x=L} = 0 \tag{a-3, 4}$$

Substitution of Eq. (10.65) into Eqs. (a-1)–(a-4) gives the corresponding boundary conditions for the modal functions,

$$W(0) = 0, \quad \vartheta(0) = 0 \tag{b-1, 2}$$

$$\vartheta'(L) = 0, \quad W'(L) - \vartheta(L) = 0 \tag{b-3, 4}$$

We next impose Eqs. (b-1)–(b-4) on the general form of the modal functions given by Eqs. (10.83) and (10.84). This gives the set of algebraic equations

$$A_1 + A_3 = 0 \tag{c-1}$$

$$g_\alpha A_2 + g_\beta A_4 = 0 \tag{c-2}$$

$$g_\alpha \alpha \left[A_1 \cosh \alpha L + A_2 \sinh \alpha L \right] - g_\beta \beta \left[A_3 \cos \beta L + A_4 \sin \beta L \right] = 0 \tag{c-3}$$

$$(\alpha - g_\alpha) \left[A_1 \sinh \alpha L + A_2 \cosh \alpha L \right] - (\beta - g_\beta) \left[A_3 \sin \beta L - A_4 \cos \beta L \right] = 0 \tag{c-4}$$

where g_α and g_β are given by Eqs. (10.85). Substituting Eqs. (c-1) and (c-2) into Eqs. (c-3) and (c-4) simplifies the calculation and gives two algebraic equations in the two constants A_1 and A_2. The equations are expressed in matrix form as

$$\begin{bmatrix} H_{11} & H_{12} \\ H_{21} & H_{22} \end{bmatrix} \begin{Bmatrix} A_1 \\ A_2 \end{Bmatrix} = \begin{Bmatrix} 0 \\ 0 \end{Bmatrix} \tag{d}$$

where

$$H_{11} = g_\alpha \alpha \cosh \alpha L + g_\beta \beta \cos \beta L \tag{e-1}$$

$$H_{12} = g_\alpha \left(\alpha \sinh \alpha L + \beta \sin \beta L \right) \tag{e-2}$$

$$H_{21} = (\alpha - g_\alpha) \sinh \alpha L + (\beta - g_\beta) \sin \beta L \tag{e-3}$$

$$H_{22} = (\alpha - g_\alpha) \cosh \alpha L - (\beta - g_\beta)(g_\alpha / g_\beta) \cos \beta L \tag{e-4}$$

Now, for nontrivial solutions the determinant of the matrix of coefficients of Eq. (d) must vanish. This results in the frequency equation,

$$\mathcal{F}(\varpi; \bar{E}) = H_{11} H_{22} - H_{12} H_{21} = 0 \tag{f}$$

which, after expanding, takes the form

$$-\left(g_\alpha + g_\beta \right) \varpi^2 \bar{E} + \frac{1}{g_\beta} \left[\beta g_\beta^2 \left(\alpha - g_\alpha \right) - \alpha g_\alpha^2 \left(\beta - g_\beta \right) \right] \cosh \alpha L \cos \beta L$$

$$- g_\alpha \left[2\alpha\beta - \left(\beta g_\alpha + \alpha g_\beta \right) \right] \sinh \alpha L \sin \beta L = 0 \tag{f'}$$

Equation (f) may be solved numerically for ϖ, for given material and geometric parameters. Then, for each root ϖ_j ($j = 1, 2, \ldots$) of the frequency equation, the natural frequencies are found from Eq. (10.49) as

$$\omega_j = \varpi_j c_a \quad (j = 1, 2, \ldots) \tag{g}$$

The first row of Eq. (d) gives A_2 in terms of A_1 as

$$A_2 = -\frac{H_{11}}{H_{12}} A_1 \tag{h}$$

Now, substituting Eq. (h) into Eq. (c-2) gives A_4 in terms of A_1. Hence,

$$A_4 = \frac{g_\alpha H_{11}}{g_\beta H_{12}} A_1 \tag{i}$$

Incorporating Eqs. (c-1), (h) and (i) into Eqs. (10.83) and (10.84) gives the modal functions for the cantilevered Timoshenko Beam in the form

$$W^{(j)}(x) = A_1^{(j)} \left[\cosh \alpha_j x - \cos \beta_j x - \frac{H_{11}^{(j)}}{H_{12}^{(j)}} \left\{ \sinh \alpha_j x - \left(g_\alpha^{(j)} / g_\beta^{(j)} \right) \sin \beta_j x \right\} \right] \tag{j-1}$$

$$\vartheta^{(j)}(x) = A_1^{(j)} \left[g_\alpha^{(j)} \sinh \alpha_j x + g_\beta^{(j)} \sin \beta_j x - g_\alpha^{(j)} \frac{H_{11}^{(j)}}{H_{12}^{(j)}} \left\{ \cosh \alpha_j x - \cos \beta_j x \right\} \right] \tag{j-2}$$

where we will choose $A_1^{(j)} = 1$ when $R > \bar{E} + 1$ and $A_1^{(j)} = i$ when $R < \bar{E} + 1$.

(b)
The frequency equation, Eq. (f), can be rewritten in a form convenient for computation by multiplying the first row of Eq. (e) by L^2 and the second row by L and noting that

$$\alpha L = \bar{\omega} \bar{r}_G \left[\tfrac{1}{2} \left\{ R - (1 + \bar{E}) \right\} \right]^{1/2} \tag{k-1}$$

$$\beta L = \bar{\omega} \bar{r}_G \left[\tfrac{1}{2} \left\{ R + (1 + \bar{E}) \right\} \right]^{1/2} \tag{k-2}$$

$$R = \sqrt{\left(\bar{E} - 1 \right)^2 + \left(4 / \bar{\omega}^2 \bar{r}_G^4 \right)} \tag{k-3}$$

where

$$\bar{\omega} = \omega / \omega_0 \tag{l-1}$$

$$\omega_0 = \sqrt{EI / mL^4} \tag{l-2}$$

and

$$\bar{r}_G = r_G / L \tag{m}$$

The first five roots of the frequency equation are computed for various values of $\bar{E}$ using the MATLAB routine "fzero." The results are summarized in Table E10.13. It is seen that the inclusion of transverse shear deformation lowers the predicted natural frequencies of the structure and that the effects are significant for frequencies other than the fundamental frequency, even for moderate values of the modulus ratio. It is also seen that the fundamental frequency predicted when the shear deformation is neglected (Rayleigh Theory) differs from that predicted by the Timoshenko Theory by 20% when the Elastic Modulus is larger than the effective shear modulus by a factor of ten. The data for the case study of a uniform cantilevered beam, contained in Tables E10.12 and E10.13, when taken together, provides a characterization of the influence of the rotatory inertia and shear deformation.

Table E10.13

The first five natural frequencies of a uniform cantilevered Timoshenko Beam with $\bar{r}_G = 0.1$, for various values of the modulus ratio

$\bar{E} = E/kG$	ω_1/ω_0	ω_2/ω_0	ω_3/ω_0	ω_4/ω_0	ω_5/ω_0
0.0*	3.437*	19.14*	46.49*	78.21*	111.8*
0.1	3.430	18.92	45.73	76.73	109.7
1.0	3.366	17.23	39.79	64.35	89.89
5.0	3.116	12.90	27.41	40.38	50.75
10.0	2.864	10.48	21.46	30.52	37.29

*Rayleigh Theory

10.7 NORMALIZATION OF THE MODAL FUNCTIONS

It was seen in Section 10.2 that the modal functions were unique to within an integration constant. In that section we implicitly set the constant equal to one. To alleviate this arbitrariness, and to allow for a uniform measure of the modal functions, we may choose to normalize the modal functions in a fashion that is analogous to that used for discrete modal vectors. That is, we shall divide the functions by their magnitudes thus rendering them unit functions. As for the discrete case, this may be done in the conventional sense or in the weighted sense as delineated below.

Conventional Scalar Product as Metric

To render a modal function a unit function (a function of unit magnitude) in the conventional sense, we divide the function by its magnitude as measured by the square root of its conventional scalar product with itself. Thus, when defined in this manner, the corresponding *normal mode* is related to the original modal function by the relation

$$\bar{U}^{(j)}(x) = \frac{U^{(j)}(x)}{\left\| U^{(j)}(x) \right\|} = \frac{U^{(j)}(x)}{\sqrt{\left\langle U^{(j)}, U^{(j)} \right\rangle}} \tag{10.88}$$

where, from Eq. (9.19), the conventional scalar product of the original modal function with itself is given by

$$\left\langle U^{(j)}, U^{(j)} \right\rangle = \int_0^L U^{(j)^2}(x)\, dx \tag{10.89}$$

It is customary to drop the over-bar when consistently using normal modes.

Weighted Scalar Product as Metric

It is often convenient to choose the weighted scalar product to provide metric (i.e., the "length scale") of a function. If this is done, then the corresponding *normal mode* is given by

$$\bar{U}^{(j)}(x) = \frac{U^{(j)}(x)}{\left\| U^{(j)} \right\|_m} = \frac{U^{(j)}(x)}{\left\langle U^{(j)}, U^{(j)} \right\rangle_m} \tag{10.90}$$

where, from Eq. (9.29), the scalar product of the modal function with itself, measured with respect to the mass operator m, is given by

$$\left\langle U^{(j)}, U^{(j)} \right\rangle_m = \int_0^L U^{(j)}(x)\, m\, U^{(j)}(x)\, dx \tag{10.91}$$

As for the conventional case discussed earlier, it is customary to drop the over-bar when consistently using normal modes.

Example 10.14

Determine the normal modes for the elastic rod of Example 10.1, (*a*) by normalizing in the conventional sense and (*b*) by normalizing with respect to the mass.

Solution
From Example 10.1, the modal functions for the rod are

$$U^{(j)}(x) = A_2^{(j)} \sin \beta_j x \quad (j = 1, 2, \ldots) \tag{a}$$

(*a*)

We first compute the magnitude of the jth modal function. Hence,

$$\left\| U^{(j)} \right\|^2 = \int_0^L U^{(j)2}(x)\,dx = A_2^{(j)2} \int_0^L \sin^2 \beta_j x\,dx = \frac{A_2^{(j)2} L}{2} \tag{b}$$

We next divide the modal function of Eq. (a) by its magnitude to obtain the jth normal mode,

$$U^{(j)}(x) = \frac{A_2^{(j)} \sin \beta_j x}{\sqrt{A_2^{(j)2} L/2}} = \sqrt{\frac{2}{L}} \sin \beta_j x \tag{◁ (c)}$$

(*b*)

Computing the magnitude of the jth modal function using the weighted scalar product gives

$$\left\| U^{(j)} \right\|_m^2 = \int_0^L U^{(j)}(x)\,m\,U^{(j)}(x)\,dx = A_2^{(j)2} \int_0^L m \sin^2 \beta_j x\,dx = \frac{A_2^{(j)2} mL}{2} \tag{d}$$

The jth normal mode is then

$$U^{(j)}(x) = \frac{A_2^{(j)} \sin \beta_j x}{\sqrt{A_2^{(j)2} mL/2}} = \sqrt{\frac{2}{mL}} \sin \beta_j x \tag{◁ (e)}$$

10.8 ORTHOGONALITY OF THE MODAL FUNCTIONS

The modal functions for a given system correspond to fundamental motions that constitute the response of that system. It will be seen that any motion, forced or free, can be described as a linear combination of the modal functions. To establish this we must first establish the mutual orthogonality, as defined in Section 9.1.2, of the modal functions for the systems of interest. In particular, we shall consider the class of one-dimensional continua that includes the mathematical models for longitudinal motion of elastic rods, torsional motion of elastic rods, transverse motion of flexible strings and cables, transverse motion of Euler-Bernoulli Beams and Beam-Columns, Rayleigh Beams and Timoshenko Beams.

10.8.1 Systems Whose Mass Operators Are Scalar Functions

In this section we establish the general condition of orthogonality of the modal functions for continuous systems whose mass operator corresponds to a single scalar function of the spatial coordinate. These systems include the mathematical models for longitudinal motion of elastic rods, torsional motion of elastic rods, transverse motion of strings and cables, and transverse motion of Euler-Bernoulli Beams.

Systems Described by a Smooth Mass Distribution

To establish the conditions for the mutual orthogonality of the modal functions for the class of systems under consideration we proceed in a fashion analogous to that which was done in Section 7.4.2 for discrete systems.

It was seen in Section 10.1 that the general free-vibration problem is reduced to finding the frequency-mode pairs that satisfy the eigenvalue problem defined by Eq. (10.3). Let us consider the ith and j^{th} natural frequencies and corresponding modal functions

$$\omega_i^2, U^{(i)}(x) \quad \text{and} \quad \omega_j^2, U^{(j)}(x)$$

Each pair represents a solution to Eq. (10.3). Therefore

$$\mathsf{k}U^{(i)}(x) = \omega_i^2 \mathsf{m}U^{(i)}(x) \tag{10.92}$$

and

$$\mathsf{k}U^{(j)}(x) = \omega_j^2 \mathsf{m}U^{(j)}(x) \tag{10.93}$$

Let us multiply Eq. (10.92) by $U^{(j)}(x)dx$ and Eq. (10.93) by $U^{(i)}(x)dx$. Integrating the resulting expressions over $[0, L]$ results in the identities

$$\int_0^L U^{(j)}\mathsf{k}U^{(i)}dx = \int_0^L \omega_i^2 U^{(j)}\mathsf{m}U^{(i)}dx \tag{10.94}$$

and

$$\int_0^L U^{(i)}\mathsf{k}U^{(j)}dx = \int_0^L \omega_j^2 U^{(i)}\mathsf{m}U^{(j)}dx \tag{10.95}$$

Subtracting Eq. (10.95) from Eq. (10.94) gives

$$\int_0^L U^{(j)}\mathsf{k}U^{(i)}dx - \int_0^L U^{(i)}\mathsf{k}U^{(j)}dx$$
$$= \omega_i^2 \int_0^L U^{(j)}\mathsf{m}U^{(i)}dx - \omega_j^2 \int_0^L U^{(i)}\mathsf{m}U^{(j)}dx \tag{10.96}$$

where

$$B_{\Bbbk}^{(ji)} \equiv \int_0^L U^{(j)} \Bbbk U^{(i)} dx - \int_0^L U^{(i)} \Bbbk U^{(j)} dx \tag{10.97}$$

will depend on the boundary conditions for the modal functions. The development to this point applies to systems with differential mass operators as well as scalar function mass operators. In the remainder of this section we restrict our attention to systems whose mass operators are single scalar functions. More complex systems are considered in Sections 10.8.4 and 10.8.5.

For systems whose mass operator is given by a smooth scalar function, $\Bbbm = m(x)$, Eq. (10.96) reduces to the form

$$B_{\Bbbk}^{(ji)} = \left(\omega_i^2 - \omega_j^2 \right) \int_0^L U^{(j)}(x) m(x) U^{(i)}(x) dx \tag{10.98}$$

where $B_{\Bbbk}^{(ji)}$ is defined by Eq. (10.97). Therefore, if

$$B_{\Bbbk}^{(jl)} = 0 \tag{10.99}$$

then, for distinct frequencies ($\omega_l^2 \neq \omega_j^2$),

$$\left\langle U^{(l)}, U^{(j)} \right\rangle_m \equiv \int_0^L U^{(l)}(x) m(x) U^{(j)}(x) dx = 0 \tag{10.100}$$

and

$$U^{(l)}(x) \underset{m}{\perp} U^{(j)}(x)$$

It follows from Eq. (10.94) that if Eq. (10.100) holds then

$$\left\langle U^{(l)}, U^{(j)} \right\rangle_{\Bbbk} \equiv \int_0^L U^{(l)} \Bbbk U^{(j)} dx = 0 \tag{10.101}$$

and

$$U^{(l)}(x) \underset{\Bbbk}{\perp} U^{(j)}(x)$$

as well. The question of mutual orthogonality of the modal functions is thus reduced to demonstrating that the boundary conditions for a given system satisfy Eq. (10.99).

Continuous Systems with One or More Concentrated Mass Points

The conditions for, and statements of, orthogonality described above hold equally well for one-dimensional continua that are described by a smooth mass distribution except at one or more isolated points, and require only minor differences in representation and interpretation. To demonstrate this consider a continuous structure such as a rod or beam whose mass distribution is smooth except at a single point, say at $x = L$, where the mass at that point is m. This situation corresponds to a smooth structure with a point mass attached to its edge (Figure 10.6). For this case the mass operator may be expressed in the form

$$\mathbb{m} = m(x) = m(x) + m\,\hat{\delta}(x - L) \tag{10.102}$$

where $m(x)$ is the mass distribution of the structure on $0 < x < L$ and $\hat{\delta}(x)$ is the Dirac Delta Function (see Chapter 4). Substitution of Eq. (10.102) into Eq. (10.100) gives the evaluation of the scalar product and corresponding statement of orthogonality for a structure possessing a point mass as

$$\begin{aligned}
\left\langle U^{(l)}, U^{(j)} \right\rangle_m &\equiv \int_0^L U^{(l)} m(x) U^{(j)} dx \\
&= \int_0^L U^{(l)} m(x) U^{(j)} dx + U^{(l)}(L)\, m\, U^{(j)}(L) = 0
\end{aligned} \tag{10.103}$$

Thus, for such situations, all prior discussions concerning the conditions of orthogonality hold, with the scalar product taken with respect to the mass distribution being interpreted in this sense. Similar statements may be made if the concentrated mass is located at an interior point of the structure.

The above development is applied to second order systems and to Euler-Bernoulli Beams and Beam-Columns in Sections 10.8.2 and 10.8.3. The conditions for, and interpretation of, the mutual orthogonality of the modes for systems with more complex mass operators, specifically Rayleigh Beams and Timoshenko Beams, are developed separately in Sections 10.8.4 and 10.8.5.

10.8.2 Second Order Systems

In this section we examine the mutual orthogonality of the modal functions corresponding to the longitudinal motion of elastic rods, torsional motion of elastic rods and transverse motion of flexible strings and cables. In each case, the mass operator is a scalar function of the axial coordinate, and the weighted scalar product with respect to the stiffness operator is of the form

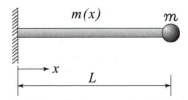

Figure 10.6 Smooth structure with attached mass.

$$\int_0^L U^{(j)} \Bbbk U^{(l)} dx = -\int_0^L U^{(j)} \frac{d}{dx} k(x) \frac{dU^{(l)}}{dx} dx \tag{10.104}$$

Integrating by parts twice gives the relation

$$\int_0^L U^{(j)} \Bbbk U^{(l)} dx = \int_0^L U^{(l)} \Bbbk U^{(j)} dx + B_{\Bbbk}^{(jl)} \tag{10.105}$$

where

$$B_{\Bbbk}^{(jl)} = \left[U^{(j)} \left\{ k \frac{dU^{(l)}}{dx} \right\} \right]_0^L - \left[\left\{ \frac{dU^{(j)}}{dx} k \right\} U^{(l)} \right]_0^L \tag{10.106}$$

It follows from Eqs. (10.99) and (10.106) that the modal functions for the continuous systems under consideration are mutually orthogonal if the boundary conditions are such that the terms in brackets of Eq. (10.106) sum to zero. The problem of orthogonality is thus reduced to properties of the boundary conditions for a given system. We next examine these conditions for individual types of motion.

Longitudinal Motion of Elastic Rods

Let us recall from Section 9.3 that, for axial motion of elastic rods, $u(x,t) \rightarrow u(x,t)$, $U(x) \rightarrow U(x)$, $k \rightarrow k_a = EA$ and $\Bbbm = m(x) = \rho A$. Let us also recall that the membrane force, N, is related to the displacement gradient as

$$N = k_a \frac{\partial u}{\partial x} = EA \frac{\partial u}{\partial x} \tag{10.107}$$

Further, from our discussions in this chapter, the free vibration response is of the form

$$u(x,t) = \sum_{j=1}^{\infty} U^{(j)}(x) e^{i\omega_j t} \tag{10.108}$$

It follows, upon substitution of Eq. (10.108) into Eq. (10.107), that

$$N(x,t) = \sum_{j=1}^{\infty} \tilde{N}^{(j)}(x) e^{i\omega_j t} \tag{10.109}$$

where

$$\tilde{N}^{(j)}(x) \equiv k_a \frac{dU^{(j)}}{dx} \tag{10.110}$$

will be referred to as *the modal membrane force*. Incorporating this identification into Eqs. (10.99) and (10.106) gives the condition for the mutual orthogonality of the modal functions for longitudinal motion of elastic rods as

$$\left[\tilde{N}^{(j)}(L)U^{(l)}(L) - \tilde{N}^{(l)}(L)U^{(j)}(L) \right]$$
$$-\left[\tilde{N}^{(j)}(0)U^{(l)}(0) - \tilde{N}^{(l)}(0)U^{(j)}(0) \right] = 0 \tag{10.111}$$

It may be seen that Eq. (10.111) will be satisfied if, at each boundary, the work of the modal membrane force of the jth mode going through the deflections of the lth mode is equal to the work of the modal membrane force of the lth mode going through the deflections of the jth mode. It is readily seen that this condition is satisfied trivially for the homogeneous boundary conditions

$$U(0) = 0 \text{ or } \tilde{N}(0) = 0 \tag{10.112}$$

and

$$U(L) = 0 \text{ or } \tilde{N}(L) = 0 \tag{10.113}$$

In summary, if the boundary conditions of an elastic rod are such that Eq. (10.111) is satisfied, then the modal functions are orthogonal as follows:

$$\left\langle U^{(l)}, U^{(j)} \right\rangle_m = \int_0^L U^{(l)}(x) m(x) U^{(j)}(x) \, dx = 0 \quad (\omega_i^2 \neq \omega_j^2) \tag{10.114}$$

$$\left\langle U^{(l)}, U^{(j)} \right\rangle_k = \int_0^L U^{(l)}(x) \frac{d}{dx} \left\{ EAU^{(j)\prime}(x) \right\} dx = 0 \quad (\omega_i^2 \neq \omega_j^2) \tag{10.115}$$

Example 10.15

Consider the elastic rod with a concentrated mass at its free end discussed in Example 10.2. Show that the modal functions for longitudinal motion of the rod are mutually orthogonal.

Solution

To establish the othogonality of the modal functions we must show that the bracketed terms of Eq. (10.111) sum to zero. For the given support condition

$$U(0) = 0 \tag{a}$$

so the second bracketed expression of Eq. (10.111) is seen to vanish identically. If we consider the concentrated mass to be part of the rod, as discussed at the end of Section 10.8.1, then the mass operator and distribution is given by Eq. (10.102). Viewing the structure in this sense, the boundary condition at the right of the point mass (at $x = L^+$) is then

$$\tilde{N}(L^+) = 0 \tag{b}$$

The first bracketed expression of Eq. (10.111) is thus seen to vanish as well. The modal functions are therefore mutually orthogonal with respect to the mass in the sense of Eq. (10.103), and are therefore orthogonal with respect to the stiffness operator in this sense as well.

Suppose, instead, that we had considered the rod and point mass as separate structures. From Eq. (10.107) and Eq. (b) of Example 10.2, the boundary condition at $x = L^-$ is now

$$\tilde{N}(L) = k_a U'(L) = k_a \beta^2 \frac{m}{m} U(L) = \omega^2 m U(L) \tag{c}$$

Substitution of Eqs. (a) and (c) into Eq. (10.111) results in the statement

$$(\omega_j^2 - \omega_i^2) U^{(j)}(L) m U^{(i)}(L) = 0 \tag{d}$$

which is not generally satisfied if the tip mass is free to move. On this basis, we might conclude that the modal functions are not mutually orthogonal. However, this simply means that the modal functions do not satisfy Eq. (10.100) on the interval $0 < x < L^-$. We would find, by direct calculation, that the modal functions do satisfy Eq. (10.103) and are thus mutually orthogonal on the domain $[0, L^+]$ in that sense.

Torsional Motion of Elastic Rods

For the case of torsional motion of rods, the vibration problem is described by the angular displacement and the internal torque. We recall from Section 9.4 that, for such motion, $k = k_T = GJ$ and $m = J_\rho$. Thus, in the conditions for general and second order systems described in Section 10.1 and at the beginning of this section, $u(x,t) \rightarrow \theta(x,t)$, $U(x) \rightarrow \Theta(x)$, and $m \rightarrow J_\rho = \rho J$. Recall from Section 9.4 that the internal torque, $T(x,t)$, is related to the gradient of the angular displacement, $\theta(x,t)$, by the relation

$$T(x,t) = k_T \frac{\partial \theta}{\partial x} = GJ \frac{\partial \theta}{\partial x} \tag{10.116}$$

Further, from earlier discussions, the free vibration response is of the general form

$$\theta(x,t) = \sum_{j=1}^{\infty} \Theta^{(j)}(x)e^{i\omega_j t} \tag{10.117}$$

It is seen, upon substitution of Eq. (10.117) into Eq. (10.116), that

$$T(x,t) = \sum_{j=1}^{\infty} \tilde{T}^{(j)}(x)e^{i\omega_j t} \tag{10.118}$$

where

$$\tilde{T}^{(j)}(x) \equiv k_T \frac{d\Theta^{(j)}}{dx} = GJ \frac{d\Theta^{(j)}}{dx} \tag{10.119}$$

will be referred to as *the modal torque*. Incorporating this identification into Eqs. (10.99) and (10.106) gives the condition for the mutual orthogonality of the modal functions for torsional motion of elastic rods as

$$\left[\tilde{T}^{(j)}(L)\Theta^{(l)}(L) - \tilde{T}^{(l)}(L)\Theta^{(j)}(L) \right]$$
$$- \left[\tilde{T}^{(j)}(0)\Theta^{(l)}(0) - \tilde{T}^{(l)}(0)\Theta^{(j)}(0) \right] = 0 \tag{10.120}$$

It is seen that Eq. (10.120) will be satisfied if, at each boundary, the work of the modal torque of the jth mode going through the rotations of the lth mode is equal to the work of the modal torque of the lth mode going through the rotations of the jth mode. The condition of orthogonality, Eq. (10.120), is satisfied trivially for the homogeneous boundary conditions

$$\Theta(0) = 0 \ \text{ or } \ \tilde{T}(0) = 0 \tag{10.121}$$

and

$$\Theta(L) = 0 \ \text{ or } \ \tilde{T}(L) = 0 \tag{10.122}$$

In summary, if the boundary conditions of an elastic rod are such that Eq. (10.120) is satisfied, then the modal functions are mutually orthogonal as follows:

$$\left\langle \Phi^{(l)}, \Phi^{(j)} \right\rangle_m = \int_0^L \Phi^{(l)}(x) \rho J \Phi^{(j)}(x)\, dx = 0 \quad (\omega_l^2 \neq \omega_j^2) \tag{10.123}$$

$$\left\langle \Phi^{(l)}, \Phi^{(j)} \right\rangle_k = \int_0^L \Phi^{(l)}(x) \frac{d}{dx}\left\{ GJ\, \Phi^{(j)\prime}(x) \right\} dx = 0 \quad (\omega_l^2 \neq \omega_j^2) \tag{10.124}$$

Example 10.16

Show that the modal functions corresponding to the system of Example 10.4 are mutually orthogonal.

Solution

From Eq. (a-1) of Example 10.4, the boundary condition at the elastic wall gives

$$\left[L\frac{\partial \theta}{\partial x} - \alpha\theta \right]_{x=0} = 0 = L\Theta'(0)e^{i\omega t} - \alpha\,\Theta(0)e^{i\omega t}$$

The corresponding modal boundary condition at the left end of the rod is thus

$$L\Theta'(0) = \alpha\,\Theta(0) \tag{a}$$

From condition (a-2) of Example 10.4, the boundary condition arising from the presence of the rigid wall at the right end of the rod follows as

$$\theta(L,t) = 0 = \Theta(L)e^{i\omega t} \quad \Rightarrow \quad \Theta(L) = 0 \tag{b}$$

Substituting Eq. (a) into Eq. (10.119) gives the relation

$$\tilde{T}^{(j)}(0) = \Theta^{(j)}(0)\,k_T\alpha/L \tag{c}$$

Finally, substituting of Eqs. (b) and (c) into the left hand side of Eq. (10.120) gives

$$\left[0 - 0\right] - \left(k_T\alpha/L\right)\left[\Theta^{(j)}(0)\Theta^{(i)}(0) - \Theta^{(i)}(0)\Theta^{(j)}(0)\right] = 0 \tag{d}$$

It is seen that Eq. (10.120) is satisfied, and thus that the modal functions are mutually orthogonal with respect to both m and k.

Transverse Motion of Strings and Cables

When considering the transverse motion of flexible strings and cables, the motion is characterized by the transverse displacement and the mass by the mass per unit length. Hence, for these systems, $u(x,t) \to w(x,t)$, $U(x) \to W(x)$ and $m = m(x)$ in the general statements of Section 10.1. We also recall from Section 9.5 that the transverse shear force, Q, is related to the displacement gradient by the relation

$$Q(x,t) = N \frac{\partial w}{\partial x} \qquad (10.125)$$

where N is the tension (tensile membrane force). Substituting the free vibration response

$$w(x,t) = \sum_{j=1}^{\infty} W^{(j)}(x) e^{i\omega t} \qquad (10.126)$$

into Eq. (10.125) gives the transverse shear force as

$$Q(x,t) = \sum_{j=1}^{\infty} \tilde{Q}^{(j)}(x) e^{i\omega_j t} \qquad (10.127)$$

where

$$\tilde{Q}^{(j)}(x) = N W^{(j)'}(x) \qquad (10.128)$$

will be referred to as *the modal shear force*. Incorporating this identification into Eqs. (10.99) and (10.106) gives the condition for the mutual orthogonality of the modal functions for transverse motion of strings and cables as

$$\begin{aligned} &\left[\tilde{Q}^{(j)}(L) W^{(l)}(L) - \tilde{Q}^{(l)}(L) W^{(j)}(L) \right] \\ &\quad - \left[\tilde{Q}^{(j)}(0) W^{(l)}(0) - \tilde{Q}^{(l)}(0) W^{(j)}(0) \right] = 0 \end{aligned} \qquad (10.129)$$

It is seen that this condition will be satisfied if, at each boundary, the work of the modal shear force of the jth mode going through the deflections of the lth mode is equal to the work of the modal shear force of the lth mode going through the deflections corresponding to the jth mode. Equation (10.129) is satisfied trivially for the homogeneous boundary conditions

$$W(0) = 0 \text{ or } \tilde{Q}(0) = 0 \qquad (10.130)$$

and

$$W(L) = 0 \text{ or } \tilde{Q}(L) = 0 \qquad (10.131)$$

In summary, if the boundary conditions of an elastic rod are such that Eq. (10.129) is satisfied, then the modal functions are mutually orthogonal as follows

$$\left\langle W^{(l)}, W^{(j)} \right\rangle_m = \int_0^L W^{(l)}(x) m(x) W^{(j)}(x) \, dx = 0 \quad (\omega_l^2 \neq \omega_j^2) \qquad (10.132)$$

$$\left\langle W^{(l)}, W^{(j)} \right\rangle_{\Bbbk} = \int_0^L W^{(l)}(x) \, N \, W^{(j)''}(x) \, dx = 0 \quad (\omega_i^2 \neq \omega_j^2) \tag{10.133}$$

Example 10.17

Show that the modal functions for the system of Example 9.6-ii are mutually orthogonal.

Solution

From Eq. (10.125) and (10.128), and Eqs. (d) and (e-ii) of Example 9.6,

$$w(0,t) = 0 = W(0)e^{i\omega t} \quad \Rightarrow \quad W(0) = 0 \tag{a}$$

$$Q(L,t) = 0 = \tilde{Q}(L)e^{i\omega t} \quad \Rightarrow \quad \tilde{Q}(L) = 0 \tag{b}$$

Conditions (a) and (b) respectively correspond to the first condition of Eq. (10.130) and the second condition of Eq. (10.131). It is readily seen that Eq. (10.129) is satisfied. This in turn shows that Eq. (10.99) is satisfied and, therefore, that Eqs. (10.100) and (10.101) are satisfied. Hence,

$$\left\langle W^{(j)}, W^{(l)} \right\rangle_m \equiv \int_0^L W^{(j)} m W^{(l)} dx = 0 \quad (l, j = 1, 2, 3, \ldots) \tag{c}$$

$$\Rightarrow \quad W^{(j)}(x) \underset{m}{\perp} W^{(l)}(x) \quad (l, j = 1, 2, 3, \ldots)$$

and

$$\left\langle W^{(j)}, W^{(l)} \right\rangle_{\Bbbk} \equiv \int_0^L W^{(j)} \Bbbk W^{(l)} dx = 0 \quad (l, j = 1, 2, 3, \ldots) \tag{d}$$

$$\Rightarrow \quad W^{(j)}(x) \underset{\Bbbk}{\perp} W^{(l)}(x) \quad (l, j = 1, 2, 3, \ldots)$$

10.8.3 Euler-Bernoulli Beams and Beam-Columns

The flexural motion of beams and beam-columns is characterized by the transverse displacement of the neutral axis of the structure. In this section we consider the mutual orthogonality of the modal functions for Euler-Bernoulli Beams and Euler-Bernoulli Beam-Columns with constant axial load. For these systems, $u(x,t) \rightarrow w(x,t)$

and $U(x) \rightarrow W(x)$ in the general statements of Section 10.1 and the mass distribution is characterized by the mass per unit length, $m(x)$.

Euler-Bernoulli Beams

For Euler-Bernoulli Beams the local stiffness corresponds to the bending stiffness $k = k_b = EI$ and the stiffness operator is the fourth order differential operator defined by

$$\mathbb{k} = \frac{\partial^2}{\partial x^2} EI \frac{\partial^2}{\partial x^2}$$

Therefore, the weighted scalar product taken with respect to the stiffness operator is given by

$$\int_0^L W^{(I)} \mathbb{k} W^{(J)} dx = \int_0^L W^{(I)} \frac{d^2}{dx^2} k_b \frac{d^2}{dx^2} W^{(J)} dx \tag{10.134}$$

Integrating Eq. (10.134) by parts four times results in the identity

$$\int_0^L W^{(I)} \mathbb{k} W^{(J)} dx = \int_0^L W^{(J)} \mathbb{k} W^{(I)} dx + B_{\mathbb{k}}^{(IJ)} \tag{10.135}$$

where

$$B_{\mathbb{k}}^{(IJ)} = \left[W^{(I)} \frac{d}{dx} \left\{ EI W^{(J)\prime\prime} \right\} \right]_0^L - \left[W^{(I)\prime} EI W^{(J)\prime\prime} \right]_0^L$$
$$+ \left[EI W^{(I)\prime\prime} W^{(J)\prime} \right]_0^L - \left[W^{(J)} \frac{d}{dx} \left\{ W^{(I)\prime\prime} EI \right\} \right]_0^L \tag{10.136}$$

From the discussion of Section 10.8.1 for systems with a scalar mass operator, we need only show that $B_{\mathbb{k}}^{(IJ)} = 0$ to establish the mutual orthogonality of the modal functions for Euler-Bernoulli Beams. Recall from Section 9.6 that for Euler-Bernoulli Beams the bending moment and transverse shear force are related to the transverse displacement by the identities

$$M(x,t) = -EI \frac{\partial^2 w}{\partial x^2} \tag{10.137}$$

$$Q(x,t) = \frac{\partial M}{\partial x} = -\frac{\partial}{\partial x} \left\{ EI \frac{\partial^2 w}{\partial x^2} \right\} \tag{10.138}$$

Further, the free vibration response is of the general form

$$w(x,t) = \sum_{j=1}^{N} W^{(j)}(x)e^{i\omega_j t} \tag{10.139}$$

where $W^{(j)}(x)$ corresponds to the jth modal function. Substitution of Eq. (10.139) into Eqs. (10.137) and (10.138) results in the relations

$$M(x,t) = \sum_{j=1}^{\infty} \tilde{M}^{(j)}(x)e^{i\omega_j t} \tag{10.140}$$

and

$$Q(x,t) = \sum_{j=1}^{\infty} \tilde{Q}^{(j)}(x)e^{i\omega_j t} \tag{10.141}$$

where

$$\tilde{M}^{(j)}(x) = -EI\,W^{(j)\prime\prime}(x) \tag{10.142}$$

is the *modal bending moment* and

$$\tilde{Q}^{(j)}(x) = \tilde{M}^{(j)\prime}(x) = -\frac{d}{dx}\left\{EI\,W^{(j)\prime\prime}(x)\right\} \tag{10.143}$$

is the *modal shear force*. Incorporating Eqs. (10.142) and (10.143) into Eqs. (10.135) and (10.136) yields the condition for orthogonality,

$$B_{lk}^{(lj)} = \left[\tilde{Q}^{(j)}W^{(i)} - \tilde{Q}^{(i)}W^{(j)}\right]_{x=L} - \left[\tilde{Q}^{(j)}W^{(i)} - \tilde{Q}^{(i)}W^{(j)}\right]_{x=0}$$
$$- \left[\tilde{M}^{(j)}W^{(i)\prime} - \tilde{M}^{(i)}W^{(j)\prime}\right]_{x=L} + \left[\tilde{M}^{(j)}W^{(i)\prime} - \tilde{M}^{(i)}W^{(j)\prime}\right]_{x=0} = 0 \tag{10.144}$$

The modal functions for an Euler-Bernoulli Beam are mutually orthogonal if Eq. (10.144) is satisfied. It may be seen that this is so if the work of the modal shear and modal moment for the lth mode going through the deflections and rotations for the jth mode is equal to the work of the modal shear and moment for the jth mode going through the deflections and rotations for the lth mode. It is observed that Eq. (10.144) is identically satisfied for the homogeneous boundary conditions

$$W(0) = 0 \quad \text{or} \quad \tilde{Q}(0) = 0 \tag{10.145}$$

$$W'(0) = 0 \quad \text{or} \quad \tilde{M}(0) = 0 \tag{10.146}$$

$$W(L) = 0 \quad \text{or} \quad \tilde{Q}(L) = 0 \tag{10.147}$$

$$W'(L) = 0 \quad \text{or} \quad \tilde{M}(L) = 0 \tag{10.148}$$

Structures with other boundary conditions must be considered individually.

In summary, if the boundary conditions of an Euler-Bernoulli Beam are such that Eq. (10.144) is satisfied, then the modal functions are mutually orthogonal as follows:

$$\left\langle W^{(i)}, W^{(j)} \right\rangle_m = \int_0^L W^{(i)}(x) m(x) W^{(j)}(x) \, dx = 0 \tag{10.149}$$

$$(\omega_i^2 \neq \omega_j^2)$$

$$\left\langle W^{(i)}, W^{(j)} \right\rangle_k = \int_0^L W^{(i)}(x) \frac{d^2}{dx^2} \left\{ EI W^{(j)''} \right\}(x) \, dx = 0 \tag{10.150}$$

Example 10.18

Consider an Euler-Bernoulli Beam that is clamped-fixed on its left edge and sits on an elastic foundation of stiffness k_L at its right edge as shown. Show that the modal functions are mutually orthogonal.

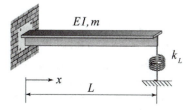

Solution
The physical boundary conditions for the beam are

$$w(0,t) = 0 , \quad \left. \frac{\partial w}{\partial x} \right|_{x=0} = 0 \tag{a-1, 2}$$

$$M(L,t) = 0 , \quad Q(L,t) = k_L w(L,t) \tag{a-3, 4}$$

The spring force imparted by the deflected spring on the right edge of the beam is equivalent to an applied shear force. The last condition then follows from the sign convention introduced in Section 9.6 (Figure 9.12). This condition also arises for the structure of Example 9.8. The boundary conditions for the modal functions are determined by substituting the assumed form of the solution, Eq. (10.28), into the physical boundary conditions established above. Hence,

$$W(0)e^{i\omega t} \quad \Rightarrow \quad W(0) = 0 \tag{b-1}$$

$$W'(0)e^{i\omega t} \quad \Rightarrow \quad W'(0) = 0 \tag{b-2}$$

$$M(L,t) = 0 = \tilde{M}(L)e^{i\omega t} \quad \Rightarrow \quad \tilde{M}(L) = 0 \tag{b-3}$$

$$\tilde{Q}(L)e^{i\omega t} = k_L W(L)e^{i\omega t} \quad \Rightarrow \quad \tilde{Q}(L) = k_L W(L) \tag{b-4}$$

Substituting Eqs. (b-1)–(b-4) into the left hand side of Eq. (10.144) gives

$$\left[\tilde{Q}^{(l)} \cdot 0 - \tilde{Q}^{(j)} \cdot 0 \right]_{x=0} - \left[\left(k_L W^{(l)} \right) W^{(j)} - \left(k_L W^{(j)} \right) W^{(l)} \right]_{x=L}$$
$$+ \left[0 \cdot W^{(j)'} - 0 \cdot W^{(l)'} \right]_{x=L} - \left[\tilde{M}^{(l)} \cdot 0 - \tilde{M}^{(j)} \cdot 0 \right]_{x=0} = 0 \tag{e}$$

The left-hand side of Eq. (e) clearly vanishes. Equation (10.144) is therefore satisfied and the corresponding modal functions are thus mutually orthogonal with respect to both m and k.

Beam-Columns with Constant Membrane Force

We next consider geometrically nonlinear beams with constant membrane force. (Structures for which the membrane force is negative are typically referred to as beam-columns.) For structures of this type, the coupling between the axial forces and transverse motion is accounted for, though the axial motion is considered much smaller than the transverse motion as discussed in Section 9.7. For these structures, the stiffness operator for the linear beam is augmented by the contribution of the component of the constant membrane force, N_0, in the transverse direction (Section 9.7). From Eq. (9.167), the stiffness operator for the geometrically nonlinear beam is

$$k = \frac{\partial^2}{\partial x^2} EI \frac{\partial^2}{\partial x^2} - N_0 \frac{\partial^2}{\partial x^2}$$

Using this operator in Eq. (10.135) and integrating by parts adds the following terms to Eq.(10.136),

$$\left[W^{(l)'} N_0 W^{(j)} - W^{(l)} N_0 W^{(j)'} \right]_0^L + \int_0^L W^{(l)''} N_0 W^{(j)} dx \tag{10.151}$$

For the nonlinear beam, the transverse shear is given by

$$Q(x,t) = -\frac{\partial}{\partial x} EI \frac{\partial^2 w}{\partial x^2} - N_0 \frac{\partial w}{\partial x} \tag{10.152}$$

(See Section 9.7.) Substitution of the free vibration response, Eq. (10.139) into Eq. (10.152) gives the corresponding modal shear force

$$\tilde{Q}^{(j)}(x) = -\frac{d}{dx}\left\{EI\,W^{(j)''}(x)\right\} - N_0 W^{(j)'}(x) \tag{10.153}$$

The modal moment remains as given by Eq. (10.140). Paralleling the remainder of the development for linear beams yields the identical form for the modal moment given by Eq. (10.142) and the corresponding condition for orthogonality, Eq. (10.144). However, it is understood that the modal shear is now given by Eq. (10.153). This also applies to the homogeneous conditions defined by Eqs. (10.145) – (10.148).

In summary, if the boundary conditions of a beam-column are such that Eq. (10.144) is satisfied, then the modal functions are mutually orthogonal as follows:

$$\left\langle W^{(l)}, W^{(j)} \right\rangle_m = \int_0^L W^{(l)}(x)\, m(x)\, W^{(j)}(x)\, dx = 0 \tag{10.154}$$

$$(\omega_l^2 \neq \omega_j^2)$$

$$\left\langle W^{(l)}, W^{(j)} \right\rangle_k = \int_0^L W^{(l)}(x)\left[\frac{d^2}{dx^2}\left\{EI\,W^{(j)''}\right\} + N_0\,W^{(j)''}\right] dx = 0 \tag{10.155}$$

Example 10.19

Let the simply supported beam-column of Example 9.12 be subjected to the static compressive load P_0 as indicated. Show that the corresponding modal functions are mutually orthogonal.

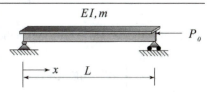

Solution
To obtain the modal boundary conditions we substitute the general form of the free vibration response of the beam column into From Eqs. (a)–(d) of Example 9.12. Hence,

$$w(0,t) = 0 = W(0)e^{i\omega t} \quad \Rightarrow \quad W(0) = 0 \tag{a}$$

$$M(0,t) = 0 = \tilde{M}(0)e^{i\omega t} \quad \Rightarrow \quad \tilde{M}(0) = 0 \tag{b}$$

$$w(L,t) = 0 = W(L)e^{i\omega t} \quad \Rightarrow \quad W(L) = 0 \tag{c}$$

$$M(L,t) = 0 = \tilde{M}(L)e^{i\omega t} \quad \Rightarrow \quad \tilde{M}(L) = 0 \tag{d}$$

Upon substitution of Eqs. (a)–(d) into Eq. (10.144) it is seen that the identity is satisfied. Therefore, the modal functions are mutually orthogonal in the sense of Eqs. (10.100) and (10.101).

10.8.4 Rayleigh Beams

In this section we establish the definitions of, and conditions for, the mutual orthogonality of the modal functions for Rayleigh Beams. Since the mass operator for these structures is a differential, rather than a scalar, operator we must also establish the explicit condition for the corresponding scalar product to be commutative. From Section 9.6.4 we have that the mass operator for Rayleigh Beams is

$$\mathbb{m} = m(x) - \frac{\partial}{\partial x} I_\rho(x) \frac{\partial}{\partial x} \tag{10.156}$$

The weighted scalar product of the lth and j^{th} modal functions taken with respect to the mass operator is then

$$\int_0^L W^{(l)} \mathbb{m} W^{(j)} dx = \int_0^L W^{(l)} m W^{(j)} dx - \int_0^L W^{(l)} \left(I_\rho W^{(j)\prime} \right)' dx \tag{10.157}$$

Integrating the second expression on the right hand side by parts gives the identity

$$\int_0^L W^{(l)} \mathbb{m} W^{(j)} dx = -\left[W^{(l)} I_\rho W^{(j)\prime} \right]_0^L$$
$$+ \int_0^L \left[W^{(l)} m W^{(j)} + W^{(l)\prime} I_\rho W^{(j)\prime} \right] dx \tag{10.158}$$

Similarly,

$$\int_0^L W^{(j)} \mathbb{m} W^{(l)} dx = -\left[W^{(j)} I_\rho W^{(l)\prime} \right]_0^L$$
$$+ \int_0^L \left[W^{(j)} m W^{(l)} + W^{(j)\prime} I_\rho W^{(l)\prime} \right] dx \tag{10.159}$$

The bending moment for a Rayleigh Beam is given by the same constitutive relation as for Euler-Bernoulli Beams, Eq. (10.137). Recall from Section 9.6.2, however, that the constitutive relation for the transverse shear force is modified due to the effects of the rotatory inertia. Rewriting Eq. (9.96) we have

$$Q(x,t) = \frac{\partial M(x,t)}{\partial x} + I_\rho \frac{\partial^2}{\partial t^2} \frac{\partial w}{\partial x} \tag{10.160}$$

Substituting the general form of the free vibration response, Eq. (10.139), into Eq. (10.160) results in a relation of the form of Eq. (10.141), where now

$$\tilde{Q}^{(j)}(x) = \tilde{M}^{(j)'}(x) - \omega_j^2 I_\rho W^{(j)'}(x) \tag{10.161}$$

and $\tilde{M}^{(j)}$ is defined by Eq. (10.142). Substituting Eqs. (10.135), (10.136), (10.158) and (10.159) into Eq. (10.96) and incorporating Eqs. (10.142) and (10.161) results in the identity

$$-\left[\tilde{Q}^{(j)} W^{(l)} - \tilde{Q}^{(l)} W^{(j)} \right]_0^L + \left[\tilde{M}^{(j)} W^{(l)'} - \tilde{M}^{(l)} W^{(j)'} \right]_0^L$$
$$= \left(\omega_j^2 - \omega_l^2 \right) \int_0^L \left[W^{(j)} m W^{(l)} + W^{(j)'} I_\rho W^{(l)'} \right] dx \tag{10.162}$$

It may be seen that if

$$\left[\tilde{M}^{(j)} W^{(l)'} - \tilde{M}^{(l)} W^{(j)'} \right]_0^L - \left[\tilde{Q}^{(j)} W^{(l)} - \tilde{Q}^{(l)} W^{(j)} \right]_0^L = 0 \tag{10.163}$$

where the modal shear force is defined by Eq. (10.161), then Eq. (10.162) reduces to the statement

$$\left(\omega_j^2 - \omega_l^2 \right) \int_0^L \left[W^{(j)} m W^{(l)} + W^{(j)'} I_\rho W^{(l)'} \right] dx = 0$$

It then follows that, for distinct natural frequencies,

$$\int_0^L \left[W^{(j)} m W^{(l)} + W^{(j)'} I_\rho W^{(l)'} \right] dx = 0 \tag{10.164}$$

The modal functions for Rayleigh Beams are mutually orthogonal with respect to the mass in this sense. It is convenient to express Eq. (10.164) in the equivalent matrix form

$$\left\langle \mathbf{W}^{(j)}, \mathbf{W}^{(l)} \right\rangle_{\mathbf{m}} = \int_0^L \mathbf{W}^{(j)\mathsf{T}} \mathbf{m} \mathbf{W}^{(l)} dx = 0 \tag{10.165}$$

where

$$\mathbf{m} = \begin{bmatrix} m & 0 \\ 0 & I_\rho \end{bmatrix} \tag{10.166}$$

and

$$\mathbf{W}^{(j)} = \begin{Bmatrix} W^{(j)}(x) \\ W^{(j)\prime}(x) \end{Bmatrix} \tag{10.167}$$

The matrix form reveals the general nature of the statement of orthogonality with respect to the mass for Rayleigh Beams. The corresponding statement of orthogonality with respect to the stiffness operator is obtained next.

The eigenvalue problem for Rayleigh Beams may be stated in the form

$$\mathbb{k} W^{(j)} = \omega_j^{\,2} \mathbb{m} W^{(j)} \tag{10.168}$$

where

$$\mathbb{k} = \frac{\partial^2}{\partial x^2} EI \frac{\partial^2}{\partial x^2} \tag{10.169}$$

and $\mathbb{m}$ is given by Eq. (10.156). Multiplying Eq. (10.168) by $W^{(l)}$ and integrating over the domain of definition of the beam gives the identity

$$\int_0^L W^{(l)} \mathbb{k} W^{(j)} dx = \omega_j^{\,2} \int_0^L W^{(l)} \mathbb{m} W^{(j)} dx \tag{10.170}$$

Integrating the left-hand side of the equation by parts twice results in the relation

$$\int_0^L W^{(l)\prime\prime} EI W^{(j)\prime\prime} dx + \tilde{B}_{\mathbb{k}}^{(lk)} = \omega_j^{\,2} \int_0^L \left[W^{(j)} m W^{(l)} + W^{(j)\prime} I_\rho W^{(l)\prime} \right] dx \tag{10.171}$$

where

$$\tilde{B}_{\mathbb{k}}^{(lj)} = \left[W^{(l)} \tilde{Q}^{(j)} + W^{(l)\prime} \tilde{M}^{(j)} \right]_0^L$$

Now, the condition for the modal functions to be orthogonal with respect to the mass, Eq. (10.163), is equivalent to the statement $\tilde{B}_{\mathbb{k}}^{(lj)} = 0$. Furthermore, when this condition holds, the right hand side of Eq. (10.171) vanishes by virtue of Eq. (10.164). Equation (10.171) therefore reduces to the statement

$$\int_0^L W^{(l)''} EI\, W^{(j)''}\, dx = 0 \qquad (10.172)$$

The modal functions of Rayleigh Beams are mutually orthogonal with respect to the stiffness in this sense. Specifically, the modal curvatures are seen to be mutually orthogonal with respect to the bending stiffness of the beam. In summary, if the boundary conditions of a given beam are such that Eq. (10.163) is satisfied, where the modal shear force is given by Eq. (10.161), then the modal functions are mutually orthogonal with respect to the mass in the sense of Eq. (10.164), or equivalently Eqs. (10.165), and with respect to the stiffness in the sense of Eq. (10.172).

Example 10.20

Consider a Rayleigh Beam that is clamped-fixed at its left edge and sits on an elastic mount of stiffness k_L at its right edge as shown. Show that the modal functions are mutually orthogonal.

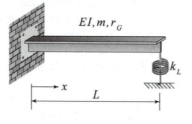

Solution
The analysis directly parallels that of Example 10.12. We first obtain the boundary conditions for the modal functions from the physical boundary conditions. Hence,

$$w(0,t) = 0 = W(0)e^{i\omega t} \quad \Rightarrow \quad W(0) = 0 \qquad (a)$$

$$\left.\frac{\partial w}{\partial x}\right|_{x=0} = 0 = W'(0)e^{i\omega t} \quad \Rightarrow \quad W'(0) = 0 \qquad (b)$$

$$M(L,t) = 0 = \tilde{M}(L)e^{i\omega t} \quad \Rightarrow \quad \tilde{M}(L) = 0 \qquad (c)$$

$$Q(L,t) = k_L w(L,t) \;\rightarrow\; \tilde{Q}(L)e^{i\omega t} = k_L W(L)e^{i\omega t} \quad \Rightarrow \quad \tilde{Q}(L) = k_L W(L) \qquad (d)$$

Substituting Eqs. (a)–(d) into the left-hand side of Eq. (10.163) gives

$$\left[\tilde{Q}^{(l)}\cdot 0-\tilde{Q}^{(j)}\cdot 0\right]_{x=0}-\left[\left(k_L W^{(l)}\right)W^{(j)}-\left(k_L W^{(j)}\right)W^{(l)}\right]_{x=L}$$
$$+\left[0\cdot W^{(j)'}-0\cdot W^{(l)'}\right]_{x=L}-\left[\tilde{M}^{(l)}\cdot 0-\tilde{M}^{(j)}\cdot 0\right]_{x=0}=0 \tag{e}$$

Equation (10.163) is clearly satisfied, therefore the corresponding modal functions for the Rayleigh Beam are mutually orthogonal in sense of Eqs. (10.164) and (10.172).

10.8.5 Timoshenko Beams

The motion of Timoshenko Beams is characterized by the displacement function matrix defined by Eq. (10.63) whose elements correspond to the transverse displacement function and the cross-section rotation function. The corresponding mass operator is the diagonal matrix operator defined by Eq. (10.61) and the stiffness operator is the differential matrix operator defined by Eq. (10.62). Establishment of the definitions and conditions for orthogonality of the modal functions therefore requires a generalization of the concepts introduced to this point, in the spirit of those introduced for Rayleigh Beams.

The eigenvalue problem for Timoshenko Beams defined by Eq. (10.67) may be written in the compact form

$$\left[\mathbf{k}-\omega^2\mathbf{m}\right]\mathbf{U}(x)=\mathbf{0} \tag{10.173}$$

where $\mathbf{m}$, $\mathbf{k}$ and $\mathbf{U}$ are defined by Eqs. (10.61), (10.62) and (10.63), respectively. Let us consider any two frequency-mode pairs for a generic Timoshenko Beam. Since they correspond to solutions of the eigenvalue problem, the lth and j^{th} frequency-mode pairs must each satisfy Eq. (10.173). Hence,

$$\mathbf{k}\mathbf{U}^{(l)}(x)=\omega_l^2\mathbf{m}\mathbf{U}^{(l)}(x) \tag{10.174}$$

and

$$\mathbf{k}\mathbf{U}^{(j)}(x)=\omega_j^2\mathbf{m}\mathbf{U}^{(j)}(x) \tag{10.175}$$

Multiplying Eq. (10.174) by $\mathbf{U}^{(j)\mathrm{T}}$, Eq. (10.175) by $\mathbf{U}^{(l)\mathrm{T}}$, and integrating the resulting expressions over $[0, L]$ results in the identities

$$\int_0^L \mathbf{U}^{(j)\mathrm{T}}\mathbf{k}\mathbf{U}^{(l)}dx=\omega_l^2\int_0^L \mathbf{U}^{(j)\mathrm{T}}\mathbf{m}\mathbf{U}^{(l)}dx \tag{10.176}$$

and

$$\int_0^L \mathbf{U}^{(j)\mathsf{T}} \mathbf{k} \mathbf{U}^{(l)} dx = \omega_j^2 \int_0^L \mathbf{U}^{(j)\mathsf{T}} \mathbf{m} \mathbf{U}^{(l)} dx \tag{10.177}$$

Note that, since **m** is a diagonal matrix it follows that

$$\mathbf{U}^{(l)\mathsf{T}} \mathbf{m} \mathbf{U}^{(j)} = \mathbf{U}^{(j)\mathsf{T}} \mathbf{m} \mathbf{U}^{(l)} \tag{10.178}$$

Subtracting Eq. (10.177) from Eq. (10.176) and incorporating Eq. (10.178) results in the identity

$$\int_0^L \mathbf{U}^{(j)\mathsf{T}} \mathbf{k} \mathbf{U}^{(l)} dx - \int_0^L \mathbf{U}^{(l)\mathsf{T}} \mathbf{k} \mathbf{U}^{(j)} dx$$

$$= \left(\omega_l^2 - \omega_j^2 \right) \int_0^L \mathbf{U}^{(j)\mathsf{T}} \mathbf{m} \mathbf{U}^{(l)} dx \tag{10.179}$$

It may be seen that if

$$\int_0^L \mathbf{U}^{(j)\mathsf{T}} \mathbf{k} \mathbf{U}^{(l)} dx = \int_0^L \mathbf{U}^{(l)\mathsf{T}} \mathbf{k} \mathbf{U}^{(j)} dx \tag{10.180}$$

then Eq. (10.179) reduces to the statement

$$\left(\omega_l^2 - \omega_j^2 \right) \int_0^L \mathbf{U}^{(j)\mathsf{T}} \mathbf{m} \mathbf{U}^{(l)} dx = 0 \tag{10.181}$$

Thus, for distinct frequencies,

$$\left\langle \mathbf{U}^{(j)}, \mathbf{U}^{(l)} \right\rangle_{\mathbf{m}} \equiv \int_0^L \mathbf{U}^{(j)\mathsf{T}} \mathbf{m} \mathbf{U}^{(l)} dx = 0 \tag{10.182}$$

and *the modal functions are mutually orthogonal with respect to the mass operator* in this sense. Substitution of Eq. (10.182) into Eq. (10.176) results in the related statement

$$\left\langle \mathbf{U}^{(j)}, \mathbf{U}^{(l)} \right\rangle_{\mathbf{k}} \equiv \int_0^L \mathbf{U}^{(j)\mathsf{T}} \mathbf{k} \mathbf{U}^{(l)} dx = 0 \tag{10.183}$$

Hence, if Eq. (10.180) is satisfied then, *the modal functions are mutually orthogonal with respect to the stiffness operator* in the above sense as well. To examine the details and implications for specific systems we next evaluate the above conditions for the pertinent mass and stiffness operators.

Substitution of Eqs. (10.61) and (10.63) into Eq. (10.182) and substituting Eqs. (10.62) and (10.63) into Eq. (10.183) and carrying through the matrix multiplication in each gives the explicit forms of the statements of orthogonality for the modal functions of Timoshenko Beams as

$$\left\langle \mathbf{U}^{(j)}, \mathbf{U}^{(i)} \right\rangle_m = \int_0^L \left[W^{(i)}(x)\, m(x)\, W^{(j)}(x) + \vartheta^{(i)}(x)\, I_\rho(x)\, \vartheta^{(j)}(x) \right] dx = 0 \qquad (10.184)$$

and

$$\int_0^L \mathbf{U}^{(i)} \mathbf{k} \mathbf{U}^{(j)}\, dx = \int_0^L W^{(i)} \left\{ k_s \vartheta^{(j)'} - \left(k_s W^{(j)'} \right)' \right\} dx$$

$$+ \int_0^L \vartheta^{(i)} \left\{ k_s \left(\vartheta^{(j)} - W^{(j)'} \right) - \left(EI\vartheta^{(j)'} \right)' \right\} dx = 0 \qquad (10.185)$$

From Eqs. (9.130, (9.131) and (9.132) the constitutive relations for the bending moment and the transverse shear force for a Timoshenko Beam are respectively given by the relations

$$M(x,t) = -EI\frac{\partial \varphi}{\partial x} \qquad (10.186)$$

and

$$Q(x,t) = k_s \left[\frac{\partial w(x,t)}{\partial x} - \varphi(x,t) \right] \qquad (10.187)$$

From Eq. (10.65), the free vibration response is of the form

$$\begin{Bmatrix} w(x,t) \\ \varphi(x,t) \end{Bmatrix} = \begin{Bmatrix} W(x) \\ \vartheta(x) \end{Bmatrix} e^{i\omega t} \qquad (10.188)$$

Substitution of the above form into Eqs. (10.186) and (10.187) gives

$$M(x,t) = \tilde{M}(x)e^{i\omega t} \qquad (10.189)$$

and

$$Q(x,t) = \tilde{Q}(x)e^{i\omega t} \qquad (10.190)$$

where

$$\tilde{M}(x) = -EI\vartheta(x) \qquad (10.191)$$

is the modal moment, and

$$\tilde{Q}(x) = k_s \left[W'(x) - \vartheta(x) \right] \qquad (10.192)$$

is the modal shear force. With the modal moment and modal shear established, we may proceed to evaluate the explicit form of the condition for the modes to be mutually orthogonal. Substitution of Eqs. (10.62) and (10.63) into Eq. (10.180), integrating the resulting expression by parts and incorporating Eqs. (10.191) and (10.192) renders the explicit condition for mutual orthogonality of the modal functions for Timoshenko Beams to the familiar form

$$\left[W^{(l)}\tilde{Q}^{(j)} - \tilde{Q}^{(l)}W^{(j)}\right]_0^L - \left[\mathcal{G}^{(l)}\tilde{M}^{(j)} - \tilde{M}^{(l)}\mathcal{G}^{(j)}\right]_0^L = 0 \qquad (10.193)$$

If the boundary conditions for a Timoshenko Beam are such that Eq. (10.193) is satisfied, then the corresponding modal functions are mutually orthogonal in the sense of Eqs. (10.182) and (10.183), or equivalently Eqs. (10.184) and (10.185).

Example 10.21

Consider a Timoshenko Beam supported as in Examples 10.18 and 10.20. Show that the modal functions for the beam are mutually orthogonal.

Solution
Proceeding identically as in Examples 10.18 and 10.20 we first state the physical boundary conditions for the beam. They are

$$w(0,t) = 0 , \quad \varphi(0,t) = 0 \qquad (a\text{-}1, 2)$$

$$M(L,t) = 0 , \quad Q(L,t) = k_L w(L,t) \qquad (a\text{-}1, 2)$$

The boundary conditions for the modal functions are obtained by substituting the assumed form of the modal functions, Eq. (10.188), into Eqs. (a-1)–(a-4). We thus have the corresponding conditions

$$W(0) = 0 , \quad \mathcal{G}(0) = 0 \qquad (b\text{-}1, 2)$$

$$\tilde{M}(L) = 0 , \quad \tilde{Q}(L) = k_L W(L) \qquad (b3, 4)$$

Substituting Eqs. (b-1)–(b-4) into the left-hand side of Eq. (10.193) results in the statement

$$\left[\tilde{Q}^{(l)}\cdot 0 - \tilde{Q}^{(j)}\cdot 0\right]_{x=0} - \left[\left(k_L W^{(l)}\right)W^{(j)} - \left(k_L W^{(j)}\right)W^{(l)}\right]_{x=L}$$
$$+ \left[0\cdot\mathcal{G}^{(j)} - 0\cdot\mathcal{G}^{(l)}\right]_{x=L} - \left[\tilde{M}^{(l)}\cdot 0 - \tilde{M}^{(j)}\cdot 0\right]_{x=0} = 0$$

Equation (10.193) is clearly satisfied. The corresponding modal functions of the Timoshenko Beam are therefore mutually orthogonal with respect to both **m** and **k**.

10.9 EVALUATION OF AMPLITUDES AND PHASE ANGLES

The free vibration response of one-dimensional continua was seen to be expressed as a series of the modal functions with harmonic time signatures. In each case the amplitudes and phase angles are a function of the specific initial conditions imposed on the particular system under consideration. In this section we establish the relations between the amplitudes and phase angles for the systems considered in this chapter. We begin by establishing the relations for systems with a single scalar mass operator. These systems include second order systems and Euler-Bernoulli Beams and geometrically nonlinear beams with constant axial loads. We then establish the conditions for Rayleigh Beams and Timoshenko Beams in separate sections.

10.9.1 Systems Possessing a Single Scalar Mass Operator

We here consider systems whose mass description corresponds to a single scalar function. These include mathematical models that describe the longitudinal and torsional motion of elastic rods, the transverse motion of strings and cables, and the flexural motion of Euler-Bernoulli Beams and geometrically nonlinear beams with constant membrane force. In each case, the free vibration response is of the general form

$$u(x,t) = \sum_{j=1}^{\infty} U^{(j)}(x) \left[A_1^{(j)} \cos \omega_j t + A_2^{(j)} \sin \omega_j t \right]$$

$$= \sum_{j=1}^{\infty} U^{(j)}(x) A^{(j)} \cos(\omega_j t - \phi_j)$$

$$(10.194)$$

where

$$A^{(j)} = \sqrt{A_1^{(j)2} + A_2^{(j)2}}, \quad \phi_j = \tan^{-1}\left(A_2^{(j)} \big/ A_1^{(j)} \right) \qquad (10.195)$$

$u(x,t)$ is the pertinent displacement measure, ω_j and $U^{(j)}(x)$ respectively correspond to the jth natural frequency and modal function, and $A^{(j)}$ and ϕ_j are the associated amplitude and phase angle.

General Initial Conditions

We wish to evaluate the amplitudes and phase angles in terms of the initial conditions

$$u(x,0) = u_0(x) \quad \text{and} \quad \left. \frac{\partial u}{\partial t} \right|_{t=0} = v_0(x) \qquad (10.196)$$

Imposing the initial conditions on the general form of the response gives the relations

$$u_0(x) = \sum_{j=1}^{\infty} U^{(j)}(x) A_1^{(j)}$$

$$v_0(x) = \sum_{j=1}^{\infty} U^{(j)}(x) \omega_j A_2^{(j)}$$

Let us next multiply the above relations by the product of the lth modal function, $U^{(l)}(x)$, and the scalar mass operator, $m(x)$, and integrate the resulting expressions over the domain of definition of the structure $[0, L]$. Doing this yields the identities

$$\int_0^L U^{(l)}(x) m(x) u_0(x) dx = \sum_{j=1}^{\infty} \left\{ \int_0^L U^{(l)}(x) m(x) U^{(j)}(x) dx \right\} A_1^{(j)}$$

$$\int_0^L U^{(l)}(x) m(x) v_0(x) dx = \sum_{j=1}^{\infty} \left\{ \int_0^L U^{(l)}(x) m(x) U^{(j)}(x) dx \right\} \omega_j A_2^{(j)}$$

In each of the above identities, the term in brackets may be recognized as the scalar product of the lth and j^{th} modal functions. If the modal functions are mutually orthogonal, that product vanishes for all terms in the series except for the term where $j = l$. The nonvanishing term is the square of the magnitude of the modal function. The above identities therefore reduce to the relations

$$A_1^{(j)} = \Lambda^{(j)} \tag{10.197}$$

$$A_2^{(j)} = X^{(j)} \tag{10.198}$$

where

$$\Lambda^{(j)} = \frac{1}{\left\| U^{(j)} \right\|_m^2} \int_0^L U^{(j)}(x) m(x) u_0(x) dx \tag{10.199}$$

$$X^{(j)} = \frac{1}{\omega_j \left\| U^{(j)} \right\|_m^2} \int_0^L U^{(j)}(x) m(x) v_0(x) dx \tag{10.200}$$

and

$$\left\| U^{(j)} \right\|_m^2 = \int_0^L U^{(j)}(x) m(x) U^{(j)}(x) dx \tag{10.201}$$

Substituting Eqs. (10.197) and (10.198) into Eqs. (10.195) gives the amplitudes and phase angles as

$$A^{(j)} = \sqrt{\Lambda^{(j)2} + X^{(j)2}} \tag{10.202}$$

and

$$\phi_j = \tan^{-1}\left(X^{(j)}/\Lambda^{(j)}\right) \tag{10.203}$$

where $\Lambda^{(j)}$ and $X^{(j)}$ are defined by Eqs. (10.199) and (10.200), and are evaluated for given initial displacements and velocities $u_0(x)$ and $v_0(x)$.

Systems Released from Rest

As a special case, let us consider systems that are initially at rest. For this case $v_0(x) = 0$ and thus, from Eq. (10.200), $X^{(j)} = 0$ ($j = 1, 2, \ldots$). We then have, from Eqs. (10.202) and (10.203) that

$$A^{(j)} = \Lambda^{(j)} = \frac{1}{\left\| U^{(j)} \right\|_m^2} \int_0^L U^{(j)}(x)\, m(x)\, u_0(x)\, dx \quad (j = 1, 2, \ldots) \tag{10.204}$$

and

$$\phi_j = 0 \tag{10.205}$$

Substituting these expressions into Eq. (10.194) gives the free vibration response of a system released from rest as

$$u(x,t) = \sum_{j=1}^{\infty} \Lambda^{(j)} U^{(j)}(x) \cos \omega_j t \tag{10.206}$$

Systems Initially in Motion at the Reference Configuration

Let us next consider the special case where the system is initially in motion while in the reference (undeformed) configuration. This may correspond to the situation where we start monitoring the motion at an instant when the body is in motion as it passes through the undeformed configuration, or when the system is "launched" from the this initial configuration. The latter may occur, for example, when a javelin is thrown, when a rocket is launched or when a vehicle is impacted and we monitor the motion of the body from the instant after it is released. In the present context, we thus consider a continuous system for which the initial displacement vanishes but the initial velocity is finite. For this case, $u_0(x) = 0$. It then follows from Eqs. (10.197) and (10.199) that $\Lambda^{(j)} = A_1^{(j)} = 0$ and hence, from Eqs. (10.194), (10.198) and (10.200), that the free vibration response of the structure is given by

$$u(x,t) = \sum_{j=1}^{\infty} U^{(j)}(x)\, X^{(j)} \sin \omega_j t = \sum_{j=1}^{\infty} U^{(j)}(x)\, X^{(j)} \cos(\omega_j t - \pi/2) \tag{10.207}$$

It is seen that, for this case, $A^{(j)} = X^{(j)}$ and $\phi_j = \pi/2 \ (j = 1, 2, ...)$.

Example 10.22

Determine the amplitudes and phase angles for the rod of Example 10.1 if it is released from rest from the configuration $u(x, 0) = \varepsilon_0 x$, where $\varepsilon_0 = P_0/EA$ is the uniform axial strain. This initial state corresponds to the deformation induced by a static tensile load of magnitude P_0 at the free end of the rod.

Solution

From Example 10.1, the modal functions for the rod are

$$U^{(j)}(x) = \sin \beta_j x \quad (j = 1, 2, ...) \tag{a-1}$$

where

$$\beta_j = (2j - 1)\pi/2L \quad (j = 1, 2, ...) \tag{a-2}$$

Further, the magnitude of the modal function was computed in Example 10.13 as

$$\left\| U^{(j)} \right\|^2 = \frac{mL}{2} \quad (j = 1, 2, ...) \tag{b}$$

Now, the initial conditions for the present case are

$$u_0(x) = \varepsilon_0 x, \quad v_0(x) = 0 \tag{c-1, 2}$$

It follows from Eqs. (10.204) and (10.205) that

$$\phi_j = 0 \quad (j = 1, 2, ...) \tag{$\triangleleft$ (d)}$$

and

$$A^{(j)} = \frac{2}{mL} \int_0^L \sin \beta_j x \ m \ \varepsilon_0 x \, dx = \frac{2\varepsilon_0}{\beta_j^2 L} \left[\sin \beta_j L - \beta_j L \cos \beta_j L \right]$$

The above expression is simplified when we recall the frequency equation for the rod, Eq. (e) of Example 10.1,

$$\cos \beta_j L = 0$$

The amplitudes of the vibrating rod are then

$$A^{(j)} = \frac{2\varepsilon_0}{\beta_j^2 L}\sin\beta_j L \quad (j=1,2,...) \qquad \triangleleft (e)$$

Example 10.23

Determine the amplitudes and phase angles for the cantilevered Euler-Bernoulli beam of Example 10.7 if it is released from rest from the configuration $w(x,0) = -\kappa_0 x^2/2$. Evaluate the resulting free vibration response. This initial configuration corresponds to the deflections produced by a bending moment, M_0, applied to the free end of the beam. The parameter $\kappa_0 = M_0/EI$ is the uniform curvature of the beam.

Solution

The modal functions for the beam were determined in Example 10.7 to be

$$W^{(j)}(x) = \cosh\beta_j x - \cos\beta_j x - Y_j\left[\sinh\beta_j x - \sin\beta_j x\right] \qquad (a)$$

where

$$Y_j = \frac{\cosh\beta_j L + \cos\beta_j L}{\sinh\beta_j L + \sin\beta_j L} \qquad (b)$$

The initial conditions are

$$w_0(x) = -\tfrac{1}{2}\kappa_0 x^2, \quad v_0(x) = 0 \qquad (c\text{-}1, 2)$$

Since the structure is initially at rest, we have from (10.205) that

$$\phi_j = 0 \qquad \triangleleft (c)$$

Further, substituting Eq. (c-1) into Eq. (10.204) gives

$$A^{(j)} = \frac{1}{\left\|W^{(j)}\right\|_m^2}\int_0^L \left\{-\tfrac{1}{2}\kappa_0 x^2\right\}mW^{(j)}(x)\,dx$$

The amplitudes of the vibrating cantilevered beam are then

$$A^{(j)} = -\frac{\kappa_0}{2} \frac{\displaystyle\int_0^L x^2\, W^{(j)}(x)\,dx}{\displaystyle\int_0^L W^{(j)\,2}(x)\,dx} \qquad (j=1,2,\ldots) \qquad \lhd \text{(d)}$$

where

$$\beta_j^{\,3} \int_0^L x^2\, W^{(j)}(x)\,dx = \beta_j^{\,3}\left\langle x^2, W^{(j)} \right\rangle$$

$$\begin{aligned}
&= \left(\beta_j L\right)^2 \left[\sinh\beta_j L - \sin\beta_j L \right] \\
&\quad - 2Y_j \left[\cosh\beta_j L - \cos\beta_j L \right] \\
&\quad - \left\{ 2\beta_j L + Y_j \right\}\left[\cosh\beta_j L + \cos\beta_j L \right] \\
&\quad + 2\left\{ \beta_j L Y_j + 1 \right\}\left[\sinh\beta_j L + \sin\beta_j L \right]
\end{aligned}$$

(e)

and

$$4\beta_j \int_0^L W^{(j)\,2}(x)\,dx = 4\beta_j \left\| W^{(j)} \right\|^2$$

$$\begin{aligned}
&= \sinh 2\beta_j L + \sin 2\beta_j L + 12\beta_j L \\
&\quad + 2Y_j \left[\cos 2\beta_j L - \cosh 2\beta_j L + 4\sinh\beta_j L \sin\beta_j L \right] \\
&\quad + Y_j^{\,2} \left[\sinh 2\beta_j L - \sin 2\beta_j L + 4\sinh 2\beta_j L \cos 2\beta_j L \right. \\
&\qquad\qquad \left. - 2\cosh 2\beta_j L \sin 2\beta_j L \right]
\end{aligned}$$

(f)

Let us now evaluate the response of the beam. To do this we recall that the first three roots of the frequency equation were computed in Example 10.7 to be

$$\beta L = 1.875,\ \ 4.695,\ \ 7.855,\ \ \ldots$$

Substitution of these values into Eqs. (e) and (f) and then substituting the resulting numbers into Eq. (d) gives the corresponding amplitudes

$$A^{(1)} = -0.1581\kappa_0 L^2,\quad A^{(2)} = 0.0039\kappa_0 L^2,\quad A^{(3)} = 0.0001881\kappa_0 L^2,\quad \ldots$$

The explicit form of the free vibration response of the beam is then

$$w(x,t) = -0.1581\kappa_0 L^2 \cos\left(3.516 t/\omega_0\right) + 0.0039\kappa_0 L^2 \cos\left(22.03 t/\omega_0\right) + \ldots \quad \lhd \text{(g)}$$

10.9.2 Rayleigh Beams

The free vibration response for Rayleigh Beams, and the associated initial conditions, are of the same general form as for the systems considered thus far. Hence,

$$w(x,t) = \sum_{j=1}^{\infty} W^{(j)}(x)\left[A_1^{(j)} \cos \omega_j t + A_2^{(j)} \sin \omega_j t \right] = \sum_{j=1}^{\infty} W^{(j)}(x) A^{(j)} \cos(\omega_j t - \phi_j)$$

where $w(x,t)$ is the transverse displacement, ω_j and $W^{(j)}(x)$ are the jth natural frequency and modal function, and $A^{(j)}$ and ϕ_j are the associated amplitude and phase angle. We wish to evaluate the amplitudes and phase angles in terms of the initial conditions

$$w(x,0) = w_0(x) \quad \text{and} \quad \left. \frac{\partial w}{\partial t} \right|_{t=0} = v_0(x)$$

For these structures, however, we have the added mass measure of the rotatory inertia as well as the mass per unit length of the beam. Because of this, it is convenient to describe the mass operator in the matrix form of Eq.(10.166), and to express the initial conditions in a similar form. Let us therefore introduce the displacement matrix

$$\mathbf{w}(x,t) = \left\{ \begin{array}{c} w(x,t) \\ \dfrac{\partial w}{\partial x} \end{array} \right\} \tag{10.208}$$

and the corresponding statement of the initial conditions

$$\mathbf{w}_0(x) = \left\{ \begin{array}{c} w_0(x) \\ w_0'(x) \end{array} \right\} \tag{10.209}$$

and

$$\mathbf{v}_0(x) = \left\{ \begin{array}{c} v_0(x) \\ v_0'(x) \end{array} \right\} \tag{10.210}$$

where $(\ )' = d(\ )/dx$. The free vibration response of the beam is expressed in matrix form as

$$\mathbf{w}(x,t) = \sum_{j=1}^{\infty} \mathbf{W}^{(j)}(x) \cos(\omega_j t - \phi_j) \tag{10.211}$$

where the matrix $\mathbf{W}^{(j)}$ is defined by Eq. (10.167). Imposing the initial conditions on the matrix form of the free vibration response gives the relations

$$\mathbf{w}_0(x) = \sum_{j=1}^{\infty} \mathbf{W}^{(j)}(x) A_1^{(j)}$$

and

$$\mathbf{v}_0(x) = \sum_{j=1}^{\infty} \mathbf{W}^{(j)}(x) A_2^{(j)}$$

Let us next multiply the above equations by the matrix product $\mathbf{W}^{(l)\mathrm{T}}(x)\,\mathbf{m}(x)$ and integrate the resulting expressions over the domain of definition of the beam. Doing this and utilizing the orthogonality relation, Eq. (10.165), results in the relations

$$A_1^{(j)} = \Lambda^{(j)} \tag{10.212}$$

$$A_2^{(j)} = \mathrm{X}^{(j)} \tag{10.213}$$

where

$$\Lambda^{(j)} = \frac{1}{\left\| \mathbf{W}^{(l)} \right\|_{\mathbf{m}}^2} \int_0^L \mathbf{W}^{(l)}(x)\,\mathbf{m}(x)\,\mathbf{w}_0(x)\,dx$$

$$= \frac{1}{\left\| \mathbf{W}^{(l)} \right\|_{\mathbf{m}}^2} \int_0^L \left\{ W^{(l)}(x)\,m(x)\,w_0(x) + W^{(l)\prime}(x)\,I_\rho(x)\,w_0{}'(x) \right\} dx \tag{10.214}$$

$$\mathrm{X}^{(l)} = \frac{1}{\left\| \mathbf{W}^{(l)} \right\|_{\mathbf{m}}^2} \int_0^L \mathbf{W}^{(l)}(x)\,\mathbf{m}(x)\,\mathbf{v}_0(x)\,dx$$

$$= \frac{1}{\left\| \mathbf{W}^{(l)} \right\|_{\mathbf{m}}^2} \int_0^L \left\{ W^{(l)}(x)\,m(x)\,v_0(x) + W^{(l)\prime}(x)\,I_\rho(x)\,v_0{}'(x) \right\} dx \tag{10.215}$$

and

$$\left\| \mathbf{W}^{(l)} \right\|_{\mathbf{m}}^2 = \int_0^L \mathbf{W}^{(l)}(x)\,\mathbf{m}(x)\,\mathbf{W}^{(l)}(x)\,dx$$

$$= \int_0^L \left\{ W^{(l)}(x)\,m(x)\,W^{(l)}(x) + W^{(l)\prime}(x)\,I_\rho(x)\,W^{(l)\prime}(x) \right\} dx \tag{10.216}$$

The amplitudes and phase angles are then given by the relations

$$A^{(j)} = \sqrt{\Lambda^{(j)2} + \mathrm{X}^{(j)2}} \tag{10.217}$$

and

$$\phi_j = \tan^{-1}\left(\mathrm{X}^{(j)} / \Lambda^{(j)} \right) \tag{10.218}$$

and may be evaluated for given initial displacements and velocities $w_0(x)$ and $v_0(x)$.

For *Rayleigh Beams that are released from rest* it follows, from the pertinent arguments of the previous section, that

$$\phi_j = 0 \quad (j = 1, 2, \ldots) \tag{10.219}$$

and

$$A^{(j)} = \frac{1}{\left\| \mathbf{W}^{(l)} \right\|_m^2} \int_0^L \left\{ W^{(l)}(x) m(x) w_0(x) + W^{(l)\prime}(x) I_\rho(x) w_0'(x) \right\} dx \tag{10.220}$$

The free vibration response for a Rayleigh Beam released from rest is then

$$w(x,t) = \sum_{j=1}^{\infty} \Lambda^{(j)} W^{(j)}(x) \cos \omega_j t \tag{10.221}$$

Example 10.24

Determine the amplitudes and phase angles for the cantilevered beam of Example 10.12 if it is released from rest from the configuration $w(x,0) = -\kappa_0 x^2/2$. As for the Euler-Bernoulli Beam of Example 10.23, this initial configuration corresponds to the deflections produced by a bending moment, M_0, applied to the free end of the beam and the parameter $\kappa_0 = M_0/EI$ corresponds to the uniform curvature of the structure.

Solution

The modal functions for the beam were determined in Example 10.12 to be

$$W^{(j)}(x) = \cosh \alpha_j x - \cos \beta_j x - Y^{(j)} \left[\beta_j \sinh \alpha_j x - \alpha_j \sin \beta_j x \right] \tag{a}$$

where

$$Y^{(j)} = \frac{1}{\beta_j} \frac{\left(\alpha_j^2 \cosh \alpha_j L + \beta_j^2 \cos \beta_j L \right)}{\left(\alpha_j^2 \sinh \alpha_j L + \alpha_j \beta_j \sin \beta_j L \right)} \tag{b}$$

The initial conditions are

$$w_0(x) = -\tfrac{1}{2}\kappa_0 x^2, \quad v_0(x) = 0 \tag{c-1, 2}$$

Since the structure is initially at rest, we have from Eq. (10.219) that

$$\phi_j = 0 \tag{◁ (d)}$$

Further, substituting Eq. (c-1) into Eq. (10.220) gives

$$A^{(j)} = \frac{1}{\left\| \mathbf{W}^{(j)} \right\|_m^2} \int_0^L \left[\left\{ -\tfrac{1}{2} \kappa_0 x^2 \right\} m W^{(j)}(x) + \int_0^L \left\{ -\kappa_0 x \right\} I_\rho W^{(j)\prime}(x)\, dx \right] dx$$

The amplitudes of the freely vibrating cantilevered Rayleigh Beam are then

$$A^{(j)} = -\kappa_0 \frac{\displaystyle\int_0^L \left[\tfrac{1}{2} x^2\, W^{(j)}(x) + r_G^2\, x\, W^{(j)\prime}(x) \right] dx}{\displaystyle\int_0^L \left[W^{(j)\,2}(x) + r_G^2\, W^{(j)\prime\,2}(x) \right] dx} \qquad (j = 1, 2, \ldots) \qquad \triangleleft \text{(e)}$$

where $W^{(j)}$ is given by Eq. (a) and r_G is the radius of gyration of the cross-section. The integrals can be evaluated analytically to give the explicit forms of the amplitudes.

10.9.3 Timoshenko Beams

The development for Timoshenko Beams is similar to that for Rayleigh Beams. However, for the present case, the motion is characterized by two displacement functions. Recall that the displacement matrix for Timoshenko Beams is

$$\mathbf{u}(x,t) = \begin{Bmatrix} w(x,t) \\ \varphi(x,t) \end{Bmatrix} \tag{10.222}$$

where $w(x,t)$ and $\varphi(x,t)$, correspond to the transverse deflection and cross-section rotation respectively. Since we now have two displacement functions, we must specify two initial conditions for each. Hence,

$$w(x,0) = w_0(x), \quad \left. \frac{\partial w}{\partial t} \right|_{t=0} = v_0(t)$$

$$\varphi(x,0) = \varphi_0(x), \quad \left. \frac{\partial \varphi}{\partial t} \right|_{t=0} = \chi_0(x)$$

which is written in matrix form as

$$\mathbf{u}(x,0) = \mathbf{u}_0(x) = \left\{ \begin{matrix} w_0(x) \\ \varphi_0(x) \end{matrix} \right\} \tag{10.223}$$

and

$$\frac{\partial \mathbf{u}}{\partial t}\bigg|_{t=0} = \mathbf{v}_0(x) = \left\{ \begin{matrix} v_0(x) \\ \chi_0(x) \end{matrix} \right\} \tag{10.224}$$

Now, the free vibration response of the beam is of the general form

$$\mathbf{u}(x,t) = \sum_{j=1}^{\infty} \mathbf{U}^{(j)}(x) \left[A_1^{(j)} \cos \omega_j t + A_2^{(j)} \sin \omega_j t \right]$$

$$= \sum_{j=1}^{\infty} \mathbf{U}^{(j)}(x) A^{(j)} \cos(\omega_j t - \phi_j) \tag{10.225}$$

Imposing the initial conditions on the time history of the response results in the identities

$$\mathbf{u}_0(x) = \sum_{j=1}^{\infty} \mathbf{U}^{(j)}(x) A_1^{(j)}$$

$$\mathbf{v}_0(x) = \sum_{j=1}^{\infty} \mathbf{U}^{(j)}(x) \omega_j A_2^{(j)}$$

where

$$\mathbf{U}^{(j)}(x) = \left\{ \begin{matrix} W^{(j)}(x) \\ \mathcal{G}^{(j)}(x) \end{matrix} \right\} \tag{10.226}$$

Paralleling the development for Rayleigh Beams with the present displacement variables, modal matrices and initial conditions gives the relations

$$A_1^{(j)} = \Lambda^{(j)} \tag{10.227}$$

$$A_2^{(j)} = \mathrm{X}^{(j)} \tag{10.228}$$

where, for Timoshenko Beams,

$$\Lambda^{(j)} = \frac{1}{\left\| \mathbf{U}^{(l)} \right\|_m^2} \int_0^L \mathbf{U}^{(l)}(x) \mathbf{m}(x) \mathbf{u}_0(x)\, dx$$

$$= \frac{1}{\left\| \mathbf{U}^{(l)} \right\|_m^2} \int_0^L \left\{ W^{(l)}(x) m(x) w_0(x) + \vartheta^{(l)}(x) I_\rho(x) \varphi_0(x) \right\} dx \tag{10.229}$$

$$\mathrm{X}^{(l)} = \frac{1}{\left\| \mathbf{U}^{(l)} \right\|_m^2} \int_0^L \mathbf{U}^{(l)}(x) \mathbf{m}(x) \mathbf{v}_0(x)\, dx$$

$$= \frac{1}{\left\| \mathbf{U}^{(l)} \right\|_m^2} \int_0^L \left\{ W^{(l)}(x) m(x) v_0(x) + \vartheta^{(l)}(x) I_\rho(x) \chi_0(x) \right\} dx \tag{10.230}$$

and

$$\left\| \mathbf{U}^{(l)} \right\|_m^2 = \int_0^L \mathbf{U}^{(l)}(x) \mathbf{m}(x) \mathbf{U}^{(l)}(x)\, dx$$

$$= \int_0^L \left\{ W^{(l)}(x) m(x) W^{(l)}(x) + \vartheta^{(l)}(x) I_\rho(x) \vartheta^{(l)}(x) \right\} dx \tag{10.231}$$

The amplitudes and phase angles are then given by

$$A^{(j)} = \sqrt{\Lambda^{(j)2} + \mathrm{X}^{(j)2}} \tag{10.232}$$

and

$$\phi_j = \tan^{-1}\left(\mathrm{X}^{(j)} / \Lambda^{(j)} \right) \tag{10.233}$$

and may be evaluated as described above for given initial displacements, rotations, velocities and rotation rates, $w_0(x)$, $\varphi_0(x)$, $v_0(x)$ and $\chi(x)$. It follows from the same reasoning as for other the structures considered in this chapter that, for Timoshenko Beams that are released from rest,

$$\phi_j = 0 \tag{10.234}$$

and

$$A^{(j)} = \frac{1}{\left\| \mathbf{U}^{(l)} \right\|_m^2} \int_0^L \left\{ W^{(l)}(x) m(x) w_0(x) + \vartheta^{(l)}(x) I_\rho(x) \varphi_0(x) \right\} dx \tag{10.235}$$

The free vibration response of a Timoshenko Beam released from rest then follows as

$$\mathbf{u}(x,t) = \sum_{j=1}^{\infty} \mathbf{U}^{(j)}(x) \Lambda^{(j)} \cos \omega_j t \tag{10.236}$$

Example 10.25

Determine the amplitudes and phase angles for the cantilevered beam of Example 10.13 if it is released from rest from the configuration $w(x,0) = -\kappa_0 x^2/2$, $\varphi(x,0) = -\kappa_0 x$. As for the beams of Examples 10.23 and 10.24, this initial configuration corresponds to the deflections produced by a bending moment, M_0, applied to the free end of the beam and the parameter $\kappa_0 = M_0/EI$ corresponds to the uniform curvature of the structure. [Note that, for this particular case $\varphi_0(x) = dw_0(x)/dx$ since there is no transverse shear in the structure initially (it is loaded in pure bending) and hence no shear deformation in the beam at that time. See Section 9.6.5, Eq. (9.130).]

Solution

The modal functions for the beam were determined in Example 10.13 to be

$$W^{(j)}(x) = \cosh \alpha_j x - \cos \beta_j x - Y^{(j)} \left[g_\beta^{(j)} \sinh \alpha_j x - g_\alpha^{(j)} \sin \beta_j x \right] \tag{a}$$

where

$$Y^{(j)} = \frac{\left(g_\alpha^{(j)} \alpha_j \cosh \alpha_j L + g_\beta^{(j)} \beta_j \cos \beta_j L \right)}{g_\alpha^{(j)} g_\beta^{(j)} \left(\alpha_j \sinh \alpha_j L + \beta_j \sin \beta_j L \right)} \tag{b}$$

and g_α and g_β are defined by Eqs. (10.85). The initial conditions are

$$\mathbf{u}_0(x) = -\kappa_0 \left\{ \begin{array}{c} \frac{1}{2}x^2 \\ x \end{array} \right\}, \quad \mathbf{v}_0(x) = \mathbf{0} \tag{c}$$

Since the beam is initially at rest we have from Eqs. (10.234) and (10.235) that

$$\phi_j = 0 \quad (j = 1, 2, \ldots) \tag{◁ (c)}$$

and

$$A^{(j)} = \frac{1}{\left\| \mathbf{U}^{(j)} \right\|_m^2} \int_0^L \left[\left\{ -\tfrac{1}{2}\kappa_0 x^2 \right\} m W^{(j)}(x) + \int_0^L \left\{ -\kappa_0 x \right\} I_\rho \vartheta^{(j)}(x)\, dx \right] dx$$

The amplitudes of the freely vibrating cantilevered Timoshenko Beam are then

$$A^{(j)} = -\kappa_0 \frac{\displaystyle\int_0^L \left[\tfrac{1}{2} x^2 W^{(j)}(x) + r_G^2 x \vartheta^{(j)}(x) \right] dx}{\displaystyle\int_0^L \left[W^{(j)2}(x) + r_G^2 \vartheta^{(j)2}(x) \right] dx} \quad (j = 1, 2, \ldots) \tag{◁ (d)}$$

where $W^{(j)}$ is given by Eq. (a) and r_G is the radius of gyration of the cross-section. The integrals can be evaluated analytically to give the explicit forms of the amplitudes.

10.10 CONCLUDING REMARKS

In most engineering systems the components are effectively continuous distributions of matter at the macroscopic level. In earlier chapters mechanical systems were treated as discrete systems, typically based on the assumption that the mass of the structure was negligible when compared with "dominant" mass concentrations attached to the body, and the focus was on the corresponding "dominant" motions associated with these conditions. The legitimacy of the assumptions and the accuracy of the predicted motions were, to this point, accepted on the basis of the aforementioned physical arguments. In the discrete models the inertia of the structure is neglected and the application of such models is limited in this regard. When the mass of the structure is comparable with other mass measures, when the assumption of vanishing mass of the structure is relaxed, or when the detailed motion of the structure itself is of interest, representation of the physical system as a continuum is warranted. Case studies for the longitudinal motion of elastic rods and the flexural motion of elastic beams were performed and the corresponding results give quantitative assessments of the limitation of the discrete model as an approximation to the first mode of the continuum. One-dimensional continua are mathematical representations of three-dimensional bodies or media for which one spatial dimension, the axial dimension, is large compared with the other two. Mathematical models for various types of motion of such structures were developed and discussed in detail in Chapter 9. In all, the thinness of the structures was central to the simplifications adopted. In the present chapter we studied the motion of various one-dimensional continua under their own volition. It was seen that the description of the system, and of the corresponding motion, is an abstraction of the parallel representations of discrete systems. In this regard it was seen that, for one-dimensional continua, the displacement and force matrices become functions of a single spatial variable and time, and the mass and stiffness matrix operators become differential operators. In one of the theories considered, "Timoshenko Beam Theory," the motion of the structure is described by two "displacement" functions that comprise a 2×1 displacement matrix and, correspondingly, a 2×1 force matrix whose elements are comprised of forcing functions. In addition, the mass and stiffness operators each take the form of 2×2 matrices of scalar function and differential operators respectively.

The free vibration problem for one-dimensional continua was seen to reduce to an eigenvalue problem, where the infinity of eigenvalues correspond to the squares of the natural frequencies of the structure and the eigenfunctions are the associated modal functions. The fundamental motions of one-dimensional continua were seen to be described by a linear combination of the infinity of modal functions. Since, as mentioned above, one-dimensional continua are mathematically one-dimensional representations of three-dimensional bodies as described in Chapter 9, the accuracy of the results predicted by these models is limited by the geometrical restrictions implicit to these models. From a vibrations perspective, these simplifications restrict the suitability of the predicted results to those modes for which the "wavelength" (the distance between nodes) is large compared with the thickness of the body. This, in turn, restricts the accuracy to the lowest modes of the system predicted by these mathematical models. To extend these models to a wider range of frequencies and modes for the case of flexural motion of beams, "corrections" are made to the basic Euler-Bernoulli Theory to account for the effects of rotatory inertia of the cross section (Rayleigh Theory) and deformation due to transverse shear in an average sense (shear beam theory) and to both transverse shear deformation and rotatory inertia (Timoshenko Theory). It was demonstrated by comparative studies for the case of the cantilevered beam that the inclusion of the shear deformation has the most pronounced contribution. If Timoshenko Beam Theory is taken to be the most accurate of those considered, the limitations of the results of the simpler models is demonstrated and quantified by the results of those examples.

The modal functions for a given structure were shown to be mutually orthogonal, with the interpretation being a clear extension of earlier definitions in terms of the scalar product of the modal functions. The basic definitions were abstracted for the Rayleigh and Timoshenko Beam Theories. Regardless of the model, in each case, the mutual orthogonality of the modal functions was seen to be a function of the boundary conditions imposed in the structure. The general response of a freely vibrating structure was shown to correspond to a linear combination of the modal functions with harmonic time signatures. The corresponding amplitudes and phase angles are computed by imposing the initial conditions on the general form of the response, and are simplified by capitalizing on the mutual orthogonality of the modal functions. In the next chapter we consider the response of one-dimensional continua to external dynamic forcing. The mutual orthogonality of the modal functions will be seen to be important in the corresponding analysis and response of any given system.

BIBLIOGRAPHY

Ginsberg, J.H., *Mechanical and Structural Vibrations: Theory and Applications*, Wiley, New York, 2001.

Hildebrand, F.B., *Methods of Applied Mathematics*, 2nd ed., Prentice-Hall, Englewood Cliffs, 1965.

Meirovitch, L., *Elements of Vibration Analysis*, 2nd ed., McGraw-Hill, New York, 1986.

Meirovitch, L., *Fundamentals of Vibrations*, McGraw-Hill, Boston, 2001.

Rayleigh, J.W.S., *The Theory of Sound*, Vol.1, Dover, New York, 1945.

Thomson, W.T., *Theory of Vibration with Applications*, 4th ed., Prentice-Hall, Englewood Cliffs, 1993.

Weaver, W. Jr., Timoshenko, S.P. and Young, D.P., *Vibration Problems in Engineering*, 5th ed., Wiley-Interscience, 1990.

PROBLEMS

10.1 Consider free longitudinal vibration of the elastic rod that is attached to an elastic wall, as shown. (a) Establish the modal boundary conditions for the structure. (See, also, Problem 9.4.) (b) Derive the frequency equation for the rod. (c) Determine the first three natural frequencies and modal functions for a system where $\bar{k} = kL/EA = 0.5$. Plot the modal functions.

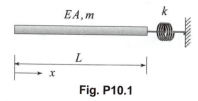

Fig. P10.1

10.2 Consider free longitudinal vibration of the elastic rod with rigid and elastic supports, as shown. (a) Establish the modal boundary conditions for the structure. (See, also, Problem 9.5.) (b) Derive the frequency equation for the rod. (c) Determine the first three natural frequencies and modal functions for a system where $\bar{k} = kL/EA = 0.5$. Plot the modal functions.

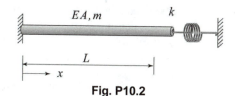

Fig. P10.2

10.3   Consider free longitudinal vibration of a uniform elastic rod of length L, membrane stiffness EA and mass per unit length m, that is constrained by elastic walls of stiffness k at each end. (a) Establish the modal boundary conditions for the structure. (Hint: See Example 9.3.) (b) Derive the frequency equation for the rod. (c) Determine the first three natural frequencies and modal functions for a structure where $\bar{k} = k_w L / EA = 0.5$. Plot the modal functions.

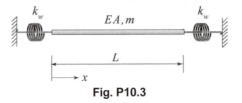

Fig. P10.3

10.4   *Case Study.* Perform a parameter study for the rod of Problem 10.3 in which you compare the first three natural frequencies for various values of the stiffness ratio $\bar{k}$.

10.5   Consider longitudinal vibrations of the free-free uniform elastic rod of length L, membrane stiffness EA and mass per unit length m shown in the figure. (a) Establish the modal boundary conditions for the structure. (b) Derive the frequency equation for the rod. (c) Determine the first three natural frequencies and modal functions. Plot the first three modes.

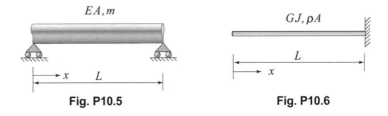

Fig. P10.5   **Fig. P10.6**

10.6   Consider free torsional vibration of a uniform circular elastic rod of length L, torsional stiffness GJ and mass per unit length m, that is free at its left end and fixed at its right end. (a) Establish the modal boundary conditions for the structure. (b) Derive the frequency equation for the rod. (c) Determine the first three natural frequencies and modal functions. Plot the first three modes.

10.7   Consider free torsional vibrations of a uniform circular elastic rod of length L, torsional stiffness GJ and mass per unit length m, that is free at its left end and embedded in an elastic wall of stiffness $k_\theta = GJ/L$ at its right end, as shown. (a) Establish the modal boundary conditions for the structure. (Hint: See Problem 9.8). (b) Derive the frequency equation for the rod. (c) Determine the first three natural frequencies and modal functions. Plot the first three modes.

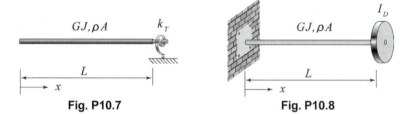

Fig. P10.7   **Fig. P10.8**

10.8   Consider free torsional vibrations of the uniform circular elastic rod of length
L, torsional stiffness GJ , mass density ρ and cross-sectional area A, that has a
rigid disk of mass moment of inertia I_D attached to its free end, as shown in
Figure P10.8. (a) Establish the modal boundary conditions for the structure.
(Hint: See Problem 9.10.) (b) Derive the frequency equation for the rod in
terms of the ratio of the inertia ratio $\alpha = I_D/I_\rho L$. (c) Determine the first three
natural frequencies and modal functions for the case where $\alpha = 2$. Plot the
first three modal functions.

10.9   *Case Study.* Conduct a comparative study of the rod of Example 10.4 and its
"equivalent" single degree of freedom model of Section 1.2.3, in the spirit of
Example 10.3. Which mode does the 1 d.o.f. system simulate? At what value
of the inertia ratio does the 1 d.o.f. system simulate the continuous system to
within 2 significant figures? To within 3 significant figures?

10.10   Consider free lateral vibrations of the chain of Example 9.6-ii. (a) Determine
the natural frequencies and modal functions for the system. (b) Plot the first
three modes.

10.11   Consider the pulley system of Problem 9.36. (a) Determine the natural
frequencies and modal functions for
the free lateral vibration of the inner
sections of the cable. (b) Plot the
first three modes.

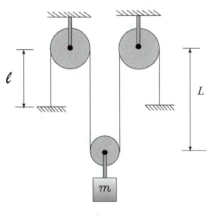

Fig. P10.11

10.12 Consider the free flexural vibrations of a uniform elastic beam of length L, bending stiffness EI and mass per unit length m that is clamped at both edges, as shown, and is represented mathematically using Euler-Bernoulli Theory. (a) Establish the modal boundary conditions for the structure. (b) Derive the frequency equation for the beam. (c) Determine the first three natural frequencies and modal functions. Plot the modal functions.

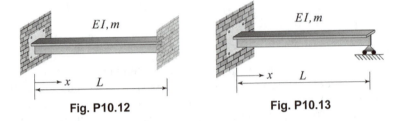

| **Fig. P10.12** | **Fig. P10.13** |

10.13 Consider the free flexural vibrations of a uniform elastic beam of length L, bending stiffness EI and mass per unit length m that is clamped at its left edge and is pin-free supported at its right edge, as shown. Let the beam be represented mathematically using Euler-Bernoulli Theory. (a) Establish the modal boundary conditions for the structure. (b) Derive the frequency equation for the beam. (c) Determine the first three natural frequencies and modal functions. Plot the modal functions.

10.14 Consider the free flexural vibrations of the uniform elastic beam of length L, bending stiffness EI and mass per unit length m that is supported by elastic clamps of (rotational) stiffness k_φ at each end, as shown. Let the beam be represented mathematically using Euler-Bernoulli Theory. (a) Establish the modal boundary conditions for the structure. (Hint: See Problem 9.15.) (b) Derive the frequency equation for the beam. (c) Determine the first three natural frequencies and modal functions for the case where $\bar{k}_\varphi = k_\varphi L/EI = 1$. Plot the modal functions.

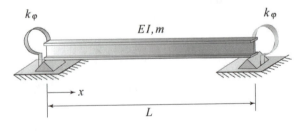

Fig. P10.14

10.15 Consider the free flexural vibrations of the uniform elastic beam of length L, bending stiffness EI and mass per unit length m that is supported by an elastic clamp of (rotational) stiffness k_φ at one end, as shown. Let the beam be represented mathematically using Euler-Bernoulli Theory. (a) Establish the modal boundary conditions for the structure. (Hint: See Problem 9.16). (b) Derive the frequency equation for the beam. (c) Determine the first three natural frequencies and modal functions for the case where $\overline{k}_\varphi = k_\varphi L/EI = 1$. Plot the modal functions.

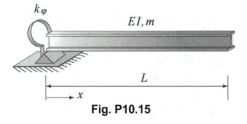

Fig. P10.15

10.16 Consider the free flexural vibrations of the uniform elastic beam of length L, bending stiffness EI and mass per unit length m that is supported by an elastic clamp at one end and an elastic mount at the other, as shown. Let the beam be represented mathematically using Euler-Bernoulli Theory. (a) Establish the modal boundary conditions for the structure. (Hint: See Problem 9.17) (b) Derive the frequency equation for the beam. (c) Determine the first three natural frequencies and modal functions if the supports are such that $\overline{k}_\varphi = k_\varphi L/EI = 1$ and $\overline{k}_w = k_w L^3/EI = 1$. Plot the modal functions.

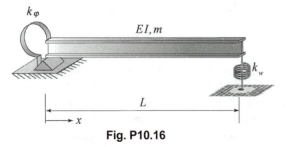

Fig. P10.16

10.17 Consider the free flexural vibrations of the uniform elastic beam of length L, bending stiffness EI and mass per unit length m that is supported by linear springs of stiffness k_w at each end, as shown. Let the beam be represented mathematically using Euler-Bernoulli Theory. (a) Establish the modal boundary conditions for the structure. (Hint: See Problem 9.18.) (b) Derive the frequency equation for the beam. (c) Determine the first three natural frequencies and modal functions for the case where $\overline{k}_w = k_w L^3/EI = 1$. Plot the modal functions.

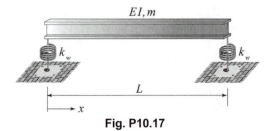

Fig. P10.17

10.18 Consider the free flexural vibrations of the cantilevered uniform elastic beam of length L, bending stiffness EI, and mass per unit length m that is supported by an elastic mount of stiffness k_w at its free end, as shown. Let the beam be represented mathematically using Euler-Bernoulli Theory. (a) Establish the modal boundary conditions for the structure. (Hint: See Problem 9.19.) (b) Derive the frequency equation for the beam. (c) Determine the first three natural frequencies and modal functions for the case where $\bar{k}_w = k_w L^3 / EI = 1$. Plot the modal functions.

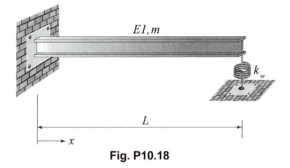

Fig. P10.18

10.19 Consider the free flexural vibrations of a simply supported uniform elastic beam-column of length L, bending stiffness EI, and mass per unit length m that is subjected to a constant compressive end load P_0. (See Example 9.12.) Determine the first three natural frequencies and modal functions for the case where the applied load is half the static buckling load. Plot the modal functions.

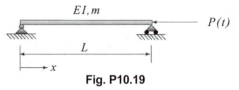

Fig. P10.19

10.20 *Case Study.* Parallel the parameter study of Example 10.10 for the simply supported structure of Problem 10.19.

10.21 Consider the free flexural vibrations of the uniform elastic beam-column of length L, bending stiffness EI, and mass per unit length m that is pinned-fixed supported at its left end and is clamped-free supported at its right end, as shown. The structure is subjected to a constant compressive end load P_0, as indicated. (a) Establish the modal boundary conditions for the structure. (Hint: See Problem 9.34.) (b) Derive the frequency equation for the beam. (c) Determine the first three natural frequencies and modal functions for the case where the applied load is half the static buckling load. Plot the modal functions.

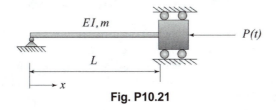

EI, m

$P(t)$

L

x

Fig. P10.21

10.22 Consider the free flexural vibrations of a simply supported uniform elastic beam of length L, bending stiffness EI, radius of gyration r_G and mass per unit length m, and let it be represented mathematically using Rayleigh Beam Theory. (a) Establish the modal boundary conditions for the structure. (b) Derive the frequency equation for the beam. (c) Determine the first three natural frequencies and modal functions for a beam with $r_G/L = 0.1$. Plot the modal functions.

10.23 *Case Study.* Parallel the study of Example 10.12 for the simply supported beam of Problem 10.22.

10.24 Solve Problem 10.12 if the beam is modeled as a Rayleigh Beam with $r_G/L = 0.1$.

10.25 Solve Problem 10.13 if the beam is modeled as a Rayleigh Beam with $r_G/L = 0.1$.

10.26 Solve Problem 10.22 if the beam is modeled as a Timoshenko Beam with $r_G/L = 0.1$ and $E/kG = 5$.

10.27 Solve Problem 10.12 if the beam is modeled as a Timoshenko Beam with $r_G/L = 0.1$ and $E/kG = 5$.

10.28 Solve Problem 10.13 if the beam is modeled as a Timoshenko Beam with $r_G/L = 0.1$ and $E/kG = 5$.

10.29 The elastic rod of Problem 10.6 is twisted and held in place such that $\theta_0(x) = ax$ where a is a constant. Determine the amplitudes and phase angles for the free vibration response if the rod is released from rest when in this configuration.

10.30 A string of mass per unit length m and length L is under tension N_0 and held in place in the configuration $w_0(x) = ax(L-x)$ where $a = q_0/2N_0$ is a constant. Determine the amplitudes and phase angles for the free vibration response of the string if the string is released from rest.

10.31 Solve problem 10.30 if the string is released from the given configuration with the velocity $v_0(x) = c_0 w_0(x)$, where c_0 is a constant.

10.32 A simply supported Euler-Bernoulli Beam of length L, bending stiffness EI and mass per unit length m is deflected by a static load and held in the configuration $w_0(x) = ax\left(L^3 - 2Lx^2 + x^3\right)$, where $a = q_0/24EI$ is a constant. Determine the amplitudes and phase angles for the free vibration response of the structure if it is released from rest when in this configuration.

10.33 Determine the amplitudes and phase angles for free vibration of the simply supported beam of Problem 10.32 if the beam is represented using Rayleigh Theory.

10.34 The cantilevered Timoshenko Beam of Example 10.12 is released from rest from the configuration

$$w_0(x) = Q_0 x^2 \left[L - \tfrac{1}{6}x\right], \quad \varphi_0(x) = Q_0 \left[\tfrac{1}{2}x^2 - Lx - k_s^{-1}\right]$$

where Q_0 is constant. Determine the amplitudes and phase angles for the free vibration response of the beam.

11

Forced Vibration of One-Dimensional Continua

The dynamic response of mechanical and structural systems to external loads is of primary importance in a variety of applications. Excessive vibrations of a structure such as an airplane wing or a highway overpass can lead to catastrophic failure, while the bowing of a violin string, the plucking of a guitar string or the pounding of a drum head can produce desirable effects. The controlled vibration of structures is germane to the performance of telephones, stereo speakers and SONAR, to name but a few examples. The behavior of continua under dynamic loading is central to the understanding and implementation in all of these applications. It is therefore of interest to study the vibrations of continua under applied dynamic forcing. In this chapter we study the forced vibration of the mathematically one-dimensional continua discussed in Chapters 9 and 10. Specifically, we discuss the forced longitudinal and torsional motions of elastic rods, the transverse motion of externally excited strings and cables, and the transverse motion of Euler-Bernoulli Beams, Rayleigh Beams and Timoshenko Beams to dynamic transverse loads and applied moments.

It was shown in Chapter 9 that the mathematical description of a continuous system is an abstraction of the description of a discrete system. This abstraction was utilized in Chapter 10 to study the free vibrations of one-dimensional continua, and the concepts of normalization and mutual orthogonality of the modes were similarly advanced. In a similar manner, the modal decomposition introduced to study the forced vibration of discrete systems can be extended to continuous systems, allowing for a fundamental and comprehensive approach to forced vibration problems for one-dimensional continua. Modal analysis of forced vibration problems is based on an expansion of the displacement function in terms of the modal functions found for the corresponding free vibration problem. The justification for such a modal expansion is discussed in the next section. The methodology and applications to the various systems considered to this point follow in subsequent sections of this chapter.

11.1 MODAL EXPANSION

In order to solve forced vibration problems it is of fundamental, as well as practical, interest to first express the displacements in terms of the fundamental motions of the system described by the modal functions. The advantage of this approach was clearly seen when we studied forced vibration of discrete systems in Chapter 8. As discussed at the beginning of Section 8.3, a vector may be expressed as a linear combination of other vectors provided those vectors are linearly independent. It was demonstrated in Chapters 9 and 10 that a function is an abstraction of a finite dimensional vector. That is, functions may be viewed as vectors with an infinite number of components or elements that are densely packed and continuously distributed. To show that a set of functions forms a basis for a given space, and hence that any other function in that space can be expressed as a linear combination of the functions of the given set, we must show that the functions that comprise the set are linearly independent. We do this for the class of modal functions described in the previous chapter.

11.1.1 Linear Independence of the Modal Functions

We wish to establish the linear independence of the modal functions for the class of continua discussed in Chapters 9 and 10. To accomplish this we must show that, for a given structure, no one modal function can be expressed as a linear combination of the other modal functions. The mutually orthogonality of the modal functions, as discussed in Section 10.8, is central to establishing this property.

The motion of each of the systems discussed in Chapters 9 and 10 was seen to be described by a displacement function or functions and the governing equation or equations by the associated mass and stiffness operators. Since the case of a single displacement function and scalar mass and stiffness differential operators is a special case of multiple displacement functions and corresponding mass and stiffness operators, we shall establish the desired condition for the more general case. Let us, therefore, consider continuous systems for which the displacements are described by the column array, $\mathbf{u}(x,t)$ with corresponding modes $\mathbf{U}^{(j)}(x)$ ($j = 1, 2, \ldots$) and associated mass and stiffness operators described by the square matrices $\mathbf{m}$ and $\mathbf{k}$, respectively. For the case of a single displacement function, the column arrays reduce to single element arrays (functions) $u(x,t)$ and $U(x)$, and the square matrices reduce to single element operators m and k.

The functions that comprise the set of modal functions $\{\mathbf{U}^{(1)}(x), \mathbf{U}^{(2)}(x), \ldots\}$ are linearly independent if no function of that set can be expressed as a linear combination of the other functions of the set. Equivalently, the modal functions for a given structure are linearly independent if the equation

$$\sum_{l=1}^{\infty} a_l \mathbf{U}^{(l)}(x) = \mathbf{0} \tag{11.1}$$

is satisfied when all $a_l = 0$ ($l = 1, 2, \ldots$). We shall show that this is the case for a set of mutually orthogonal modal functions. To do this, we first multiply the above equation by $\mathbf{U}^{(l)\mathsf{T}}\mathbf{m}\,dx$ and integrate the resulting expression over the domain $[0, L]$. Hence,

$$\int_0^L \sum_{l=1}^{\infty} a_l \mathbf{U}^{(j)\mathsf{T}}(x)\mathbf{m}\mathbf{U}^{(l)}(x)\,dx = 0$$

We next interchange the order of summation and integration in the above expression to obtain

$$\sum_{l=1}^{\infty} a_l \int_0^L \mathbf{U}^{(j)\mathsf{T}}(x)\mathbf{m}\mathbf{U}^{(l)}(x)\,dx = 0 \tag{11.2}$$

Now, if the modal functions are mutually orthogonal then

$$\int_0^L \mathbf{U}^{(j)\mathsf{T}}(x)\mathbf{m}\mathbf{U}^{(l)}(x)\,dx = \left\langle \mathbf{U}^{(j)}, \mathbf{U}^{(l)} \right\rangle_{\mathbf{m}} = \begin{cases} 0 & (l \neq j) \\ \left\| \mathbf{U}^{(j)} \right\|_{\mathbf{m}}^2 & (l = j) \end{cases}$$

Accounting for the mutual orthogonality of the modal functions in Eq. (11.2) reduces that statement to the form

$$a_j \left\| \mathbf{U}^{(j)} \right\|_{\mathbf{m}}^2 = 0 \quad (j = 1, 2, \ldots) \tag{11.3}$$

Since the square of the magnitude of a modal function does not vanish we have that

$$a_j = 0 \quad (j = 1, 2, \ldots) \tag{11.4}$$

which is what we set out to show. The modal functions for a given structure are thus linearly independent if they are mutually orthogonal.

11.1.2 Generalized Fourier Series

Since the mutually orthogonal modal functions of a given set are linearly independent, any function $\mathbf{u}(x,t)$ defined on the domain of definition of the modal functions can be expressed as a linear combination of these functions. That is,

$$\mathbf{u}(x,t) = \sum_{j=1}^{\infty} \eta_j(t)\mathbf{U}^{(j)}(x) \qquad (11.5)$$

where the time dependent coefficients, η_j ($j = 1, 2, \ldots$) depend upon the particular loading applied to the system and the parameters that define the system. Note that if the set of functions corresponds to a sequence of harmonic functions whose wave numbers differ by multiples of 2π then Eq. (11.5) reduces to a standard Fourier series, Eq. (3.145). A general expansion of the form of Eq. (11.5) is thus referred to as a *generalized Fourier Series*. The modal expansion tells us that the response of a given system to external forcing is comprised of a linear combination of the responses of the individual modes.

11.2 DECOMPOSITION OF THE FORCED VIBRATION PROBLEM

The solution of forced vibration problems for one-dimensional continua is facilitated by the results of the last section. It was seen therein that the response of a given structure may be expressed as a linear combination of the modal functions. It will be shown in the present section that the governing equations for a given structure can be decomposed into the governing equations for an infinite system of uncoupled single degree of freedom systems whose displacements correspond to the modal coordinates. As in the previous section, we carry out the development for the general case where several displacement functions may characterize the motion of the system. Structures for which a single displacement function characterizes the motion are interpreted as a special case.

The governing equation for the class of structures under consideration takes the general form

$$\mathbf{m}\frac{\partial^2 \mathbf{u}}{\partial t^2} + \mathbf{k}\mathbf{u} = \mathbf{F} \qquad (11.6)$$

where $\mathbf{F} = \mathbf{F}(x,t)$ represents the distributed external loads applied to the structure (see Chapter 9). Let us express the array of displacement functions as a linear combination of the arrays of the corresponding modal functions. Upon substituting Eq. (11.5) into Eq. (11.6) we have

$$\sum_{j=1}^{\infty}\left[\mathbf{m}\mathbf{U}^{(j)}(x)\ddot{\eta}_j(t) + \mathbf{k}\mathbf{U}^{(j)}(x)\eta_j(t)\right] = \mathbf{F}(x,t) \qquad (11.7)$$

Multiplying (11.7) on the left by $\mathbf{U}^{(l)\mathrm{T}}dx$ and integrating over $[0, L]$ then results in the identity

$$\int_0^L \sum_{j=1}^{\infty} \left[\mathbf{U}^{(l)\mathsf{T}}(x)\mathbf{m}\mathbf{U}^{(j)}(x)\ddot{\eta}_j(t) + \mathbf{U}^{(l)\mathsf{T}}(x)\mathbf{k}\mathbf{U}^{(j)}(x)\eta_j(t) \right] dx$$

$$= \int_0^L \mathbf{U}^{(l)\mathsf{T}}(x)\mathbf{F}(x,t)\, dx \tag{11.8}$$

Let us now interchange the order of integration and summation in Eq. (11.8) and note the relations for the mutual orthogonality of the modal functions,

$$\left\langle \mathbf{U}^{(l)}, \mathbf{U}^{(j)} \right\rangle_{\mathbf{m}} \equiv \int_0^L \mathbf{U}^{(l)\mathsf{T}}(x)\mathbf{m}\mathbf{U}^{(j)}(x)\, dx = 0 \quad (j \neq l)$$

$$\left\langle \mathbf{U}^{(l)}, \mathbf{U}^{(j)} \right\rangle_{\mathbf{k}} \equiv \int_0^L \mathbf{U}^{(l)\mathsf{T}}(x)\mathbf{k}\mathbf{U}^{(j)}(x)\, dx = 0 \quad (j \neq l)$$

After doing this, Eq. (11.8) reduces to the system of uncoupled ordinary differential equations in η_j,

$$\tilde{m}_j \ddot{\eta}_j(t) + \tilde{k}_j \eta_j(t) = \tilde{F}_j(t) \quad (j = 1, 2, \ldots) \tag{11.9}$$

where

$$\tilde{m}_j = \left\| \mathbf{U}^{(j)} \right\|_{\mathbf{m}}^2 = \int_0^L \mathbf{U}^{(j)\mathsf{T}}(x)\mathbf{m}\mathbf{U}^{(j)}(x)\, dx \quad (j = 1, 2, \ldots) \tag{11.10}$$

$$\tilde{k}_j = \left\| \mathbf{U}^{(j)} \right\|_{\mathbf{k}}^2 = \int_0^L \mathbf{U}^{(j)\mathsf{T}}(x)\mathbf{k}\mathbf{U}^{(j)}(x)\, dx \quad (j = 1, 2, \ldots) \tag{11.11}$$

and

$$\tilde{F}_j(t) = \left\langle \mathbf{U}^{(j)}, \mathbf{F} \right\rangle = \int_0^L \mathbf{U}^{(j)\mathsf{T}}(x)\mathbf{F}(x,t)\, dx \quad (j = 1, 2, \ldots) \tag{11.12}$$

respectively correspond to the modal mass, modal stiffness and modal force for the jth mode. Note that if we choose to normalize the modal functions with respect to the mass, then the modal masses are all equal to one. Equations (11.9) are seen to correspond to the equations of motion for an infinite number of uncoupled single degree of freedom systems whose displacements are the modal coordinates $\eta_j(t)$ (Figure 11.1). Each equation describes the amplitude of an individual mode. The corresponding masses and stiffnesses of these single degree of freedom systems are the modal masses and modal stiffnesses defined by Eqs. (11.10) and (11.11). The modal forces (the forces acting on the modal masses) defined by Eq. (11.12) correspond to the portion of the applied force distributed to the individual mode. Before proceeding it is useful to identify the natural frequencies for each of the equivalent single degree of freedom systems.

Recall from Chapter 10 that natural frequencies and modal functions for a given structure each satisfy the relation

$$\mathbf{k}\mathbf{U}^{(j)} - \omega_j^2\,\mathbf{m}\mathbf{U}^{(j)} = \mathbf{0} \quad (j = 1, 2, \ldots) \tag{11.13}$$

where ω_j is the jth natural frequency of the structure. If we multiply Eq. (11.13) on the left by $\mathbf{U}^{(i)\mathrm{T}}dx$ and integrate the resulting expression over the domain of definition of the structure $[0, L]$ we arrive at the identity

$$\tilde{k}_j - \omega_j^2\tilde{m}_j = 0 \quad (j = 1, 2, \ldots) \tag{11.14}$$

With the relation between the frequencies of the equivalent single degree of freedom systems seen, perhaps not surprisingly, to correspond to the natural frequencies of a given structure we return to the problem defined by Eq. (11.9).

Dividing Eq. (11.9) by $\tilde{m}_j$ and incorporating Eq. (11.14) puts the uncoupled system of equations for the modal coordinates in the standard form (Chapter 3)

$$\ddot{\eta}_j(t) + \omega_j^2\eta_j(t) = \omega_j^2\tilde{f}_j(t) \quad (j = 1, 2, \ldots) \tag{11.15}$$

where

$$\tilde{f}_j(t) \equiv \frac{\tilde{F}_j(t)}{\tilde{k}_j} = \frac{\tilde{F}_j(t)}{\omega_j^2\tilde{m}_j} \tag{11.16}$$

The equations are seen to be a system of uncoupled forced harmonic equations which may be solved for $\eta_j(t)$ using the methods of Chapters 3–5. The solutions correspond to the displacements of effective single degree of freedom systems and are the modal coordinates. Once these modal displacements are determined for a given system and applied forces, they may be substituted into Eq. (11.5) giving the forced vibration response in the form

$$\mathbf{u}(x,t) = \sum_{j=1}^{\infty}\eta_j(t)\mathbf{U}^{(j)}(x)$$

The above development may be applied to the various one-dimensional continua discussed in Chapter 10. In particular, the modal analysis described above may be applied directly to problems concerning the longitudinal motion of elastic rods, torsional motion of elastic rods, transverse vibration of strings and cables, flexural motion of Euler-Bernoulli Beams and Beam-Columns, and Timoshenko Beams, with proper identification of the mass and stiffness operators and structural parameters for the system of interest. The general procedure holds for Rayleigh Beams as well, though the definitions for modal mass and modal stiffness differ slightly from those stated above and will be introduced in Section 11.3.5. In the next section we use modal analysis to solve forced vibration problems for various one-dimensional continua.

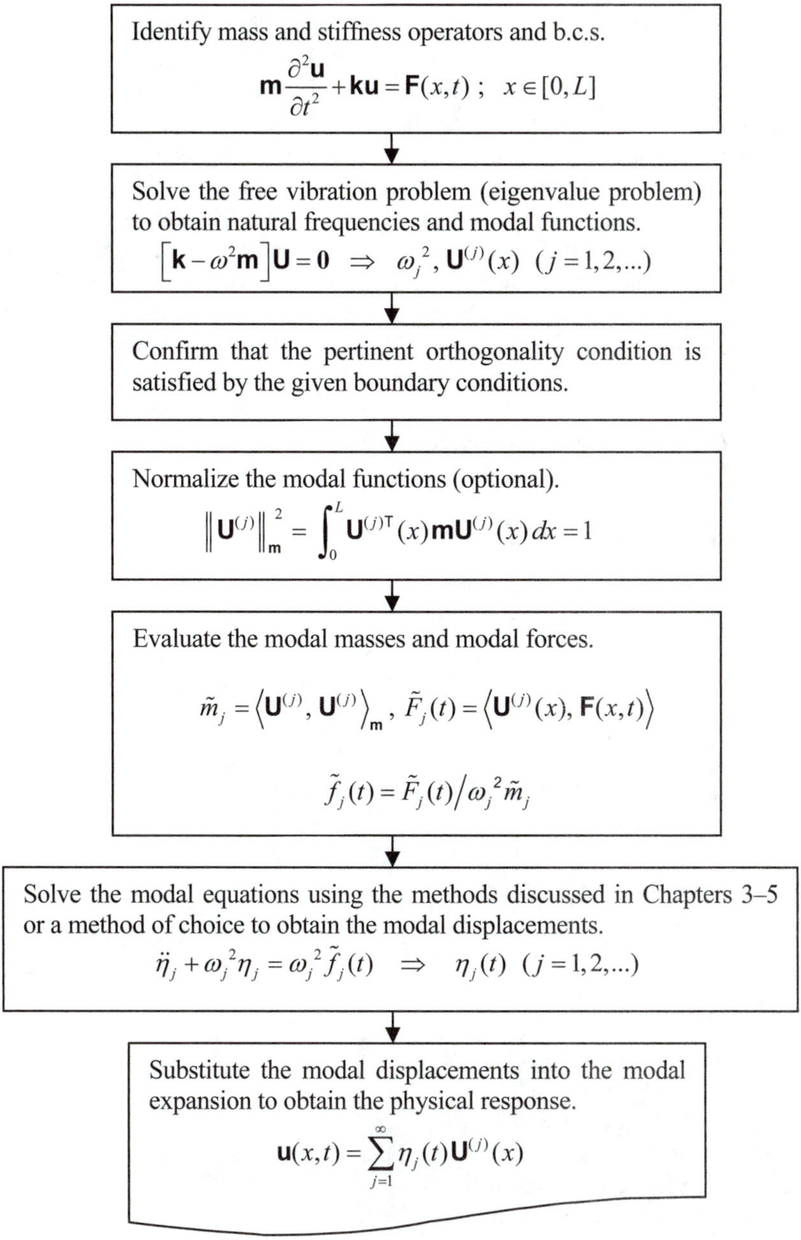

Identify mass and stiffness operators and b.c.s.

$$\mathbf{m}\frac{\partial^2 \mathbf{u}}{\partial t^2} + \mathbf{k}\mathbf{u} = \mathbf{F}(x,t) \; ; \quad x \in [0,L]$$

Solve the free vibration problem (eigenvalue problem) to obtain natural frequencies and modal functions.

$$\left[\mathbf{k} - \omega^2 \mathbf{m}\right]\mathbf{U} = \mathbf{0} \;\; \Rightarrow \;\; \omega_j^2, \, \mathbf{U}^{(j)}(x) \;\; (j = 1,2,...)$$

Confirm that the pertinent orthogonality condition is satisfied by the given boundary conditions.

Normalize the modal functions (optional).

$$\left\| \mathbf{U}^{(j)} \right\|_{\mathbf{m}}^2 = \int_0^L \mathbf{U}^{(j)\mathsf{T}}(x)\mathbf{m}\mathbf{U}^{(j)}(x)\,dx = 1$$

Evaluate the modal masses and modal forces.

$$\tilde{m}_j = \left\langle \mathbf{U}^{(j)}, \mathbf{U}^{(j)} \right\rangle_{\mathbf{m}}, \;\; \tilde{F}_j(t) = \left\langle \mathbf{U}^{(j)}(x), \mathbf{F}(x,t) \right\rangle$$

$$\tilde{f}_j(t) = \tilde{F}_j(t) \big/ \omega_j^2 \tilde{m}_j$$

Solve the modal equations using the methods discussed in Chapters 3–5 or a method of choice to obtain the modal displacements.

$$\ddot{\eta}_j + \omega_j^2 \eta_j = \omega_j^2 \tilde{f}_j(t) \;\; \Rightarrow \;\; \eta_j(t) \;\; (j = 1,2,...)$$

Substitute the modal displacements into the modal expansion to obtain the physical response.

$$\mathbf{u}(x,t) = \sum_{j=1}^{\infty} \eta_j(t)\mathbf{U}^{(j)}(x)$$

Figure 11.1 Recipe for modal analysis of 1-D continua subjected to external forcing.

11.3 SOLUTION OF FORCED VIBRATION PROBLEMS

The forced vibration problem for any of the one-dimensional continua described in Chapters 9 and 10 can be solved by way of the modal decomposition discussed in the previous section. This procedure is referred to as modal analysis. To perform such an analysis for a given structure we must first solve the corresponding free vibration problem, confirm that the modal functions are mutually orthogonal by checking that the boundary conditions satisfy the requisite conditions defined in Section 10.8, and then computing the modal masses and/or modal stiffnesses for the system and the modal forces for the particular forces applied to the structure. We then solve the system of differential equations for the modal displacements. Once the modal displacements are determined we substitute them, along with the corresponding modal functions, into the modal expansion defined by Eq. (11.5) to obtain the forced response of the structure. The general procedure is outlined in Figure 11.1.

Examples pertaining to each of the structures discussed in Chapters 9 and 10 are presented in the remainder of this section.

11.3.1 Axially Loaded Elastic Rods

We first consider the longitudinal vibrations of elastic rods due to applied forces. Recall the equation of longitudinal motion, Eq. 9.44,

$$m\frac{\partial^2 u}{\partial t^2} - \frac{\partial}{\partial x}EA\frac{\partial}{\partial x}u = p(x,t)$$

where $u(x,t)$ is the axial displacement of the cross-section originally at x and $p(x,t)$ is the distributed axial load applied to the rod. Thus, for the present type of structure, $\mathbf{u}(x,t) \to u(x,t)$, $\mathbf{F}(x,t) \to p(x,t)$ and $\mathbf{m} \to m(x)$ in the development and formulation presented in Sections 11.1–11.2.

Example 11.1

A uniform elastic rod of mass density ρ, Young's Modulus E, cross-sectional area A and length L is fixed at its left end and free to translate at its right end as shown. Determine the steady state response of the rod if a harmonic force $P(t) = P_0\sin\Omega t$ is applied at its free end as indicated.

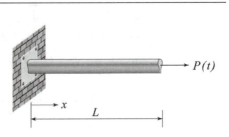

Solution

The natural frequencies and modal functions for this structure were determined in Example 10.1 to be

$$\omega_j = (2j-1)\frac{\pi}{2}\omega_0 \quad (j=1,2,...) \tag{a-1}$$

where

$$\omega_0 = \sqrt{\frac{EA}{mL^2}} = c_a/L \tag{a-2}$$

$$U^{(j)}(x) = \sin\beta_j x \quad (j=1,2,...) \tag{b-1}$$

where

$$\beta_j = (2j-1)\pi/2L \tag{b-2}$$

The applied force may be represented as a distributed axial force with the aid of the Dirac Delta Function (Chapter 4) as

$$p(x,t) = P_0 \sin\Omega t \,\hat{\delta}(x-L) \tag{c}$$

To determine the response of the rod, $u(x,t)$, in the form of Eq. (11.5) we must determine the modal displacements, $\eta_j(t)$, which are solutions to Eq. (11.15) evaluated for the present system. To determine the modal displacements, we must first calculate the modal masses (or modal stiffnesses) and the modal forces. Substituting Eq. (b-1) into Eq. (11.10) and performing the indicated integration gives the modal masses as

$$\tilde{m}_j = \int_0^L U^{(j)}m(x)U^{(j)}(x)\,dx = m\int_0^L \sin^2\beta_j x\,dx = \tfrac{1}{2}mL \tag{d}$$

The corresponding modal forces are determined by substituting Eqs. (c) and (b-1) into Eq. (11.12). Doing this we find that

$$\tilde{F}_j(t) = \int_0^L U^{(j)}(x)p(x,t)\,dx$$

$$= P_0 \sin\Omega t \int_0^L \sin\beta_j x\,\hat{\delta}(x-L)\,dx = P_0 \sin\Omega t \,\sin\beta_j L$$

Thus,

$$\tilde{F}_j(t) = \tilde{F}_j^0 \sin\Omega t \tag{e-1}$$

where

$$\tilde{F}_j^0 = P_0 \sin\beta_j L = (-1)^{j+1} P_0 \tag{e-2}$$

Hence, for the present system, Eq. (11.15) takes the specific form

$$\ddot{\eta}_j(t) + \omega_j^2 \eta_j(t) = \omega_j^2 \tilde{f}_j^0 \sin\Omega t \quad (j = 1, 2, \ldots) \tag{f}$$

where

$$\tilde{f}_j^0 = \frac{\tilde{F}_j^0}{\omega_j^2 \tilde{m}_j} = \frac{8}{\pi^2} \frac{P_0 L}{EA} \frac{(-1)^{j+1}}{(2j-1)^2} \tag{g}$$

The solutions to Eq. (f) may be written directly from Eq. (3.28). In this way, the jth modal displacement is found to be

$$\eta_j(t) = \frac{\tilde{f}_j^0}{1 - (\Omega/\omega_j)^2} \sin\Omega t = \frac{(-1)^{j+1}}{\left[\pi^2 (2j-1)^2 - 4(\Omega/\omega_0)^2\right]} \frac{8 P_0 L}{EA} \sin\Omega t \tag{h}$$

where ω_0 is given by Eq. (a-2). Substitution of Eq. (h) into Eq. (11.5) gives the steady state response of the rod to the harmonic edge load as

$$u(x,t) = \frac{8 P_0 L}{EA} \sin\Omega t \sum_{j=1}^{\infty} \frac{(-1)^{j+1}}{\left[\pi^2 (2j-1)^2 - 4(\Omega/\omega_0)^2\right]} \sin\left\{(2j-1)\pi x/2L\right\} \quad \triangleleft (i)$$

11.3.2 Torsion of Elastic Rods

We next consider the vibrations of elastic rods due to applied torques. Recall that for a rod of circular cross section with polar moment of inertia $J(x)$, the equation of torsional motion, Eq. (9.59), is

$$J_\rho \frac{\partial^2 \theta}{\partial t^2} - \frac{\partial}{\partial x} GJ \frac{\partial}{\partial x} \theta(x,t) = \mu(x,t)$$

where $\theta(x,t)$ is the rotational displacement of the cross section at coordinate x. Correspondingly $\mu(x,t)$ represents the distributed (body) torque applied along the axis of the rod. Thus, for the present type of structure, $\mathbf{u}(x,t) \rightarrow \theta(x,t)$, $\mathbf{F}(x,t) \rightarrow \mu(x,t)$ and $\mathbf{m}(x) \rightarrow J_\rho(x) = \rho(x) J(x)$ in the development and formulation presented in Sections 11.1–11.2.

Example 11.2

A sleeve of negligible mass fits around a circular elastic rod of length L, radius R, shear modulus G and mass density ρ. The structure is fixed at its right end as shown. At a certain instant the sleeve is quickly rotated such that it exerts a sudden uniformly distributed torque of magnitude μ_0 along the shaft. Determine the response of the shaft and the reaction at the support.

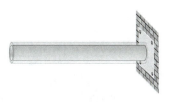

Solution

The frequency equation, natural frequencies and modal functions for a rod fixed at its right end are determined by solving the corresponding free vibration problem (Problem 10.6). These quantities may also be determined from the corresponding expressions of Example 10.4 by setting $\alpha = 0$. The natural frequencies and modal functions for the shaft are thus, respectively,

$$\omega_j = c_T (2j-1)\pi/2L \quad (j=1,2,...) \tag{a-1}$$

and

$$\Theta^{(j)}(x) = \cos \beta_j x \quad (j=1,2,...) \tag{a-2}$$

where

$$\beta_j = (2j-1)\pi/2L \quad (j=1,2,...) \tag{a-3}$$

and $c_T = \sqrt{G/\rho}$. The modal masses are then calculated as

$$\begin{aligned}
\tilde{m}_j &= \int_0^L \Theta^{(j)}(x) J_\rho \Theta^{(j)}(x)\, dx \\
&= \rho J \int_0^L \cos^2 \beta_j x\, dx = \tfrac{1}{2}\rho JL \quad (j=1,2,...)
\end{aligned} \tag{b}$$

For the given problem, the applied torque takes the form

$$\mu(x,t) = \mu_0 \mathcal{H}(t) \tag{c}$$

The modal forces are then

$$\tilde{F}_j(t) = \int_0^L \Theta^{(j)}(x)\, \mu(x,t)\, dx = \frac{2\mu_0 L}{\pi}\mathcal{H}(t)\frac{(-1)^j}{(2j-1)} \tag{d}$$

Hence,

$$\tilde{f}_j(t) = \frac{\tilde{F}_j(t)}{\omega_j^2 \tilde{m}_j} = \tilde{f}_j^0 \frac{(-1)^{j+1}}{(2j-1)^3} \mathcal{H}(t) \tag{e}$$

where

$$\tilde{f}_j^0 = \frac{16L^2 \mu_0}{\pi^3 GJ} \frac{(-1)^{j+1}}{(2j-1)^3} = \frac{32L^2 \mu_0}{\pi R^4 G} \frac{(-1)^{j+1}}{(2j-1)^3} \tag{f}$$

Thus, for the present problem, Eq. (11.15) takes the form

$$\ddot{\eta}_j + \omega_j^2 \eta_j = \omega_j^2 \tilde{f}_j^0 \mathcal{H}(t) \tag{g}$$

The solution of Eq. (g) follows directly from Eqs. (4.31) and (4.32) as

$$\eta_j(t) = \tilde{f}_j^0 S_j(t) = \frac{32L^2 \mu_0}{\pi R^4 G} \frac{(-1)^{j+1}}{(2j-1)^3} \cos \omega_j t \tag{h}$$

The response is then obtained by summing the contributions of each of the modes, as per Eq. (11.5). We thus have

$$\theta(x,t) = \sum_{j=1}^{\infty} \eta_j(t) \Theta^{(j)}(x) = \frac{32L^2 \mu_0}{\pi R^4 G} \sum_{j=1}^{\infty} \frac{(-1)^{j+1}}{(2j-1)^3} \cos \beta_j x \cos \omega_j t \qquad \triangleleft (i)$$

The reaction at the support is equal to the internal torque at $x = L$. Hence, from Eq. (9.55) and Eq. (i),

$$T_L = T(L,t) = GJ \frac{\partial \theta}{\partial x}\bigg|_{x=L} = -\frac{16L^2 \mu_0}{\pi^3} \sum_{j=1}^{\infty} \frac{1}{(2j-1)^2} \cos \omega_j t \qquad \triangleleft (j)$$

11.3.3 Strings and Cables

We now apply the formulation presented in Sections 9.1 and 9.2 to the forced vibration of strings and cables. The corresponding equation of motion, Eq. (9.74), is

$$m \frac{\partial^2 w}{\partial t^2} - N \frac{\partial^2 w}{\partial x^2} = q(x,t)$$

where $w(x,t)$ is the transverse displacement the point on the axis of the string originally at x and $q(x,t)$ represents a distributed transverse load. Thus, for the present
class of structure, $\mathbf{u}(x,t) \rightarrow w(x,t)$, $\mathbf{F}(x,t) \rightarrow q(x,t)$ and $\mathbf{m} \rightarrow m(x)$ in the development and formulation presented in Sections 11.1–11.2.

Example 11.3

The cable and sign system of Example
9.6-(i) undergoes a sudden wind load that
varies parabolically along the length as
$q(x,t) = \hat{q}(t) \cdot (x/L)^2$ and then abruptly
dies off. Determine the motion of the cable if the time dependent amplitude of the
wind load may be represented by a half
sine wave of magnitude q_0 and duration
$t*$.

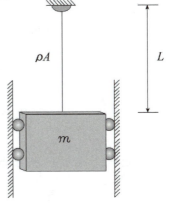

Solution
The wind load is represented mathematically as

$$q(x,t) = \hat{q}(t) \cdot (x/L)^2 \tag{a-1}$$

where

$$\hat{q}(t) = q_0 \sin(\pi t/t*)\left[\mathcal{H}(t) - \mathcal{H}(t - t*)\right] \tag{a-2}$$

The natural frequencies and modal functions for a cable under uniform tension
and fixed at its ends were determined in Example 10.5. Setting $N = mg$ in
those results gives the frequencies and modal functions for the present system
as

$$\omega_j = j\pi \sqrt{\frac{mg}{mL^2}} \tag{b-1}$$

and

$$W^{(j)}(x) = \sin\left(j\pi x/L\right) \tag{b-2}$$

The modal mass is then

$$\tilde{m}_j = \int_0^L W^{(j)}(x) m W^{(j)}(x)\,dx = m\int_0^L \cos^2\left(j\pi x/L\right)dx = \frac{mL}{2} \tag{c}$$

The corresponding modal forces are calculated as

$$\tilde{F}_j(t) = \int_0^L W^{(j)}(x)q(x,t)\,dx$$

$$= \int_0^L \sin(j\pi x/L)q_0[x/L]^2 \sin(\pi t/t^*)[\mathcal{H}(t) - \mathcal{H}(t-t^*)]\,dx \qquad \text{(d)}$$

$$= -\frac{q_0 L}{(j\pi)^3}\left[2 + \{(j\pi)^2 - 2\}(-1)^j\right]\sin(\pi t/t^*)[\mathcal{H}(t) - \mathcal{H}(t-t^*)]$$

from which it follows that

$$\tilde{f}_j(t) = \tilde{f}_j^0 \sin(\pi t/t^*)[\mathcal{H}(t) - \mathcal{H}(t-t^*)] \qquad \text{(e-1)}$$

where

$$\tilde{f}_j^0 = \frac{2q_0 L^2}{mg}\frac{\left[2 + \{(j\pi)^2 - 2\}(-1)^j\right]}{(j\pi)^5} \qquad \text{(e-2)}$$

Therefore, for the present problem, Eq. (11.15) takes the form

$$\ddot{\eta}_j(t) + \omega_j^2 \eta_j(t) = \omega_j^2 \tilde{f}_j^0 \sin(\pi t/t^*)[\mathcal{H}(t) - \mathcal{H}(t-t^*)] \qquad \text{(f)}$$

The solution to Eq. (f) is obtained by direct application of Eq. (3.28) with $\Omega = \pi/t^*$ giving

$$\eta_j(t) = \frac{\tilde{f}_j^0}{1 - \left(mL^2/j^2\pi mg\right)^2}\sin(\pi t/t^*)[\mathcal{H}(t) - \mathcal{H}(t-t^*)] \qquad \text{(g)}$$

The response of the cable is then

$$w(x,t) = \sum_{j=1}^{\infty} \eta_j(t)W^{(j)}(x)$$

$$= \frac{2q_0 L^2}{mg}\sin(\pi t/t^*)\sum_{j=1}^{\infty}\frac{\sin(j\pi x/L)}{(j\pi)^5}\frac{\left[2 + \{(j\pi)^2 - 2\}(-1)^j\right]}{\left[1 - \left(mL^2/j^2\pi \bar{m}g\right)^2\right]}[\mathcal{H}(t) - \mathcal{H}(t-t^*)]$$

$$\text{(h)}$$

Hence,

$$w(x,t) = 0 \quad (t < 0) \qquad \lhd$$

$$w(x,t) = \frac{2q_0 L^2}{mg}\sin(\pi t/t^*)\sum_{j=1}^{\infty}\frac{\sin(j\pi x/L)}{(j\pi)^5}\frac{\left[2 + \{(j\pi)^2 - 2\}(-1)^j\right]}{\left[1 - \left(mL^2/j^2\pi \bar{m}g\right)^2\right]} \quad (0 \le t \le t^*) \qquad \lhd$$

$$w(x,t) = 0 \quad (t > t^*)$$ ◁

11.3.4 Euler-Bernoulli Beams

We now apply the formulation presented in Sections 9.1 and 9.2 to determine the dynamic response of Euler-Bernoulli Beams to externally applied forces. The corresponding equation of motion, Eq. (9.104), is

$$\frac{\partial^2}{\partial x^2} EI \frac{\partial^2 w}{\partial x^2} + m \frac{\partial^2 w}{\partial t^2} = q(x,t) - \frac{\partial b}{\partial x}$$

where $w(x,t)$ is the transverse displacement of the point on the axis of the beam originally at x, $q(x,t)$ represents a distributed transverse load and $b(x,t)$ is a distributed couple. Thus, for the present type of structure, $\mathbf{u}(x,t) \rightarrow w(x,t)$, $\mathbf{F}(x,t) \rightarrow q(x,t) - \partial b/\partial x$ and $\mathbf{m}(x) \rightarrow m(x)$ in the development and formulation of Sections 11.1 and 11.2.

Example 11.4

Consider the simply supported uniform Euler-Bernoulli Beam shown in the figure. Determine the response of the beam if a distributed transverse load that varies linearly along the length of the beam is suddenly applied, maintained at a constant level over a time duration t^*, and is then suddenly unloaded. Also, compute the reactions of the supports during the time that the distributed load is present.

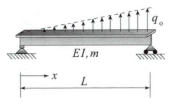

Solution

The free vibration problem for a simply supported beam was considered in Example 10.6. The corresponding natural frequencies and normal modes are found to be

$$\omega_j = (j\pi)^2 \sqrt{\frac{EI}{mL^4}} \qquad (a\text{-}1)$$

and

$$W^{(j)}(x) = \sqrt{\frac{2}{mL}} \sin\left(j\pi x/L\right) \qquad (a\text{-}2)$$

respectively, where we have normalized the latter as described in Section 10.7. To solve the problem using modal analysis we seek a solution in the form

$$w(x,t) = \sum_{j=1}^{\infty} \eta_j(t) W^{(j)}(x)$$

where the modal displacements $\eta_j(t)$ are solutions to the problems

$$\ddot{\eta}_j + \omega_j^2 \eta_j = \omega_j^2 \tilde{f}_j(t)$$

To solve for the modal displacements we must first determine the modal forces. The applied loading may be expressed mathematically as

$$q(x,t) = q_0 \frac{x}{L} g(t) \tag{b-1}$$

where,

$$g(t) = \mathcal{H}(t) - \mathcal{H}(t - t^*) \tag{b-2}$$

is a rectangular pulse (Section 4.6.1) and $\mathcal{H}(t)$ is the Heaviside Step Function (Section 4.1.2). The modal force is then determined using Eq. (11.12). Hence,

$$\tilde{F}_j(t) = \int_0^L W^{(j)}(x) q(x,t) \, dx = \int_0^L \sqrt{\frac{2}{mL}} \sin(j\pi x/L) q_0 \frac{x}{L} g(t) \, dx$$

Performing the integration gives the jth modal force as

$$\tilde{F}_j(t) = -q_0 g(t) \sqrt{\frac{2}{mL}} \cdot \frac{L}{2\pi} \cos(j\pi) = \frac{(-1)^{j+1}}{j\pi} q_0 \sqrt{\frac{2L}{m}} g(t) \tag{c}$$

Since the modal functions have been normalized with respect to m, the modal masses are equal to one. Therefore

$$\omega_j^2 = \frac{\tilde{k}_j}{\tilde{m}_j} = \frac{\tilde{k}_j}{1} \Rightarrow \tilde{k}_j = \omega_j^2 = (j\pi)^2 \sqrt{\frac{EI}{mL^4}} \tag{d}$$

It then follows that,

$$\tilde{f}_j(t) \equiv \frac{\tilde{F}_j(t)}{\tilde{k}_j} = \tilde{f}_j^0 g(t) \tag{e-1}$$

where

$$\tilde{f}_j^0 = \frac{(-1)^{j\pi}}{(j\pi)^3} q_0 \sqrt{\frac{2L}{m}} \qquad \text{(e-2)}$$

To determine the modal displacements, we must solve the problem

$$\ddot{\eta}_j + \omega_j^2 \eta_j = \omega_j^2 \tilde{f}_j^0 \left[\mathcal{H}(t) - \mathcal{H}(t - t^*) \right] \qquad \text{(f)}$$

Each mode is seen to behave as a single degree of freedom system subjected to a step load (Section 4.6.1). The solution of Eq. (f) follows directly from Eqs. (4.38)–(4.40) giving the modal displacements as

$$\eta_j(t) = f_j^0 \left[1 - \cos \omega_j t \right] \mathcal{H}(t) - f_j^0 \left[1 - \cos \omega_j (t - t_1) \right] \mathcal{H}(t - t^*) \qquad \text{(g)}$$

Hence,

$$\eta_j(t) = q_0 \sqrt{\frac{2L}{m}} \frac{(-1)^{j+1}}{j\pi \omega_j^2} \left[1 - \cos \omega_j t \right] \qquad (0 \le t < t^*) \qquad \text{(h-1)}$$

$$\eta_j(t) = q_0 \sqrt{\frac{2L}{m}} \frac{(-1)^{j+1}}{j\pi \omega_j^2} \left[\cos \omega_j (t - t^*) - \cos \omega_j t \right] \qquad (t \ge t^*) \qquad \text{(h-2)}$$

The forced response of the beam is then

$$w(x,t) = q_0 \frac{mL^4}{EI} \sum_{j=1}^{\infty} \frac{(-1)^{j+1}}{(j\pi)^5} \left[1 - \cos \omega_j t \right] \sin \left(j\pi x / L \right) \quad (0 \le t < t^*) \quad \triangleleft \text{(i-1)}$$

$$w(x,t) = q_0 \frac{mL^4}{EI} \sum_{j=1}^{\infty} \frac{(-1)^{j+1}}{(j\pi)^5} \left[\cos \omega_j (t - t^*) - \cos \omega_j t \right] \sin \left(j\pi x / L \right)$$

$$(t \ge t^*) \quad \triangleleft \text{(i-2)}$$

To determine the reactions we must first determine the transverse shear in the beam. This is accomplished by substituting Eqs. (i-1) and (i-2) into Eqs. (9.102) and (9.97). Hence,

$$Q(x,t) = -EI \frac{\partial^3 w}{\partial x^3}$$

$$= q_0 \frac{mL}{EI} \sum_{j=1}^{\infty} \frac{(-1)^{j+1}}{(j\pi)^2} \left[1 - \cos \omega_j t \right] \cos \left(j\pi x / L \right) \quad (0 \le t < t^*) \qquad \text{(j-1)}$$

$$Q(x,t) = -EI\frac{\partial^3 w}{\partial x^3}$$

$$= q_0 \frac{mL}{EI} \sum_{j=1}^{\infty} \frac{(-1)^{j+1}}{(j\pi)^2} \left[\cos \omega_j (t-t^*) - \cos \omega_j t\right] \cos\left(j\pi x/L\right) \quad (t \geq t^*) \tag{j-2}$$

We can compute the reactions during the first phase of the loading using Eq. (j-1) as follows. Employing the convention shown in the figure we obtain

$$R_0(t) = Q(0,t) = q_0 \frac{mL}{EI} \sum_{j=1}^{\infty} \frac{(-1)^{j+1}}{(j\pi)^2} \left[1 - \cos \omega_j t\right] \quad (0 \leq t < t^*) \qquad \triangleleft \text{(k-1)}$$

$$R_L(t) = -Q(L,t) = q_0 \frac{mL}{EI} \sum_{j=1}^{\infty} \frac{(-1)^{j+3}}{(j\pi)^2} \left[1 - \cos \omega_j t\right] \quad (0 \leq t < t^*) \qquad \triangleleft \text{(k-2)}$$

Example 11.5

A simply supported Euler-Bernoulli Beam is impacted from above at a point located a distance a from the left support as shown. Determine the response of the beam if the magnitude of the impact is $\mathcal{I}_0$.

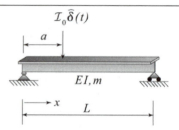

Solution

The impact force may be expressed as an equivalent distributed load by using the Dirac Delta Function (Section 4.4.1) in space as well as in time. The applied force is then

$$q(x,t) = -\mathcal{I}_0 \widehat{\delta}(t) \widehat{\delta}(x-a) \tag{a}$$

The frequencies and normal modes were presented in Eqs. (a-1) and (a-2) of Example 11.4. Since the modes were normalized with respect to the mass, the modal masses are all equal to one. The modal forces for the present problem are computed as follows:

$$\tilde{F}_j(t) = \int_0^L W^{(j)}(x) q(x,t)\, dx = -\sqrt{\frac{2}{mL}} \int_0^L \sin\left(j\pi x/L\right) \mathcal{I}_0 \widehat{\delta}(t)\widehat{\delta}(x-a)\, dx \tag{b}$$

Hence,

$$\tilde{F}_j(t) = \mathcal{I}_j\,\hat{\delta}(t) \tag{c-1}$$

where

$$\mathcal{I}_j = -\mathcal{I}_0\sqrt{\frac{2}{mL}}\,\sin\left(j\pi a/L\right) \tag{c-2}$$

Since the modal masses are unity, Eq. (11.9) takes the form

$$\overset{1}{\tilde{m}_j}\,\ddot{\eta}_j + \tilde{k}_j\eta_j = \mathcal{I}_j\,\hat{\delta}(t) \tag{d}$$

The modal response follows directly from Eq. (4.20) as

$$\eta_j(t) = \frac{\mathcal{I}_j}{\tilde{m}_j\,\omega_j}\sin\omega_j t\,\mathcal{H}(t) = -\mathcal{I}_0\sqrt{\frac{2L}{EI}}\,\sin\left(j\pi a/L\right)\sin\omega_j t\,\mathcal{H}(t) \tag{e}$$

The response of the beam is then

$$w(x,t) = \sum_{j=1}^{\infty}\eta_j(t)W^{(j)}(x)$$

$$= -\frac{2\mathcal{I}_0}{\sqrt{mEI}}\sum_{j=1}^{\infty}\sin\left(j\pi a/L\right)\sin\left(j\pi x/L\right)\sin(\omega_j t)\,\mathcal{H}(t) \qquad \triangleleft\text{(f)}$$

Example 11.6

Consider a simply supported Euler-Bernoulli Beam subjected to a moving point load of magnitude Q_0 that is directed downward as shown. Determine the response of the beam if the load moves with velocity c_Q.

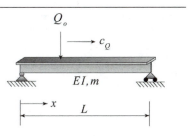

Solution

The point load may be represented as a distributed load by using the Dirac Delta Function in space, as in Example 11.5, and letting $a = c_Q t$. Hence,

$$q(x,t) = -Q_0\tilde{\delta}(x - c_Q t) \tag{a}$$

The modal forces are then,

$$\tilde{F}_j(t) = -Q_0 \int_0^L W^{(j)}(x)\hat{\delta}(x - c_Q t)\,dx = -Q_0 W^{(j)}(c_Q t) \tag{b}$$

The natural frequencies and normalized modal functions are, from Example 11.4,

$$W^{(j)}(x) = \sqrt{\frac{2}{mL}}\sin(j\pi x/L), \quad \omega_j = (j\pi)^2 \sqrt{\frac{EI}{mL^4}} \tag{c-1, 2}$$

Thus,

$$W^{(j)}(c_Q t) = \sqrt{\frac{2}{mL}}\sin\Omega_j t \tag{d}$$

where

$$\Omega_j = j\pi c_Q/L \tag{e}$$

Then, for the present problem, Eq. (11.15) takes the form

$$\ddot{\eta}_j + \omega_j^2 \eta_j = \omega_j^2 \tilde{f}_j^0 \sin\Omega_j t \tag{f}$$

where

$$\tilde{f}_j^0 = -\frac{Q_0 L^3}{EI}\frac{\sqrt{2mL}}{(j\pi)^4} \tag{g}$$

The solution to Eq. (f) is written directly from Eq. (3.28) as

$$\eta_j(t) = \tilde{f}_j^0 \Gamma_j \sin\Omega_j t = \tilde{f}_j^0 \Gamma_j \sin(j\pi c_Q t/L) \tag{h}$$

where

$$\Gamma_j = \frac{1}{1 - (\Omega_j/\omega_j)^2} = \frac{(j\pi)^2}{(j\pi)^2 - (\bar{c}_Q/\bar{r}_G)^2} \tag{i}$$

$$\bar{c}_Q = c_Q/c_a \tag{j}$$

$$\bar{r}_G = r_G/L \tag{k}$$

and

$$r_G^2 = I/A \tag{l}$$

The response of the beam to the moving point load is then

$$w(x,t) = -\frac{2Q_0 L^3}{EI}\sum_{j=1}^{\infty}\frac{1}{(j\pi)^2\left[(j\pi)^2-\left(\bar{c}_Q/\bar{r}_G\right)^2\right]}\sin\left(j\pi x/L\right)\sin\left(j\pi c_Q t/L\right)$$ ◁

It is seen that a resonance condition exists when $c_Q = j\pi\omega_0 L$ or, equivalently, when $\bar{c}_Q = j\pi\bar{r}_G$.

Example 11.7

A cantilevered Euler-Bernoulli Beam is loaded by a time dependent moment, $M_L(t)$, at its free end. Determine the steady state response of the beam if the applied moment is of the form $M_L(t) = M_0 \cos \Omega t$.

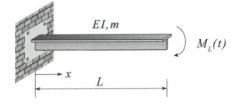

Solution

The frequencies for a uniform cantilevered beam are, from Example 10.7,

$$\omega_1 = 3.516\,\omega_0, \quad \omega_2 = 22.03\,\omega_0, \quad \omega_3 = 61.70\,\omega_0, \ ... \tag{a}$$

where

$$\omega_0 = \sqrt{\frac{EI}{mL^4}} = \frac{c_a r_G}{L^2} \tag{b}$$

The corresponding modal functions were found to be

$$W^{(j)}(x) = \cosh \beta_j x - \cos \beta_j x - Y_j\left[\sinh \beta_j x - \sin \beta_j x\right] \tag{c}$$

where

$$Y_j = \left(\cosh \beta_j L + \cos \beta_j L\right)\big/\left(\sinh \beta_j L + \sin \beta_j L\right) \tag{d}$$

and

$$\beta_1 L = 1.875, \ \beta_2 L = 4.695, \ \beta_3 L = 7.855, \ ... \tag{e}$$

The loading for the present problem, the applied moment at $x = L$, may be expressed as a distributed moment with the aid of the Dirac Delta Function. The corresponding body couple is then

$$b(x,t) = M_L(t)\widehat{\delta}(x-L) \tag{f}$$

Since there is no distributed transverse load, we have that

$$F(x,t) = \cancel{q(x,t)} - \frac{\partial b}{\partial x} = -\frac{\partial b}{\partial x} \tag{g}$$

The modal forces for the cantilever beam are then found as

$$\tilde{F}_j(x,t) = -\int_0^L W^{(j)}(x) \frac{\partial b(x,t)}{\partial x} dx = -M_0 \cos\Omega t \int_0^L W^{(j)}(x)\hat{\delta}'(x-L)\,dx$$

The integral in the above expression is evaluated using Eq. (4.8) to give

$$\tilde{F}_j(t) = -M_0 \cos\Omega t\, W^{(j)'}(L) = -2M_0 \cos\Omega t \left[\frac{\beta_j \sinh\beta_j L \sin\beta_j L}{\sinh\beta_j L + \sin\beta_j L} \right] \tag{h}$$

Hence,

$$\tilde{f}_j(t) = \frac{\tilde{F}_j(t)}{\omega_j^2 \tilde{m}_j} = \tilde{f}_j^0 \cos\Omega t \tag{i-1}$$

where

$$\tilde{f}_j^0 = -2\frac{M_0}{EI}\frac{1}{\beta_j^3\left(\tilde{m}_j/m\right)}\left[\frac{\sinh\beta_j L \sin\beta_j L}{\sinh\beta_j L + \sin\beta_j L} \right] \tag{i-2}$$

and

$$\tilde{m}_j = \int_0^L W^{(j)}(x)\, m\, W^{(j)}(x)\, dx = m\left\| W^{(j)} \right\|^2$$
$$= m \int_0^L \left[\cosh\beta_j x - \cos\beta_j x - g_j\left(\sinh\beta_j x - \sin\beta_j x\right) \right]^2 dx \tag{j}$$

Thus, for the present problem, Eq. (11.15) takes the form

$$\ddot{\eta}_j(t) + \omega_j^2 \eta(t) = \omega_j^2 \tilde{f}_j^0 \cos\Omega t \tag{k}$$

where $\tilde{f}_j^0$ is given by Eq. (i-2). The solution is then written directly from Eq. (3.27) as

$$\eta_j(t) = -\frac{2M_0 L^3}{EI}\frac{\tilde{G}(\beta_j L)}{\left[1-\left(\Omega/\omega_j\right)^2\right]}\cos\Omega t \tag{l}$$

where

$$\tilde{G}(\beta_j L) = \frac{\sinh \beta_j L \sin \beta_j L}{\left\| W^{(j)} \right\|^2 (\beta_j L)^3 \left(\sinh \beta_j L + \sin \beta_j L \right)} \tag{m}$$

and

$$4\beta_j \left\| W^{(j)} \right\|^2 = \sinh 2\beta_j L + \sin 2\beta_j L + 12\beta_j L$$
$$+ 2Y_j \left[\cos 2\beta_j L - \cosh 2\beta_j L + 4\sinh \beta_j L \sin \beta_j L \right]$$
$$+ Y_j^2 \left[\sinh 2\beta_j L - \sin 2\beta_j L + 4\sinh 2\beta_j L \cos 2\beta_j L \right. \tag{n}$$
$$\left. - 2\cosh 2\beta_j L \sin 2\beta_j L \right]$$

The steady state response of the beam is then

$$w(x,t) = -2\frac{M_0 L^2}{EI} \cos \Omega t \sum_{j=1}^{\infty} \frac{\tilde{G}(\beta_j L)}{\left[1 - \left(\Omega/\omega_j \right)^2 \right]} \cdot$$
$$\left[\cosh \beta_j x - \cos \beta_j x - Y_j \left(\sinh \beta_j x - \sin \beta_j x \right) \right] \tag{$\triangleleft$ (o)}$$

We now compute $\tilde{G}(\beta L)$ for the first three modes using Eqs. (m) and (n) to get

$$\tilde{G}(\beta_1 L) = \tilde{G}(1.875) = 0.0527 \tag{p-1}$$
$$\tilde{G}(\beta_2 L) = \tilde{G}(4.695) = 7.210 \times 10^{-6} \tag{p-2}$$
$$\tilde{G}(\beta_3 L) = \tilde{G}(7.855) = -4.887 \times 10^{-7} \tag{p-3}$$

from which it is seen that the first mode is dominant (except near resonance conditions other than $\Omega = \omega_1$). Substitution of Eqs. (p) into Eq. (o) gives the explicit steady state response

$$w(x,t) = -\frac{M_0 L^2}{EI} \cos \Omega t \frac{0.1055}{\left[1 - 0.08089 \left(\Omega/\omega_0 \right)^2 \right]} \cdot$$
$$\left[\cosh(1.875x/L) - \cos(1.875x/L) \right.$$
$$\left. - 0.7351 \{ \sinh(1.875x/L) - \sin(1.875x/L) \} \right] + \dots \tag{$\triangleleft$ (q)}$$

Example 11.8

The support of a uniform cantilevered elastic beam of bending stiffness *EI*, length *L*, and mass per unit length *m* undergoes the prescribed lateral motion $h(t) = h_0 \sin \Omega t$, where h_0 is a constant. Determine the steady state motion of the beam when modeled using Euler-Bernoulli Theory.

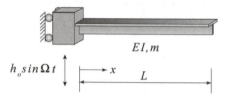

EI, m

$h_o \sin \Omega t$ x *L*

Solution

The equation of motion for the beam can be written directly from Eq. (9.184) with $\chi_z(t) = h(t)$. For the present problem, $N = \chi_x = p = 0$. We also have no distributed transverse load or body couple as well ($q = 6 = 0$), so that the equation of transverse motion takes the form

$$EI\frac{\partial^4 w}{\partial x^4} + m\frac{\partial^2 w}{\partial t^2} = -m\ddot{h}(t) \tag{a}$$

where $w(x,t)$ represents the displacement *relative to the support*. It should be pointed out that if we neglect geometric nonlinearities, as is done for Euler-Bernoulli Theory (presently) and for the Rayleigh and Timoshenko theories, then the nonlinear terms appearing in Eqs. (9.184) and (9.185) are neglected even when they do not vanish, and the equation of motion takes the form of Eq. (a), but with nonvanishing *q* when appropriate. That is, the equations for transverse and longitudinal motion decouple. Equation (a) could also be derived directly for this relatively simple problem by replacing $w(x,t)$ with the total deflection $\xi(x,t) = h(t) + w(x,t)$ in Eq. (9.104) (which was restated in the beginning of the current section). That is, Eq. (9.104) is derived with respect to a fixed reference frame. Thus, if $\xi_z(x,t)$ measures the absolute displacement then Eq. (9.104) takes the form

$$\frac{\partial^2}{\partial x^2}EI\frac{\partial^2 \xi_z}{\partial x^2} + m\frac{\partial^2 \xi_z}{\partial t^2} = q - \frac{\partial \phi}{\partial x}$$

Substituting $\xi_z(x,t) = h(t) + w(x,t)$ into the above equation, with *EI* = constant, results in Eq. (a).

Now, for the given excitation,

$$F(x,t) = q - \frac{\partial \phi}{\partial x} - m\ddot{h} = F_0 \sin \Omega t \tag{b-1}$$

where

$$F_0 = \Omega^2 m h_0 \tag{b-2}$$

From this point on, the problem proceeds as for a fixed beam subjected to a uniformly distributed harmonic excitation. The natural frequencies and modal functions for the structure were established in Example 10.7 and employed in Example 11.7. We employ these same functions for the present problem. The modal forces are then

$$\tilde{F}^{(j)}(t) = \int_0^L W^{(j)}(x) F_0 \sin \Omega t \, dx$$

$$= \frac{F_0 \sin \Omega t}{\beta_j} \left[\sinh \beta_j x - \sin \beta_j x - Y_j \left\{ \cosh \beta_j x + \cos \beta_j x \right\} \right] \tag{c}$$

where

$$\mathfrak{J}(\beta_j x) = \beta_j^{-1} \left[\sinh \beta_j x - \sin \beta_j x - Y_j \left\{ \cosh \beta_j x + \cos \beta_j x \right\} \right] \tag{d}$$

It follows that

$$\tilde{f}^{(j)}(t) = \tilde{f}_j^0 \sin \Omega t \tag{e-1}$$

where

$$\tilde{f}_j^0 = \frac{\tilde{F}_j^0}{\omega_j^2 \tilde{m}} = \frac{F_0 \mathfrak{J}(\beta_j L)}{EI \beta_j^4 \left\| W^{(j)} \right\|^2} \tag{e-2}$$

and $\left\| W^{(j)} \right\|^2$ is given by Eq. (n) of Example 11.7. For the present system, Eq. (11.15) takes the form

$$\ddot{\eta}_j(t) + \omega_j^2 \eta_j(t) = \omega_j^2 \tilde{f}_j^0 \sin \Omega t \quad (j = 1, 2, \ldots) \tag{f}$$

The modal displacements are written directly from Eq. (3.28) as

$$\eta_j(t) = \frac{\tilde{f}_j^0}{1 - \left(\Omega / \omega_j \right)^2} \sin \Omega t \tag{g}$$

The steady state response of the beam then follows from Eq. (11.5) as

$$w(x,t) = \frac{mL^4}{EI} \Omega^2 h_0 \sin \Omega t \sum_{j=1}^{\infty} \frac{\mathfrak{J}(\beta_j L)}{\left[1 - \left(\Omega / \omega_j \right)^2 \right] \beta_j^4 \left\| W^{(j)} \right\|^2} \quad \triangleleft \text{(h)}$$

Recall that $w(x,t)$ is measured relative to the support. The motion of the beam measured with respect to a fixed coordinate system is then

$$\xi_z(x,t) = h(t) + w(x,t)$$

$$= h_0 \sin \Omega t \left\{ 1 + \frac{mL^4}{EI} \Omega^2 \sum_{j=1}^{\infty} \frac{\mathcal{J}(\beta_j L)}{\left[1 - \left(\Omega/\omega_j \right)^2 \right] \beta_j^4 \left\| W^{(j)} \right\|^2} \right\} \quad \triangleleft (i)$$

11.3.5 Rayleigh Beams

In this section we apply the formulation presented in Sections 9.1 and 9.2 to determine the dynamic response of Rayleigh Beams to externally applied forces. For this model there are slight differences in the form of the statements for orthogonality of the modal functions, and hence for the definitions of the modal mass and modal stiffness, from those pertaining to the other systems considered in this chapter. We therefore parallel the development and formulation of Section 11.1 in detail for the particular case of Rayleigh Beams.

The equation of motion for these structures, Eq. (9.115), is

$$m \frac{\partial^2 w}{\partial t^2} - I_\rho \frac{\partial^4 w}{\partial x^2 \partial t^2} + \frac{\partial^2}{\partial x^2} EI \frac{\partial^2 w}{\partial x^2} = q(x,t) - \frac{\partial \mathfrak{b}}{\partial x}$$

where I_ρ is the rotatory inertia and, as for Euler-Bernoulli Beams, $w(x,t)$ is the transverse displacement of the point on the axis of the beam originally at x, $q(x,t)$ represents a distributed transverse load, $\mathfrak{b}(x,t)$ is a distributed couple and $m(x)$ is the mass per unit length.

We seek a solution in the form of an expansion in terms of the modal functions. For the present case, Eq. (11.5) takes the explicit form

$$w(x,t) = \sum_{j=1}^{\infty} \eta_j(t) W^{(j)}(x)$$

Substituting the above modal expansion into the equation of motion, multiplying the resulting expression by $W^{(l)} dx$ and integrating over the domain of definition of the beam gives

$$\sum_{j=1}^{\infty} \ddot{\eta}_j(t) \int_0^L \left[W^{(l)} m W^{(j)} + W^{(l)'} I_\rho W^{(j)'} \right] dx$$

$$+ \sum_{j=1}^{\infty} \eta_j(t) \int_0^L W^{(l)''} I_\rho W^{(j)''} dx + B_l^* = \int_0^L W^{(l)}(x) F(x,t) dx \tag{11.17}$$

where

$$F(x,t) = q(x,t) - \frac{\partial b}{\partial x} \tag{11.18}$$

and

$$B_l^* = \left[W^{(l)}(x) Q(x,t) + W^{(l)'}(x) M(x,t) \right]_0^L$$

If the boundary conditions are such that

$$B_l^* = 0 \tag{11.19}$$

which is consistent with the conditions for orthogonality, Eq. (10.159), and we exploit the mutual orthogonality of the modal functions as per Eqs. (10.160) and (10.168), Eq. (11.17) reduces to the familiar system of uncoupled differential equations for the modal displacements

$$\tilde{m}_j \ddot{\eta}_j(t) + \tilde{k}_j \eta_j(t) = \tilde{F}_j(t) \quad (j = 1, 2, ...) \tag{11.20}$$

where, for Rayleigh Beams,

$$\tilde{m}_j = \int_0^L m(x) W^{(j)2}(x) dx + \int_0^L I_\rho(x) W^{(j)'2}(x) dx \quad (j = 1, 2, ...) \tag{11.21}$$

$$\tilde{k}_j = \int_0^L W^{(j)''} EI W^{(j)''} dx \quad (j = 1, 2, ...) \tag{11.22}$$

and

$$\tilde{F}_j(t) = \int_0^L W^{(j)}(x) \left[q(x,t) - \partial b / \partial x \right] dx \quad (j = 1, 2, ...) \tag{11.23}$$

Now, it follows from Eq. (10.166), (11.21) and (11.22) that

$$\tilde{k}_j = \omega_j^2 \tilde{m}_j \quad (j = 1, 2, ...) \tag{11.24}$$

Equations (11.20) can thus be rewritten in the standard form of Eqs. (11.15) and (11.16),

$$\ddot{\eta}_j + \omega_j^2 \eta_j = \omega_j^2 \tilde{f}_j(t) \quad (j = 1, 2, ...)$$

where

$$\tilde{f}_j(t) = \frac{\tilde{F}_j(t)}{\tilde{k}_j} = \frac{\tilde{F}_j(t)}{\omega_j^2 \tilde{m}_j} \quad (j = 1, 2, \ldots)$$

At this stage the analysis proceeds as for previous systems.

Example 11.9

Set up the solution for a uniform cantilevered Rayleigh Beam of bending stiffness EI, mass m, rotatory inertia I_ρ and length L, that is loaded by a moment on its free edge as in Example 11.7.

Solution

From Example 10.12, the modal functions for the beam are of the form

$$W^{(j)}(x) = \left\{ \cosh \alpha_j x - \cos \beta_j x \right\} - h^{(j)} \left\{ \sinh \alpha_j x - \left(\alpha_j / \beta_j \right) \sin \beta_j x \right\} \tag{a}$$

where

$$h^{(j)} = \frac{\alpha_j^2 \cosh \alpha_j L + \beta_j^2 \cos \beta_j L}{\alpha_j^2 \sinh \alpha_j L + \alpha_j \beta_j \sin \beta_j L} \tag{b}$$

In the above expressions, α_j and β_j correspond to the values of the parameters defined by Eqs. (10.51) and (10.52) evaluated at the jth natural frequency ω_j, which corresponds to the jth root of Eq. (e) of Example 10.12. Next, substituting Eq. (a) into Eq. (11.21) allows for the evaluation of the jth modal mass. Hence,

$$\tilde{m}_j = m \int_0^L \left[\left\{ \cosh \alpha_j x - \cos \beta_j x \right\} - h^{(j)} \left\{ \sinh \alpha_j x - \left(\alpha_j / \beta_j \right) \sin \beta_j x \right\} \right]^2 dx$$

$$+ I_\rho \int_0^L \left[\left\{ \alpha_j \sinh \alpha_j x + \beta_j \sin \beta_j x \right\} - h^{(j)} \alpha_j \left\{ \cosh \alpha_j x - \cos \beta_j x \right\} \right]^2 dx \tag{c}$$

The corresponding modal force for the cantilever beam is calculated as

$$\tilde{F}_j(x,t) = -\int_0^L W^{(j)}(x) \frac{\partial \mathfrak{b}(x,t)}{\partial x} dx$$

$$= -M_0 \cos \Omega t \int_0^L W^{(j)}(x) \hat{\delta}'(x - L) dx = -M_0 \cos \Omega t \, W^{(j)\prime}(L)$$

Hence,

$$\tilde{F}_j(t) = -M_0 \cos\Omega t \left[\left\{ \alpha_j \sinh\alpha_j x + \beta_j \sin\beta_j x \right\} - h^{(j)}\alpha_j \left\{ \cosh\alpha_j x - \cos\beta_j x \right\} \right]$$

(d)

It follows that

$$\tilde{f}_j(t) = \tilde{f}_j^0 \cos\Omega t \tag{e-1}$$

where

$$\tilde{f}_j^0 = -\frac{M_0\left[\left\{\alpha_j \sinh\alpha_j L + \beta_j \sin\beta_j L\right\} - h^{(j)}\alpha_j\left\{\cosh\alpha_j L - \cos\beta_j L\right\}\right]}{\omega_j^2 \tilde{m}_j} \tag{e-2}$$

Thus, for the present problem, Eq. (11.15) takes the form

$$\ddot{\eta}_j(t) + \omega_j^2 \eta(t) = \omega_j^2 \tilde{f}_j^0 \cos\Omega t \tag{f}$$

where $\tilde{f}_j^0$ is given by Eq. (e-2). The solution to this equation is then written directly from Eq. (3.27) as

$$\eta_j(t) = M_0 \cos\Omega t \frac{\left[\left\{\alpha_j \sinh\alpha_j L + \beta_j \sin\beta_j L\right\} - h^{(j)}\alpha_j\left\{\cosh\alpha_j L - \cos\beta_j L\right\}\right]}{\tilde{m}_j\left(\Omega^2 - \omega_j^2\right)}$$

(g)

The steady state response of the beam then follows as

$$w(x,t) = M_0 \cos\Omega t \cdot$$

$$\sum_{j=1}^{\infty} \frac{\left[\left\{\alpha_j \sinh\alpha_j L + \beta_j \sin\beta_j L\right\} - h^{(j)}\alpha_j\left\{\cosh\alpha_j L - \cos\beta_j L\right\}\right]}{\tilde{m}_j\left(\Omega^2 - \omega_j^2\right)} W^{(j)}(x)$$

◁ (h)

where $W^{(j)}(x)$ is given by Eq. (a).

11.3.6 Timoshenko Beams

We now apply the formulation presented in Sections 9.1 and 9.2 to determine the dynamic response of Timoshenko Beams to externally applied forces. The corresponding equation of motion is, from Eq. (9.135),

$$
\begin{bmatrix} m & 0 \\ 0 & I_\rho \end{bmatrix} \begin{Bmatrix} \partial^2 w/\partial t^2 \\ \partial^2 \varphi/\partial t^2 \end{Bmatrix} + \begin{bmatrix} -\dfrac{\partial}{\partial x} k_s \dfrac{\partial}{\partial x} & \dfrac{\partial}{\partial x} k_s \\ -k_s \dfrac{\partial}{\partial x} & k_s - \dfrac{\partial}{\partial x} EI \dfrac{\partial}{\partial x} \end{bmatrix} \begin{Bmatrix} w \\ \varphi \end{Bmatrix} = \begin{Bmatrix} q(x,t) \\ 6(x,t) \end{Bmatrix}
$$

where $w(x,t)$ is the transverse displacement of the point on the axis of the beam originally at x, $\varphi(x,t)$ is the rotation of the cross section at x, $q(x,t)$ represents a distributed
transverse load and $6(x,t)$ is a distributed couple. Thus, for the present type of struc-
ture, $\mathbf{u} \rightarrow [w\ \varphi]^T$, $\mathbf{F} \rightarrow [q\ 6]^T$, and $\mathbf{m}$ and $\mathbf{k}$ are given by the square matrices in the
above equation of motion.

Inserting the mass and displacement matrices for the Timoshenko Beam into
Eq. (11.10) gives the explicit expression for the modal masses as

$$
\tilde{m}_j = \int_0^L m(x) W^{(j)\,2}(x)\, dx + \int_0^L I_\rho(x) \varphi^{(j)\,2}(x)\, dx \quad (j = 1,2,\ldots) \qquad (11.25)
$$

Note that if we normalize with respect to the mass operator this quantity is one for
each mode. Likewise, inserting the force and displacement matrices for the Timoshenko Beam into Eq. (11.12) gives the explicit expression for the modal forces as

$$
\tilde{F}_j(t) = \int_0^L W^{(j)}(x) q(x,t)\, dx + \int_0^L \varphi^{(j)}(x) 6(x,t)\, dx \qquad (11.26)
$$

Example 11.10

Set up the solution for a uniform cantilevered Timoshenko Beam of bending
stiffness EI, modulus ratio E/kG, mass per unit length m, rotatory inertia I_ρ and
length L, that is loaded by a moment on its free edge as in Examples 11.7 and
11.9.

Solution

From Example 10.13, the modal functions for the beam are of the form

$$
\mathbf{U}^{(j)}(x) = \begin{Bmatrix} W^{(j)}(x) \\ \vartheta^{(j)}(x) \end{Bmatrix}
$$

$$
= \begin{Bmatrix} H_{12}^{(j)}\left[\cosh \alpha_j x - \cos \beta_j x\right] - H_{11}^{(j)}\left[\sinh \alpha_j x - \left(g_\alpha^{(j)}/g_\beta^{(j)}\right)\sin \beta_j x\right] \\ H_{12}^{(j)}\left[g_\alpha^{(j)} \sinh \alpha_j x + g_\beta^{(j)} \sin \beta_j x\right] - g_\alpha^{(j)} H_{11}^{(j)}\left[\cosh \alpha_j x - \cos \beta_j x\right] \end{Bmatrix}
$$

(a)

where $g_\alpha^{(j)}(\omega_j)$ and $g_\beta^{(j)}(\omega_j)$ are defined by Eqs. (10.81), the natural frequencies are the roots of Eq. (f) of Example 10.8 and the values of $H_{nl}^{(j)}(\omega_j)$ ($n,l =$

1, 2, ...) are defined by Eqs. (e-1)–(e-4) of that example. The modal masses for the uniform beam are then, from Eq. (11.25),

$$\tilde{m}_j = m\left\|W^{(j)}\right\|^2 + I_\rho\left\|\mathcal{G}^{(j)}\right\|^2 \quad (j = 1,2,...) \tag{b}$$

where $W^{(j)}(x)$ and $\mathcal{G}^{(j)}(x)$ are given by the first and second rows of Eq. (a), respectively.

For the problem at hand, $q(x,t) = 0$ and $6(x,t) = M_0 \sin\Omega t\, \hat{\delta}(x-L)$. Hence, for a Timoshenko Beam loaded in this manner,

$$\mathbf{F}(x,t) = \begin{Bmatrix} 0 \\ M_0 \sin\Omega t\, \hat{\delta}(t-L) \end{Bmatrix} \tag{c}$$

Substituting Eq. (c) into Eq. (11.26) gives the modal force for the jth mode as

$$\tilde{F}_j(t) = M_0 \sin\Omega t \int_0^L \mathcal{G}^{(j)}(x)\hat{\delta}(x-L)\,dx = M_0 \sin\Omega t\, \mathcal{G}^{(j)}(L) \tag{d}$$

It follows that

$$\tilde{f}_j(t) = \frac{\tilde{F}_j(t)}{\tilde{m}_j\omega_j^2} = \tilde{f}_j^0 \sin\Omega t \tag{e-1}$$

where

$$\tilde{f}_j^0 = \frac{M_0\mathcal{G}^{(j)}(L)\big/\omega_j^2}{m\left\|W^{(j)}\right\|^2 + I_\rho\left\|\mathcal{G}^{(j)}\right\|^2} \tag{e-2}$$

For the present problem, Eq. (11.15) thus takes the form

$$\ddot{\eta}_j + \omega_j^2\eta = \omega_j^2\tilde{f}_j^0 \sin\Omega t \tag{f}$$

The particular solution to Eq. (f), and hence the steady state amplitude for the jth mode, follows directly from Eq. (3.28) as

$$\eta_j(t) = \frac{\tilde{f}_j^0}{1-\left(\Omega\big/\omega_j\right)^2}\sin\Omega t \tag{g}$$

Substituting Eq. (g) into Eq. (11.5) gives the steady state response of the cantilevered Timoshenko Beam subjected to a harmonic edge moment as

$$\begin{Bmatrix} w(x,t) \\ \varphi(x,t) \end{Bmatrix} = \sum_{j=1}^{\infty} \eta_j(t) \mathbf{U}^{(j)}(x)$$

$$= M_0 \sin \Omega t \sum_{j=1}^{\infty} \frac{\vartheta^{(j)}(L)}{\left(\omega_j^2 - \Omega^2\right)\left[m \left\| W^{(j)} \right\|^2 + I_\rho \left\| \vartheta^{(j)} \right\|^2 \right]} \begin{Bmatrix} W^{(j)}(x) \\ \vartheta^{(j)}(x) \end{Bmatrix} \quad \triangleleft \text{(h)}$$

where $W^{(j)}(x)$ and $\varphi^{(j)}(x)$ are respectively given by the first and second rows of Eq. (a).

11.4 CONCLUDING REMARKS

In this chapter the response of one-dimensional continua to time dependent forcing was studied. It was shown that the response of such systems is comprised of the sum of the modal functions with time dependent amplitudes. In this context, the modal expansion corresponds to a generalized Fourier Series. It was further shown that, when expressed in terms of such an expansion, the equations of motion reduce to a system of uncoupled forced harmonic equations in time for each mode. The dependent variables in these equations correspond to the modal displacements which, in turn, correspond to the coefficients of the modal expansion. The decomposition for continua was seen to be a generalization of the *modal analysis* introduced in Chapter 8 for discrete systems and offers the same interpretation. Namely, that each of the uncoupled harmonic equations corresponds to an effective single degree of freedom system where the modal displacements are the displacements of the effective modal mass and the modal force measures the degree and manner in which the applied forces are distributed to the individual modes. The procedure applies to all continua considered in Chapters 9 and 10 and may be extended to more general continua.

BIBLIOGRAPHY

Ginsberg, J.H., *Mechanical and Structural Vibrations: Theory and Applications*, Wiley, New York, 2001.

Hildebrand, F.B., *Methods of Applied Mathematics*, 2nd ed., Prentice-Hall, Englewood Cliffs, 1965.

Meirovitch, L., *Elements of Vibration Analysis*, 2nd ed., McGraw-Hill, New York, 1986.

Meirovitch, L., *Fundamentals of Vibrations*, McGraw-Hill, Boston, 2001.

Thomson, W.T., *Theory of Vibration with Applications*, 4th ed., Prentice-Hall, Englewood Cliffs, 1993.

Weaver, W. Jr., Timoshenko, S.P. and Young, D.P., *Vibration Problems in Engineering*, 5th ed., Wiley-Interscience, 1990.

PROBLEMS

11.1 Solve the problem of Example 11.1 using the normal modes computed in Example 10.14(b).

11.2 Compare the magnitude of the edge deflection computed in Example 11.1 and Problem 11.1 with the corresponding edge deflection due to a static load of the same magnitude as discussed in Section 1.4.

11.3 Determine the response of the rod of Example 11.1 if it is struck on its right end by an impulse of magnitude $\mathcal{I}_0$.

11.4 A rocket lies at rest on a frictionless bed when the thruster is activated. The resultant thrust acts at a distance a from the nozzle, as shown, and the geometrical and material properties of the structure may be treated as approximately uniform. The casing is of effective length L, effective area A, and is made of elastic material of Young's modulus E and mass density ρ. Determine the motion of the rocket if the engine is ramped up to a constant thrust of magnitude P_0 over the time interval τ.

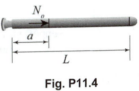

Fig. P11.4

11.5 An external field exerts the body force $p(x,t) = mg(L-x)\sin\Omega t$ on the uniform rod of Problem 10.2. Determine the steady state response of the structure.

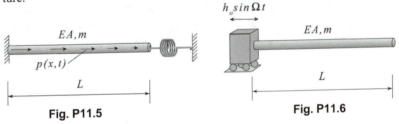

Fig. P11.5   **Fig. P11.6**

11.6 A uniform elastic rod of length L, membrane stiffness EA and mass per unit length m is attached to a rigid base at its left end, as shown. Determine the steady state motion of the rod if a motor causes the base to undergo the prescribed motion $\chi_x(t) = h_0 \sin\Omega t$.

11.7 A torque of magnitude $\mathcal{T}_0$ is suddenly applied to the free end of the elastic rod shown. Determine the response of the rod.

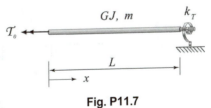

Fig. P11.7

11.8   Determine the response of the fixed-free rod shown in Figure P11.8 if the base suddenly undergoes a rotation of magnitude θ_0 as shown.

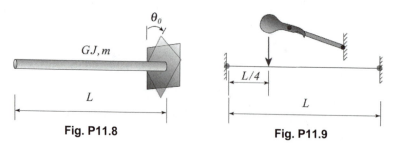

Fig. P11.8   **Fig. P11.9**

11.9   A piano wire of length L, cross-sectional area A and mass density ρ is tuned to a tension of magnitude N_0. Determine the response of the wire if the string is struck at its quarter point by a hammer that imparts an impulse of magnitude $\mathcal{I}_0$.

11.10   A simply supported uniform beam of length L, bending stiffness EI and mass per unit length m is subjected to the uniform distributed load $p(x,t) = q_0\sin\Omega t$. Determine the bending moment at the center of the span if the behavior of the beam is predicted using Euler-Bernoulli Theory.

11.11   The end of a seven foot diving board is bolted to the support as shown. The board is 2 feet wide and 1.5 inches thick, has a specific weight of 36 lb/ft³ and an elastic modulus of 10^6 psi. A 200 lb man stands at the edge of the board preparing for a dive, and then jumps once before leaving the board. If the time history of the jump is as described in the figure, use Euler-Bernoulli Theory to determine the response of the board after the dive has ensued.

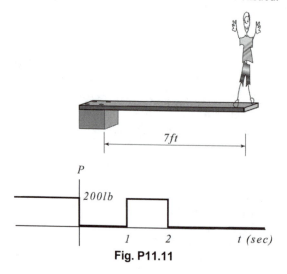

Fig. P11.11

11.12 The clamped-clamped beam shown is subjected to a concentrated load $P(t) = P_0$ that is suddenly applied to the center of the span. Determine the response of the beam after the load is suddenly removed.

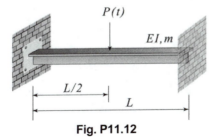

Fig. P11.12

11.13 The beam shown is subjected to the time dependent moment $\mathcal{M}(t) = M_0\sin\Omega t$ applied at its right end. Determine the steady state response of the beam.

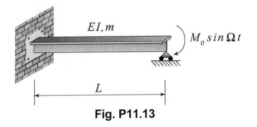

Fig. P11.13

11.14 The beam shown in Figure P11.14 is subjected to the time dependent body couple distribution whose magnitude varies linearly over the span from 0 to b_0, as shown. Determine the steady state response of the structure if the body couple distribution varies harmonically in time with an excitation frequency that is twice the fundamental frequency of the beam.

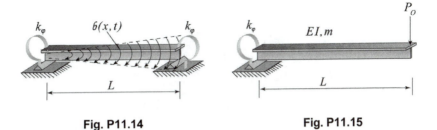

Fig. P11.14   **Fig. P11.15**

11.15 A concentrated load of magnitude P_0 is suddenly applied to the right edge of the beam of Figure P11.15. Determine the response of the structure.

11.16 The beam shown in Figure P11.16 is subjected to a concentrated harmonic load at its right end. If the magnitude of the applied load is P_0 and its frequency is equal to half of the fundamental frequency of the structure, determine the transverse shear at the left edge of the beam.

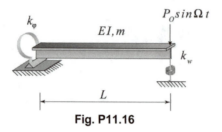

Fig. P11.16

11.17 The beam shown in Figure P11.17 is impacted at its left end. If the magnitude of the impulse imparted to the structure is $\mathcal{I}_0$, determine the response of the beam.

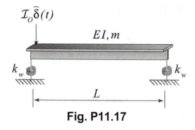

Fig. P11.17

11.18 The beam shown in Figure 11.18 is impacted at its right end. If the magnitude of the impulse imparted to the structure is $\mathcal{I}_0$, determine the reactions of the beam.

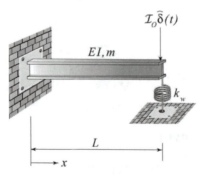

Fig. P11.18

11.19 A simply supported uniform beam of length L, bending stiffness EI and mass per unit length m is subjected to the uniform distributed load $p(x,t) = q_0\sin\Omega t$. Determine the bending moment at the center of the span if the behavior of the beam is predicted using Rayleigh Beam Theory with $r_G/L = 0.1$.

11.20 Let the clamped-clamped beam of Problem 11.12 possess radius of gyration $r_G/L = 0.1$. Determine the response of the beam if its behavior is now represented using Rayleigh Beam Theory.

11.21 Let the clamped-pinned beam of Problem 11.13 possess radius of gyration $r_G/L = 0.1$. Determine the response of the beam if its behavior is now represented using Rayleigh Beam Theory.

11.22 A simply supported uniform beam of length L, bending stiffness EI and mass per unit length m is subjected to the uniform distributed load $p(x,t) = q_0\sin\Omega t$. Determine the bending moment at the center of the span if the behavior of the beam is predicted using Timoshenko Beam Theory for a structure where $r_G/L = 0.1$ and $E/kG = 5$.

11.23 Let the clamped-clamped beam of Problem 11.12 possess radius of gyration $r_G/L = 0.1$ and modulus ratio $E/kG = 5$. Determine the response of the beam if its behavior is now represented using Timoshenko Beam Theory.

11.24 Let the clamped-pinned beam of Problem 11.13 possess radius of gyration $r_G/L = 0.1$ and modulus ratio $E/kG = 5$. Determine the response of the beam if its behavior is now represented using Timoshenko Beam Theory.

Index

Other Related Titles Include:

Mechanical Vibration: Analysis, Uncertainities, and Control, Second Edition
Haym Benaroya, Rutgers University, Piscataway, New Jersey
ISBN: 0824753801

Vibrations of Shells and Plates, Third Edition
Werner Soedel, Prudue University, West Lafayette, Indiana
ISBN: 0824756290

Human Response to Vibration
Neil J. Mansfield, Loughborough University, Loughborough, England
ISBN: 041528239X